Mathematics with A
Custom Version

Margaret L. Lial
American River College

Thomas W. Hungerford
Cleveland State University

Charles D. Miller

ADDISON-WESLEY

An Imprint of Addison Wesley Longman, Inc.

Reading, Massachusetts • Menlo Park, California • New York • Harlow, England
Don Mills, Ontario • Sydney • Mexico City • Madrid • Amsterdam

Manager of Addison Wesley Longman Custom Publishing: Lynn Colgin
Production Administrator: Rohnda Barnes

Mathematics with Applications, Custom Version includes selected sections from *Mathematics with Applications*, 7ed. (0-321-02294-7) by Margaret L. Lial, Thomas W. Hungerford, and Charles D. Miller.

ISBN: 0-201-43465-2

99 00 9 8 7 6 5 4 3

Contents

CHAPTER 11
Introduction to Statistics 506

Preface

Mathematics with Applications, Seventh Edition, is designed to provide the mathematical topics needed by students in the fields of business, management, social science, and natural science. We have done our best to present sound mathematics in an informal manner that stresses meaningful motivation, careful explanations, and numerous examples, with an ongoing focus on real-world problem solving.

This book is written at a level appropriate for the intended audience. Topics are presented by proceeding from what is already known to new material, from concrete examples to general rules and formulas. Almost every section includes pertinent applications. The only prerequisite for using this book is a course in algebra. Chapters 1 and 2 provide a thorough review of algebra for those students who need it.

 ## GRAPHING TECHNOLOGY

The use of graphing technology is an optional feature of this book. Throughout the text, there are a number of *Technology Tips* to inform students of various features of their graphing calculators and to guide them in implementing these features. Some exercises are designed especially for the use of graphing technology.

Instructors who want to make graphing technology an integral part of the course rather than an optional add-on should consider our new text, *Mathematics with Applications: Graphing Technology Version,* which is available from your Addison Wesley Longman representative. It integrates graphing technology into the course without losing sight of the fact that the underlying mathematics is the crucial issue.

 ## NEW CONTENT HIGHLIGHTS

Several topics are treated differently than they were in the sixth edition:

- The introductory algebraic material that formerly constituted a very long first chapter has been reorganized into two shorter chapters. Elementary review material is in Chapter 1. Basic graphing is introduced at the beginning of Chapter 2 so that it can be used later in the chapter to provide an explanation for the algebraic techniques used in solving polynomial and rational inequalities.
- The treatment of standard deviation in Chapter 11 has been rewritten so that the distinction between the population and the sample standard deviations is more fully explained. In addition, the treatment of various statistical graphs has been expanded in Chapters 10 and 11.
- Chapter 13, Applications of the Derivative, has been significantly reordered to produce a better organized, more cohesive discussion of the first and second derivatives and their use in optimization applications.

PEDAGOGICAL FEATURES

We have retained popular features from earlier editions of *Mathematics with Applications:* extensive examples; exercises keyed to the text (many of them new); conceptual and writing exercises; realistic and timely applications, many based on real data; end-of-chapter case studies; margin problems; highlighted rules, definitions, and summaries; and Connection exercises (marked with a ◀▷) that involve material from earlier sections.

For the benefit of students who use graphing calculators, there is a Program Appendix. It contains programs of two types: programs to update older calculators (e.g., a table program for TI-85 and a root finder program for TI-81 and older Casio models) and programs to do specific tasks discussed in the text (e.g., programs to create amortization tables, to carry out the simplex method, and to approximate a definite integral by using the areas of rectangles). Depending on which parts of the text an instructor covers, some of these programs are likely to prove extremely useful.

COURSE FLEXIBILITY

This book can be used for a variety of courses, including the following:

Finite Mathematics and Calculus (one year or less). Use the entire book; cover topics from Chapters 1–5 as needed before proceeding to further topics.

Finite Mathematics (one semester or two quarters). Use as much of Chapters 1–5 as needed, then go into Chapters 6–11 as time permits and local needs require.

Calculus (one semester or quarter). Cover the precalculus topics in Chapters 1–5 as necessary, and then use Chapters 12–15.

College Algebra with Applications (one semester or quarter). Use Chapters 1–9, with the topics in Chapters 8 and 9 being optional.

Chapter interdependence is as follows:

	Chapter	Prerequisite
1	Fundamentals of Algebra	None
2	Graphs, Equations, and Inequalities	Chapter 1
3	Functions and Graphs	Chapters 1 and 2
4	Polynomial and Rational Functions	Chapter 3
5	Exponential and Logarithmic Functions	Chapter 3
6	Mathematics of Finance	Chapter 5
7	Systems of Linear Equations and Matrices	Chapters 1 and 2
8	Linear Programming	Chapters 3 and 7
9	Sets and Probability	None
10	Further Topics in Probability	Chapter 9
11	Introduction to Statistics	Chapter 9
12	Differential Calculus	Chapters 1–5
13	Applications of the Derivative	Chapter 12
14	Integral Calculus	Chapters 12 and 13
15	Multivariate Calculus	Chapters 12–14

Contact your Addison Wesley Longman sales representative to order a customized version of this text.

SUPPLEMENTS

FOR THE INSTRUCTOR

Instructor's Resource Guide and Solutions Manual (ISBN 0-321-03954-8) The Manual contains detailed solutions to all text problems and test items with their answers.

Answer Book (ISBN 0-321-03955-6)   The Answer Book contains the answers to all the problems in the text.

TestGen-EQ with QuizMaster (ISBN for Windows: 0-321-03561-5; ISBN for Macintosh: 0-321-03562-3)   TestGen-EQ is a computerized test generator with algorithmically defined problems organized specifically for this textbook. Its user-friendly graphical interface enables instructors to select, view, edit, and add test items, then print tests in a variety of fonts and forms. Seven question types are available, and search and sort features let the instructor quickly locate questions and arrange them in a preferred order. A built-in question editor gives the user the power to create graphs, import graphics, insert mathematical symbols and templates, and insert variable numbers or text. An "Export to HTML" feature lets instructors create practice tests that can be posted to a Web site. Tests created with TestGen-EQ can be used with QuizMaster-EQ, which enables students to take exams on a computer network. QuizMaster-EQ automatically grades the exams, stores results on disk, and allows the instructor to view or print a variety of reports for individual students, classes, or courses. This program, available in Windows and Macintosh formats, is free to adopters of the text.

Printed Test Bank (ISBN 0-321-03956-4)   The Test Bank provides prepared tests for each chapter.

FOR THE STUDENT

Student's Solutions Manual (ISBN 0-321-03953-X)   The Student's Solutions Manual contains detailed, carefully worked-out solutions to all odd-numbered section and chapter review exercises, and all case study exercises. Also included are the *Explorations in Finite Mathematics* and *Visual Calculus* software programs, by David Schneider of the University of Maryland. These are Windows/ DOS programs that contain several routines directly tied to this text. Many routines allow for application and discovery.

Graphing Calculator Manual (ISBN 0-321-03959-9)   The Graphing Calculator Manual provides step-by-step instructions for using graphing calculators to work through examples from the main text.

Web Site   A Web site to accompany this text contains additional application problems (and their answers) and TI-83 graphing calculator programs that can be downloaded to a computer and transferred to a TI-83 calculator using TI-Graph Link. http://hepg.awl.com   Keyword: CalcZone or FiniteZone

ACKNOWLEDGMENTS

We thank the following instructors who reviewed the manuscript and made many helpful suggestions for improvement.

Jean Davis, *Southwest Texas State University*
J. Franklin Fitzgerald, *Boston University*
Leland J. Fry, *Kirkwood Community College*
Joseph A. Guthrie, *University of Texas–El Paso*
Alec Ingraham, *New Hampshire College*
Jeffrey Lee, *Texas Tech University*
Arthur M. Lieberman, *Cleveland State University*
Norman Lindquist, *Western Washington University*
C.G. Mendez, *Metropolitan State College of Denver*
Kandasamy Muthuvel, *University of Wisconsin–Oshkosh*
Michael I. Ratliff, *Northern Arizona University*
Bhushan Wadhwa, *Cleveland State University*

We also wish to thank our accuracy checkers who did an excellent job of checking all of the exercise answers: Jean Davis, *Southwest Texas State University,* and Michael I. Ratliff, *Northern Arizona University.*

Special thanks go to Laurel Technical Services and Paula Young, who did an outstanding job preparing the print supplements; to Terry McGinnis, whose careful proofreading has helped eliminate errors from the text; to Paul Van Erden, who created an accurate and complete index for us; and to Becky Troutman, who carefully compiled the Index of Applications.

We want to thank the staff of Addison Wesley Longman for their assistance with and contributions to this book, particularly Greg Tobin, Christine O'Brien, and Becky Malone. Finally we express our deep appreciation to freelance editor Sandi Goldstein, who kept us on schedule and on target, and to Cathy Wacaser of Elm Street Publishing Services, who did her usual fine production work.

Margaret L. Lial
Thomas W. Hungerford

To the Student

Several features of the text are designed to assist you in understanding the concepts and learning the mathematical procedures involved. To help you learn new concepts and reinforce your understanding, there are numerous *side problems* in the margin. They are referred to in the text by numbers in colored squares, such as $\boxed{2}$. When you see that symbol, you should work the indicated problem in the margin before going on.

Graphing calculators are not required to use this book. However, for those students who have them, there are *Technology Tips* throughout the text. These Tips describe the proper menus or keys to be used on specific calculators in order to carry out a particular procedure. When these Tips are not sufficient, consult the instruction manual for your calculator. There are also some optional exercises that require a graphing calculator.

The key to succeeding in this course is to remember that

mathematics is not a spectator sport.

You can't expect to learn mathematics without *doing* mathematics any more than you could learn to swim without getting wet. You have to take an active role, making use of all the resources at your disposal: your instructor, your fellow students, and this book.

There is no way that your instructor can possibly cover every aspect of a topic during class time. You simply won't develop the level of understanding you need to succeed unless you read the text carefully. In particular, you should read the text *before* starting the exercises. However, you can't read a math book the way you read a novel. You should have pencil, paper, and calculator handy to do the side problems, work out the statements you don't understand, and make notes on things to ask your fellow students and/or your instructor.

Finally, remember the words of the great Hillel: "The bashful do not learn." There is no such thing as a "dumb question" (assuming, of course, that you have read the book and your class notes and attempted the homework). Your instructor will welcome questions that arise from a serious effort on your part. So get your money's worth: Ask questions.

CHAPTER 1

Fundamentals of Algebra

This book deals with the application of mathematics to business, social science, and biology. Because algebra is vital to understanding these applications, we begin with a review of the fundamental ideas of algebra. If you have not used your algebraic skills for some time, it is important for you to study the review material in this chapter carefully; your success in covering the material that follows will depend on your algebraic skills.

1.1 THE REAL NUMBERS

Only real numbers will be used in this book.* The names of the most common types of real numbers are as follows.

THE REAL NUMBERS	
Natural (counting) numbers	$1, 2, 3, 4, \ldots$
Whole numbers	$0, 1, 2, 3, 4, \ldots$
Integers	$\ldots, -3, -2, -1, 0, 1, 2, 3, \ldots$
Rational numbers	All numbers of the form p/q, where p and q are integers, with $q \neq 0$
Irrational numbers	Real numbers that are not rational

The relationships among these types of numbers are shown in Figure 1.1 on the next page. Notice, for example, that the integers are also rational numbers and real numbers, but the integers are not irrational numbers.

---

*Not all numbers are real numbers. An example of a number that is not a real number is $\sqrt{-1}$.

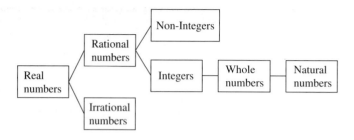

FIGURE 1.1

One example of an irrational number is π, the ratio of the circumference of a circle to its diameter. The number π can be approximated by writing $\pi \approx 3.14159$ ($\approx$ means "is approximately equal to"), but there is no rational number that is exactly equal to π.

1 Name all the types of numbers that apply to the following.

(a) -2

(b) $-5/8$

(c) $\pi/5$

Answers:

(a) Integer, rational, real

(b) Rational, real

(c) Irrational, real

EXAMPLE 1   What kind of number is each of the following?

(a) 6

The number 6 is a natural number, whole number, integer, rational number, and real number.

(b) 3/4

This number is rational and real.

(c) 3π

Because π is not a rational number, 3π is irrational and real.   ■   **1***

All real numbers can be written in decimal form. A rational number, when written in decimal form, is either a terminating decimal, such as .5 or .128, or a repeating decimal in which some block of digits eventually repeats forever, such as 1.3333. . . or 4.7234234234. . . .† Irrational numbers are decimals that neither terminate nor repeat.

The only real numbers that can be entered exactly into a calculator are those rational numbers that are terminating decimals of no more than 10 or 12 digits (depending on the calculator). Similarly, the answers produced by a calculator are often 10–12 digit decimal *approximations*—accurate enough for most applications. As a general rule, *you should not round off any numbers during a long calculator computation,* so that your final answer will be as accurate as possible. For convenience, however, we usually round off the final answer to three or four decimal places.

The important basic properties of the real numbers are as follows.

---

*The use of margin problems is explained in the "To the Student" section preceding this chapter.

†Some calculators have a FRAC key that automatically converts some repeating decimals to fraction form.

PROPERTIES OF THE REAL NUMBERS

For all real numbers, a, b, and c, the following properties hold true.

Commutative properties $a + b = b + a$ $ab = ba$

Associative properties $(a + b) + c = a + (b + c)$ $(ab)c = a(bc)$

Identity properties   There exists a unique real number 0, called the **additive identity,** such that

$$a + 0 = a \quad \text{and} \quad 0 + a = a.$$

There exists a unique real number 1, called the **multiplicative identity,** such that

$$a \cdot 1 = a \quad \text{and} \quad 1 \cdot a = a.$$

Inverse properties   For each real number a, there exists a unique real number $-a$, called the **additive inverse** of a, such that

$$a + (-a) = 0 \quad \text{and} \quad (-a) + a = 0.$$

If $a \neq 0$, there exists a unique real number $1/a$, called the **multiplicative inverse** of a, such that

$$a \cdot \frac{1}{a} = 1 \quad \text{and} \quad \frac{1}{a} \cdot a = 1.$$

Distributive property $a(b + c) = ab + ac$

2 Name the property illustrated in each of the following examples.

(a) $(2 + 3) + 9$
$= (3 + 2) + 9$

(b) $(2 + 3) + 9$
$= 2 + (3 + 9)$

(c) $(2 + 3) + 9$
$= 9 + (2 + 3)$

(d) $(4 \cdot 6)p = (6 \cdot 4)p$

(e) $4(6p) = (4 \cdot 6)p$

Answers:
(a) Commutative property
(b) Associative property
(c) Commutative property
(d) Commutative property
(e) Associative property

EXAMPLE 2   The following statements are examples of the commutative properties. Notice that the order of the numbers changes from one side of the equals sign to the other, so that the order in which two numbers are added or multiplied is unimportant.

(a) $(6 + x) + 9 = (x + 6) + 9$

(b) $(6 + x) + 9 = 9 + (6 + x)$

(c) $5 \cdot (9 \cdot 8) = (9 \cdot 8) \cdot 5$

(d) $5 \cdot (9 \cdot 8) = 5 \cdot (8 \cdot 9)$ ■

EXAMPLE 3   The following statements are examples of the associative properties. Here the order of the numbers does not change, but the placement of the parentheses does change. This means that when three numbers are to be added, the sum of any two can be found first, then that result is added to the remaining number.

(a) $4 + (9 + 8) = (4 + 9) + 8$

(b) $3(9x) = (3 \cdot 9)x$ ■ **2**

3 Name the property illustrated in each of the following examples.

(a) $2 + 0 = 2$

(b) $-\frac{1}{4} \cdot (-4) = 1$

(c) $-\frac{1}{4} + \frac{1}{4} = 0$

(d) $1 \cdot \frac{2}{3} = \frac{2}{3}$

Answers:
(a) Identity property
(b) Inverse property
(c) Inverse property
(d) Identity property

4 Use the distributive property to complete each of the following.

(a) $4(-2 + 5)$

(b) $2(a + b)$

(c) $-3(p + 1)$

(d) $(8 - k)m$

(e) $5x + 3x$

Answers:
(a) $4(-2) + 4(5) = 12$
(b) $2a + 2b$
(c) $-3p - 3$
(d) $8m - km$
(e) $(5 + 3)x = 8x$

EXAMPLE 4   By the identity properties,

(a) $-8 + 0 = -8$,

(b) $(-9)1 = -9$. ∎

EXAMPLE 5   By the inverse properties, the statements in parts (a) through (d) are true.

(a) $9 + (-9) = 0$

(b) $-15 + 15 = 0$

(c) $-8 \cdot \left(\frac{1}{-8}\right) = 1$

(d) $\frac{1}{\sqrt{5}} \cdot \sqrt{5} = 1$ ∎

N O T E   There is no real number x such that $0 \cdot x = 1$, so 0 has no inverse for multiplication. ◆   3

One of the most important properties of the real numbers, and the only one that involves both addition and multiplication, is the distributive property. The next example shows how this property is applied.

EXAMPLE 6   By the distributive property,

(a) $9(6 + 4) = 9 \cdot 6 + 9 \cdot 4$

(b) $3(x + y) = 3x + 3y$

(c) $-8(m + 2) = (-8)(m) + (-8)(2) = -8m - 16$

(d) $(5 + x)y = 5y + xy$. ∎

N O T E   As shown in Example 6(d), by the commutative property, the distributive property can also be written as $(a + b)c = ac + bc$. ◆   4

ORDER OF OPERATIONS   We avoid possible ambiguity when working problems with real numbers by using the following *order of operations,* which has been agreed on as the most useful. This order of operations is used by computers and graphing calculators.

ORDER OF OPERATIONS

If parentheses or square brackets are present:

1. Work separately above and below any fraction bar.
2. Use the rules below within each set of parentheses or square brackets. Start with the innermost set and work outward.

If no parentheses or square brackets are present:

1. Find all powers and roots, working from left to right.
2. Do any multiplications or divisions in the order in which they occur, working from left to right.
3. Do any additions or subtractions in the order in which they occur, working from left to right.

⁵ Evaluate the following if $m = -5$ and $n = 8$.

(a) $-2mn - 2m^2$

(b) $\dfrac{4(n-5)^2 - m}{m+n}$

Answers:

(a) 30

(b) $\dfrac{41}{3}$

⁶ Simplify the following.

(a) $4^2 \div 8 + 3^2 \div 3$

(b) $[-7 + (-9)](-4) - 8(3)$

(c) $\dfrac{-11 - (-12) - 4 \cdot 5}{4(-2) - (-6)(-5)}$

(d) $\dfrac{36 \div 4 \cdot 3 \div 9 + 1}{9 \div (-6) \cdot 8 - 4}$

Answers:

(a) 5

(b) 40

(c) $\dfrac{19}{38} = \dfrac{1}{2} = .5$

(d) $-\dfrac{1}{4} = -.25$

EXAMPLE 7   Use the order of operations to evaluate each expression if $x = -2$, $y = 5$, and $z = -3$.

(a) $-4x^2 - 7y + 4z$

Use parentheses when replacing letters with numbers.

$$
\begin{aligned}
-4x^2 - 7y + 4z &= -4(-2)^2 - 7(5) + 4(-3) \\
&= -4(4) - 7(5) + 4(-3) \\
&= -16 - 35 - 12 \\
&= -63
\end{aligned}
$$

(b) $\dfrac{2(x-y)^2 + 4y}{z+4} = \dfrac{2(-2-5)^2 + 4(5)}{-3+4}$

$$
\begin{aligned}
&= \dfrac{2(-7)^2 + 20}{1} \\
&= 2(49) + 20 \\
&= 118 \quad \blacksquare \quad \boxed{5} \quad \boxed{6}
\end{aligned}
$$

SQUARE ROOTS   There are two numbers whose square is 16, namely, 4 and -4. The positive one, 4, is called the *square root* of 16. Similarly, the square root of a nonnegative number d is defined to be the *nonnegative* number whose square is d; it is denoted $\sqrt{d}$. For instance,

$$
\sqrt{36} = 6 \text{ because } 6^2 = 36 \quad \text{and} \quad \sqrt{1.44} = 1.2 \text{ because } (1.2)^2 = 1.44.
$$

No negative number has a square root in the real numbers. For instance, there is no real number whose square is -4, so -4 has no square root.

Every nonnegative real number has a square root. Unless an integer is a perfect square (such as $64 = 8^2$), its square root is an irrational number. A calculator can be used to obtain a rational approximation of these square roots.

⁷ Estimate each of the following.

(a) $\sqrt{73}$

(b) $\sqrt{22} + 3$

(c) Confirm your estimates in parts (a) and (b) with a calculator.

Answers:

(a) Between 8 and 9

(b) Between 7 and 8

(c) 8.5440; 7.6904

EXAMPLE 8   Estimate each of the following quantities. Verify your estimate with a calculator.

(a) $\sqrt{40}$

Since $6^2 = 36$ and $7^2 = 49$, $\sqrt{40}$ must be a number between 6 and 7. A typical calculator shows that $\sqrt{40} \approx 6.32455532$.

(b) $5\sqrt{7}$

$\sqrt{7}$ is between 2 and 3 because $2^2 = 4$ and $3^2 = 9$, so $5\sqrt{7}$ must be a number between $5 \cdot 2 = 10$ and $5 \cdot 3 = 15$. A calculator shows that $5\sqrt{7} \approx 13.22875656.$ $\blacksquare$ $\boxed{7}$

CAUTION   If c and d are positive real numbers, then $\sqrt{c+d}$ is *not* equal to $\sqrt{c} + \sqrt{d}$. For example, $\sqrt{9+16} = \sqrt{25} = 5$, but $\sqrt{9} + \sqrt{16} = 3 + 4 = 7$. ◆

8 Draw a number line and graph the numbers -4, -1, 0, 1, 2.5, and 13/4 on it.

Answer:

$-4 \quad -1 \; 0 \; 1 \; 2.5 \; \frac{13}{4}$
$-4 \; -3 \; -2 \; -1 \; 0 \; 1 \; 2 \; 3 \; 4$

THE NUMBER LINE The real numbers can be illustrated geometrically with a diagram called a **number line.** Each real number corresponds to exactly one point on the line and vice-versa. A number line with several sample numbers located (or **graphed**) on it is shown in Figure 1.2. ■ **8**

FIGURE 1.2

Comparing two real numbers requires symbols that indicate their order on the number line. The following symbols are used to indicate that one number is greater than or less than another number.

> $<$ means *is less than* $\qquad$ $\leq$ means *is less than or equal to*
>
> $>$ means *is greater than* $\qquad$ $\geq$ means *is greater than or equal to*

The following definitions show how the number line is used to decide which of two given numbers is the greater.

> For real numbers a and b,
>
> if a is to the left of b on a number line, then $a < b$;
>
> if a is to the right of b on a number line, then $a > b$.

EXAMPLE 9 Write *true* or *false* for each of the following.

(a) $8 < 12$

This statement says that 8 is less than 12, which is true.

(b) $-6 > -3$

The graph of Figure 1.3 shows that -6 is to the *left* of -3. Thus, $-6 < -3$, and the given statement is false.

9 Write *true* or *false* for the following.

(a) $-9 \leq -2$

(b) $8 > -3$

(c) $-14 \leq -20$

Answers:
(a) True
(b) True
(c) False

FIGURE 1.3

(c) $-2 \leq -2$

Because $-2 = -2$, this statement is true. ■ **9**

10 Graph all integers x such that

(a) $-3 < x < 5$

(b) $1 \le x \le 5$.

Answers:

(a)

$-3\ -2\ -1\ 0\ 1\ 2\ 3\ 4\ 5$

(b)

$0\ 1\ 2\ 3\ 4\ 5\ 6$

11 Graph all real numbers x such that

(a) $-5 < x < 1$

(b) $4 < x < 7$.

Answers:

(a)

$-5 \qquad 1$

(b)

$4 \qquad 7$

A number line can be used to draw the graph of a set of numbers, as shown in the next few examples.

EXAMPLE 10 Graph all integers x such that $1 < x < 5$.

The only integers between 1 and 5 are 2, 3, and 4. These integers are graphed on the number line in Figure 1.4. ■ **10**

$-3 \quad -2 \quad -1 \quad 0 \quad 1 \quad 2 \quad 3 \quad 4 \quad 5$

FIGURE 1.4

EXAMPLE 11 Graph all real numbers x such that $1 < x < 5$.

The graph includes all the real numbers between 1 and 5 and not just the integers. Graph these numbers by drawing a heavy line from 1 to 5 on the number line, as in Figure 1.5. Open circles at 1 and 5 show that neither of these points belongs to the graph. ■ **11**

$-1 \quad 0 \quad 1 \quad 2 \quad 3 \quad 4 \quad 5 \quad 6 \quad 7$

FIGURE 1.5

A set that consists of all the real numbers between two points, such as $1 < x < 5$ in Example 11, is called an **interval.** A special notation called **interval notation** is used to indicate an interval on the number line. For example, the interval including all numbers x, where $-2 < x < 3$, is written as $(-2, 3)$. The parentheses indicate that the numbers -2 and 3 are *not* included. If -2 and 3 are to be included in the interval, square brackets are used, as in $[-2, 3]$. The chart below shows several typical intervals, where $a < b$.

Inequality	Interval Notation	Explanation
$a \le x \le b$	$[a, b]$	Both a and b are included.
$a \le x < b$	$[a, b)$	a is included, b is not.
$a < x \le b$	$(a, b]$	b is included, a is not.
$a < x < b$	(a, b)	Neither a nor b is included.

Interval notation is also used to describe sets such as the set of all numbers x, with $x \ge -2$. This interval is written $[-2, \infty)$.

[12] Graph all real numbers x in the interval.

(a) $[4, \infty)$

(b) $[-2, 1]$

Answers:

(a)

(b)

EXAMPLE 12 Graph the interval $[-2, \infty)$.

Start at -2 and draw a heavy line to the right, as in Figure 1.6. Use a solid circle at -2 to show that -2 itself is part of the graph. The symbol, ∞, read "infinity," *does not* represent a number. This notation simply indicates that *all* numbers greater than -2 are in the interval. Similarly, the notation $(-\infty, 2)$ indicates the set of all numbers x with $x < 2$. ■ [12]

FIGURE 1.6

ABSOLUTE VALUE Distance is always a nonnegative number. For example, the distance from 0 to -2 on a number line is 2, the same as the distance from 0 to 2. The **absolute value** of a number a gives the distance on the number line from a to 0. Thus, the absolute value of both 2 and -2 is 2. We write the absolute value of the real number a as $|a|$. For example, the distance on the number line from 9 to 0 is 9, as is the distance from -9 to 0. (See Figure 1.7.) By definition, $|9| = 9$ and $|-9| = 9$.

FIGURE 1.7

The facts that $|9| = 9$ and that $|-9| = 9 = -(-9)$ suggest the following algebraic definition of absolute value.

[13] Find the following.

(a) $|-6|$

(b) $-|7|$

(c) $-|-2|$

(d) $|-3 - 4|$

(e) $|2 - 7|$

Answers:

(a) 6

(b) -7

(c) -2

(d) 7

(e) 5

ABSOLUTE VALUE

For any real number a,

$$|a| = a \quad \text{if } a \geq 0$$
$$|a| = -a \quad \text{if } a < 0.$$

Note the second part of the definition: for a negative number, say -5, the negative of -5 is the positive number $-(-5) = 5$. Similarly, if a is any negative number, then $-a$ is a *positive* number. Thus, *for every real number a, $|a|$ is nonnegative.*

EXAMPLE 13 To evaluate $|8 - 9|$, you should first simplify the expression within the absolute value bars:

$$|8 - 9| = |-1| = 1.$$

Similarly, $-|-5 - 8| = -|-13| = -13$. ■ [13]

1.1 EXERCISES

Label each of the following as true *or* false.

1. Every integer is a rational number.

2. Every integer is a whole number.

3. Every whole number is an integer.

4. No whole numbers are rational.

Identify the properties that are illustrated in each of the following. Some may require more than one property. Assume all variables represent real numbers. (See Examples 2–6.)

5. $-7 + 0 = -7$

6. $3 + (-3) = (-3) + 3$

7. $0 + (-7) = -7 + 0$

8. $8 + (12 + 6) = (8 + 12) + 6$

9. $[5(-8)](-3) = 5[(-8)(-3)]$

10. $8(m + 4) = 8m + 8 \cdot 4$

11. $x(y + 2) = xy + 2x$

12. $8(4 + 2) = (2 + 4)8$

13. How is the additive inverse property related to the additive identity property? the multiplicative inverse property to the multiplicative identity property?

14. Explain the distinction between the commutative and associative properties.

Evaluate each of the following if $p = -2$, $q = 4$, and $r = -5$. (See Example 7.)

15. $-3(p + 5q)$

16. $2(q - r)$

17. $\dfrac{q + r}{q + p}$

18. $\dfrac{3q}{3p - 2r}$

19. $\dfrac{\dfrac{q}{r} - \dfrac{r}{5}}{\dfrac{p}{2} + \dfrac{q}{2}}$

20. $\dfrac{\dfrac{3r}{10} - \dfrac{5p}{2}}{q + \dfrac{2r}{5}}$

Evaluate each expression, using the order of operations given in the text.

21. $8 - 4^2 - (-12)$

22. $8 - (-4)^2 - (-12)$

23. $-(3 - 5) - [2 - (3^2 - 13)]$

24. $\dfrac{2(3 - 7) + 4(8)}{4(-3) + (-3)(-2)}$

25. $\dfrac{2(-3) + 3/(-2) - 2/(-\sqrt{16})}{\sqrt{64} - 1}$

26. $\dfrac{6^2 - 3\sqrt{25}}{\sqrt{6^2 + 13}}$

State whether each of the following numbers is rational *or* irrational. *If the number is irrational, approximate it to four decimal places.*

27. 3π

28. $2/\pi$

29. $\sqrt{3}$

30. $\sqrt{4^2 - 3^2}$

31. $-\sqrt{4 + 5^2 - 13}$

32. $\dfrac{\sqrt{4 - 6}}{3^2 + 7}$

Express each of the following statements in symbols, using $<$, $>$, $\leq$, or $\geq$.

33. 5 is less than 7.

34. -4 is greater than -9.

35. y is less than or equal to 8.3.

36. z is greater than or equal to -3.

37. t is positive.

38. c is at most 14.

Graph each of the following on a number line. (See Examples 10 and 11.)

39. All integers x such that $-4 < x < 4$

40. All integers x such that $-4 \leq x < 2$

41. All natural numbers x such that $-2 < x < 5$

42. All natural numbers x such that $x \leq 2$

43. All real numbers x such that $-2 < x \leq 3$

44. All real numbers x such that $x \geq -3$

Graph the following intervals. (See Example 12.)

45. $(3, \infty)$

46. $(-\infty, 5)$

47. $(-\infty, -2]$

48. $[-4, \infty)$

49. $(-8, -1)$

50. $[-1, 10]$

51. $[-2, 2)$

52. $(3, 7]$

Physical Science   *The wind-chill factor is a measure of the cooling effect that the wind has on a person's skin. It calculates the equivalent cooling temperature if there were no wind. The table gives the wind-chill factor for various wind speeds and temperatures.* *

°F\Wind	5 mph	10 mph	15 mph	20 mph	25 mph	30 mph	35 mph	40 mph
40°	37	28	22	18	16	13	11	10
30°	27	16	9	4	0	−2	−4	−6
20°	16	4	−5	−10	−15	−18	−20	−21
10°	6	−9	−18	−25	−29	−33	−35	−37
0°	−5	−21	−36	−39	−44	−48	−49	−53
−10°	−15	−33	−45	−53	−59	−63	−67	−69
−20°	−26	−46	−58	−67	−74	−79	−82	−85
−30°	−36	−58	−72	−82	−88	−94	−98	−100
−40°	−47	−70	−85	−96	−104	−109	−113	−116
−50°	−57	−83	−99	−110	−118	−125	−129	−132

Suppose that we wish to determine the difference between two of these entries, and we are interested only in the magnitude, or absolute value, of this difference. Then we subtract the two entries and find the absolute value. For example, the difference in wind-chill factors for wind at 20 miles per hour with a 20° temperature and wind at 30 miles per hour with a 40° temperature is $|-10° - 13°| = 23°$, *or equivalently,* $|13° - (-10°)| = 23°$.

Find the absolute value of the difference of the two indicated wind-chill factors.

53. Wind at 15 miles per hour with a 30° temperature and wind at 10 miles per hour with a −10° temperature

54. Wind at 20 miles per hour with a −20° temperature and wind at 5 miles per hour with a 30° temperature

55. Wind at 30 miles per hour with a −30° temperature and wind at 15 miles per hour with a −20° temperature

56. Wind at 40 miles per hour with a 40° temperature and wind at 25 miles per hour with a −30° temperature

Evaluate each of the following. (See Example 13.)

57. $|8| - |-4|$

58. $|-9| - |-12|$

59. $-|-4| - |-1 - 14|$

60. $-|6| - |-12 - 4|$

In each of the following problems, fill in the blank with either =, <, or >, so that the resulting statement is true.

61. $|5|$ _____ $|-5|$

62. $-|-4|$ _____ $|4|$

63. $|10 - 3|$ _____ $|3 - 10|$

64. $|6 - (-4)|$ _____ $|-4 - 6|$

65. $|-2 + 8|$ _____ $|2 - 8|$

66. $|3 + 1|$ _____ $|-3 - 1|$

67. $|3| \cdot |-5|$ _____ $|3(-5)|$

68. $|3| \cdot |2|$ _____ $|3(2)|$

69. $|3 - 2|$ _____ $|3| - |2|$

70. $|5 - 1|$ _____ $|5| - |1|$

Write the expression without using absolute value.

71. $|a - 7|$ if $a < 7$

72. $|b - c|$ if $b \geq c$

73. In general, if a and b are any real numbers having the same sign (both negative or both positive), is it always true that $|a + b| = |a| + |b|$? Explain your answer.

74. If a and b are any two real numbers, is it always true that $|a - b| = |b - a|$? Explain your answer.

75. If a and b are any real numbers, is it always true that $|a + b| = |a| + |b|$? Explain your answer.

76. For which real numbers b does $|2 - b| = |2 + b|$? Explain your answer.

Social Science   *Use inequality symbols to rewrite each of the following statements, which are based on an article in* The Sacramento Bee *newspaper.* * *Using x as the variable, describe what x represents in each exercise and then write an inequality. Example: At least 4000 foreign students attend the University of Southern California (USC). Let x represent the number of foreign students attending USC. Then x ≥ 4000.*

77. Foreign students contribute more than $1 billion annually to the California economy.

---

*Miller, A. and J. Thompson, *Elements of Meteorology*, 2e, Charles E. Merrill Publishing Co., 1975.

*"State colleges drawing foreigners despite cuts" by Lisa Lapin from *The Sacramento Bee*, December 2, 1992. Copyright, The Sacramento Bee, 1994. Reprinted by permission.

78. More than 60% of the international students come from Asian countries.

79. Less than 7.5% of foreign students in the United States now come from Middle Eastern countries.

80. No more than 10% of the foreign students in the United States originate in Japan.

81. California has more than 13% of all foreign students in the United States.

82. Foreign students must prove they have at least $22,000 in cash to spend here each year.

Social Science   *Sociologists measure the status of an individual within a society by evaluating for that individual the number x, which gives the percentage of the population with less income than the given person, and the number y, the percentage of the population with less education. The average status is defined as (x + y)/2, while the individual's status incongruity is defined by | (x − y)/2 |. People with high status incongruities would include unemployed Ph.D.'s (low x, high y) and millionaires who didn't make it past the second grade (high x, low y).*

83. What is the highest possible average status for an individual? The lowest?

84. What is the highest possible status incongruity for an individual? The lowest?

85. Jolene Rizzo makes more money than 56% of the population and has more education than 78%. Find her average status and status incongruity.

86. A popular movie star makes more money than 97% of the population and is better educated than 12%. Find the average status and status incongruity for this individual.

1.2   FIRST-DEGREE EQUATIONS

One of the main uses of algebra is to solve equations. An **equation** is a statement that two mathematical expressions are equal; for example,

$$3x^2 - 2x + 4 = 7x - 2, \quad 4y^3 + 8 = 12, \quad 2z + 6 = -9.$$

The letter in each equation is called the **variable.**

In this section we concentrate on **first-degree equations,** which are equations that can be written in the form $ax + b = c$, where a, b, and c are constants (real numbers) and $a \neq 0$. Examples of first-degree equations are

$$5x - 3 = 13, \quad 8y = 4, \quad -3p + 5 = -8.$$

Examples of equations that are *not* first-degree include $x^3 = 15$, $2x^2 = 5x + 6$ and $\sqrt{x + 2} = 4$ (because of the radical).

A **solution** of an equation is a number that can be substituted for the variable in the equation to produce a true statement. For example, substituting the number 9 for x in the equation $2x + 1 = 19$ gives

$$2x + 1 = 19$$
$$2(\mathbf{9}) + 1 = 19 \qquad \text{Let } x = 9.$$
$$18 + 1 = 19. \qquad \text{True}$$

This true statement indicates that 9 is a solution of $2x + 1 = 19$. ☐**1**

1 Is -4 a solution of the following equations?

(a) $3x + 5 = -7$

(b) $2x - 3 = 5$

(c) Is there more than one solution of the equation in part (a)?

Answers:

(a) Yes

(b) No

(c) No

The following properties are used to solve equations.

PROPERTIES OF EQUALITY

1. The same number may be added to or subtracted from both sides of an equation:

$$\text{If } a = b, \text{ then } a + c = b + c \quad \text{and} \quad a - c = b - c.$$

2. Both sides of an equation may be multiplied or divided by the same nonzero number:

$$\text{If } a = b \text{ and } c \neq 0, \text{ then } ac = bc \quad \text{and} \quad \frac{a}{c} = \frac{b}{c}.$$

EXAMPLE 1   Solve the equation $5x - 3 = 12$.

Using the first property of equality, add 3 to both sides. This isolates the term containing the variable on one side of the equals sign.

$$5x - 3 = 12$$
$$5x - 3 + 3 = 12 + 3 \qquad \text{Add 3 to both sides.}$$
$$5x = 15$$

Now arrange for the coefficient of x to be 1 by using the second property of equality.

$$5x = 15$$
$$\frac{5x}{5} = \frac{15}{5} \qquad \text{Divide both sides by 5.}$$
$$x = 3$$

The solution of the original equation, $5x - 3 = 12$, is 3. Check the solution by substituting 3 for x in the original equation. ■ [2]

2 Solve the following.

(a) $3p - 5 = 19$

(b) $4y + 3 = -5$

(c) $-2k + 6 = 2$

Answers:

(a) 8

(b) -2

(c) 2

EXAMPLE 2   Solve $2k + 3(k - 4) = 2(k - 3)$.

First simplify the equation by using the distributive property on the left-side term $3(k - 4)$ and the right-side term $2(k - 3)$:

$$2k + 3(k - 4) = 2(k - 3)$$
$$2k + 3k - 12 = 2k - 6.$$

On the left, $2k + 3k = (2 + 3)k = 5k$, again by the distributive property, which gives

$$5k - 12 = 2k - 6.$$

One way to proceed is to add $-2k$ to both sides.

$$5k - 12 + (-2k) = 2k - 6 + (-2k) \quad \text{Add } -2k \text{ to both sides.}$$

$$3k - 12 = -6$$

$$3k - 12 + 12 = -6 + 12 \quad \text{Add 12 to both sides.}$$

$$3k = 6$$

$$\frac{1}{3}(3k) = \frac{1}{3}(6) \quad \text{Multiply both sides by } \frac{1}{3}.$$

$$k = 2$$

The solution is 2. Check this result by substituting 2 for k in the original equation. ■ [3]

[3] Solve the following.
(a) $3(m - 6) + 2(m + 4) = 4m - 2$
(b) $-2(y + 3) + 4y = 3(y + 1) - 6$

Answers:
(a) 8
(b) -3

EXAMPLE 3   Use a calculator to solve $42.19x + 121.34 = 16.83x + 19.15$.
To avoid round-off errors in the intermediate steps, do all the algebra first, without using the calculator.

$$42.19x = 16.83x + 19.15 - 121.34 \quad \begin{array}{l}\text{Subtract 121.34}\\\text{from both sides.}\end{array}$$

$$42.19x - 16.83x = 19.15 - 121.34 \quad \begin{array}{l}\text{Subtract } 16.83x\\\text{from both sides.}\end{array}$$

$$(42.19 - 16.83)x = 19.15 - 121.34 \quad \text{Distributive property}$$

$$x = \frac{19.15 - 121.34}{42.19 - 16.83} \quad \begin{array}{l}\text{Divide both sides by}\\(42.19 - 16.83).\end{array}$$

Now use the calculator and determine that $x \approx -4.0296$. Because this answer is approximate, it may not check exactly when substituted in the original equation. ■

The next three examples show how to simplify the solution of first-degree equations involving fractions. We solve these equations by multiplying both sides of the equation by a **common denominator,** a number that can be divided (with remainder 0) by each denominator in the equation. This step will eliminate the fractions.

CAUTION Because this is an application of the multiplication property of *equality,* it can be done *only in an equation.* ◆

EXAMPLE 4   Solve $\dfrac{r}{10} - \dfrac{2}{15} = \dfrac{3r}{20} - \dfrac{1}{5}$.
Here the denominators are 10, 15, 20, and 5. Each of these numbers can be divided into 60; therefore, 60 is a common denominator. Multiply both sides of the equation by 60.

$$60\left(\frac{r}{10} - \frac{2}{15}\right) = 60\left(\frac{3r}{20} - \frac{1}{5}\right)$$

Use the distributive property to eliminate the denominators.

$$60\left(\frac{r}{10}\right) - 60\left(\frac{2}{15}\right) = 60\left(\frac{3r}{20}\right) - 60\left(\frac{1}{5}\right)$$

$$6r - 8 = 9r - 12$$

Add $-6r$ and 12 to both sides.

$$6r - 8 + (-6r) + 12 = 9r - 12 + (-6r) + 12$$

$$4 = 3r$$

Multiply both sides by 1/3 to get the solution.

$$r = \frac{4}{3}$$

Check this solution in the original equation. ■ 4

4 Solve the following.

(a) $\dfrac{x}{2} - \dfrac{x}{4} = 6$

(b) $\dfrac{2x}{3} + \dfrac{1}{2} = \dfrac{x}{4} - \dfrac{9}{2}$

Answers:
(a) 24
(b) -12

EXAMPLE 5   Solve $\dfrac{4}{3(k+2)} - \dfrac{k}{3(k+2)} = \dfrac{5}{3}$.

Multiply both sides of the equation by the common denominator $3(k+2)$. Here $k \neq -2$, since $k = -2$ would give a 0 denominator, making the fraction undefined.

$$3(k+2) \cdot \frac{4}{3(k+2)} - 3(k+2) \cdot \frac{k}{3(k+2)} = 3(k+2) \cdot \frac{5}{3}$$

Simplify each side and solve for k.

$$4 - k = 5(k+2)$$

$$4 - k = 5k + 10 \qquad \text{Distributive property}$$

$$4 - k + k = 5k + 10 + k \qquad \text{Add } k \text{ to both sides.}$$

$$4 = 6k + 10$$

$$4 + (-10) = 6k + 10 + (-10) \qquad \text{Add } -10 \text{ to both sides.}$$

$$-6 = 6k$$

$$-1 = k \qquad \text{Multiply by } \frac{1}{6}.$$

The solution is -1. Substitute -1 for k as a check.

5 Solve the equation

$$\frac{5p+1}{3(p+1)} = \frac{3p-3}{3(p+1)}$$

$$+ \frac{9p-3}{3(p+1)}.$$

Answer:
1

$$\frac{4}{3(-1+2)} - \frac{-1}{3(-1+2)} \overset{?}{=} \frac{5}{3}$$

$$\frac{4}{3} - \frac{-1}{3} \overset{?}{=} \frac{5}{3}$$

$$\frac{5}{3} = \frac{5}{3}$$

The check shows that -1 is the solution. ■ 5

CAUTION Because the equation in Example 5 has a restriction on k, it is *essential* to check the solution. ◆

EXAMPLE 6   Solve $\dfrac{x}{x-2} = \dfrac{2}{x-2} + 2$.

Multiply both sides of the equation by $x - 2$, assuming that $x - 2 \neq 0$. This gives

$$x = 2 + 2(x - 2)$$
$$x = 2 + 2x - 4$$
$$x = 2.$$

Recall the assumption that $x - 2 \neq 0$. Because $x = 2$, we have $x - 2 = 0$, and the multiplication property of equality does not apply. To see this, substitute 2 for x in the original equation—this substitution produces a 0 denominator. Since division by zero is not defined, there is no solution for the given equation.   ■   6

Sometimes an equation with several variables must be solved for one of the variables. This process is called **solving for a specified variable.**

EXAMPLE 7   Solve for x: $3(ax - 5a) + 4b = 4x - 2$.

Use the distributive property to get

$$3ax - 15a + 4b = 4x - 2.$$

Treat x as the variable, the other letters as constants. Get all terms with x on one side of the equals sign, and all terms without x on the other side.

$3ax - 4x = 15a - 4b - 2$	Isolate terms with x on the left.
$(3a - 4)x = 15a - 4b - 2$	Distributive property
$x = \dfrac{15a - 4b - 2}{3a - 4}$	Multiply by $\dfrac{1}{3a - 4}$.

The final equation is solved for x, as required.   ■   7

Recall from Section 1.1 that the absolute value of the number a, written $|a|$, gives the distance on a number line from a to 0. For example, $|4| = 4$, and $|-7| = 7$.

EXAMPLE 8   Solve the equation $|x| = 3$.

There are two numbers whose absolute value is 3, namely 3 and -3. The solutions of the given equation are 3 and -3.   ■

---

6 Solve each equation.

(a) $\dfrac{3p}{p+1} = 1 - \dfrac{3}{p+1}$

(b) $\dfrac{8y}{y-4} = \dfrac{32}{y-4} - 3$

Answer:
Neither equation has a solution.

7 Solve for x.

(a) $2x - 7y = 3xk$

(b) $8(4 - x) + 6p = -5k - 11yx$

Answers:

(a) $x = \dfrac{7y}{2 - 3k}$

(b) $x = \dfrac{5k + 32 + 6p}{8 - 11y}$

8 Solve each equation.

(a) $|y| = 9$

(b) $|r + 3| = 1$

(c) $|2k - 3| = 7$

Answers:
(a) 9, −9

(b) −2, −4

(c) 5, −2

EXAMPLE 9   Solve $|p - 4| = 2$.

The only numbers with absolute value 2 are 2 and −2. So this equation will be satisfied if the expression inside the absolute value bars, $p - 4$, equals either 2 or −2:

$$p - 4 = 2 \quad \text{or} \quad p - 4 = -2.$$

Solving these two equations produces

$$p = 6 \quad \text{or} \quad p = 2,$$

so that 6 and 2 are solutions for the original equation. As before, check by substituting in the original equation. ■ **8**

EXAMPLE 10   Solve $|4m - 3| = |m + 6|$.

The quantities in absolute value bars must either be equal or be negatives of one another to satisfy the equation. That is,

$$4m - 3 = m + 6 \quad \text{or} \quad 4m - 3 = -(m + 6)$$
$$3m = 9 \qquad\qquad 4m - 3 = -m - 6$$
$$m = 3 \qquad\qquad\quad 5m = -3$$
$$m = -\frac{3}{5}.$$

9 Solve each equation.

(a) $|r + 6| = |2r + 1|$

(b) $|5k - 7| = |10k - 2|$

Answers:
(a) 5, −7/3

(b) −1, 3/5

Check that the solutions for the original equation are 3 and −3/5. ■ **9**

APPLIED PROBLEMS   One of the main reasons for learning mathematics is to be able to use it to solve practical problems. There are no hard and fast rules for dealing with real-world applications, except perhaps to use common sense. However, you will find it much easier to deal with such problems if you don't try to do everything at once. After reading the problem carefully, attack it in stages, as suggested in the following guidelines.

SOLVING APPLIED PROBLEMS

1. Decide on the unknown. Name it with some variable that you *write down*. Many students try to skip this step. They are eager to get on with the writing of the equation. But this is an important step. If you don't know what the variable represents, how can you write a meaningful equation or interpret a result?

2. Draw a sketch or make a chart, if appropriate, showing the information given in the problem.

3. Decide on a variable expression to represent any other unknowns in the problem. For example, if x represents the width of a rectangle, and you know that the length is one more than twice the width, then *write down* that the length is $1 + 2x$.

continued

> **4.** Using the results of Steps 1–3, write an equation that expresses a condition that must be satisfied.
> **5.** Solve the equation.
> **6.** Check the solution in the words of the *original problem*, not just in the equation you have written.

The following examples illustrate this approach.

EXAMPLE 11   If the length of a side of a square is increased by 3 centimeters, the new perimeter is 40 centimeters more than twice the length of the side of the original square. Find the length of a side of the original square.

Step 1   What should the variable represent? To find the length of a side of the original square, let

$$x = \text{length of a side of the original square.}$$

Step 2   Draw a sketch, as in Figure 1.8.

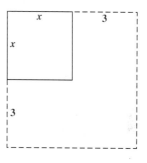

FIGURE 1.8

Step 3   The length of a side of the new square is 3 centimeters more than the length of a side of the old square, so

$$x + 3 = \text{length of a side of the new square.}$$

Now write a variable expression for the new perimeter. Since the perimeter of a square is four times the length of a side,

$$4(x + 3) = \text{the perimeter of the new square.}$$

Step 4   Write an equation by looking again at the information in the problem. The new perimeter is 40 more than twice the length of a side of the original square, so the equation is

$$\left(\begin{array}{c}\text{the new} \\ \text{perimeter}\end{array}\right) \quad \text{is} \quad 40 \quad \left(\begin{array}{c}\text{more} \\ \text{than}\end{array}\right) \quad \left(\begin{array}{c}\text{twice the side of} \\ \text{the original square}\end{array}\right)$$
$$4(x + 3) \quad = \quad 40 \quad + \quad 2x.$$

10 **(a)** A triangle has a perimeter of 45 centimeters. Two of the sides of the triangle are equal in length, with the third side 9 centimeters longer than either of the two equal sides. Find the lengths of the sides of the triangle.

(b) A rectangle has a perimeter which is five times its width. The length is 4 more than the width. Find the length and width of the rectangle.

Answers:
(a) 12 centimeters, 12 centimeters, 21 centimeters
(b) Length is 12, width is 8

Step 5   Solve the equation.

$$4(x + 3) = 40 + 2x$$
$$4x + 12 = 40 + 2x$$
$$2x = 28$$
$$x = 14$$

Step 6   Check the solution using the wording of the original problem. The length of a side of the new square would be $14 + 3 = 17$ centimeters; its perimeter would be $4(17) = 68$ centimeters. Twice the length of the side of the original square is $2(14) = 28$ centimeters. Because $40 + 28 = 68$ centimeters, the solution checks with the words of the original problem. ■   **10**

EXAMPLE 12   Chuck travels 80 kilometers in the same time that Mary travels 180 kilometers. Mary travels 50 kilometers per hour faster than Chuck. Find the speed of each person.

Use the steps given earlier.

Step 1   Use x to represent Chuck's speed and $x + 50$ to represent Mary's speed, which is 50 kilometers per hour faster than Chuck's.

Steps 2 and 3   Constant rate problems of this kind require the distance formula

$$d = rt,$$

where d is the distance traveled in t hours at a constant rate of speed r. The distance traveled by each person is given, along with the fact that the time traveled by each person is the same. Solve the formula $d = rt$ for t.

$$d = rt$$
$$\frac{1}{r} \cdot d = \frac{1}{r} \cdot rt$$
$$\frac{d}{r} = t$$

For Chuck, $d = 80$ and $r = x$, giving $t = 80/x$. For Mary, $d = 180$, $r = x + 50$, and $t = 180/(x + 50)$. Use these facts to complete a chart, which organizes the information given in the problem.

	d	r	t
Chuck	80	x	$\dfrac{80}{x}$
Mary	180	$x + 50$	$\dfrac{180}{x + 50}$

Step 4   Because both people traveled for the *same time,* the equation is

$$\frac{80}{x} = \frac{180}{x + 50}.$$

11 (a) Tom and Dick are in a run for charity. Tom runs at 7 mph and Dick runs at 5 mph. If they start at the same time, how long will it be until they are 1/2 mile apart?

(b) In part (a), suppose the run has a staggered start. If Dick starts first, and Tom starts 10 minutes later, how long will it be until they are neck and neck?

Answers:
(a) 15 minutes (1/4 hour)
(b) After Tom runs 25 minutes.

Step 5   Multiply both sides of the equation by $x(x + 50)$.

$$x(x + 50)\frac{80}{x} = x(x + 50)\frac{180}{x + 50}$$

$$80(x + 50) = 180x$$
$$80x + 4000 = 180x$$
$$4000 = 100x$$
$$40 = x$$

Step 6   Since x represents Chuck's speed, Chuck went 40 kilometers per hour. Mary's speed is $x + 50$, or $40 + 50 = 90$ kilometers per hour. Check these results in the words of the original problem. ■ **11**

EXAMPLE 13   A financial manager has $14,000 to invest for her company. She plans to invest part of the money in tax-free bonds at 6% interest and the remainder in taxable bonds at 9%. She wants to earn $1005 per year in interest from the investments. Find the amount she should invest at each rate.

Let x represent the amount to be invested at 6%, so that $14,000 - x$ is the amount to be invested at 9%. Interest is given by the product of principal, rate, and time in years ($i = prt$). Summarize this information in a chart.

Investment	Amount Invested	Interest Rate	Interest Earned in 1 Year
Tax-free Bonds	x	6% = .06	.06x
Taxable Bonds	$14,000 - x$	9% = .09	$.09(14,000 - x)$
Totals	14,000		1005

Because the total interest is to be $1005,

$$.06x + .09(14,000 - x) = 1005.$$

12 An investor owns two pieces of property. One, worth twice as much as the other, returns 6% in annual interest, while the other returns 4%. Find the value of each piece of property if the total annual interest earned is $8000.

Answer:
6% return: $100,000; 4% return: $50,000

Solve this equation.

$$.06x + 1260 - .09x = 1005$$
$$-.03x = -255$$
$$x = 8500$$

The manager should invest $8500 at 6%, and $14,000 - $8500 = $5500 at 9%. ■ **12**

1.2 EXERCISES

Solve each equation. (See Examples 1–6.)

1. $3x + 5 = 20$

2. $4 - 5y = 9$

3. $.6k - .3 = .5k + .4$

4. $2.5 + 5.04m = 8.5 - .06m$

5. $\frac{2}{5}r + \frac{1}{4} - 3r = \frac{6}{5}$

6. $\frac{2}{3} - \frac{1}{4}p = \frac{3}{2} + \frac{1}{3}p$

7. $2a - 1 = 3(a + 1) + 7a + 5$

8. $3(k - 2) - 6 = 4k - (3k - 1)$

9. $2[x - (3 + 2x) + 9] = 2x + 4$

10. $-2[4(k + 2) - 3(k + 1)] = 14 + 2k$

11. $\frac{3x}{5} - \frac{4}{5}(x + 1) = 2 - \frac{3}{10}(3x - 4)$

12. $\frac{4}{3}(x - 2) - \frac{1}{2} = 2\left(\frac{3}{4}x - 1\right)$

13. $\frac{5y}{6} - 8 = 5 - \frac{2y}{3}$

14. $\frac{x}{2} - 3 = \frac{3x}{5} + 1$

15. $\frac{m}{2} - \frac{1}{m} = \frac{6m + 5}{12}$

16. $-\frac{3k}{2} + \frac{9k - 5}{6} = \frac{11k + 8}{k}$

17. $\frac{4}{x - 3} - \frac{8}{2x + 5} + \frac{3}{x - 3} = 0$

18. $\frac{5}{2p + 3} - \frac{3}{p - 2} = \frac{4}{2p + 3}$

19. $\frac{3}{2m + 4} = \frac{1}{m + 2} - 2$

20. $\frac{8}{3k - 9} - \frac{5}{k - 3} = 4$

Use a calculator to solve each of the following equations. Round your answer to the nearest hundredth. (See Example 3.)

21. $9.06x + 3.59(8x - 5) = 12.07x + .5612$

22. $-5.74(3.1 - 2.7p) = 1.09p + 5.2588$

23. $\frac{2.63r - 8.99}{1.25} - \frac{3.90r - 1.77}{2.45} = r$

24. $\frac{8.19m + 2.55}{4.34} - \frac{8.17m - 9.94}{1.04} = 4m$

Solve each equation for x. (See Example 7. In Exercises 29 and 30, recall that $a^2 = a \cdot a$.)

25. $4(a - x) = b - a + 2x$

26. $(3a + b) - bx = a(x - 2)$

27. $5(b - x) = 2b + ax$

28. $bx - 2b = 2a - ax$

29. $x = a^2x + ax - 3a + 3$

30. $a^2x - 2a^2 = 3x$

Solve each equation for the specified variable. Assume all denominators are nonzero. (See Example 7.)

31. $PV = k$ for V

32. $i = prt$ for p

33. $V = V_0 + gt$ for g

34. $S = S_0 + gt^2 + k$ for g

35. $A = \dfrac{1}{2}(B + b)h$ for B

36. $C = \dfrac{5}{9}(F - 32)$ for F

37. $\dfrac{1}{R} = \dfrac{1}{r_1} + \dfrac{1}{r_2}$ for R

38. $m = \dfrac{Ft}{v_1 - v_2}$ for v_2

Solve each equation. (See Examples 8–10.)

39. $|2h + 1| = 5$

40. $|4m - 3| = 12$

41. $|6 - 2p| = 10$

42. $|-5x + 7| = 15$

43. $\left|\dfrac{5}{r - 3}\right| = 10$

44. $\left|\dfrac{3}{2h - 1}\right| = 4$

45. $\left|\dfrac{6y + 1}{y - 1}\right| = 3$

46. $\left|\dfrac{3a - 4}{2a + 3}\right| = 1$

47. $|3y - 2| = |4y + 5|$

48. $|1 - 3z| = |z + 2|$

49. Natural Science The excess lifetime cancer risk R is a measure of the likelihood that an individual will develop cancer from a particular pollutant. For example, if $R = .01$, then a person has a 1% increased chance of developing cancer during a lifetime. The value of R for formaldehyde can be calculated using the linear equation $R = kd$, where k is a constant and d is the daily dose in parts per million. The constant k for formaldehyde can be calculated using the formula $k = .132(\frac{B}{W})$, where B is the total number of cubic meters of air a person breathes in one day and W is a person's weight in kilograms.*

(a) Find k for a person that breathes in 20 cubic meters of air per day and weighs 75 kg.

(b) Mobile homes in Minnesota were found to have a mean daily dose d of .42 parts per million. Calculate R.†

---

*Hines, A., Ghosh, T., Layalka, S., and Warder, R., *Indoor Air Quality & Control,* Prentice Hall, 1993. (TD 883.1.I476 1993)

†Ritchie, I. and R. Lehnen, "An Analysis of Formaldehyde Concentration in Mobile and Conventional Homes." *J. Env. Health* 47: 300–305.

(c) For every 5000 people, how many cases of cancer could be expected each year from these levels of formaldehyde? Assume an average life expectancy of 72 years.

50. Refer to Exercise 29. Suppose someone tells you that there is no reason to solve for x, because the left side of the equation is already equal to x. Is this correct? Explain.

Management *The approximate annual interest rate of a loan paid off with monthly payments is given by*

$$A = \frac{24f}{b(p + 1)},$$

where f is the finance charge on the loan, p is the total number of payments, and b is the original balance of the loan. Use the formula to find the requested value in the following. Round A to the nearest percent and round other variables to the nearest whole numbers. (This formula is not accurate enough for the requirements of federal law.)

51. $f = \$800$, $b = \$4000$, $p = 36$; find A

52. $A = 5\%$, $b = \$1500$, $p = 24$; find f

53. $A = 6\%$, $f = \$370$, $p = 36$; find b

54. $A = 10\%$, $f = \$490$, $p = 48$; find b.

Management *When a loan is paid off early, a portion of the finance charge must be returned to the borrower. By one method of calculating finance charge (called the rule of 78), the amount of unearned interest (finance charge to be returned) is given by*

$$u = f \cdot \frac{n(n + 1)}{q(q + 1)},$$

where u represents unearned interest, f is the original finance charge, n is the number of payments remaining when the loan is paid off, and q is the original number of payments. Find the amount of the unearned interest in each of the following.

55. Original finance charge = \$800, loan scheduled to run 36 months, paid off with 18 payments remaining

56. Original finance charge = \$1400, loan scheduled to run 48 months, paid off with 12 payments remaining

Solve each applied problem. (See Examples 11–13.)

57. A closed recycling bin is in the shape of a rectangular box. Find the height of the bin if its length is 18 feet, its width is 8 feet, and its surface area is 496 square feet.

(a) Choose a variable and write down what it represents.

(b) Write an equation relating the height, length, and width of a box to its surface area.

(c) Solve the equation and check the solution in the wording of the original problem.

58. The length of a rectangular label is 3 centimeters less than twice the width. The perimeter is 54 centimeters. Find the width. Follow the steps outlined in Exercise 57.

59. A puzzle piece in the shape of a triangle has a perimeter of 30 centimeters. Two sides of the triangle are each twice as long as the shortest side. Find the length of the shortest side.

60. A triangle has a perimeter of 27 centimeters. One side is twice as long as the shortest side. The third side is seven centimeters longer than the shortest side. Find the length of the shortest side.

61. A plane flies nonstop from New York to London, which are about 3500 miles apart. After one hour and six minutes in the air, the plane passes over Halifax, Nova Scotia, which is 600 miles from New York. Estimate the flying time from New York to London.

62. On vacation, Le Hong averaged 50 mph traveling from Denver to Minneapolis. Returning by a different route that covered the same number of miles, he averaged 55 mph. What is the distance between the two cities, if his total traveling time was 32 hours?

63. Russ and Janet are running in the Apple Hill Fun Run. Russ runs at 7 mph, Janet at 5 mph. If they start at the same time, how long will it be before they are 2/3 mile apart?

64. If the run in Exercise 63 has a staggered start and Janet starts first, with Russ starting 15 minutes later, how long will it be before he catches up with her?

65. Joe Gonzalvez received $52,000 profit from the sale of some land. He invested part at 5% interest, and the rest at 4% interest. He earned a total of $2290 interest per year. How much did he invest at 5%?

66. Weijen Luan invests $20,000 received from an insurance settlement in two ways: some at 6%, and some at 4%.

Altogether, she makes $1040 per year interest. How much is invested at 4%?

67. Maria Martinelli bought two plots of land for a total of $120,000. On the first plot, she made a profit of 15%. On the second, she lost 10%. Her total profit was $5500. How much did she pay for each piece of land?

68. Suppose $20,000 is invested at 5%. How much additional money must be invested at 4% to produce a yield of 4.8% on the entire amount invested?

69. Cathy Wacaser earns take-home pay of $592 a week. If her deductions for taxes, retirement, union dues, and medical plan amount to 26% of her wages, what is her weekly pay before deductions?

70. Barbara Dalton gives 10% of her net income to the church. This amounts to $80 a month. In addition, her paycheck deductions are 24% of her gross monthly income. What is her gross monthly income?

Natural Science *Exercises 71 and 72 depend on the idea of the octane rating of gasoline, a measure of its antiknock qualities. Actual gasoline blends are compared to standard fuels. In one measure of octane, a standard fuel is made with only two ingredients: heptane and isooctane. For this fuel, the octane rating is the percent of isooctane; i.e., a gasoline with an octane rating of 98 has the same antiknock properties as a standard fuel that is 98% isooctane.*

71. How many liters of 94 octane gasoline should be mixed with 200 liters of 99 octane gasoline to get a mixture that is 97 octane?

72. A service station has 92 octane and 98 octane gasoline. How many liters of each gasoline should be mixed to provide 12 liters of 96 octane gasoline for a chemistry experiment?

1.3 POLYNOMIALS

We begin with exponents, whose properties are essential for understanding polynomials. You are familiar with the usual notation for squares and cubes:

$$5^2 = 5 \cdot 5 \quad \text{and} \quad 6^3 = 6 \cdot 6 \cdot 6.$$

We now extend this convenient notation to other cases.

If n is a natural number and a is any real number, then

a^n denotes the product $a \cdot a \cdot a \cdots a$ (n factors).

The number a is the **base** and n is the **exponent**.

EXAMPLE 1 4^6, which is read "four to the sixth" is the number

$$4 \cdot 4 \cdot 4 \cdot 4 \cdot 4 \cdot 4 = 4096.$$

Similarly, $(-5)^3 = (-5)(-5)(-5) = -125$ and

$$\left(\frac{3}{2}\right)^4 = \frac{3}{2} \cdot \frac{3}{2} \cdot \frac{3}{2} \cdot \frac{3}{2} = \frac{81}{16}. \quad \blacksquare$$

EXAMPLE 2 Use a calculator to approximate the following.

(a) $(1.2)^8$

Key in 1.2, then use the x^y key (labeled $\wedge$ on some calculators), and finally key in the exponent 8. The calculator displays the (exact) answer 4.29981696.

(b) $\left(\frac{12}{7}\right)^{23}$

Don't compute 12/7 separately. Use parentheses and key in (12/7), followed by the x^y key and the exponent 23 to obtain the approximate answer 242,054.822. $\quad \blacksquare$ 　 $\boxed{1}$

CAUTION A common error in using exponents occurs with expressions such as $4 \cdot 3^2$. The exponent 2 applies only to the base 3, so that

$$4 \cdot \mathbf{3^2} = 4 \cdot \mathbf{3} \cdot \mathbf{3} = 36.$$

On the other hand,

$$(\mathbf{4 \cdot 3})^2 = (\mathbf{4 \cdot 3})(\mathbf{4 \cdot 3}) = 12 \cdot 12 = 144,$$

and so

$$4 \cdot 3^2 \neq (4 \cdot 3)^2.$$

Be careful to distinguish between expressions like -2^4 and $(-2)^4$.

$$-2^4 = -(2^4) = -(2 \cdot 2 \cdot 2 \cdot 2) = -16$$
$$(\mathbf{-2})^4 = (\mathbf{-2})(\mathbf{-2})(\mathbf{-2})(\mathbf{-2}) = 16$$

and so

$$-2^4 \neq (-2)^4. \blacklozenge \quad \boxed{2}$$

By the definition of an exponent,

$$3^4 \cdot 3^2 = (3 \cdot 3 \cdot 3 \cdot 3)(3 \cdot 3) = 3^6.$$

This suggests the following property for the product of two powers of a number.

> If m and n are natural numbers and a is a real number, then
>
> $$a^m \cdot a^n = a^{m+n}.$$

---

1 Evaluate the following.

(a) 6^3

(b) 5^{12}

(c) 1^9

(d) $\left(\frac{7}{5}\right)^8$

Answers:

(a) 216

(b) 244,140,625

(c) 1

(d) 14.75789056

---

2 Evaluate the following.

(a) $3 \cdot 6^2$

(b) $5 \cdot 4^3$

(c) -3^6

(d) $(-3)^6$

(e) $-2 \cdot (-3)^5$

Answers:

(a) 108

(b) 320

(c) -729

(d) 729

(e) 486

3 Simplify the following.

(a) $5^3 \cdot 5^6$

(b) $(-3)^4 \cdot (-3)^{10}$

(c) $(5p)^2 \cdot (5p)^8$

Answers:

(a) 5^9

(b) $(-3)^{14}$

(c) $(5p)^{10}$

EXAMPLE 3   Simplify the following.

(a) $7^4 \cdot 7^6 = 7^{4+6} = 7^{10}$

(b) $(-2)^3 \cdot (-2)^5 = (-2)^{3+5} = (-2)^8$

(c) $(3k)^2 \cdot (3k)^3 = (3k)^5$

(d) $(m+n)^2 \cdot (m+n)^5 = (m+n)^7$ ■   **3**

POLYNOMIALS   A polynomial is an algebraic expression like

$$5x^4 + 2x^3 + 6x, \quad 8m^3 + 9m^2 - 6m + 3, \quad 10p, \quad \text{or} \quad -9.$$

More formally, a **polynomial** in one variable is an expression of the form

$$a_n x^n + a_{n-1} x^{n-1} + \cdots + a_1 x + a_0, \tag{1}$$

where n is a whole number, x is a **variable,** and $a_0, a_1, a_2, \ldots, a_n$ are real numbers (called the **coefficients** of the polynomial). For example, the polynomial

$$8x^3 + 9x^2 - 6x + 3$$

is of the form

$$a_n x^n + a_{n-1} x^{n-1} + \cdots + a_1 x + a_0,$$

with $n = 3$, $a_n = a_3 = 8$, $a_{n-1} = a_2 = 9$, $a_1 = -6$, and $a_0 = 3$. Each of the expressions $8x^3$, $9x^2$, $-6x$, and 3 is called a **term** of the polynomial $8x^3 + 9x^2 - 6x + 3$. The coefficient a_0 of a polynomial (for instance, 3 in the polynomial $8x^3 + 9x^2 - 6x + 3$) is called the **constant term.** Letters other than x may be used for the variable of a polynomial.

Only expressions that can be put in the form (1) are polynomials. Consequently, the following expressions are *not* polynomials:

$$8x^3 + \frac{6}{x}, \quad \frac{9+x}{2-x}, \quad \text{and} \quad \frac{-p^2 + 5p + 3}{2p - 1}.$$

The **degree of a nonzero term** with only one variable is the exponent on the variable. For example, the term $9p^4$ has degree 4. The **degree of a polynomial** is the highest degree of any of its nonzero terms. Thus, the degree of $-p^2 + 5p + 3$ is 2. The **zero polynomial** consists of the constant term 0 and no other terms. No degree is assigned to the zero polynomial because it has no nonzero terms. A polynomial with two terms, such as $5x + 2$ or $x^3 + 7$, is called a **binomial** and a polynomial with three terms, such as $3x^2 - 4x + 7$, is called a **trinomial.**

ADDITION AND SUBTRACTION   Two terms having the same variable with the same exponent are called **like terms;** other terms are called **unlike terms.** Polynomials can be added or subtracted by using the distributive property to combine like terms. Only like terms can be combined. For example,

$$12y^4 + 6y^4 = (12 + 6)y^4 = 18y^4$$

and

$$-2m^2 + 8m^2 = (-2 + 8)m^2 = 6m^2.$$

The polynomial $8y^4 + 2y^5$ has unlike terms, so it cannot be further simplified. Polynomials are subtracted using the fact that $a - b = a + (-b)$. The next example shows how to add and subtract polynomials by combining terms.

EXAMPLE 4   Add or subtract as indicated.

(a) $(8x^3 - 4x^2 + 6x) + (3x^3 + 5x^2 - 9x + 8)$
Combine like terms.

$(8x^3 - 4x^2 + 6x) + (3x^3 + 5x^2 - 9x + 8)$ Commutative property

$= (8x^3 + 3x^3) + (-4x^2 + 5x^2) + (6x - 9x) + 8$ and associative property

$= 11x^3 + x^2 - 3x + 8$ Distributive property

(b) $(-4x^4 + 6x^3 - 9x^2 - 12) + (-3x^3 + 8x^2 - 11x + 7)$
$= -4x^4 + 3x^3 - x^2 - 11x - 5$

(c) $(2x^2 - 11x + 8) - (7x^2 - 6x + 2)$
Use the definition of subtraction: $a - b = a + (-b)$. Here a and b are polynomials, and $-b$ is

$$-(7x^2 - 6x + 2) = -7x^2 + 6x - 2.$$

Now perform the subtraction.

$(2x^2 - 11x + 8) - (7x^2 - 6x + 2)$
$= (2x^2 - 11x + 8) + (-7x^2 + 6x - 2)$
$= -5x^2 - 5x + 6$ ■   4

MULTIPLICATION   The distributive property is also used to multiply polynomials. For example, the product of $8x$ and $6x - 4$ is found as follows.

$8x(6x - 4) = 8x(6x) - 8x(4)$ Distributive property
$= 48x^2 - 32x$ $x \cdot x = x^2$

EXAMPLE 5   Find each product.

(a) $2p^3(3p^2 - 2p + 5) = 2p^3(3p^2) + 2p^3(-2p) + 2p^3(5)$
$= 6p^5 - 4p^4 + 10p^3$

(b) $(3k - 2)(k^2 + 5k - 4) = 3k(k^2 + 5k - 4) - 2(k^2 + 5k - 4)$
$= 3k^3 + 15k^2 - 12k - 2k^2 - 10k + 8$
$= 3k^3 + 13k^2 - 22k + 8$ ■   5

---

4 Add or subtract.

(a) $(-2x^2 + 7x + 9)$
$+ (3x^2 + 2x - 7)$

(b) $(4x + 6) - (13x - 9)$

(c) $(9x^3 - 8x^2 + 2x)$
$- (9x^3 - 2x^2 - 10)$

Answers:
(a) $x^2 + 9x + 2$
(b) $-9x + 15$
(c) $-6x^2 + 2x + 10$

5 Find the following products.

(a) $-6r(2r - 5)$

(b) $(8m + 3)(m^4 - 2m^2 + 6m)$

Answers:
(a) $-12r^2 + 30r$
(b) $8m^5 + 3m^4 - 16m^3 + 42m^2 + 18m$

EXAMPLE 6 The product $(2x - 5)(3x + 4)$ can be found by using the distributive property twice.

$$(2x - 5)(3x + 4) = 2x(3x + 4) - 5(3x + 4)$$
$$= 2x \cdot 3x + 2x \cdot 4 + (-5) \cdot 3x + (-5) \cdot 4$$
$$= 6x^2 + \underbrace{8x - 15x} - 20$$
$$= 6x^2 - \quad 7x \quad - 20 \quad \blacksquare$$

Observe the pattern in the second line of Example 6 and its relationship to the terms being multiplied.

$$(\mathbf{2x} - 5)(\mathbf{3x} + 4) = 2x \cdot 3x + 2x \cdot 4 + (-5) \cdot 3x + (-5) \cdot 4$$

First terms

$(\mathbf{2x} - 5)(3x + \mathbf{4})$

Outside terms

$(2x \quad 5)(\mathbf{3x} + 4)$

Inside terms

$(2x \quad 5)(3x + \mathbf{4})$

Last terms

This pattern is easy to remember by using the acronym **FOIL** (**F**irst, **O**utside, **I**nside, **L**ast). The FOIL method makes it easy to find products such as this one mentally, without the necessity of writing out the intermediate steps.

EXAMPLE 7

$$(3x + 2)(x + 5) = \quad 3x^2 \quad + \quad 15x \quad + \quad 2x \quad + \quad 10 \quad = 3x^2 + 17x + 10$$
$$\uparrow \qquad \uparrow \qquad \uparrow \qquad \uparrow$$
$$\text{First} \quad \text{Outside} \quad \text{Inside} \quad \text{Last} \quad \blacksquare \quad \boxed{6}$$

The profit from the sales of an item is the difference between the revenue R received from the sales and the cost C to sell the item. Thus, profit P is found from the equation $P = R - C$.

EXAMPLE 8 Suppose the cost to sell x compact discs is $2x^2 - 2x + 10$ and the revenue from the sale of x discs is $5x^2 + 12x - 1$. Write an expression for the profit. Because profit equals revenue less cost, or $P = R - C$, the profit is

$$P = (5x^2 + 12x - 1) - (2x^2 - 2x + 10)$$
$$= 3x^2 + 14x - 11. \quad \blacksquare \quad \boxed{7}$$

**Side notes (left column):**

$\boxed{6}$ Use FOIL to find these products.

(a) $(5k - 1)(2k + 3)$

(b) $(7z - 3)(2z + 5)$

Answers:
(a) $10k^2 + 13k - 3$

(b) $14z^2 + 29z - 15$

$\boxed{7}$ Write an expression for profit if revenue is $7x^2 - 3x + 8$ and cost is $3x^2 + 5x - 2$.

Answer:
$4x^2 - 8x + 10$

1.3 EXERCISES

Use a calculator to evaluate the following, approximating when necessary. (See Examples 1 and 2.)

1. 17^9

2. $(-6.54)^{11}$

3. $(-18/7)^6$

4. $(7/9)^8$

5. Explain how the value of -3^2 differs from $(-3)^2$. Do -3^3 and $(-3)^3$ differ in the same way? Why or why not?

6. Describe the steps used to multiply 4^3 and 4^5. Is the product of 4^3 and 3^4 found in the same way? Explain.

Simplify each of the following. Leave answers with exponents. (See Example 3.)

7. $2^4 \cdot 2^3$

8. $3^8 \cdot 3^3$

9. $(-5)^2 \cdot (-5)^5$

10. $(-4)^4 \cdot (-4)^6$

11. $(-3)^5 \cdot 3^4$

12. $8^2 \cdot (-8)^3$

13. $(2z)^5 \cdot (2z)^6$

14. $(6y)^3 \cdot (6y)^5$

Add or subtract as indicated. (See Example 4.)

15. $(3x^3 + 2x^2 - 5x) + (-4x^3 - x^2 + 8x)$

16. $(-2p^3 - 5p + 7) + (-4p^2 + 8p + 2)$

17. $(-4y^2 - 3y + 8) - (2y^2 - 6y - 2)$

18. $(7b^2 + 2b - 5) - (3b^2 + 2b - 6)$

19. $(2x^3 - 2x^2 + 4x - 3) - (2x^3 + 8x^2 - 1)$

20. $(3y^3 + 9y^2 - 11y + 8) - (-4y^2 + 10y - 6)$

Find each of the following products and sums. (See Examples 5–7.)

21. $-9m(2m^2 + 3m - 1)$

22. $2a(4a^2 - 6a + 3)$

23. $(3z + 5)(4z^2 - 2z + 1)$

24. $(2k + 3)(4k^3 - 3k^2 + k)$

25. $(6k - 1)(2k - 3)$

26. $(8r + 3)(r - 1)$

27. $(3y + 5)(2y - 1)$

28. $(5r - 3s)(5r + 4s)$

29. $(9k + q)(2k - q)$

30. $(.012x - .17)(.3x + .54)$

31. $(6.2m - 3.4)(.7m + 1.3)$

32. $2p - 3[4p - (3p + 1)]$

33. $5k - [k + (-3 + 5k)]$

34. $(3x - 1)(x + 2) - (2x + 5)^2$

35. $(4x + 3)(2x - 1) - (x + 3)^2$

Management The accompanying bar graph depicts the number of North American users, in millions, of on-line services, as reported by Forrester Research, Inc. Using these figures, it can be determined that the polynomial

$$.035x^4 - .266x^3 + 1.005x^2 + .509x + 2.986$$

gives a good approximation of the number of users in the year x, where x = 0 corresponds to 1992, x = 1 corresponds to 1993, and so on. For the given year (a) use the bar graph to determine the number of users and then (b) use the polynomial to determine the number of users.

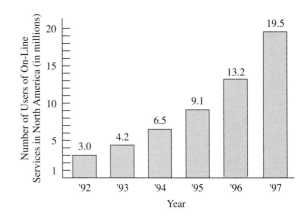

36. 1992

37. 1993

38. 1994

39. 1997

40. **Physical Science** One of the most amazing formulas in all of ancient mathematics is the formula discovered by the Egyptians to find the volume of the frustum of a square pyramid shown in the figure. Its volume is given by $(1/3)h(a^2 + ab + b^2)$, where b is the length of the base, a is the length of the top, and h is the height.*

(a) When the Great Pyramid in Egypt was partially completed to a height h of 200 feet, b was 756 feet, and a was 314 feet. Calculate its volume at this stage of construction.

(b) Try to visualize the figure if a = b. What is the resulting shape? Find its volume.

(c) Let a = b in the Egyptian formula and simplify. Are the results the same?

*Freebury, H. A., *A History of Mathematics*. MacMillan Company, New York, 1968.

41. Physical Science Refer to the formula and the discussion in Exercise 40.

(a) Use the expression $(1/3)h(a^2 + ab + b^2)$ to determine a formula for the volume of a pyramid with a square base b and height h by letting $a = 0$.

(b) The Great Pyramid in Egypt had a square base of 756 feet and a height of 481 feet. Find the volume of the Great Pyramid. Compare it with the 273-foot-tall Superdome in New Orleans, which has an approximate volume of 100 million cubic feet.*

(c) The Superdome covers an area of 13 acres. How many acres does the Great Pyramid cover? (*Hint:* 1 acre = 43,560 ft^2.)

42. Suppose one polynomial has degree 3 and another also has degree 3. Find all possible values for the degree of their

(a) sum
(b) difference
(c) product.

43. If one polynomial has degree 3 and another has degree 4, find all possible values for the degree of their

The Guinness Book of Records 1995.

(a) sum
(b) difference
(c) product.

44. Generalize the results of Exercise 43: Suppose one polynomial has degree m and another has degree n, where m and n are natural numbers with $n < m$. Find all possible values for the degree of their

(a) sum
(b) difference
(c) product.

45. Find $(a + b)(a - b)$. Then state in words a formula for the result of the product $(3x + 2y)(3x - 2y)$.

46. Find $(a - b)^2$. Then state in words a formula for finding $(2x - 5)^2$.

Management *Write an expression for profit given the following expressions for revenue and cost. (See Example 8.)*

47. Revenue: $5x^3 - 3x + 1$; cost: $4x^2 + 5x$

48. Revenue: $3x^3 + 2x^2$; cost: $x^3 - x^2 + x + 10$

49. Revenue: $2x^2 - 4x + 50$; cost: $x^2 + 3x + 10$

50. Revenue: $10x^2 + 8x + 12$; cost: $2x^2 - 3x + 20$

1.4 FACTORING

The number 18 can be written as a product in several ways: $9 \cdot 2$, $(-3)(-6)$, $1 \cdot 18$, etc. The numbers in each product (9, 2, -3, etc.) are called **factors** and the process of writing 18 as a product of factors is called **factoring.** Thus, factoring is the reverse of multiplication.

Factoring of polynomials is also important. It provides a means to simplify many situations and to solve certain types of equations. As is the usual custom, factoring of polynomials in this book will be restricted to finding factors with *integer* coefficients (otherwise there are an infinite number of possible factors).

GREATEST COMMON FACTOR The algebraic expression $15m + 45$ is made up of two terms, $15m$ and 45. Each of these terms can be divided by 15. In fact, $15m = 15 \cdot m$ and $45 = 15 \cdot 3$. By the distributive property,

$$15m + 45 = \mathbf{15} \cdot m + \mathbf{15} \cdot 3 = \mathbf{15}(m + 3).$$

Both 15 and $m + 3$ are factors of $15m + 45$. Since 15 divides into all terms of $15m + 45$ and is the largest number that will do so, it is called the **greatest common factor** for the polynomial $15m + 45$. The process of writing $15m + 45$ as $15(m + 3)$ is called **factoring out** the greatest common factor.

EXAMPLE 1 Factor out the greatest common factor.

(a) $12p - 18q$

Both $12p$ and $18q$ are divisible by 6, so

$$12p - 18q = 6 \cdot 2p - 6 \cdot 3q$$
$$= 6(2p - 3q).$$

(b) $8x^3 - 9x^2 + 15x$

Each of these terms is divisible by x.

$$8x^3 - 9x^2 + 15x = (8x^2) \cdot x - (9x) \cdot x + 15 \cdot x$$
$$= x(8x^2 - 9x + 15)$$

(c) $5(4x - 3)^3 + 2(4x - 3)^2$

The quantity $(4x - 3)^2$ is a common factor. Factoring it out gives

$$5(4x - 3)^3 + 2(4x - 3)^2 = (4x - 3)^2[5(4x - 3) + 2]$$
$$= (4x - 3)^2(20x - 15 + 2)$$
$$= (4x - 3)^2(20x - 13). \quad \blacksquare \quad \boxed{1}$$

$\boxed{1}$ Factor out the greatest common factor.

(a) $12r + 9k$

(b) $75m^2 + 100n^2$

(c) $6m^4 - 9m^3 + 12m^2$

(d) $3(2k + 1)^3 + 4(2k + 1)^4$

Answers:

(a) $3(4r + 3k)$

(b) $25(3m^2 + 4n^2)$

(c) $3m^2(2m^2 - 3m + 4)$

(d) $(2k + 1)^3(7 + 8k)$

FACTORING TRINOMIALS Factoring is the opposite of multiplication. Because the product of two binomials is usually a trinomial, we can expect factorable trinomials (that have terms with no common factor) to have two binomial factors. Thus, factoring trinomials requires using FOIL backwards.

EXAMPLE 2 Factor each trinomial.

(a) $4y^2 - 11y + 6$

To factor this trinomial, we must find integers a, b, c, and d such that

$$4y^2 - 11y + 6 = (ay + b)(cy + d)$$
$$= acy^2 + ady + bcy + bd$$
$$= acy^2 + (ad + bc)y + bd.$$

Since the coefficients of y^2 must be the same on both sides, we see that $ac = 4$. Similarly, the constant terms show that $bd = 6$. The positive factors of 4 are 4 and 1 or 2 and 2. Since the middle term is negative, we consider only negative factors of 6. The possibilities are -2 and -3 or -1 and -6. Now we try various arrangements of these factors until we find one that gives the correct coefficient of y.

$$(2y - 1)(2y - 6) = 4y^2 - \mathbf{14y} + 6 \qquad \text{Incorrect}$$
$$(2y - 2)(2y - 3) = 4y^2 - \mathbf{10y} + 6 \qquad \text{Incorrect}$$
$$(y - 2)(4y - 3) = 4y^2 - \mathbf{11y} + 6 \qquad \text{Correct}$$

The last trial gives the correct factorization.

[2] Factor the following.

(a) $r^2 - 5r - 14$

(b) $3m^2 + 5m - 2$

(c) $6p^2 + 13pq - 5q^2$

Answers:

(a) $(r - 7)(r + 2)$

(b) $(3m - 1)(m + 2)$

(c) $(2p + 5q)(3p - q)$

(b) $6p^2 - 7pq - 5q^2$

Again, we try various possibilities. The positive factors of 6 could be 2 and 3 or 1 and 6. As factors of -5 we have only -1 and 5 or -5 and 1. Try different combinations of these factors until the correct one is found.

$$(2p - 5q)(3p + q) = 6p^2 - \mathbf{13pq} - 5q^2 \quad \text{Incorrect}$$
$$(3p - 5q)(2p + q) = 6p^2 - \mathbf{7pq} - 5q^2 \quad \text{Correct}$$

Finally, $6p^2 - 7pq - 5q^2$ factors as $(3p - 5q)(2p + q)$. ■ [2]

N O T E In Example 2, we chose positive factors of the positive first term. Of course, we could have used two negative factors, but the work is easier if positive factors are used. ◆

The method shown above can be used to factor a **perfect square trinomial,** one that is the square of a binomial. The binomial can be predicted by observing the following patterns that always apply to a perfect square trinomial.

$$\left. \begin{array}{l} x^2 + 2xy + y^2 = (x + y)^2 \\ x^2 - 2xy + y^2 = (x - y)^2 \end{array} \right\} \quad \text{Perfect square trinomials}$$

EXAMPLE 3 Factor each trinomial.

(a) $16p^2 - 40pq + 25q^2$

Because $16p^2 = (4p)^2$ and $25q^2 = (5q)^2$, use the second pattern shown above with $4p$ replacing x and $5q$ replacing y to get

$$16p^2 - 40pq + 25q^2 = (\mathbf{4p})^2 - 2(\mathbf{4p})(\mathbf{5q}) + (\mathbf{5q})^2$$
$$= (4p - 5q)^2.$$

[3] Factor each trinomial.

(a) $4m^2 + 4m + 1$

(b) $25z^2 - 80zt + 64t^2$

Answers:

(a) $(2m + 1)^2$

(b) $(5z - 8t)^2$

Make sure that the middle term of the trinomial being factored, $-40pq$ here, is twice the product of the two terms in the binomial $4p - 5q$.

$$-40pq = 2(\mathbf{4p})(-\mathbf{5q})$$

(b) $169x^2 + 104xy^2 + 16y^4 = (13x + 4y^2)^2$, since $2(13x)(4y^2) = 104xy^2$. ■ [3]

FACTORING BINOMIALS Three special factoring patterns are listed below. Each can be verified by multiplying on the right side of the equation. These formulas should be memorized.

$$x^2 - y^2 = (x + y)(x - y) \quad \text{Difference of two squares}$$
$$x^3 - y^3 = (x - y)(x^2 + xy + y^2) \quad \text{Difference of two cubes}$$
$$x^3 + y^3 = (x + y)(x^2 - xy + y^2) \quad \text{Sum of two cubes}$$

EXAMPLE 4   Factor each of the following.

(a) $4m^2 - 9$

Notice that $4m^2 - 9$ is the difference of two squares, since $4m^2 = (2m)^2$ and $9 = 3^2$. Use the pattern for the difference of two squares, letting $2m$ replace x and 3 replace y. Then the pattern $x^2 - y^2 = (x + y)(x - y)$ becomes

$$4m^2 - 9 = (2m)^2 - 3^2$$
$$= (2m + 3)(2m - 3).$$

(b) $128p^2 - 98q^2$

First factor out the common factor of 2.

$$128p^2 - 98q^2 = 2(64p^2 - 49q^2)$$
$$= 2[(8p)^2 - (7q)^2]$$
$$= 2(8p + 7q)(8p - 7q)$$

(c) $x^2 + 36$

The *sum* of two squares usually cannot be factored. To see this, check some possibilities.

$$(x + 6)(x + 6) = (x + 6)^2 = x^2 + 12x + 36$$
$$(x + 4)(x + 9) = x^2 + 13x + 36$$

Any product of two binomials will always have a middle term unless it is the *difference* of two squares.

(d) $4z^2 + 12z + 9 - w^2$

Notice that the first three terms can be factored as a perfect square.

$$4z^2 + 12z + 9 - w^2 = (2z + 3)^2 - w^2$$

Written in this form, the expression is the difference of squares, which can be factored as

$$(2z + 3)^2 - w^2 = [(2z + 3) + w][(2z + 3) - w]$$
$$= (2z + 3 + w)(2z + 3 - w).$$

(e) $256k^4 - 625m^4$

Use the difference of two squares pattern twice, as follows:

$$256k^4 - 625m^4 = (16k^2)^2 - (25m^2)^2$$
$$= (16k^2 + 25m^2)(16k^2 - 25m^2)$$
$$= (16k^2 + 25m^2)(4k + 5m)(4k - 5m).$$ ∎ 4

4 Factor the following.

(a) $9p^2 - 49$

(b) $y^2 + 100$

(c) $9r^2 + 12r + 4 - t^2$

(d) $81x^4 - 16y^4$

Answers:

(a) $(3p + 7)(3p - 7)$

(b) Cannot be factored

(c) $(3r + 2 + t)(3r + 2 - t)$

(d) $(9x^2 + 4y^2)(3x + 2y) \cdot (3x - 2y)$

EXAMPLE 5   Factor each of the following.

(a) $k^3 - 8$

Use the pattern for the difference of two cubes, since $k^3 = (k)^3$ and $8 = (2)^3$, to get

$$k^3 - 8 = k^3 - 2^3 = (k - 2)(k^2 + 2k + 4).$$

⑤ Factor the following.

(a) $a^3 + 1000$

(b) $z^3 - 64$

(c) $100m^3 - 27z^3$

Answers:

(a) $(a + 10)(a^2 - 10a + 100)$

(b) $(z - 4)(z^2 + 4z + 16)$

(c) $(10m - 3z) \cdot$
$(100m^2 + 30mz + 9z^2)$

⑥ Factor.

(a) $6x^2 - 27x - 15$

(b) $18 - 8xy - 2y^2 - 8x^2$

Answers:

(a) $3(2x + 1)(x - 5)$

(b) $2(3 - 2x - y)(3 + 2x + y)$

(b) $m^3 + 125 = m^3 + 5^3 = (m + 5)(m^2 - 5m + 25)$

(c) $8k^3 - 27z^3 = (2k)^3 - (3z)^3 = (2k - 3z)(4k^2 + 6kz + 9z^2)$ ■ ⑤

EXAMPLE 6 Factor each of the following.

(a) $12x^2 - 26x - 10$

Look first for a common factor. Here there is a common factor of 2: $12x^2 - 26x - 10 = 2(6x^2 - 13x - 5)$. Now try to factor $6x^2 - 13x - 5$. Possible factors of 6 are 3 and 2 or 6 and 1. The only factors of -5 are -5 and 1 or 5 and -1. Try various combinations. You should find the trinomial factors as $(3x + 1)(2x - 5)$. Thus,

$$12x^2 - 26x - 10 = 2(3x + 1)(2x - 5).$$

(b) $16a^2 - 100 - 48ac + 36c^2$

Factor out the common factor of 4 first.

$$16a^2 - 100 - 48ac + 36c^2 = 4[4a^2 - 25 - 12ac + 9c^2]$$
$$= 4[(4a^2 - 12ac + 9c^2) - 25] \quad \text{Rearrange terms and group.}$$
$$= 4[(2a - 3c)^2 - 25] \quad \text{Factor the trinomial.}$$
$$= 4(2a - 3c + 5)(2a - 3c - 5) \quad \text{Factor the difference of squares.} ■$$

CAUTION Remember always to look first for a common factor. ◆ ⑥

1.4 EXERCISES

Factor out the greatest common factor in each of the following. (See Example 1.)

1. $12x^2 - 24x$

2. $5y - 25xy$

3. $r^3 - 5r^2 + r$

4. $t^3 + 3t^2 + 8t$

5. $2m - 5n + p$

6. $4k + 6h - 5c$

7. $6z^3 - 12z^2 + 18z$

8. $5x^3 + 35x^2 + 10x$

9. $25p^4 - 20p^3q + 100p^2q^2$

10. $60m^4 - 120m^3n + 50m^2n^2$

11. $3(2y - 1)^2 + 5(2y - 1)^3$

12. $(3x + 7)^5 - 2(3x + 7)^3$

13. $3(x + 5)^4 + (x + 5)^6$

14. $3(x + 6)^2 + 2(x + 6)^4$

Factor each of the following completely. Factor out the greatest common factor as necessary. (See Examples 2–4 and 6.)

15. $2a^2 + 3a - 5$

16. $6a^2 - 48a - 120$

17. $x^2 - 64$

18. $x^2 + 17xy + 72y^2$

19. $9p^2 - 24p + 16$

20. $3r^2 - r - 2$

21. $r^2 - 3rt - 10t^2$

22. $2a^2 + ab - 6b^2$

23. $m^2 - 6mn + 9n^2$

24. $8k^2 - 16k - 10$

25. $4p^2 - 9$

26. $8r^2 + r + 6$

27. $3x^2 - 24xz + 48z^2$

28. $9m^2 - 25$

29. $a^2 + 4ab + 5b^2$

30. $6y^2 - 11y - 7$

31. $-x^2 + 7x - 12$

32. $4y^2 + y - 3$

33. $3a^2 - 13a - 30$

34. $3k^2 + 2k - 8$

35. $21m^2 + 13mn + 2n^2$

36. $81y^2 - 100$

37. $20y^2 + 39yx - 11x^2$

38. $12s^2 + 11st - 5t^2$

39. $64z^2 + 25$

40. $p^2q^2 - 10 - 2q^2 + 5p^2$

41. $y^2 - 4yz - 21z^2$

42. $49a^2 + 9$

43. $3n^2 - 4m^2 + m^2n^2 - 12$

44. $y^2 + 20yx + 100x^2$

45. $121x^2 - 64$

46. $4z^2 + 56zy + 196y^2$

47. $24a^4 + 10a^3b - 4a^2b^2$

48. $10x^2 + 34x + 12$

49. $18x^5 + 15x^4z - 75x^3z^2$

50. $16m^2 + 40m + 25$

51. $5m^3(m^3 - 1)^2 - 3m^5(m^3 - 1)^3$

52. $9(x - 4)^5 - (x - 4)^3$

53. When asked to factor $6x^4 - 3x^2 - 3$ completely, a student gave the following result:

$$6x^4 - 3x^2 - 3 = (2x^2 + 1)(3x^2 - 3).$$

Is this answer correct? Explain why.

54. When can the sum of two squares be factored? Give examples.

Factor each of the following. (See Example 5.)

55. $a^3 - 216$ **56.** $b^3 + 125$

57. $8r^3 - 27s^3$
58. $1000p^3 + 27q^3$
59. $64m^3 + 125$
60. $216y^3 - 343$
61. $1000y^3 - z^3$
62. $125p^3 + 8q^3$

63. Explain why $(x + 2)^3$ is not the correct factorization of $x^3 + 8$ and give the correct factorization.

64. Describe how factoring and multiplication are related. Give examples.

1.5 RATIONAL EXPRESSIONS

We now consider **rational expressions,** such as

$$\frac{8}{x - 1}, \quad \frac{3x^2 + 4x}{5x - 6}, \quad \text{and} \quad \frac{2 + \dfrac{1}{y}}{y}.$$

Because rational expressions involve quotients, it is important to keep in mind values of the variables that make denominators 0. For example, 1 cannot be used as a replacement for x in the first rational expression above, and 6/5 cannot be used in the second one, since these values make the respective denominators equal 0. ⬛1

OPERATIONS WITH RATIONAL EXPRESSIONS The rules for operations with rational expressions are the usual rules for fractions.

⬛1 What values of the variable make each denominator equal 0?

(a) $\dfrac{5}{x - 3}$

(b) $\dfrac{2x - 3}{4x - 1}$

(c) $\dfrac{x + 2}{x}$

(d) Why do we need to determine these values?

Answers:
(a) 3
(b) 1/4
(c) 0
(d) Because division by 0 is undefined.

OPERATIONS WITH RATIONAL EXPRESSIONS

For all mathematical expressions P, $Q \neq 0$, R, and $S \neq 0$

(a) $\dfrac{P}{Q} = \dfrac{PS}{QS}$ Fundamental property

(b) $\dfrac{P}{Q} \cdot \dfrac{R}{S} = \dfrac{PR}{QS}$ Multiplication

(c) $\dfrac{P}{Q} + \dfrac{R}{Q} = \dfrac{P + R}{Q}$ Addition

(d) $\dfrac{P}{Q} - \dfrac{R}{Q} = \dfrac{P - R}{Q}$ Subtraction

(e) $\dfrac{P}{Q} \div \dfrac{R}{S} = \dfrac{P}{Q} \cdot \dfrac{S}{R}, R \neq 0.$ Division

---

The following examples illustrate these operations.

EXAMPLE 1 Write each of the following rational expressions in lowest terms (so that the numerator and denominator have no common factor with integer coefficients except 1 or -1).

(a) $\dfrac{12m}{-18}$

2 Write each of the following in lowest terms.

(a) $\dfrac{12k + 36}{18}$

(b) $\dfrac{15m + 30m^2}{5m}$

(c) $\dfrac{2p^2 + 3p + 1}{p^2 + 3p + 2}$

Answers:

(a) $\dfrac{2(k + 3)}{3}$ or $\dfrac{2k + 6}{3}$

(b) $3(1 + 2m)$ or $3 + 6m$

(c) $\dfrac{2p + 1}{p + 2}$

Both $12m$ and -18 are divisible by 6. By operation (a) above,

$$\frac{12m}{-18} = \frac{2m \cdot 6}{-3 \cdot 6}$$
$$= \frac{2m}{-3}$$
$$= -\frac{2m}{3}.$$

(b) $\dfrac{8x + 16}{4} = \dfrac{8(x + 2)}{4} = \dfrac{4 \cdot 2(x + 2)}{4} = \dfrac{2(x + 2)}{1} = 2(x + 2)$

The numerator, $8x + 16$, was factored so that the common factor could be identified. The answer could also be written as $2x + 4$, if desired.

(c) $\dfrac{k^2 + 7k + 12}{k^2 + 2k - 3} = \dfrac{(k + 4)(k + 3)}{(k - 1)(k + 3)} = \dfrac{k + 4}{k - 1}$ ■   **2**

The values of k in Example 1(c) are restricted to $k \neq 1$ and $k \neq -3$. From now on, such restrictions will be assumed when working with rational expressions.

EXAMPLE 2   (a) Multiply $\dfrac{2}{3} \cdot \dfrac{y}{5}$.

Multiply the numerators and then the denominators.

$$\frac{2}{3} \cdot \frac{y}{5} = \frac{2 \cdot y}{3 \cdot 5} = \frac{2y}{15}$$

The result, $2y/15$, is in lowest terms.

(b) $\dfrac{3y + 9}{6} \cdot \dfrac{18}{5y + 15}$

Factor where possible.

$$\frac{3y + 9}{6} \cdot \frac{18}{5y + 15} = \frac{3(y + 3)}{6} \cdot \frac{18}{5(y + 3)}$$
$$= \frac{3 \cdot 18(y + 3)}{6 \cdot 5(y + 3)} \quad \text{Multiply numerators and denominators.}$$
$$= \frac{3 \cdot 6 \cdot 3(y + 3)}{6 \cdot 5(y + 3)} \quad 18 = 6 \cdot 3$$
$$= \frac{3 \cdot 3}{5} \quad \text{Write in lowest terms.}$$
$$= \frac{9}{5}$$

(c) $\dfrac{m^2 + 5m + 6}{m + 3} \cdot \dfrac{m^2 + m - 6}{m^2 + 3m + 2}$

$$= \frac{(m + 2)(m + 3)}{m + 3} \cdot \frac{(m - 2)(m + 3)}{(m + 2)(m + 1)} \quad \text{Factor.}$$

3 Multiply.

(a) $\dfrac{3r^2}{5} \cdot \dfrac{20}{9r}$

(b) $\dfrac{y-4}{y^2-2y-8} \cdot \dfrac{y^2-4}{3y}$

Answers:

(a) $\dfrac{4r}{3}$

(b) $\dfrac{y-2}{3y}$

$$= \frac{(m+2)(m+3)(m-2)(m+3)}{(m+3)(m+2)(m+1)} \qquad \text{Multiply.}$$

$$= \frac{(m-2)(m+3)}{m+1} \qquad \text{Lowest terms}$$

$$= \frac{m^2+m-6}{m+1} \qquad \blacksquare \quad \boxed{3}$$

EXAMPLE 3   (a) Divide $\dfrac{8x}{5} \div \dfrac{11x^2}{20}$.

Invert the second expression and multiply.

$$\frac{8x}{5} \div \frac{11x^2}{20} = \frac{8x}{5} \cdot \frac{20}{11x^2} \qquad \text{Invert and multiply.}$$

$$= \frac{8x \cdot 20}{5 \cdot 11x^2} \qquad \text{Multiply.}$$

$$= \frac{32}{11x} \qquad \text{Lowest terms}$$

4 Divide.

(a) $\dfrac{5m}{16} \div \dfrac{m^2}{10}$

(b) $\dfrac{2y-8}{6} \div \dfrac{5y-20}{3}$

(c) $\dfrac{m^2-2m-3}{m(m+1)} \div \dfrac{m+4}{5m}$

Answers:

(a) $\dfrac{25}{8m}$

(b) $\dfrac{1}{5}$

(c) $\dfrac{5(m-3)}{m+4}$

(b) $\dfrac{9p-36}{12} \div \dfrac{5(p-4)}{18}$

$$= \frac{9p-36}{12} \cdot \frac{18}{5(p-4)} \qquad \text{Invert and multiply.}$$

$$= \frac{9(p-4)}{12} \cdot \frac{18}{5(p-4)} \qquad \text{Factor.}$$

$$= \frac{27}{10} \qquad \text{Multiply and write in lowest terms.} \quad \blacksquare \quad \boxed{4}$$

EXAMPLE 4   Add or subtract as indicated.

(a) $\dfrac{4}{5k} - \dfrac{11}{5k}$

When two rational expressions have the same denominators, subtract by subtracting the numerators and keeping the common denominator.

$$\frac{4}{5k} - \frac{11}{5k} = \frac{4-11}{5k} = -\frac{7}{5k}$$

(b) $\dfrac{7}{p} + \dfrac{9}{2p} + \dfrac{1}{3p}$

These three denominators are different; addition requires the same denominators. Find a common denominator, one which can be divided by p, $2p$, and $3p$. A common denominator here is $6p$. Rewrite each rational expression, using operation (a), with a denominator of $6p$. Then, using operation (c), add the numerators and keep the common denominator.

$$\frac{7}{p} + \frac{9}{2p} + \frac{1}{3p} = \frac{6 \cdot 7}{6 \cdot p} + \frac{3 \cdot 9}{3 \cdot 2p} + \frac{2 \cdot 1}{2 \cdot 3p} \qquad \text{Operation (a)}$$

$$= \frac{42}{6p} + \frac{27}{6p} + \frac{2}{6p}$$

$$= \frac{42 + 27 + 2}{6p} \qquad \text{Operation (c)}$$

$$= \frac{71}{6p}$$

(c) $\dfrac{k^2}{k^2 - 1} - \dfrac{2k^2 - k - 3}{k^2 + 3k + 2}$

Factor the denominators to find a common denominator.

$$\frac{k^2}{k^2 - 1} - \frac{2k^2 - k - 3}{k^2 + 3k + 2} = \frac{k^2}{(k+1)(k-1)} - \frac{2k^2 - k - 3}{(k+1)(k+2)}$$

The common denominator is $(k + 1)(k - 1)(k + 2)$. Write each fraction with the common denominator.

$$\frac{k^2}{(k+1)(k-1)} - \frac{2k^2 - k - 3}{(k+1)(k+2)}$$

$$= \frac{k^2(k+2)}{(k+1)(k-1)(k+2)} - \frac{(2k^2 - k - 3)(k-1)}{(k+1)(k-1)(k+2)}$$

$$= \frac{k^3 + 2k^2 - (2k^2 - k - 3)(k-1)}{(k+1)(k-1)(k+2)} \qquad \text{Subtract fractions.}$$

$$= \frac{k^3 + 2k^2 - (2k^3 - 3k^2 - 2k + 3)}{(k+1)(k-1)(k+2)} \qquad \begin{array}{l}\text{Multiply} \\ (2k^2 - k - 3)(k - 1).\end{array}$$

$$= \frac{k^3 + 2k^2 - 2k^3 + 3k^2 + 2k - 3}{(k+1)(k-1)(k+2)} \qquad \text{Polynomial subtraction}$$

$$= \frac{-k^3 + 5k^2 + 2k - 3}{(k+1)(k-1)(k+2)} \qquad \text{Combine terms.} \qquad \blacksquare \quad \boxed{5}$$

COMPLEX FRACTIONS Any quotient of two rational expressions is called a **complex fraction.** Complex fractions can be simplified by the methods shown in the following examples.

EXAMPLE 5 Simplify each complex fraction.

(a) $\dfrac{6 - \dfrac{5}{k}}{1 + \dfrac{5}{k}}$

$\boxed{5}$ Add or subtract.

(a) $\dfrac{3}{4r} + \dfrac{8}{3r}$

(b) $\dfrac{1}{m - 2} - \dfrac{3}{2(m - 2)}$

(c) $\dfrac{p + 1}{p^2 - p} - \dfrac{p^2 - 1}{p^2 + p - 2}$

Answers:

(a) $\dfrac{41}{12r}$

(b) $\dfrac{-1}{2(m - 2)}$

(c) $\dfrac{-p^3 + p^2 + 4p + 2}{p(p - 1)(p + 2)}$

Multiply both numerator and denominator by the common denominator k.

$$\frac{6 - \dfrac{5}{k}}{1 + \dfrac{5}{k}} = \frac{k\left(6 - \dfrac{5}{k}\right)}{k\left(1 + \dfrac{5}{k}\right)} \qquad \text{Multiply by } \dfrac{k}{k}.$$

$$= \frac{6k - k\left(\dfrac{5}{k}\right)}{k + k\left(\dfrac{5}{k}\right)} \qquad \text{Distributive property}$$

$$= \frac{6k - 5}{k + 5} \qquad \text{Simplify.}$$

(b) $\dfrac{\dfrac{a}{a + 1} + \dfrac{1}{a}}{\dfrac{1}{a} + \dfrac{1}{a + 1}}$

Multiply both numerator and denominator by the common denominator of all the fractions, in this case $a(a + 1)$. Doing so gives

| 6 | Simplify each complex fraction.

(a) $\dfrac{t - \dfrac{1}{t}}{2t + \dfrac{3}{t}}$

(b) $\dfrac{\dfrac{m}{m + 2} + \dfrac{1}{m}}{\dfrac{1}{m} - \dfrac{1}{m + 2}}$

Answers:

(a) $\dfrac{t^2 - 1}{2t^2 + 3}$

(b) $\dfrac{m^2 + m + 2}{2}$

$$\frac{\dfrac{a}{a + 1} + \dfrac{1}{a}}{\dfrac{1}{a} + \dfrac{1}{a + 1}} = \frac{\left(\dfrac{a}{a + 1} + \dfrac{1}{a}\right)a(a + 1)}{\left(\dfrac{1}{a} + \dfrac{1}{a + 1}\right)a(a + 1)}$$

$$= \frac{a^2 + (a + 1)}{(a + 1) + a} = \frac{a^2 + a + 1}{2a + 1}.$$

As an alternative method of solution, first perform the indicated additions in the numerator and denominator, and then divide.

$$\frac{\dfrac{a}{a + 1} + \dfrac{1}{a}}{\dfrac{1}{a} + \dfrac{1}{a + 1}} = \frac{\dfrac{a^2 + 1(a + 1)}{a(a + 1)}}{\dfrac{1(a + 1) + 1(a)}{a(a + 1)}} = \frac{\dfrac{a^2 + a + 1}{a(a + 1)}}{\dfrac{2a + 1}{a(a + 1)}}$$

$$= \frac{a^2 + a + 1}{a(a + 1)} \cdot \frac{a(a + 1)}{2a + 1} = \frac{a^2 + a + 1}{2a + 1} \qquad \blacksquare \quad \boxed{6}$$

1.5 EXERCISES

Write each of the following in lowest terms. Factor as necessary. (See Example 1.)

1. $\dfrac{8x^2}{40x}$

2. $\dfrac{27m}{81m^3}$

3. $\dfrac{20p^2}{35p^3}$

4. $\dfrac{18y^4}{27y^2}$

5. $\dfrac{5m + 15}{4m + 12}$

6. $\dfrac{10z + 5}{20z + 10}$

7. $\dfrac{4(w - 3)}{(w - 3)(w + 3)}$

8. $\dfrac{-6(x + 2)}{(x - 4)(x + 2)}$

9. $\dfrac{3y^2 - 12y}{9y^3}$

10. $\dfrac{15k^2 + 45k}{9k^2}$

11. $\dfrac{8x^2 + 16x}{4x^2}$

12. $\dfrac{36y^2 + 72y}{9y}$

13. $\dfrac{m^2 - 4m + 4}{m^2 + m - 6}$

14. $\dfrac{r^2 - r - 6}{r^2 + r - 12}$

15. $\dfrac{x^2 + 3x - 4}{x^2 - 1}$

16. $\dfrac{z^2 - 5z + 6}{z^2 - 4}$

Multiply or divide as indicated in each of the following. Write all answers in lowest terms. (See Examples 2 and 3.)

17. $\dfrac{4p^3}{49} \cdot \dfrac{7}{2p^2}$

18. $\dfrac{24n^4}{6n^2} \cdot \dfrac{18n^2}{9n}$

19. $\dfrac{21a^5}{14a^3} \div \dfrac{8a}{12a^2}$

20. $\dfrac{2x^3}{6x^2} \div \dfrac{10x^2}{15x}$

21. $\dfrac{2a + b}{2c} \cdot \dfrac{15}{4(2a + b)}$

22. $\dfrac{4(x + 2)}{w} \cdot \dfrac{3w}{8(x + 2)}$

23. $\dfrac{15p - 3}{6} \div \dfrac{10p - 2}{3}$

24. $\dfrac{6m - 18}{18} \cdot \dfrac{20}{4m - 12}$

25. $\dfrac{2k + 8}{6} \div \dfrac{3k + 12}{2}$

26. $\dfrac{5m + 25}{10} \cdot \dfrac{12}{6m + 30}$

27. $\dfrac{9y - 18}{6y + 12} \cdot \dfrac{3y + 6}{15y - 30}$

28. $\dfrac{12r + 24}{36r - 36} \div \dfrac{6r + 12}{8r - 8}$

29. $\dfrac{4a + 12}{2a - 10} \div \dfrac{a^2 - 9}{a^2 - a - 20}$

30. $\dfrac{6r - 18}{9r^2 + 6r - 24} \cdot \dfrac{12r - 16}{4r - 12}$

31. $\dfrac{k^2 - k - 6}{k^2 + k - 12} \cdot \dfrac{k^2 + 3k - 4}{k^2 + 2k - 3}$

32. $\dfrac{n^2 - n - 6}{n^2 - 2n - 8} \div \dfrac{n^2 - 9}{n^2 + 7n + 12}$

33. In your own words, explain how to find the least common denominator for two fractions.

34. Describe the steps required to add three rational expressions. You may use an example to illustrate.

Add or subtract as indicated in each of the following. Write all answers in lowest terms. (See Example 4.)

35. $\dfrac{3}{5z} - \dfrac{2}{3z}$

36. $\dfrac{7}{4z} - \dfrac{5}{3z}$

37. $\dfrac{r + 2}{3} - \dfrac{r - 2}{3}$

38. $\dfrac{3y - 1}{8} - \dfrac{3y + 1}{8}$

39. $\dfrac{4}{x} + \dfrac{1}{3}$

40. $\dfrac{6}{r} - \dfrac{3}{4}$

41. $\dfrac{2}{y} - \dfrac{1}{4}$

42. $\dfrac{6}{11} + \dfrac{3}{a}$

43. $\dfrac{1}{6m} + \dfrac{2}{5m} + \dfrac{4}{m}$

44. $\dfrac{8}{3p} + \dfrac{5}{4p} + \dfrac{9}{2p}$

45. $\dfrac{1}{m - 1} + \dfrac{2}{m}$

46. $\dfrac{8}{y + 2} - \dfrac{3}{y}$

47. $\dfrac{8}{3(a - 1)} + \dfrac{2}{a - 1}$

48. $\dfrac{5}{2(k + 3)} + \dfrac{2}{k + 3}$

49. $\dfrac{2}{5(k - 2)} + \dfrac{3}{4(k - 2)}$

50. $\dfrac{11}{3(p + 4)} - \dfrac{5}{6(p + 4)}$

51. $\dfrac{2}{x^2 - 2x - 3} + \dfrac{5}{x^2 - x - 6}$

52. $\dfrac{3}{m^2 - 3m - 10} + \dfrac{5}{m^2 - m - 20}$

53. $\dfrac{2y}{y^2 + 7y + 12} - \dfrac{y}{y^2 + 5y + 6}$

54. $\dfrac{-r}{r^2 - 10r + 16} - \dfrac{3r}{r^2 + 2r - 8}$

55. $\dfrac{3k}{2k^2 + 3k - 2} - \dfrac{2k}{2k^2 - 7k + 3}$

56. $\dfrac{4m}{3m^2 + 7m - 6} - \dfrac{m}{3m^2 - 14m + 8}$

In each of the following exercises, simplify the complex fraction. (See Example 5.)

57. $\dfrac{1 + \dfrac{1}{x}}{1 - \dfrac{1}{x}}$

58. $\dfrac{2 - \dfrac{2}{y}}{2 + \dfrac{2}{y}}$

59. $\dfrac{\dfrac{1}{x + h} - \dfrac{1}{x}}{h}$

60. $\dfrac{\dfrac{1}{(x + h)^2} - \dfrac{1}{x^2}}{h}$

61. $\dfrac{1 + \dfrac{1}{1 - b}}{1 - \dfrac{1}{1 + b}}$

62. $\dfrac{m - \dfrac{1}{m^2 - 4}}{\dfrac{1}{m + 2}}$

1.6 EXPONENTS AND RADICALS

Exponents were introduced in Section 1.3. In this section the definition of exponents will be extended to include negative exponents and rational number exponents, such as 1/2 and 7/3.

INTEGER EXPONENTS   In Section 1.3 we defined positive integer exponents and showed that $a^m \cdot a^n = a^{m+n}$ for positive integer values of m and n. Now we develop an analogous property for quotients. By definition,

$$\frac{6^5}{6^2} = \frac{6 \cdot 6 \cdot 6 \cdot 6 \cdot 6}{6 \cdot 6}$$
$$= 6 \cdot 6 \cdot 6$$
$$= 6^3.$$

Because there are 5 factors of 6 in the numerator and 2 factors of 6 in the denominator, the quotient has $5 - 2 = 3$ factors of 6. In general,

> If a is a nonzero real number and m, n are positive integers with $m > n$, then
> $$\frac{a^m}{a^n} = a^{m-n}.$$

Next, we want to give a meaning to expressions such as 3^0. If the quotient property in the preceding box is to continue to be valid, we must define 3^0 in such a way that

$$\frac{3^5}{3^5} = 3^{5-5}$$
$$= 3^0.$$

Since $3^5/3^5 = 1$, it is reasonable to define $3^0 = 1$, and similarly in the general case. The symbol 0^0 is undefined.

[1] Evaluate the following.

(a) 17^0

(b) 30^0

(c) $(-10)^0$

(d) $-(12)^0$

Answers:

(a) 1

(b) 1

(c) 1

(d) -1

> **ZERO EXPONENT**
> If a is any nonzero real number, then
> $$a^0 = 1.$$

EXAMPLE 1   Evaluate the following.

(a) $6^0 = 1$

(b) $(-9)^0 = 1$

(c) $-(4)^0 = -(1) = -1$ ■ [1]

The next step is to define negative integer exponents. If they are to be defined in such a way that the quotient rule given above remains valid, then we must have, for example,

$$\frac{3^2}{3^4} = 3^{2-4} = 3^{-2}.$$

However,

$$\frac{3^2}{3^4} = \frac{3 \cdot 3}{3 \cdot 3 \cdot 3 \cdot 3} = \frac{1}{3^2},$$

which suggests that 3^{-2} should be defined to be $1/3^2$. Thus, we have the following definition of a negative exponent.

NEGATIVE EXPONENT

If n is a natural number, and if $a \neq 0$, then

$$a^{-n} = \frac{1}{a^n}.$$

EXAMPLE 2 Evaluate the following.

(a) $3^{-2} = \dfrac{1}{3^2} = \dfrac{1}{9}$

(b) $5^{-4} = \dfrac{1}{5^4} = \dfrac{1}{625}$

(c) $9^{-1} = \dfrac{1}{9^1} = \dfrac{1}{9}$

(d) $-4^{-2} = -\dfrac{1}{4^2} = -\dfrac{1}{16}$

(e) $\left(\dfrac{3}{4}\right)^{-1} = \dfrac{1}{\left(\dfrac{3}{4}\right)^1} = \dfrac{1}{\dfrac{3}{4}} = \dfrac{4}{3}$

(f) $\left(\dfrac{2}{3}\right)^{-3} = \dfrac{1}{\left(\dfrac{2}{3}\right)^3} = \dfrac{1}{\left(\dfrac{2^3}{3^3}\right)} = 1 \cdot \dfrac{3^3}{2^3} = \dfrac{3^3}{2^3} = \dfrac{27}{8}$ ■

2 Evaluate the following.

(a) 6^{-2}

(b) -6^{-3}

(c) -3^{-4}

(d) $\left(\dfrac{5}{8}\right)^{-1}$

(e) $\left(\dfrac{1}{2}\right)^{-4}$

(f) $\left(\dfrac{7}{3}\right)^{-2}$

Answers:
(a) 1/36
(b) −1/216
(c) −1/81
(d) 8/5
(e) 16
(f) 9/49

Parts (e) and (f) of Example 2 involve work with fractions that can lead to error. For a useful shortcut with such fractions, use the properties of division of rational numbers and the definition of a negative exponent to get

$$\left(\frac{a}{b}\right)^{-n} = \frac{1}{\left(\dfrac{a}{b}\right)^n} = \frac{1}{\left(\dfrac{a^n}{b^n}\right)} = 1 \cdot \frac{b^n}{a^n} = \frac{b^n}{a^n} = \left(\frac{b}{a}\right)^n. \quad \boxed{2}$$

ROOTS AND RATIONAL EXPONENTS The definition of a^n will now be extended to include rational values of n, such as 1/2 and 7/3. In order to do this, we need some terminology.

There are two numbers whose square is 16, 4 and -4. The positive one, 4, is called the **square root** (or second root) of 16.* Similarly, there are two numbers whose fourth power is 16, 2 and -2. We call 2 the **fourth root** of 16. This suggests the following generalization.

> If n is even, the **nth root of a** is the positive real number whose nth power is a.

All nonnegative numbers have nth roots for every natural number n, but *no negative number has a real, even nth root.* For example there is no real number whose square is -16, so -16 has no square root.

We say that the **cube root** (or third root) of 8 is 2 because $2^3 = 8$. Similarly, since $(-2)^3 = -8$, we say -2 is the cube root of -8. Again, we can generalize.

> If n is odd, the **nth root of a** is the real number whose nth power is a.

Every real number has an nth root for every *odd* natural number n.

We can now define rational exponents. If they are to have the same properties as integer exponents, we want $a^{1/2}$ to be a number such that

$$(a^{1/2})^2 = a^{1/2} \cdot a^{1/2} = a^{1/2 + 1/2} = a^1 = a.$$

Thus, $a^{1/2}$ should be a number whose square is a and it is reasonable to *define* $a^{1/2}$ to be the square root of a (if it exists). Similarly, $a^{1/3}$ is defined to be the cube root of a and we have the following definition.

> If a is a real number and n is a positive integer, then
>
> $$a^{1/n} \text{ is defined to be the } n\text{th root of } a \text{ (if it exists).}$$

3 Evaluate the following.

(a) $16^{1/2}$

(b) $16^{1/4}$

(c) $-256^{1/2}$

(d) $(-256)^{1/2}$

(e) $-8^{1/3}$

(f) $243^{1/5}$

Answers:

(a) 4

(b) 2

(c) -16

(d) Not a real number

(e) -2

(f) 3

EXAMPLE 3   Evaluate the following roots.

(a) $36^{1/2} = 6$ because $6^2 = 36$.

(b) $-100^{1/2} = -10$

(c) $-(225^{1/2}) = -15$

(d) $625^{1/4} = 5$ because $5^4 = 625$.

(e) $(-1296)^{1/4}$ is not a real number, but $-1296^{1/4} = -6$ because $6^4 = 1296$.

(f) $(-27)^{1/3} = -3$

(g) $-32^{1/5} = -2$ ■   **3**

A calculator can be used to evaluate expressions with fractional exponents. Whenever it's easy to do so, enter the fractional exponents in their equivalent decimal form. For instance, to find $625^{1/4}$, enter $625^{.25}$ on the calculator. When the deci-

---

*Sometimes the positive square root is called the *principal* square root.

mal equivalent of a fraction is an infinitely repeating decimal, however, it is best to enter the fractional exponent directly using parentheses. For example, $17^{1/3}$ is entered as $17^{(1 \div 3)}$. If you omit the parentheses or use a shortened decimal approximation (such as .33 for 1/3), you will *not* get the correct answer.

For more general rational exponents, the symbol $a^{m/n}$ should be defined so that the properties for exponents still hold. For example,

$$(a^{1/n})^m \text{ must equal } a^{m/n}.$$

This suggests the following definition.

For all integers m and all positive integers n, and for all real numbers a for which $a^{1/n}$ is a real number,

$$a^{m/n} = (a^{1/n})^m.$$

---

4 Evaluate the following.

(a) $16^{3/4}$

(b) $25^{5/2}$

(c) $32^{7/5}$

(d) $100^{3/2}$

Answers:

(a) 8

(b) 3125

(c) 128

(d) 1000

EXAMPLE 4   Evaluate the following.

(a) $27^{2/3} = (27^{1/3})^2$
$\qquad = 3^2 = 9$

(b) $32^{2/5} = (32^{1/5})^2$
$\qquad = 2^2 = 4$

(c) $64^{4/3} = (64^{1/3})^4$
$\qquad = 4^4 = 256$

(d) $25^{3/2} = (25^{1/2})^3$
$\qquad = 5^3 = 125$ ■   **4**

The definitions and properties discussed above are summarized below, together with three power rules that follow from the definition of an exponent. These rules and definitions should be memorized.

DEFINITIONS AND PROPERTIES OF EXPONENTS

For any rational numbers m and n, and any real numbers a and b for which the following exist,

(a) $a^m \cdot a^n = a^{m+n}$ Product property

(b) $\dfrac{a^m}{a^n} = a^{m-n}$ Quotient property

(c) $(a^m)^n = a^{mn}$

(d) $(ab)^m = a^m \cdot b^m$ Power properties

(e) $\left(\dfrac{a}{b}\right)^m = \dfrac{a^m}{b^m}$

(f) $a^0 = 1$

(g) $a^{-n} = \dfrac{1}{a^n}$

(h) $\left(\dfrac{a}{b}\right)^{-n} = \left(\dfrac{b}{a}\right)^n.$

5 Simplify the following.

(a) $9^6 \cdot 9^{-4}$

(b) $\dfrac{8^7}{8^{-3}}$

(c) $(13^4)^{-3}$

(d) $6^{2/5} \cdot 6^{3/5}$

(e) $\dfrac{8^{2/3} \cdot 8^{-4/3}}{8^2}$

Answers:

(a) 9^2

(b) 8^{10}

(c) $1/13^{12}$

(d) 6

(e) $1/8^{8/3}$ or $1/2^8$

6 Simplify the following. Give answers with only positive exponents. Assume all variables represent positive real numbers.

(a) $\dfrac{(t^{-1})^2}{t^{-5}}$

(b) $\dfrac{(3z)^{-1}z^4}{z^2}$

(c) $3x^{1/4} \cdot 5x^{5/4}$

(d) $\left(\dfrac{2k^{1/3}}{p^{5/4}}\right)^2 \cdot \left(\dfrac{4k^{-2}}{p^5}\right)^{3/2}$

(e) $a^{5/8}(2a^{3/8} + a^{-1/8})$

Answers:

(a) t^3

(b) $z/3$

(c) $15x^{3/2}$

(d) $32/(p^{10}k^{7/3})$

(e) $2a + a^{1/2}$

7 Simplify.

(a) $\sqrt[3]{27}$

(b) $\sqrt[4]{625}$

(c) $\sqrt[6]{64}$

(d) $\sqrt[3]{\dfrac{64}{125}}$

Answers:

(a) 3

(b) 5

(c) 2

(d) $4/5$

EXAMPLE 5 Use the properties of exponents to simplify each of the following. Write answers with positive exponents.

(a) $7^{-4} \cdot 7^6 = 7^2$ Property (a)

(b) $\dfrac{9^{14}}{9^{-6}} = 9^{14-(-6)} = 9^{20}$ Property (b)

(c) $(2^{-3})^{-4} = 2^{(-3)(-4)} = 2^{12}$ Property (c)

(d) $\dfrac{27^{1/3} \cdot 27^{5/3}}{27^3} = \dfrac{27^{1/3+5/3}}{27^3}$ Product property

$\qquad\qquad = \dfrac{27^2}{27^3} = 27^{2-3}$ Quotient property

$\qquad\qquad = 27^{-1} = \dfrac{1}{27}$ Definition of negative exponent   ■   **5**

You can use a calculator to check computations, such as those in Example 5, by computing the left and right sides separately and confirming that the answers are the same in each case.

EXAMPLE 6 Simplify each expression. Give answers with only positive exponents. Assume all variables represent positive real numbers.

(a) $\dfrac{(m^3)^{-2}}{m^4} = \dfrac{m^{-6}}{m^4} = m^{-6-4} = m^{-10} = \dfrac{1}{m^{10}}$

(b) $6y^{2/3} \cdot 2y^{-1/2} = 12y^{2/3-1/2} = 12y^{1/6}$

(c) $\left(\dfrac{3m^{5/6}}{y^{3/4}}\right)^2 = \dfrac{3^2 m^{5/3}}{y^{3/2}} = \dfrac{9m^{5/3}}{y^{3/2}}$

(d) $m^{2/3}(m^{7/3} + 2m^{1/3}) = (m^{2/3+7/3} + 2m^{2/3+1/3}) = m^3 + 2m$ ■   **6**

RADICALS The nth root of a was denoted above as $a^{1/n}$. An alternative notation for nth roots uses **radicals.**

If n is an even natural number and $a \geq 0$, or n is an odd natural number,

$$a^{1/n} = \sqrt[n]{a}.$$

In the radical expression $\sqrt[n]{a}$, a is called the **radicand** and n is called the **index.** When $n = 2$, the familiar square root symbol $\sqrt{a}$ is used instead of $\sqrt[2]{a}$.

EXAMPLE 7 Simplify the following.

(a) $\sqrt[4]{16} = 16^{1/4} = 2$

(b) $\sqrt[5]{-32} = -2$

(c) $\sqrt[3]{1000} = 10$

(d) $\sqrt[6]{\dfrac{64}{729}} = \dfrac{2}{3}$ ■   **7**

The symbol $a^{m/n}$ also can be written in an alternative notation using radicals.

> For all rational numbers m/n and all real numbers a for which $\sqrt[n]{a}$ exists,
>
> $$a^{m/n} = (\sqrt[n]{a})^m \quad \text{or} \quad a^{m/n} = \sqrt[n]{a^m}.$$

Notice that $\sqrt[n]{x^n}$ cannot be written simply as x when n is even. For example, if $x = -5$,

$$\sqrt{x^2} = \sqrt{(-5)^2} = \sqrt{25} = 5 \neq x.$$

However, $|-5| = 5$, so that $\sqrt{x^2} = |x|$ when x is -5. This is true in general.

> For any real number a and any natural number n,
>
> $$\sqrt[n]{a^n} = |a| \text{ if } n \text{ is even}$$
>
> and
>
> $$\sqrt[n]{a^n} = a \text{ if } n \text{ is odd.}$$

To avoid this difficulty that $\sqrt[n]{a^n}$ is not necessarily equal to a, we shall assume that all variables in radicands represent only nonnegative numbers, as they usually do in applications.

The properties of exponents can be written with radicals as shown below.

> For all real numbers a and b and positive integers m and n for which all indicated roots exist,
>
> **(a)** $\sqrt[n]{a} \cdot \sqrt[n]{b} = \sqrt[n]{ab}$ **(b)** $\dfrac{\sqrt[n]{a}}{\sqrt[n]{b}} = \sqrt[n]{\dfrac{a}{b}}$ $(b \neq 0)$

8 Simplify.

(a) $\sqrt{3} \cdot \sqrt{27}$

(b) $\sqrt{\dfrac{3}{49}}$

(c) $\sqrt{25 - 4}$

(d) $\sqrt{25} - \sqrt{4}$

Answers:

(a) 9

(b) $\dfrac{\sqrt{3}}{7}$

(c) $\sqrt{21}$

(d) 3

EXAMPLE 8 Simplify the following.

(a) $\sqrt{6} \cdot \sqrt{54} = \sqrt{6 \cdot 54} = \sqrt{324} = 18$

Alternatively, simplify $\sqrt{54}$ first.

$$\sqrt{6} \cdot \sqrt{54} = \sqrt{6} \cdot \sqrt{9 \cdot 6}$$
$$= \sqrt{6} \cdot 3\sqrt{6} = 3 \cdot 6 = 18$$

(b) $\sqrt{\dfrac{7}{64}} = \dfrac{\sqrt{7}}{\sqrt{64}} = \dfrac{\sqrt{7}}{8}$ ∎

CAUTION $\sqrt[n]{a + b} \neq \sqrt[n]{a} + \sqrt[n]{b}$. For example, $\sqrt{9 + 16} = \sqrt{25} = 5$, but $\sqrt{9} + \sqrt{16} = 3 + 4 = 7.$ ◆ **8**

Multiplying radical expressions is much like multiplying polynomials.

EXAMPLE 9 Multiply the following.

(a) $(\sqrt{2} + 3)(\sqrt{8} - 5) = \sqrt{2}(\sqrt{8}) - \sqrt{2}(5) + 3\sqrt{8} - 3(5)$ FOIL

$$= \sqrt{16} - 5\sqrt{2} + 3(2\sqrt{2}) - 15$$

$$= 4 - 5\sqrt{2} + 6\sqrt{2} - 15$$

$$= -11 + \sqrt{2}$$

(b) $(\sqrt{7} - \sqrt{10})(\sqrt{7} + \sqrt{10}) = (\sqrt{7})^2 - (\sqrt{10})^2$

$$= 7 - 10 = -3 \quad \blacksquare \quad \boxed{9}$$

RATIONALIZING THE DENOMINATOR Before calculators were easily available, it was useful to **rationalize denominators** (write denominators with no radicals) because it was easy to calculate results like $\sqrt{2}/2 \approx 1.414/2 = .707$ mentally, but more difficult to compute $1/\sqrt{2} \approx 1/1.414$ without pencil and paper. However, there are other good reasons to rationalize denominators (and sometimes numerators). The process of rationalizing the denominator is explained in the next example.

EXAMPLE 10 Rationalize each denominator.

(a) $\dfrac{4}{\sqrt{3}}$

To rationalize the denominator, multiply by 1 in the form $\sqrt{3}/\sqrt{3}$ so that the denominator is $\sqrt{3} \cdot \sqrt{3} = 3$, a rational number.

$$\frac{4}{\sqrt{3}} \cdot \frac{\sqrt{3}}{\sqrt{3}} = \frac{4\sqrt{3}}{3}$$

(b) $\dfrac{1}{1 - \sqrt{2}}$

A useful approach here is to multiply both the numerator and denominator by the **conjugate*** of the denominator, in this case $1 + \sqrt{2}$. As suggested by Example 9(b), the product $(1 - \sqrt{2})(1 + \sqrt{2})$ is rational.

$$\frac{1}{1 - \sqrt{2}} = \frac{1(1 + \sqrt{2})}{(1 - \sqrt{2})(1 + \sqrt{2})} = \frac{1 + \sqrt{2}}{1 - 2}$$

$$= \frac{1 + \sqrt{2}}{-1} = -1 - \sqrt{2} \quad \blacksquare \quad \boxed{10}$$

---

*The conjugate of $a\sqrt{m} + b\sqrt{n}$ is $a\sqrt{m} - b\sqrt{n}$.

Margin notes:

$\boxed{9}$ Multiply.
(a) $(\sqrt{5} - \sqrt{2})(3 + \sqrt{2})$
(b) $(\sqrt{3} + \sqrt{7})(\sqrt{3} - \sqrt{7})$

Answers:
(a) $3\sqrt{5} + \sqrt{10} - 3\sqrt{2} - 2$
(b) -4

$\boxed{10}$ Rationalize the denominator.

(a) $\dfrac{2}{\sqrt{5}}$

(b) $\dfrac{1}{2 + \sqrt{3}}$

Answers:
(a) $\dfrac{2\sqrt{5}}{5}$
(b) $2 - \sqrt{3}$

1.6 EXERCISES

Evaluate each expression. Write all answers without exponents. (See Examples 1 and 2.)

1. 5^0

2. 8^0

3. 6^{-1}

4. 10^{-3}

5. 2^{-5}

6. 5^{-2}

7. -4^{-3}

8. -7^{-4}

9. $(7.94)^{-3}$

10. $(12.5)^{-2}$

11. $\left(\dfrac{1}{3}\right)^{-2}$

12. $\left(\dfrac{1}{6}\right)^{-3}$

13. $\left(\dfrac{2}{5}\right)^{-4}$

14. $\left(\dfrac{4}{3}\right)^{-2}$

15. Explain why $-2^{-4} = -1/16$, but $(-2)^{-4} = 1/16$.

16. Explain the reason a negative exponent is defined as a reciprocal: $a^{-n} = 1/a^n$.

Evaluate each expression. Write all answers without exponents. Write decimal answers to the nearest tenth. (See Examples 3 and 4.)

17. $49^{1/2}$

18. $8^{1/3}$

19. $(7.51)^{1/4}$

20. $(68.93)^{1/5}$

21. $27^{2/3}$

22. $24^{3/2}$

23. $(947)^{2/5}$

24. $(58.1)^{3/4}$

25. $-64^{2/3}$

26. $-64^{3/2}$

27. $(9/25)^{1/2}$

28. $(2401/16)^{1/4}$

29. $(16/9)^{-3/2}$

30. $(8/27)^{-4/3}$

31. $(27/64)^{-1/3}$

Simplify each expression. Write all answers using only positive exponents. (See Example 5.)

32. $\dfrac{4^{-2}}{4^3}$

33. $\dfrac{9^{-4}}{9^{-3}}$

34. $4^{-3} \cdot 4^6$

35. $5^{-9} \cdot 5^{10}$

36. $8^{2/3} \cdot 8^{-1/3}$

37. $12^{-3/4} \cdot 12^{1/4}$

38. $\dfrac{8^9 \cdot 8^{-7}}{8^{-3}}$

39. $\dfrac{5^{-4} \cdot 5^6}{5^{-1}}$

40. $\dfrac{9^{-5/3}}{9^{2/3} \cdot 9^{-1/5}}$

41. $\dfrac{3^{5/3} \cdot 3^{-3/4}}{3^{-1/4}}$

Simplify each expression. Assume all variables represent positive real numbers. Write answers with only positive exponents. (See Example 6.)

42. $\dfrac{z^5 \cdot z^2}{z^4}$

43. $\dfrac{k^6 \cdot k^9}{k^{12}}$

44. $\dfrac{2^{-1}(p^{-1})^3}{2p^{-4}}$

45. $\dfrac{(5x^3)^{-2}}{x^4}$

46. $(q^{-5}r^2)^{-1}$

47. $(2y^2z^{-2})^{-3}$

48. $(2p^{-1})^3 \cdot (5p^2)^{-2}$

49. $(5^{-1}m^2)^{-3} \cdot (3m^{-2})^4$

50. $(2p)^{1/2} \cdot (2p^3)^{1/3}$

51. $(5k^2)^{3/2} \cdot (5k^{1/3})^{3/4}$

52. $p^{2/3}(2p^{1/3} + 5p)$

53. $2z^{1/2}(3z^{-1/2} + z^{1/2})$

Match the rational exponent expression in Column I with the equivalent radical expression in Column II. Assume that x is not zero.

I	**II**
54. $(-3x)^{1/3}$	**(a)** $\dfrac{3}{\sqrt[3]{x}}$
55. $-3x^{1/3}$	**(b)** $-3\sqrt[3]{x}$
56. $(-3x)^{-1/3}$	**(c)** $\dfrac{1}{\sqrt[3]{3x}}$
57. $-3x^{-1/3}$	**(d)** $\dfrac{-3}{\sqrt[3]{x}}$
58. $(3x)^{1/3}$	**(e)** $3\sqrt[3]{x}$
59. $3x^{-1/3}$	**(f)** $\sqrt[3]{-3x}$
60. $(3x)^{-1/3}$	**(g)** $\sqrt[3]{3x}$
61. $3x^{1/3}$	**(h)** $\dfrac{1}{\sqrt[3]{-3x}}$

62. Some calculators will not compute a value for expressions like $(-8)^{2/3}$, having a negative base and a rational exponent with an odd denominator and an even numerator. Check to see whether your model is one of these. If it is, use the fact that $(-8)^{2/3} = [(-8)^{1/3}]^2$ to calculate it. What rule for exponents applies here?

Simplify each of the following. (See Examples 7–9.)

63. $\sqrt[3]{64}$

64. $\sqrt{64}$

65. $\sqrt[4]{625}$

66. $\sqrt[5]{-243}$

67. $\sqrt[7]{-128}$

68. $\sqrt{44} \cdot \sqrt{11}$

69. $\sqrt[3]{81} \cdot \sqrt[3]{9}$

70. $\sqrt{49 - 16}$

71. $\sqrt{81 - 4}$

72. $\sqrt{49} - \sqrt{16}$

73. $\sqrt{81} - \sqrt{4}$

74. $(\sqrt{2} + 3)(\sqrt{2} - 3)$

75. $(\sqrt{5} + \sqrt{2})(\sqrt{5} - \sqrt{2})$

76. $(3\sqrt{2} + \sqrt{3})(2\sqrt{3} - \sqrt{2})$

77. $(4\sqrt{5} - 1)(3\sqrt{5} + 2)$

Rationalize the denominator of each of the following. (See Example 10.)

78. $\dfrac{3}{1 - \sqrt{2}}$

79. $\dfrac{2}{1 + \sqrt{5}}$

80. $\dfrac{4 - \sqrt{2}}{2 - \sqrt{2}}$

81. $\dfrac{\sqrt{3} - 1}{\sqrt{3} - 2}$

82. What is wrong with the statement $\sqrt[3]{4} \cdot \sqrt[3]{4} = 4$?

The following exercises are applications of exponentiation and radicals.

83. Management The theory of economic lot size shows that, under certain conditions, the number of units to order to minimize total cost is

$$x = \sqrt{\dfrac{kM}{f}}.$$

Here k is the cost to store one unit for one year, f is the (constant) setup cost to manufacture the product, and M is the total number of units produced annually. (See Section 13.3.) Find x for the following values of f, k, and M.
(a) $k = \$1$, $f = \$500$, $M = 100{,}000$
(b) $k = \$3$, $f = \$7$, $M = 16{,}700$
(c) $k = \$1$, $f = \$5$, $M = 16{,}800$

84. Social Science The threshold weight T for a person is the weight above which the risk of death increases greatly. One researcher found that the threshold weight in pounds for men aged 40–49 is related to height in inches by the equation $h = 12.3T^{1/3}$. What height corresponds to a threshold of 216 pounds for a man in this age group?

85. Social Science The length L of an animal is related to its surface area S by the equation

$$L = \left(\frac{S}{a}\right)^{1/2},$$

where a is a constant that depends on the type of animal. Find the length of an animal with a surface area of 1000 square centimeters if $a = 2/5$.

86. Management An eccentric billionaire offers you a job for the month of September. You may choose either to be paid $300,000 per day or to be paid 2 cents on the first day, 4 cents on the second day, 8 cents on the third day, and so on, with your pay doubling on each successive day.
   (a) Write the equation that gives your salary (in cents) on day t if you choose the second option.
   (b) Which pay option will give you the larger total income for the month? (*Hint:* Under the second option, how much will you be paid on September 30?)

Natural Science *The wind-chill factor is a measure of the cooling effect that the wind has on a person's skin. It calculates the equivalent cooling temperature if there were no wind. The table gives the wind-chill factor for various wind speeds and temperatures.** *

87. One model of the wind-chill factor is

$$C = T - \left(\frac{v}{4} + 7\sqrt{v}\right)\left(1 - \frac{T}{90}\right),$$

where T represents the temperature and v represents the wind velocity. Evaluate C for **(a)** $T = -10$ and $v = 30$, and **(b)** $T = -40$ and $v = 5$.

88. Another model of the wind-chill factor is

$$C = 91.4 - (91.4 - T)(.478 + .301\sqrt{v} - .02v).$$

*Miller, A. and J. Thompson, *Elements of Meteorology,* 2e, Charles E. Merrill Publishing Co., 1975.

Repeat parts (a) and (b) of Exercise 87 using this model.

°F\Wind	5 mph	10 mph	15 mph	20 mph	25 mph	30 mph
40°	37	28	22	18	16	13
30°	27	16	9	4	0	−2
20°	16	4	−5	−10	−15	−18
10°	6	−9	−18	−25	−29	−33
0°	−5	−21	−36	−39	−44	−48
−10°	−15	−33	−45	−53	−59	−63
−20°	−26	−46	−58	−67	−74	−79
−30°	−36	−58	−72	−82	−88	−94
−40°	−47	−70	−85	−96	−104	−109
−50°	−57	−83	−99	−110	−118	−125

89. Use the table above to find the wind-chill factor for the information in parts (a) and (b) of Exercise 87.
   (c) Which expression in Exercises 87 and 88 models the wind-chill factor better?

90. Management Using a technique from statistics called *exponential regression,* it can be shown that the equation $y = 386(1.18)^x$ provides a fairly good model for product sales generated by infomercials, where y is in millions of dollars and $x = 0$ corresponds to the year 1988, $x = 1$ corresponds to 1989, and so on through the year 1993.* Use a calculator to estimate, to the nearest million dollars, the amount generated during the following years.
   (a) 1988
   (b) 1990
   (c) 1992

*National Infomercial Marketing Association

CHAPTER **1** S U M M A R Y

Key Terms and Symbols

1.1
$\approx$ is approximately equal to
π pi
$|a|$ absolute value of a
real number
natural (counting) number
whole number

integer
rational number
irrational number
properties of real numbers
additive inverse
multiplicative inverse

order of operations
number line
interval
interval notation
absolute value
1.2 variable
first-degree equation
properties of equality
solving for a specified variable
1.3 a^n a to the power n
exponent or power
base
polynomial
coefficient
term
constant term
degree of a nonzero term
degree of a polynomial
zero polynomial
binomial
trinomial

like terms
FOIL
1.4 factor
factoring
greatest common factor
perfect square trinomial
difference of two squares
sum and difference of two cubes
1.5 rational expression
operations with rational expressions
complex fraction
1.6 $a^{1/n}$ nth root of a
$\sqrt{a}$ square root of a
$\sqrt[n]{a}$ nth root of a
zero exponent
properties of exponents
radical
radicand
index
rationalizing the denominator

Key Concepts

Absolute Value

Assume a and b are real numbers with $b > 0$.

The solutions of $|a| = b$ or $|a| = |b|$ are $a = b$ or $a = -b$.

The solutions of $|a| < b$ are $-b < a < b$.

The solutions of $|a| > b$ are $a < -b$ or $a > b$.

Factoring

$x^2 + 2xy + y^2 = (x + y)^2$ $x^3 - y^3 = (x - y)(x^2 + xy + y^2)$

$x^2 - 2xy + y^2 = (x - y)^2$ $x^3 + y^3 = (x + y)(x^2 - xy + y^2)$

$x^2 - y^2 = (x + y)(x - y)$

Rules for Radicals

Let a and b be real numbers, n be a positive integer, and m be any integer for which the following exist.

$a^{m/n} = \sqrt[n]{a^m} = (\sqrt[n]{a})^m$ $\sqrt[n]{a^n} = |a|$ if n is even $\sqrt[n]{a^n} = a$ if n is odd

$\sqrt[n]{a} \cdot \sqrt[n]{b} = \sqrt[n]{ab}$ $\dfrac{\sqrt[n]{a}}{\sqrt[n]{b}} = \sqrt[n]{\dfrac{a}{b}}$ $(b \neq 0)$

Rules for Exponents

Let a, b, r, and s be any real numbers for which the following exist.

$a^{-r} = \dfrac{1}{a^r}$ $a^0 = 1$ $\left(\dfrac{a}{b}\right)^r = \dfrac{a^r}{b^r}$

$a^r \cdot a^s = a^{r+s}$ $(a^r)^s = a^{rs}$ $a^{1/r} = \sqrt[r]{a}$

$\dfrac{a^r}{a^s} = a^{r-s}$ $(ab)^r = a^r b^r$ $\left(\dfrac{a}{b}\right)^{-r} = \left(\dfrac{b}{a}\right)^r$

Chapter 1 Review Exercises

Name the numbers from the list $-12, -6, -9/10, -\sqrt{7}, -\sqrt{4},$
$0, 1/8, \pi/4, 6, \sqrt{11}$ *that are*

1. whole numbers;

2. integers;

3. rational numbers;

4. irrational numbers.

Identify the properties of real numbers that are illustrated in each of the following.

5. $7[(-3)4] = 7[4(-3)]$

6. $9(4 + 5) = (4 + 5)9$

7. $6(x + y - 3) = 6x + 6y + 6(-3)$

8. $11 + (8 + 3) = (11 + 8) + 3$

Express each statement in symbols.

9. x is at least 6.

10. x is negative.

Write the following numbers in numerical order from smallest to largest.

11. $-7, -3, 8, \pi, -2, 0$

12. $\dfrac{5}{6}, \dfrac{1}{2}, -\dfrac{2}{3}, -\dfrac{5}{4}, -\dfrac{3}{8}$

13. $|6 - 4|, -|-2|, |8 + 1|, -|3 - (-2)|$

14. $\sqrt{7}, -\sqrt{8}, -|\sqrt{16}|, |-\sqrt{12}|$

Write without absolute value bars.

15. $-|-6| + |3|$

16. $|-5| + |-9|$

17. $7 - |-8|$

18. $|-2| - |-7 + 3|$

Graph each of the following on a number line.

19. $x \geq -3$

20. $-4 < x \leq 6$

21. $x < -2$

22. $x \leq 1$

Use the order of operations to simplify.

23. $(-6 + 2 \cdot 5)(-2)$

24. $-4(-7 - 9 \div 3)$

25. $\dfrac{-8 + (-6)(-3) \div 9}{6 - (-2)}$

26. $\dfrac{20 \div 4 \cdot 2 \div 5 - 1}{-9 - (-3) - 12 \div 3}$

Solve each equation.

27. $2x - 5(x - 4) = 3x + 2$

28. $5y + 6 = -2(1 - 3y) + 4$

29. $\dfrac{2z}{5} - \dfrac{4z - 3}{10} = \dfrac{-z + 1}{10}$

30. $\dfrac{p}{p + 2} - \dfrac{3}{4} = \dfrac{2}{p + 2}$

31. $\dfrac{2m}{m - 3} = \dfrac{6}{m - 3} + 4$

32. $\dfrac{15}{k + 5} = 4 - \dfrac{3k}{k + 5}$

Solve for x.

33. $5ax - 1 = x$

34. $6x - 3y = 4bx$

35. $\dfrac{2x}{3 - c} = ax + 1$

36. $b^2x - 2x = 4b^2$

Solve each equation.

37. $|m - 3| = 9$

38. $|6 - x| = 12$

39. $\left|\dfrac{2 - y}{5}\right| = 8$

40. $\left|\dfrac{3m + 5}{7}\right| = 15$

41. $|4k + 1| = |6k - 3|$

Management   *Solve each of the following problems.*

42. A computer printer is on sale for 15% off. The sale price is $425. What was the original price?

43. To make a special mix for Valentine's Day, the owner of a candy store wants to combine chocolate hearts which sell for $5 per pound with candy kisses which sell for $3.50 per pound. How many pounds of each kind should be used to get 30 pounds of a mix that can be sold for $4.50 per pound?

44. A real estate firm invests the $100,000 proceeds from a sale in two ways. The first portion in invested in a shopping center that provides an annual return of 8%. The rest is invested in a small apartment building with an annual return of 5%. The firm wants an annual income of $6800 from these investments. How much should be put into each investment?

Perform each of the indicated operations.

45. $(3x^4 - x^2 + 5x) - (-x^4 + 3x^2 - 8x)$

46. $(-8y^3 + 8y^2 - 3y) - (2y^3 - 4y^2 - 10)$

47. $-2(q^4 - 3q^3 + 4q^2) + 4(q^4 + 2q^3 + q^2)$

48. $5(3y^4 - 4y^5 + y^6) - 3(2y^4 + y^5 - 3y^6)$

49. $(5z + 2)(3z - 2)$

50. $(8p - 4)(5p + 3)$

51. $(4k - 3h)(4k + 3h)$

52. $(2r - 5y)(2r + 5y)$

53. $(6x + 3y)^2$

54. $(2a - 5b)^2$

Factor as completely as possible.

55. $2kh^2 - 4kh + 5k$

56. $2m^2n^2 + 6mn^2 + 16n^2$

57. $3a^4 + 13a^3 + 4a^2$

58. $24x^3 + 4x^2 - 4x$

59. $10y^2 - 11y + 3$

60. $8q^2 + 3m + 4qm + 6q$

61. $4a^2 - 20a + 25$

62. $36p^2 + 12p + 1$

63. $144p^2 - 169q^2$

64. $81z^2 - 25x^2$

65. $8y^3 - 1$

66. $125a^3 + 216$

Perform each operation.

67. $\dfrac{4x}{5} \cdot \dfrac{35x}{12}$

68. $\dfrac{5k^2}{24} - \dfrac{75k}{36}$

69. $\dfrac{c^2 - 3c + 2}{2c(c - 1)} \div \dfrac{c - 2}{8c}$

70. $\dfrac{p^3 - 2p^2 - 8p}{3p(p^2 - 16)} \div \dfrac{p^2 + 4p + 4}{9p^2}$

71. $\dfrac{2y - 10}{5y} \cdot \dfrac{20y - 25}{12}$

72. $\dfrac{m^2 - 2m}{15m^3} \cdot \dfrac{5}{m^2 - 4}$

73. $\dfrac{2m^2 - 4m + 2}{m^2 - 1} \div \dfrac{6m + 18}{m^2 + 2m - 3}$

74. $\dfrac{x^2 + 6x + 5}{4(x^2 + 1)} \cdot \dfrac{2x(x + 1)}{x^2 - 25}$

75. $\dfrac{6}{15z} + \dfrac{2}{3z} - \dfrac{9}{10z}$

76. $\dfrac{5}{y - 2} - \dfrac{4}{y}$

77. $\dfrac{2}{5q} + \dfrac{10}{7q}$

78. $\dfrac{\dfrac{1}{x} - \dfrac{2}{y}}{3 - \dfrac{1}{xy}}$

79. Describe the steps needed to find the sum of the following expression:

$$\frac{2a + b}{4a^2 - b^2} + \frac{5a}{2a - b}.$$

80. Give some examples of corresponding rules for exponents and radicals, and explain how they are related.

81. Give two ways to evaluate $125^{2/3}$ and then compare them. Which do you prefer? Why?

Simplify each of the following. In Exercises 82–103, write all answers without negative exponents. Assume all variables represent positive real numbers.

82. 5^{-3}

83. 10^{-2}

84. -7^0

85. -3^{-1}

86. $\left(-\dfrac{6}{5}\right)^{-2}$

87. $\left(\dfrac{2}{3}\right)^{-3}$

88. $4^6 \cdot 4^{-3}$

89. $7^{-5} \cdot 7^{-1}$

90. $\dfrac{9^{-4}}{9^{-3}}$

91. $\dfrac{6^{-2}}{6^3}$

92. $\dfrac{9^4 \cdot 9^{-5}}{(9^{-2})^2}$

93. $\dfrac{k^4 \cdot k^{-3}}{(k^{-2})^{-3}}$

94. $4^{-1} + 2^{-1}$

95. $3^{-2} + 3^{-1}$

96. $125^{2/3}$

97. $128^{3/7}$

98. $9^{-5/2}$

99. $\left(\dfrac{144}{49}\right)^{-1/2}$

100. $\dfrac{5^{1/3} \cdot 5^{1/2}}{5^{3/2}}$

101. $\dfrac{2^{3/4} \cdot 2^{-1/2}}{2^{1/4}}$

102. $(3a^2)^{1/2} \cdot (3^2a)^{3/2}$

103. $(4p)^{2/3} \cdot (2p^3)^{3/2}$

104. $\sqrt[3]{27}$

105. $\sqrt[4]{625}$

106. $\sqrt[5]{-32}$

107. $\sqrt[6]{-64}$

108. $\sqrt{24}$

109. $\sqrt{63}$

110. $\sqrt[3]{54p^3q^5}$

111. $\sqrt[4]{64a^5b^3}$

112. $\sqrt{\dfrac{5n^2}{6m}}$

113. $\sqrt{\dfrac{3x^3}{2z}}$

114. $2\sqrt{3} - 5\sqrt{12}$

115. $8\sqrt{7} + 2\sqrt{28}$

116. $(\sqrt{5} - 1)(\sqrt{5} + 1)$

117. $(\sqrt{7} - \sqrt{3})(\sqrt{7} + \sqrt{3})$

118. $(2\sqrt{5} - \sqrt{3})(\sqrt{5} + 2\sqrt{3})$

119. $(4\sqrt{7} + \sqrt{2})(3\sqrt{7} - \sqrt{2})$

120. $\dfrac{\sqrt{2}}{1 + \sqrt{3}}$

121. $\dfrac{4 + \sqrt{2}}{4 - \sqrt{5}}$

Social Science   *In our system of government, the president is elected by the electoral college, and not by individual voters. Because of this, smaller states have a greater voice in the selection of a president than they otherwise would have. Two political scientists have studied the problems of campaigning for president under the current system and have concluded that candidates should allot their money according to the formula*

$$\text{Amount for large state} = \left(\frac{E_{large}}{E_{small}}\right)^{3/2} \times \text{amount for small state.}$$

Here E_{large} represents the electoral vote of the large state, and E_{small} represents the electoral vote of the small state. Find the amount that should be spent in each of the following larger states if \$1,000,000 is spent in the small state and the following statements are true.

122. The large state has 48 electoral votes, and the small state has 3.

123. The large state has 36 electoral votes, and the small state has 4.

124. 6 votes in a small state; 28 in a large

125. 9 votes in a small state; 32 in a large

CASE 1

Consumers Often Defy Common Sense*

Imagine two refrigerators in the appliance section of a department store. One sells for \$700 and uses \$85 worth of electricity a year. The other is \$100 more expensive but costs only \$25 a year to run. Given that either refrigerator should last at least 10 years without repair, consumers would overwhelmingly buy the second model, right?

Well, not exactly. Many studies by economists have shown that in a wide range of decisions about money—from paying taxes to buying major appliances—consumers consistently make decisions that defy common sense.

In some cases—as in the refrigerator example—this means that people are generally unwilling to pay a little more money up front to save a lot of money in the long run. At times, psychological studies have shown, consumers appear to assign entirely whimsical values to money, values that change depending on time and circumstances.

In recent years, these apparently irrational patterns of human behavior have become a subject of intense interest among economists and psychologists, both for what they say about the way the human mind works and because of their implications for public policy.

How, for example, can the United States move toward a more efficient use of electricity if so many consumers refuse to buy energy-efficient appliances even when such a move is in their own best interest?

At the heart of research into the economic behavior of consumers is a concept known as the discount rate. It is a measure of how consumers compare the value of a dollar received today with one received tomorrow.

Consider, for example, if you won \$1000 in a lottery. How much more money would officials have to give you before you would agree to postpone cashing the check for a year?

Some people might insist on at least another \$100, or 10 percent, since that is roughly how much it would take to make up for the combined effects of a year's worth of inflation and lost interest.

But the studies show that someone who wants immediate gratification might not be willing to postpone receiving the \$1000 for 20 percent or 30 percent or even 40 percent more money.

In the language of economists, this type of person has a high discount rate: He or she discounts the value of \$1000 so much over a year that it would take hundreds of extra dollars to make waiting as attractive as getting the money immediately.

Of the two alternatives, waiting a year for more money is clearly more rational than taking the check now. Why would people turn down \$1400 next year in favor of \$1000 today? Even if they needed the \$1000 immediately, they would be better off borrowing it from a bank, even at 20 percent or 30

*"Consumers often defy common sense" by Malcolm Gladwell from *The Washington Post.* Copyright ©1990, The Washington Post. Reprinted with permission.

percent interest. Then, a year later, they could pay off the loan—including the interest—with the $1400 and pocket the difference.

The fact is, however, that economists find numerous examples of such high discount rates implicit in consumer behavior.

While consumers were very much aware of savings to be made at the point of purchase, they so heavily discounted the value of monthly electrical costs that they would pay over the lifetime of their dryer or freezer that they were oblivious of the potential for greater savings.

Gas water heaters, for example, were found to carry an implicit discount rate of 100 percent. This means that in deciding which model was cheapest over the long run, consumers acted as if they valued a $100 gas bill for the first year as if it were really $50. Then, in the second year, they would value the next $100 gas bill as if it were really worth $25, and so on through the life of the appliance.

Few consumers actually make this formal calculation, of course. But there are clearly bizarre behavioral patterns in evidence.

Some experiments, for example, have shown that the way in which consumers make decisions about money depends a great deal on how much is at stake. Few people are willing to give up $10 now for $15 next year. But they are if the choice is between $100 now and $150 next year, a fact that would explain why consumers appear to care less about many small electricity bills—even if they add up to a lot—than one big initial outlay.

EXERCISES

1. Suppose a refrigerator that sells for $700 costs $85 a year for electricity. Write an expression for the cost to buy and run the refrigerator for x years.

2. Suppose another refrigerator costs $1000 and $25 a year for electricity. Write an expression for the total cost for this refrigerator over x years.

3. Over 10 years which refrigerator costs the most? By how much?

4. In how many years will the total costs for the two refrigerators be equal?

CHAPTER 2

Graphs, Equations, and Inequalities

The solutions of many applied problems involve equations and inequalities. This chapter presents both algebraic and graphical methods of dealing with such situations.

2.1 GRAPHS

Just as the number line associates the points on a line with real numbers, a similar construction in two dimensions associates points in the plane with *ordered pairs* of real numbers. A **Cartesian coordinate system,** as shown in Figure 2.1, consists of a horizontal number line (usually called the **x-axis**) and a vertical number line (usually called the **y-axis**). The point where the number lines meet is called the **origin.** Each point in a Cartesian coordinate system is labeled with an **ordered pair** of real numbers, such as $(-2, 4)$ or $(3, 2)$. Several points and their corresponding ordered pairs are shown in Figure 2.1.

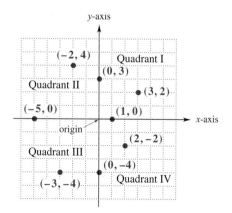

FIGURE 2.1

1 Locate $(-1, 6)$, $(-3, -5)$, $(4, -3)$, $(0, 2)$, and $(-5, 0)$ on a coordinate system.

Answer:

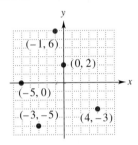

For the point $(-2, 4)$, for example, -2 is the **x-coordinate** and 4 is the **y-coordinate.** You can think of these coordinates as directions telling you how to move to this point from the origin: you go 2 horizontal units to the left (x-coordinate) and 4 vertical units upward (y-coordinate). From now on, instead of referring to "the point labeled by the ordered pair $(-2, 4)$," we will say "the point $(-2, 4)$." **1**

The x-axis and the y-axis divide the plane into four parts, or **quadrants,** which are numbered as shown in Figure 2.1. The points on the coordinate axes belong to no quadrant.

EQUATIONS AND GRAPHS A **solution of an equation** in two variables, such as

$$y = -2x + 3$$

or

$$y = x^2 + 7x - 2,$$

is an ordered pair of numbers such that the substitution of the first number for x and the second number for y produces a true statement.

2 Which of the following are solutions of $y = x^2 + 7x - 2$?

(a) $(1, 6)$

(b) $(-2, -20)$

(c) $(-1, -8)$

Answer:
Parts (a) and (c)

EXAMPLE 1 Which of the following are solutions of $y = -2x + 3$?

(a) $(2, -1)$
This is a solution of $y = -2x + 3$ because "$-1 = -2 \cdot 2 + 3$" is a true statement.

(b) $(4, 7)$
Since $-2 \cdot 4 + 3 = -5$ and not 7, the ordered pair $(4, 7)$ is not a solution of $y = -2x + 3$. ■ **2**

Equations in two variables, such as $y = -2x + 3$, typically have an infinite number of solutions. To find one, choose a number for x and then compute the value of y that produces a solution. For instance, if $x = 5$, then $y = -2 \cdot 5 + 3 = -7$, so that $(5, -7)$ is a solution of $y = -2x + 3$. Similarly, if $x = 0$, then $y = -2 \cdot 0 + 3 = 3$, so that $(0, 3)$ is also a solution.

3 Graph $x = 5y$.

Answer:

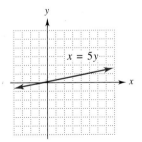

The **graph** of an equation in two variables is the set of points in the plane whose coordinates (ordered pairs) are solutions of the equation. Thus, the graph of an equation is a picture of its solutions. Since a typical equation has infinitely many solutions, its graph has infinitely many points.

EXAMPLE 2 Sketch the graph of $y = -2x + 5$.
Since we cannot plot infinitely many points, we construct a table of y-values for a reasonable number of x-values, plot the corresponding points, and make an "educated guess" about the rest. The table of values and points in Figure 2.2 suggest that the graph is a straight line, as shown in Figure 2.3. ■ **3**

x	−2x + 5
−1	7
0	5
2	1
4	−3
5	−5

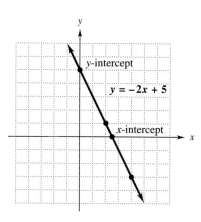

FIGURE 2.2

FIGURE 2.3

An **x-intercept** of a graph is the x-coordinate of a point where the graph meets the x-axis (the y-coordinate of this point is 0 since it's on the axis). Consequently, to find the x-intercepts of the graph of an equation, set $y = 0$ and solve for x. For instance, in Example 2, the x-intercept of the graph of $y = -2x + 5$ (see Figure 2.3) is found by setting $y = 0$ and solving for x.

$$0 = -2x + 5$$
$$2x = 5$$
$$x = \frac{5}{2}$$

4 Find the x- and y-intercepts of the graphs of these equations.

(a) $3x + 4y = 12$

(b) $5x - 2y = 8$

Answers:

(a) x-intercept 4, y-intercept 3

(b) x-intercept 8/5, y-intercept −4

Similarly, a **y-intercept** of a graph is the y-coordinate of a point where the graph meets the y-axis (the x-coordinate of this point is 0—why?). The y-intercepts are found by setting $x = 0$ and solving for y. For example, the graph of $y = -2x + 5$ in Figure 2.3 has y-intercept 5.   **4**

EXAMPLE 3   Find the x- and y-intercepts of the graph of $y = 4 - x^2$ and sketch the graph.

Setting $x = 0$ in $y = 4 - x^2$, we see that the y-intercept is $y = 4$. Similarly, setting $y = 0$ gives $x^2 = 4$. Thus the x-intercepts are $x = 2$ and $x = -2$. Now make a table, being sure to take both positive and negative values for x, and plot the corresponding points, as in Figure 2.4. These points suggest that the entire graph looks like Figure 2.5.   ■

x	4 − x²
−3	−5
−2	0
−1	3
0	4
1	3
2	0
3	−5

FIGURE 2.4

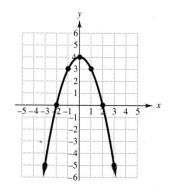

FIGURE 2.5

EXAMPLE 4   Sketch the graph of $y = \sqrt{x}$.

Notice that x can only take nonnegative values (since the square root of a negative number is not defined) and that the corresponding value of y is also nonnegative. Hence all the points on the graph will lie in the first quadrant. Computing some typical values, we obtain Figure 2.6.   ■

x	$\sqrt{x}$
0	0
1	1
4	2
6	$\sqrt{6} \approx 2.45$
9	3
11	$\sqrt{11} \approx 3.32$

FIGURE 2.6

A graphing calculator or computer graphing program follows essentially the same procedure used in the preceding examples: the calculator selects a large number of x-values (95 or more), equally spaced along the x-axis, and plots the corresponding points, simultaneously connecting them with line segments. Calculator-generated graphs are generally quite accurate, although they may not appear as smooth as hand-drawn ones. Figure 2.7, for example, shows the graph of $y = x^3 - 5x + 1$. If you have a graphing calculator, learn how to use it.

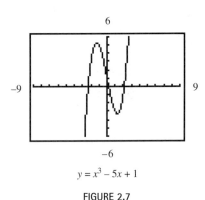

$$y = x^3 - 5x + 1$$

FIGURE 2.7

GRAPH READING   Information often is given in graphical form, so you must be able to read and interpret graphs, that is, to translate graphical information into statements in English.

EXAMPLE 5   A device at the weather bureau records the temperature over a 24-hour period in the form of a graph (Figure 2.8). The first coordinate of each point on the graph represents the time (measured in hours after midnight) and the second coordinate the temperature at that time.

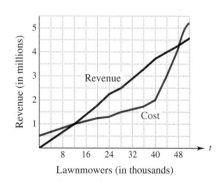

FIGURE 2.8

(a) What was the temperature at 6 A.M. and at 6 P.M.?

The point (6, 40) is on the graph, which means that the temperature at 6 A.M. was 40°. Now 6 P.M. is 18 hours after midnight and the point on the graph with first coordinate 18 is (18, 60). So the temperature at 6 P.M. was 60°.

(b) At what times during the day was the temperature below 50°?

Look for the points whose second coordinates are less than 50, that is the points that lie below the horizontal line through 50°. The first coordinates of these points are the times when the temperature was below 50°. Figure 2.8 shows that these are the points with first coordinates less than 10 or greater than 20. Since 20 hours corresponds to 8 P.M., we see that the temperature was below 50° from midnight to 10 A.M. and from 8 P.M. to midnight.

(c) During what time period *before* 4 P.M. was the temperature at least 60°?

Since 4 P.M. is 16 hours after midnight, we look for points with first coordinate less than 16 and second coordinate greater than 60. Figure 2.8 shows that these are the points with first coordinates between 13 and 16. So the temperature was at least 60° from 1 P.M. to 4 P.M. ■ 5

5 In Example 5, what are the highest and lowest temperatures during the day? When do they occur?

Answer:
Highest is about 64° near 4 P.M.; lowest is about 37° at midnight when the day begins.

EXAMPLE 6 Monthly revenue and costs for the Lange Lawnmower Company are determined by the number t of lawnmowers produced, as shown in Figure 2.9.

FIGURE 2.9

(a) How many lawnmowers should be produced each month if the company is to make a profit?

Profit is Revenue − Cost, so the company makes a profit whenever revenue is greater than cost, that is, when the revenue graph is above the cost graph. Figure 2.9 shows that this occurs between $t = 12$ and $t = 48$, that is, when 12,000 to 48,000 lawnmowers are produced. If the company makes fewer than 12,000 mowers, it will lose money (costs will be greater than revenue). They also lose money by making more than 48,000 mowers (one reason might be that high production levels require large amounts of overtime pay, which drives costs up too much).

(b) Is it more profitable to make 40,000 or 44,000 mowers?

On the revenue graph, the point with first coordinate 40 has second coordinate approximately 3.7, meaning that the revenue from 40,000 mowers is about 3.7 million dollars. The point with first coordinate 40 on the cost graph is (40, 2), meaning that the cost of producing 40,000 mowers is 2 million dollars. Therefore, the profit on 40,000 mowers is about $3.7 - 2 = 1.7$ million dollars. For 44,000 mowers, we have the approximate points (44, 4) on the revenue graph and (44, 3) on the cost graph. So the profit on 44,000 mowers is $4 - 3 = 1$ million dollars. Consequently, it is more profitable to make 40,000 mowers. ■ 6

6 In Example 6, find the profit from making

(a) 32,000 lawnmowers;

(b) 4000 lawnmowers.

Answers:

(a) About $1,000,000 (rounded)

(b) About −$500,000 (that is, a loss of $500,000)

EXAMPLE 7 The graph in Figure 2.10, which appeared in the March 1997 issue of *Scientific American,* shows actual and projected population figures for various parts of the world over a two hundred year period. Two projections are given for the United States; graph A assumes current immigration and fertility rates will continue, while graph B assumes that one or both will decrease.

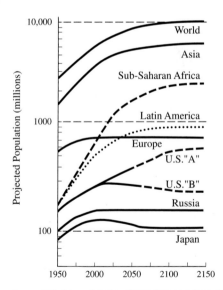

Source: U.S. data are based on Census Bureau series projections through 2050. Data for all other areas are from Eduard Bos et al., *World Population Projections, 1994–95 Edition.* (Johns Hopkins University Press for the World Bank, 1994.)

FIGURE 2.10

(a) Approximately when will the population of Latin America exceed that of Europe?

In approximately 2030, because Figure 2.10 shows that the Latin America graph lies below the Europe graph until then, and above it thereafter

(b) Which region will grow the fastest between now and 2050?

Sub-saharan Africa, because its graph rises more steeply than any of the others between 2000 and 2050

(c) Which regions are projected to have the same number or fewer people in 2150 than in 2000?

These are the regions whose graphs do not rise between 2000 and 2150, namely, Japan, Russia, Europe, and possibly the United States (if graph B is accurate). ∎

2.1 EXERCISES

Determine whether the given ordered pair is a solution of the given equation. (See Example 1.)

1. $(1, -2)$; $3x - y - 5 = 0$ ~Yes $3(1) + 2 - 5 = 0$
$3 + 2 - 5 = 0$
$0 = 0$

2. $(2, -1)$; $x^2 + y^2 - 6x + 8y = -15$

3. $(3, 4)$; $(x - 2)^2 + (y + 5)^2 = 4$

4. $(1, -1)$; $\dfrac{x^2}{2} + \dfrac{y^2}{3} = 1$ ~Yes $\dfrac{1}{2} - \dfrac{1}{3} = 1$

$\dfrac{3 - 2}{}= 1$

Sketch the graph of each of these equations. (See Example 2.)

5. $4y + 3x = 12$ **6.** $2x + 7y = 14$

7. $8x + 3y = 12$ **8.** $9y - 4x = 12$

9. $x = 2y + 3$ **10.** $x - 3y = 0$

List the x-intercepts and y-intercepts of each graph.

11.
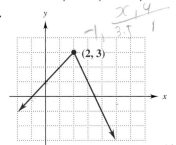
$\dfrac{x \mid y}{3.5 \mid}$
-1 1
(2, 3)

12.

$\dfrac{x \mid y}{1 \mid 2}$

13.

x-intercept
$-2, 0 \ 2$
y-intercept o
$\dfrac{x}{-2, 0, 2}$ $\dfrac{y}{0}$

14.
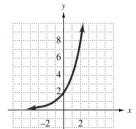

Find the x-intercepts and y-intercepts of the graph of each equation. You need not sketch the graph. (See Example 3.)

15. $3x + 4y = 24$ → $x = 8, \ y = 6$ $8, 0$ $0, 6$

16. $x - 2y = 3$

17. $2x - 3y = 6$ → $x = 3, \ y = -2$

18. $3x + y = 4$

19. $y = x^2 - 9$

20. $y = x^2 + 4$

Sketch the graph of the equation. (See Examples 2–4.)

21. $y = x^2$

22. $y = x^2 + 2$ $y = x^2$

23. $y = x^2 - 3$

$\dfrac{x \mid y}{\begin{array}{c|c} -2 & 4 \\ -1 & 1 \\ 0 & 0 \\ 1 & 1 \end{array}}$

24. $y = 2x^2$

25. $y = x^3$

26. $y = x^3 - 3$

27. $y = x^3 + 1$

28. $y = x^3/2$

✳ **29.** $y = \sqrt{x + 2}$ ✳

30. $y = \sqrt{x - 2}$

✳ **31.** $y = \sqrt{4 - x^2}$

32. $y = \sqrt{9 - x^2}$

Physical Science *According to an article in the December 1994 issue of* Scientific American, *the coast-down time for a typical 1993 car as it drops 10 miles per hour from an initial speed depends on variations from the standard condition (automobile in neutral; average drag and tire pressure). The accompanying graph illustrates some of these conditions with coast-down time in seconds and initial speed in miles per hour.*

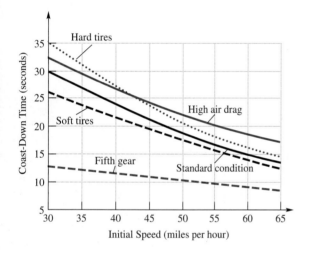

Initial Speed (miles per hour)

Use the graph to answer the following questions.

33. What is the approximate coast-down time in fifth gear if the initial speed is 40 miles per hour?

34. For what speed is the coast-down time the same for the conditions of high air drag and hard tires?

Physical Science *The temperature graphs for Fargo and Seattle on the same day are shown in the figure at the top of the next column. Use them to do the following exercises. (See Example 5.)*

Time (hrs after midnight)

35. Approximately when did the temperature first reach 60° in Fargo? In Seattle?

36. At what times during the day did the two cities have the same temperature?

37. At what times during the day was it warmer in Fargo than in Seattle?

38. Was there any time when it was at least 10° warmer in Seattle than in Fargo?

Management *Use the revenue and cost graphs for the Lange Lawnmower Company (Figure 2.9) to do Exercises 39–42.*

39. Find the approximate cost of manufacturing the given number of lawnmowers.
(a) 20,000   (b) 36,000
(c) 48,000

40. Find the approximate revenue from selling the given number of lawnmowers.
(a) 12,000   (b) 24,000
(c) 36,000

41. Use the fact that profit = revenue − cost to find the approximate profit from manufacturing the given number of lawnmowers.
(a) 20,000   (b) 28,000
(c) 36,000

42. The company must replace its aging machinery with better, but much more expensive machines. In addition, raw material prices increase, so that monthly costs go up by $250,000. Owing to competitive pressure, lawnmower prices cannot be increased, so revenue remains the same. Under these new circumstances, find the approximate profit from manufacturing the given number of lawnmowers.
(a) 20,000   (b) 36,000
(c) 40,000

43. Management Sandi Goldstein takes out a 30-year mortgage on which her monthly payment is $850. During the early years of the mortgage, most of each payment is for interest and the rather small remainder for principal. As time goes on the portion of each payment that goes for interest decreases, while the portion for principal increases, as shown in the accompanying graph.

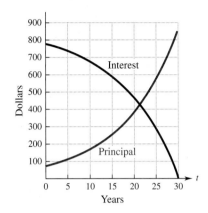

(a) Approximately how much of the $850 monthly payment goes for interest in year 5? In year 15? In year 25?

(b) In what year will the monthly payment be equally divided between interest and principal?

Management *The accompanying graph, which shows non-agricultural employment in the United States, appeared in a recent* economic review (U.S. Territory 1997, published by the Economics Department of U.S. Bancorp). Use the graph to answer the following questions.

44. How many people were employed at the beginning of 1990? At the beginning of 1995?

45. During what period was employment below 109,000,000?

46. Approximately when was employment the lowest?

U.S. Nonagricultural Employment, seasonally adjusted (in thousands)

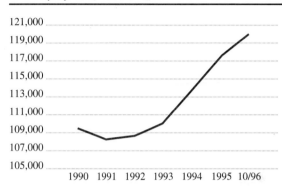

Source: Department of Labor, Bureau of Labor Statistics.

2.2 SLOPE AND THE EQUATIONS OF A LINE

Straight lines, which are the simplest graphs, play an important role in a wide variety of applications. They are considered here from both a geometric and an algebraic point of view.

The key geometric feature of a nonvertical straight line is how steeply it rises or falls as you move from left to right. The "steepness" of a line can be represented numerically by a number called the *slope* of the line.

To see how slope is defined, start with Figure 2.11 on the next page, which shows a line passing through the two different points $(x_1, y_1) = (-3, 5)$ and $(x_2, y_2) = (2, -4)$. The difference in the two x-values,

$$x_2 - x_1 = 2 - (-3) = 5$$

in this example, is called the **change in x.** The Greek letter Δ (delta) is used to denote change. The symbol Δx (read "delta x") represents the change in x. In the same way, Δy represents the **change in y.** In this example,

$$\Delta y = y_2 - y_1 = -4 - 5 = -9.$$

The **slope** of the line through the two points (x_1, y_1) and (x_2, y_2), where $x_1 \neq x_2$, is defined as the quotient of the change in y and the change in x, or

$$\text{slope} = \frac{\text{change in } y}{\text{change in } x} = \frac{\Delta y}{\Delta x} = \frac{y_2 - y_1}{x_2 - x_1}.$$

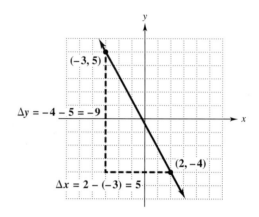

FIGURE 2.11

The slope of the line in Figure 2.11 is

$$\text{slope} = \frac{\Delta y}{\Delta x} = \frac{-4 - 5}{2 - (-3)} = -\frac{9}{5}.$$

Using similar triangles from geometry, it can be shown that the slope is independent of the choice of points on the line. That is, the same value of the slope will be obtained for *any* choice of two different points on the line.

EXAMPLE 1   Find the slope of the line through the points $(-7, 6)$ and $(4, 5)$.

Let $(x_1, y_1) = (-7, 6)$. Then $(x_2, y_2) = (4, 5)$. Use the definition of slope.

$$\text{slope} = \frac{\Delta y}{\Delta x} = \frac{5 - 6}{4 - (-7)} = -\frac{1}{11}$$

The slope can also be found by letting $(x_1, y_1) = (4, 5)$ and $(x_2, y_2) = (-7, 6)$. In that case,

$$\text{slope} = \frac{6 - 5}{-7 - 4} = \frac{1}{-11} = -\frac{1}{11},$$

the same answer.   ■   ☐1

☐1 Find the slope of the line through

(a) $(6, 11), (-4, -3)$;

(b) $(-3, 5), (-2, 8)$.

Answers:

(a) $7/5$

(b) 3

CAUTION   When finding the slope of a line, be careful to subtract the *x*-values and the *y*-values in the same order. For example, with the points (4, 3) and (2, 9), if you use $9 - 3$ for the numerator, you must use $2 - 4$ (*not* $4 - 2$) for the denominator. ◆

EXAMPLE 2   Find the slope of the horizontal line in Figure 2.12.

Every point on the line has the same *y*-coordinate, -5. Choose any two of them to compute the slope, say $(x_1, y_1) = (-3, -5)$ and $(x_2, y_2) = (2, -5)$:

$$\text{slope} = \frac{-5 - (-5)}{2 - (-3)}$$

$$= \frac{0}{5}$$

$$= 0. \quad\blacksquare$$

FIGURE 2.12   FIGURE 2.13

EXAMPLE 3   What is the slope of the vertical line in Figure 2.13?

Every point on the line has the same *x*-coordinate, 4. If you attempt to compute the slope with two of these points, say $(x_1, y_1) = (4, -2)$ and $(x_2, y_2) = (4, 1)$, you obtain

$$\text{slope} = \frac{1 - (-2)}{4 - 4}$$

$$= \frac{3}{0}.$$

Division by 0 is not defined, so the slope of this line is undefined.   ∎

The arguments used in Examples 2 and 3 work in the general case and lead to the following conclusion.

The slope of every horizontal line is 0.
The slope of every vertical line is undefined.

SLOPE-INTERCEPT FORM   Slope can be used to develop an algebraic description of nonvertical straight lines. Assume that a line with slope m has y-intercept b, so that it goes through the point $(0, b)$ (see Figure 2.14). Let (x, y) be any point on the line other than $(0, b)$. Using the definition of slope with the points $(0, b)$ and (x, y) gives

$$m = \frac{y - b}{x - 0}$$

$$m = \frac{y - b}{x}$$

$$mx = y - b \qquad \text{Multiply by } x.$$

from which

$$y = mx + b \qquad \text{Add } b. \text{ Reverse the equation.}$$

In other words, the coordinates of any point on the line satisfy the equation $y = mx + b$.

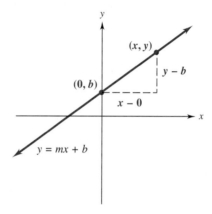

FIGURE 2.14

SLOPE–INTERCEPT FORM

If a line has slope m and y-intercept b, then it is the graph of the equation

$$y = mx + b.$$

This equation is called the **slope-intercept form** of the equation of the line.

EXAMPLE 4   Find an equation for the line with y-intercept 7/2 and slope $-5/2$. Use the slope-intercept form with $b = 7/2$ and $m = -5/2$.

$$y = mx + b$$

$$y = -\frac{5}{2}x + \frac{7}{2} \qquad \blacksquare$$

2 Find an equation for the line with

(a) y-intercept -3 and slope 2/3;

(b) y-intercept 1/4 and slope $-3/2$.

Answers:

(a) $y = \dfrac{2}{3}x - 3$

$-y = Mx + b$

$y = \dfrac{2}{3}x - 3$

(b) $y = -\dfrac{3}{2}x + \dfrac{1}{4}$

EXAMPLE 5 Find the equation of the horizontal line with y-intercept 3.

The slope of the line is 0 (why?) and its y-intercept is 3, so its equation is

$$y = mx + b$$
$$y = 0x + 3$$
$$y = 3 \quad \blacksquare \quad \boxed{2}$$

The argument in Example 5 also works in the general case.

---

3 Find the slope and y-intercept for

(a) $x + 4y = 6$;

(b) $3x - 2y = 1$.

$y = Mx + b$

$4y = -x + b$

$y = -\dfrac{x}{4} + \dfrac{3}{2}$

Answers:

(a) Slope $-1/4$; y-intercept 3/2

(b) Slope 3/2; y-intercept $-1/2$

If k is a constant, then the graph of the equation $y = k$ is the horizontal line with y-intercept k.

EXAMPLE 6 Find the slope and y-intercept for each of the following lines.

(a) $5x - 3y = 1$

Solve for y.

$-2y = -3x + 1$

$y = \dfrac{3}{2}x - \dfrac{1}{2}$

$$5x - 3y = 1$$
$$-3y = -5x + 1$$
$$y = \dfrac{5}{3}x - \dfrac{1}{3}$$

4 **(a)** List the slopes of the following lines:

E: $y = -.3x$, F: $y = -x$,

G: $y = -2x$, H: $y = -5x$.

(b) Graph all four lines on the same set of axes.

(c) How are the slopes of the lines related to their steepness?

Answers:

(a) Slope E $= -.3$; slope F $= -1$; slope G $= -2$; slope H $= -5$.

(b)

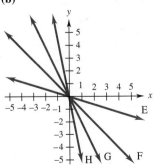

(c) The larger the slope in absolute value, the more steeply the line falls from left to right.

This equation is in the form $y = mx + b$, with $m = 5/3$ and $b = -1/3$. So the slope is 5/3 and the y-intercept is $-1/3$.

(b) $-9x + 6y = 2$

Solve for y.

$$-9x + 6y = 2$$
$$6y = 9x + 2$$
$$y = \dfrac{3}{2}x + \dfrac{1}{3}$$

The slope is 3/2 (coefficient of x) and the y-intercept is 1/3. $\blacksquare$ $\boxed{3}$

The slope-intercept form can be used to show how the slope measures the steepness of a line. Consider the straight lines A, B, C, and D given by the following equations; each has y-intercept 0 and slope as indicated.

A: $y = .5x$,   B: $y = x$,   C: $y = 3x$,   D: $y = 7x$,
Slope .5   Slope 1   Slope 3   Slope 7

For these lines, Figure 2.15 on the next page shows that the bigger the slope, the more steeply the line rises from left to right. $\boxed{4}$

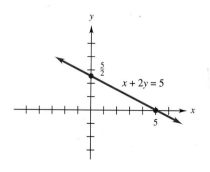

FIGURE 2.15

5 Graph the lines and label the intercepts.

(a) $3x + 4y = 12$

(b) $5x - 2y = 8$

Answers:

(a)

(b)

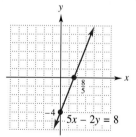

The preceding discussion and Problem 4 in the margin may be summarized as follows.

Direction of Line (moving from left to right)	Slope
Upward	**Positive** (larger for steeper lines)
Horizontal	**0**
Downward	**Negative** (larger in absolute value for steeper lines)
Vertical	**Undefined**

EXAMPLE 7   Sketch the graph of $x + 2y = 5$ and label the intercepts.
Find the x-intercept by setting $y = 0$ and solving for x.

$$x + 2 \cdot 0 = 5$$
$$x = 5$$

The x-intercept is 5 and (5, 0) is on the graph. The y-intercept is found similarly, by setting $x = 0$ and solving for y.

$$0 + 2y = 5$$
$$y = 5/2$$

The y-intercept is 5/2 and (0, 5/2) is on the graph. The points (5, 0) and (0, 5/2) can be used to sketch the graph (Figure 2.16).   ■   **5**

FIGURE 2.16

TECHNOLOGY TIP To graph a linear equation on a graphing calculator, you must first put the equation in slope-intercept form $y = mx + b$ so that it can be entered in the equation memory (called the Y= list on some calculators). Vertical lines cannot be graphed on most calculators. ✔

PARALLEL AND PERPENDICULAR LINES We shall assume the following facts without proof. The first one is a consequence of the fact that slope measures steepness and parallel lines have the same steepness.

> Two nonvertical lines are parallel whenever they have the same slope.
>
> Two nonvertical lines are perpendicular whenever the product of their slopes is -1.

EXAMPLE 8 Determine whether each of the following pairs of lines are *parallel*, *perpendicular*, or *neither*.

(a) $2x + 3y = 5$ and $4x + 5 = -6y$

Put each equation in slope-intercept form by solving for y.

$$3y = -2x + 5 \qquad -6y = 4x + 5$$
$$y = -\frac{2}{3}x + \frac{5}{3} \qquad y = -\frac{2}{3}x - \frac{5}{6}$$

In each case the slope (coefficient of x) is $-2/3$, so the lines are parallel.

(b) $3x = y + 7$ and $x + 3y = 4$

The slope of $3x = y + 7$ is 3 (why?). Verify that the slope of $x + 3y = 4$ is $-1/3$. Since $3(-1/3) = -1$, these lines are perpendicular.

(c) $x + y = 4$ and $x - 2y = 3$

Verify that the slope of the first line is -1 and the slope of the second is $1/2$. The slopes are not equal and their product is not -1, so the lines are neither parallel nor perpendicular. ■ ⑥

⑥ Tell if the lines in each of the following pairs are *parallel*, *perpendicular*, or *neither*.

(a) $x - 2y = 6$ and $2x + y = 5$

(b) $3x + 4y = 8$ and $x + 3y = 2$

(c) $2x - y = 7$ and $2y = 4x - 5$

Answers:
(a) Perpendicular
(b) Neither
(c) Parallel

TECHNOLOGY TIP Perpendicular lines may not appear perpendicular on a graphing calculator, unless you use a *square window*—one in which a one-unit segment on the y-axis is the same length as a one-unit segment on the x-axis. To obtain such a window on most calculators, use a viewing window in which the y-axis is about 2/3 as long as the x-axis. ✔

POINT-SLOPE FORM The slope-intercept form of the equation of a line involves the slope and the y-intercept. Sometimes, however, the slope of a line is known, together with one point on the line (perhaps *not* the y-intercept). The *point-slope form* of the equation of a line is used to find an equation in this case. Let (x_1, y_1) be any fixed point on the line and let (x, y) represent any other point on the line. If m is the slope of the line, then, by the definition of slope,

$$\frac{y - y_1}{x - x_1} = m.$$

Multiplying both sides by $x - x_1$ shows that

$$y - y_1 = m(x - x_1).$$

POINT-SLOPE FORM

If a line has slope m and passes through the point (x_1, y_1), then

$$y - y_1 = m(x - x_1)$$

is the **point-slope form** of the equation of the line.

EXAMPLE 9 Find an equation of the line with the given slope that passes through the given point.

(a) $(-4, 1), m = -3$

Use the point-slope form because a point on the line, together with the slope of the line, is known. Substitute the values $x_1 = -4$, $y_1 = 1$, and $m = -3$ into the point-slope form.

$$y - y_1 = m(x - x_1)$$
$$y - 1 = -3[x - (-4)] \qquad \text{Point-slope form}$$

Using algebra we obtain the slope-intercept form of this equation.

$$y - 1 = -3(x + 4)$$
$$y - 1 = -3x - 12 \qquad \text{Distributive property}$$
$$y = -3x - 11 \qquad \text{Slope-intercept form}$$

(b) $(3, -7), m = 5/4$

$$y - y_1 = m(x - x_1)$$
$$y - (-7) = \frac{5}{4}(x - 3) \qquad \text{Let } y_1 = -7, m = \frac{5}{4}, x_1 = 3.$$
$$y + 7 = \frac{5}{4}(x - 3) \qquad \text{Point-slope form}$$
$$y + 7 = \frac{5}{4}x - \frac{15}{4}$$
$$y = \frac{5}{4}x - \frac{43}{4} \qquad \text{Slope-intercept form} \quad ■ \quad \boxed{7}$$

$\boxed{7}$ Find both the point-slope and the slope-intercept forms of the equation of the line having the given slope and passing through the given point.

(a) $m = -3/5, (5, -2)$

(b) $m = 1/3, (6, 8)$

Answers:

(a) $y + 2 = -\dfrac{3}{5}(x - 5)$;

$y = -\dfrac{3}{5}x + 1$

(b) $y - 8 = \dfrac{1}{3}(x - 6)$;

$y = \dfrac{1}{3}x + 6$

The point-slope form can also be used to find an equation of a line given two different points on the line. The procedure for doing this is shown in the next example.

EXAMPLE 10   Find an equation of the line through $(5, 4)$ and $(-10, -2)$.

Begin by using the definition of slope to find the slope of the line that passes through the two points.

$$\text{slope} = m = \frac{-2 - 4}{-10 - 5} = \frac{-6}{-15} = \frac{2}{5}$$

Use $m = 2/5$ and either of the given points in the point-slope form. If $(x_1, y_1) = (5, 4)$, then

$$y - y_1 = m(x - x_1)$$

$$y - 4 = \frac{2}{5}(x - 5) \qquad \text{Let } y_1 = 4, m = \frac{2}{5}, x_1 = 5.$$

$$5(y - 4) = 2(x - 5) \qquad \text{Multiply both sides by 5.}$$

$$5y - 20 = 2x - 10 \qquad \text{Distributive property}$$

$$5y = 2x + 10.$$

Check that the result is the same when $(x_1, y_1) = (-10, -2)$. ■   8

8 Find an equation of the line through

(a) $(2, 3)$ and $(-4, 6)$;

(b) $(-8, 2)$ and $(3, -6)$.

Answers:

(a) $2y = -x + 8$

(b) $11y = -8x - 42$

EXAMPLE 11   In an experiment testing a person's reaction time y (in seconds) after undergoing x hours of stressful activity, the linear equation $y = .1957x + .1243$ was found to be a good approximation of the relationship between stress and reaction time during the first five hours.

(a) Assuming that reaction times continue to follow this pattern, what would be the approximate reaction time after eight and a half hours?

Substitute $x = 8.5$ in the equation and use a calculator to compute y.

$$y = .1957x + .1243$$

$$y = .1957(8.5) + .1243 = 1.78775$$

So the reaction time is approximately 1.8 seconds.

(b) When the reaction time is 1.5 seconds, how long has the subject been undergoing stressful activity?

Substitute $y = 1.5$ in the equation and solve for x.

$$1.5 = .1957x + .1243$$

$$1.5 - .1243 = .1957x$$

$$x = \frac{1.5 - .1243}{.1957} \approx 7.0296$$

Hence the stressful activity has been going on for slightly more than 7 hours. ■

VERTICAL LINES   The equation forms developed above do not apply to vertical lines because slope is not defined for such lines. However, vertical lines can easily be described as graphs of equations.

EXAMPLE 12   Find the equation whose graph is the vertical line in Figure 2.17.

FIGURE 2.17

Every point on the line has x-coordinate -1 and hence has the form $(-1, y)$. Thus every point is a solution of the equation $x + 0y = -1$, which is usually written simply as $x = -1$. Note that -1 is the x-intercept of the line.   ■

The argument in Example 12 also works in the general case.

If k is a constant, then the graph of the equation $x = k$ is the vertical line with x-intercept k.

A **linear equation** is an equation in two variables that can be put in the form $ax + by = c$ for some constants $a, b, c,$ with at least one of a, b nonzero. These equations are so named because their graph is always a straight line. For instance, when $b \neq 0$, then $ax + by = c$ can be written

$$by = -ax + c$$

$$y = -\frac{a}{b}x + \frac{c}{b},$$

which is the equation of a line with slope $-a/b$ and y-intercept c/b. Similarly, when $b = 0$, then $a \neq 0$ and $ax + by = c$ becomes $x = c/a$, whose graph is a vertical line. Conversely, the equation of every line is a linear equation. For example, the line with slope -3 and y-intercept 5 has the equation $y = -3x + 5$ (why?), which can be written $3x + 1y = 5$. Here is a summary of the basic facts about linear equations and their graphs.

Equation	Description
$ax + by = c$	If $a \neq 0$ and $b \neq 0$, the line has x-intercept c/a and y-intercept c/b.
$x = k$	**Vertical line,** x-intercept k, no y-intercept, undefined slope
$y = k$	**Horizontal line,** y-intercept k, no x-intercept, slope 0
$y = mx + b$	**Slope-intercept form,** slope m, y-intercept b
$y - y_1 = m(x - x_1)$	**Point-slope form,** slope m, the line passes through (x_1, y_1).

2.2 EXERCISES

Find the slope of the line, if it is defined. (See Examples 1–3.)

1. Through $(2, 5)$ and $(0, 6)$

2. Through $(9, 0)$ and $(12, 15)$

3. Through $(-4, 7)$ and $(3, 0)$

4. Through $(-5, -2)$ and $(-4, 11)$

5. Through the origin and $(-4, 6)$

6. Through the origin and $(8, -2)$

7. Through $(-1, 4)$ and $(-1, 8)$

8. Through $(-3, 5)$ and $(2, 5)$

Find an equation of the line with the given y-intercept and slope m. (See Examples 4 and 5.)

9. $5, m = 3$

10. $-3, m = -7$

11. $1.5, m = -2.3$

12. $-4.5, m = 1.5$

13. $4, m = -3/4$

14. $-3, m = 2/3$

Find the slope and the y-intercept of the line whose equation is given. (See Example 6.)

15. $2x - y = 7$

16. $x + 2y = 7$

17. $6x = 2y + 4$

18. $4x + 3y = 12$

19. $6x - 9y = 14$

20. $4x + 2y = 0$

21. $2x - 3y = 0$

22. $y = 5$

23. $x = y - 5$

24. On one graph, sketch six straight lines that meet at a single point and satisfy this condition: one line has slope 0, two lines have positive slope, two lines have negative slope, and one line has undefined slope.

25. For which of the line segments in the figure at the top of the next column is the slope
    (a) largest? (b) smallest?
    (c) largest in absolute value? (d) closest to 0?

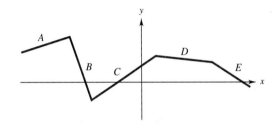

26. Match each equation with the line that most closely resembles its graph. (*Hint:* Consider the signs of m and b in the slope-intercept form.)
    (a) $y = 3x + 2$ (b) $y = -3x + 2$
    (c) $y = 3x - 2$ (d) $y = -3x - 2$

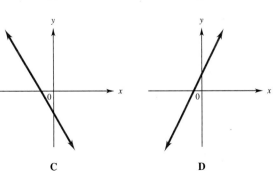

Sketch the graph of the equation and label its intercepts. (See Example 7.)

27. $2x - y = -2$

28. $2y + x = 4$

29. $2x + 3y = 4$

30. $-5x + 4y = 3$

31. $4x - 5y = 2$

32. $3x + 2y = 8$

Determine whether each pair of lines is parallel, perpendicular, or neither. (See Example 8.)

33. $4x - 3y = 6$ and $3x + 4y = 8$

34. $2x - 5y = 7$ and $15y - 5 = 6x$

35. $3x + 2y = 8$ and $6y = 5 - 9x$

36. $x - 3y = 4$ and $y = 1 - 3x$

37. $4x = 2y + 3$ and $2y = 2x + 3$

38. $2x - y = 6$ and $x - 2y = 4$

39. **(a)** Find the slope of each side of the triangle with vertices $(9, 6)$, $(-1, 2)$, and $(1, -3)$.

   (b) Is this triangle a right triangle? (*Hint:* Are two sides perpendicular?)

40. **(a)** Find the slope of each side of the quadrilateral with vertices $(-5, -2)$, $(-3, 1)$, $(3, 0)$, and $(1, -3)$.

   (b) Is this quadrilateral a parallelogram?

41. **Physical Science** The graph shows the winning times (in minutes) at the Olympic Games for the 5000-meter run, together with a linear approximation of the data.*

Olympic times for 5000-meter run
(minutes)

(a) The equation for the linear approximation is $y = -.0221x + 57.14$. What does the slope of this line represent? Why is the slope negative?

(b) Can you think of any reason why there are no data points for the years 1940 and 1944?

42. **Management** The graph shows the number of U.S. radio stations on the air along with a linear approximation of the data.*

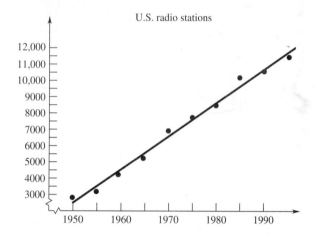

U.S. radio stations

(a) Discuss the accuracy of the linear approximation.

(b) Use the two data points $(1950, 2773)$ and $(1994, 11600)$ to find the approximate slope of the line shown. Interpret this number.

Find an equation of the line that passes through the given point and has the given slope. (See Examples 9 and 12.)

43. $(-1, 2)$, $m = -2/3$

44. $(-4, -3)$, $m = 5/4$

45. $(-2, -2)$, $m = 2$

46. $(-2, 3)$, $m = -1/2$

47. $(8, 2)$, $m = 0$

48. $(2, -4)$, $m = 0$

49. $(6, -5)$, undefined slope

50. $(-8, 9)$, undefined slope

Find an equation of the line that passes through the given points. (See Example 10.)

51. $(-1, 1)$ and $(2, 5)$

52. $(2, 5)$ and $(0, 6)$

---

*United States Olympic Committee

*National Association of Broadcasters

53. (1, 2) and (3, 7)

54. (−1, −2) and (2, −1)

Find an equation of the line satisfying the given conditions.

55. Through the origin with slope 7

56. Through the origin and horizontal

57. Through (5, 8) and vertical

58. Through (7, 11) and parallel to $y = 6$

59. Through (3, 4) and parallel to $4x − 2y = 5$

60. Through (6, 8) and perpendicular to $y = 2x − 3$

61. x-intercept 5; y-intercept −5

62. Through (−5, 2) and parallel to the line through (1, 2) and (4, 3)

63. Through (−1, 3) and perpendicular to the line through (0, 1) and (2, 3)

64. y-intercept 3 and perpendicular to $2x − y + 6 = 0$

65. Management Ral Corp. has an incentive compensation plan under which a branch manager receives 10% of the branch's income after deduction of the bonus but before deduction of income tax.* Branch income for 1988 before the bonus and income tax was $165,000. The tax rate was 30%. The 1988 bonus amounted to
(a) $12,600 (b) $15,000
(c) $16,500 (d) $18,000.

Management *The lost value of equipment over a period of time is called* **depreciation.** *The simplest method for calculating depreciation is* **straight-line depreciation.** *The annual straight-line depreciation on an item that cost x dollars with a useful life of n years is D = (1/n)x. Find the depreciation for items with the following characteristics.*

66. Cost: $12,482; life 10 yr

67. Cost: $39,700; life 12 yr

68. Cost: $145,000; life 28 yr

69. Management In a recent issue of *Business Week,* the president of InstaTune, a chain of franchised automobile tune-up shops, says that people who buy a franchise and open a shop pay a weekly fee (in dollars) of

$$y = .07x + \$135$$

to company headquarters. Here y is the fee and x is the total amount of money taken in during the week by the tune-up center. Find the weekly fee if x is
(a) $0; (b) $1000; (c) $2000;
(d) $3000, (e) Graph y.

70. Management In a recent issue of *The Wall Street Journal,* we are told that the relationship between the amount of money that an average family spends on food eaten at home, x, and the amount of money it spends on eating out, y, is approximated by the model $y = .36x$. Find y if x is
(a) $40; (b) $80; (c) $120. (d) Graph y.

Use a calculator to answer the following. (See Example 11.)

71. Physical Science Kitchen gas ranges are a source of indoor pollutants such as carbon monoxide and nitrogen dioxide. One of the most effective ways of removing contaminants from the air while cooking is to use a *vented* range hood. If a range hood removes F liters of air per second, the percentage P of contaminants that are also removed from the surrounding air is given by $P = 1.06F + 7.18$, where $10 \le F \le 75$.* For instance, if $F = 10$, then $1.06(10) + 7.18 = 17.78$ percent of the contaminants are removed. To the nearest liter, what rate F is needed to remove at least 58 percent of the contaminants from the air?

72. Management An insurance company determines that the amount of damage y a building sustains in a fire is related to the distance x from the building to the nearest fire station and is given by the equation $y = 9.879x + 5.117$, where x is in miles and y is in thousands of dollars. To the nearest tenth of a mile, how far from the fire station is a building that sustains more than $60,000 worth of damage?

73. Social Science The capital outlay for education by state and local governments during the period from 1985 to 1992 can be approximated by the equation $y = 2480x + 12,726$, where $x = 0$ corresponds to 1985 and y is in millions of dollars.† Based on this model, in what year did this amount first exceed 22,600 million dollars?

74. Social Science The number of farms with milk cows in the United States has steadily declined since 1985. The equation $y = −13.5x + 261.5$ gives a good approximation of the number of such farms in each year from 1985 to 1992, with $x = 0$ representing 1985 and with y in thousands.‡ If these trends continued, in what year would the number of such farms first fall below 150 thousand?

*Uniform CPA Examination, May, 1989, American Institute of Certified Public Accountants.

*Rezvan, R.L., "Effectiveness of Local Ventilation in Removing Simulated Pollutants from Point Sources" in *Proceedings of the Third International Conference on Indoor Air Quality and Climate,* 1984.

†U.S. Bureau of the Census, *Historical Statistics on Governmental Finances and Employment,* and *Government Finance,* Series GF, annual.

‡U.S. Dept. of Agriculture, National Agricultural Statistics Service, *Dairy Products,* annual; and *Milk: Production, Disposition, and Income,* annual.

75. Management The linear equation $y = 1.082x + 16.882$ provides the approximate average monthly rate for basic cable television subscribers between the years 1990 and 1993, where $x = 0$ corresponds to 1990, $x = 1$ corresponds to 1991, and so on, and y is in dollars. Use this equation to answer the following questions.

(a) What was the approximate average monthly rate in 1991?

(b) What was the approximate average monthly rate in 1993?

(c) During 1994, the rates dropped substantially to $18.86. The model above is based on data from 1990 through 1993. If you were to use the model for 1994, what would the rate be?

(d) Why do you think there is such a discrepancy between the actual rate and the rate based on the model in part (c)? Discuss the pitfalls of using the model to predict for years following 1993.

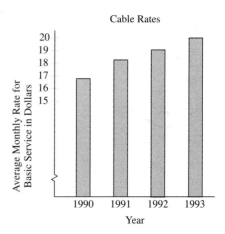

Cable Rates

Source: National Cable Television Association; Nations Bank; Paul Kagan Associates

2.3 APPLICATIONS OF LINEAR EQUATIONS

When mathematics is used to solve real-world problems, there are no hard and fast rules that can be used in every case. You must use the given information, your common sense, and the mathematics available to you to build an appropriate *mathematical model* (such as an equation or inequality) that describes the situation. The examples in this section, which are restricted to models given by linear equations, illustrate this process in several situations.

In most cases, linear (or any other reasonably simple) equations provide only *approximations* to the real-world situations. Nevertheless, these approximations are often remarkably useful. Furthermore, the concepts and procedures presented here are applicable to more complicated situations, as we shall see later.

We begin with a common relationship found in everyday situations. Recall that water freezes at a temperature of 32° Fahrenheit, which is 0° Celsius, while it boils at 212° Fahrenheit (100° Celsius).

EXAMPLE 1 The relationship between the Celsius and Fahrenheit temperature scales is known to be linear.

(a) Find the equation that relates the Celsius temperature y to the corresponding Fahrenheit temperature x and display its graph.

The fact that 32° Fahrenheit corresponds to 0° Celsius means that the point (32, 0) is on the graph. Similarly, (212, 100) is also on the graph. The graph is a straight line whose slope can be found from these two points.

$$\text{slope} = m = \frac{100 - 0}{212 - 32} = \frac{100}{180} = \frac{5}{9}$$

Using $m = 5/9$ and $(x_1, y_1) = (32, 0)$ in the point-slope form, we find that the equation of this line is

$$y - y_1 = m(x - x_1)$$

$$y - 0 = \frac{5}{9}(x - 32)$$

$$y = \frac{5}{9}(x - 32).$$

The graph of this equation is in Figure 2.18.

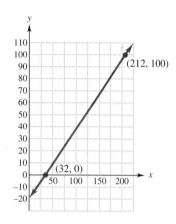

FIGURE 2.18

(b) Use the graph to find the Celsius temperature corresponding to 50° Fahrenheit.

Figure 2.18 shows that the point on the graph with first coordinate 50 is (50, 10), which means that 50° Fahrenheit is 10° Celsius.

(c) Find the Celsius temperature corresponding to 75° Fahrenheit.

The graph in Figure 2.18 suggests that the second coordinate of the point with first coordinate 75 is a bit smaller than 25. We can confirm this algebraically and obtain an accurate answer by substituting $x = 75$ in the equation $y = \frac{5}{9}(x - 32)$ and solving for y.

$$y = \frac{5}{9}(75 - 32)$$

$$= \frac{5}{9} \cdot 43$$

$$\approx 23.89$$

Therefore 75° Fahrenheit is equivalent to 23.89° Celsius. ∎ ☐1

1 Find the Celsius temperature corresponding to 23° Fahrenheit.

Answer:
−5° Celsius

EXAMPLE 2 According to an article in *The New York Times* of November 15, 1995, sales per square foot at discount stores in New England were $142 in 1991 and $175 in 1994 (in constant 1994 dollars). Records for several years indicate that sales have increased in a linear pattern.

(a) Find a linear equation that describes the sales y in year x.

For computational convenience, let 1990 correspond to $x = 0$, 1991 to $x = 1$, and so on, so that 1994 corresponds to $x = 4$. If y represents sales in year x, then the points $(1, 142)$ and $(4, 175)$ are on the graph. The slope of the line through these points is

$$m = \frac{175 - 142}{4 - 1} = \frac{33}{3} = 11.$$

Using the point-slope form of the equation of a line, with $m = 11$ and $(x_1, y_1) = (1, 142)$, we obtain the following.

$$y - 142 = 11(x - 1)$$
$$y - 142 = 11x - 11$$
$$y = 11x + 131$$

(b) Assuming that the equation in part (a) remains valid in later years, find the sales in the year 2000.

Since 2000 corresponds to $x = 10$, substitute $x = 10$ in the equation.

$$y = 11x + 131 = 11(10) + 131 = 241$$

Therefore sales in 2000 are $241 per square foot. ■ $\boxed{2}$

$\boxed{2}$ The article mentioned in Example 2 also stated that sales per square foot at discount stores in the Mid-Atlantic region were $131 in 1991 and $197 in 1994.

(a) Find a linear equation describing the Mid-Atlantic sales y in year x.

(b) What were the sales in 1996?

Answers:
(a) $y = 22x + 109$
(b) $241

BREAK-EVEN ANALYSIS In a manufacturing and sales situation the basic relationship is

$$\textbf{Profit = Revenue - Cost.}$$

Typically, revenue and cost can be described in terms of equations. Here, as in many applications, we use as variables letters that suggest what they represent, rather than the letters x or y: r for revenue, c for cost, and p for profit.

EXAMPLE 3 The cost c (in dollars) of making x leaf-blowers is given by the equation $c = 45x + 6000$. Each leaf-blower can be sold for $60.

(a) Find an equation that expresses the revenue r from selling x leaf-blowers.

The revenue r from selling x leaf-blowers is the product of the price per item, $60, and the number of units sold (demand), x. So $r = 60x$.

(b) What is the revenue from selling 500 leaf-blowers?

Using the revenue equation of part (a) with $x = 500$, we have

$$r = 60 \cdot 500 = \$30,000.$$

(c) Find an equation that expresses the profit p from selling x leaf-blowers.

The profit is the difference between revenue and cost, that is,

$$p = r - c = 60x - (45x + 6000) = 15x - 6000.$$

(d) What is the profit from selling 500 leaf-blowers?

Using the profit equation of part (c) with $x = 500$, we have

$$p = 15 \cdot 500 - 6000 = 7500 - 6000 = \$1500.$$ ∎

A company can make a profit only if the revenue received from its customers exceeds the cost of producing its goods and services. The number of units at which revenue equals cost (that is, profit is 0) is the **break-even point.**

EXAMPLE 4   A firm producing poultry feed finds that the total cost c of producing x units is given by

$$c = 20x + 100.$$

Management plans to charge \$24 per unit for the feed. How many units must be sold for the firm to break even?

The revenue equation is $r = 24x$ (price per unit times number of units). The firm will break even (0 profit) as long as revenue equals cost, that is, when

$$r = c.$$
$$24x = 20x + 100$$
$$4x = 100$$
$$x = 25$$

The firm breaks even by selling 25 units. The graphs of the revenue and cost equations are shown in Figure 2.19. The break-even point (where $x = 25$) is shown on the graph. If the company produces more than 25 units ($x > 25$), it makes a profit. If it produces less than 25 units, it loses money (that is, profit is negative). ∎ $\boxed{3}$

$\boxed{3}$ For a certain magazine, the cost equation is $c = .70x + 1200$, where x is the number of magazines sold. The magazine sells for \$1 per copy. Find the break-even point.

Answer:

4000 magazines

$r = 1x$

$r = c$

$1x = .70x + 1700$

$.3\phi x = 120\,000\phi$

$x = 4000$

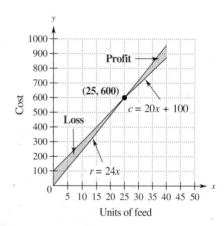

FIGURE 2.19

TECHNOLOGY TIP The break-even point can be found on a graphing calculator by graphing the cost and revenue equations on the same screen and using the calculator's intersection finder to locate their point of intersection. Depending on the calculator, the intersection finder is on the CALC or MATH or FCN or G-SOLVE or JUMP menu; check your instruction manual to see how to use it. ✔

SUPPLY AND DEMAND   The supply of and demand for an item are usually related to its price. Producers will supply large numbers of the item at a high price, but consumer demand will be low. As the price of the item decreases, consumer demand increases, but producers are less willing to supply large numbers of the item. The curves that show the quantity that will be supplied at a given price and the quantity that will be demanded at a given price are called **supply and demand curves.** Straight lines often appear as supply and demand curves, as in the next example. In supply and demand problems, we use p for price and q for quantity. We will discuss the economic concepts of supply and demand in more detail in later chapters.

EXAMPLE 5   Bill Cornett, an economist, has studied the supply and demand for aluminum siding and has determined that price per unit,* p, and the quantity demanded, q, are related by the linear equation

$$p = 60 - \frac{3}{4}q.$$

(a) Find the demand at a price of $40 per unit.
　　Let $p = 40$.

$$p = 60 - \frac{3}{4}q$$

$$\mathbf{40} = 60 - \frac{3}{4}q \qquad \text{Let } p = 40.$$

$$-20 = -\frac{3}{4}q \qquad \text{Add } -60 \text{ on both sides.}$$

$$\frac{80}{3} = q \qquad \text{Multiply both sides by } -\frac{4}{3}.$$

At a price of $40 per unit, 80/3 (or $26\frac{2}{3}$) units will be demanded.

(b) Find the price if the demand is 32 units.
　　Let $q = 32$.

$$p = 60 - \frac{3}{4}q$$

$$p = 60 - \frac{3}{4}(\mathbf{32}) \qquad \text{Let } q = 32.$$

$$p = 60 - 24$$

$$p = 36$$

With a demand of 32 units, the price is $36.

---

*An appropriate unit here might be, for example, one thousand square feet of siding.

4 Suppose price and demand are related by $p = 100 - 4q$.

(a) Find the price if the demand is 10 units.

(b) Find the demand if the price is $80.

(c) Write the corresponding ordered pairs.

Answers:

(a) $60

(b) 5 units

(c) (10, 60); (5, 80)

(c) Graph $p = 60 - \dfrac{3}{4}q$.

It is customary to use the horizontal axis for quantity q and the vertical axis for price p. In part (a) we saw that 80/3 units would be demanded at a price of $40 per unit; this gives the ordered pair (80/3, 40). Part (b) shows that with a demand of 32 units, the price is $36, which gives the ordered pair (32, 36). Use the points (80/3, 40) and (32, 36) to get the demand graph shown in Figure 2.20. Only the portion of the graph in Quadrant I is shown, because supply and demand are meaningful only for positive values of p and q.   **4**

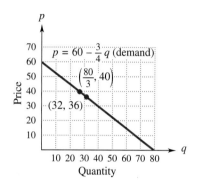

FIGURE 2.20

(d) From Figure 2.20, at a price of $30, what quantity is demanded?

Price is located on the vertical axis. Look for 30 on the p-axis and read across to where the line $p = 30$ crosses the demand graph. As the graph shows, this occurs where the demand is 40.

(e) At what price will 60 units be demanded?

Quantity is located on the horizontal axis. Find 60 on the q-axis and read up to where the vertical line $q = 60$ crosses the demand graph. This occurs where the price is about $15 per unit.

(f) What quantity is demanded at a price of $60 per unit?

The point (0, 60) on the demand graph shows that the demand is 0 at a price of $60 (that is, there is no demand at such a high price). ■

EXAMPLE 6 Suppose the economist of Example 5 concludes that the supply q of siding is related to its price p by the equation

$$p = .85q.$$

(a) Find the supply if the price is $51 per unit.

$$51 = .85q \qquad \text{Let } p = 51.$$
$$60 = q$$

If the price is $51 per unit, then 60 units will be supplied to the marketplace.

(b) Find the price per unit if the supply is 20 units.

$$p = .85(20) = 17 \qquad \text{Let } q = 20.$$

If the supply is 20 units, then the price is $17 per unit.

(c) Graph the supply equation $p = .85q$.

As with demand, each point on the graph has quantity q as its first coordinate and the corresponding price p as its second coordinate. Part (a) shows that the ordered pair $(60, 51)$ is on the graph of the supply equation and part (b) shows that $(20, 17)$ is on the graph. Using these points we obtain the supply graph in Figure 2.21.

FIGURE 2.21

(d) Use the graph in Figure 2.21 to find the approximate price at which 35 units will be supplied. Then use algebra to find the exact price.

The point on the graph with first coordinate $q = 35$ is approximately $(35, 30)$. Therefore, 35 units will be supplied when the price is approximately $30. To determine the exact price algebraically, substitute $q = 35$ in the supply equation:

$$p = .85q = .85(35) = \$29.75. \quad \blacksquare$$

EXAMPLE 7 The supply and demand curves of Examples 5 and 6 are shown in Figure 2.22. Graphically determine whether there is a surplus or a shortage of supply at a price of $40 per unit.

FIGURE 2.22

Find 40 on the vertical axis in Figure 2.22 and read across to the point where the horizontal line $p = 40$ crosses the supply graph (that is, the point corresponding to a price of $40). This point lies above the demand graph, so supply is greater than demand at a price of $40 and there is a surplus of supply. ■

Supply and demand are equal at the point where the supply curve intersects the demand curve. This is the **equilibrium point.** Its second coordinate is the **equilibrium price,** the price at which the same quantity will be supplied as is demanded. Its first coordinate is the quantity that will be demanded and supplied at the equilibrium price; this number is called either the **equilibrium demand** or the **equilibrium supply.**

EXAMPLE 8   In the situation described in Examples 5–7, what is the equilibrium demand? What is the equilibrium price?

The equilibrium point is where the supply and demand curves in Figure 2.22 intersect. To find the quantity q at which the price given by the supply equation $p = 60 - .75q$ (Example 5) is the same as that given by the demand equation $p = .85q$ (Example 6), set these two expressions for p equal and solve the resulting equation.

$$60 - .75q = .85q$$
$$60 = 1.6q$$
$$37.5 = q$$

5 The demand for a certain commodity is related to the price by $p = 80 - (2/3)q$. The supply is related to the price by $p = (4/3)q$. Find

(a) the equilibrium demand;

(b) the equilibrium price.

Answers:
(a) 40
(b) $160/3 \approx \$53.33$

Therefore, the equilibrium demand is 37.5 units, the number of units for which supply will equal demand. Substituting $q = 37.5$ in either the demand or supply equation shows that

$$p = 60 - .75(37.5) = 31.875 \quad \text{or} \quad p = .85(37.5) = 31.875.$$

So the equilibrium price is $31.875 (or $31.88 rounded). (To avoid error, it's a good idea to substitute in both equations, as we did here, to be sure the same value of p results; if it doesn't, a mistake has been made.) In this case, the equilibrium point—the point whose coordinates are the equilibrium demand and price—is (37.5, 31.875). ■ **5**

TECHNOLOGY TIP The equilibrium point (37.5, 31.875) can be found on a graphing calculator by graphing the supply and demand curves on the same screen and using the calculator's intersection finder to locate their point of intersection. ✔

2.3 EXERCISES

Physical Science   *Use the equation derived in Example 1 for conversion between Fahrenheit and Celsius temperatures.*

1. Convert each temperature.
   (a) 58°F to Celsius   **(b)** 50°C to Fahrenheit
   (c) −10°C to Fahrenheit   **(d)** −20°F to Celsius

2. According to the *1995 Guinness Book of Records,* Venus is the hottest planet, with a surface temperature of 864° Fahrenheit. What is this temperature in Celsius?

3. Find the temperature at which Celsius and Fahrenheit temperatures are numerically equal.

4. You may have heard that the average temperature of the human body is 98.6°. Recent experiments show that the actual figure is closer to 98.2°.* The figure of 98.6 comes from experiments done by Carl Wunderlich in 1868. But Wunderlich measured the temperatures in degrees Celsius and rounded the average to the nearest degree, giving 37°C as the average temperature.
   (a) What is the Fahrenheit equivalent of 37°C?
   (b) Given that Wunderlich rounded to the nearest degree Celsius, his experiments tell us that the actual average human temperature is somewhere between 36.5°C and 37.5°C. Find what this range corresponds to in degrees Fahrenheit.

5. **Management** Assume that the sales of a certain appliance dealer are approximated by a linear equation. Suppose that sales were $850,000 in 1987 and $1,262,500 in 1992. Let $x = 0$ represent 1987.
   (a) Find an equation giving the dealer's yearly sales.
   (b) Use this equation to approximate the sales in 1999.
   (c) The dealer estimates that a new store will be necessary once sales exceed $2,170,000. When is this expected to occur?

6. **Management** Assume that the sales of a certain automobile parts company are approximated by a linear equation. Suppose that sales were $200,000 in 1985 and $1,000,000 in 1992. Let $x = 0$ represent 1985 and $x = 7$ represent 1992.
   (a) Find the equation giving the company's yearly sales.
   (b) Use this equation to approximate the sales in 1996.
   (c) The company wants to negotiate a new contract once sales reach $2,000,000. When is this expected to occur?

In each of the following problems, assume that the data can be closely approximated by a straight line. Find the equation of the line and then answer the question. (See Examples 1–2.)

7. **Social Science** The graph, in which $x = 0$ corresponds to 1980, shows an idealized linear relationship for the average monthly family payment to families with dependent children in 1994 dollars.† Based on this information what was the average payment in 1987? (*Hint:* Use the graph to find two points, from which the equation can be found.)

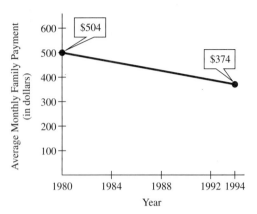

Source: Office of Financial Management, Administration for Children and Families.

8. **Management** Consumer prices in the United States at the beginning of each year, measured as a percent of the 1967 average, have produced a graph that is approximately linear. In 1984 the consumer price index was 300% and in 1987 it was 333%. Let y represent the consumer price index in year x, where $x = 0$ corresponds to 1980. In what year did the consumer price index reach 350%?

9. **Physical Science** Suppose a baseball is thrown at 85 miles per hour. The ball will travel 320 feet when hit by a bat swung at 50 miles per hour and will travel 440 feet when hit by a bat swung at 80 miles per hour. Let y be the number of feet traveled by the ball when hit by a bat swung at x miles per hour. (*Note:* This is valid for $50 \le x \le 90$, where the bat is 35 inches long, weighs 32 ounces, and strikes a waist-high pitch so the place of the swing lies at 10° from the diagonal.*) How much farther will a ball travel for each mile per hour increase in the speed of the bat?

10. **Natural Science** The amount of tropical rain forests in Central America decreased from 130,000 square miles to about 80,000 square miles from 1969 to 1985. Let y be the amount (in ten thousands of square miles) x years after 1965. How large were the rain forests in the year 1997?

11. **Physical Science** Ski resorts require large amounts of water in order to make snow. Snowmass Ski Area in Colorado plans to pump at least 1120 gallons of water per minute for at least 12 hours a day from Snowmass Creek between mid-October and late December.† Environmentalists are concerned about the effects on the ecosystem.

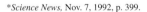

Science News, Nov. 7, 1992, p. 399.

†Office of Financial Management, Administration for Children and Families.

*Adair, Robert K. *The Physics of Baseball;* Harper & Row 1990.

†York Snow Incorporated.

Find the minimum amount of water pumped in 30 days. (*Hint:* Let y be the total number of gallons pumped x days after pumping begins. Note that $(0, 0)$ is on the graph of the equation.)

12. Physical Science Suppose that instead of making snow in Exercise 11, the water being pumped from Snowmass Creek was used to fill swimming pools. If the average backyard pool holds 20,000 gallons of water, how many pools could be filled each day? In how many days could at least 1000 pools be filled? (*Hint:* Let y be the number of pools filled after x days.)

13. Physical Science In 1994 Leroy Burrell (USA) set a world record in the 100-meter dash with a time of 9.85 seconds.* If this pace could be maintained for an entire 26 mile 385-yard marathon, how would this time compare to the fastest time for a marathon, 2 hours, 6 minutes, 50 seconds? (*Hint:* Let x be the distance in meters and y the time in seconds. Use the point $(0, 0)$ and the point obtained from the data in the problem to obtain the equation of a line. To answer the question, remember that 1 meter ≈ 3.281 ft.)

14. Natural Science In 1990, the Intergovernmental Panel on Climate Change predicted that the average temperature on the earth would rise $.3°C$ per decade in the absence of international controls on greenhouse emissions.† The average global temperature was $15°C$ in 1970. Let y be the average global temperature t years after 1970. Scientists have estimated that the sea level will rise by 65 centimeters, if the average global temperature rises to $19°C$. From your equation, when will this occur?

15. Physical Science Ventilation is an effective method of removing air pollutants. According to the American Society of Heating, Refrigerating and Air-Conditioning Engineers, Inc., a classroom should have a ventilation rate of 15 cubic feet per minute for each person in the classroom. How much ventilation y (in cubic feet per hour) is needed for a classroom with 30 people in it?

16. Physical Science A common unit of ventilation is an air exchange per hour (ach). 1 ach is equivalent to replacing all of the air in the room every hour. If a classroom with a volume of 15,000 cubic feet has 40 people in it, how many air exchanges per hour are necessary to keep the room properly ventilated? (*Hint:* Let x be the number of people and y the number of ach. Use the information developed in Exercise 15.)

17. Management The U.S. is China's largest export market. Imports from China have grown from about 8 billion dollars in 1988 to 39 billion dollars in 1995.* This growth has been approximately linear. Use the given data pairs to write a linear equation that describes this growth in imports over the years. Let $x = 88$ represent 1988 and $x = 95$ represent 1995.

18. Management U.S. exports to China have grown (although at a slower rate than imports) since 1988. In 1988, about 8 billion dollars of goods were exported to China. By 1995, this amount had grown to 15.9 billion dollars.* Write a linear equation describing the number of exports each year, with $x = 88$ representing 1988 and $x = 95$ representing 1995.

Use algebra to find the intersection points of the graphs of the given equations. (See Examples 4 and 8.)

19. $2x - y = 7$ and $y = 8 - 3x$

20. $6x - y = 2$ and $y = 4x + 7$

21. $y = 3x - 7$ and $y = 7x + 4$

22. $y = 3x + 5$ and $y = 12 - 2x$

Use a graphing calculator to find the intersection points of the graphs of the given equations.

23. $y = x^2 + x - 3$ and $y = x^3$

24. $y = .2x^2 + .5x - 3$ and $y = -x^5 + 2x + 1$

25. $y = .2x^2 + .5x - 3$ and $9x - 2y = 6$

26. $y = \sqrt{x^2 + 1}$ and $y = x^3 - 15x^2 + .8x - 6$

Management *Work the following problems. (See Examples 3–4.)*

27. For x thousand policies, an insurance company claims that their monthly revenue in dollars is given by $R = 125x$ and their monthly cost in dollars is given by $C = 100x + 5000$.
(a) Find the break-even point.
(b) Graph the revenue and cost equations on the same axes.
(c) From the graph, estimate the revenue and cost when $x = 100$ (100 thousand policies).

28. The owners of a parking lot have determined that their weekly revenue and cost in dollars are given by $R = 80x$ and $C = 50x + 2400$, where x is the number of long-term parkers.
(a) Find the break-even point.
(b) Graph R and C on the same axes.
(c) From the graph, estimate the revenue and cost when there are 60 long-term parkers.

29. The revenue (in millions of dollars) from the sale of x units at a home supply outlet is given by $r = .21x$. The profit (in

*International Amateur Athletic Association.

†*Science News,* June 23, 1990, p. 391.

*Problems 17 and 18 from *Economist,* U.S.-China Business Council, *China Business Review,* U.S. Commerce Department.

millions of dollars) from the sale of x units is given by $p = .084x - 1.5$.
(a) Find the cost equation.
(b) What is the cost of producing 7 units?
(c) What is the break-even point?

30. The profit (in millions of dollars) from the sale of x million units of Blue Glue is given by $p = .7x - 25.5$. The cost is given by $c = .9x + 25.5$.
(a) Find the revenue equation.
(b) What is the revenue from selling 10 million units?
(c) What is the break-even point?

Management *Suppose you are the manager of a firm. The accounting department has provided cost estimates and the sales department sales estimates on a new product. You must analyze the data they give you, determine what it will take to break even, and decide whether or not to go ahead with production of the new product. (See Examples 3 and 4.)*

31. Cost estimate is given by $c = 80x + 7000$ and revenue estimate by $r = 95x$; no more than 400 units can be sold.

32. Cost is $c = 140x + 3000$ and revenue is $r = 125x$.

33. Cost is $c = 125x + 42,000$ and revenue is $r = 165.5x$; no more than 2000 units can be sold.

34. Cost is $c = 1750x + 95,000$ and revenue is $r = 1975x$; no more than 600 units can be sold.

35. Management The graph shows the productivity of U.S. and Japanese workers in appropriate units over a 35-year period. Estimate the break-even point (the point at which workers in the two countries produced the same amounts).*

MANUFACTURING PRODUCTIVITY INDEX
■ Japan
■ U.S
1977 equals 100
Source: U.S. Bureau of Labor Statistics and Salomon Brothers Inc.

36. Management The graph gives U.S. imports and exports in billions of dollars over a five-year period. Estimate the break-even point.

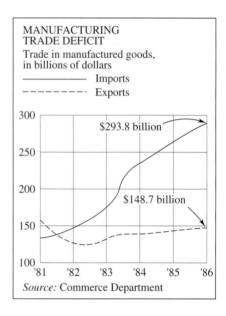

MANUFACTURING TRADE DEFICIT
Trade in manufactured goods, in billions of dollars
——— Imports
- - - - Exports
$293.8 billion
$148.7 billion
Source: Commerce Department

37. Management Canadian and Japanese investment in the United States in billions of dollars in 1980 and 1990 are shown in the chart.*

	1980	*1990*
Canada	12.1	27.7
Japan	4.7	108.1

(a) Assuming the change in investment in each case is linear, write an equation giving the investment in year x for each country. Let x be the number of years since 1980.
(b) Graph the equations from part (a) on the same coordinate axes.
(c) Find the intersection point where the graphs in part (b) cross and interpret your answer.

38. Social Science The median family income (in thousands of dollars) for the white population in the United States is given by $y = (4/3)x + 8$, where x is the number of years since 1973. The median family income (in thousands of dollars) for the black population in the United States is given by $y = (2/3)x + 6$.

*The figures for Exercises 35 and 36 appeared in *The Sacramento Bee,* December 21, 1987. Reprinted by permission of The Associated Press.

*Data reprinted with permission from *The World Almanac and Book of Facts,* 1992. Copyright © 1991. All rights reserved. The World Almanac is an imprint of Funk & Wagnalls Corporation.

(a) Graph both equations on the same coordinate axes.

(b) Do the graphs intersect? If so, in what year was median family income the same for white and black populations?

(c) What can you infer from the two graphs in part (a)?

Management *Use the supply and demand curves graphed below to answer Exercises 39–42. (See Examples 5–8.)*

39. At what price are 20 items supplied?

40. At what price are 20 items demanded?

41. Find the equilibrium supply and the equilibrium demand.

42. Find the equilibrium price.

Management *Work the following exercises. (See Examples 5–8.)*

43. Suppose that the demand and price for a certain brand of shampoo are related by

$$p = 16 - \frac{5}{4}q,$$

where p is price, in dollars, and q is demand. Find the price for a demand of

(a) 0 units; **(b)** 4 units; **(c)** 8 units.

Find the demand for the shampoo at a price of

(d) \$6; **(e)** \$11; **(f)** \$16.

(g) Graph $p = 16 - (5/4)q$. Suppose the price and supply of the shampoo are related by

$$p = \frac{3}{4}q,$$

where q represents the supply, and p the price. Find the supply when the price is

(h) \$0; **(i)** \$10; **(j)** \$20.

(k) Graph $p = (3/4)q$ on the same axes used for part (g).

(l) Find the equilibrium supply.

(m) Find the equilibrium price.

44. Let the supply and demand for radial tires in dollars be given by

$$\text{supply: } p = \frac{3}{2}q \quad \text{and} \quad \text{demand: } p = 81 - \frac{3}{4}q.$$

(a) Graph these on the same axes.

(b) Find the equilibrium demand.

(c) Find the equilibrium price.

45. Let the supply and demand for bananas in cents per pound be given by

$$\text{supply: } p = \frac{2}{5}q \quad \text{and} \quad \text{demand: } p = 100 - \frac{2}{5}q.$$

(a) Graph these on the same axes.

(b) Find the equilibrium demand.

(c) Find the equilibrium price.

(d) On what interval does demand exceed supply?

46. Let the supply and demand for sugar be given by

$$\text{supply: } p = 1.4q - .6$$

and

$$\text{demand: } p = -2q + 3.2,$$

where p is in dollars.

(a) Graph these on the same axes.

(b) Find the equilibrium demand.

(c) Find the equilibrium price.

(d) On what interval does supply exceed demand?

2.4 QUADRATIC EQUATIONS

In the first part of this chapter we studied graphs of equations in two variables, such as $y = x^2$ and $y = 2x - 5$. In the remainder of the chapter we consider equations and inequalities in one variable, such as

$$9x^2 - 12x = 1, \quad 3x + 5 > 11, \quad \text{and} \quad z^3 - 4z < 0.$$

Many applied problems reduce to solving such equations and inequalities. Although the emphasis will be on algebraic solution methods, we shall use some graphical facts about equations in two variables in order to understand why these algebraic methods work.

In Section 1.2, we considered first-degree equations, such as $3x + 7 = 15$. In this section we consider second-degree or quadratic equations. An equation that can be put in the form

$$ax^2 + bx + c = 0,$$

where a, b, and c are real numbers with $a \neq 0$, is called a **quadratic equation.** For example, each of

$$2x^2 + 3x + 4 = 0, \quad x^2 = 6x - 9, \quad 3x^2 + x = 6, \quad \text{and} \quad x^2 = 5$$

is a quadratic equation. A solution of an equation that is a real number is said to be a **real solution** of the equation.

One method of solving quadratic equations is based on the following property of real numbers.

ZERO–FACTOR PROPERTY

If a and b are real numbers, with $ab = 0$, then $a = 0$ or $b = 0$ or both.

EXAMPLE 1 Solve the equation $(x - 4)(3x + 7) = 0$.

By the zero-factor property, the product $(x - 4)(3x + 7)$ can equal 0 only if at least one of the factors equals 0. That is, the product equals zero only if $x - 4 = 0$ or $3x + 7 = 0$. Solving each of these equations separately will give the solutions of the original equation.

$$
\begin{aligned}
x - 4 &= 0 \quad \text{or} \quad 3x + 7 = 0 \\
x &= 4 \quad \text{or} \qquad 3x = -7 \\
&\qquad\qquad\qquad\qquad x = -\frac{7}{3}
\end{aligned}
$$

The solutions of the equation $(x - 4)(3x + 7) = 0$ are 4 and $-7/3$. Check these solutions by substitution in the original equation. ■ $\boxed{1}$

$\boxed{1}$ Solve the following equations.

(a) $(y - 6)(y + 2) = 0$

(b) $(5k - 3)(k + 5) = 0$

(c) $(2r - 9)(3r + 5)(r + 3) = 0$

Answers:

(a) $6, -2$

(b) $3/5, -5$

(c) $9/2, -5/3, -3$

EXAMPLE 2 Solve $6r^2 + 7r = 3$.

Rewrite the equation as

$$6r^2 + 7r - 3 = 0.$$

Now factor $6r^2 + 7r - 3$ to get

$$(3r - 1)(2r + 3) = 0.$$

By the zero-factor property, the product $(3r - 1)(2r + 3)$ can equal 0 only if

$$3r - 1 = 0 \quad \text{or} \quad 2r + 3 = 0.$$

2 Solve each equation by factoring.

(a) $y^2 + 3y = 10$

(b) $2r^2 + 9r = 5$

(c) $4k^2 = 9k$

Answers:

(a) $2, -5$

(b) $1/2, -5$

(c) $9/4, 0$

Solving each of these equations separately gives the solutions of the original equation.

$$3r = 1 \quad \text{or} \quad 2r = -3$$

$$r = \frac{1}{3} \qquad r = -\frac{3}{2}$$

Verify that both $1/3$ and $-3/2$ are solutions by substituting them in the original equation. ■   **2**

An equation such as $x^2 = 5$ has two solutions, $\sqrt{5}$ and $-\sqrt{5}$. The same idea is true in general.

SQUARE-ROOT PROPERTY

If $b > 0$, then the solutions of $x^2 = b$ are $\sqrt{b}$ and $-\sqrt{b}$.

The two solutions are sometimes abbreviated $\pm\sqrt{b}$.

EXAMPLE 3   Solve each equation.

(a) $m^2 = 17$

By the square-root property, the solutions are $\sqrt{17}$ and $-\sqrt{17}$, abbreviated $\pm\sqrt{17}$.

3 Solve each equation by using the square-root property.

(a) $p^2 = 21$

(b) $(m + 7)^2 = 15$

(c) $(2k - 3)^2 = 5$

Answers:

(a) $\pm\sqrt{21}$

(b) $-7 \pm \sqrt{15}$

(c) $(3 \pm \sqrt{5})/2$

(b) $(y - 4)^2 = 11$

Use a generalization of the square-root property, working as follows:

$$(y - 4)^2 = 11$$
$$y - 4 = \sqrt{11} \quad \text{or} \quad y - 4 = -\sqrt{11}$$
$$y = 4 + \sqrt{11} \qquad y = 4 - \sqrt{11}.$$

Abbreviate the solutions as $4 \pm \sqrt{11}$. ■   **3**

As suggested by Example 3(b), any quadratic equation can be solved using the square-root property if the equation can be written in the form $(x + n)^2 = k$. For instance, to write $4x^2 - 24x + 19 = 0$ in this form, we need to write the left side as a perfect square, $(x + n)^2$. To get a perfect square trinomial, first add -19 to both sides of the equation, then multiply both sides by $1/4$, so that x^2 has a coefficient of 1.

$$4x^2 - 24x + 19 = 0$$
$$4x^2 - 24x = -19$$
$$x^2 - 6x = -\frac{19}{4} \qquad \text{(*)}$$

Adding 9 to both sides will make the left side a perfect square, so the equation can be written in the desired form.

$$x^2 - 6x + 9 = -\frac{19}{4} + 9$$

$$(x - 3)^2 = \frac{17}{4}$$

Use the square-root property to complete the solution.

$$x - 3 = \pm\sqrt{\frac{17}{4}} = \pm\frac{\sqrt{17}}{2}$$

$$x = 3 \pm \frac{\sqrt{17}}{2} = \frac{6 \pm \sqrt{17}}{2}$$

The two solutions are $\dfrac{6 + \sqrt{17}}{2}$ and $\dfrac{6 - \sqrt{17}}{2}$.

Notice that the number 9 added to both sides of equation (*) is found by taking $(1/2)(6) = 3$ and squaring it: $3^2 = 9$. This always works because

$$\left(x + \frac{b}{2}\right)^2 = x^2 + 2\left(\frac{b}{2}\right)x + \left(\frac{b}{2}\right)^2 = x^2 + bx + \left(\frac{b}{2}\right)^2.$$

The process of changing $4x^2 - 24x + 19 = 0$ to $(x - 3)^2 = 17/4$ is called **completing the square.**†

The method of completing the square can be used on the general quadratic equation,

$$ax^2 + bx + c = 0 \quad (a \neq 0),$$

to convert it to one whose solutions can be found by the square-root property. This will give a general formula for solving any quadratic equation. Going through the necessary algebra produces the following important result.

QUADRATIC FORMULA

The solutions of the quadratic equation $ax^2 + bx + c = 0$, where $a \neq 0$, are given by

$$x = \frac{-b \pm \sqrt{b^2 - 4ac}}{2a}.$$

CAUTION When using the quadratic formula, remember the equation must be in the form $ax^2 + bx + c = 0$. Also, notice that the fraction in the quadratic formula extends under *both* terms in the numerator. Be sure to add $-b$ to $\pm\sqrt{b^2 - 4ac}$ *before* dividing by $2a$. ◆

---

†Completing the square is discussed further in Section 4.1.

EXAMPLE 4   Solve $x^2 + 1 = 4x$.

First add $-4x$ to both sides, to get 0 alone on the right side.

$$x^2 - 4x + 1 = 0$$

Now identify the letters a, b, and c. Here $a = 1$, $b = -4$, and $c = 1$. Substitute these numbers into the quadratic formula

$$x = \frac{-(-4) \pm \sqrt{(-4)^2 - 4(1)(1)}}{2(1)}$$

$$= \frac{4 \pm \sqrt{16 - 4}}{2}$$

$$= \frac{4 \pm \sqrt{12}}{2}$$

$$= \frac{4 \pm 2\sqrt{3}}{2} \qquad \sqrt{12} = \sqrt{4 \cdot 3} = \sqrt{4} \cdot \sqrt{3} = 2\sqrt{3}$$

$$= \frac{2(2 \pm \sqrt{3})}{2} \qquad \text{Factor } 4 \pm 2\sqrt{3}.$$

$$x = 2 \pm \sqrt{3}.$$

The $\pm$ sign represents the two solutions of the equation. First use $+$ and then use $-$ to find each of the solutions: $2 + \sqrt{3}$ and $2 - \sqrt{3}$. ■   4

4 Use the quadratic formula to solve each equation.

(a) $x^2 - 2x = 2$

(b) $u^2 - 6u + 4 = 0$

Answers:

(a) $x = 1 + \sqrt{3}$ or $1 - \sqrt{3}$

(b) $u = 3 + \sqrt{5}$ or $3 - \sqrt{5}$

Example 4 shows that the quadratic formula produces exact solutions of quadratic equations that do not readily factor. In many real-world situations, however, accurate decimal approximations of solutions are needed. A calculator is ideal for this. In Example 4, for instance, we can use a calculator to determine that approximate solutions (accurate to seven decimal places) are

$$x = 2 + \sqrt{3} \approx 3.7320508 \quad \text{and} \quad x = 2 - \sqrt{3} \approx .2679492.$$

Experiment with your calculator to find the most efficient keystroke sequence for using the quadratic formula.

EXAMPLE 5   Use the quadratic formula and a calculator to solve

$$3.2x^2 + 15.93x - 7.1 = 0.$$

Compute $\sqrt{b^2 - 4ac} = \sqrt{15.93^2 - 4(3.2)(-7.1)} \approx 18.56461419$ and store the result in memory. This number may be inserted in any computation by using the memory recall key, which we shall denote here by MR (it may be labeled differently on your calculator). The solutions of the equation are given by

$$x = (-15.93 + \text{MR})/(2 \cdot 3.2) \approx .411658467$$

and

$$x = (-15.93 - \text{MR})/(2 \cdot 3.2) \approx -5.38978347.$$

5 Find approximate solutions for $5.1x^2 - 3.3x - 240.624 = 0$.

Answer:
$x = 7.2$ or $x \approx -6.5529$

Note the use of parentheses in these calculations—omitting the parentheses leads to wrong answers. Also remember that these answers are *approximations,* so they may not check exactly when substituted in the original equation. ■ **5**

TECHNOLOGY TIP You can approximate the solutions of quadratic equations on a graphing calculator by using a quadratic formula program (see the Program Appendix) or using a built-in quadratic equation solver if your calculator has one. Then you need only enter the coefficients a, b, c to obtain the approximate solutions. See your instruction manual for details. ✔

EXAMPLE 6 Solve $9x^2 - 30x + 25 = 0$.

Applying the quadratic formula with $a = 9$, $b = -30$, and $c = 25$, we have

$$x = \frac{-(-30) \pm \sqrt{(-30)^2 - 4(9)(25)}}{2(9)}$$

$$= \frac{30 \pm \sqrt{900 - 900}}{18} = \frac{30 \pm 0}{18} = \frac{30}{18} = \frac{5}{3}.$$

Therefore, the given equation has only one real solution. The fact that the solution is a rational number indicates that this equation could have been solved by factoring. ■

EXAMPLE 7 Solve $x^2 - 6x + 10 = 0$.

Apply the quadratic formula with $a = 1$, $b = -6$, and $c = 10$.

$$x = \frac{-(-6) \pm \sqrt{(-6)^2 - 4(1)(10)}}{2(1)}$$

$$= \frac{6 \pm \sqrt{36 - 40}}{2}$$

$$= \frac{6 \pm \sqrt{-4}}{2}$$

6 Solve each equation.

(a) $9k^2 - 6k + 1 = 0$

(b) $4m^2 + 28m + 49 = 0$

(c) $2x^2 - 5x + 5 = 0$

Answers:
(a) 1/3

(b) −7/2

(c) No real number solutions

Since no negative number has a square root in the real number system, $\sqrt{-4}$ is not a real number. Hence the equation has no real solutions. ■ **6**

As illustrated in Examples 4–7,

7 Use the discriminant to determine the number of real solutions of each equation.

(a) $x^2 + 8x + 3 = 0$

(b) $2x^2 + x + 3 = 0$

(c) $x^2 - 194x + 9409 = 0$

Answers:
(a) 2

(b) 0

(c) 1

Every quadratic equation has either two or one or no real solutions.

These examples also show that $b^2 - 4ac$, the quantity under the radical in the quadratic formula, determines the number of real solutions of the equation as follows.

If $b^2 - 4ac > 0$ (as in Examples 4 and 5), there are two real solutions.

If $b^2 - 4ac = 0$ (as in Example 6), there is one real solution.

If $b^2 - 4ac < 0$ (as in Example 7), there are no real solutions.

The number $b^2 - 4ac$ is called the **discriminant** of the quadratic equation $ax^2 + bx + c = 0$. **7**

TECHNOLOGY TIP The real number solutions of a one-variable equation such as $x^2 - 6x + 10 = 0$ are the x-intercepts of the graph of the two-variable equation $y = x^2 - 6x + 10$. So you can use a graphing calculator to determine the number of solutions of the equation by graphing $y = x^2 - 6x + 10$ and counting the number of x-intercepts. You can also find these solutions graphically by using either zoom-in or a graphical root finder (if your calculator has one) to locate the x-intercepts of the graph. See your instruction manual for details. ✔

EXAMPLE 8   A landscape architect wants to make an exposed gravel border of uniform width around a small shed behind a company plant. The shed is 10 feet by 6 feet. He has enough gravel to cover 36 square feet. How wide should the border be?

Follow the steps given in Section 1.2 for solving applied problems. A sketch of the shed with border is given in Figure 2.23. Let x represent the width of the border.

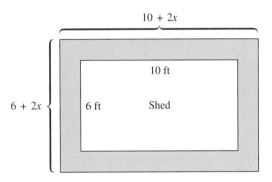

FIGURE 2.23

Then the width of the large rectangle is $6 + 2x$ and its length is $10 + 2x$. We must write an equation relating the given areas and dimensions. The area of the large rectangle is $(6 + 2x)(10 + 2x)$. The area occupied by the shed is $6 \cdot 10 = 60$. The area of the border is found by subtracting the area of the shed from the area of the large rectangle. This difference should be 36 square feet, giving the equation

$$(6 + 2x)(10 + 2x) - 60 = 36.$$

Solve this equation with the following sequence of steps.

$$60 + 32x + 4x^2 - 60 = 36$$
$$4x^2 + 32x - 36 = 0$$
$$x^2 + 8x - 9 = 0$$
$$(x + 9)(x - 1) = 0$$

The solutions are -9 and 1. The number -9 cannot be the width of the border, so the solution is to make the border 1 foot wide. ■ 8

8 The length of a picture is 2 inches more than the width. It is mounted on a mat that extends 2 inches beyond the picture on all sides. What are the dimensions of the picture if the area of the mat is 99 square inches?

Answer:
5 inches by 7 inches

Some equations that are not quadratic can be solved as quadratic equations by making a suitable substitution. Such equations are called **quadratic in form.**

EXAMPLE 9 Solve $4m^4 - 9m^2 + 2 = 0$.

Use the substitutions

$$x = m^2 \quad \text{and} \quad x^2 = m^4$$

to rewrite the equation as

$$4x^2 - 9x + 2 = 0.$$

This quadratic equation can be solved by factoring.

$$4x^2 - 9x + 2 = 0$$
$$(x - 2)(4x - 1) = 0$$
$$x - 2 = 0 \quad \text{or} \quad 4x - 1 = 0$$
$$x = 2 \quad \text{or} \quad 4x = 1$$
$$x = \frac{1}{4}$$

Because $x = m^2$,

$$m^2 = 2 \qquad \text{or} \quad m^2 = \frac{1}{4}$$

$$m = \pm\sqrt{2} \quad \text{or} \quad m = \pm\frac{1}{2}.$$

There are four solutions: $-\sqrt{2}, \sqrt{2}, -1/2,$ and $1/2$. ■ 9

9 Solve each equation.

(a) $9x^4 - 23x^2 + 10 = 0$

(b) $4x^4 = 7x^2 - 3$

Answers:

(a) $-\sqrt{5}/3, \sqrt{5}/3, -\sqrt{2}, \sqrt{2}$

(b) $-\sqrt{3}/2, \sqrt{3}/2, -1, 1$

EXAMPLE 10 Solve $4 + \dfrac{1}{z - 2} = \dfrac{3}{2(z - 2)^2}$.

First, substitute u for $z - 2$ to get

$$4 + \frac{1}{z - 2} = \frac{3}{2(z - 2)^2}$$

$$4 + \frac{1}{u} = \frac{3}{2u^2}.$$

Now multiply both sides of the equation by the common denominator $2u^2$; then solve the resulting quadratic equation.

$$2u^2\left(4 + \frac{1}{u}\right) = 2u^2\left(\frac{3}{2u^2}\right)$$

$$8u^2 + 2u = 3$$

$$8u^2 + 2u - 3 = 0$$

$$(4u + 3)(2u - 1) = 0$$

10 Solve each equation.

(a) $\dfrac{13}{3(p+1)} + \dfrac{5}{3(p+1)^2} = 2$

(b) $1 + \dfrac{1}{a} = \dfrac{5}{a^2}$

Answers:
(a) $-4/3,\ 3/2$
(b) $(-1+\sqrt{21})/2,$
$(-1-\sqrt{21})/2$

$$4u + 3 = 0 \qquad \text{or} \quad 2u - 1 = 0$$

$$u = -\frac{3}{4} \qquad \text{or} \qquad u = \frac{1}{2}$$

$$z - 2 = -\frac{3}{4} \qquad \text{or} \quad z - 2 = \frac{1}{2} \qquad \text{Replace } u \text{ with } z - 2.$$

$$z = -\frac{3}{4} + 2 \qquad\qquad z = \frac{1}{2} + 2$$

$$z = \frac{5}{4} \qquad \text{or} \qquad z = \frac{5}{2}.$$

The solutions are 5/4 and 5/2. Check both solutions in the original equation. ■ **10**

The next example shows how to solve an equation for a specified variable when the equation is quadratic in that variable.

11 Solve each of the following equations for the indicated variable. Assume all variables are positive.

(a) $k = mp^2 - bp$ for p

(b) $r = \dfrac{APk^2}{3}$ for k

Answers:
(a) $p = \dfrac{b \pm \sqrt{b^2 + 4mk}}{2m}$

(b) $k = \pm\sqrt{\dfrac{3r}{AP}}$ or $\dfrac{\pm\sqrt{3rAP}}{AP}$

EXAMPLE 11 Solve $v = mx^2 + x$ for x. (Assume $m \neq 0$.)

The equation is quadratic in x because of the x^2 term. Use the quadratic formula, first writing the equation in standard form.

$$v = mx^2 + x$$
$$0 = mx^2 + x - v$$

Let $a = m$, $b = 1$, and $c = -v$. Then the quadratic formula gives

$$x = \frac{-1 \pm \sqrt{1^2 - 4(m)(-v)}}{2m}$$

$$x = \frac{-1 \pm \sqrt{1 + 4mv}}{2m}. \qquad ■ \quad \boxed{11}$$

2.4 EXERCISES

Use factoring to solve each equation. (See Examples 1 and 2.)

1. $(x + 3)(x - 12) = 0$
2. $(p - 16)(p - 5) = 0$
3. $x(x + 5) = 0$
4. $x^2 - 2x = 0$
5. $3z^2 = 6z$
6. $x^2 - 81 = 0$
7. $y^2 + 15y + 56 = 0$
8. $k^2 - 4k - 5 = 0$
9. $2x^2 = 5x - 3$
10. $2 = 12z^2 + 5z$
11. $6r^2 + r = 1$
12. $3y^2 = 16y - 5$

13. $2m^2 + 20 = 13m$
14. $10a^2 + 17a + 3 = 0$
15. $m(m - 7) = -10$
16. $z(2z + 7) = 4$
17. $9x^2 - 16 = 0$
18. $25y^2 - 64 = 0$
19. $16x^2 - 16x = 0$
20. $12y^2 - 48y = 0$

Solve each equation by the square-root property. (See Example 3.)

21. $(r - 2)^2 = 7$
22. $(b + 5)^2 = 8$
23. $(4x - 1)^2 = 20$
24. $(3t + 5)^2 = 11$

Use the quadratic formula to solve each equation. If the solutions involve square roots, give both the exact and approximate solutions. (See Examples 4–7.)

25. $2x^2 + 5x + 1 = 0$

26. $3x^2 - x - 7 = 0$

27. $4k^2 + 2k = 1$

28. $r^2 = 3r + 5$

29. $5y^2 + 6y = 2$

30. $2z^2 + 3 = 8z$

31. $6x^2 + 6x + 5 = 0$

32. $3a^2 - 2a + 2 = 0$

33. $2r^2 - 7r + 5 = 0$

34. $8x^2 = 8x - 3$

35. $6k^2 - 11k + 4 = 0$

36. $8m^2 - 10m + 3 = 0$

37. $2x^2 - 7x + 30 = 0$

38. $3k^2 + k = 6$

Use the discriminant to determine the number of real solutions of each equation. You need not solve the equations.

39. $25t^2 + 49 = 70t$

40. $9z^2 - 12z = 1$

41. $13x^2 + 24x - 6 = 0$

42. $22x^2 + 19x + 5 = 0$

Use a calculator and the quadratic formula to find approximate solutions of the equation. (See Example 5.)

43. $4.42x^2 - 10.14x + 3.79 = 0$

44. $3x^2 - 82.74x + 570.4923 = 0$

45. $7.63x^2 + 2.79x = 5.32$

46. $8.06x^2 + 25.8726x = 25.047256$

Give all real number solutions of the following equations. (See Examples 9 and 10. Hint: In Exercise 51, let $u = p - 3$.)

47. $z^4 - 2z^2 = 15$

48. $6p^4 = p^2 + 2$

49. $2q^4 + 3q^2 - 9 = 0$

50. $4a^4 = 2 - 7a^2$

51. $6(p - 3)^2 + 5(p - 3) - 6 = 0$

52. $12(q + 4)^2 - 13(q + 4) - 4 = 0$

53. $1 + \dfrac{7}{2a} = \dfrac{15}{2a^2}$

54. $5 - \dfrac{4}{k} - \dfrac{1}{k^2} = 0$

55. $-\dfrac{2}{3z^2} + \dfrac{1}{3} + \dfrac{8}{3z} = 0$

56. $2 + \dfrac{5}{x} + \dfrac{1}{x^2} = 0$

Students often confuse expressions with equations. In part (a) add or subtract the expressions as indicated. In part (b) solve the equation. Compare the results.

57. (a) $\dfrac{6}{r} - \dfrac{5}{r - 2} - 1$ **(b)** $\dfrac{6}{r} = \dfrac{5}{r - 2} - 1$

58. (a) $\dfrac{8}{z - 1} - \dfrac{5}{z} - \dfrac{2z}{z - 1}$ **(b)** $\dfrac{8}{z - 1} = \dfrac{5}{z} + \dfrac{2z}{z - 1}$

Solve the following problems. (See Example 8.)

59. An ecology center wants to set up an experimental garden. It has 300 meters of fencing to enclose a rectangular area of 5000 square meters. Find the length and width of the rectangle.
 (a) Let $x =$ the length and write an expression for the width.
 (b) Write an equation relating the length, width, and area, using the result of part (a).
 (c) Solve the problem.

60. A shopping center has a rectangular area of 40,000 square yards enclosed on three sides for a parking lot. The length is 200 yards more than twice the width. Find the length and width of the lot. Let x equal the width and follow steps similar to those in Exercise 59.

61. A landscape architect has included a rectangular flower bed measuring 9 ft by 5 ft in her plans for a new building. She wants to use two colors of flowers in the bed, one in the center and the other for a border of the same width on all four sides. If she can get just enough plants to cover 24 sq ft for the border, how wide can the border be?

62. Joan wants to buy a rug for a room that is 12 feet by 15 feet. She wants to leave a uniform strip of floor around the rug. She can afford 108 square feet of carpeting. What dimensions should the rug have?

63. In 1991 Rick Mears won the (500 mi) Indianapolis 500 race. His speed (rate) was 100 mph (to the nearest mph) faster than that of the 1911 winner, Ray Harroun. Mears completed the race in 3.74 hours less time than Harroun. Find Mears's rate to the nearest whole number.

64. Chris and Josh have received walkie-talkies for Christmas. If they leave from the same point at the same time, Chris walking north at 2.5 mph and Josh walking east at 3 mph, how long will they be able to talk to each other if the range of the walkie-talkies is 4 mi? Round your answer to the nearest minute.

65. Management The manager of a bicycle shop knows that the cost of selling x bicycles is $C = 20x + 60$ and the revenue from selling x bicycles is $R = x^2 - 8x$. Find the break-even value of x.

66. Management A company that produces breakfast cereal has found that its operating cost in dollars is $C = 40x + 150$ and its revenue in dollars is $R = 65x - x^2$. For what value(s) of x will the company break even?

67. Physical Science If a ball is thrown upward with an initial velocity of 64 feet per second, then its height after t seconds is $h = 64t - 16t^2$. In how many seconds will the ball reach
(a) 64 ft?   (b) 28 ft?
(c) Why are two answers possible?

68. Physical Science A particle moves horizontally with its distance from a starting point in centimeters at t seconds given by $d = 11t^2 - 10t$.
(a) How long will it take before the particle returns to the starting point?
(b) When will the particle be 100 centimeters from the starting point?

Solve each of the following equations for the indicated variable. Assume all denominators are nonzero, and that all variables represent positive real numbers. (See Example 11.)

69. $S = \dfrac{1}{2}gt^2$ for t

70. $a = \pi r^2$ for r

71. $L = \dfrac{d^4 k}{h^2}$ for h

72. $F = \dfrac{kMv^2}{r}$ for v

73. $P = \dfrac{E^2 R}{(r + R)^2}$ for R

74. $S = 2\pi rh + 2\pi r^2$ for r

2.5 LINEAR INEQUALITIES

An **inequality** is a statement that one mathematical expression is greater than (or less than) another. Inequalities are very important in applications. For example, a company wants revenue to be *greater than* costs and must use *no more than* the total amount of capital or labor available.

Inequalities may be solved using algebraic or geometric methods. In this section we shall concentrate on algebraic methods of solving **linear inequalities,** such as

$$4 - 3x \le 7 + 2x \quad \text{and} \quad -2 < 5 + 3m < 20,$$

and absolute value inequalities, such as $|x - 2| < 5$. The following properties are the basic algebraic tools for working with inequalities.

PROPERTIES OF INEQUALITY

For real numbers a, b, and c,

(a) if $a < b$, then $a + c < b + c$
(b) if $a < b$, and if $c > 0$, then $ac < bc$
(c) if $a < b$, and if $c < 0$, then $ac > bc$.

Throughout this section, definitions are given only for $<$; but they are equally valid for $>$, $\le$, or $\ge$.

1 **(a)** First multiply both sides of $-6 < -1$ by 4, and then multiply both sides of $-6 < -1$ by -7.

(b) First multiply both sides of $9 \geq -4$ by 2, and then multiply both sides of $9 \geq -4$ by -5.

(c) First add 4 to both sides of $-3 < -1$, and then add -6 to both sides of $-3 < -1$.

Answers:
(a) $-24 < -4;\ 42 > 7$
(b) $18 \geq -8;\ -45 \leq 20$
(c) $1 < 3;\ -9 < -7$

CAUTION Pay careful attention to part (c): if both sides of an inequality are multiplied by a negative number, the direction of the inequality symbol must be reversed. For example, starting with the true statement $-3 < 5$ and multiplying both sides by the positive number 2 gives

$$-3 \cdot \mathbf{2} < 5 \cdot \mathbf{2},$$

or

$$-6 < 10,$$

still a true statement. On the other hand, starting with $-3 < 5$ and multiplying both sides by the negative number -2 gives a true result only if the direction of the inequality symbol is reversed:

$$-3(\mathbf{-2}) > 5(\mathbf{-2})$$
$$6 > -10. \quad \blacklozenge \quad \boxed{1}$$

EXAMPLE 1 Solve $3x + 5 > 11$. Graph the solution.

First, add -5 to both sides.

$$3x + 5 + (\mathbf{-5}) > 11 + (\mathbf{-5})$$
$$3x > 6$$

Now multiply both sides by 1/3.

$$\frac{1}{3}(3x) > \frac{1}{3}(6)$$
$$x > 2$$

2 Solve these inequalities. Graph each solution.

(a) $5z - 11 < 14$

(b) $-3k \leq -12$

(c) $-8y \geq 32$

Answers:
(a) $z < 5$

(b) $k \geq 4$

(c) $y \leq -4$

(Why was the direction of the inequality symbol not changed?) As a check, note that 0, which is not part of the solution, makes the inequality false, while 3, which is part of the solution, makes it true.

$$3(0) + 5 > 11 \qquad\qquad 3(3) + 5 > 11$$
$$5 > 11 \quad \text{False} \qquad\qquad 14 > 11 \quad \text{True}$$

In interval notation (introduced in Section 1.1), the solution is the interval $(2, \infty)$, which is graphed on the number line in Figure 2.24. An open circle at 2 shows that 2 is not included. ■ **2**

FIGURE 2.24

EXAMPLE 2 Solve $4 - 3x \leq 7 + 2x$.

$$4 - 3x \leq 7 + 2x$$
$$4 - 3x + (\mathbf{-4}) \leq 7 + 2x + (\mathbf{-4})$$
$$-3x \leq 3 + 2x$$

Add $-2x$ to both sides. (Remember that *adding* to both sides never changes the direction of the inequality symbol.)

$$-3x + (-2x) \le 3 + 2x + (-2x)$$
$$-5x \le 3$$

3 Solve these inequalities. Graph each solution.

(a) $8 - 6t \ge 2t + 24$

(b) $-4r + 3(r + 1) < 2r$

Answers:

(a) $t \le -2$

(b) $r > 1$

Multiply both sides by $-1/5$. Since $-1/5$ is negative, change the direction of the inequality symbol.

$$-\frac{1}{5}(-5x) \ge -\frac{1}{5}(3)$$

$$x \ge -\frac{3}{5}$$

Figure 2.25 shows a graph of the solution $[-3/5, \infty)$. The solid circle in Figure 2.25 shows that $-3/5$ is included in the solution. ■ **3**

FIGURE 2.25

4 Solve each of the following. Graph each solution.

(a) $9 < k + 5 < 13$

(b) $-6 \le 2z + 4 \le 12$

Answers:

(a) $4 < k < 8$

(b) $-5 \le z \le 4$

EXAMPLE 3 Solve $-2 < 5 + 3m < 20$. Graph the solution.

The inequality $-2 < 5 + 3m < 20$ says that $5 + 3m$ is *between* -2 and 20. We can solve this inequality with an extension of the properties given above. Work as follows, first adding -5 to each part.

$$-2 + (-5) < 5 + 3m + (-5) < 20 + (-5)$$
$$-7 < 3m < 15$$

Now multiply each part by $1/3$.

$$-\frac{7}{3} < m < 5$$

A graph of the solution, $(-7/3, 5)$, is given in Figure 2.26. ■ **4**

FIGURE 2.26

EXAMPLE 4 The formula for converting from Celsius to Fahrenheit temperature is

$$F = \frac{9}{5}C + 32.$$

What Celsius temperature range corresponds to the range from 32°F to 77°F? The Fahrenheit temperature range is $32 < F < 77$. Since $F = (9/5)C + 32$,

$$32 < \frac{9}{5}C + 32 < 77.$$

Solve the inequality for C.

$$32 < \frac{9}{5}C + 32 < 77$$

$$0 < \frac{9}{5}C < 45$$

5 In Example 4, what Celsius temperatures correspond to 5°F to 95°F?

$$0 < C < \frac{5}{9} \cdot 45$$

$$0 < C < 25$$

Answer:
$-15°C$ to $35°C$

The corresponding Celsius temperature range is 0°C to 25°C. ■ **5**

A product will break even, or produce a profit, only if the revenue R from selling the product at least equals the cost C of producing it, that is if $R \geq C$.

EXAMPLE 5 A company analyst has determined that the cost to produce and sell x units of a certain product is $C = 20x + 1000$. The revenue for that product is $R = 70x$. Find the values of x for which the company will break even or make a profit on the product.

Solve the inequality $R \geq C$.

$$R \geq C$$
$$70x \geq 20x + 1000 \qquad \text{Let } R = 70x, C = 20x + 1000.$$
$$50x \geq 1000$$
$$x \geq 20$$

The company must produce and sell 20 items to break even and more than 20 to make a profit. ■

The next examples show how to solve inequalities with absolute value.

EXAMPLE 6 Solve each inequality.

(a) $|x| < 5$

Because absolute value gives the distance from a number to 0, the inequality $|x| < 5$ is true for all real numbers whose distance from 0 is less than 5. This includes all numbers from -5 to 5, or numbers in the interval $(-5, 5)$. A graph of the solution is shown in Figure 2.27.

FIGURE 2.27

Solve each inequality. Graph each solution.

(a) $|x| \leq 1$

(b) $|y| \geq 3$

Answers:

(a) $[-1, 1]$

(b) All numbers in

$(-\infty, -3]$ or $[3, \infty)$

(b) $|x| > 5$

In a similar way, the solution of $|x| > 5$ is given by all those numbers whose distance from 0 is *greater* than 5. This includes the numbers satisfying $x < -5$ or $x > 5$. A graph of the solution, all numbers in

$$(-\infty, -5) \quad \text{or} \quad (5, \infty),$$

is shown in Figure 2.28. ■

FIGURE 2.28

The examples above suggest the following generalizations.

Assume a and b are real numbers with b positive.

1. Solve $|a| = |b|$ by solving $a = b$ or $a = -b$.

2. Solve $|a| < b$ by solving $-b < a < b$.

3. Solve $|a| > b$ by solving $a < -b$ or $a > b$.

 Solve each inequality. Graph each solution.

(a) $|p + 3| < 4$

(b) $|2k - 1| \leq 7$

Answers:

(a) $(-7, 1)$

(b) $[-3, 4]$

EXAMPLE 7 Solve $|x - 2| < 5$.

Replace a with $x - 2$ and b with 5 in property (2) above. Now solve $|x - 2| < 5$ by solving the inequality

$$-5 < x - 2 < 5.$$

Add 2 to each part, getting the solution

$$-3 < x < 7,$$

which is graphed in Figure 2.29. ■

FIGURE 2.29

EXAMPLE 8 Solve $|2 - 7m| - 1 > 4$.

First add 1 on both sides.

$$|2 - 7m| > 5$$

Now use property (3) from above to solve $|2 - 7m| > 5$ by solving the inequality

$$2 - 7m < -5 \quad \text{or} \quad 2 - 7m > 5.$$

8 Solve each inequality. Graph each solution.

(a) $|y - 2| > 5$

(b) $|3k - 1| \geq 2$

(c) $|2 + 5r| - 4 \geq 1$

Answers:

(a) All numbers in $(-\infty, -3)$ or $(7, \infty)$

(b) All numbers in $\left(-\infty, -\dfrac{1}{3}\right]$ or $[1, \infty)$

(c) All numbers in $\left(-\infty, -\dfrac{7}{5}\right]$ or $\left[\dfrac{3}{5}, \infty\right)$

9 Solve each inequality.

(a) $|5m - 2| > -1$

(b) $|2 + 3a| < -3$

(c) $|6 + r| > 0$

Answers:

(a) All real numbers

(b) No solution

(c) All real numbers except -6

Solve each part separately.

$$-7m < -7 \quad \text{or} \quad -7m > 3$$

$$m > 1 \quad \text{or} \quad m < -\frac{3}{7}$$

The solution, all numbers in $\left(-\infty, -\dfrac{3}{7}\right)$ or $(1, \infty)$, is graphed in Figure 2.30. ■ **8**

FIGURE 2.30

EXAMPLE 9 Solve $|2 - 5x| \geq -4$.

The absolute value of a number is always nonnegative. Therefore, $|2 - 5x| \geq -4$ is always true, so that the solution is the set of all real numbers. Note that the inequality $|2 - 5x| < -4$ has no solution, because the absolute value of a quantity can never be less than a negative number. ■ **9**

Absolute value inequalities can be used to indicate how far a number may be from a given number. The next example illustrates this use of absolute value.

EXAMPLE 10 Write each statement using absolute value.

(a) k is at least 4 units from 1.

If k is at least 4 units from 1, then the distance from k to 1 is greater than or equal to 4. See Figure 2.31(a). Since k may be on either side of 1 on the number line, k may be less than -3 or greater than 5. Write this statement using absolute value as follows:

$$|k - 1| \geq 4.$$

FIGURE 2.31

10 Write each statement using absolute value.

(a) m is at least 3 units from 5.

(b) t is within .01 of 4.

Answers:

(a) $|m - 5| \geq 3$

(b) $|t - 4| \leq .01$

(b) p is within 2 units of 5.

This statement means that the distance between p and 5 must be less than or equal to 2. See Figure 2.31(b). Using absolute value notation, the statement is written as

$$|p - 5| \leq 2.$$ ■ **10**

2.5 EXERCISES

1. Explain how to determine whether an open circle or a solid circle is used when graphing the solution of a linear inequality.

2. The three-part inequality $p < x < q$ means "p is less than x and x is less than q." Which one of the following inequalities is not satisfied by any real number x? Explain why.
   (a) $-3 < x < 5$ (b) $0 < x < 4$
   (c) $-7 < x < -10$ (d) $-3 < x < -2$

Solve each inequality. Graph each solution in Exercises 3–32. (See Examples 1–3.)

3. $-8k \le 32$

4. $-6a \le 36$

5. $-2b > 0$

6. $6 - 6z < 0$

7. $3x + 4 \le 12$

8. $2y - 5 < 9$

9. $-4 - p \ge 3$

10. $5 - 3r \le -4$

11. $7m - 5 < 2m + 10$

12. $6x - 2 > 4x - 8$

13. $m - (4 + 2m) + 3 < 2m + 2$

14. $2p - (3 - p) \le -7p - 2$

15. $-2(3y - 8) \ge 5(4y - 2)$

16. $5r - (r + 2) \ge 3(r - 1) + 5$

17. $3p - 1 < 6p + 2(p - 1)$

18. $x + 5(x + 1) > 4(2 - x) + x$

19. $-7 < y - 2 < 4$

20. $-3 < m + 6 < 2$

21. $8 \le 3r + 1 \le 13$

22. $-6 < 2p - 3 \le 5$

23. $-4 \le \dfrac{2k - 1}{3} \le 2$

24. $-1 \le \dfrac{5y + 2}{3} \le 4$

25. $\dfrac{3}{5}(2p + 3) \ge \dfrac{1}{10}(5p + 1)$

26. $\dfrac{8}{3}(z - 4) \le \dfrac{2}{9}(3z + 2)$

27. $42.75x > 7.460$

28. $15.79y < 6.054$

29. $8.04z - 9.72 < 1.72z - .25$

30. $3.25 + 5.08k > .76k + 6.28$

31. $-(1.42m + 7.63) + 3(3.7m - 1.12) \le 4.81m - 8.55$

32. $3(8.14a - 6.32) - (4.31a - 4.84) > .34a + 9.49$

33. **Natural Science** Federal guidelines require drinking water to have fewer than .050 milligram per liter of lead. A test using 21 samples of water in a midwestern city found that the average amount of lead in the samples was .040 milligram per liter. All samples had lead content within 5% of the average. Did all the samples meet the federal requirement?
   (a) Select a variable and write down what it represents.
   (b) Write a three-part inequality to express the sample results.
   (c) Answer the question.

34. **Management** The federal income tax for an income of $24,651 to $59,750 is 28% times (net income − 24,650) + $3697.50.
   (a) State what the variables represent.
   (b) Write this income bracket as an inequality.
   (c) Write the tax range in dollars for this income bracket as an inequality.

35. **Natural Science** Exposure to radon gas is a known lung cancer risk. According to the Environmental Protection Agency (EPA), the individual lifetime excess cancer risk R for radon exposure is between .0015 and .006, where $R = .01$ represents a 1% increase in risk of developing cancer.*
   (a) Write the preceding information as an inequality.
   (b) Determine the range of individual annual risk by dividing R by an average life expectancy of 75 years.

Solve each inequality. Graph each solution. (See Examples 6–9.)

36. $|p| > 7$

37. $|m| < 1$

38. $|r| \le 4$

39. $|a| < -2$

40. $|b| > -5$

41. $|2x + 5| < 3$

42. $\left| x - \dfrac{1}{2} \right| < 2$

43. $|3z + 1| \ge 7$

Indoor-Air-Assessment: A Review of Indoor Air Quality Risk Characterization Studies, Report No. EPA/600/8-90/044, Environmental Protection Agency, 1991.

44. $|8b + 5| \geq 7$

45. $\left|5x + \dfrac{1}{2}\right| - 2 < 5$

46. $\left|x + \dfrac{2}{3}\right| + 1 < 4$

47. Physical Science The temperatures on the surface of Mars in degrees Celsius approximately satisfy the inequality $|C - 84| \leq 56$. What range of temperatures corresponds to this inequality?

48. Natural Science Dr. Tydings has found that, over the years, 95% of the babies he has delivered have weighed y pounds, where $|y - 8.0| \leq 1.5$. What range of weights corresponds to this inequality?

49. Physical Science The industrial process that is used to convert methanol to gasoline is carried out at a temperature range of 680°F to 780°F. Using F as the variable, write an absolute value inequality that corresponds to this range.

50. Physical Science When a model kite was flown across the wind in tests to determine its limits of power extraction, it attained speeds of 98 to 148 feet per second in winds of 16 to 26 feet per second. Using x as the variable in each case, write absolute value inequalities that correspond to these ranges.

51. Natural Science Human beings emit carbon dioxide when they breathe. In one study, the emission rates of carbon dioxide by college students were measured both during lectures and exams. The average individual rate R_L (in grams per hour) during a lecture class satisfied the inequality $|R_L - 26.75| \leq 1.42$, whereas during an exam the rate R_E satisfied the inequality $|R_E - 38.75| \leq 2.17$.*
(a) Find the range of values for R_L and R_E.
(b) The class had 225 students. If T_L and T_E represent the total amounts of carbon dioxide (in grams) emitted during a one-hour lecture and one-hour exam respectively, write inequalities that describe the ranges for T_L and T_E.

52. Social Science When administering a standard intelligence test, we expect about 1/3 of the scores to be more than 12 units above 100 or more than 12 units below 100. Describe this situation by writing an absolute value inequality.

Management *Work the following problems. (See Example 4.)*

53. In 1993 Pacific Bell charged $.15 for the first minute plus $.14 for each additional minute (or part of a minute) for a dial-direct call from Sacramento to North Highland, California.* How many minutes could a person talk for no more than $2.00?

54. A student has a total of 970 points before the final exam in her algebra class. She must have 81% of the 1300 points possible in order to get a B. What is the lowest score she can earn on the 100-point final to get a B in the class?

55. Bill Bradkin went to a conference in Montreal, Canada for a week. He decided to rent a car and checked with two rental firms. Avery wanted $56 per day, with no mileage fee. Hart wanted $216 per week and $.28 per mile (or part of a mile). How many miles must Bradkin drive before the Avery car is the better deal?

56. After considering the situation in Exercise 55, Bradkin contacted Lowcost Rental Cars and was offered a car for $198 per week plus $.30 a mile (or part of a mile). With this offer, can he travel more miles for the same price as Avery charges? How many more or fewer miles?

Management *In Exercises 57–62, find all values of x where the following products will at least break even. (See Example 5.)*

57. The cost to produce x units of wire is $C = 50x + 5000$, while the revenue is $R = 60x$.

58. The cost to produce x units of squash is $C = 100x + 6000$, while the revenue is $R = 500x$.

59. $C = 85x + 900$; $R = 105x$

60. $C = 70x + 500$; $R = 60x$

61. $C = 1000x + 5000$; $R = 900x$

62. $C = 2500x + 10{,}000$; $R = 102{,}500x$

Write each of the following statements using absolute value. (See Example 10.)

63. x is within 4 units of 2.

64. m is no more than 8 units from 9.

65. z is no less than 2 units from 12.

66. p is at least 5 units from 9.

67. If x is within .0004 unit of 2, then y is within .00001 unit of 7.

68. y is within .001 unit of 10 whenever x is within .02 unit of 5.

---

*Pacific Bell White Pages, January 1996–January 1997.

---

*Wang, T. C., *ASHRAE Transactions* 81 (Part 1), 32 (1975).

2.6 POLYNOMIAL AND RATIONAL INEQUALITIES

This section deals with the solution of polynomial and rational inequalities, such as

$$r^2 + 3r - 4 \geq 0, \quad x^3 - x \leq 0, \quad \text{and} \quad \frac{2x - 1}{3x + 4} < 5.$$

We shall concentrate on algebraic solution methods, but to understand why these methods work, we must first look at such inequalities from a graphical point of view.

Consider the inequality $6x^2 + x - 15 < 0$ and suppose that we know what the graph of the two-variable equation $y = 6x^2 + x - 15$ looks like (see Figure 2.32). A typical point on the graph has coordinates $(x, 6x^2 + x - 15)$. The number x is a solution of the inequality $6x^2 + x - 15 < 0$ precisely when the second coordinate of this point is *negative,* that is, when this point lies *below* the x-axis. So to solve the inequality, we need only find the first coordinates of points on the graph that are below the x-axis. This information can be read from the graph in Figure 2.32: the graph is below the x-axis when x lies between the two x-intercepts c and d.

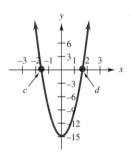

FIGURE 2.32

1 **(a)** Determine the x-intercepts c and d of $y = 6x^2 + x - 15$ in Figure 2.32 by setting $y = 0$ and solving for x; in other words, solve

$$6x^2 + x - 15 = 0.$$

(b) Find the solutions of $6x^2 + x - 15 < 0$.

(c) Find the solutions of $6x^2 + x - 15 > 0$.

Answers:

(a) $c = -5/3$ and $d = 3/2$

(b) The numbers x between the intercepts, that is,

$$-5/3 < x < 3/2.$$

(c) The numbers x to the left of c or to the right of d, that is,

$$x < -5/3 \text{ or } x > 3/2.$$

The graph in Figure 2.32 also allows us to solve the inequality

$$6x^2 + x - 15 > 0.$$

The solutions of this inequality are the first coordinates of the points on the graph that lie *above* the x-axis—that is, to the left of the x-intercept c or to the right of the x-intercept d. To complete the solution of these two inequalities, you need only determine the x-intercepts c and d. **1**

The next example shows how the preceding solution method can be carried out even if you can't see the graph. It depends on the fact that the graph is a smoot' unbroken curve—the only way it can pass from above to below the x-axis is to pas. through an x-intercept.

EXAMPLE 1 Solve each of these quadratic inequalities.

(a) $x^2 - x < 12$

First, rewrite the inequality as $x^2 - x - 12 < 0$. Now we don't know what the graph of $y = x^2 - x - 12$ looks like, but we can still find its x-intercepts by solving the equation

$$x^2 - x - 12 = 0$$
$$(x + 3)(x - 4) = 0.$$
$$x + 3 = 0 \quad \text{or} \quad x - 4 = 0$$
$$x = -3 \quad \text{or} \quad x = 4$$

These numbers divide the x-axis (number line) into three regions, as indicated in Figure 2.33.

FIGURE 2.33

In each region, the graph of $y = x^2 - x - 12$ is an unbroken curve, so it will be entirely on one side of the x-axis, that is, entirely above or entirely below the axis. It can only pass from above to below the x-axis at the x-intercepts. To see whether the graph is above or below the x-axis when x is in region A, choose a value of x in region A, say $x = -5$, and substitute this in the equation.

$$y = x^2 - x - 12 = (-5)^2 - (-5) - 12 = 18$$

Therefore, the point $(-5, 18)$ is on the graph. Since its y-coordinate 18 is positive, this point lies above the x-axis and hence the entire graph lies above the x-axis in region A.

Similarly, we can choose a value of x in region B, say $x = 0$. Then

$$y = x^2 - x - 12 = 0^2 - 0 - 12 = -12,$$

so that $(0, -12)$ is on the graph. Since this point lies below the x-axis (why?), the entire graph in region B must be below the x-axis. Finally, in region C, let $x = 5$. Then $y = 5^2 - 5 - 12 = 8$, so that $(5, 8)$ is on the graph and the entire graph in region C lies above the x-axis. We can summarize the results as follows.

Region	Graph	Conclusion
A: $x < -3$	Above x-axis	$x^2 - x - 12 > 0$
B: $-3 < x < 4$	Below x-axis	$x^2 - x - 12 < 0$
C: $x > 4$	Above x-axis	$x^2 - x - 12 > 0$

Solve each inequality. Graph the solution on the number line.

(a) $x^2 + 2x - 3 < 0$

(b) $2p^2 + 3p - 2 < 0$

Answers:
(a) $(-3, 1)$

(b) $(-2, 1/2)$

The last column shows that the only region where $x^2 - x - 12 < 0$ is region B, so the solutions of the inequality are all numbers x with $-3 < x < 4$, that is, the interval $(-3, 4)$, as shown in the number line graph in Figure 2.34.

FIGURE 2.34

(b) $x^2 - x - 12 > 0$

Use the chart in part (a). The last column shows that $x^2 - x - 12 > 0$ only when x is in region A or region C. Hence, the solutions of the inequality are all numbers x with $x < -3$ or $x > 4$, that is, all numbers in the intervals $(-\infty, -3)$ or $(4, \infty)$. ■ [2]

TECHNOLOGY TIP If you have a graphing calculator, you need not choose a number in each region to solve the inequality. Once you have found the x-intercepts by solving the appropriate equation, graph the corresponding two-variable equation and visually check the regions where the graph is above or below the x-axis, as in the introductory example preceding Example 1. ✔

EXAMPLE 2 Solve the quadratic inequality $r^2 + 3r \geq 4$.

First rewrite the inequality so that one side is 0.

$$r^2 + 3r \geq 4$$
$$r^2 + 3r - 4 \geq 0 \qquad \text{Add } -4 \text{ to both sides.}$$

Now solve the corresponding equation.

$$r^2 + 3r - 4 = 0$$
$$(r - 1)(r + 4) = 0$$
$$r = 1 \quad \text{or} \quad r = -4$$

[3] Solve each inequality. Graph each solution.

(a) $k^2 + 2k - 15 \geq 0$

(b) $3m^2 + 7m \geq 6$

Answers:
(a) All numbers in $(-\infty, -5]$ or $[3, \infty)$

(b) All numbers in $(-\infty, -3]$ or $[2/3, \infty)$

These numbers separate the number line into three regions, as shown in Figure 2.35. Test a number from each region.

Let $x = -5$ from region **A:** $(-5)^2 + 3(-5) - 4 = 6 > 0$.
Let $x = 0$ from region B: $(0)^2 + 3(0) - 4 = -4 < 0$.
Let $x = 2$ from region **C:** $(2)^2 + 3(2) - 4 = 6 > 0$.

We want the inequality to be ≥ 0, that is, positive or 0. The solution includes numbers in region A and in region C, as well as -4 and 1, the endpoints. The solution, which includes all numbers in the intervals $(-\infty, -4]$ or $[1, \infty)$, is graphed in Figure 2.35. ■ [3]

FIGURE 2.35

EXAMPLE 3 Solve $q^3 - 4q > 0$.

Solve the corresponding equation by factoring.

$$q^3 - 4q = 0$$
$$q(q^2 - 4) = 0$$
$$q(q + 2)(q - 2) = 0$$
$$q = 0 \quad \text{or} \quad q + 2 = 0 \quad \text{or} \quad q - 2 = 0$$
$$q = 0 \qquad\qquad q = -2 \qquad\qquad q = 2$$

These three numbers separate the number line into the four regions shown in Figure 2.36.

FIGURE 2.36

---

4 Solve each inequality. Graph each solution.

(a) $m^3 - 9m > 0$

(b) $2k^3 - 50k \leq 0$

Answers:
(a) All numbers in $(-3, 0)$ or $(3, \infty)$

(b) All numbers in $(-\infty, -5]$ or $[0, 5]$

Test a number from each region.

A: If $q = -3$, $(-3)^3 - 4(-3) = -15 < 0$.

B: If $q = -1$, $(-1)^3 - 4(-1) = 3 > 0$.

C: If $q = 1$, $(1)^3 - 4(1) = -3 < 0$.

D: If $q = 3$, $(3)^3 - 4(3) = 15 > 0$.

The numbers that make the polynomial > 0, or positive, are in the intervals

$$(-2, 0) \quad \text{or} \quad (2, \infty),$$

as graphed in Figure 2.36. ■ **4**

RATIONAL INEQUALITIES Inequalities with quotients of algebraic expressions are called **rational inequalities.** These inequalities can be solved in much the same way as polynomial inequalities.

EXAMPLE 4 Solve the rational inequality $\dfrac{5}{x + 4} \geq 1$.

Write an equivalent inequality with one side equal to 0.

$$\frac{5}{x + 4} \geq 1$$

$$\frac{5}{x + 4} - 1 \geq 0$$

Write the left side as a single fraction.

$$\frac{5}{x + 4} - \frac{x + 4}{x + 4} \geq 0 \qquad \text{Get a common denominator.}$$

$$\frac{5 - (x + 4)}{x + 4} \geq 0 \qquad \text{Subtract fractions.}$$

$$\frac{5 - x - 4}{x + 4} \geq 0 \qquad \text{Distributive property}$$

$$\frac{1 - x}{x + 4} \geq 0$$

The quotient can change sign only when the denominator is 0 or when the numerator is 0. (In graphical terms, these are the only places where the graph of $y = \dfrac{1 - x}{x + 4}$ can change from above the x-axis to below.) This happens when

$$1 - x = 0 \quad \text{or} \quad x + 4 = 0$$
$$x = 1 \quad \text{or} \qquad x = -4.$$

As in the earlier examples, test a number from each of the regions determined by 1 and -4 in the new form of the inequality.

Let $x = -5$: $\quad \dfrac{1 - (-5)}{-5 + 4} = -6 \leq 0.$

Let $x = 0$: $\quad \dfrac{1 - 0}{0 + 4} = \dfrac{1}{4} \geq \mathbf{0}.$

Let $x = 2$: $\quad \dfrac{1 - 2}{2 + 4} = -\dfrac{1}{6} \leq 0.$

The test shows that numbers in $(-4, 1)$ satisfy the inequality. With a quotient, the endpoints must be considered individually to make sure that no denominator is 0. In this inequality, -4 makes the denominator 0, while 1 satisfies the given inequality. Write the solution as $(-4, 1]$. ■

CAUTION As suggested by Example 4, be very careful with the endpoints of the intervals in the solution of rational inequalities. ◆ 5

EXAMPLE 5 Solve $\dfrac{2x - 1}{3x + 4} < 5.$

Write an equivalent inequality with 0 on one side. Begin by subtracting 5 on both sides and combining the terms on the left into a single fraction.

$$\frac{2x - 1}{3x + 4} < 5$$

$$\frac{2x - 1}{3x + 4} - 5 < 0 \qquad \text{Get 0 on one side.}$$

$$\frac{2x - 1 - 5(3x + 4)}{3x + 4} < 0 \qquad \text{Subtract.}$$

$$\frac{-13x - 21}{3x + 4} < 0 \qquad \text{Combine terms.}$$

5 Solve each inequality.

(a) $\dfrac{3}{x - 2} \geq 4$

(b) $\dfrac{p}{1 - p} < 3$

(c) Why is 2 excluded from the solution in part (a)?

Answers:

(a) $(2, 11/4]$

(b) All numbers in $(-\infty, 3/4)$ or $(1, \infty)$

(c) When $x = 2$, the fraction is undefined.

Set the numerator and denominator each equal to 0 and solve the two equations.

$$-13x - 21 = 0 \quad \text{or} \quad 3x + 4 = 0$$

$$x = -\frac{21}{13} \quad \text{or} \quad x = -\frac{4}{3}$$

6 Solve each rational inequality.

(a) $\dfrac{3y - 2}{2y + 5} < 1$

(b) $\dfrac{3c - 4}{2 - c} \geq -5$

Answers:
(a) $(-5/2, 7)$

(b) All numbers in $(-\infty, 2)$ or $[3, \infty)$

Use the values $-21/13$ and $-4/3$ to divide the number line into three intervals. Test a number from each interval in the inequality. The quotient is negative for numbers in $(-\infty, -21/13)$ or $(-4/3, \infty)$. Neither endpoint satisfies the given inequality. ■

CAUTION In problems like Examples 4 and 5, we cannot begin by simply multiplying both sides by the denominator to simplify the inequality, because we do not know whether the variable denominator is positive or negative. ◆ **6**

2.6 EXERCISES

Solve each of these quadratic inequalities. Graph the solutions on the number line. (See Examples 1 and 2.)

1. $(x + 5)(2x - 3) \leq 0$

2. $(5y - 1)(y + 4) > 0$

3. $r^2 + 4r > -3$

4. $z^2 + 6z < -8$

5. $4m^2 + 7m - 2 \leq 0$

6. $6p^2 - 11p + 3 \geq 0$

7. $4x^2 + 3x - 1 > 0$

8. $3x^2 - 5x > 2$

9. $x^2 \leq 25$

10. $y^2 \geq 4$

11. $p^2 - 16p > 0$

12. $r^2 - 9r < 0$

Solve these inequalities. (See Example 3.)

13. $x^3 - 9x \geq 0$

14. $p^3 - 25p \leq 0$

15. $(x + 6)(x + 1)(x - 4) \geq 0$

16. $(2x + 5)(x^2 - 1) \leq 0$

17. $(x + 4)(x^2 - 2x - 3) < 0$

18. $x^3 - 2x^2 - 3x \geq 0$

19. $6k^3 - 5k^2 < 4k$

20. $2m^3 + 7m^2 > 4m$

21. A student solved the inequality $p^2 < 16$ by taking the square root of both sides to get $p < 4$. She wrote the solution as $(-\infty, 4)$. Is her solution correct?

Solve the following rational inequalities. (See Examples 4 and 5.)

22. $\dfrac{y + 2}{y - 4} \leq 0$

23. $\dfrac{r - 3}{r - 1} \geq 0$

24. $\dfrac{z + 6}{z + 3} > 1$

25. $\dfrac{a - 2}{a - 5} < -1$

26. $\dfrac{1}{3k - 5} < \dfrac{1}{3}$

27. $\dfrac{1}{p - 2} < \dfrac{1}{3}$

28. $\dfrac{7}{k + 2} \geq \dfrac{1}{k + 2}$

29. $\dfrac{5}{p + 1} > \dfrac{12}{p + 1}$

30. $\dfrac{x^2 - 4}{x} > 0$

31. $\dfrac{x^2 - x - 6}{x} < 0$

32. $\dfrac{x^2 + x - 2}{x^2 - 2x - 3} < 0$

Use a graphing calculator to solve these inequalities. (See the Technology Tip after Example 1. You may have to approximate the x-intercepts of the graph by zooming in or by using a graphical root finder.)

33. $3x^2 + 4x > 5$

34. $3x + 7 < 2x^2$

35. $.5x^2 - 1.2x < .1$

36. $3.1x^2 - 7.4x + 3.2 > 0$

37. $x^3 - 2x^2 - 5x + 7 \geq 2x + 1$

38. $x^4 - 6x^3 + 2x^2 < 5x - 2$

39. $2x^4 + 3x^3 < 2x^2 + 4x - 2$

40. $x^5 + 5x^4 > 4x^3 - 3x^2 - 2$

41. $\dfrac{2x^2 + x - 1}{x^2 - 4x + 4} \le 0$

42. $\dfrac{x^3 - 3x^2 + 5x - 29}{x^2 - 7} > 3$

Work these problems.

Management   *A product will break even or produce a profit only if the revenue from selling the product at least equals the cost of producing it. Find all values of x for which the product will at least break even.*

43. The cost to produce x souvenir plates is $C = 5x + 350$; the revenue is $R = 50x - x^2$.

44. The cost to produce x file cabinets is $C = 10x + 600$; the revenue is $R = 80x - x^2$.

45. Management   An analyst has found that his company's profits, in hundreds of thousands of dollars, are given by $P = 3x^2 - 35x + 50$, where x is the amount, in hundreds of dollars, spent on advertising. For what values of x does the company make a profit?

46. Management   The commodities market is very unstable; money can be made or lost quickly on investments in soybeans, wheat, and so on. Suppose that an investor kept track of her total profit, P, at time t, in months, after she began investing, and found that $P = 4t^2 - 29t + 30$. Find the time intervals where she has been ahead.

47. Management   The manager of a large apartment complex has found that the profit is given by $P = -x^2 + 250x - 15,000$, where x is the number of apartments rented. For what values of x does the complex produce a profit?

48. Physical Science   A physicist has found that the velocity of a moving particle is given by $2t^2 - 5t - 12$, where t is time in seconds since he began his observations. (Here t can be positive or negative; think of t seconds before his observations began.) Find the time intervals in which the velocity has been negative.

49. Physical Science   A projectile is fired from ground level. After t seconds its height above the ground is $220t - 16t^2$ feet. For what time period is the projectile at least 624 feet above the ground?

CHAPTER 2   SUMMARY

Key Terms and Symbols

2.1
Cartesian coordinate system
x-axis
y-axis
origin
ordered pair
x-coordinate
y-coordinate
quadrant
solution of an equation
graph
x-intercept
y-intercept
graph reading

2.2
Δx change in x
Δy change in y
slope
slope-intercept form
parallel and perpendicular lines
point-slope form
linear equations

2.3
mathematical models
revenue
profit

break-even point
supply and demand curves
equilibrium point
equilibrium price
equilibrium supply/demand

2.4
quadratic equation
real solution
zero-factor property
square-root property
completing the square
quadratic formula
discriminant

2.5
$<$ is less than
$\le$ is less than or equal to
$>$ is greater than
$\ge$ is greater than or equal to
linear inequality
properties of inequality
absolute value inequality

2.6
polynomial inequality
graphical solution methods
algebraic solution methods
rational inequality

Key Concepts

The **slope** of the line through the points (x_1, y_1) and (x_2, y_2), where $x_1 \neq x_2$ is $m = \dfrac{y_2 - y_1}{x_2 - x_1}$.

Nonvertical **parallel lines** have the same slope and **perpendicular lines,** if neither is vertical, have slopes with a product of -1.

The line with equation $ax + by = c$ (with $a \neq 0, b \neq 0$) has x-intercept c/a and y-intercept c/b.
The line with equation $x = k$ is vertical with x-intercept k, no y-intercept, and undefined slope.
The line with equation $y = k$ is horizontal, with y-intercept k, no x-intercept, and slope 0.
The line with equation $y = mx + b$ has slope m and y-intercept b.
The line with equation $y - y_1 = m(x - x_1)$ has slope m and goes through (x_1, y_1).

If $p = f(q)$ gives the price per unit when x units can be supplied, and $p = g(q)$ gives the price per unit when q units are demanded, then the **equilibrium price, supply,** and **demand** occur at the q-value such that $f(q) = g(q)$.

Facts needed to solve quadratic equations (in which a, b, and c are real numbers):
Factoring: If $ab = 0$, then $a = 0$ or $b = 0$ or both.
Square-Root Property: If $b > 0$, then the solutions of $x^2 = b$ are $\sqrt{b}$ and $-\sqrt{b}$.
Quadratic Formula: The solutions of $ax^2 + bx + c$ ($a \neq 0$) are

$$x = \frac{-b \pm \sqrt{b^2 - 4ac}}{2a}.$$

Discriminant: There are two real solutions if $b^2 - 4ac > 0$, one real solution if $b^2 - 4ac = 0$, and no real solutions if $b^2 - 4ac < 0$.

Chapter 2 Review Exercises

Which of the ordered pairs $(-2, 3)$, $(0, -5)$, $(2, -3)$, $(3, -2)$, $(4, 3)$, $(7, 2)$ *are solutions of the given equation?*

1. $y = x^2 - 2x - 5$

2. $x - y = 5$

Sketch the graph of each equation.

3. $5x - 3y = 15$

4. $2x + 7y - 21 = 0$

5. $y + 3 = 0$

6. $y - 2x = 0$

7. $y = .25x^2 + 1$

8. $y = \sqrt{x} + 4$

9. The following temperature graph was recorded in Bratenahl, Ohio.

(a) At what times during the day was the temperature over 55°?

(b) When was the temperature below 40°?

10. Greenville, South Carolina, is 500 miles south of Bratenahl, Ohio, and its temperature is 7° higher all day long (see the graph in Exercise 9). At what time was the temperature in Greenville the same as the temperature at noon in Bratenahl?

11. In your own words, define the slope of a line.

In Exercises 12–21, find the slope of the line.

12. Through $(-1, 4)$ and $(2, 3)$

13. Through $(5, -3)$ and $(-1, 2)$

14. Through $(7, -2)$ and the origin

15. Through $(8, 5)$ and $(0, 3)$

16. $2x + 3y = 30$

17. $4x - y = 7$

18. $x + 5 = 0$

19. $y = 3$

20. Parallel to $3x + 8y = 0$

21. Perpendicular to $x = 3y$

22. Graph the line through $(0, 5)$ with $m = -2/3$.

23. Graph the line through $(-4, 1)$ with $m = 3$.

24. What information is needed to determine the equation of a line?

Find an equation for each of the following lines.

25. Through $(5, -1)$, slope $2/3$

26. Through $(8, 0)$, slope $-1/4$

27. Through $(5, -2)$ and $(1, 3)$

28. Through $(2, -3)$ and $(-3, 4)$

29. Undefined slope, through $(-1, 4)$

30. Slope 0, through $(-2, 5)$

31. x-intercept -3, y-intercept 5

32. x-intercept $-2/3$, y-intercept $1/2$

33. Social Science The percent of children living with a never-married parent was 4.2 in 1960 and 30.6 in 1990.
(a) Assuming that this growth was linear, write an equation that relates the percent y and the year x. Let 1960 correspond to $x = 0$, 1961 to $x = 1$, and so on.
(b) Graph the equation for the years from 1960 to 2000.
(c) Is the slope of the line positive or negative? Why?

34. Social Science In 1960 the percent of children living with two parents was 87.7 and in 1990 it was 72.5.
(a) Assuming the decrease was linear, write an equation that relates the percent y and the year x, with 1960 corresponding to $x = 0$.
(b) Graph the equation for the years from 1960 to 2000.
(c) Is the slope of the line positive or negative? Why?

35. Management The cost c of producing x units of a product is given by the equation $c = 20x + 100$. The product sells for $40 per unit.
(a) What is the revenue equation?
(b) Find the break-even point.
(c) If the company sells exactly the number of units needed to break even, what will its revenue be?
(d) Find the equation that relates the profit p gained from selling x units.

36. Management A product can be sold for $25 per unit. The cost c of producing x units in the Zanesville plant is given by $c = 24x + 5000$.
(a) What is the revenue equation?
(b) How many units must be sold for the company to break even?

37. Management For producing x tons of crushed rock, a quarry has average monthly costs given by the equation $c = 56.75x + 192.44$. The rock can be sold for $102.50 per ton. How many tons must be sold each month for the quarry to break even?

38. Natural Science Lead is a neurotoxin found in drinking water, old paint, and the air. It is particularly hazardous to people because it is not easily eliminated from the body. As directed by the "Safe Drinking Water Act" of 1974, the EPA proposed a maximum lead level in public drinking water of .05 milligram per liter. This standard assumes that an individual consumes two liters of water per day.*
(a) If EPA guidelines are followed, write an equation that expresses the maximum amount y of lead that could be ingested in x years. Assume that there are 365.25 days in a year.
(b) If the average life expectancy is 75 years, what is the EPA maximum lead intake from water over an average lifetime?

39. Natural Science Ground level ozone is toxic to both plants and animals and causes respiratory problems and eye irritation in humans. Automobiles are a major source of this type of ozone, which often occurs when smog levels are significant. Ozone from outside air can enter buildings through ventilation systems. Guideline levels for indoor ozone recommend less than 50 parts per billion (ppb). In a scientific study, a purafil air filter was found to remove 43% of the ozone.†
(a) Write an equation that expresses the amount y of ozone remaining from an initial concentration of x ppb when this filter is used.
(b) If the initial concentration is 140 ppb, does this type of filter reduce the ozone to acceptable levels?
(c) What is the maximum initial concentration of ozone that this filter will reduce to an acceptable level?

---

*Nemerow, N. and Dasgupta, A., *Industrial and Hazardous Waste Treatment,* New York: Van Nostrand Reinhold, 1991.

†Parmar and Grosjean, *Removal of Air Pollutants from Museum Display Cases,* Marina del Rey, CA: Getty Conservation Institute, 1989.

40. Management   The supply and demand for a certain commodity are related by

supply: $p = 6q + 3$ demand: $p = 19 - 2q$,

where p represents the price at a supply or demand of q units. Find the supply and the demand when the price is as follows.
(a) $10
(b) $15
(c) $18
(d) Find the equilibrium price.
(e) Find the equilibrium quantity (supply/demand).

41. Management   For a particular product, 72 units will be supplied at a price of $34, while 16 units will be supplied at a price of $6.
(a) Write a linear supply equation for this product.
(b) The demand for this product is given by the equation $p = 12 - .5q$. Find the equilibrium price and quantity.

Determine the number of real solutions of the quadratic equation.

42. $x^2 - 6x = 4$

43. $-3x^2 + 5x + 2 = 0$

44. $4x^2 - 12x + 9 = 0$

45. $5x^2 + 2x + 1 = 0$

46. $x^2 + 3x + 5 = 0$

Find all real solutions of the equation.

47. $(b + 7)^2 = 5$

48. $(2p + 1)^2 = 7$

49. $2p^2 + 3p = 2$

50. $2y^2 = 15 + y$

51. $x^2 - 2x = 2$

52. $r^2 + 4r = 1$

53. $2m^2 - 12m = 11$

54. $9k^2 + 6k = 2$

55. $2a^2 + a - 15 = 0$

56. $12x^2 = 8x - 1$

57. $2q^2 - 11q = 21$

58. $3x^2 + 2x = 16$

59. $6k^4 + k^2 = 1$

60. $21p^4 = 2 + p^2$

61. $2x^4 = 7x^2 + 15$

62. $3m^4 + 20m^2 = 7$

63. $3 = \dfrac{13}{z} + \dfrac{10}{z^2}$

64. $1 + \dfrac{13}{p} + \dfrac{40}{p^2} = 0$

65. $\dfrac{15}{x - 1} + \dfrac{18}{(x - 1)^2} = -2$

66. $\dfrac{5}{(2t + 1)^2} = 2 - \dfrac{9}{2t + 1}$

Solve each equation for the specified variable.

67. $p = \dfrac{E^2 R}{(r + R)^2}$ for r

68. $p = \dfrac{E^2 R}{(r + R)^2}$ for E

69. $K = s(s - a)$ for s

70. $kz^2 - hz - t = 0$ for z

71. Management   A landscaper wants to put a cement walk of uniform width around a rectangular garden that measures 24 by 40 feet. She has enough cement to cover 740 square feet. To the nearest tenth of a foot, how wide should the walk be in order to use up all the cement?

72. Management   A corner lot measures 25 by 40 yards. The city plans to take a strip of uniform width along the two sides bordering the streets in order to widen these roads. To the nearest tenth of a yard, how wide should the strip be if the remainder of the lot is to have an area of 814 square yards?

73. Management   A recreation director wants to fence off a rectangular playground beside an apartment building. The building forms one boundary, so she needs to fence only the other three sides. The area of the playground is to be 11,250 square meters. She has enough material to build 325 meters of fence. Find the length and width of the playground.

74. Two cars leave an intersection at the same time. One travels north and the other heads west traveling 10 mph faster. After 1 hour they are 50 miles apart. What were their speeds?

Solve each inequality.

75. $-6x + 3 < 2x$

76. $12z \geq 5z - 7$

77. $2(3 - 2m) \geq 8m + 3$

78. $6p - 5 > -(2p + 3)$

79. $-3 \leq 4x - 1 \leq 7$

80. $0 \leq 3 - 2a \leq 15$

Solve each inequality.

81. $|b| \leq 8$

82. $|a| > 7$

83. $|2x - 7| \geq 3$

84. $|4m + 9| \leq 16$

85. $|5k + 2| - 3 \leq 4$

86. $|3z - 5| + 2 \geq 10$

87. Natural Science Dr. Ryan has found that, over the years, 95% of the babies he has delivered have weighed y pounds, where $|\,y - 7.5\,| \le 2$. What range of weights corresponds to this inequality?

88. Natural Science The number of milligrams of a certain substance per liter in drinking water samples all tested within .05 of 40 milligrams per liter. Write this information as an inequality, using absolute value.

89. Management One automobile rental firm charges $75 for a weekend rental (Friday afternoon through Monday morning) with unlimited mileage. A second firm charges $50 plus 5 cents per mile. For what range of miles driven is the second firm cheaper?

90. Management A nearby business college charges an annual tuition of $6400. Jack plans to save $150 from each monthly paycheck, which he receives at the end of each month. What is the least number of months it will take him to save enough for one year's tuition?

Solve each inequality.

91. $r^2 + r - 6 < 0$

92. $y^2 + 4y - 5 \ge 0$

93. $2z^2 + 7z \ge 15$

94. $3k^2 \le k + 14$

95. $(x - 3)(x^2 + 7x + 10) \le 0$

96. $(x + 4)(x^2 - 1) \ge 0$

97. $\dfrac{m + 2}{m} \le 0$

98. $\dfrac{q - 4}{q + 3} > 0$

99. $\dfrac{5}{p + 1} > 2$

100. $\dfrac{6}{a - 2} \le -3$

101. $\dfrac{2}{r + 5} \le \dfrac{3}{r - 2}$

102. $\dfrac{1}{z - 1} > \dfrac{2}{z + 1}$

CASE 2

Depreciation

Straight-Line (Linear) Depreciation Because machines and equipment wear out or become obsolete over time, business firms must take into account the value that equipment has lost during each year of its useful life. This lost value, called **depreciation,** may be calculated in several ways.

Historically, to find the annual straight-line depreciation, the net cost was first found by deducting any estimated residual or **salvage value.** The depreciation was then divided by the number of years of useful life. For example, a $10,000 item, with a salvage value of $2000 and a useful life of 4 years would have an annual depreciation of

$$\frac{\$10,000 - \$2000}{4} = \$2000.$$

The $8000 difference in the numerator is called the **depreciable basis.** However, from a practical viewpoint, it is often difficult to determine the salvage value when the item is new. Also, for many items, such as computers, there is no residual dollar value at the end of the useful life due to obsolescence.

The tendency now is to focus away from salvage value. In actual practice, it is customary to determine the annual straight-line depreciation by simply dividing the cost by the number of years of useful life.* In the example above, then, the annual straight-line depreciation would be

$$\frac{\$10,000}{4} = \$2500.$$

In general, the annual straight-line depreciation is given by

$$D = \frac{1}{n}x.$$

Suppose a copier costs $4500 and has a useful life of 3 years. Because the useful life is 3 years, 1/3 of the cost of the copier is depreciated each year.

$$D = \frac{1}{3}(\$4500) = \$1500$$

---

*Joel E. Halle, CPA.

Straight-line depreciation is the easiest method of depreciation to use, but it often does not accurately reflect the rate at which assets actually lose value. Some assets, such as new cars, lose more value annually at the beginning of their useful life than at the end. For this reason, two other methods of depreciation, the *sum-of-the-years'-digits* method and the *double declining balance* method (a nonlinear method) are permitted by the Internal Revenue Service.

Sum-of-the-Years'-Digits Depreciation

With sum-of-the-years'-digits depreciation, a successively smaller fraction is applied each year to the cost of the asset. The denominator of the fraction, which remains constant, is the "sum of the years' digits" that gives the method its name. For example, if the asset has a 5-year life, the denominator is $5 + 4 + 3 + 2 + 1 = 15$. The numerator of the fraction, which changes each year, is the number of years of life that remain. For the first year the numerator is 5, for the second year it is 4, and so on. For example, for an asset with a cost of \$15,000 and a 5-year life, the depreciation in each year is shown in the following table.

Year	Fraction	Depreciation	Accumulated Depreciation
1	5/15	\$5000	\$ 5,000
2	4/15	\$4000	\$ 9,000
3	3/15	\$3000	\$12,000
4	2/15	\$2000	\$14,000
5	1/15	\$1000	\$15,000

Now, let us develop a model (or formula) for the depreciation D in year j of an asset with a cost of x dollars and a life of n years. The constant denominator will be the sum

$$n + (n - 1) + (n - 2) + \cdots + 2 + 1.$$

We can simplify this expression for the sum as follows.

$$
\begin{aligned}
S &= 1 + 2 + 3 + \cdots + (n - 1) + n \\
S &= n + (n - 1) + \cdots + 3 + 2 + 1 \\
\hline
2S &= (n + 1) + (n + 1) + \cdots + (n + 1)
\end{aligned}
$$

There are n terms, so adding term by term gives n sums of $n + 1$.

$$2S = n(n + 1)$$

$$S = \frac{n(n + 1)}{2}$$

Since the numerator is the number of years of life that remain, in the first year (when $j = 1$), the numerator is n; when $j = 2$, the numerator is $n - 1$; when $j = 3$, the numerator is $n - 2$; and so on. In each case, the value of j plus the numerator equals $n + 1$,

so the numerator can be written as $n + 1 - j$. Thus, the fraction for the year j is

$$\frac{n + 1 - j}{\dfrac{n(n + 1)}{2}} = \frac{2(n + 1 - j)}{n(n + 1)},$$

and the depreciation in year j (denoted D_j) is

$$D_j = \frac{2(n + 1 - j)}{n(n + 1)} x.$$

Double Declining Balance Depreciation

Another common method is double declining balance depreciation. In the first year, the depreciation is found by multiplying the cost x by $2/n$, double the amount of value lost each year. Thus, the depreciation in year 1 is

$$\frac{2}{n} x.$$

The depreciation in later years of an asset's life can be found by multiplying the depreciation from the previous year by $1 - 2/n$. For example, an asset costing \$5000 with a life of 5 years would lead to a depreciation of \$5000(2/5), or \$2000, during the first year of its life. To find the depreciation in year 2, multiply the depreciation in year 1 by $1 - 2/5$, as follows.

$$
\begin{aligned}
\text{depreciation in year 2} &= (\text{depreciation in year 1}) \times \left(1 - \frac{2}{n}\right) \\
&= 2000\left(1 - \frac{2}{5}\right) \qquad \text{Let } n = 5. \\
&= 2000\left(\frac{3}{5}\right) \\
&= 1200,
\end{aligned}
$$

or \$1200. To find the depreciation in year 3, multiply this result by $1 - 2/5$, or 3/5, to get \$720.

The depreciation by the double declining balance method in each of the first four years of the life of an asset is shown in the following table.

Year	Amount of Depreciation
1	$\dfrac{2}{n} \cdot x$
2	$\dfrac{2}{n} \cdot x\left(1 - \dfrac{2}{n}\right)$
3	$\dfrac{2}{n} \cdot x\left(1 - \dfrac{2}{n}\right)^2$
4	$\dfrac{2}{n} \cdot x\left(1 - \dfrac{2}{n}\right)^3$

As the table suggests, each entry is found by multiplying the preceding entry by $1 - 2/n$. Based on this, the depreciation in year j, written D_j, is the amount

$$D_j = \frac{2}{n} \cdot x \cdot \left(1 - \frac{2}{n}\right)^{j-1}.$$

This result is a general formula for the entries in the table above. It is a mathematical model for double declining balance depreciation. If double declining balance depreciation were to be used for each year of the life of an asset, then the total depreciation would be less than the net cost of the asset. For this reason, it is permissible to switch to straight-line depreciation toward the end of the useful life of the asset.

In the example above, the total amount depreciated after 3 years is $2000 + 1200 + 720 = 3920$, leaving an undepreciated balance of $5000 - 3920 = 1080$. If we switch to straight-line depreciation after 3 years, we would depreciate $1080/2 = 540$, or $540, in each of the last two years.

EXERCISES

1. A machine tool costs $55,000 and has a useful life of 5 years. For each of the methods discussed, find the depreciation in each year. The total depreciation at the end of the 5 years should be $55,000 for each method. For the double declining balance method, switch to straight-line depreciation in the last 2 years.

2. A new airplane costs $600,000 and has a useful life of 10 years. Find the depreciation in years 1 and 4 by each of the three methods discussed.

3. Which method gives the largest deductions in the early years?

Functions and Graphs

Functions are an extremely useful way of describing many real-world situations in which the value of one quantity varies with, depends on, or determines the value of another. In this chapter you will be introduced to functions, learn how to use functional notation, develop skill in constructing and interpreting the graphs of functions, and, finally, learn to apply this knowledge in a variety of situations.

3.1 FUNCTIONS

To understand the origin of the concept of function, we consider some "real life" situations in which one numerical quantity depends on, corresponds to, or determines another.

EXAMPLE 1   The amount of income tax you pay depends on the amount of your income. The way in which the income determines the tax is given by the tax law.   ■

EXAMPLE 2   The weather bureau records the temperature over a 24-hour period in the form of a graph (see Figure 3.1). The graph shows the temperature that corresponds to each given time.   ■

FIGURE 3.1

EXAMPLE 3   Suppose a rock is dropped straight down from a high point. From physics we know that the distance traveled by the rock in t seconds is $16t^2$ feet. So the distance depends on the time.   ■

The first common feature shared by these examples is that each involves two sets of numbers, which we can think of as a set of inputs and a set of outputs.

	Set of Inputs	*Set of Outputs*
Example 1	All incomes	All tax amounts
Example 2	Hours since midnight	Temperatures during the day
Example 3	Seconds elapsed after dropping the rock	Distances the rock travels

The second common feature is that in each example there is a definite *rule* by which each input determines an output. In Example 1 the rule is given by the tax law, which specifies how each income (input) determines a tax amount (output). Similarly, the rule is given by the time/temperature graph in Example 2 and by the formula (distance $= 16t^2$) in Example 3.

Each of these examples could be represented by an idealized calculator that has a single operation key and can receive or display any real number. When a number is entered (*input*), and the "rule key" is pressed, an answer is displayed (*output*). (See Figure 3.2.) The formal definition of function has these same common features (input/rule/output), with a slight change of terminology.

INPUT — OUTPUT

FIGURE 3.2

A **function** consists of a set of input numbers called the **domain,** a set of output numbers called the **range,** and a rule by which each input (number in the domain) determines exactly one output (number in the range).

The domain in Example 1 consists of all possible income amounts; the rule is given by the tax law, and the range consists of all possible tax amounts. In Example

2 the domain is the set of hours in the day (that is, all real numbers from 0 to 24); the rule is given by the time/temperature graph, which shows the temperature at each time. The graph also shows that the range (the temperatures that actually occur during the day) includes all the numbers from 38 to 63.

In Example 2, for each time of day (number in the domain) there is one and only one temperature (number in the range). Notice, however, that it is possible to have the same temperature (number in the range) corresponding to two different times (numbers in the domain).

> By the rule of a function, each number in the domain determines *one and only one* number in the range. But several different numbers in the domain may determine the same number in the range.

In other words, exactly one output is produced for each input, but different inputs may produce the same output.

EXAMPLE 4 Which of the following rules describe functions?

(a) Use the optical reader at the checkout counter of the supermarket to convert codes to prices.

For each code, the reader produces exactly one price, so this is a function.

(b) Enter a number in a calculator and press the x^2 key.

This is a function because the calculator produces just one number x^2 for each number x that is entered.

(c) Assign to each number x the number y given by this table.

x	1	1	2	2	3	3
y	3	−3	5	−5	8	−8

Since at least one x-value corresponds to more than one y-value, this table does not define a function.

(d) Assign to each number x the number y given by this equation: $y = 3x - 5$.

Because the equation determines a unique value of y for each value of x, it defines a function. ■ 1

The preceding examples show that there are many ways to define a function. Almost all of the functions used in this book will be defined by equations, as in part (d) of Example 4, and *x will be assumed to represent the input variable.*

The domain and range may or may not be the same set. For instance, in the function given by the x^2 key on a calculator, the domain consists of all numbers (positive, negative, or 0) that can be entered in the calculator, but the range consists only of *nonnegative* numbers (since $x^2 \geq 0$ for every x). In the function given by the equation $y = 3x - 5$, both the domain and range are the set of all real numbers because of the following *agreement on domains.*

1 Do the following define functions?

(a) The correspondence defined by the rule $y = x^2 + 5$

(b) The $\boxed{1/x}$ key on a calculator

(c) The correspondence between a computer, x, and several users of the computer, y

Answers:
(a) Yes
(b) Yes
(c) No

Unless otherwise stated, assume that the domain of any function defined by an equation is the largest set of real numbers that are meaningful replacements for the input variable.

For example, suppose

$$y = \frac{-4x}{2x - 3}.$$

Any real number can be used for x except $x = 3/2$, which makes the denominator equal 0. By the agreement on domains, the domain of this function is the set of all real numbers except $3/2$.

2 Do the following define y as a function of x?

(a) $y = -6x + 1$

(b) $y = x^2$

(c) $x = y^2 - 1$

(d) $y < x + 2$

Answers:

(a) Yes

(b) Yes

(c) No

(d) No

EXAMPLE 5   Decide whether each of the following equations defines y as a function of x. Give the domain of any functions.

(a) $y = -4x + 11$

For a given value of x, calculating $-4x + 11$ produces exactly one value of y. (For example, if $x = -7$, then $y = -4(-7) + 11 = 39$.) Because one value of the input variable leads to exactly one value of the output variable, $y = -4x + 11$ defines a function. Since x may take on any real-number values, the domain is the set of all real numbers, written $(-\infty, \infty)$.

(b) $y^2 = x$

Suppose $x = 36$. Then $y^2 = x$ becomes $y^2 = 36$, from which $y = 6$ or $y = -6$. Since one value of x can lead to two values of y, $y^2 = x$ does not define y as a function of x. ■   **2**

EXAMPLE 6   Find the domain of each of the following functions.

(a) $y = x^4$

Any number may be raised to the fourth power so the domain is $(-\infty, \infty)$.

(b) $y = \sqrt{6 - x}$

For y to be a real number, $6 - x$ must be nonnegative. This happens only when $6 - x \geq 0$, or $6 \geq x$, making the domain the interval $(-\infty, 6]$.

(c) $y = \dfrac{1}{x + 3}$

Because the denominator cannot be 0, $x \neq -3$ and the domain consists of all numbers in the intervals

$$(-\infty, -3) \quad \text{or} \quad (-3, \infty). \quad ■ \quad \boxed{3}$$

3 Give the domain.

(a) $y = 3x + 1$

(b) $y = x^2$

(c) $y = \sqrt{-x}$

Answers:

(a) $(-\infty, \infty)$

(b) $(-\infty, \infty)$

(c) $(-\infty, 0]$

CAUTION   Notice the difference between the rule $y = \sqrt{6 - x}$ in Example 6(b) and the rule $y^2 = 6 - x$. For a particular x-value, say 2, the radical expression in the first rule represents a single positive number. The second rule, however, when $x = 2$, produces *two* numbers, $y = 2$ or $y = -2$. ◆

FUNCTIONAL NOTATION In actual practice, functions are seldom presented in the style of domain, rule, range, as they have been here. Functions are usually denoted by a letter (f is frequently used). If x is an input (number in the domain), then $f(x)$ denotes the output number that the function f produces from the input x. The symbol $f(x)$ is read "f of x." The rule is usually given by a formula, such as $f(x) = \sqrt{x^2 + 1}$. This formula can be thought of as a set of directions.

Name of function   Input number

$$\underbrace{f(x)}_{\text{Output number}} = \underbrace{\sqrt{x^2 + 1}}_{\begin{array}{l}\text{Directions that tell you what to do with input } x \\ \text{in order to produce the corresponding output} \\ f(x); \text{ namely, "square it, add 1, and take the} \\ \text{square root of the result."}\end{array}}$$

For example, to find $f(3)$ (the output number produced by the input 3), simply replace x by 3 in the formula:

$$f(3) = \sqrt{3^2 + 1}$$
$$= \sqrt{10}.$$

Similarly, replacing x by -5 and 0 shows that

$$f(-5) = \sqrt{(-5)^2 + 1} \qquad \text{and} \qquad f(0) = \sqrt{0^2 + 1}$$
$$= \sqrt{26} \qquad\qquad\qquad\qquad = 1.$$

These directions can be applied to any quantities, such as $a + b$ or c^4 (where a, b, c are real numbers). Thus, to compute $f(a + b)$, the output corresponding to input $a + b$, we square the input [obtaining $(a + b)^2$], add 1 [obtaining $(a + b)^2 + 1$] and take the square root of the result:

$$f(a + b) = \sqrt{(a + b)^2 + 1}$$
$$= \sqrt{a^2 + 2ab + b^2 + 1}.$$

Similarly, the output $f(c^4)$ corresponding to the input c^4 is computed by squaring the input [$(c^4)^2$], adding 1 [$(c^4)^2 + 1$], and taking the square root of the result:

$$f(c^4) = \sqrt{(c^4)^2 + 1}$$
$$= \sqrt{c^8 + 1}.$$

$x - 1$

EXAMPLE 7   Let $g(x) = -x^2 + 4x - 5$. Find each of the following.

(a) $g(-2)$

Replace x with -2.

$$g(-2) = -(-2)^2 + 4(-2) - 5$$
$$= -4 - 8 - 5$$
$$= -17$$

(b) $g(x + h)$

Replace x by the quantity $x + h$ in the rule of g.

$$g(x + h) = -(x + h)^2 + 4(x + h) - 5$$
$$= -(x^2 + 2xh + h^2) + (4x + 4h) - 5$$
$$= -x^2 - 2xh - h^2 + 4x + 4h - 5$$

(c) $g(x + h) - g(x)$

Use the result from part (b) and the rule for $g(x)$.

$$g(x + h) - g(x) = (-x^2 - 2xh - h^2 + 4x + 4h - 5) - (-x^2 + 4x - 5)$$
$$= -2xh - h^2 + 4h$$

(d) $\dfrac{g(x + h) - g(x)}{h}$ (assuming $h \neq 0$)

The numerator was found in part (c). Divide it by h as follows.

$$\frac{g(x + h) - g(x)}{h} = \frac{-2xh - h^2 + 4h}{h}$$
$$= \frac{h(-2x - h + 4)}{h}$$
$$= -2x - h + 4 \quad \blacksquare$$

The quotient found in Example 7(d)

$$\frac{g(x + h) - g(x)}{h}$$

is important in calculus, and we will see it again in Chapter 12. ☐4

4 Let $f(x) = 5x^2 - 2x + 1$.
Find the following.

(a) $f(1)$

(b) $f(3)$

(c) $f(1 + 3)$

(d) $f(1) + f(3)$

(e) $f(m)$

(f) $f(x + h) - f(x)$

Answers:

(a) 4

(b) 40

(c) 73

(d) 44

(e) $5m^2 - 2m + 1$

(f) $10xh + 5h^2 - 2h$

CAUTION Functional notation is *not* the same as ordinary algebraic notation. You cannot simplify an expression such as $f(x + h)$ by writing $f(x) + f(h)$. To see why, consider the answers to Problems 4(c) and (d) at the side, that show that

$$f(1 + 3) \neq f(1) + f(3). \blacklozenge$$

EXAMPLE 8 Suppose the projected sales (in thousands of dollars) of a small company over the next ten years are approximated by the function

$$S(x) = .08x^4 - .04x^3 + x^2 + 9x + 54.$$

(a) What are the projected sales this year?

The current year corresponds to $x = 0$ and the sales this year are given by $S(0)$. Substituting 0 for x in the rule of S, we see that $S(0) = 54$. So the projected sales are $54,000.

5 A developer estimates that the total cost of building x large apartment complexes in a year is approximated by

$$A(x) = x^2 + 80x + 60,$$

where $A(x)$ represents the cost in hundred thousands of dollars. Find the cost of building

(a) 4 complexes;

(b) 10 complexes.

Answers:
(a) $39,600,000
(b) $96,000,000

(b) What will sales be in three years?

The sales three years from now are given by $S(3)$, which can be computed by hand or with a calculator:

$$S(x) = .08x^4 - .04x^3 + x^2 + 9x + 54$$

$$S(3) = .08(3)^4 - .04(3)^3 + (3)^2 + 9(3) + 54 \qquad \text{Let } x = 3.$$

$$= 95.4,$$

which means that sales are projected to be $95,400. ■   **5**

TECHNOLOGY TIP Many graphing calculators have a table feature that displays a table of values for a function. Some, but not all, graphing calculators also allow you to use functional notation. Check your instruction manual to see if your calculator has these features, and if it does, learn how to use them. ✔

3.1 EXERCISES

Which of the following rules define y as a function of x? (See Examples 1–5.)

1.

x	3	2	1	0	−1	−2	−3
y	9	4	1	0	1	4	9

2.

x	9	4	1	0	1	4	9
y	3	2	1	0	−1	−2	−3

3. $y = x^3$ ~yes~

4. $y = \sqrt{x - 1}$ ~yes~

5. $x = |y + 2|$

6. $x = y^2 + 3$ ~no~

7. $y = \dfrac{-1}{x - 1}$

8. $y = \dfrac{4}{2x + 3}$ ~yes~

State the domain of each function. (See Examples 5 and 6.)

9. $f(x) = 4x - 1$

10. $f(x) = 2x + 7$ ~All real #'s~

11. $f(x) = x^4 - 1$ ~All real #'s~

12. $f(x) = (2x + 5)^2$ ~All real #'s~

13. $f(x) = \sqrt{-x} + 3$

14. $f(x) = \sqrt{4 - x}$ ~Nothing higher then~ ~x≤4~ 4

15. $f(x) = \dfrac{1}{x - 1}$

16. $f(x) = \dfrac{1}{x - 3}$ ~All #'s except 3~

17. $f(x) = |5 - 4x|$

18. $f(x) = |-x - 6|$ ~All real #'s~

For each of the following functions, find
(a) $f(4)$, ~only~
(b) $f(-3)$, ~A&B done~
(c) $f(2.7)$,
(d) $f(-4.9)$.
(See Example 7.)

19. $f(x) = 6$

20. $f(x) = 0$ ~A○ constant B○ function~

21. $f(x) = 2x^2 + 4x$

22. $f(x) = x^2 - 2x$ ~A&B not a function~

23. $f(x) = \sqrt{x + 3}$

24. $f(x) = \sqrt{5 - x}$

25. $f(x) = x^3 - 6.2x^2 + 4.5x - 1$

26. $f(x) = x^4 + 5.5x^2 - .3x$

27. $f(x) = |x^2 - 6x - 4|$

28. $f(x) = |x^3 - x^2 + x - 1|$

29. $f(x) = \dfrac{\sqrt{x - 1}}{x^2 - 1}$

30. $f(x) = \sqrt{-x} + \dfrac{2}{x + 1}$

If you have a graphing calculator with table making ability, display a table showing the (approximate) values of the given function at 3.5, 3.9, 4.3, 4.7, 5.1, and 5.5.

31. $g(x) = 3x^4 - x^3 + 2x$

32. $f(x) = \sqrt{x^2 - 2.4x + 8}$

For each of the following functions, find
(a) $f(p)$,
(b) $f(-r)$, and
(c) $f(m + 3)$.
(See Example 7.)

33. $f(x) = 5 - x$

34. $f(x) = 3x + 7$

35. $f(x) = \sqrt{4 - x}$

36. $f(x) = \sqrt{-2x}$

37. $f(x) = x^3 + 1$

38. $f(x) = 2 - x^3$

39. $f(x) = \dfrac{3}{x - 1}$

40. $f(x) = \dfrac{-1}{2 + x}$

For each of the following find
$$\frac{f(x + h) - f(x)}{h}.$$

41. $f(x) = 2x - 4$

42. $f(x) = 2 - 3x$

43. $f(x) = x^2 + 1$

44. $f(x) = x^2 - x$

Use a calculator to work these exercises. (See Example 8.)

45. Natural Science The table contains incidence ratios by age for death from coronary heart disease (CHD) and lung cancer (LC) when comparing smokers (21–39 cigarettes per day) to nonsmokers.*

Age	CHD	LC
55–64	1.9	10
65–74	1.7	9

The incidence ratio of 10 means that smokers are 10 times more likely than nonsmokers to die of lung cancer between

---

*Walker, A., *Observations and Inference: An Introduction to the Methods of Epidemiology*, Newton Lower Falls, MA: Epidemiology Resources, Inc., 1991.

the ages of 55 and 64. If the incidence ratio is x, then the percent P (expressed as a decimal) of deaths caused by smoking is given by the function $P(x) = \dfrac{x - 1}{x}$.

(a) What is the percent of lung cancer deaths that can be attributed to smoking between the ages of 65 and 74?

(b) What is the percent of coronary heart disease deaths that can be attributed to smoking between the ages of 55 and 64?

46. Management The number of fliers on commuter airlines (10- to 30-seat planes) between 1975 and 2010 is approximated by the function $g(x) = .0138x^2 - .172x + 1.4$ (where $x = 0$ corresponds to 1975 and $g(x)$ is in millions).*

(a) How many fliers were there in 1975 and 1994?

(b) How many fliers are projected for 2006?

47. Social Science The number of Americans (in thousands) that are or are expected to be over 100 years old in year x can be approximated by the function $h(x) = .4018x^2 + 2.039x + 50$ (where $x = 0$ corresponds to 1994).†

(a) How many Americans were over 100 in 1994? In 1996?

(b) Predict the number of Americans that will be over 100 years old in the year 2008.

48. Management The total annual amount (in millions of dollars) of government guaranteed student loans from 1986 to 1994 is approximated by the function $f(x) = 1.088(x - 1986) + 8.6$.‡

(a) How much was spent in 1986? In 1989?

(b) If this function remains valid in later years, how much will be spent in 1999?

Work these exercises.

49. Physical Science The distance from Chicago to Seattle is approximately 2000 miles. A plane flying directly to Seattle passes over Chicago at noon. If the plane travels at 475 mph, find the rule of the function $f(t)$ that gives the distance of the plane from Seattle at time t hours (with $t = 0$ corresponding to noon).

50. Management A pretzel factory has daily fixed costs of $1800. In addition, it costs 50 cents to produce each bag of pretzels. A bag of pretzels sells for $1.20.

(a) Find the rule of the cost function $c(x)$ that gives the total daily cost of producing x bags of pretzels.

(b) Find the rule of the revenue function $r(x)$ that gives the daily revenue from selling x bags of pretzels.

(c) Find the rule of the profit function $p(x)$ that gives the daily profit from x bags of pretzels.

---

*Based on data from the Federal Aviation Administration (in *USA Today,* March 27, 1995).

†Based on data from the U.S. Census Bureau.

‡Based on data in *USA Today.*

3.2 GRAPHS OF FUNCTIONS

The **graph** of a function $f(x)$ is defined to be the graph of the *equation* $y = f(x)$. For example, the graph of $f(x) = x^3 + x + 5$ is the graph of the equation $y = x^3 + x + 5$. Thus, the graph consists of all points in the coordinate plane whose coordinates are of the form $(x, x^3 + x + 5)$, that is, all points $(x, f(x))$. The same thing is true in the general case: Each point on the graph of a function f is an ordered pair whose first coordinate is an input number from the domain of f and whose second coordinate is the corresponding output number.

We begin with the simplest function graphs. In Section 2.2 we saw that the graph of an equation of the form $y = ax + b$ is a straight line. This equation also defines y as a function of x, which leads to the following terminology.

A **linear function** is a function whose rule can be written in the form

$$f(x) = ax + b$$

for some constants a and b.

EXAMPLE 1   The graph of the linear function $g(x) = .5x - 3$ is the graph of the equation $y = .5x - 3$. So the graph is a straight line with slope .5 and y-intercept -3, as shown in Figure 3.3. ∎

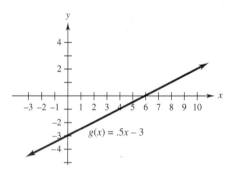

FIGURE 3.3

PIECEWISE LINEAR FUNCTIONS   We now consider functions whose graphs consist of straight line segments. Such functions are called **piecewise linear functions** and are typically defined with different equations for different parts of the domain.

EXAMPLE 2   Graph the following function.

$$f(x) = \begin{cases} x + 1 & \text{if } x \leq 2 \\ -2x + 7 & \text{if } x > 2 \end{cases}$$

When $x \leq 2$, the graph consists of the part of the line $y = x + 1$ that lies to the left of $x = 2$. When $x > 2$, the graph consists of the part of the line $y = -2x + 7$ that lies to the right of $x = 2$. These line segments can be graphed by plotting the points determined by the following tables.

$x \leq 2$			
x	-2	0	2
$y = x + 1$	-1	1	3

$x > 2$			
x	2	3	4
$y = -2x + 7$	3	1	-1

Note that even though 2 is not in the interval $x > 2$, we found the ordered pair for that endpoint because the graph will extend right up to that point. Because this endpoint $(2, 3)$ agrees with the endpoint for the interval $x \leq 2$, the two parts of the graph are joined at that point as shown in Figure 3.4. ◼

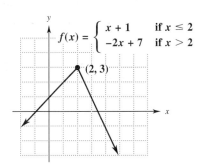

$$f(x) = \begin{cases} x + 1 & \text{if } x \leq 2 \\ -2x + 7 & \text{if } x > 2 \end{cases}$$

$(2, 3)$

FIGURE 3.4

1 Graph each function.

(a) $f(x) = 2 - |x|$

(b) $f(x) = |5x - 7|$

Answers:

(a)

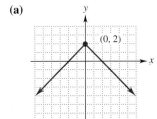

$(0, 2)$

(b)

$\left(\frac{7}{5}, 0\right)$

CAUTION In Example 2, notice that we did not graph the entire lines but only those portions with domain as given. Graphs of these functions should *not* be two intersecting lines. ◆

EXAMPLE 3 Graph the **absolute value function,** whose rule is $f(x) = |x|$. The function f is a piecewise linear function because from the definition of $|x|$,

$$f(x) = \begin{cases} x & \text{if } x \geq 0 \\ -x & \text{if } x < 0. \end{cases}$$

So the right half of the graph (that is, where $x \geq 0$) will consist of a portion of the line $y = x$. It can be graphed by plotting two points, say $(0, 0)$ and $(1, 1)$. The left half of the graph (where $x < 0$) will consist of a portion of the line $y = -x$, which can be graphed by plotting $(-2, 2)$ and $(-1, 1)$, as shown in Figure 3.5 on the next page. ◼ **1**

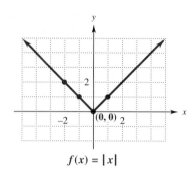

$$f(x) = |x|$$

FIGURE 3.5

 Graph $f(x)$, where

$$f(x) = \begin{cases} -2x + 5 & \text{if } x < 2 \\ x - 4 & \text{if } x \geq 2. \end{cases}$$

Answer:

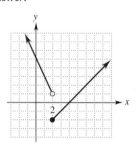

EXAMPLE 4 Graph the function

$$f(x) = \begin{cases} x - 2 & \text{if } x \leq 3 \\ -x + 8 & \text{if } x > 3. \end{cases}$$

The ordered pairs $(-2, -4)$, $(0, -2)$, and $(3, 1)$ satisfy $y = x - 2$ and the pairs $(4, 4)$ and $(6, 2)$ satisfy $y = -x + 8$. Use them to graph the two straight line segments that make up the graph of the function. The point $(3, 5)$, where the right half of the graph begins is an endpoint, but is *not* part of the graph, as indicated by the open circle in Figure 3.6. ■ 2

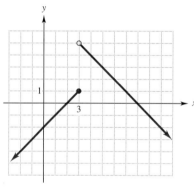

$$f(x) = \begin{cases} x - 2 & \text{if } x \leq 3 \\ -x + 8 & \text{if } x > 3 \end{cases}$$

FIGURE 3.6

TECHNOLOGY TIP In order to graph most piecewise linear functions on a graphing calculator, you must use a special syntax. For example, on TI and HP-38 calculators, the best way to obtain the graph in Example 4 is to graph two separate equations on the same screen:

$$y_1 = (x - 2)/(x \leq 3) \quad \text{and} \quad y_2 = (-x + 8)/(x > 3);$$

the inequality symbols are in the TEST (or CHAR) menu. However, most calculators will graph absolute value functions directly. To graph $f(x) = |x + 2|$, for instance, graph the equation $y = \text{abs}(x + 2)$. "Abs" (for absolute value) is on the keyboard or in the MATH menu. ✔

Graphs used in business and economics are often graphs of piecewise linear functions. It is just as important to know how to *read* such graphs as it is to construct them.

EXAMPLE 5   The graph in Figure 3.7 shows the per capita personal income in the greater Sacramento area from 1980 to 1992. What does it say about the rate of income growth?

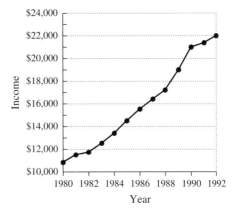

Graph showing per capita personal income in the greater Sacramento area, *Economic Profile Greater Sacramento Area* from Sacramento Area Commerce and Trade Organization, Summer, 1992. Reprinted by permission.

FIGURE 3.7

After increasing slightly from 1980 to 1982, income rose at about the same steady rate from 1982 to 1988, which is indicated by the fact that a straight line connecting these two points would be virtually identical to the graph for this period. Income increased quite sharply from 1988 to 1990, then tended to level off, increasing much more slowly from 1990 to 1992. ■

EXAMPLE 6   China's economy boomed during the late 80s and early 90s. Figure 3.8 shows the graphs of two piecewise linear functions representing China's exports and imports (in billions of dollars) during this period.* The rules of these functions are

$$E(x) = \text{total exports in year } x \text{ (in billions of dollars)}$$
$$I(x) = \text{total imports in year } x \text{ (in billions of dollars).}$$

Growing Trade Surplus

FIGURE 3.8

(a) Find the function values $E(1988)$ and $E(1991)$ and interpret them.

$E(1988) = 50$ and $E(1991) \approx 70$, because the points (1988, 50) and (1991, 70) are on the graph of E. This means that exports totaled 50 billion dollars in 1988 and 70 billion in 1991.

(b) Find the function values $I(1986)$ and $I(1991)$ and interpret them.

$I(1986) \approx 44$ and $I(1991) \approx 57$, because the (approximate) points (1986, 44) and (1991, 57) are on the graph of I. This means that total imports were 44 billion dollars in 1986 and 57 billion in 1991.

(c) What was the difference between exports and imports in 1991? What does this mean?

In functional notation, we must find $E(1991) - I(1991)$. Using parts (a) and (b), we have $E(1991) - I(1991) = 70 - 57 = 13$. Therefore, China had a positive net balance of trade of 13 billion dollars. ■

STEP FUNCTIONS The **greatest integer function,** usually written $f(x) = [x]$, is defined by saying that $[x]$ denotes the largest integer that is less than or equal to x. For example, $[8] = 8$, $[7.45] = 7$, $[\pi] = 3$, $[-1] = -1$, $[-2.6] = -3$, and so on.

EXAMPLE 7 Graph $f(x) = [x]$.

If $-1 \le x < 0$, then $[x] = -1$. If $0 \le x < 1$, then $[x] = 0$. If $1 \le x < 2$, then $[x] = 1$, and so on. Thus, the graph, as shown in Figure 3.9, consists of a series of

---

*The graph is from "China's Booming Economy" in *U.S. News and World Report,* October 19, 1992. Copyright © 1992 by U.S. News and World Report. Reprinted by permission.

3 Graph $y = [\tfrac{1}{2}x + 1]$.

Answer:

4 Assume that the post office charges 30¢ per ounce, or fraction of an ounce, to mail a letter. Graph the ordered pairs (ounces, cost).

Answer:

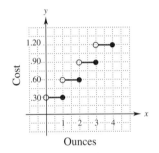

horizontal line segments. In each one, the left endpoint is included and the right endpoint is excluded. The shape of the graph is the reason that this function is called a **step function.** ■ **3**

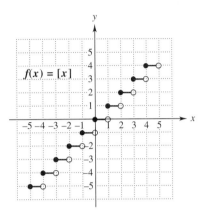

FIGURE 3.9

EXAMPLE 8 An overnight delivery service charges $25 for a package weighing up to 2 pounds. For each additional pound or fraction of a pound there is an additional charge of $3. Let $D(x)$ represent the cost to send a package weighing x pounds. Graph $D(x)$ for x in the interval (0, 6].

For x in the interval (0, 2], $y = 25$. For x in (2, 3], $y = 25 + 3 = 28$. For x in (3, 4], $y = 28 + 3 = 31$, and so on. The graph, which is that of a step function, is shown in Figure 3.10. ■ **4**

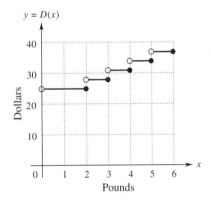

FIGURE 3.10

TECHNOLOGY TIP On most graphing calculators the greatest integer function is denoted INT or FLOOR (look on the MATH menu or its NUM submenu). Casio calculators use INTG for the greatest integer function and INT for a different function. When graphing these functions, put your calculator in "dot" graphing mode rather than the usual "connected" mode to avoid erroneous vertical line segments in the graph. ✔

OTHER FUNCTIONS The graphs of many functions do not consist only of straight line segments. As a general rule when graphing functions by hand, you should follow the procedure introduced in Section 2.1.

GRAPHING A FUNCTION BY POINT PLOTTING

1. Determine the domain of the function.
2. Select a few numbers in the domain of f (include both negative and positive ones when possible) and compute the corresponding values of $f(x)$.
3. Plot the points $(x, f(x))$ computed in Step two. Use these points and any other information you may have about the function to make an "educated guess" about the shape of the entire graph.
4. Unless you have information to the contrary, assume that the graph is continuous (unbroken) wherever it is defined.

---

5 Graph $f(x) = \sqrt{4 - x}$.

Answer:

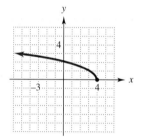

This method was used to find the graphs of the functions $f(x) = 4 - x^2$ and $g(x) = \sqrt{x}$ in Examples 3 and 4 of Section 2.1. Here are some more examples.

EXAMPLE 9 Graph $g(x) = \sqrt{x + 1}$.

Because the rule of the function is defined only when $x + 1 \geq 0$ (that is, when $x \geq -1$), the domain of g is the interval $[-1, \infty)$. Use a calculator to get a table of ordered pairs, such as the one in Figure 3.11. Plot the points and connect them in order (as x increases) to get the graph in Figure 3.11. ■ **5**

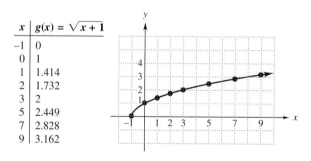

x	$g(x) = \sqrt{x + 1}$
-1	0
0	1
1	1.414
2	1.732
3	2
5	2.449
7	2.828
9	3.162

FIGURE 3.11

EXAMPLE 10 Graph the function whose rule is $f(x) = 2 - x^3/5$.

Make a table of values and plot the corresponding points. They suggest the graph in Figure 3.12. ■

x	$f(x) = 2 - x^3/5$
-4	14.8
-3	7.4
-2	3.6
-1	2.2
0	2
1	1.8
2	.4
3	-3.4
4	-10.8

FIGURE 3.12

The method used in Examples 9 and 10 is satisfactory for functions with simple algebraic rules. In later chapters we will consider graphing functions with more complicated rules. The following fact, which distinguishes function graphs from other graphs, is sometimes useful for identifying graphs of functions.

VERTICAL LINE TEST

No vertical line intersects the graph of a function more than once.

In other words, if a vertical line intersects a graph at more than one point, the graph is not the graph of a function. To see why this is true, consider the graph in Figure 3.13. The vertical line $x = 3$ intersects the graph at two points. If this were the graph of a function f, this would mean that $f(3) = 2$ (because $(3, 2)$ is on the graph) *and* that $f(3) = -1$ (because $(3, -1)$ is on the graph). This is impossible because a *function* can have only one value when $x = 3$ (each input determines exactly one output). Therefore, this cannot be the graph of a function. A similar argument works in the general case.

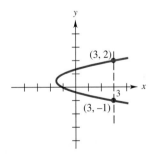

FIGURE 3.13

EXAMPLE 11 Which of the graphs in Figure 3.14 are the graphs of functions?

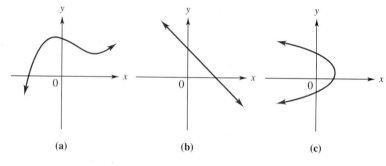

(a) (b) (c)

FIGURE 3.14

6 Does this graph represent a function? How can you tell?

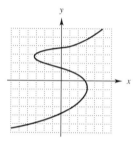

Answer:
No; a vertical line crosses the graph at more than one point.

(a) Every vertical line intersects this graph in at most one point, so this is the graph of a function.

(b) Again, each vertical line intersects the graph in at most one point, showing that this is the graph of a function.

(c) It is possible for a vertical line to intersect the graph in part (c) twice. This is not the graph of a function. ■ 6

3.2 EXERCISES

Graph each function. (See Examples 1–4.)

1. $f(x) = -.5x + 2$

2. $g(x) = 3 - x$

3. $f(x) = \begin{cases} x + 2 & \text{if } x \le 1 \\ 3 & \text{if } x > 1 \end{cases}$

4. $g(x) = \begin{cases} 2x - 1 & \text{if } x < 0 \\ -1 & \text{if } x \ge 0 \end{cases}$

5. $y = \begin{cases} 3 - x & \text{if } x \le 0 \\ 2x + 3 & \text{if } x > 0 \end{cases}$

6. $y = \begin{cases} x + 5 & \text{if } x \le 1 \\ 2 - 3x & \text{if } x > 1 \end{cases}$

7. $f(x) = \begin{cases} |x| & \text{if } x \le 2 \\ -x & \text{if } x > 2 \end{cases}$

8. $g(x) = \begin{cases} -|x| & \text{if } x \le 1 \\ 2x & \text{if } x > 1 \end{cases}$

9. $f(x) = |x - 4|$

10. $g(x) = |4 - x|$

11. $f(x) = |3 - 4x|$

12. $g(x) = -|x|$

13. $y = -|x - 1|$

14. $f(x) = |x| - 2$

15. $y = |x| + 3$

16. $|x| + |y| = 1$ (*Hint:* This is not the graph of a function, but is made up of four straight line segments. Find them by using the definition of absolute value in these four cases: $x \ge 0$ and $y \ge 0$; $x \ge 0$ and $y < 0$; $x < 0$ and $y \ge 0$; $x < 0$ and $y < 0$.)

Graph each of the following functions. (See Examples 7 and 8.)

17. $f(x) = [x - 3]$

18. $g(x) = [x + 2]$

19. $g(x) = [-x]$

20. $f(x) = -[x]$

21. $f(x) = [x] + [-x]$ (The graph contains horizontal segments, but is *not* a horizontal line.)

22. Assume that postage rates are 32¢ for the first ounce, plus 23¢ for each additional ounce and that each letter carries one 32¢ stamp and as many 23¢ stamps as necessary. Graph the *postage stamp function* whose rule is

$p(x)$ = the number of stamps on a letter weighing x ounces.

Graph each function. (See Examples 9 and 10.)

23. $f(x) = 3 - 2x^2$

24. $g(x) = 2 - x^2$

25. $h(x) = x^3/10 + 2$

26. $f(x) = x^3/20 - 3$

27. $g(x) = \sqrt{-x}$

28. $h(x) = \sqrt{x} - 1$

29. $f(x) = \sqrt[3]{x}$

30. $g(x) = \sqrt[3]{x - 4}$

Which of these are graphs of functions? (See Example 11.)

31.

32.

33.

34.

35.

36.

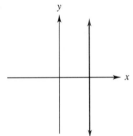

Work the following problems. (See Examples 2–5.)

37. Social Science   United States health care costs as a percent of the gross national product from 1960 to 1992 are given by

$$y = \begin{cases} .22x + 5.5 & \text{for 1960 to 1985} \\ .29x + 3.75 & \text{for 1985 to 1992.} \end{cases}$$

Let $x = 0$ represent 1960 and graph this function. What does the graph suggest about health care costs?

38. Social Science   Personal taxes in the United States from 1960 to 1990 are approximated by

$$f(x) = \begin{cases} 7.9x + 50.4 & \text{from 1960 to 1975} \\ 35.4x - 361.6 & \text{from 1975 to 1990.} \end{cases}$$

Let $x = 0$ represent 1960 and graph the function. What happened to personal taxes in 1975?

39. Natural Science   The snow depth in Michigan's Isle Royale National Park varies throughout the winter. In a typical winter, the snow depth in inches is approximated by the following function.

$$f(x) = \begin{cases} 6.5x & \text{if } 0 \le x \le 4 \\ -5.5x + 48 & \text{if } 4 < x \le 6 \\ -30x + 195 & \text{if } 6 < x \le 6.5 \end{cases}$$

Here, x represents the time in months with $x = 0$ representing the beginning of October, $x = 1$ representing the beginning of November, and so on.
(a) Graph $f(x)$.
(b) In what month is the snow deepest? What is the deepest snow depth?
(c) In what months does the snow begin and end?

40. **Natural Science** A factory begins emitting particulate matter into the atmosphere at 8 A.M. each workday, with the emissions continuing until 4 P.M. The level of pollutants, $P(t)$, measured by a monitoring station 1/2 mile away is approximated as follows, where t represents the number of hours since 8 A.M.

$$P(t) = \begin{cases} 75t + 100 & \text{if } 0 \le t \le 4 \\ 400 & \text{if } 4 < t < 8 \\ -100t + 1200 & \text{if } 8 \le t \le 10 \\ -\dfrac{50}{7}t + \dfrac{1900}{7} & \text{if } 10 < t < 24 \end{cases}$$

Find the level of pollution at
(a) 9 A.M.   (b) 11 A.M.   (c) 5 P.M.
(d) 7 P.M.   (e) Midnight.
(f) Graph $y = P(t)$.
(g) From the graph in part (f), at what time(s) is the pollution level highest? lowest?

41. **Natural Science** The table shows the percentage of babies delivered by Caesarean section in 1970, 1980, and 1990.*

Year	Percentage
1970	5%
1980	17%
1990	23%

(a) Let $x = 0$ correspond to 1970 and $x = 20$ to 1990. Find the rule of a piecewise linear function f that models this

data, that is, a piecewise linear function with $f(0) = .05$, $f(10) = .17$, and $f(20) = .23$.
(b) Graph the function f when $0 \le x \le 20$.
(c) Use the function f to estimate the percentage of Caesarean deliveries in 1975 and 1982.

42. **Management** The graph (from the Office of Management and Budget) shows the total federal debt from 1982 to 1994 (in billions of dollars). Note that the straight line segment joining the endpoints of this piecewise linear graph is a very good approximation of the graph.

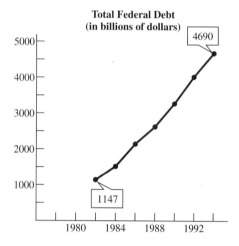

Total Federal Debt (in billions of dollars)

(a) Let $x = 0$ correspond to 1982 and $x = 12$ to 1994. Find the rule of the function g whose graph is the straight line that approximates the federal debt graph when $0 \le x \le 12$.
(b) Use the function g to find the approximate federal debt in 1990. How does this compare with the actual debt that year of 3190 billion dollars?

Work the following problems. (See Example 6.)

43. **Management** The graph at the top of the next page shows two functions of time.* The first, which we will designate $I(t)$, shows the energy intensity of the United States, which is defined as the thousands of BTUs of energy consumed per dollar of Gross National Product (GNP) in year t (in 1982 dollars). The second, which we will designate $S(t)$, represents the national savings per year (also in 1982 dollars) due to a decreased use of energy.

*Teutsch, S. and R. Churchill, *Principles and Practice of Public Health Surveillance*, New York: Oxford University Press, 1994.

*Graph "Energy Use Down, Savings Up" from "The Greenhouse Effect" by the Union of Concerned Scientists. Reprinted by permission.

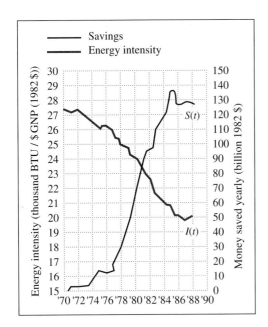

Savings

Energy intensity

(a) Find $I(1976)$.   **(b)** Find $S(1984)$.

(c) Find $I(t)$ and $S(t)$ at the time when the two graphs cross. In what year does this occur?

(d) What significance, if any, is there to the point at which the two graphs cross?

44. Management   A chain-saw rental firm charges \$7 per day or fraction of a day to rent a saw, plus a fixed fee of \$4 for resharpening the blade. Let $S(x)$ represent the cost of renting a saw for x days. Find each of the following.

(a) $S\left(\dfrac{1}{2}\right)$ **(b)** $S(1)$ **(c)** $S\left(1\dfrac{1}{4}\right)$ **(d)** $S\left(3\dfrac{1}{2}\right)$

(e) What does it cost to rent a saw for $4\frac{9}{10}$ days?

(f) A portion of the graph of $y = S(x)$ is shown here. Explain how the graph could be continued.

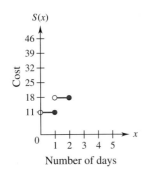

Number of days

(g) What is the domain variable?

(h) What is the range variable?

(i) Write a sentence or two explaining what part (c) and its answer represent.

(j) We have left $x = 0$ out of the graph. Discuss why it should or shouldn't be included. If it were included, how would you define $S(0)$?

45. Natural Science   The table gives estimates of the percent of ozone change from 1985 for several years in the future, if chlorofluorocarbon production is reduced 80% globally.

Year	Percent
1985	0
2005	1.5
2025	3
2065	4
2085	5

(a) Plot these ordered pairs on a grid.

(b) The points from part (a) should lie approximately on a straight line. Use the pairs (2005, 1.5) and (2085, 5) to write an equation of the line.

(c) Letting $f(x)$ represent the percent of ozone change and x represent the year, write your equation from part (b) as a rule that defines a function.

(d) Find $f(2065)$. Does it agree fairly closely with the number in the table that corresponds to 2065? Do you think the expression from part (c) describes this function adequately?

46. Management   The graph below, based on data given in *Business Week* magazine, shows the number of aircraft with airfones since 1984.*

Airfones Everywhere

Number of aircraft with airfones

(a) Is this graph that of a function?

(b) What does the domain represent?

(c) Estimate the range.

· *"Airfones Everywhere" reprinted from February 15, 1990 issue of *Business Week* by special permission, copyright © 1990 by McGraw-Hill, Inc.

47. Natural Science A laboratory culture contains about one million bacteria at midnight. The culture grows very rapidly until noon, when a bactericide is introduced and the bacteria population plunges. By 4 P.M. the bacteria have adapted to the bactericide and the culture slowly increases in population until 9 P.M. when the culture is accidentally destroyed by the clean-up crew. Let $g(t)$ denote the bacteria population at time t (with $t = 0$ corresponding to midnight) and draw a plausible graph of the function g. (Many correct answers are possible.)

48. Management A plane flies from Austin, Texas, to Cleveland, Ohio, a distance of 1200 miles. Let f be the function whose rule is:

$f(t) =$ distance (in miles) from Austin at time t hours,

with $t = 0$ corresponding to the 4 P.M. takeoff. In each of the following, draw a plausible graph of f under the given circumstances. (There are many correct answers for each part.)

(a) The flight is nonstop and takes between 3.5 and 4 hours.

(b) Bad weather forces the plane to land in Dallas (about 200 miles from Austin) at 5 P.M., remain overnight, and leave at 8 A.M. the next morning, flying nonstop to Cleveland.

(c) The plane flies nonstop, but due to heavy traffic it must fly in a holding pattern for an hour over Cincinnati (about 200 miles from Cleveland), then go on to Cleveland.

Management *Work the following problems. (See Examples 7 and 8.)*

49. The charge to rent a Haul-It-Yourself Trailer is $25 plus $2 per hour or portion of an hour. Find the cost to rent a trailer for
(a) 2 hours; **(b)** 1.5 hours; **(c)** 4 hours;
(d) 3.7 hours.
(e) Graph the ordered pairs (hours, cost).

50. A delivery company charges $3 plus 50¢ per mile or part of a mile. Find the cost for a trip of
(a) 3 miles; **(b)** 3.4 miles; **(c)** 3.9 miles;
(d) 5 miles.
(e) Graph the ordered pairs (miles, cost).
(f) Is this a function?

51. A college typing service charges $3 plus $7 per hour or fraction of an hour. Graph the ordered pairs (hours, cost).

52. A parking garage charges $1 plus 50¢ per hour or fraction of an hour. Graph the ordered pairs (hours, cost).

53. A car rental costs $37 for 1 day, which includes 50 free miles. Each additional 25 miles, or portion, costs $10. Graph the ordered pairs (miles, cost).

54. For a lift truck rental of no more than 3 days, the charge is $300. An additional charge of $75 is made for each day or portion of a day after 3. Graph the ordered pairs (days, cost).

3.3 APPLICATIONS OF LINEAR FUNCTIONS

In this section linear functions are applied to a variety of real-world situations. One of the most common occurs when discrete data are used to construct a linear function that approximates the data. Such a function provides a **linear model** of the situation that can be used (within limits) to predict future behavior.

EXAMPLE 1 The average annual cost of tuition and fees at public four-year colleges has been rising steadily, as illustrated in the following table from the College Board.

Year	1981	1983	1985	1987	1989	1991	1993	1995
Cost	$909	$1148	$1318	$1537	$1781	$2137	$2527	$2686

(a) Display this information graphically.

Let $x = 0$ represent the year 1980, $x = 1$ represent 1981, and so on. Then plot the points given by the table: (1, 909), (3, 1148), (5, 1318), etc. (Figure 3.15).

FIGURE 3.15

(b) The points in Figure 3.15 do not lie on a straight line, but they come close. Find a linear model for these data.

There are a variety of sophisticated methods of finding a straight line that "best fits" data that is almost linear. We shall use a simpler approach. Choose two of the points in Figure 3.15, say (1, 909) and (11, 2137), and draw the straight line they determine. All the data points lie reasonably close to this line (Figure 3.16).

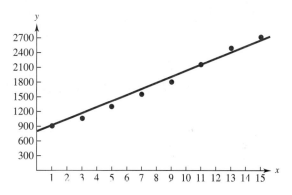

FIGURE 3.16

The line through (1, 909) and (11, 2137) has slope

$$\frac{y_2 - y_1}{x_2 - x_1} = \frac{2137 - 909}{11 - 1} = \frac{1228}{10} = 122.8.$$

Its equation can be found by using the point-slope form (see Section 2.2).

$$y - y_1 = m(x - x_1)$$
$$y - 909 = 122.8(x - 1)$$
$$y = 122.8x + 786.2$$

Therefore, the function $f(x) = 122.8x + 786.2$ provides a linear model of the situation.

(c) Use the linear model in part (b) to estimate the average annual cost of tuition and fees in 1984 and 2001.

According to the model, the average cost in 1984 (that is, $x = 4$) was

$$f(4) = 122.8(4) + 786.2$$
$$= 1277.4$$

or $1277.40, which is a bit higher than the actual cost of $1228 that year. Since 2001 corresponds to $x = 21$, the average cost in 2001 is given by

$$f(21) = 122.8(21) + 786.2$$
$$= \$3365. \quad \blacksquare$$

CAUTION A linear model need not be accurate for all values of x. The model in Example 1, for instance, suggests that the average cost in 1970 (corresponding to $x = -10$) is the negative number $f(-10) = -441.8$, which is obviously nonsense. Similarly, this model may not accurately predict costs far in the future. ◆

COMPARING RATES OF CHANGE One way to compare the change in some quantity in two or more situations is to compare the rates at which the quantity changes in the two cases. If the increase or decrease in a quantity can be approximated (or modeled) by a linear function, we can use the work in Section 2.2 to find the rate of change of the quantity with respect to time.

EXAMPLE 2 The chart below shows sales in two different years for two stores of a major chain of discount stores.

Store	Sales in 1992	Sales in 1995
A	$100,000	$160,000
B	50,000	140,000

A study of company records suggests that the sales of both stores have increased linearly (that is, the sales can be closely approximated by a linear function). Find a linear equation describing the sales for Store A.

To find a linear equation describing the sales, let $x = 0$ represent 1992 so that 1995 corresponds to $x = 3$. Then, by the chart above, the line representing the sales for Store A passes through the points $(0, 100000)$ and $(3, 160000)$. The slope of the line through these points is

$$\frac{160,000 - 100,000}{3 - 0} = 20,000.$$

1 Find a linear equation describing the sales for store B from Example 2.

Answer:
$y = 30,000x + 50,000$

Using the point-slope form of the equation of a line gives

$$y - 100,000 = 20,000(x - 0)$$
$$y = 20,000x + 100,000$$

as an equation describing the sales of Store A. ■ **1**

Point-Slope equation
$$y - y_1 = m(x - x_1)$$

AVERAGE RATE OF CHANGE Notice that the sales for Store A in Example 2 increased from $100,000 to $160,000 over the period from 1992 to 1995 representing a total increase of $60,000 in 3 years.

$$\text{Average rate of change in sales} = \frac{\$60,000}{3} = \$20,000 \text{ per year}$$

This is the same as the slope found in Example 2. Verify that the average annual rate of change in sales for store B over the 3-year period also agrees with the slope of the equation found in Problem 1 at the side. Management needs to watch the rate of change in sales closely in order to be aware of any unfavorable trends. If the rate of change is decreasing, then sales growth is slowing down, and this trend may require some response.

Example 2 illustrates a useful fact about linear functions. If $f(x) = mx + b$ is a *linear* function, then the **average rate of change** in y with respect to x (the change in y divided by the corresponding change in x) is the slope of the line $y = mx + b$. In particular, the average rate of change of a linear function is constant. (See Exercises 12 and 13 in this section for examples of functions where the average rate of change is not constant.)

2 A certain new anticholesterol drug is related to the blood cholesterol level by the linear model

$$y = 280 - 3x,$$

where x is the dosage of the drug (in grams) and y is the blood cholesterol level.

(a) Find the blood cholesterol level if 12 grams of the drug are administered.

(b) In general, an increase of 1 gram in the dose causes what change in the blood cholesterol level?

Answers:
(a) 244
(b) A decrease of 3 units

EXAMPLE 3 Average family health care costs in dollars y over the 10-year period of the 90s are expected to increase each year as given by

$$y = 510x + 4300,$$

where x is the number of years since 1990. By this estimate, how much did health care cost the average family in 1996? What is the average rate of increase?

Let $x = 1996 - 1990 = 6$. The health care costs were

$$y = 510(6) + 4300 = 7360 \quad \text{or} \quad \$7360.$$

The average rate of increase in cost is given by the slope of the line. The slope is 510, so this model indicates that each year costs will increase $510. ■ **2**

COST ANALYSIS The cost of manufacturing an item commonly consists of two parts. The first is a **fixed cost** for designing the product, setting up a factory, training workers, and so on. Within broad limits, the fixed cost is constant for a particular product and does not change as more items are made. The second part is a *cost per item* for labor, materials, packing, shipping, and so on. The total value of this second cost *does* depend on the number of items made.

(a) $y = 280 - 3(12)$
$y = 280 - 36 =$
244

(b) $280 - 3(13)^3$
39
241 $244 - 241$
$= 3$ units

EXAMPLE 4 Suppose that the cost of producing clock-radios can be approximated by the linear model

$$C(x) = 12x + 100,$$

where $C(x)$ is the cost in dollars to produce x radios. The cost to produce 0 radios is

$$C(0) = 12(0) + 100 = 100,$$

or $100. This amount, $100, is the fixed cost.

Once the company has invested the fixed cost into the clock-radio project, what is the additional cost per radio? To find out, first find the cost of a total of 5 radios:

$$C(5) = 12(5) + 100 = 160,$$

or $160. The cost of 6 radios is

$$C(6) = 12(6) + 100 = 172,$$

or $172. The sixth radio costs $172 − $160 = $12 to produce. In the same way, the 81st radio costs $C(81) − C(80) = \$1072 − \$1060 = \$12$ to produce. In fact, the $(n + 1)$st radio costs

$$C(n + 1) − C(n) = [12(n + 1) + 100] − [12n + 100] = 12,$$

or $12 to produce. Because each additional radio costs $12 to produce, $12 is the variable cost per radio. The number 12 is also the slope of the cost function, $C(x) = 12x + 100$. ■

Example 4 can easily be generalized. Suppose the total cost to make x items is given by the linear cost function $C(x) = mx + b$. Then the fixed cost (the cost which occurs even if no items are produced) is found by letting $x = 0$.

$$C(0) = m \cdot 0 + b = b$$

Thus, the fixed cost is the y-intercept of the cost function.

In economics, **marginal cost** is the rate of change of cost. Marginal cost is important to management in making decisions in areas such as cost control, pricing, and production planning. If the cost function is $C(x) = mx + b$, then its graph is a straight line with slope m. Since the slope represents the average rate of change, the marginal cost is the number m. ③

In Example 4, the marginal cost (slope) 12 was also the cost of producing one more radio. We now show that the same thing is true for any linear cost function $C(x) = mx + b$. If n items have been produced, the cost of the $(n + 1)$st item is the difference between the cost of producing $n + 1$ items and the cost of producing n items, namely, $C(n + 1) − C(n)$. Because $C(x) = mx + b$,

$$\text{Cost of one more item} = C(n + 1) − C(n)$$
$$= [m(n + 1) + b] − [mn + b]$$
$$= mn + m + b − mn − b = m$$
$$= \text{marginal cost.}$$

③ The cost in dollars to produce x kilograms of chocolate candy is given by $C(x)$, where in dollars

$$C(x) = 3.5x + 800.$$

Find each of the following.

(a) The fixed cost

(b) The total cost for 12 kilograms

(c) The marginal cost per kilogram

(d) The marginal cost of the 40th kilogram

Answers:
(a) $800

(b) $842

(c) $3.50

(d) $3.50

This discussion is summarized as follows.

> In a **linear cost function** $C(x) = mx + b$, m represents the marginal cost and b the fixed cost. The marginal cost is the cost of producing one more item.
> Conversely, if the fixed cost is b and the marginal cost is always the same constant m, then the cost function for producing x items is $C(x) = mx + b$.

EXAMPLE 5 The marginal cost to produce an anticlot drug is $10 per unit, while the cost to produce 100 units is $1500. Find the cost function $C(x)$, given that it is linear.

Since the cost function is linear, it can be written in the form $C(x) = mx + b$. The marginal cost is $10 per unit, which gives the value for m, leading to $C(x) = 10x + b$. To find b, use the fact that the cost of producing 100 units of the drug is $1500, or $C(100) = 1500$. Substituting $x = 100$ and $C(x) = 1500$ into $C(x) = 10x + b$ gives

$$C(x) = 10x + b$$
$$1500 = 10(100) + b$$
$$1500 = 1000 + b$$
$$500 = b.$$

4 The total cost of producing 10 units of a business calculator is $100. The marginal cost per calculator is $4. Find the cost function, $C(x)$, if it is linear.

Answer:
$C(x) = 4x + 60$

The cost function is $C(x) = 10x + 500$, where the fixed cost is $500. ■ **4**

If $C(x)$ is the total cost to manufacture x items, then the **average cost** per item is given by

$$\overline{C}(x) = \frac{C(x)}{x}.$$

In Example 4, the average cost per clock-radio is

$$\overline{C}(x) = \frac{C(x)}{x} = \frac{12x + 100}{x} = 12 + \frac{100}{x}.$$

Note what happens to the term $100/x$ as x gets larger.

$$\overline{C}(100) = 12 + \frac{100}{100} = 12 + 1 = \$13$$

$$\overline{C}(500) = 12 + \frac{100}{500} = 12 + \frac{1}{5} = \$12.20$$

$$\overline{C}(1000) = 12 + \frac{100}{1000} = 12 + \frac{1}{10} = \$12.10$$

As more and more items are produced, $100/x$ gets smaller and the average cost per item decreases. However, the average cost is never less than $12.

EXAMPLE 6   Find the average cost per unit to produce 50 units and 500 units of the anticlot drug in Example 5.

The cost function from Example 5 is $C(x) = 10x + 500$, so the average cost per unit is

$$\overline{C}(x) = \frac{C(x)}{x} = \frac{10x + 500}{x} = 10 + \frac{500}{x}.$$

If 50 units of the drug are produced, the average cost is

$$\overline{C}(50) = 10 + \frac{500}{50} = 20,$$

or $20 per unit. Producing 500 units of the drug will lead to an average cost of

$$\overline{C}(500) = 10 + \frac{500}{500} = 11,$$

or $11 per unit.   ■   5

5 In Problem 4, at the side, find the average cost per calculator to produce 100 calculators.

Answer:
$4.60

3.3 EXERCISES

1. Social Science   The number of children in the U.S. from 5 to 13 years old decreased from 31.2 million in 1980 to 30.3 million in 1986.
(a) Write a linear function describing this population y in terms of year x for the given period.
(b) What was the average rate of change in this population over the period from 1980 to 1986?

2. Social Science   According to the U.S. Census Bureau, the poverty level income for a family of four was $5500 in 1975, $8414 in 1980, and $13,359 in 1990.
(a) Let $x = 0$ correspond to 1975 and use the points $(0, 5500)$ and $(15, 13359)$ to find a linear model for these data.
(b) Compare the poverty level given by the model for 1985 with the actual level of $10,989. How accurate is the model?
(c) How accurate is the level given by the model for 1970, when the actual poverty level was $3968?
(d) According to this model, what will the poverty level income be in 2000?

3. Physical Science   The table lists the distances (in megaparsecs) and velocities (in km/sec) of four galaxies moving rapidly away from Earth.*

Galaxy	Distance	Velocity
Virgo	15	1600
Ursa Minor	200	15,000
Corona Borealis	290	24,000
Bootes	520	40,000

(a) Plot the data using distance for the x-values and velocity for the y-values. What type of relationship seems to hold for these data?
(b) Use the points $(0, 0)$ and $(520, 40000)$ to find a linear function of the form $f(x) = mx$ that models these data.
(c) The galaxy Hydra has a velocity of 60,000 km/sec. About how far away is it?
(d) The constant m in the rule of the function is called the **Hubble constant.** The Hubble constant can be used to estimate the age of the universe A (in years), with the formula

$$A = \frac{9.5 \times 10^{11}}{m}.$$

Approximate A using your value of m.

4. Social Science   The table lists the average annual cost (in dollars) of tuition and fees at private four-year colleges for selected years.*

*Acker, A. and C. Jaschek, *Astronomical Methods and Calculations,* John Wiley & Sons, 1986. Karttunen, H. (editor), *Fundamental Astronomy,* Springer-Verlag, 1994.

*The College Board.

Year	Tuition and Fees
1981	4,113
1983	5,093
1985	6,121
1987	7,116
1989	8,446
1991	10,017
1993	11,025

Let $x = 1$ correspond to 1981 and let $f(x)$ be the tuition and fees in year x.

(a) Determine a linear function $f(x) = mx + b$ that models the data, using the points (1, 4113) and (13, 11025).

(b) Graph f and the data on the same coordinate axes.

(c) What does the slope of the graph of f indicate?

(d) Use this function to approximate the tuition and fees in the year 1990. Compare it with the true value of $9340.

5. Natural Science In one study of HIV patients who were infected by intravenous drug use, it was found that after four years 17% of the patients had AIDS and after seven years 33% had AIDS.*

(a) Use the points (4, .17) and (7, .33) to find a linear function that models the relationship between the time interval and the percentage of patients with AIDS.

(b) Predict the number of years until half of these patients have AIDS.

6. Management The table lists the total federal debt (in billions of dollars) from 1985 to 1989.†

Year	Federal Debt
1985	1828
1986	2130
1987	2354
1988	2615
1989	2881

(a) Plot the data by letting $x = 0$ correspond to 1985. Discuss any trends of the federal debt over this time period.

(b) Find a linear function $f(x) = mx + b$ that approximates the data, using the points (0, 1828) and (4, 2881). What does the slope of the graph of f represent?

(c) Use f to predict the federal debt in the years 1984 and 1990. Compare your results to the true values of 1577 and 3191 billion dollars.

(d) Now use f to predict the federal debt in the years 1980 and 1994. Compare your results to the true values of 914 and 4690 billion dollars.

7. Natural Science To achieve the maximum benefit for the heart when exercising, your heart rate (in beats per minute) should be in the Target Heart Rate Zone. The lower limit of this zone is found by taking 70% of the difference between 220 and your age. The upper limit is found by using 85%.*

(a) Find formulas for the upper and lower limits ($u(x)$ and $l(x)$) as a function of age x.

(b) What is the Target Heart Rate Zone for a 20-year-old?

(c) What is the Target Heart Rate Zone for a 40-year-old?

(d) Two women in an aerobics class stop to take their pulse, and are surprised to find that they have the same pulse. One woman is 36 years older than the other and is working at the upper limit of her Target Heart Rate Zone. The younger woman is working at the lower limit of her Target Heart Rate Zone. What are the ages of the two women, and what is their pulse?

(e) Run for 10 minutes, take your pulse, and see if it is in your Target Heart Rate Zone. (After all, this is listed as an exercise!)

8. Management In deciding whether to set up a new manufacturing plant, company analysts have established that a reasonable function for the total cost to produce x items is

$$C(x) = 500{,}000 + 4.75x.$$

(a) Find the total cost to produce 100,000 items.

(b) Find the rate of change of the cost of the items to be produced in this plant.

9. Management Suppose the sales of a particular brand of electric guitar satisfy the relationship

$$S(x) = 300x + 2000,$$

where $S(x)$ represents the number of guitars sold in year x, with $x = 0$ corresponding to 1987.
Find the sales in each of the following years.

(a) 1987 **(b)** 1990 **(c)** 1991

(d) The manufacturer needed sales to reach 4000 guitars by 1996 to pay off a loan. Did sales reach that goal?

(e) Find the annual rate of change of sales.

*Alcabes, P., A. Munoz, D. Vlahov, and G. Friedland, "Incubation Period of Human Immunodeficiency Virus," *Epidemiologic Review,* Vol. 15, No. 2, The Johns Hopkins University School of Hygiene and Public Health, 1993.

†U.S. Office of Management and Budget.

*The New York Times, August 17, 1994, p. B9.

10. Management In the early 1990s, the Office of Management and Budget made the estimates for future Medicare costs (in billions of dollars) shown in the table.

Year	Costs
1995	157
1996	178
1997	194
1998	211
1999	229
2000	247

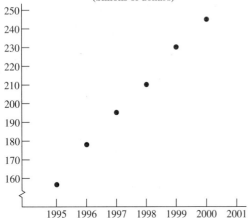

Projected Medicare Spending
(billions of dollars)

(a) The graph shows that the data are approximately linear. Use the data from 1995 and 2000 to find a linear function $f(x)$ that approximates the cost in year x.

(b) According to this function f, at what rate are costs increasing? How is this number related to the graph of the function?

11. Management Used car "superstores" that compete in the used car market with new car dealers are a recent phenomenon.*

(a) The figure shows the total number of new car dealers selling used cars in thousands. Describe what the graph tells us about the number of dealers selling used cars over this time period. From the data in the figure, what was the average rate of change in dealers from 1990 to 1995?

*"A Revolution in Car Buying," *Chicago Tribune,* February 18, 1996.

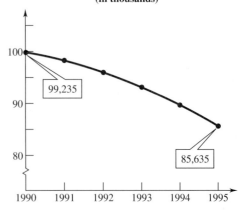

Number of Dealers Selling Used Cars
(in thousands)

99,235

85,635

(b) The sales graph showing the number of used vehicles (in millions) sold by used car "superstores" is approximately linear, as shown in the figure. What does the slope of this line indicate about the sales of these superstores?

Number of Used Vehicles Sold
by Superstores (in millions)

14.2

15.7

(c) Compare the graphs in parts (a) and (b). Describe a possible connection between the number of new car dealers selling used cars and the number of used cars sold in superstores.

12. Management The graph at the top of the next page shows the total sales, y, in thousands of dollars from the distribution of x thousand catalogs. Find the average rate of change of sales with respect to the number of catalogs distributed for the following changes in x.

(a) 10 to 20   **(b)** 10 to 40
(c) 20 to 30   **(d)** 30 to 40

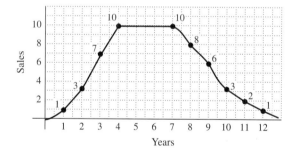

Number of Catalogs

13. Management The graph below shows annual sales (in units) of a typical product. Sales increase slowly at first to some peak, hold steady for a while, and then decline as the product goes out of style. Find the average annual rate of change in sales for the following changes in years.

(a) 1 to 3   (b) 2 to 4   (c) 3 to 6
(d) 5 to 7   (e) 7 to 9   (f) 8 to 11
(g) 9 to 10   (h) 10 to 12

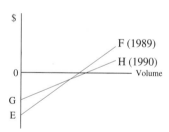

Years

14. What is meant by the marginal cost of a product? the fixed cost?

Management *Write a cost function for each of the following. Identify all variables used. (See Example 5.)*

15. A chain-saw rental firm charges $12 plus $1 per hour.

16. A trailer-hauling service charges $45 plus $2 per mile.

17. A parking garage charges 35¢ plus 30¢ per half hour.

18. For a 1-day rental, a car rental firm charges $14 plus 6¢ per mile.

Management *Assume that each of the following can be expressed as a linear cost function. Find the appropriate cost function in each case. (See Example 5.)*

19. Fixed cost, $100; 50 items cost $1600 to produce.

20. Fixed cost, $1000; 40 items cost $2000 to produce.

21. Marginal cost, $120; 100 items cost $15,800 to produce.

22. Marginal cost, $90; 150 items cost $16,000 to produce.

23. Management The total cost (in dollars) to produce x algebra books is $C(x) = 5.25x + 40,000$.
(a) What is the marginal cost per book?
(b) What is the total cost to produce 5000 books?

24. Management In Exercise 8, we were given the following function for the total cost in dollars to produce x items.

$$C(x) = 500,000 + 4.75x$$

(a) Find the marginal cost per item of the items to be produced in this plant.
(b) Find the average cost per item.

▷ **Management** *Assume that each of the following can be expressed as a linear cost function. Find* (a) *the cost function;* (b) *the revenue function.*

	Fixed Cost	Marginal Cost per Item	Item Sells for
25.	$500	$10	$35
26.	$180	$11	$20
27.	$250	$18	$28
28.	$1500	$30	$80

29. Management In the profit-volume chart below, EF and GH represent the profit-volume graphs of a single-product company for 1989 and 1990, respectively.*

$ F (1989)
                                 H (1990)
0 ────────────────────── Volume
G
E

If 1989 and 1990 unit sales prices are identical, how did total fixed costs and unit variable costs of 1990 change compared to 1989?

	1990 Total Fixed Costs	1990 Unit Variable Costs
(a)	Decreased	Increased
(b)	Decreased	Decreased
(c)	Increased	Increased
(d)	Increased	Decreased

*Uniform CPA Examination, May, 1991, American Institute of Certified Public Accountants.

CHAPTER 3 SUMMARY

Key Terms and Symbols

3.1 function
domain
range
functional notation

3.2 graph
linear model
piecewise linear function
absolute value function
greatest integer function
step function
vertical line test

3.3 linear function
average rate of change
fixed cost
marginal cost
linear cost function
average cost

Key Concepts

A **function** consists of a set of input numbers called the **domain,** a set of output numbers called the **range,** and a rule by which each number in the domain determines exactly one number in the range.

If a vertical line intersects a graph in more than one point, the graph is not that of a function.

A **linear cost function** has equation $C(x) = mx + b$ where m is the **marginal cost** (the cost of producing one more item) and b is the **fixed cost.**

Chapter 3 Review Exercises

Which of the following rules defines a function?

1.

x	3	2	1	0	1	2
y	8	5	2	0	-2	-5

2.

x	2	1	0	-1	-2
y	5	3	1	-1	-3

3. $y = \sqrt{x}$

4. $x = |y|$

5. $x = y^2 + 1$

6. $y = 5x - 2$

For each function, find
(a) $f(6)$,
(b) $f(-2)$,
(c) $f(p)$,
(d) $f(r + 1)$.

7. $f(x) = 4x - 1$

8. $f(x) = 3 - 4x$

9. $f(x) = -x^2 + 2x - 4$

10. $f(x) = 8 - x - x^2$

11. Let $f(x) = 5x - 3$ and $g(x) = -x^2 + 4x$. Find each of the following.
(a) $f(-2)$
(b) $g(3)$
(c) $g(-k)$
(d) $g(3m)$
(e) $g(k - 5)$
(f) $f(3 - p)$

12. Let $f(x) = x^2 + x + 1$. Find each of the following.
(a) $f(3)$
(b) $f(1)$
(c) $f(4)$
(d) Based on your answers in parts (a)–(c), is it true that
$f(a + b) = f(a) + f(b)$ for all real numbers a, b?

Graph each function.

13. $f(x) = |x| - 3$

14. $f(x) = -|x| - 2$

15. $f(x) = -|x + 1| + 3$

16. $f(x) = 2|x - 3| - 4$

17. $f(x) = [x - 3]$

18. $f(x) = \left[\dfrac{1}{2}x - 2\right]$

19. $f(x) = \begin{cases} -4x + 2 & \text{if } x \le 1 \\ 3x - 5 & \text{if } x > 1 \end{cases}$

20. $f(x) = \begin{cases} 3x + 1 & \text{if } x < 2 \\ -x + 4 & \text{if } x \ge 2 \end{cases}$

21. $f(x) = \begin{cases} |x| & \text{if } x < 3 \\ 6 - x & \text{if } x \ge 3 \end{cases}$

22. $f(x) = \sqrt{x^2}$

23. $g(x) = x^2/8 - 3$

24. $h(x) = \sqrt{x} + 2$

Management   *In Exercises 25–28, find the following.*
(a) *the linear cost function*
(b) *the marginal cost*
(c) *the average cost per unit to produce 100 units*

25. Eight units cost $300; fixed cost is $60.

26. Fixed cost is $2000; 36 units cost $8480.

27. Twelve units cost $445; 50 units cost $1585.

28. Thirty units cost $1500; 120 units cost $5640.

29. Management   The graph shows the percent of car owners who own foreign cars since 1975. Estimate the average rate of change in the percent over the following intervals.
(a) 1975 to 1983   **(b)** 1983 to 1987
(c) 1975 to 1991   **(d)** 1987 to 1991

Foreign Car Ownership

30. Management   Let f be a function that gives the cost to rent a floor polisher for x hours. The cost is a flat $3 for renting the polisher plus $4 per day or fraction of a day for using the polisher.
(a) Graph f.
(b) Give the domain and range of f.
(c) David Fleming wants to rent a polisher, but he can spend no more than $15. At most how many days can he use it?

31. Management   A trailer hauling service charges $45, plus $2 per mile or part of a mile.
(a) Is $90 enough for a 20-mile haul?
(b) Graph the ordered pairs (miles, cost).
(c) Give the domain and range.

32. Social Sciences   The birth rate per thousand in developing countries, for 1775–1977, can be approximated by

$$f(x) = \begin{cases} 42 & \text{from 1775 to 1925} \\ 67.5 - .17x & \text{from 1925 to 1977.} \end{cases}$$

Graph the function. Let $x = 0$ correspond to 1775. What does the graph suggest about the birth rate in developing countries?

33. Social Science   The per capita consumption of red meat in the U.S. has decreased from 131.7 pounds in 1970 to 114.1 pounds in 1992.* Assume a linear function describes the decrease. Write a linear equation defining the function. Let x represent the number of years since 1900 and y represent the number of pounds of red meat consumed.

34. Social Science   More people are staying single longer in the United States. In 1970, the number of never-married adults, ages 18 and over, was 21.4 million. By 1993, it was 42.3 million.† Assume the data increase linearly, and write an equation that defines a linear function for the data. Let x represent the number of years since 1900.

35. Social Science   The birth rate in the United States was 14.0 (per thousand) in 1975 and 16.7 in 1990.‡ Assume that the birth rate changes linearly.
(a) Find an equation for the birth rate in the United States as a linear function of time t, where t is measured in years since 1975.
(b) Israel's birth rate in 1990 was 22.2. In what year would the U.S. rate be at least that big (assuming the linear trend continues)?

36. Social Science   The percentage of college students who are ages 35 and older has been increasing at roughly a linear rate. In 1972 the percentage was 9%, and in 1992 it was 17%.§
(a) Find the percentage of college students age 35 and older as a function of time t, where t represents the number of years since 1970.
(b) If this linear trend continues, what percentage of college students will be 35 and over in 2010?
(c) If this linear trend continues, in what year will the percentage of college students 35 and over reach 31%?

---

*U.S. Department of Agriculture, Economic Research Service, *Food Consumption, Price, and Expenditures,* annual.

†U.S. Bureau of the Census, *1970 Census of Population,* vol. 1, part 1, and *Current Population Reports,* pp. 20, 450.

‡*Statistical Abstract of the United States 1994,* U.S. Department of Commerce, Economics and Statistics Division, Bureau of the Census.

§*Physical Fitness: The Pathway to Healthful Living,* Robert V. Hockey; Times Mirror/Mosby College Publishing, 1989, pp. 85–87.

CASE 3

Marginal Cost—Booz, Allen and Hamilton*

Booz, Allen and Hamilton is a large management consulting firm. One of the services it provides to client companies is profitability studies, which show ways in which the client can increase profit levels. The client company requesting the analysis presented in this case is a large producer of a staple food. The company buys from farmers, and then processes the food in its mills, resulting in a finished product. The company sells both at retail and under its own brands, and in bulk to other companies who use the product in the manufacture of convenience foods.

The client company has been reasonably profitable in recent years, but the management retained Booz, Allen and Hamilton to see whether its consultants could suggest ways of increasing company profits. The management of the company had long operated with the philosophy of trying to process and sell as much of its product as possible, since they felt this would lower the average processing cost per unit sold. However, the consultants found that the client's fixed mill costs were quite low, and that, in fact, processing extra units made the cost per unit start to increase. (There are several reasons for this: the company must run three shifts, machines break down more often, and so on.)

In this case, we shall discuss the marginal cost of two of the company's products. The marginal cost (cost of producing an extra unit) of production for product A was found by the consultants to be approximated by the linear function

$$y = .133x + 10.09,$$

where x is the number of units produced (in millions) and y is the marginal cost.

For example, at a level of production of 3.1 million units, an additional unit of product A would cost about

$$y = .133(3.1) + 10.09$$
$$\approx \$10.50.†$$

At a level of production of 5.7 million units, an extra unit costs $10.85. Figure 1 shows a graph of the marginal cost function from $x = 3.1$ to $x = 5.7$, the domain to which the function above was found to apply.

The selling price for product A is $10.73 per unit, so that, as shown on the graph of Figure 1, the company was losing money on many units of the product that it sold. Since the selling price could not be raised if the company was to remain competitive, the consultants recommended that production of product A be cut.

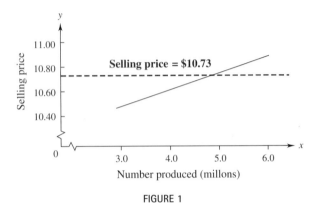

FIGURE 1

For product B, the Booz, Allen and Hamilton consultants found a marginal cost function given by

$$y = .0667x + 10.29,$$

with x and y as defined above. Verify that at a production level of 3.1 million units, the marginal cost is $10.50, while at a production level of 5.7 million units, the marginal cost is $10.67. Since the selling price of this product is $9.65, the consultants again recommended a cutback in production.

The consultants ran similar cost analyses of other products made by the company, and then issued their recommendations: the company should reduce total production by 2.1 million units. The analysts predicted that this would raise profits for the products under discussion from $8.3 million annually to $9.6 million, which is very close to what actually happened when the client took this advice.

EXERCISES

1. At what level of production, x, was the marginal cost of a unit of product A equal to the selling price?

2. Graph the marginal cost function for product B from $x = 3.1$ million units to $x = 5.7$ million units.

3. Find the number of units for which marginal cost equals the selling price for product B.

4. For product C, the marginal cost of production is

$$y = .133x + 9.46.$$

   (a) Find the marginal cost at a level of production of 3.1 million units; of 5.7 million units.
   (b) Graph the marginal cost function.
   (c) For a selling price of $9.57, find the level of production for which the cost equals the selling price.

---

*Case study, "Marginal Cost—Booz, Allen and Hamilton" supplied by John R. Dowdle of Booz, Allen & Hamilton, Inc. Reprinted by permission.

†The symbol "≈" means *is approximately equal to.*

Polynomial and Rational Functions

Polynomial functions (that is, functions defined by polynomial expressions) arise naturally in many real world applications. They are also useful for approximating many complicated functions in applied mathematics. In Chapters 2 and 3 we considered linear polynomial functions. In this chapter we examine polynomial functions of higher degree, as well as rational functions (whose rules are given by quotients of polynomials).

4.1   QUADRATIC FUNCTIONS

A **quadratic function** is a function whose rule is given by a quadratic polynomial, such as

$$f(x) = x^2, \quad g(x) = 3x^2 + 30x + 67, \quad \text{and} \quad h(x) = -x^2 + 4x.$$

Thus, a quadratic function is one whose rule can be written in the form

$$f(x) = ax^2 + bx + c$$

for some constants a, b, c, with $a \neq 0$.

The quadratic function $f(x) = 4 - x^2$ was graphed in Figure 2.5. Here are some other quadratic graphs.

EXAMPLE 1   Graph each of these quadratic functions.

(a) $f(x) = x^2$

Choose several negative, zero, and positive values of x, find the values of $f(x)$, and plot the corresponding points. Connecting these points with a smooth curve, we obtain Figure 4.1 on the next page.

1 Graph each of the following quadratic functions.

(a) $f(x) = x^2 - 4$

(b) $f(x) = -(x - 3)^2$

Answers:

(a)

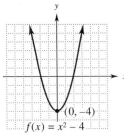

$f(x) = x^2 - 4$

(b)

$f(x) = -(x - 3)^2$

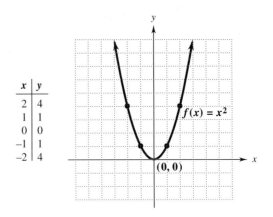

x	y
2	4
1	1
0	0
-1	1
-2	4

$f(x) = x^2$

(0, 0)

FIGURE 4.1

(b) $h(x) = -(x + 2)^2$

Since $h(x) = -(x + 2)^2 = -(x^2 + 4x + 4) = -x^2 - 4x - 4$, we see that h actually is a quadratic function. When $x = -2$, then

$$h(x) = h(-2) = -(-2 + 2)^2 = 0.$$

Therefore $(-2, 0)$ is on the graph. When $x \neq -2$, then $x + 2 \neq 0$, so that $(x + 2)^2$ is positive, and hence, $h(x) = -(x + 2)^2$ is negative. Thus, the graph lies below the x-axis whenever $x \neq -2$. Using these facts and plotting some points leads to Figure 4.2. ■ **1**

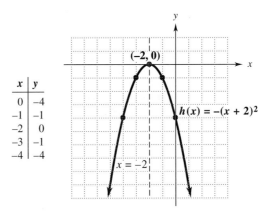

x	y
0	-4
-1	-1
-2	0
-3	-1
-4	-4

(-2, 0)

$h(x) = -(x + 2)^2$

$x = -2$

FIGURE 4.2

The curves in Figures 2.5, 4.1, 4.2 and in Problem 1 in the margin are called **parabolas.** It can be shown that the graph of every quadratic function is a parabola. All parabolas have the same basic "cup" shape, though the cup may be broad or narrow. The parabola opens upward when the coefficient of x^2 is positive (as in Figure 4.1) and downward when this coefficient is negative (as in Figure 4.2).

Parabolas have many useful properties. Cross sections of radar dishes and spot-lights form parabolas. Discs often visible on the sidelines of televised football games are microphones having reflectors with parabolic cross sections. These microphones are used by the television networks to pick up the shouted signals of the quarterbacks.

When a parabola opens upward (as in Figure 4.1), its lowest point is called the **vertex.** When a parabola opens downward (as in Figure 4.2), its highest point is called the **vertex.** The vertical line through the vertex of a parabola is called the **axis of the parabola.** For example, (0, 0) is the vertex of the parabola in Figure 4.1 and its axis is the y-axis. If you were to fold this graph along its axis, the two halves of the parabola would match exactly. This means that a parabola is *symmetric* about its axis.

The vertex of a parabola can be roughly approximated by using the trace feature or accurately approximated by using the minimum or maximum finder on a graphing calculator. However, there are algebraic techniques for finding the vertex precisely.

EXAMPLE 2   The function $g(x) = 2(x - 3)^2 + 1$ is quadratic because its rule can be written in the required form:

$$g(x) = 2(x - 3)^2 + 1 = 2(x^2 - 6x + 9) + 1 = 2x^2 - 12x + 19.$$

Its graph will be an upward-opening parabola. Note that

$$g(3) = 2(3 - 3)^2 + 1 = 1,$$

so that (3, 1) is on the graph. We claim that (3, 1) is the vertex—the lowest point on the graph. To see why this is true, note that when $x \neq 3$, the quantity $2(x - 3)^2$ is positive and hence

$$g(x) = 2(x - 3)^2 + 1 = \text{(a positive number)} + 1,$$

so that $g(x) > 1$. Therefore, every point $(x, g(x))$ on the graph with $x \neq 3$ has second coordinate $g(x)$ larger than 1 and hence lies *above* the point (3, 1). In other words, (3, 1) is the lowest point on the graph, the vertex of the parabola. Knowing this fact makes it easy to plot some points on both sides of the vertex and obtain the graph in Figure 4.3. Since the vertex is (3, 1), the vertical line $x = 3$ is the axis of the parabola. ∎

x	y
1	9
2	3
3	1
4	3
5	9

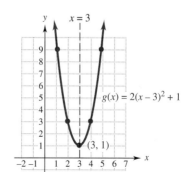

FIGURE 4.3

Example 2 illustrates some of the following facts, whose proofs are omitted.

2 Determine the vertex of the parabola algebraically and graph it.

(a) $f(x) = (x + 4)^2 - 3$

(b) $f(x) = -2(x - 3)^2 + 1$

Answers:

(a)

$f(x) = (x + 4)^2 - 3$

(b)

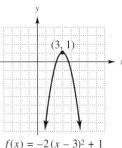

$f(x) = -2(x - 3)^2 + 1$

> If f is a quadratic function defined by $y = a(x - h)^2 + k$, then the graph of the function f is a parabola having its vertex at (h, k) and axis of symmetry $x = h$.
>
> If $a > 0$, the parabola opens upward; if $a < 0$, it opens downward.
>
> If $0 < |a| < 1$, the parabola is "broader" than $y = x^2$, while if $|a| > 1$, the parabola is "narrower" than $y = x^2$.

EXAMPLE 3   Determine algebraically whether the parabola opens upward or downward and find its vertex.

(a) $f(x) = -3(x - 4)^2 - 7$

The rule of the function is in the form $f(x) = a(x - h)^2 + k$ (with $a = -3, h = 4$, and $k = -7$). The parabola opens downward ($a < 0$) and its vertex is $(h, k) = (4, -7)$.

(b) $g(x) = 2(x + 3)^2 + 5$

Be careful here; the vertex is *not* $(3, 5)$. To put the rule of $g(x)$ in the form $a(x - h)^2 + k$, we must rewrite it so that there is a minus sign inside the parentheses:

$$g(x) = 2(x + 3)^2 + 5$$
$$= 2(x - (-3))^2 + 5.$$

This is the required form with $a = 2, h = -3$, and $k = 5$. The parabola opens upward and its vertex is $(-3, 5)$. ■ **2**

EXAMPLE 4   Find the rule of a quadratic function whose graph has vertex $(3, 4)$ and passes through the point $(6, 22)$.

The graph of $f(x) = a(x - h)^2 + k$ has vertex (h, k). We want $h = 3$ and $k = 4$, so that $f(x) = a(x - 3)^2 + 4$. Since $(6, 22)$ is on the graph, we must have $f(6) = 22$. Therefore

$$f(x) = a(x - 3)^2 + 4$$
$$f(6) = a(6 - 3)^2 + 4$$
$$22 = a(3)^2 + 4$$
$$9a = 18$$
$$a = 2.$$

Thus, the graph of $f(x) = 2(x - 3)^2 + 4$ is a parabola with vertex $(3, 4)$ that passes through $(6, 22)$. ■

The vertex of each parabola in Examples 2 and 3 was easily determined because the rule of the function had the form

$$f(x) = a(x - h)^2 + k.$$

The rule of *any* quadratic function can be put in this form by using the technique of completing the square, which was discussed in Section 2.4.

EXAMPLE 5   By completing the square, determine the vertex of the graph of $f(x) = x^2 - 2x + 3$. Then graph the parabola.

Rewrite the equation in the form $f(x) = a(x - h)^2 + k$ by first writing $f(x) = x^2 - 2x + 3$ as

$$f(x) = (x^2 - 2x \quad) + 3.$$

Now complete the square for the expression in parentheses. Take half the coefficient of x, namely $(\frac{1}{2})(-2) = -1$, and square the result: $(-1)^2 = 1$. In order to complete the square we must add 1 inside the parentheses, but in order not to change the rule of the function we must also subtract 1:

$$f(x) = (x^2 - 2x + 1 - 1) + 3.$$

By the associative property,

$$f(x) = (x^2 - 2x + 1) + (-1 + 3).$$

Factor $x^2 - 2x + 1$ as $(x - 1)^2$, to get

$$f(x) = (x - 1)^2 + 2.$$

With the rule in this form, we can see that the graph is an upward-opening parabola with vertex $(1, 2)$, as shown in Figure 4.4.   ■   3

3 Rewrite the rule of the function by completing the square and use this form to find the vertex of the graph.

(a) $f(x) = x^2 - 6x + 11$

(b) $g(x) = x^2 + 8x + 18$

Answers:

(a) $f(x) = (x - 3)^2 + 2; (3, 2)$

(b) $g(x) = (x + 4)^2 + 2; (-4, 2)$

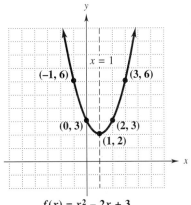

$$f(x) = x^2 - 2x + 3$$
$$f(x) = (x - 1)^2 + 2$$

FIGURE 4.4

4 Complete the square, find the vertex, and graph the following.

(a) $f(x) = 3x^2 - 12x + 14$

(b) $f(x) = -x^2 + 6x - 12$

Answers:

(a) $f(x) = 3(x - 2)^2 + 2$

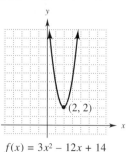

(2, 2)

$f(x) = 3x^2 - 12x + 14$

(b) $f(x) = -(x - 3)^2 - 3$

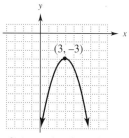

(3, −3)

$f(x) = -x^2 + 6x - 12$

EXAMPLE 6   Find the vertex of the graph of $f(x) = -2x^2 + 12x - 19$ and graph the parabola.

Get $f(x)$ in the form $f(x) = a(x - h)^2 + k$ by first factoring -2 from $-2x^2 + 12x$.

$$f(x) = -2x^2 + 12x - 19 = -2(x^2 - 6x) - 19$$

Working inside the parentheses, take half of -6 (the coefficient of x): $(1/2)(-6)$ $= -3$. Square this result: $(-3)^2 = 9$. Add and subtract 9 inside the parentheses.

$$y = -2(x^2 - 6x + 9 - 9) - 19$$
$$y = \mathbf{-2}(x^2 - 6x + 9) + (\mathbf{-2})(-9) - 19 \qquad \text{Distributive property}$$

Simplify and factor to get

$$y = -2(x - 3)^2 - 1.$$

Since $a = -2$ is negative, the parabola opens downward. The vertex is at $(3, -1)$, and the parabola is narrower than $y = x^2$. Use these results and plot additional ordered pairs as needed to get the graph in Figure 4.5.   ■   **4**

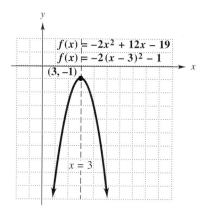

$f(x) = -2x^2 + 12x - 19$
$f(x) = -2(x - 3)^2 - 1$
(3, −1)

$x = 3$

FIGURE 4.5

The technique of completing the square can be used to rewrite the general equation $f(x) = ax^2 + bx + c$ in the form $f(x) = a(x - h)^2 + k$, as is shown in Exercise 40. When this is done we obtain a formula for the coordinates of the vertex.

The graph of $f(x) = ax^2 + bx + c$ is a parabola with vertex (h, k), where

$$h = \frac{-b}{2a} \quad \text{and} \quad k = f(h).$$

EXAMPLE 7 Find the vertex, the axis, and the x- and y-intercepts of the graph of $f(x) = x^2 - x - 6$.

Since $a = 1$ and $b = -1$, the x-value of the vertex is

$$\frac{-b}{2a} = \frac{-(-1)}{2 \cdot 1} = \frac{1}{2}.$$

The y-value of the vertex is

$$f\left(\frac{1}{2}\right) = \left(\frac{1}{2}\right)^2 - \frac{1}{2} - 6 = -\frac{25}{4}.$$

The vertex is $(1/2, -25/4)$ and the axis of the parabola is $x = 1/2$, as shown in Figure 4.6. The intercepts are found by setting each variable equal to 0.

Set $f(x) = y = 0$:	Set $x = 0$:
$0 = x^2 - x - 6$	$f(x) = y = 0^2 - 0 - 6 = -6$
$0 = (x + 2)(x - 3)$	The y-intercept is -6.
$x + 2 = 0$ or $x - 3 = 0$	
$x = -2$ $x = 3$	

The x-intercepts are -2 and 3. ■ 5

5 Find the vertex of the graph.

(a) $f(x) = 2x^2 - 7x + 12$

(b) $k(x) = -.4x^2 + 1.6x + 5$

Answers:

(a) $(7/4, 47/8)$

(b) $(2, 6.6)$

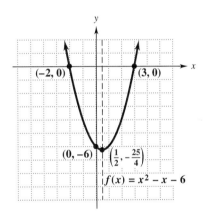

FIGURE 4.6

4.1 EXERCISES

Without graphing, determine the vertex of each of the following parabolas and state whether it opens upward or downward. (See Examples 2, 3, 5–7.)

1. $f(x) = 3(x - 5)^2 + 2$

2. $g(x) = -6(x - 2)^2 - 5$

3. $f(x) = -(x - 1)^2 + 2$

4. $g(x) = x^2 + 1$

5. $f(x) = x^2 + 12x + 1$

6. $g(x) = x^2 - 10x + 3$

7. $f(x) = 2x^2 + 4x + 1$

8. $g(x) = -3x^2 + 6x + 5$

Without graphing, determine the x- and y-intercepts of each of the following parabolas. (See Example 7.)

9. $f(x) = 3(x - 2)^2 - 3$

10. $f(x) = x^2 - 4x - 1$

11. $g(x) = 2x^2 + 8x + 6$

12. $g(x) = x^2 - 10x + 20$

Find the rule of a quadratic function whose graph has the given vertex and passes through the given point. (See Example 4.)

13. Vertex $(3, -5)$; point $(5, -9)$

14. Vertex $(2, 4)$; point $(0, -8)$

15. Vertex $(1, 2)$; point $(5, 6)$

16. Vertex $(-3, 2)$; point $(2, 1)$

17. Vertex $(-1, -2)$; point $(1, 2)$

18. Vertex $(2, -4)$; point $(5, 2)$

Graph each of the following parabolas. Find the vertex and axis of symmetry of each. (See Examples 1–7.)

19. $f(x) = (x + 2)^2$

20. $f(x) = -(x + 5)^2$

21. $f(x) = (x - 1)^2 - 3$

22. $f(x) = (x - 2)^2 + 1$

23. $f(x) = x^2 - 4x + 6$

24. $f(x) = x^2 + 6x + 3$

25. $f(x) = 2x^2 - 4x + 5$

26. $f(x) = -3x^2 + 24x - 46$

27. $f(x) = -x^2 + 6x - 6$

28. $f(x) = -x^2 + 2x + 5$

Determine the vertex of the parabola.

29. $f(x) = 5x^2 - 72x + 271.2$

30. $f(x) = 3x^2 - 30x + 27$

31. $f(x) = .3x^2 - 57x + 12$

32. $f(x) = .005x^2 - .002x + .16$

In Exercises 33–38, graph the functions in parts (a)–(d) on the same set of axes; then answer part (e).

33. (a) $k(x) = x^2$
   (b) $f(x) = 2x^2$
   (c) $g(x) = 3x^2$
   (d) $h(x) = 3.5x^2$
   (e) Explain how the coefficient a in the function given by $f(x) = ax^2$ affects the shape of the graph, in comparison with the graph of $k(x) = x^2$, when $a > 1$.

34. (a) $k(x) = x^2$
   (b) $f(x) = .8x^2$
   (c) $g(x) = .5x^2$
   (d) $h(x) = .3x^2$
   (e) Explain how the coefficient a in the function given by $f(x) = ax^2$ affects the shape of the graph, in comparison with the graph of $k(x) = x^2$, when $0 < a < 1$.

35. (a) $k(x) = -x^2$
   (b) $f(x) = -2x^2$
   (c) $g(x) = -3x^2$
   (d) $h(x) = -3.5x^2$
   (e) Compare these graphs to the ones in Exercise 33 and explain how changing the sign of the coefficient a in the function given by $f(x) = ax^2$ affects the graph.

36. (a) $k(x) = x^2$
   (b) $f(x) = x^2 + 2$
   (c) $g(x) = x^2 + 3$
   (d) $h(x) = x^2 + 5$
   (e) Explain how the graph of $f(x) = x^2 + c$ (where c is a positive constant) can be obtained from the graph of $k(x) = x^2$.

37. (a) $k(x) = x^2$
   (b) $f(x) = x^2 - 1$
   (c) $g(x) = x^2 - 2$
   (d) $h(x) = x^2 - 4$
   (e) Explain how the graph of $f(x) = x^2 - c$ (where c is a positive constant) can be obtained from the graph of $k(x) = x^2$.

38. (a) $k(x) = x^2$
   (b) $f(x) = (x + 2)^2$
   (c) $g(x) = (x - 2)^2$
   (d) $h(x) = (x - 4)^2$
   (e) Explain how the graph of $f(x) = (x + c)^2$ and $f(x) = (x - c)^2$ (where c is a positive constant) can be obtained from the graph of $k(x) = x^2$.

39. Find the rule of a quadratic function whose graph is a parabola with vertex $(0, 0)$ that includes the point $(2, 12)$.

40. Verify that the right side of the equation $ax^2 + bx + c = a\left(x - \left(\dfrac{-b}{2a}\right)\right)^2 + \left(c - \dfrac{b^2}{4a}\right)$ equals the left side. Since the right side is in the form $a(x - h)^2 + k$, we conclude that the vertex of the parabola $f(x) = ax^2 + bx + c$ has x-coordinate $h = -b/2a$.

4.2 APPLICATIONS OF QUADRATIC FUNCTIONS

The fact that the vertex of a parabola $y = ax^2 + bx + c$ is the highest or lowest point on the graph can be used in applications to find a maximum or a minimum value. When $a > 0$, the graph opens upward, so the function has a minimum. When $a < 0$, the graph opens downward, producing a maximum.

EXAMPLE 1   Leslie Lahr owns and operates Aunt Emma's Blueberry Pies. She has hired a consultant to analyze her business operations. The consultant tells her that her profits, $P(x)$, from the sale of x units of pies, are given by

$$P(x) = 120x - x^2.$$

How many units of pies should she sell in order to maximize profit? What is the maximum profit?

The profit function can be rewritten as $P(x) = -x^2 + 120x$. Its graph is a downward-opening parabola (Figure 4.7). Its vertex, which may be found as in Section 4.1, is (60, 3600). For each point on the graph,

> the x-coordinate is the number of units of pies;
>
> the y-coordinate is the profit on that number of units.

Only the portion of the graph in Quadrant I (where both coordinates are positive) is relevant here because she cannot sell a negative number of pies and she is not interested in a negative profit. *Maximum* profit occurs at the point with the largest y-coordinate, namely, the vertex, as shown in Figure 4.7. The maximum profit of $3600 is obtained when 60 units of pies are sold.   ■   ☐1

1 When a company sells x units of a product, its profit is $P(x) = -2x^2 + 40x + 280$. Find

(a) the number of units that should be sold so that maximum profit is received;

(b) the maximum profit.

Answers:
(a) 10 units
(b) $480

FIGURE 4.7

Supply and demand curves were introduced in Section 2.3. Here is a quadratic example.

EXAMPLE 2 Suppose that the price and demand for an item are related by

$$p = 150 - 6q^2, \qquad \text{Demand function}$$

where p is the price (in dollars) and q is the number of items demanded (in hundreds). The price and supply are related by

$$p = 10q^2 + 2q, \qquad \text{Supply function}$$

where q is the number of items supplied (in hundreds). Find the equilibrium demand (and supply) and the equilibrium price.

The graphs of both of these equations are parabolas (Figure 4.8). Only those portions of the graphs that lie in the first quadrant are included, because neither supply, demand, nor price can be negative.

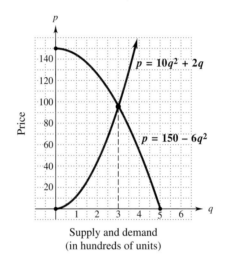

Supply and demand
(in hundreds of units)

FIGURE 4.8

The point where the demand and supply curves intersect is the equilibrium point. Its first coordinate is the equilibrium demand (and supply) and its second coordinate is the equilibrium price. These coordinates may be found as follows. At the equilibrium point, the second coordinate of the demand curve must be the same as the second coordinate of the supply curve so that

$$150 - 6q^2 = 10q^2 + 2q.$$

Write this quadratic equation in standard form as follows.

$$0 = 16q^2 + 2q - 150 \qquad \text{Add } -150 \text{ and } 6q^2 \text{ to both sides.}$$
$$0 = 8q^2 + q - 75 \qquad \text{Multiply both sides by } \frac{1}{2}.$$

This equation can be solved by the quadratic formula given in Section 2.4. Here $a = 8$, $b = 1$, and $c = -75$.

$$q = \frac{-1 \pm \sqrt{1 - 4(8)(-75)}}{2(8)}$$

$$= \frac{-1 \pm \sqrt{1 + 2400}}{16} \qquad -4(8)(-75) = 2400$$

$$= \frac{-1 \pm 49}{16} \qquad \sqrt{1 + 2400} = \sqrt{2401} = 49$$

$$q = \frac{-1 + 49}{16} = \frac{48}{16} = 3 \quad \text{or} \quad q = \frac{-1 - 49}{16} = -\frac{50}{16} = -\frac{25}{8}$$

2 The price and demand for an item are related by $p = 32 - x^2$, while price and supply are related by $p = x^2$. Find

(a) the equilibrium supply;

(b) the equilibrium price.

Answers:

(a) 4

(b) $16

It is not possible to make $-25/8$ units, so discard that answer and use only $q = 3$. Hence the equilibrium demand (and supply) is 300. Find the equilibrium price by substituting 3 for q in either the supply or the demand function (and check your answer by using the other one). Using the supply function gives

$$p = 10q^2 + 2q$$
$$p = 10 \cdot 3^2 + 2 \cdot 3 \qquad \text{Let } q = 3.$$
$$= 10 \cdot 9 + 6$$
$$p = 96. \quad \blacksquare \quad \boxed{2}$$

EXAMPLE 3 The rental manager of a small apartment complex with 16 units has found from experience that each $40 increase in the monthly rent results in an empty apartment. All 16 apartments will be rented at a monthly rent of $500. How many $40 increases will produce maximum monthly income for the complex?

Let x represent the number of $40 increases. Then the number of apartments rented will be $16 - x$. Also, the monthly rent per apartment will be $500 + 40x$. (There are x increases of $40 for a total increase of $40x$.) The monthly income, $I(x)$, is given by the number of apartments rented times the rent per apartment, so

$$I(x) = (16 - x)(500 + 40x)$$
$$= 8000 + 640x - 500x - 40x^2$$
$$= 8000 + 140x - 40x^2.$$

Since x represents the number of $40 increases and each $40 increase causes one empty apartment, x must be a whole number. Since there are only 16 apartments, $0 \le x \le 16$. Because there is a small number of possibilities, the value of x that produces maximum income may be found in two ways.

Brute Force Method Find the sixteen values of $I(x)$ when $x = 1, 2, \ldots, 16$ and determine the largest one. (If you have a graphing calculator, use the table feature to do this quickly.)

Algebraic Method   The graph of $I(x) = 8000 + 140x - 40x^2$ is a downward-opening parabola (why?) and the value of x that produces maximum income occurs at the vertex. The methods of Section 4.1 show that the vertex is (1.75, 8122.50). Since x must be a whole number, evaluate $I(x)$ at $x = 1$ and $x = 2$ to see which one gives the best result.

$$\text{If } x = 1, \text{ then } I(1) = -40(1)^2 + 140(1) + 8000 = 8100.$$
$$\text{If } x = 2, \text{ then } I(2) = -40(2)^2 + 140(2) + 8000 = 8120.$$

So maximum income occurs when $x = 2$. The manager should charge a rent of $500 + 2(40) = \$580$ and leave 2 apartments vacant.   ■

QUADRATIC MODELS   Real-world data can sometimes be used to construct a quadratic function that approximates the data. Such **quadratic models** can then be used (subject to limitations) to predict future behavior.

EXAMPLE 4   The table below lists the cumulative number of AIDS cases diagnosed in the United States during 1982–1993.* It shows, for example, that 22,620 cases were diagnosed in 1982–1985.

Year	AIDS Cases	Year	AIDS Cases
1982	1,563	1988	105,489
1983	4,647	1989	147,170
1984	10,845	1990	193,245
1985	22,620	1991	248,023
1986	41,662	1992	315,329
1987	70,222	1993	361,509

(a) Display this information graphically.

Let $x = 0$ represent the year 1980, $x = 1$ represent 1981, and so on. Then plot the points given by the table: (2, 1563), (3, 4647), etc. (Figure 4.9).

FIGURE 4.9

FIGURE 4.10

*U.S. Dept. of Health and Human Services, Centers for Disease Control and Prevention, *HIV/AIDS Surveillance,* March 1994.

(b) The shape of the data points in Figure 4.9 resembles the right half of an upward-opening parabola. Find a quadratic model $f(x) = a(x - h)^2 + k$ for these data.

Based on Figure 4.9 we let $(2, 1563)$ be the vertex of the parabola, so that $f(x) = a(x - 2)^2 + 1563$. Use another point, say $(13, 361509)$, to find a (another choice here would lead to a different quadratic model).

$$f(x) = a(x - 2)^2 + 1563$$
$$361{,}509 = a(13 - 2)^2 + 1563$$
$$121a = 359{,}946$$
$$a \approx 2974.76$$

3 Find another quadratic model in Example 4(b) by using $(2, 1563)$ as vertex and $(11, 248023)$ as the other point.

Answer:
$f(x) = 3042.72(x - 2)^2 + 1563$

Therefore $f(x) = 2974.76(x - 2)^2 + 1563$ is a quadratic model for the data. The graph of $f(x)$ in Figure 4.10 appears to be a reasonable approximation of the data.

(c) Use the quadratic model in part (b) to estimate the total number of AIDS cases diagnosed by 2000.

The year 2000 corresponds to $x = 20$, so the number is approximately

$$f(20) = 2974.76(20 - 2)^2 + 1563 \approx 965{,}385. \quad \blacksquare \quad \boxed{3}$$

TECHNOLOGY TIP The maximum or minimum finder on a graphing calculator can approximate the vertex of a parabola with a high degree of accuracy. Similarly, the calculator's graphical root finder can approximate the x-intercepts. Consult your instruction manual for details. ✔

4.2 EXERCISES

Work these problems. (See Example 1.)

1. Management   Shannise Cole makes and sells candy. She has found that the cost per box for making x boxes of candy is given by

$$C(x) = x^2 - 10x + 32.$$

(a) How much does it cost per box to make 2 boxes? 4 boxes? 10 boxes?

(b) Graph the cost function $C(x)$ and mark the points corresponding to 2, 4, and 10 boxes.

(c) What point on the graph corresponds to the number of boxes that will make the cost per box as small as possible?

(d) How many boxes should she make in order to keep the cost per box at a minimum? What is the minimum cost per box?

2. Management   Greg Tobin sells bottled water. He has found that the average amount of time he spends with each customer is related to his weekly sales volume by the function

$$f(x) = x(60 - x),$$

where x is the number of minutes per customer and $f(x)$ is the number of cases sold per week.

(a) How many cases does he sell if he spends 10 minutes with each customer? 20 minutes? 45 minutes?

(b) Choose an appropriate scale for the axes and sketch the graph of $f(x)$. Mark the points on the graph corresponding to 10, 20, and 45 minutes.

(c) Explain what the vertex of the graph represents.

(d) How long should Greg spend with each customer in order to sell as many cases per week as possible? In this case, how many cases will he sell?

3. Management   French fries produce a tremendous profit (150% to 300%) for many fast food restaurants. Management, therefore, desires to maximize the number of bags sold. Suppose that a mathematical model connecting p, the profit per day from french fries (in tens of dollars), and x, the price per bag (in dimes), is $p = -2x^2 + 24x + 8$.

(a) Find the price per bag that leads to maximum profit.

(b) What is the maximum profit?

4. Natural Science A researcher in physiology has decided that a good mathematical model for the number of impulses fired after a nerve has been stimulated is given by $y = -x^2 + 20x - 60$, where y is the number of responses per millisecond and x is the number of milliseconds since the nerve was stimulated.

(a) When will the maximum firing rate be reached?

(b) What is the maximum firing rate?

5. Physical Science If an object is thrown upward with an initial velocity of 32 feet per second, then its height, in feet, above the ground after t seconds is given by

$$h = 32t - 16t^2.$$

Find the maximum height attained by the object. Find the number of seconds it takes for the object to hit the ground.

6. Management Colleen Davis owns a factory that manufactures souvenir key chains. Her weekly profit (in hundreds of dollars) is given by $P(x) = -2x^2 + 60x - 120$, where x is the number of cases of key chains sold.

(a) What is the largest number of cases she can sell and still make a profit?

(b) Explain how it is possible for her to lose money if she sells more cases than your answer in part (a).

(c) How many cases should she make and sell in order to maximize her profits?

7. Management The manager of a bicycle shop has found that, at a price (in dollars) of $p(x) = 150 - \dfrac{x}{4}$ per bicycle, x bicycles will be sold.

(a) Find an expression for the total revenue from the sale of x bicycles. (*Hint:* revenue = demand × price.)

(b) Find the number of bicycle sales that leads to maximum revenue.

(c) Find the maximum revenue.

8. Management If the price (in dollars) of a videotape is $p(x) = 40 - \dfrac{x}{10}$, then x tapes will be sold.

(a) Find an expression for the total revenue from the sale of x tapes.

(b) Find the number of tapes that will produce maximum revenue.

(c) Find the maximum revenue.

9. Management The demand for a certain type of cosmetic is given by

$$p = 500 - x,$$

where p is the price when x units are demanded.

(a) Find the revenue, $R(x)$, that would be obtained at a demand of x.

(b) Graph the revenue function, $R(x)$.

(c) From the graph of the revenue function, estimate the price that will produce the maximum revenue.

(d) What is the maximum revenue?

Work the following problems. (See Example 2.)

10. Management Suppose the price p of widgets is related to the quantity q that is demanded by

$$p = 640 - 5q^2,$$

where q is measured in hundreds of widgets. Find the price when the number of widgets demanded is

(a) 0;   (b) 5;   (c) 10.

Suppose the supply function for widgets is given by $p = 5q^2$, where q is the number of widgets (in hundreds) that are supplied at price p.

(d) Graph the demand function $p = 640 - 5q^2$ and the supply function $p = 5q^2$ on the same axes.

(e) Find the equilibrium supply.

(f) Find the equilibrium price.

Management *Suppose the supply and demand for a certain textbook are given by:*

$$\text{supply: } p = \frac{1}{5}q^2; \quad \text{demand: } p = -\frac{1}{5}q^2 + 40,$$

where p is price and q is quantity.

11. How many books are demanded at a price of

(a) 10?   (b) 20?

(c) 30?   (d) 40?

How many books are supplied at a price of

(e) 5?   (f) 10?

(g) 20?   (h) 30?

(i) Graph the supply and demand functions on the same axes.

12. Find the equilibrium demand and the equilibrium price in Exercise 11.

Management *Work each problem. (See Example 3.)*

13. A charter flight charges a fare of $200 per person, plus $4 per person for each unsold seat on the plane. If the plane holds 100 passengers and if x represents the number of unsold seats, find the following.

(a) An expression for the total revenue received for the flight (*Hint:* Multiply the number of people flying, $100 - x$, by the price per ticket.)

(b) The graph for the expression of part (a)

(c) The number of unsold seats that will produce the maximum revenue

(d) The maximum revenue

14. The revenue of a charter bus company depends on the number of unsold seats. If 100 seats are sold, the price is $50. Each unsold seat increases the price per seat by $1. Let x represent the number of unsold seats.
  (a) Write an expression for the number of seats that are sold.
  (b) Write an expression for the price per seat.
  (c) Write an expression for the revenue.
  (d) Find the number of unsold seats that will produce the maximum revenue.
  (e) Find the maximum revenue.

15. Farmer Linton wants to find the best time to take her hogs to market. The current price is 88 cents per pound and her hogs weigh an average of 90 pounds. The hogs gain 5 pounds per week and the market price for hogs is falling each week by 2 cents per pound. How many weeks should Ms. Linton wait before taking her hogs to market in order to receive as much money as possible? At this time, how much money (per hog) will she get?

Work these problems. (See Example 4.)

16. Management   Federal government revenues (in billions of dollars) for selected years are shown in the table.*

Year	Revenue
1960	92.5
1965	116.8
1970	192.8
1975	279.1
1980	517.1
1985	734.1

  (a) Let $x = 0$ correspond to 1960. Use $(0, 92.5)$ as the vertex and the data from 1985 to find a quadratic function $f(x) = a(x - h)^2 + k$ that models this data.
  (b) Use the quadratic model to estimate government revenue in the years 1981, 1988, and 1991. How well do the estimates compare to the actual revenues of 599.3, 909, and 1119.1? Does this model seem reasonable?
  (c) Estimate government revenue in 1999.

*Economic Reports of the President, 1960–1991.

17. Natural Science   The table lists the total cumulative number of deaths in the United States to date known to have been caused by AIDS.*

Year	Deaths
1982	620
1983	2,122
1984	5,600
1985	12,529
1986	24,550
1987	40,820
1988	61,723
1989	89,172
1990	119,821
1991	154,567
1992	191,508
1993	220,592

  (a) Let $x = 0$ correspond to 1980 and plot the data points.
  (b) Use $(2, 620)$ as the vertex and $(13, 220592)$ as the other point to find a quadratic model $g(x) = a(x - h)^2 + k$ for these data.
  (c) Use g to estimate the number of deaths by the year 2000.

18. Management   The gross national product (GNP) of the United States in billions of 1982 dollars was 2416.2 in 1970, 3187.1 in 1980, and 4155.8 in 1990.†
  (a) Use the points corresponding to 1970 (vertex) and 1990 to construct a quadratic model for these data, with $x = 0$ corresponding to 1970.
  (b) Estimate the GNP in 1980, 1985, 1995, and 2000. How good is the 1980 estimate?

Work these problems.

19. A field bounded on one side by a river is to be fenced on three sides to form a rectangular enclosure. There are 320 ft of fencing available. What should the dimensions be to have an enclosure with the maximum possible area?

20. A rectangular garden bounded on one side by a river is to be fenced on the other three sides. Fencing material for the side parallel to the river costs $30 per foot and material for the other two sides costs $10 per foot. What are the dimensions of the garden of largest possible area, if $1200 is to be spent for fencing material?

*U.S. Dept. of Health and Human Services, Centers for Disease Control and Prevention, *HIV/AIDS Surveillance,* March 1994.

†Economic Reports of the President, 1970–1991.

21. A trough is to be made by bending a long, flat piece of metal 10 inches wide into a rectangular shape. What depth should the trough be to have maximum possible cross-sectional area?

22. A culvert is shaped like a parabola, 18 centimeters across the top and 12 centimeters deep. How wide is the culvert 8 centimeters from the top?

◀▷ **Management**   *Recall that profit equals revenue minus cost. In Exercises 23 and 24 find the following.*

(a) *The break-even point (to the nearest tenth)*
(b) *The x-value that makes revenue a maximum*
(c) *The x-value that makes profit a maximum*
(d) *The maximum profit*
(e) *For what x-values will a loss occur?*
(f) *For what x-values will a profit occur?*

23. $R(x) = 60x - 2x^2$; $C(x) = 20x + 80$

24. $R(x) = 75x - 3x^2$; $C(x) = 21x + 65$

4.3 POLYNOMIAL FUNCTIONS

A **polynomial function of degree n** is a function whose rule is given by a polynomial of degree n.* For example, linear functions, such as

$$f(x) = 3x - 2 \quad \text{and} \quad g(x) = 7x + 5,$$

are polynomial functions of degree 1 and quadratic functions, such as

$$f(x) = 3x^2 + 4x - 6 \quad \text{and} \quad g(x) = -2x^2 + 9$$

are polynomial functions of degree 2. Similarly, the function given by

$$f(x) = 2x^4 + 5x^3 - 6x^2 + x - 3$$

is a polynomial function of degree 4.

The simplest polynomial functions are those whose rules are of the form $f(x) = ax^n$ (where a is a constant).

1 Graph $f(x) = x^3 - 5$.

Answer:

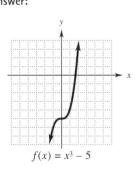

$f(x) = x^3 - 5$

EXAMPLE 1   Graph $f(x) = x^3$.

First, find several ordered pairs belonging to the graph. Be sure to choose some negative x-values, $x = 0$, and some positive x-values to get representative ordered pairs. Find as many ordered pairs as you need in order to see the shape of the graph. Then plot the ordered pairs and draw a smooth curve through them, getting the graph in Figure 4.11. ■   **1**

───────────────

*The degree of a polynomial was defined on page 24.

2 Graph these functions.

(a) $f(x) = -.25x^5 - 2$

(b) $f(x) = -x^4 + 3$

Answers:

(a)

(b)

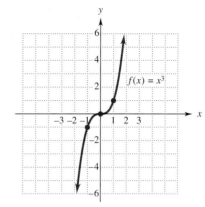

x	y
2	8
1	1
0	0
−1	−1
−2	−8

FIGURE 4.11

EXAMPLE 2   Graph $f(x) = (3/2)x^4$.

The following table gives some typical ordered pairs.

x	-2	-1	0	1	2
y	24	3/2	0	3/2	24

The graph is shown in Figure 4.12.   ■   2

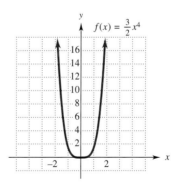

FIGURE 4.12

The graph of $f(x) = ax^n$ has the same general shape as one of the graphs in Figures 4.11 or 4.12 or Problem 2 in the margin.

> ### GRAPH OF $f(x) = ax^n$
>
> If the exponent n is even, then the graph of $f(x) = ax^n$ is cup-shaped with the bottom of the cup at the origin and the y-axis through the middle; it opens upward when $a > 0$ and downward when $a < 0$.
>
> If the exponent n is odd, then the graph of $f(x) = ax^n$ moves upward from left to right when $a > 0$ and downward when $a < 0$, with a single bend at the origin.

The domain of every polynomial function is the set of all real numbers. The range of a polynomial function of odd degree (1, 3, 5, 7, and so on) is also the set of all real numbers. Graphs typical of polynomial functions of odd degree are shown in Figure 4.13.

FIGURE 4.13

A polynomial function of even degree has a range that takes the form $y \leq k$ or $y \geq k$, for some real number k. Figure 4.14 shows two graphs of typical polynomial functions of even degree.

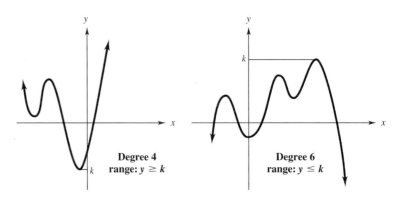

FIGURE 4.14

As illustrated in the preceding figures, the graph of a typical polynomial function may have several "peaks" and "valleys." For instance, the right-hand graph in Figure 4.14 has 3 peaks and 2 valleys. Note that this is the graph of a polynomial of degree 6 and that there are a total of 5 peaks and valleys. The location of peaks and valleys on a graph can be approximated with a high degree of accuracy by a maximum or minimum finder on a graphing calculator. Calculus is needed to determine their exact locations.

Since the domain of a polynomial function is the set of all real numbers, every polynomial graph extends forever to the left and to the right. As shown in Figures 4.13 and 4.14, the graph moves sharply away from the x-axis at the far left and far right. Note, for example, that the graph of the fourth degree polynomial function on the left in Figure 4.14 moves upward on both ends, just as the graph of $f(x) = (3/2)x^4$ does in Figure 4.12.

The preceding discussion illustrates the following facts, which require calculus for their proof.

 3 Identify the x-intercepts of each graph.

(a)

(b)

Answers:

(a) $-2, 1, 2$

(b) -3

■ **POLYNOMIAL GRAPHS**

The graph of a polynomial function $f(x)$ is a smooth, unbroken curve that extends forever to the left and right.

When $|x|$ is large, the graph of $f(x)$ resembles the graph of its highest degree term and moves sharply away from the x-axis.

If $f(x)$ has degree n, then

the number of x-intercepts of its graph is $\leq n$ and

the total number of peaks and valleys on its graph is $\leq n - 1$.

When a polynomial can be completely factored, the general shape of its graph can be determined by using the facts in the box above and some algebra. Although determining the exact location of the peaks and valleys requires calculus, this approach is a reasonable way to graph polynomial functions by hand. This method requires you to find the x-intercepts of the graph.   **3**

EXAMPLE 3   Graph $f(x) = (2x + 3)(x - 1)(x + 2)$.

Multiplying out the expression on the right shows that $f(x)$ is the cubic (third-degree) polynomial

$$f(x) = 2x^3 + 5x^2 - x - 6.$$

Begin by using the factored form of f to find any x-intercepts; do this by setting $f(x) = 0$.

$$f(x) = 0$$
$$(2x + 3)(x - 1)(x + 2) = 0$$

Solve this equation by placing each of the three factors equal to 0.

$$2x + 3 = 0 \quad \text{or} \quad x - 1 = 0 \quad \text{or} \quad x + 2 = 0$$

$$x = -\frac{3}{2} \qquad\qquad x = 1 \qquad\qquad x = -2$$

The three numbers, $-3/2$, 1, and -2, divide the x-axis into four regions:

$$x < -2, \quad -2 < x < -\frac{3}{2}, \quad -\frac{3}{2} < x < 1, \quad \text{and} \quad 1 < x.$$

These regions are shown in Figure 4.15.

FIGURE 4.15

Since the graph is an unbroken curve, it can only change from above the x-axis to below it by passing through the x-axis. As we have seen, this occurs only at the x-intercepts $x = -2$, $-3/2$, and 1. Consequently, in the region between two intercepts (or to the left of $x = -2$ or to the right of $x = 1$), the graph of $f(x)$ must lie entirely above or entirely below the x-axis.

We can determine where the graph lies over a region, by evaluating $f(x) = (2x + 3)(x - 1)(x + 2)$ at a number in that region. For example, $x = -3$ is in the region where $x < -2$ and

$$f(-3) = (2(-3) + 3)(-3 - 1)(-3 + 2)$$
$$= -12.$$

Therefore, $(-3, -12)$ is on the graph. Since this point lies below the x-axis, all points in this region (that is, all points with $x < -2$) must lie below the x-axis. By testing numbers in the other intervals, we obtain this chart.

Region	Test Number	Value of $f(x)$	Sign of $f(x)$	Graph
$x < -2$	-3	-12	negative	below x-axis
$-2 < x < -3/2$	$-7/4$	$11/32$	positive	above x-axis
$-3/2 < x < 1$	0	-6	negative	below x-axis
$1 < x$	2	28	positive	above x-axis

4 Graph
$f(x) = .5(x - 1)(x + 2)(x - 3)$.

Answer:

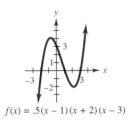

$f(x) = .5(x - 1)(x + 2)(x - 3)$

Since the graph touches the x-axis at the intercepts $x = -2$ and $x = -3/2$ and is above the x-axis between these intercepts, there must be at least one peak there. Similarly, there must be at least one valley between $x = -3/2$ and $x = 1$ because the graph is below the x-axis there. However, a polynomial function of degree 3 can have a total of at most $3 - 1 = 2$ peaks and valleys (as noted in the box before the example). So there must be exactly one peak and exactly one valley on this graph.

Furthermore, when $|x|$ is large, the graph must resemble the graph of $y = 2x^3$ (the highest degree term). The graph of $y = 2x^3$, like the graph of $y = x^3$ in Figure 4.11, moves upward to the right and downward to the left. Using these facts and plotting the x-intercepts shows that the graph must have the general shape shown in Figure 4.16. Plotting additional points leads to the reasonably accurate graph in Figure 4.17. We say "reasonably accurate" because we cannot be sure of the exact locations of the peaks and valleys on the graph without using calculus. ■ **4**

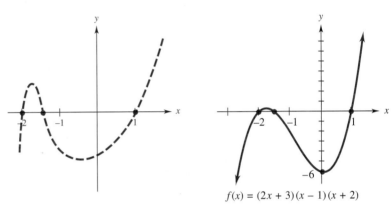

$f(x) = (2x + 3)(x - 1)(x + 2)$

FIGURE 4.16   FIGURE 4.17

EXAMPLE 4   Sketch the graph of $f(x) = 3x^4 + x^3 - 2x^2$.
The polynomial can be factored as follows:

$$3x^4 + x^3 - 2x^2 = x^2(3x^2 + x - 2)$$
$$= x^2(3x - 2)(x + 1).$$

By letting $f(x) = 0$, we find that the x-intercepts are at $x = 0$, $x = 2/3$, and $x = -1$. Between any two adjacent intercepts, the graph must be entirely above or entirely below the x-axis. We can determine whether the graph is above or below the x-axis in each interval by evaluating the function at a "test number" in that interval, as summarized in the following chart.

5 Graph
$f(x) = x^3 - 4x$.

Answer:

$f(x) = x^3 - 4x$

Region	Test Number	Value of $f(x)$	Sign of $f(x)$	Graph
$x < -1$	-2	32	positive	above x-axis
$-1 < x < 0$	$-\dfrac{1}{2}$	$-\dfrac{7}{16}$	negative	below x-axis
$0 < x < \dfrac{2}{3}$	$\dfrac{1}{2}$	$-\dfrac{3}{16}$	negative	below x-axis
$\dfrac{2}{3} < x$	1	2	positive	above x-axis

By plotting the x-intercepts and the points determined by the test numbers and using the facts that there can be a total of at most 3 peaks and valleys (why?) and that the graph must resemble the graph of $y = 3x^4$ when $|x|$ is large, we obtain the graph in Figure 4.18.   ■   **5**

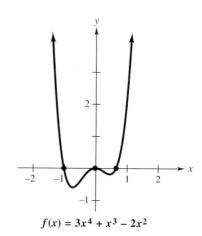

$f(x) = 3x^4 + x^3 - 2x^2$

FIGURE 4.18

4.3 EXERCISES

Graph each of the following polynomial functions. (See Examples 1 and 2.)

1. $f(x) = x^4$

2. $g(x) = -.5x^6$

3. $h(x) = -.2x^5$

4. $f(x) = x^7$

In Exercises 5–12, state whether the graph could possibly be the graph of (a) *some polynomial function;* (b) *a polynomial function of degree 3;* (c) *a polynomial function of degree 4;* (d) *a polynomial function of degree 5. (See the box preceding Example 3.)*

5.

6.

7.

8.

9.

10.

11.

12.

Graph each of the following polynomial functions. (See Examples 3–4.)

13. $f(x) = (x + 2)(x - 3)(x + 4)$

14. $f(x) = (x - 3)(x - 1)(x + 1)$

15. $f(x) = x^2(x - 2)(x + 3)$

16. $f(x) = x^2(x + 1)(x - 1)$

17. $f(x) = x^3 + x^2 - 6x$

18. $f(x) = x^3 - 2x^2 - 8x$

19. $f(x) = x^3 + 3x^2 - 4x$

20. $f(x) = x^4 - 5x^2$

21. $f(x) = x^4 - 2x^2$

22. $f(x) = x^4 - 7x^2 + 12$

Exercises 23–26 require a graphing calculator. Find a viewing window that shows a complete graph of the polynomial function (that is, a graph that includes all the peaks and valleys and indicates how the curve moves away from the x-axis at the far left and far right). There are many possible correct answers. Consider your answer correct if it shows all the features that appear in the window given in the answers. (See the box preceding Example 3.)

23. $g(x) = x^3 - 3x^2 - 4x - 5$

24. $f(x) = x^4 - 10x^3 + 35x^2 - 50x + 24$

25. $f(x) = 2x^5 - 3.5x^4 - 10x^3 + 5x^2 + 12x + 6$

26. $g(x) = x^5 + 8x^4 + 20x^3 + 9x^2 - 27x - 7$

In Exercises 27–31, use a calculator to evaluate the functions. Graph each function either by plotting points or by using a graphing calculator.

27. Natural Science The polynomial function defined by

$$A(x) = -.015x^3 + 1.058x$$

gives the approximate alcohol concentration (in tenths of a percent) in an average person's bloodstream x hours after drinking about eight ounces of 100 proof whiskey. The function is approximately valid for $0 \le x < 8$. Find the following values.

(a) $A(1)$

(b) $A(2)$

(c) $A(4)$

(d) $A(6)$

(e) $A(8)$

(f) Graph $A(x)$.

(g) Using the graph you drew for part (f), estimate the time of maximum alcohol concentration.

(h) In one state, a person is legally drunk if the blood alcohol concentration exceeds .15%. Use the graph of part (f) to estimate the period in which this average person is legally drunk.

28. Natural Science A technique for measuring cardiac output depends on the concentration of a dye after a known amount is injected into a vein near the heart. In a normal heart, the concentration of the dye at time x (in seconds) is given by the function defined by

$$g(x) = -.006x^4 + .140x^3 - .053x^2 + 1.79x.$$

(a) Find the following: $g(0)$; $g(1)$; $g(2)$; $g(3)$.

(b) Graph $g(x)$ for $x \ge 0$.

29. Natural Science The pressure of the oil in a reservoir tends to drop with time. By taking sample pressure readings for a particular oil reservoir, petroleum engineers have found that the change in pressure is given by

$$P(t) = t^3 - 18t^2 + 81t,$$

where t is time in years from the date of the first reading.

(a) Find the following: $P(0)$; $P(3)$; $P(7)$; $P(10)$.

(b) Graph $P(t)$.

(c) For what time period is the change in pressure (drop) increasing? decreasing?

30. Natural Science During the early part of the 20th century, the deer population of the Kaibab Plateau in Arizona experienced a rapid increase because hunters had reduced the number of natural predators. The increase in population depleted the food resources and eventually caused the population to decline. For the period from 1905 to 1930, the deer population was approximated by

$$D(x) = -.125x^5 + 3.125x^4 + 4000,$$

where x is time in years from 1905.

(a) Find the following: $D(0)$; $D(5)$; $D(10)$; $D(15)$; $D(20)$; $D(25)$.

(b) Graph $D(x)$.

(c) From the graph, over what period of time (from 1905 to 1930) was the population increasing? relatively stable? decreasing?

31. Management According to data for the years 1983–1994 in *Insider Flyer Magazine,* the number of frequent flier miles (in billions) earned by customers of various airlines, but not yet redeemed in year x, could be approximated by the polynomial function

$$f(x) = .015x^4 - .68x^3 + 11.33x^2 - 20.15x \ (3 \le x \le 14),$$

where $x = 0$ corresponds to 1980.

(a) How many unredeemed miles were there in 1985? In 1990?

(b) Assume this model remains valid through 2003 ($x = 23$). How many unredeemed frequent flier miles are there in 1998? In 2000?

(c) If all the unredeemed miles in 2002 are redeemed for jumbo jet flights from New York to Los Angeles (each requiring 30,000 frequent flier miles) and each plane holds 400 people, how many planes would be needed?

4.4 RATIONAL FUNCTIONS

A **rational function** is a function whose rule is the quotient of two polynomials, such as

$$f(x) = \frac{2}{1 + x}, \qquad g(x) = \frac{3x + 2}{2x + 4}, \qquad h(x) = \frac{x^2 - 2x - 4}{x^3 - 2x^2 + x}.$$

Thus a rational function is one whose rule can be written in the form

$$f(x) = \frac{P(x)}{Q(x)},$$

where $P(x)$ and $Q(x)$ are polynomials, with $Q(x) \neq 0$. The function is undefined for any values of x that make $Q(x) = 0$, so there are breaks in the graph at these numbers.

LINEAR RATIONAL FUNCTIONS   We begin with rational functions in which both numerator and denominator are first-degree or constant polynomials. Such functions are sometimes called **linear rational functions.**

EXAMPLE 1   Graph the rational function defined by $y = \dfrac{2}{1+x}$.

   This function is undefined for $x = -1$, since -1 leads to a 0 denominator. For this reason, the graph of this function will not intersect the vertical line $x = -1$. Since x can take on any value except -1, the values of x can approach -1 as closely as desired from either side of -1, as shown in the following table of values.

x approaches −1 ↓

x	−1.5	−1.2	−1.1	−1.01	−.99	−.9	−.8	−.5
$1+x$	−.5	−.2	−.1	−.01	.01	.1	.2	.5
$\dfrac{2}{1+x}$	−4	−10	−20	−200	200	20	10	4

↑ |$f(x)$| gets larger and larger

   The table above suggests that as x gets closer and closer to -1 from either side, the denominator $1 + x$ gets closer and closer to 0, and $|2/(1 + x)|$ gets larger and larger. The part of the graph near $x = -1$ in Figure 4.19 on the next page shows this behavior. The vertical line $x = -1$ that is approached by the curve is called a *vertical asymptote*. For convenience, the vertical asymptote is indicated by a dashed line in Figure 4.19, but this line is *not* part of the graph of the function.
   As $|x|$ gets larger and larger so does the absolute value of the denominator $1 + x$. Hence $y = 2/(1 + x)$ gets closer and closer to 0, as shown in the table below.

x	−101	−11	−2	0	9	99
$1+x$	−100	−10	−1	1	10	100
$\dfrac{2}{1+x}$	−.02	−.2	−2	2	.2	.02

☐1 Graph the following.

(a) $f(x) = \dfrac{3}{5 - x}$

(b) $f(x) = \dfrac{-4}{x + 4}$

Answers:

(a)

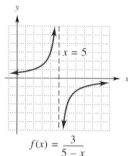

$$f(x) = \frac{3}{5 - x}$$

(b)

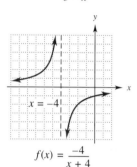

$$f(x) = \frac{-4}{x + 4}$$

The horizontal line $y = 0$ is called a *horizontal asymptote* for this graph. Using the asymptotes and plotting the intercept and other points gives the graph of Figure 4.19. ■ ☐1

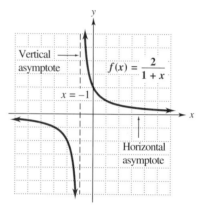

FIGURE 4.19

Example 1 suggests the following conclusion.

> If a number c makes the denominator 0 but the numerator nonzero in the expression defining a rational function, then the line $x = c$ is a **vertical asymptote** for the graph of the function.
>
> Also, wherever the values of y approach but do not equal some number k as $|x|$ gets larger and larger, the line $y = k$ is a **horizontal asymptote** for the graph.

EXAMPLE 2 Graph $f(x) = \dfrac{3x + 2}{2x + 4}$.

Find the vertical asymptote by setting the denominator equal to 0, then solving for x.

$$2x + 4 = 0$$
$$x = -2$$

In order to see what the graph looks like when $|x|$ is very large, we rewrite the rule of the function. When $x \neq 0$, dividing both numerator and denominator by x does not change the value of the function:

Graph the following.

(a) $f(x) = \dfrac{3x + 5}{x - 3}$

(b) $f(x) = \dfrac{2 - x}{x + 3}$

Answers:

(a)

$$f(x) = \frac{3x + 5}{x - 3}$$

(b)

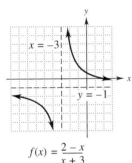

$$f(x) = \frac{2 - x}{x + 3}$$

$$f(x) = \frac{3x + 2}{2x + 4} = \frac{\dfrac{3x + 2}{x}}{\dfrac{2x + 4}{x}}$$

$$= \frac{\dfrac{3x}{x} + \dfrac{2}{x}}{\dfrac{2x}{x} + \dfrac{4}{x}} = \frac{3 + \dfrac{2}{x}}{2 + \dfrac{4}{x}}.$$

Now when $|x|$ is very large, the fractions $2/x$ and $4/x$ are very close to 0 (for instance, when $x = 200$, $4/x = 4/200 = .02$). Therefore the numerator of $f(x)$ is very close to $3 + 0 = 3$ and the denominator is very close to $2 + 0 = 2$. Hence $f(x)$ is very close to 3/2 when $|x|$ is large, so the line $y = 3/2$ is the horizontal asymptote of the graph, as shown in Figure 4.20. ■

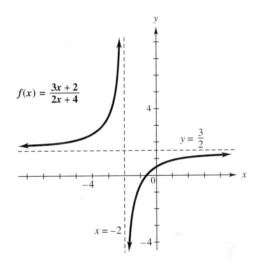

FIGURE 4.20

TECHNOLOGY TIP Depending on the viewing window, a graphing calculator may not accurately represent the graph of a rational function, particularly near a vertical asymptote. This problem can often be avoided by using a window that has the vertical asymptote at the center of the x-axis. ✔

In Examples 1 and 2, each graph has a single vertical asymptote, determined by the root of the denominator, and a horizontal asymptote determined by the coefficients of x. In Example 2, for instance, $f(x) = \dfrac{3x + 2}{2x + 4}$ has horizontal asymptote $y = \dfrac{3}{2}$, and in Example 1, $f(x) = \dfrac{2}{1 + x} = \dfrac{0x + 2}{1x + 1}$ has horizontal asymptote $y = \dfrac{0}{1} = 0$ (the x-axis). Similar arguments work for any linear rational function.

The graph of $f(x) = \dfrac{ax + b}{cx + d}$ (where $c \neq 0$ and $ad \neq bc$) has a vertical asymptote at the root of the denominator and horizontal asymptote $y = \dfrac{a}{c}$.

OTHER RATIONAL FUNCTIONS When the numerator or denominator of a rational function has degree greater than 1, its graph can be more complicated than those in Examples 1 and 2. The graph may have several vertical asymptotes, as well as peaks and valleys.

EXAMPLE 3   Graph $f(x) = \dfrac{2x^2}{x^2 - 4}$.

Find the vertical asymptotes by setting the denominator equal to 0 and solving for x.

$$x^2 - 4 = 0$$
$$(x + 2)(x - 2) = 0$$
$$x + 2 = 0 \quad \text{or} \quad x - 2 = 0$$
$$x = -2 \qquad\qquad x = 2$$

Since neither of these numbers make the numerator 0, the lines $x = -2$ and $x = 2$ are vertical asymptotes of the graph. The horizontal asymptote can be determined by dividing both the numerator and denominator of $f(x)$ by x^2 (the highest power of x that appears in either one).

$$f(x) = \frac{2x^2}{x^2 - 4}$$

$$= \frac{\dfrac{2x^2}{x^2}}{\dfrac{x^2 - 4}{x^2}}$$

$$= \frac{\dfrac{2x^2}{x^2}}{\dfrac{x^2}{x^2} - \dfrac{4}{x^2}}$$

$$= \frac{2}{1 - \dfrac{4}{x^2}}$$

3 List the vertical and horizontal asymptotes of the function.

(a) $f(x) = \dfrac{3x + 5}{x + 5}$

(b) $g(x) = \dfrac{2 - x^2}{x^2 - 4}$

Answers:

(a) Vertical $x = -5$; horizontal $y = 3$.

(b) Vertical $x = -2$ and $x = 2$; horizontal $y = -1$.

When $|x|$ is very large, the fraction $4/x^2$ is very close to 0, so that the denominator is very close to 1 and $f(x)$ is very close to 2. Hence the line $y = 2$ is the horizontal asymptote of the graph. Using this information and plotting several points in each of the three regions determined by the vertical asymptotes, we obtain Figure 4.21. ∎ **3**

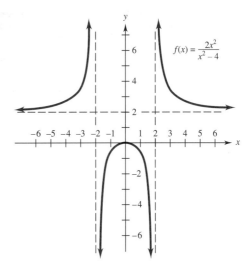

$$f(x) = \frac{2x^2}{x^2 - 4}$$

FIGURE 4.21

The arguments used to find the horizontal asymptotes in Examples 1–3 work in the general case and lead to this conclusion.

> If the numerator of the rational function $f(x)$ has *smaller* degree than the denominator, then the x-axis (the line $y = 0$) is the horizontal asymptote of the graph. If the numerator and denominator have the *same* degree, say
> $$f(x) = \frac{ax^n + \cdots}{cx^n + \cdots},$$
> then the line $y = \frac{a}{c}$ is the horizontal asymptote.*

APPLICATIONS Rational functions have a variety of applications, some of which are explored here.

EXAMPLE 4 In many situations involving environmental pollution, much of the pollutant can be removed from the air or water at a fairly reasonable cost, but the last, small part of the pollutant can be very expensive to remove.

Cost as a function of the percentage of pollutant removed from the environment can be calculated for various percentages of removal, with a curve fitted through the resulting data points. This curve then leads to a function that approximates the situation. Rational functions often are a good choice for these **cost-benefit functions.**

For example, suppose a cost-benefit function is given by

$$f(x) = \frac{18x}{106 - x},$$

where $f(x)$ or y is the cost (in thousands of dollars) of removing x percent of a certain pollutant. The domain of x is the set of all numbers from 0 to 100, inclusive; any

---

*When the numerator has larger degree than the denominator, the graph has no horizontal asymptote, but may have nonhorizontal lines or other curves as asymptotes; see Exercises 30 and 31 for examples.

4 Using the function of Example 4, find the cost to remove the following percents of pollutants.

(a) 70%

(b) 85%

(c) 98%

Answers:

(a) $35,000

(b) About $73,000

(c) About $221,000

amount of pollutant from 0% to 100% can be removed. To remove 100% of the pollutant here would cost

$$y = \frac{18(100)}{106 - 100} = 300,$$

or $300,000. Check that 95% of the pollutant can be removed for $155,000, 90% for $101,000, and 80% for $55,000, as shown in Figure 4.22. ■ **4**

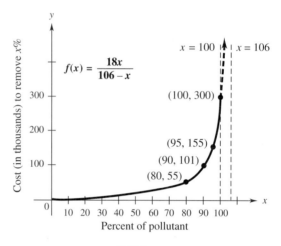

FIGURE 4.22

In management, **product-exchange functions** give the relationship between quantities of two items that can be produced by the same machine or factory. For example, an oil refinery can produce gasoline, heating oil, or a combination of the two; a winery can produce red wine, white wine, or a combination of the two. The next example discusses a product-exchange function.

EXAMPLE 5 The product-exchange function for the Golden Grape Winery for red wine x and white wine y, in tons, is

$$y = \frac{100,000 - 50x}{1000 + x}.$$

Graph the function and find the maximum quantity of each kind of wine that can be produced.

5 Rework Example 5 with the product-exchange function

$$y = \frac{60,000 - 10x}{60 + x}$$

to find the maximum amount of each wine that can be produced.

Answer:

6000 tons of red, 1000 tons of white

Only nonnegative values of x and y make sense in this situation, so we graph the function in the first quadrant (Figure 4.23). Note that the y-intercept of the graph (found by setting $x = 0$) is 100, and the x-intercept (found by setting $y = 0$ and solving for x) is 2000. Since we are interested only in the Quadrant I portion of the graph, we can find a few more points in Quadrant I and complete the graph as shown in Figure 4.23.

The maximum value of y occurs when $x = 0$, so the maximum amount of white wine that can be produced is 100 tons, given by the y-intercept. The x-intercept gives the maximum amount of red wine that can be produced, 2000 tons. ■ **5**

FIGURE 4.23

4.4 EXERCISES

Graph each function. Give the equations of the vertical and horizontal asymptotes (see Examples 1–3).

1. $f(x) = \dfrac{1}{x + 5}$

2. $g(x) = \dfrac{-7}{x - 6}$

3. $f(x) = \dfrac{-3}{2x + 5}$

4. $h(x) = \dfrac{-4}{2 - x}$

5. $f(x) = \dfrac{3x}{x - 1}$

6. $g(x) = \dfrac{x - 2}{x}$

7. $f(x) = \dfrac{x + 1}{x - 4}$

8. $f(x) = \dfrac{x - 3}{x + 5}$

9. $f(x) = \dfrac{2 - x}{x - 3}$

10. $g(x) = \dfrac{3x - 2}{x + 3}$

11. $f(x) = \dfrac{2x - 1}{4x + 2}$

12. $f(x) = \dfrac{3x - 6}{6x - 1}$

13. $h(x) = \dfrac{x + 1}{x^2 + 3x - 4}$

14. $g(x) = \dfrac{1}{x(x + 1)^2}$

15. $f(x) = \dfrac{x^2 + 1}{x^2 - 1}$

16. $f(x) = \dfrac{x - 1}{x^2 - x - 6}$

Find the equations of the vertical asymptotes of each of the following rational functions.

17. $f(x) = \dfrac{x - 3}{x^2 + x - 2}$

18. $g(x) = \dfrac{x + 2}{x^2 - 1}$

19. $g(x) = \dfrac{x^2 + 2x}{x^2 - 4x - 5}$

20. $f(x) = \dfrac{x^2 - 2x - 4}{x^3 - 2x^2 + x}$

Work these problems. (See Example 4.)

21. Natural Science Suppose a cost-benefit model is given by

$$f(x) = \frac{4.3x}{100 - x},$$

where $f(x)$ is the cost in thousands of dollars of removing x percent of a given pollutant. Find the cost of removing each of the following percents of pollutants.

(exercise continues)

(a) 50%   (b) 70%
(c) 80%   (d) 90%
(e) 95%   (f) 98%
(g) 99%
(h) Is it possible, according to this model to remove *all* the pollutant?
(i) Graph the function.

22. Natural Science Suppose a cost-benefit model is given by

$$f(x) = \frac{4.5x}{101 - x},$$

where $f(x)$ is the cost in thousands of dollars of removing x percent of a certain pollutant. Find the cost of removing the following percents of pollutants.

(a) 0%   (b) 50%
(c) 80%   (d) 90%
(e) 95%   (f) 99%
(g) 100%
(h) Graph the function.

23. Management SuperStar Cablevision Company recently began service to the city of Megapolis. Based on past experience, it estimates that the number $N(x)$ of subscribers (in thousands) at the end of x months is

$$N(x) = \frac{250x}{x + 6}.$$

Find the number of subscribers at the end of

(a) 6 months
(b) 18 months
(c) two years.
(d) Graph $N(x)$.
(e) What part of the graph is relevant to this situation?
(f) What is the horizontal asymptote of the graph? What does this suggest that the maximum possible number of subscribers will be?

24. Social Science The average waiting time in a line (or queue) before getting served is given by

$$W = \frac{S(S - A)}{A},$$

where A is the average rate that people arrive at the line and S is the average service time. At a certain fast food restaurant, the average service time is 3 minutes. Find W for each of the following average arrival times.

(a) 1 minute
(b) 2 minutes
(c) 2.5 minutes
(d) What is the vertical asymptote?
(e) Graph the equation on the interval (0, 3].
(f) What happens to W when $A > 3$? What does this mean?

25. Management At Ewing's Clothing Store, daily sales of shirts (in dollars) after x days of newspaper ads are given by

$$S = \frac{600x + 3800}{x + 1}.$$

(a) Graph the first quadrant portion of the sales function.
(b) Assuming that the ads keep running, will sales tend to level off? What is the minimum amount of sales to be expected? What feature of the graph gives this information?
(c) If newspaper ads cost $1000 per day, at what point should they be discontinued? Why?

Management *Sketch the Quadrant I portion of the graph of each of the functions defined as follows, and then estimate the maximum quantities of each product that can be produced. (See Example 5.)*

26. The product-exchange function for gasoline, x, and heating oil, y, in hundreds of gallons per day, is

$$y = \frac{125,000 - 25x}{125 + 2x}.$$

27. A drug factory found the product-exchange function for a red tranquilizer, x, and a blue tranquilizer, y, is

$$y = \frac{900,000,000 - 30,000x}{x + 90,000}.$$

28. Physical Science The failure of several O-rings in field joints was the cause of the fatal crash of the *Challenger* space shuttle in 1986. NASA data from 24 successful launches prior to *Challenger* suggested that O-ring failure was related to launch temperature by a function similar to

$$N(t) = \frac{600 - 7t}{4t - 100} \quad (50 \le t \le 85),$$

where t is the temperature (in °F) at launch and N is the approximate number of O-rings that fail. Assume that this function accurately models the number of O-ring failures that would occur at lower launch temperatures (an assumption NASA did not make).

(a) Does $N(t)$ have a vertical asymptote? At what value of t does it occur?
(b) Without graphing, what would you conjecture that the graph would look like just to the right of the vertical asymptote? What does this suggest about the number of O-ring failures that might be expected near that temperature? (The temperature at the *Challenger* launching was 31°.)
(c) Confirm your conjecture by graphing $N(t)$ between the vertical asymptote and $t = 85$.

29. Management   A company has fixed costs of $40,000 and marginal cost of $2.60 per unit.
   (a) Find the linear cost function.
   (b) Find the average cost function. (Average cost was defined after Example 5 of Section 3.3.)
   (c) Find the horizontal asymptote of the graph of the average cost function. Explain what the asymptote means in this situation (how low can the average cost possibly be?).

Use a graphing calculator to do Exercises 30 and 31.

30. (a) Graph $f(x) = \dfrac{x^3 + 3x^2 + x + 1}{x^2 + 2x + 1}$.

   (b) Does the graph appear to have a horizontal asymptote? Does the graph appear to have some nonhorizontal straight line as an asymptote?
   (c) Graph $f(x)$ and the line $y = x + 1$ on the same screen. Does this line appear to be an asymptote of the graph of $f(x)$?

31. (a) Graph $g(x) = \dfrac{x^3 - 2}{x - 1}$ in the window with $-4 \leq x \leq 6$ and $-6 \leq y \leq 12$.

   (b) Graph $g(x)$ and the parabola $y = x^2 + x + 1$ on the same screen. How do the two graphs compare when $|x| \geq 2$?

CHAPTER 4   SUMMARY

Key Terms and Symbols

4.1   quadratic function
   parabola
   vertex
   axis

4.2   quadratic model

4.3   polynomial function
   graph of $f(x) = ax^n$
   properties of polynomial graphs

4.4   rational function
   linear rational function
   vertical asymptote
   horizontal asymptote

Key Concepts

The **quadratic function** defined by $y = a(x - h)^2 + k$ has a graph that is a **parabola** with vertex (h, k) and axis of symmetry $x = h$. The parabola opens upward if $a > 0$, downward if $a < 0$.
If the equation is in the form $f(x) = ax^2 + bx + c$, the vertex is $\left(-\dfrac{b}{2a}, f\left(-\dfrac{b}{2a}\right)\right)$.

If a number c makes the denominator of a **rational function** 0, but the numerator nonzero, then the line $x = c$ is a **vertical asymptote** for the graph.

Whenever the values of y approach, but never equal, some number k as $|x|$ gets larger and larger, the line $y = k$ is a **horizontal asymptote** for the graph.

Chapter 4 Review Exercises

Without graphing, determine whether each of the following parabolas opens upward or downward and find its vertex.

1. $f(x) = 3(x - 2)^2 + 6$

2. $f(x) = 2(x + 3)^2 - 5$

3. $g(x) = -4(x + 1)^2 + 8$

4. $g(x) = -5(x - 4)^2 - 6$

Graph each of the following and label its vertex.

5. $f(x) = x^2 - 4$

6. $f(x) = 6 - x^2$

7. $f(x) = x^2 + 2x - 3$

8. $f(x) = -x^2 + 6x - 3$

9. $f(x) = -x^2 - 4x + 1$

10. $f(x) = 4x^2 - 8x + 3$

11. $f(x) = 2x^2 + 4x - 3$

12. $f(x) = -3x^2 - 12x - 8$

Determine whether each of the following functions has a minimum or a maximum value and find that value.

13. $f(x) = x^2 + 6x - 2$

14. $f(x) = x^2 + 4x + 5$

15. $g(x) = -4x^2 + 8x + 3$

16. $g(x) = -3x^2 - 6x + 3$

Solve each problem.

17. Management The commodity market is very unstable; money can be made or lost quickly when investing in soybeans, wheat, pork bellies, and the like. Suppose that an investor kept track of her total profit, P, at time t, measured in months, after she began investing and found that $P = -4t^2 + 32t - 20$. At what time is her profit largest? (*Hint:* $t > 0$ in this case.)

18. Physical Science The height h (in feet) of a rocket at time t seconds is given by $h = -16t^2 + 800t$.
(a) How long does it take the rocket to reach 3200 feet?
(b) What is the maximum height of the rocket?

19. Management The Hopkins Company's profit in thousands of dollars is given by $P = -4x^2 + 88x - 259$, where x is the number of hundreds of cases produced. How many cases should be produced if the company is to make a profit?

20. Management The manager of a large apartment complex has found that the profit is given by $P = -x^2 + 250x - 15,000$, where x is the number of units rented. For what value of x does the complex produce the largest profit?

21. Management A rectangular enclosure is to be built with three sides made out of redwood fencing at a cost of $15 per running foot and the fourth side made out of cement blocks at a cost of $30 per running foot. $900 is available for the project. What are the dimensions of the enclosure with maximum possible area and what is this area?

22. Find the dimensions of the rectangular region of maximum area that can be enclosed with 200 meters of fencing.

23. Management The table lists expenditures by the federal government (in billions of dollars) for Medicare in selected years.

Year	1981	1985	1989	1993
Expenditures	40	70	90	140

(a) Let $x = 0$ correspond to 1980. Assuming the data points (1, 40), (5, 70), etc. fit a quadratic model, find a quadratic function $f(x) = a(x - h)^2 + k$ that models these data by using (1, 40) as the vertex and (13, 140) to determine a.
(b) Assuming this model remains valid, what are the Medicare expenditures (to the nearest billion) in 1999 and 2000?

Graph each of the following polynomial functions.

24. $f(x) = x^4 - 2$

25. $g(x) = x^3 - x$

26. $g(x) = x^4 - x^2$

27. $f(x) = x(x - 2)(x + 3)$

28. $f(x) = (x - 1)(x + 2)(x - 3)$

29. $f(x) = x(2x - 1)(x + 2)$

30. $f(x) = 3x(3x + 2)(x - 1)$

31. $f(x) = 2x^3 - 3x^2 - 2x$

32. $f(x) = x^3 - 3x^2 - 4x$

33. $f(x) = x^4 - 5x^2 - 6$

34. $f(x) = x^4 - 7x^2 - 8$

List the vertical and horizontal asymptotes of each function and sketch its graph.

35. $f(x) = \dfrac{1}{x - 3}$

36. $f(x) = \dfrac{-2}{x + 4}$

37. $f(x) = \dfrac{-3}{2x - 4}$

38. $f(x) = \dfrac{5}{3x + 7}$

39. $g(x) = \dfrac{5x - 2}{4x^2 - 4x - 3}$

40. $g(x) = \dfrac{x^2}{x^2 - 1}$

41. Management   The average cost per unit of producing x cartons of cocoa is given by

$$C(x) = \frac{650}{2x + 40}.$$

Find the average cost per carton to make the following number of cartons.
(a) 10 cartons
(b) 50 cartons
(c) 70 cartons
(d) 100 cartons
(e) Graph $C(x)$.

42. Management   The cost and revenue functions (in dollars) for a frozen yogurt shop are given by

$$C(x) = \frac{400x + 400}{x + 4} \quad \text{and} \quad R(x) = 100x,$$

where x is measured in hundreds of units.
(a) Graph $C(x)$ and $R(x)$ on the same axes.
(b) What is the break-even point for this shop?
(c) If the profit function is given by $P(x)$, does $P(1)$ represent a profit or a loss?
(d) Does $P(4)$ represent a profit or a loss?

43. Management   The supply and demand functions for the yogurt shop in Exercise 42 are as follows:

$$\text{supply: } p = \frac{q^2}{4} + 25; \quad \text{demand: } p = \frac{500}{q},$$

where p is the price in dollars for q hundred units of yogurt.
(a) Graph both functions on the same axes, and from the graph, estimate the equilibrium point.
(b) Give the q-intervals where supply exceeds demand.
(c) Give the q-intervals where demand exceeds supply.

44. Management   A cost-benefit curve for pollution control is given by

$$y = \frac{9.2x}{106 - x},$$

where y is the cost in thousands of dollars of removing x percent of a specific industrial pollutant. Find y for each of the following values of x.
(a) $x = 50$
(b) $x = 98$
(c) What percent of the pollutant can be removed for $22,000?

CASE 4

Error-Correcting Codes*

Both the *Voyager* satellite and musical CDs send data over "noisy" channels, so that error-correcting techniques are needed to get the original information. To see how these techniques work, consider a simple 2-word message coded as the numbers 2.6 and 5.7. This information will be interpreted as the ordered pairs (1, 2.6) and (2, 5.7), where the first numbers indicate the position of the word in the message.

These two pairs determine a straight line, whose equation can be found to be $y = 3.1x - .5$. This equation can be used to find more pairs: (3, 8.8), (4, 11.9), (5, 15.0), and (6, 18.1), for example. Now the message is sent encoded as (2.6, 5.7, 8.8, 11.9, 15.0, 18.1). This establishes a strong pattern that makes it possible to recover any data that has been incorrectly received. For instance, if the data was received as (5.7, 2.6, 8.8, 11.9, 15.0, 18.1), after plotting the points, the four collinear points could be used to determine the equation of the line containing them. Then, by substituting the x-coordinates of the points that don't fit, the correct y-values can be determined.

*Linda Kurz, "Error-Correcting Codes," given in "Snapshots of Applications in Mathematics" edited by Dennis Callas and David J. Hildreth in the *AMATYC Review*, Fall 1995.

Messages with more words require higher-order polynomials, but the same procedure can be used. A graphing calculator with a polynomial regression feature can be used to find a polynomial through the points. (For the correct syntax, look for "regression" in the index of your instruction manual.)

EXERCISES

1. Use the method discussed here to determine the correct 2-word message if (2.9, 14.1, 8.5, 11.3, 5.7, 16.9) is received.

2. A 3-word message requires a quadratic equation. Suppose the message to be sent is (1.2, 2.5, 3.7). Use quadratic regression on a calculator to find a quadratic equation that satisfies the corresponding ordered pairs, (1, 1.2), (2, 2.5), (3, 3.7). Find the points $(4, y)$, $(5, y)$, $(6, y)$, and $(7, y)$. What message would be sent?

3. If your calculator can do cubic regression, create a strong pattern with eight points for sending the following message: (2.3, 1.1, 1.7, 2.4).

CHAPTER 5

Exponential and Logarithmic Functions

Exponential and logarithmic functions play a key role in management, economics, the social and physical sciences, and engineering. They are used to study the growth of money and organizations; learning curves; the growth of human, animal, and bacterial populations; the spread of disease; and radioactive decay. Specific examples of applications that involve expressions with variable exponents are given in the exercises throughout the chapter.

5.1 EXPONENTIAL FUNCTIONS

In polynomial functions, the variable is raised to various constant exponents; for instance, $f(x) = x^2 + 3x - 5$. In *exponential functions,* such as

$$f(x) = 10^x \quad \text{or} \quad g(x) = 2^{-x^2} \quad \text{or} \quad h(x) = 3^{.6x},$$

the variable is in the exponent and the base is a positive constant. Exponential functions can be used to model situations such as vehicle theft in the United States as we shall see in Exercise 53. We begin with the simplest type of exponential function.

> If a is a positive constant other than 1, then the function defined by
>
> $$f(x) = a^x$$
>
> is called an **exponential function with base a.**

The function $f(x) = 1^x$ is the constant function $f(x) = 1^x = 1$. Exponential functions with negative bases are not of interest because when a is negative, a^x may not be defined for some values of x; for instance, $(-4)^{1/2} = \sqrt{-4}$ is not a real number.

EXAMPLE 1 Graph each exponential function and discuss its characteristics.
(a) $f(x) = 2^x$

Begin by making a table of values of x and y, as shown in Figure 5.1. Plot the points and draw a smooth curve through them. From the graph we can see that $f(x)$

(or y) gets larger as x increases. However, there is no vertical asymptote. The graph approaches the negative x-axis, but will never touch it because *every* power of 2 is positive. Thus, the x-axis is a horizontal asymptote. Since x may take any real number value, the domain is the set of all real numbers. Because 2^x is always positive, the range is the set of positive real numbers.

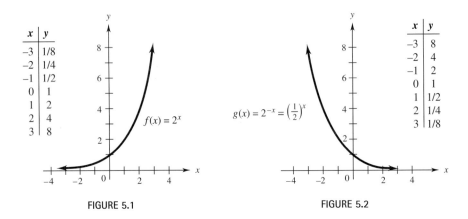

x	y
-3	1/8
-2	1/4
-1	1/2
0	1
1	2
2	4
3	8

$f(x) = 2^x$

FIGURE 5.1

$g(x) = 2^{-x} = \left(\frac{1}{2}\right)^x$

x	y
-3	8
-2	4
-1	2
0	1
1	1/2
2	1/4
3	1/8

FIGURE 5.2

1 Graph $f(x) = \left(\frac{1}{3}\right)^x$.

Answer:

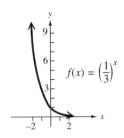

$f(x) = \left(\frac{1}{3}\right)^x$

(b) $g(x) = 2^{-x}$

Again we construct a table of values and draw a smooth curve through the resulting points. This graph, in Figure 5.2, approaches the positive x-axis as x increases, so the x-axis is again a horizontal asymptote. The graph indicates that $g(x)$ decreases as x increases. Notice that

$$2^{-x} = \frac{1}{2^x} = \left(\frac{1}{2}\right)^x,$$

by properties of exponents. The graph of $g(x) = (1/2)^x$ is the mirror image of the graph of $f(x) = 2^x$, with the y-axis as the mirror. The domain and range are the same as for $f(x)$. ■ **1**

The graph of $f(x) = 2^x$ in Figure 5.1 illustrates **exponential growth,** which is far more explosive than polynomial growth. For example, if the graph in Figure 5.1 were extended on the same scale to include the point $(50, 2^{50})$, the graph would be more than 1.5 billion *miles* high! Similarly, the graph of $g(x) = 2^{-x} = (1/2)^x$ illustrates **exponential decay.**

When $a > 1$, the graph of the exponential function $h(x) = a^x$ has the same basic shape as the graph of $f(x) = 2^x$ in Figure 5.1. It has the negative x-axis as a horizontal asymptote toward the left, crosses the y-axis at 1, and rises steeply to the right. The larger the base a is, the more steeply the graph rises, as illustrated in Figure 5.3(a).

When $0 < a < 1$, the graph of $h(x) = a^x$ has the same basic shape as the graph of $g(x) = (1/2)^x$ in Figure 5.2. Graphs of some typical exponential functions with $0 < a < 1$ are shown in Figure 5.3(b).

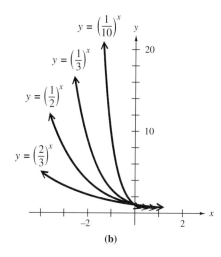

FIGURE 5.3

The graphs of exponential functions such as $f(x) = 3^{1-x}$ or $g(x) = 2^{.6x}$ or $h(x) = 3 \cdot 10^{2x+1}$ have the same general shape as the exponential graphs above. The only differences are that the graph may rise or fall at a different rate or the entire graph may be shifted vertically or horizontally.

EXAMPLE 2 Graph $f(x)$ and $g(x)$ as defined below and explain how the two graphs are related.

(a) $f(x) = 3^{1-x}$ and $g(x) = 3^{-x}$

Choose values of x that make the exponent positive, zero, and negative, and plot the corresponding points. The graphs are shown in Figure 5.4. The graph of $f(x) = 3^{1-x}$ has the same shape as the graph of $g(x) = 3^{-x}$, but is shifted 1 unit to the right, making the y-intercept $(0, 3)$ rather than $(0, 1)$.

FIGURE 5.4

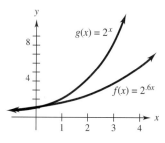

FIGURE 5.5

(b) $f(x) = 2^{.6x}$ and $g(x) = 2^x$

Comparing the graphs of $f(x) = 2^{.6x}$ and $g(x) = 2^x$ in Figure 5.5, we see that the graphs are both increasing, but the graph of $f(x)$ rises at a slower rate. This happens

2 Graph $f(x) = 2^{x+1}$.

Answer:

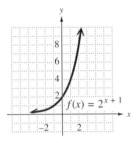

$f(x) = 2^{x+1}$

3 Graph $f(x) = \left(\dfrac{1}{2}\right)^{-x^2}$.

Answer:

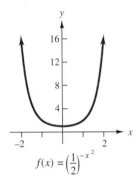

$f(x) = \left(\dfrac{1}{2}\right)^{-x^2}$

because of the .6 in the exponent. If the coefficient of x were greater than 1, the graph would rise at a faster rate than the graph of $f(x) = 2^x$. ■ **2**

When the exponent involves a nonlinear expression in x, an exponential function graph may have a much different shape than the preceding ones.

EXAMPLE 3 Graph $f(x) = 2^{-x^2}$.

Find several ordered pairs and plot them. You should get a graph like the one in Figure 5.6. The graph is symmetric about the y-axis—that is, if the figure were folded on the y-axis, the two halves would match. Like the graphs above, this graph has the x-axis as a horizontal asymptote. The domain is still all real numbers, but here the range is $0 < y \leq 1$. Graphs such as this are important in probability, where the normal curve has an equation similar to $f(x)$ in this example. ■ **3**

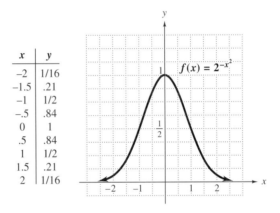

x	y
-2	1/16
-1.5	.21
-1	1/2
$-.5$	.84
0	1
.5	.84
1	1/2
1.5	.21
2	1/16

$f(x) = 2^{-x^2}$

FIGURE 5.6

The graphs of typical exponential functions $y = a^x$ in Figure 5.3 suggest that a given value of x leads to exactly one value of a^x and each value of a^x corresponds to exactly one value of x. Because of this, an equation with a variable in the exponent, called an **exponential equation,** can often be solved using the following property.

If $a > 0$, $a \neq 1$, and $a^x = a^y$, then $x = y$.

4 Solve each equation.

(a) $6^x = 6^4$

(b) $3^{2x} = 3^9$

(c) $5^{-4x} = 5^3$

Answers:
(a) 4

(b) 9/2

(c) $-3/4$

CAUTION Both bases must be the same. The value $a = 1$ is excluded because, for example, $1^2 = 1^3$ even though $2 \neq 3$. ◆

As an example, we solve $2^{3x} = 2^7$ using this property as follows:

$$2^{3x} = 2^7$$

$$3x = 7$$

$$x = \frac{7}{3}. \quad \boxed{4}$$

5 Solve each equation.

(a) $8^{2x} = 4$

(b) $5^{3x} = 25^4$

(c) $36^{-2x} = 6$

Answers:

(a) 1/3

(b) 8/3

(c) −1/4

EXAMPLE 4   Solve $9^x = 27$.

First rewrite both sides of the equation so the bases are the same. Since $9 = 3^2$ and $27 = 3^3$,

$$(3^2)^x = 3^3$$
$$3^{2x} = 3^3$$
$$2x = 3$$
$$x = \frac{3}{2}.\ \blacksquare\quad \boxed{5}$$

Exponential functions have many practical applications. For example, in situations that involve growth or decay of a quantity, the amount of the quantity present at a given time t often is determined by an exponential function of t.

EXAMPLE 5   The International Panel on Climate Change (IPCC) in 1990 published its finding that if current trends of burning fossil fuel and deforestation continue, then future amounts of atmospheric carbon dioxide in parts per million (ppm) will increase as shown in the table.

Year	Carbon Dioxide
1990	353
2000	375
2075	590
2175	1090
2275	2000

(a) Plot the data. Do the carbon dioxide levels appear to grow exponentially (follow an exponential curve)?

We show a calculator-generated plot of the data in Figure 5.7(a). The data do appear to have the shape of the graph of an increasing exponential function.

(a)

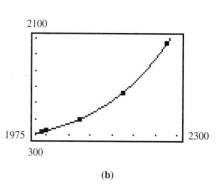

(b)

FIGURE 5.7

6 The number of organisms present at time t is given by $f(t) = 75(2.5)^{.5t}$.

(a) Is this a growth function or a decay function?

Find the number of organisms present at

(b) $t = 0$;

(c) $t = 2$;

(d) $t = 4$.

Answers:
(a) A growth function
(b) 75
(c) About 188
(d) About 469

(b) A good model for the data is given by

$$y = 353(1.0061)^{t-1990}.$$

The calculator graph of this function in Figure 5.7(b) shows that it closely fits the data points. Use the model to find the level of carbon dioxide in the year 2005. Let $t = 2005$ and evaluate y.

$$y = 353(1.0061)^{t-1990}$$
$$y = 353(1.0061)^{2005-1990}$$
$$y = 353(1.0061)^{15} \approx 387$$

In 2005 the model shows the carbon dioxide level will be 387 ppm. ■ **6**

5.1 EXERCISES

Classify each function as linear, quadratic, or exponential.

1. $f(x) = 2x^2 + x - 3$ **2.** $g(x) = 5^{x-3}$

3. $h(x) = 3 \cdot 6^{2x^2+4}$ **4.** $f(x) = 3x + 1$

Without graphing, **(a)** *describe the shape of the graph of each function, and* **(b)** *complete the ordered pairs* $(0, \ )$ *and* $(1, \ )$ *for each function. (See Examples 1–2.)*

5. $f(x) = .8^x$ **6.** $g(x) = 6^{-x}$

7. $f(x) = 5^{.4x}$ **8.** $g(x) = -(2^x)$

Graph each function. (See Examples 1–2.)

9. $f(x) = 3^x$ **10.** $g(x) = 3^{.5x}$

11. $f(x) = 2^{x/2}$ **12.** $g(x) = 4^{-x}$

13. $f(x) = (1/5)^x$ **14.** $g(x) = 2^{3x}$

15. Graph these functions on the same axes.
 (a) $f(x) = 2^x$ **(b)** $g(x) = 2^{x+3}$
 (c) $h(x) = 2^{x-4}$
 (d) If c is a positive constant, explain how the graphs of $y = 2^{x+c}$ and $y = 2^{x-c}$ are related to the graph of $f(x) = 2^x$.

16. Graph these functions on the same axes.
 (a) $f(x) = 3^x$ **(b)** $g(x) = 3^x + 2$
 (c) $h(x) = 3^x - 4$
 (d) If c is a positive constant, explain how the graphs of $y = 3^x + c$ and $y = 3^x - c$ are related to the graph of $f(x) = 3^x$.

The figure shows the graphs of $y = a^x$ *for* $a = 1.8, 2.3, 3.2, .4, .75,$ *and* $.31.$ *They are identified by letter, but not necessarily in the same order as the values of a just given. Use your knowledge of how the exponential function behaves for various powers of a to match each lettered graph with the correct value of a.*

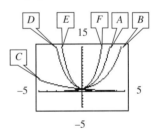

17. A **18.** B **19.** C **20.** D **21.** E **22.** F

In Exercises 23 and 24, the graph of an exponential function with base a is given. Follow the directions in parts (a)–(f) in each exercise.

23.

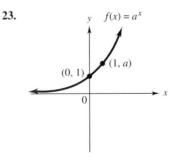

(a) Is $a > 1$ or is $0 < a < 1$?
(b) Give the domain and range of f.
(c) Sketch the graph of $g(x) = -a^x$.
(d) Give the domain and range of g.
(e) Sketch the graph of $h(x) = a^{-x}$.
(f) Give the domain and range of h.

24.

(a) Is $a > 1$ or is $0 < a < 1$?
(b) Give the domain and range of f.
(c) Sketch the graph of $g(x) = a^x + 2$.
(d) Give the domain and range of g.
(e) Sketch the graph of $h(x) = a^{x+2}$.
(f) Give the domain and range of h.

25. If $f(x) = a^x$ and $f(3) = 27$, find the following values of $f(x)$.
(a) $f(1)$
(b) $f(-1)$
(c) $f(2)$
(d) $f(0)$

26. Give an equation of the form $f(x) = a^x$ to define the exponential function whose graph contains the given point.
(a) $(3, 8)$
(b) $(-3, 64)$

Graph each function. (See Example 3.)

27. $f(x) = 2^{-x^2+2}$

28. $g(x) = 2^{x^2-2}$

29. $f(x) = x \cdot 2^x$

30. $f(x) = x^2 \cdot 2^x$

Solve each equation. (See Example 4.)

31. $5^x = 25$

32. $3^x = \dfrac{1}{9}$

33. $4^x = 64$

34. $a^x = a^2 \quad (a > 0)$

35. $16^x = 64$

36. $\left(\dfrac{3}{4}\right)^x = \dfrac{16}{9}$

37. $5^{-2x} = \dfrac{1}{25}$

38. $3^{x-1} = 9$

39. $16^{-x+1} = 8$

40. $25^{-2x} = 3125$

41. $81^{-2x} = 3^{x-1}$

42. $7^{-x} = 49^{x+3}$

43. $2^{|x|} = 16$

44. $5^{-|x|} = \dfrac{1}{25}$

45. $2^{x^2-4x} = \dfrac{1}{16}$

46. $5^{x^2+x} = 1$

47. Use a calculator and the graph of $f(x) = 2^x$ to explain why the solution of $2^x = 12$ must be a number between 3.5 and 3.6.

48. Explain why the equation $4^{x^2+1} = 2$ has no solutions.

Work the following exercises.

49. Management If \$1 is deposited into an account paying 6% per year compounded annually, then after t years the account will contain
$$y = (1 + .06)^t = (1.06)^t$$
dollars.
(a) Use a calculator to complete the following table.

t	0	1	2	3	4	5	6	7	8	9	10
y	1					1.34					1.79

(b) Graph $y = (1.06)^t$.

50. Management If money loses value at the rate of 3% per year, the value of \$1 in t years is given by
$$y = (1 - .03)^t = (.97)^t.$$
(a) Use a calculator to complete the following table.

t	0	1	2	3	4	5	6	7	8	9	10
y	1					.86					.74

(b) Graph $y = (.97)^t$.

51. Management If money loses value, then it takes more dollars to buy the same item. Use the results of Exercise 50(a) to answer the following questions.
(a) Suppose a house costs \$105,000 today. Estimate the cost of a similar house in 10 years. (*Hint:* Solve the equation $.74t = \$105,000$.)
(b) Estimate the cost of a \$50 textbook in 8 years.

52. **Natural Science** Biologists studying salmon have found that the oxygen consumption of yearling salmon is given by $100(3)^{.6x}$, where x is the speed in feet per second. Find each of the following.
    (a) The oxygen consumption when the fish are still.
    (b) The oxygen consumption at a speed of 2 feet per second.

53. **Social Science** Vehicle theft in the United States has risen exponentially since 1972. The number of stolen vehicles (in millions) can be approximated by $f(x) = .88(1.03)^x$, where $x = 0$ represents the year 1972. Use this function to estimate the number of vehicles stolen in the following years.
    (a) 1980
    (b) 1990
    (c) 2000

54. **Natural Science** The amount (in cubic centimeters) of a tracer dye injected into the bloodstream decreases exponentially as described by the function with $f(x) = 6(3^{-.03x})$, where x is the number of minutes since the dye was injected. How much of the dye remains after 10 minutes?

55. **Social Science** The number of people in a company who have heard a rumor after t days is given by $f(t) = (2.7)^{.8t}$. How many people have heard the rumor after
    (a) 3 days?
    (b) 5 days?
    (c) 8 days?

56. **Social Science** If the world population continues to grow at the present rate, the population in billions in year t will be given by the function $P(t) = 4.2(2^{.0285(t-1980)})$.
    (a) There were fewer than a billion people on earth when Thomas Jefferson died in 1826. What was the world population in 1980? If this model is accurate, what will the population of the world be in the year
    (b) 2000?
    (c) 2010?
    (d) 2040?

57. **Social Science** By using the function in Exercise 56 and experimenting with a calculator, estimate how many years it will take for the world population in 2000 to triple. Is it likely that you will be alive then?

58. **Natural Science** The amount of plutonium remaining from one kilogram after x years is given by the function $W(x) = 2^{-x/24360}$. How much will be left after
    (a) 1000 years?
    (b) 10,000 years?
    (c) 15,000 years?
    (d) Estimate how long it will take for the one kilogram to decay to half its original weight. This may help to explain why nuclear waste disposal is a serious problem.

Management *The scrap value of a machine is the value of the machine at the end of its useful life. By one method of calculating scrap value, where it is assumed a constant percentage of value is lost annually, the scrap value S is given by*

$$S = C(1 - r)^n,$$

where C is the original cost, n is the useful life of the machine in years, and r is the constant annual percentage of value lost. Find the scrap value for each of the following machines.

59. Original cost, \$54,000; life, 8 years; annual rate of value loss, 12%

60. Original cost, \$178,000; life, 11 years; annual rate of value loss, 14%

61. Use the graphs of $f(x) = 2^x$ and $g(x) = 2^{-x}$ (not a calculator) to explain why $2^x + 2^{-x}$ is approximately equal to 2^x when x is very large.

62. **Management** The number of Internet users was estimated to be 1.6 million in October of 1989 and 39 million in October of 1994.* This growth can be approximated by an exponential function.
    (a) Write two ordered pairs that satisfy the function. Let 0 represent October 1989, 1 represent October 1990, and so on.
    (b) Find an exponential function of the form $f(x) = b \cdot a^x$. (*Hint:* Write two equations using the ordered pairs from part (a) and use them to determine b and a.)
    (c) Use the result of part (b) to determine $f(3)$. Interpret your result.

Use a graphing calculator for Exercises 63 and 64.

63. **Natural Science** The number of fruit flies present after t days of a laboratory experiment is given by $p(t) = 100 \cdot 3^{t/10}$. How many flies will be present after
    (a) 15 days?   (b) 25 days?
    (c) When will the fly population reach 3500? (*Hint:* Where does the graph of p meet the horizontal line $y = 3500$?)

64. **Natural Science** The population of fish in a certain lake at time t months is given by the function

$$p(t) = \frac{20,000}{1 + 24(2^{-.36t})}.$$

    (a) Graph the population function from $t = 0$ to $t = 48$ (a four-year period).
    (b) What was the population at the beginning?
    (c) Use the graph to estimate the one-year period in which the population grew most rapidly.
    (d) When do you think the population will reach 25,000? What factors in nature might explain your answer?

---

*Genesis Corporation.

5.2 APPLICATIONS OF EXPONENTIAL FUNCTIONS

1 Evaluate the following powers of e.

(a) $e^{.06}$

(b) $e^{-.06}$

(c) $e^{2.30}$

(d) $e^{-2.30}$

Answers:
(a) 1.06184
(b) .94176
(c) 9.97418
(d) .10026

A certain irrational number, denoted e, arises naturally in a variety of mathematical situations (much the same way that the number π appears when you consider the area of a circle). To nine decimal places,

$$e = 2.718281828.$$

Perhaps the single most useful exponential function is the function defined by $f(x) = e^x$.

TECHNOLOGY TIP To evaluate powers of e with a calculator, use the $\boxed{e^x}$ key. On some calculators, you will need to use the two keys $\boxed{\text{INV}}$ $\boxed{\text{LN}}$ or $\boxed{\text{2nd}}$ $\boxed{\text{LN}}$. For example, your calculator should show that $e^{.14} = 1.150273799$ to 8 decimal places. If these keys do not work, consult the instruction manual for your calculator. ✔ 1

N O T E To display the decimal expansion of e on your calculator screen, calculate e^1. ◆

In Figure 5.8 the functions defined by

$$g(x) = 2^x, \quad f(x) = e^x, \quad \text{and} \quad h(x) = 3^x$$

are graphed for comparison.

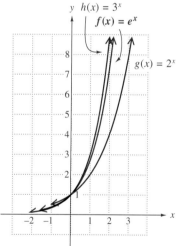

FIGURE 5.8

In many situations in biology, economics, and the social sciences, a quantity changes at a rate proportional to the quantity present. In such cases, the amount present at time t is a function of t called the *exponential growth function*.

EXPONENTIAL GROWTH FUNCTION

Under normal conditions, growth is described by the function

$$f(t) = y_0 e^{kt},$$

where $f(t)$ is the amount or quantity present at time t, y_0 is the amount present at time $t = 0$, and k is a constant that depends on the rate of growth.

It is understood in this context that "growth" can involve either growing larger or growing smaller. Here is an example of growing larger.

EXAMPLE 1 Chlorofluorocarbons (CFCs) are greenhouse gases that were produced by various products after 1930. CFCs are found in refrigeration units, foaming agents, and aerosols. They have great potential for destroying the ozone layer. As a result, governments have agreed to phase out their production by the year 2000. CFC-11 is a CFC which has increased faster than any other greenhouse gas. The exponential function

$$f(x) = .05e^{.03922(x-1950)}$$

models the concentration of atmospheric CFC-11 in parts per billion (ppb) from 1950 to 2000, where x is the year.* This is an example of an exponential growth function with $k = .03922$ and $y_0 = .05$.

(a) What was the concentration of atmospheric CFC-11 in 1950?
Evaluating $f(1950)$ gives

$$f(1950) = .05e^{.03922(1950-1950)}$$
$$= .05e^0 = .05 \text{ ppb.}$$

(b) What is the concentration in 2000?

$$f(2000) = .05e^{.03922(2000-1950)}$$
$$\approx .355 \text{ ppb} \quad \blacksquare \quad \boxed{2}$$

When the constant k in the exponential growth function is negative, the quantity will *decrease* over time, that is, grow smaller.

EXAMPLE 2 Radioactive lead-210 decays to polonium-210. The amount y of radioactive lead-210 at time t is given by $y = y_0 e^{-.032t}$, where t is time in years. How much of an initial 500 grams of lead-210 will remain after 5 years?
Since the initial amount is 500 grams, $y_0 = 500$. Substitute 500 for y_0 and 5 for t in the definition of the function.

$$y = 500e^{-.032(5)} \approx 426$$

About 426 grams will remain. $\blacksquare$ $\boxed{3}$

$\boxed{2}$ In Example 1, what was the atmospheric concentration of CFC-11 in 1990?

Answer:
About .24 ppb

$\boxed{3}$ Suppose the number of bacteria in a culture at time t is
$$y = 500e^{.4t},$$
where t is measured in hours.

(a) How many bacteria are present initially?

(b) How many bacteria are present after 10 hours?

Answers:
(a) 500

(b) About 27,300

---

*Nilsson, A., *Greenhouse Earth*, John Wiley & Sons, New York, 1992.

The exponential growth function deals with quantities that eventually either grow very large or decrease practically to 0. Other exponential functions are needed to deal with different patterns of growth.

4 Suppose the value of the assets (in thousands of dollars) of a certain company after t years is given by

$$V(t) = 100 - 75e^{-.2t}.$$

(a) What is the initial value of the assets?

(b) What is the limiting value of the assets?

(c) Find the value after 10 years.

(d) Graph $V(t)$.

Answers:
(a) $25,000
(b) $100,000
(c) $89,850
(d)

EXAMPLE 3   Sales of a new product often grow rapidly at first and then begin to level off with time. For example, suppose the sales, $S(x)$, in some appropriate unit, of a new model calculator are approximated by

$$S(x) = 1000 - 800e^{-x},$$

where x represents the number of years the calculator has been on the market. Calculate $S(0)$, $S(1)$, $S(2)$, and $S(4)$. Graph $S(x)$.

Find $S(0)$ by letting $x = 0$.

$$S(0) = 1000 - 800 \cdot 1 = 200 \qquad e^{-x} = e^{-0} = 1$$

Using one of these calculators,

$$S(1) = 1000 - 800e^{-1} \approx 1000 - 294.3 = 705.7,$$

which rounds to 706.

In the same way, verify that $S(2) \approx 892$ and $S(4) \approx 985$. Plotting several such points leads to the graph shown in Figure 5.9. Notice that as x increases, e^{-x} decreases and approaches 0. Thus, $S(x) = 1000 - 800e^{-x}$ approaches $S(x) = 1000 - 0 = 1000$. This means that the graph approaches the horizontal asymptote $y = 1000$. As the graph suggests, sales will tend to level off with time and gradually approach a level of 1000 units. ■ **4**

FIGURE 5.9

The sales graph in Figure 5.9 is typical of the graphs of *limited growth functions.* Another example of a limited growth function is a **learning curve.** Certain types of skills involving the repetitive performance of the same task characteristically improve rapidly, but then the learning tapers off and approaches some upper limit. (See Exercises 12 and 13.) The next example shows a **forgetting curve.**

EXAMPLE 4 Psychologists have measured people's ability to remember facts that they have memorized. In one such experiment, it was found that the number of facts $N(t)$ remembered after t days was given by

$$N(t) = 10\left(\frac{1 + e^{-8}}{1 + e^{.8t-8}}\right).$$

At the beginning of the experiment, $t = 0$ and

$$N(0) = 10\left(\frac{1 + e^{-8}}{1 + e^{.8(0)-8}}\right) = 10\left(\frac{1 + e^{-8}}{1 + e^{-8}}\right) = 10.$$

5 In Example 4,

(a) find the number of facts remembered after 10 days.

(b) use the graph to estimate when just 1 fact will be remembered.

Answers:

(a) 5

(b) After about 12 days

So 10 facts were known at the beginning. After 7 days, the number remembered was

$$N(7) = 10\left(\frac{1 + e^{-8}}{1 + e^{(.8)(7)-8}}\right) = 10\left(\frac{1 + e^{-8}}{1 + e^{-2.4}}\right) \approx 9.17.$$

Plotting several such points (see the problem at the side) leads to Figure 5.10. In the expression for $N(t)$, as t gets larger, the denominator increases and the fraction decreases, so fewer facts are remembered as time goes on. The graph of $N(t)$ illustrates the situation.   ■   **5**

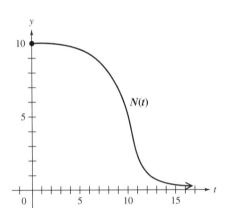

FIGURE 5.10

5.2 EXERCISES

1. **Social Science** A recent report by the U.S. Census Bureau predicts that the Latino-American population will increase from 26.7 million in 1995 to 96.5 million in 2050.* Assuming an exponential growth pattern, the population can be approximated by $f(t) = 26.7e^{.023t}$, where t is the number of years since 1995. What is this population in the year 2000?

2. **Social Science** (Refer to Exercise 1.) The report also predicted that the African-American population of the U.S. would increase from 31.4 million in 1995 to 53.6 million in 2050.* Find the population in the year 2010, if the population growth is approximated by $f(t) = 31.4e^{.009t}$.

*Population Projections of the U.S. by Age, Race, and Hispanic Origin: 1995 to 2050, U.S. Census Bureau.

3. **Natural Science** A sealed box contains radium. The number of grams present at time t is given by

$$Q(t) = 100e^{-.00043t},$$

where t is measured in years. Find the amount of radium in the box at the following times.
(a) $t = 0$
(b) $t = 800$
(c) $t = 1600$
(d) $t = 5000$

4. **Natural Science** The amount of a chemical in grams that will dissolve in a solution is given by

$$C(t) = 10e^{.02t},$$

where t is the temperature of the solution. Find each of the following.
(a) $C(0°)$
(b) $C(10°)$
(c) $C(30°)$
(d) $C(100°)$

5. **Natural Science** When a bactericide is introduced into a certain culture, the number of bacteria present, $D(t)$, is given by

$$D(t) = 50,000e^{-.01t},$$

where t is time measured in hours. Find the number of bacteria present at each of the following times.
(a) $t = 0$
(b) $t = 5$
(c) $t = 20$
(d) $t = 50$

6. **Management** The estimated number of units that will be sold by the Goldstein Widget Works t months from now is given by

$$N(t) = 100,000e^{-.09t}.$$

(a) What are current sales ($t = 0$)?
(b) What will sales be in 2 months? In 6 months?
(c) Will sales ever regain their present level? (What does the graph of $N(t)$ look like?)

Natural Science *The pressure of the atmosphere, $p(h)$, in pounds per square inch, is given by*

$$p(h) = p_0e^{-kh},$$

where h is the height above sea level and p_0 and k are constants. The pressure at sea level is 15 pounds per square inch and the pressure is 9 pounds per square inch at a height of 12,000 feet.

7. Find the pressure at an altitude of 6000 feet.

8. What would be the pressure encountered by a spaceship at an altitude of 150,000 feet?

9. **Management** In Chapter 6, it is shown that P dollars compounded continuously (every instant) at an annual rate of interest i would amount to

$$A = Pe^{ni}$$

at the end of n years. How much would $20,000 amount to at 8% compounded continuously for the following number of years?
(a) 1 year
(b) 5 years
(c) 10 years
(d) By experimenting with your calculator, estimate how long it will take for the initial $20,000 to triple.

10. **Natural Science** The graph shows how the risk of chromosomal abnormality in a child rises with the age of the mother.*
(a) Read from the graph the risk of chromosomal abnormality (per 1000) at ages 20, 35, 42, and 49.
(b) Verify by substitution that the exponential equation $y = .590e^{.061t}$ "fits" the graph for ages 20 and 35.
(c) Does the equation in part (b) also fit the graph for ages 42 and 49? What does this mean?

Maternity

Source: American College of Obstetricians and Gynecologists.

11. **Natural Science** The number (in hundreds) of fish in a small commercial pond is given by

$$F(t) = 27 - 15e^{-.8t},$$

where t is in years. Find each of the following.
(a) $F(0)$ (b) $F(1)$ (c) $F(5)$ (d) $F(10)$

*The New York Times, Feb. 5, 1994, p. 24.

12. Social Science The number of words per minute that an average typist can type is given by

$$W(t) = 60 - 30e^{-.5t},$$

where t is time in months after the beginning of a typing class. Find each of the following.

(a) $W(0)$
(b) $W(1)$
(c) $W(4)$
(d) $W(6)$

13. Management Assembly line operations tend to have a high turnover of employees, forcing the companies involved to spend much time and effort in training new workers. It has been found that a worker new to the operation of a certain task on the assembly line will produce $P(t)$ items on day t, where

$$P(t) = 25 - 25e^{-.3t}.$$

(a) How many items will be produced on the 1st day?
(b) How many items will be produced on the 8th day?
(c) What is the maximum number of items, according to the function, the worker can produce?

14. Management Sales of a new model can opener are approximated by

$$S(x) = 5000 - 4000e^{-x},$$

where x represents the number of years that the can opener has been on the market, and $S(x)$ represents sales in thousands. Find each of the following.

(a) $S(0)$
(b) $S(1)$
(c) $S(2)$
(d) $S(5)$
(e) $S(10)$
(f) Find the horizontal asymptote for the graph.
(g) Graph $y = S(x)$.

Management *The national debt has grown from 65 million dollars at the beginning of the Civil War to more than 4 trillion dollars currently. The graph of the debt, in which the horizontal axis represents time and the vertical axis dollars, has the approximate shape of an exponential growth function.* * *Over the past thirty years, the amount of debt (in billions of dollars) has been roughly approximated by the function*

$$D(t) = e^{.1293(t+36)} + 150,$$

where t is measured in years with t = 0 representing 1964. Use this function in Exercises 15 and 16.

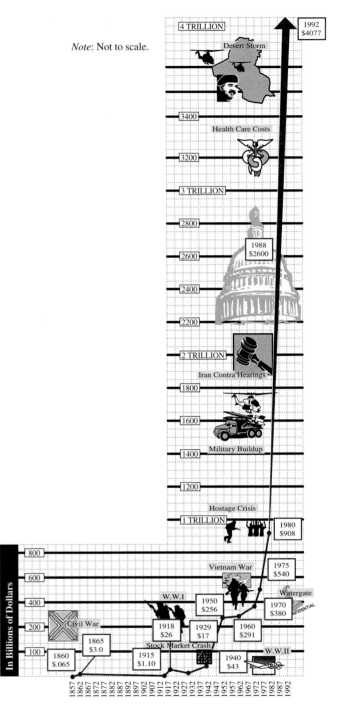

Note: Not to scale.

*Graph from "Debt Dwarfs Deficit" by Sam Hodges as appeared in *The Mobile Register*. Reprinted by permission of Newhouse News Service.

15. Use the function D to estimate the debt in each of the following years. Compare your results with the graph.
    (a) 1972
    (b) 1982
    (c) 1992

16. (a) The Congressional Budget Office projected that the national debt would be less than 6 trillion dollars at the turn of the century. How does this figure compare with the one predicted by the function D? What might explain the difference?
    (b) Experiment with your calculator and various exponential functions to see if you can find a model for the national debt for the decade 1992–2002. Your model should indicate a debt of approximately 4077 billion in 1992 and approximately 6000 billion in 2002.

Natural Science *Newton's Law of Cooling says that the rate at which a body cools is proportional to the difference in temperature between the body and an environment into which it is introduced. Using calculus, the temperature F(t) of the body at time t after being introduced into an environment having constant temperature T_0 is*

$$F(t) = T_0 + Ce^{-kt},$$

where C and k are constants. Use this result in Exercises 17 and 18.

17. Boiling water, at $100°$ Celsius, is placed in a freezer at $0°$ Celsius. The temperature of the water is $50°$ Celsius after 24 minutes. Find the temperature of the water after 96 minutes.

18. Paisley refuses to drink coffee cooler than $95°F$. She makes coffee with a temperature of $170°F$ in a room with a temperature of $70°F$. The coffee cools to $120°F$ in 10 minutes. What is the longest time she can let the coffee sit before she drinks the coffee?

19. **Natural Science** The beaver population in a certain lake in year t is approximately

$$p(t) = \frac{2000}{1 + e^{-.5544t}}.$$

    (a) What is the population now ($t = 0$)?
    (b) What will the population be in five years?

20. **Social Science** A class in elementary Tibetan is given a final exam worth 100 points. Then as part of an experiment they are tested weekly thereafter to see how much they remember. The average score on the test taken after t weeks is given by

$$T(x) = 77\left(\frac{1 + e^{-1}}{1 + e^{.08x-1}}\right).$$

(a) What was the average score on the original exam?
(b) What was the average score on the exam after 3 weeks? after 8 weeks?
(c) Do the students remember much Tibetan after a year?

21. **Social Science** A sociologist has shown that the fraction $y(t)$ of people in a group who have heard a rumor after t days is approximated by

$$y(t) = \frac{y_0 e^{kt}}{1 - y_0(1 - e^{kt})},$$

where y_0 is the fraction of people who have heard the rumor at time $t = 0$, and k is a constant. A graph of $y(t)$ for a particular value of k is shown in the figure.
(a) If $k = .1$ and $y_0 = .05$, find $y(10)$.
(b) If $k = .2$ and $y_0 = .10$, find $y(5)$.

22. **Social Science** Data from the National Highway Traffic Safety Administration for the period 1982–1992 indicates that the approximate percentage of people who use seat belts when driving is given by

$$f(t) = \frac{880}{11 + 69e^{-.3(t-1982)}}.$$

What percentage used seat belts in
(a) 1982?
(b) 1989?
(c) 1992?
Assuming this function is accurate after 1992, what percentage of people use seat belts in
(d) 1997?
(e) 2000?
(f) 2005?
(g) If this function remains accurate in the future, will there ever be a time when 95% of people use seat belts?

Use a graphing calculator for Exercises 23–26.

23. Social Science The probability P percent of having an automobile accident is related to the alcohol level of the driver's blood t by the function $P(t) = e^{21.459t}$.
(a) Graph $P(t)$ in a viewing window with $0 \le t \le .2$ and $0 \le P(t) \le 100$.
(b) At what blood alcohol level is the probability of an accident at least 50 percent? What is the legal blood alcohol level in your state?

The logistic growth model,

$$y = \frac{N}{1 + be^{-kx}},$$

describes growth that is limited (by environmental factors, for example) to a maximum size N.

24. Management The data in the chart reflect the sales y of a new model of car, with x representing the number of months since the car was introduced.

x	0	2	4	6	8	10	12
y	1000	7000	35,000	80,000	97,000	99,500	99,900

(a) Plot the data points.
(b) Graph the function $y = \dfrac{100,060}{1 + 101e^{-x}}$ in the same window as the points from part (a). Does this function model the data well? Is this a logistic model?
(c) Use the function from part (b) to find sales 5 months after the model was introduced.

(d) Does the graph indicate a limit on sales? If so, what is the limit?

25. Natural Science The chart shows the number of mites y in a population being studied in a laboratory after t weeks.

t	0	40	80	120	160	200
y	99	690	3530	8010	9680	9950

(a) Plot the ordered pairs and decide whether or not they fit a logistic model.
(b) Graph the function defined by $y = \dfrac{10,000}{1 + 100e^{-.05t}}$ in the same window as the points in part (a). Is the function an appropriate model for the data?
(c) Use the function to find the number of mites after 100 weeks.
(d) Is there a limit on the population? If so, what is it?

26. Management Midwest Creations finds that its total sales $T(x)$, in thousands, from the distribution of x catalogs, where x is measured in thousands, is approximated by

$$T(x) = \frac{2500}{1 + 24 \cdot 2^{-x/4}}.$$

(a) Graph $T(x)$ on the interval $[0, 50]$ by $[0, 3000]$.
(b) Use the graph to find the total sales if 0 catalogs are distributed.
(c) Find the total sales if 20,000 catalogs are distributed.
(d) From the graph, what happens to the total sales as the number of catalogs approaches 50,000 ($x = 50$)?

5.3 LOGARITHMIC FUNCTIONS

Until the development of computers and calculators, logarithms were the only effective tool for large-scale numerical computations. They are no longer needed for this, but logarithmic functions still play a crucial role in calculus and in many applications.

Logarithms are simply a *new language for old ideas*—essentially a special case of exponents.

DEFINITION OF COMMON (BASE 10) LOGARITHMS

$y = \log x$ means $10^y = x.$

"Log x," which is read "the logarithm of x," is the answer to the question

To what exponent must 10 be raised to produce x?

| 1 | Find each common logarithm.

(a) log 100

(b) log 1000

(c) log .1

Answers:

(a) 2

(b) 3

(c) −1

EXAMPLE 1   To find log 10,000 ask yourself, "To what exponent must 10 be raised to produce 10,000?" Since $10^4 = 10{,}000$, we see that log 10,000 $= 4$. Similarly,

$$\log 1 = 0 \quad \text{because} \quad 10^0 = 1;$$

$$\log .01 = -2 \quad \text{because} \quad 10^{-2} = \frac{1}{10^2} = \frac{1}{100} = .01;$$

$$\log \sqrt{10} = 1/2 \quad \text{because} \quad 10^{1/2} = \sqrt{10}. \quad \blacksquare \quad \boxed{1}$$

EXAMPLE 2   Log(-25) is the exponent to which 10 must be raised to produce -25. But every power of 10 is positive! So there is no exponent that will produce -25. *Logarithms of negative numbers and 0 are not defined.* $\blacksquare$

| 2 | Find each common logarithm.

(a) log 27

(b) log 1089

(c) log .00426

Answers:

(a) 1.4314

(b) 3.0370

(c) −2.3706

EXAMPLE 3   **(a)** We know that log 359 must be a number between 2 and 3 because $10^2 = 100$ and $10^3 = 1000$. By using the log key, we find that log 359 (to four decimal places) is 2.5551. You can verify this by computing $10^{2.5551}$; the result (rounded) is 359.

(b) When 10 is raised to a negative exponent, the result is a number less than 1. Consequently, the logarithms of numbers between 0 and 1 are negative. For instance, log .026 $= -1.5850$. $\blacksquare$ $\boxed{2}$

Although common logarithms still have some uses (one of which is discussed in Section 5.4), the most widely used logarithms today are defined in terms of the number e (whose decimal expansion begins 2.71828 . . .) rather than 10. They have a special name and notation.

DEFINITION OF NATURAL (BASE e) LOGARITHMS

$$y = \ln x \quad \text{means} \quad e^y = x.$$

Thus the number **ln x** (which is sometimes read "el-en x") is the exponent to which e must be raised to produce the number x. For instance, ln 1 $= 0$ because $e^0 = 1$. Although logarithms to base e may not seem as "natural" as common

3 Find the following.

(a) ln 6.1

(b) ln 20

(c) ln .8

(d) ln .1

Answers:

(a) 1.8083

(b) 2.9957

(c) −.2231

(d) −2.3026

logarithms, there are several reasons for using them, some of which are discussed in Section 5.4.

EXAMPLE 4 **(a)** To find ln 85 use the $\boxed{\text{LN}}$ key of your calculator. The result is 4.4427. Thus, 4.4427 is the exponent (to four decimal places) to which e must be raised to produce 85. You can verify this by computing $e^{4.4427}$; the answer (rounded) is 85.

(b) A calculator shows that ln .38 = −.9676 (rounded), which means that $e^{-.9676} \approx .38$. ■ **3**

EXAMPLE 5 You don't need a calculator to find ln e^8. Just ask yourself, "To what exponent must e be raised to produce e^8?" The answer, obviously, is 8. So ln $e^8 = 8$. ■

Example 5 is an illustration of the following fact.

$$\ln e^k = k \text{ for every real number } k.$$

The procedure used to define common and natural logarithms can be carried out with any positive number $a \neq 1$ as the base (in place of 10 or e).

DEFINITION OF LOGARITHMS TO THE BASE a

$$y = \log_a x \quad \text{means} \quad a^y = x.$$

Read $y = \log_a x$ as "y is the logarithm of x to the base a." For example, the exponential statement $2^4 = 16$ can be translated into the equivalent logarithmic statement $4 = \log_2 16$. Thus, $\log_a x$ is an *exponent;* it is the answer to the question

To what power must a be raised to produce x?

This key definition should be memorized. It is important to remember the location of the base and exponent in each part of the definition.

Exponent
↓
Logarithmic form: $y = \log_a x$
↑
Base

Exponent
↓
Exponential form: $a^y = x$
↑
Base

4 Write the logarithmic form of
(a) $5^3 = 125$;
(b) $3^{-4} = 1/81$;
(c) $8^{2/3} = 4$.

Answers:
(a) $\log_5 125 = 3$
(b) $\log_3(1/81) = -4$
(c) $\log_8 4 = 2/3$

5 Write the exponential form of
(a) $\log_{16} 4 = 1/2$;
(b) $\log_3(1/9) = -2$;
(c) $\log_{16} 8 = 3/4$.

Answers:
(a) $16^{1/2} = 4$
(b) $3^{-2} = 1/9$
(c) $16^{3/4} = 8$

Common and natural logarithms are the special cases when $a = 10$ and when $a = e$ respectively. Both $\log u$ and $\log_{10} u$ mean the same thing. Similarly, $\ln u$ and $\log_e u$ mean the same thing.

EXAMPLE 6 This example shows several statements written in both exponential and logarithmic forms.

Exponential Form	Logarithmic Form
(a) $3^2 = 9$	$\log_3 9 = 2$
(b) $(1/5)^{-2} = 25$	$\log_{1/5} 25 = -2$
(c) $10^5 = 100,000$	$\log_{10} 100,000$ (or $\log 100,000) = 5$
(d) $4^{-3} = 1/64$	$\log_4(1/64) = -3$
(e) $2^{-4} = 1/16$	$\log_2(1/16) = -4$
(f) $e^0 = 1$	$\log_e 1$ (or $\ln 1) = 0$

4 **5**

PROPERTIES OF LOGARITHMS The usefulness of logarithmic functions depends in large part on the following *properties of logarithms*. These properties will be needed to solve exponential and logarithmic equations in the next section.

PROPERTIES OF LOGARITHMS

Let x and y be any positive real numbers and r be any real number. Let a be a positive real number, $a \neq 1$. Then

(a) $\log_a xy = \log_a x + \log_a y$;   **(b)** $\log_a \dfrac{x}{y} = \log_a x - \log_a y$;

(c) $\log_a x^r = r \log_a x$;   **(d)** $\log_a a = 1$;

(e) $\log_a 1 = 0$;   **(f)** $\log_a a^r = r$;

(g) $a^{\log_a x} = x$.

N O T E Because these properties are so useful, they should be memorized. ◆

To prove property (a), let

$$m = \log_a x \quad \text{and} \quad n = \log_a y.$$

Then, by the definition of logarithm,

$$a^m = x \quad \text{and} \quad a^n = y.$$

Multiply to get

$$a^m \cdot a^n = x \cdot y,$$

or, by a property of exponents,

$$a^{m+n} = xy.$$

Use the definition of logarithm to rewrite this last statement as

$$\log_a xy = m + n.$$

Replace m with $\log_a x$ and n with $\log_a y$ to get

$$\log_a xy = \log_a x + \log_a y.$$

Properties (b) and (c) can be proven in a similar way. Since $a^1 = a$ and $a^0 = 1$, properties (d) and (e) come from the definition of logarithm.

6 Rewrite, using the properties of logarithms.

(a) $\log_a 5x + \log_a 3x^4$

(b) $\log_a 3p - \log_a 5q$

(c) $4 \log_a k - 3 \log_a m$

Answers:

(a) $\log_a 15x^5$

(b) $\log_a(3p/5q)$

(c) $\log_a(k^4/m^3)$

EXAMPLE 7   If all the following variable expressions represent positive numbers, then for $a > 0$, $a \neq 1$,

(a) $\log_a x + \log_a(x - 1) = \log_a x(x - 1)$;

(b) $\log_a \dfrac{x^2 + 4}{x + 6} = \log_a(x^2 + 4) - \log_a(x + 6)$;

(c) $\log_a 9x^5 = \log_a 9 + \log_a x^5 = \log_a 9 + 5 \log_a x.$ ■ 6

CAUTION   There is no logarithm property that allows you to simplify the logarithm of a sum, such as $\log_a(x^2 + 4)$. In particular, $\log_a(x^2 + 4)$ is *not* equal to $\log_a x^2 + \log_a 4$. Property (a) of logarithms in the box above shows that $\log_a x^2 + \log_a 4 = \log_a 4x^2.$ ◆

EXAMPLE 8   Using the properties of logarithms, if $\log_6 7 \approx 1.09$ and $\log_6 5 \approx .90$,

(a) $\log_6 35 = \log_6(7 \cdot 5) = \log_6 7 + \log_6 5 \approx 1.09 + .90 = 1.99$;

(b) $\log_6 5/7 = \log_6 5 - \log_6 7 \approx .90 - 1.09 = -.19$;

(c) $\log_6 5^3 = 3 \log_6 5 \approx 3(.90) = 2.70$;

7 Use the properties of logarithms to rewrite and evaluate each of the following, given $\log_3 7 \approx 1.77$ and $\log_3 5 \approx 1.46$.

(a) $\log_3 35$

(b) $\log_3 7/5$

(c) $\log_3 25$

(d) $\log_3 3$

(e) $\log_3 1$

Answers:
(a) 3.23
(b) .31
(c) 2.92
(d) 1
(e) 0

(d) $\log_6 6 = 1$;

(e) $\log_6 1 = 0$. ■ **7**

In Example 8 several logarithms to base 6 were given. However, they could have been found by using a calculator and the following formula.

CHANGE OF BASE THEOREM

For any positive numbers a and x (with $a \neq 1$),

$$\log_a x = \frac{\ln x}{\ln a}.$$

EXAMPLE 9   To find $\log_7 3$, use the theorem with $a = 7$ and $x = 3$:

$$\log_7 3 = \frac{\ln 3}{\ln 7} \approx \frac{1.0986}{1.9459} \approx .5646.$$

You can check this on your calculator by verifying that $7^{.5646} \approx 3$. ■

LOGARITHMIC EQUATIONS   Equations involving logarithms are often solved by using the fact that a logarithmic equation can be rewritten (with the definition of logarithm) as an exponential equation. In other cases, the properties of logarithms may be useful in simplifying an equation involving logarithms.

EXAMPLE 10   Solve each of the following equations.

(a) $\log_x 8/27 = 3$

First, use the definition of logarithm and write the equation in exponential form.

$$x^3 = \frac{8}{27} \qquad \text{Definition of logarithm}$$

$$x^3 = \left(\frac{2}{3}\right)^3 \qquad \text{Write } \frac{8}{27} \text{ as a cube.}$$

$$x = \frac{2}{3} \qquad \text{Set the bases equal.}$$

The solution is 2/3.

8 Solve each equation.

(a) $\log_x 6 = 1$

(b) $\log_5 25 = m$

(c) $\log_{27} x = 2/3$

Answers:

(a) 6

(b) 2

(c) 9

(b) $\log_4 x = 5/2$

In exponential form, the given statement becomes

$$4^{5/2} = x \qquad \text{Definition of logarithm}$$
$$(4^{1/2})^5 = x \qquad \text{Definition of rational exponent}$$
$$2^5 = x$$
$$32 = x.$$

The solution is 32. ∎ **8**

In the next example, properties of logarithms are needed to solve equations.

EXAMPLE 11 Solve each equation.

(a) $\log_2 x - \log_2(x - 1) = 1$

By a property of logarithms, the left-hand side can be simplified as

$$\log_2 x - \log_2(x - 1) = \log_2 \frac{x}{x - 1}.$$

The original equation then becomes

$$\log_2 \frac{x}{x - 1} = 1.$$

Use the definition of logarithm to write this last result in exponential form.

$$\frac{x}{x - 1} = 2^1 = 2$$

Solve this equation.

$$\frac{x}{x - 1} \cdot (x - 1) = 2(x - 1)$$
$$x = 2(x - 1)$$
$$x = 2x - 2$$
$$-x = -2$$
$$x = 2$$

The domain of logarithmic functions includes only positive real numbers, so it is *necessary* to check this proposed solution in the original equation.

$$\log_2 x - \log_2(x - 1) \overset{?}{=} 1$$
$$\log_2 2 - \log_2(2 - 1) \overset{?}{=} 1 \qquad \text{Let } x = 2.$$
$$1 - 0 = 1 \qquad \text{Definition of logarithm}$$

The solution, 2, checks.

(b) $\log x - \log 8 = 1$

Use a property of logarithms to combine the terms on the left side.

$$\log x - \log 8 = \log \frac{x}{8}$$

The equation becomes

$$\log \frac{x}{8} = 1.$$

Recall $\log x$ means $\log_{10} x$, so

$$\log \frac{x}{8} = \log_{10} \frac{x}{8} = 1$$

$$\frac{x}{8} = 10^1 = 10 \qquad \text{Definition of logarithm}$$

$$x = 80.$$

Since $x = 80$ is in the domain of $\log x$, the solution is 80.

(c) Solve $\log x + \log(x + 1) = \log 2$.

First use a property of logarithms to rewrite the left side of the equation.

$$\log x(x + 1) = \log 2$$

$$\log x(x + 1) - \log 2 = 0$$

$$\log \frac{x(x + 1)}{2} = 0$$

$$\frac{x(x + 1)}{2} = 10^0 \qquad \text{Write in exponential form.}$$

$$x^2 + x = 2(1) \qquad 10^0 = 1$$

$$x^2 + x - 2 = 0$$

$$(x + 2)(x - 1) = 0$$

$$x = -2 \quad \text{or} \quad x = 1$$

Since $x = -2$ makes $\log x$ undefined (as well as $\log(x + 1)$), the only solution is $x = 1$. Verify that $x = 1$ satisfies the given equation. ■ 9

9 Solve each equation.

(a) $\log_5 x + 2 \log_5 x = 3$

(b) $\log_6(a + 2)$
$- \log_6 \dfrac{a - 7}{5} = 1$

Answers:
(a) 5
(b) 52

LOGARITHMIC FUNCTIONS For a given *positive* value of x, the definition of logarithm leads to exactly one value of y, so that $y = \log_a x$ defines a logarithmic function of base a. (The base a must be positive, with $a \neq 1$.)

If $a > 0$ and $a \neq 1$, the **logarithmic function** with base a is defined as

$$f(x) = \log_a x.$$

The most important logarithmic function is the natural logarithmic function.

EXAMPLE 12   Graph $f(x) = \ln x$ and $g(x) = e^x$ on the same axes.

For each function, use a calculator to compute some ordered pairs. Then plot the corresponding points and connect them with a curve to obtain the graphs in Figure 5.11.

10 Graph $f(x) = \log x$ and $g(x) = 10^x$ on the same axes.

Answer:

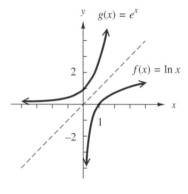

FIGURE 5.11

In Figure 5.11, the dashed straight line is the graph of $y = x$. Observe that the graph of $f(x) = \ln x$ is the mirror image of the graph of $g(x) = e^x$, with the line $y = x$ being the mirror.   ■   **10**

When the base $a > 1$, the graph of $f(x) = \log_a x$ has the same basic shape as the graph of the natural logarithmic function in Figure 5.11, as summarized below.

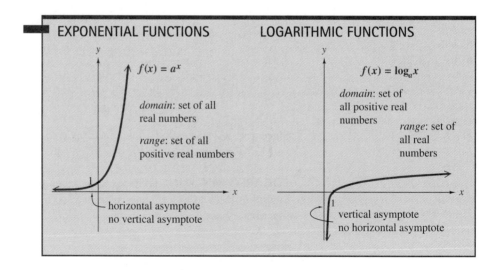

Finally, the graph of $f(x) = \log_a x$ is the mirror image of the graph of $g(x) = a^x$, with the line $y = x$ being the mirror. Functions whose graphs are related in this way are said to be **inverses** of each other. A more complete discussion of inverse functions is given in most standard college algebra books.

5.3 EXERCISES

Complete each statement in Exercises 1–4.

1. $y = \log_a x$ means $x =$ _____.

2. The statement $\log_5 125 = 3$ tells us that _____ is the power of _____ that equals _____.

3. What is wrong with the expression $y = \log_b$?

4. Logarithms of negative numbers are not defined because _____.

Translate each logarithmic statement into an equivalent exponential statement. (See Examples 1, 5, and 6.)

5. $\log 100{,}000 = 5$

6. $\log .001 = -3$

7. $\log_3 81 = 4$

8. $\log_2(1/4) = -2$

Translate each exponential statement into an equivalent logarithmic statement. (See Examples 5–6.)

9. $10^{1.8751} = 75$

10. $e^{3.2189} = 25$

11. $3^{-2} = 1/9$

12. $16^{1/2} = 4$

Without using a calculator, evaluate each of the following. (See Examples 1, 5, and 6.)

13. $\log 1000$

14. $\log .0001$

15. $\log_5 25$

16. $\log_9 81$

17. $\log_4 64$

18. $\log_6 216$

19. $\log_2 \dfrac{1}{4}$

20. $\log_3 \dfrac{1}{27}$

21. $\ln \sqrt{e}$

22. $\ln(1/e)$

23. $\ln e^{3.78}$

24. $\log 10^{56.9}$

Use a calculator to evaluate each logarithm to three decimal places. (See Examples 3 and 4.)

25. $\log 47$

26. $\log .004$

27. $\ln .351$

28. $\ln 2160$

29. Why does $\log_a 1$ always equal 0 for any valid base a?

Write each expression as the logarithm of a single number or expression. Assume all variables represent positive numbers. (See Example 7.)

30. $\log 15 - \log 3$

31. $\log 4 + \log 8 - \log 2$

32. $3 \ln 2 + 2 \ln 3$

33. $2 \ln 5 - \dfrac{1}{2} \ln 25$

34. $3 \log x - 2 \log y$

35. $2 \log u + 3 \log w - 6 \log v$

36. $\ln(3x + 2) + \ln(x + 4)$

37. $2 \ln(x + 1) - \ln(x + 2)$

Write each expression as a sum and/or a difference of logarithms with all variables to the first degree.

38. $\log 5x^2 y^3$

39. $\ln \sqrt{6m^4 n^2}$

40. $\ln \dfrac{3x}{5y}$

41. $\log \dfrac{\sqrt{xz}}{z^3}$

42. The calculator-generated table in the figure is for $y_1 = \log(4 - x)$. Why do the values in the y_1 column show ERROR for $x \geq 4$?

Express each of the following in terms of u and v, where $u = \ln x$ and $v = \ln y$. For example, $\ln x^3 = 3(\ln x) = 3u$.

43. $\ln(x^2 y^5)$

44. $\ln(\sqrt{x} \cdot y^2)$

45. $\ln(x^3/y^2)$

46. $\ln(\sqrt{x}/y)$

Evaluate each expression. (See Example 9.)

47. $\log_6 543$

48. $\log_{20} 97$

49. $\log_{35} 6874$

50. $\log_5 50 - \log_{50} 5$

Find numerical values for b and c for which the given statement is false.

51. $\log(b + c) = \log b + \log c$

52. $\dfrac{\ln b}{\ln c} = \ln\left(\dfrac{b}{c}\right)$

Solve each equation. (See Examples 10 and 11.)

53. $\log_x 25 = -2$

54. $\log_x \dfrac{1}{16} = -2$

55. $\log_9 27 = m$

56. $\log_8 4 = z$

57. $\log_y 8 = \dfrac{3}{4}$

58. $\log_r 7 = \dfrac{1}{2}$

59. $\log_3(5x + 1) = 2$

60. $\log_5(9x - 4) = 1$

61. $\log x - \log(x + 3) = -1$

62. $\log m - \log(m - 4) = -2$

63. $\log_3(y + 2) = \log_3(y - 7) + \log_3 4$

64. $\log_8(z - 6) = 2 - \log_8(z + 15)$

65. $\ln(x + 9) - \ln x = 1$

66. $\ln(2x + 1) - 1 = \ln(x - 2)$

67. $\log x + \log(x - 3) = 1$

68. $\log(x - 1) + \log(x + 2) = 1$

69. Suppose you overhear the following statement: "I must reject any negative answer when I solve an equation involving logarithms." Is this correct? Write an explanation of why it is or is not correct.

70. What values of x could not possibly be solutions of the following equation?

$$\log_a(4x - 7) + \log_a(x^2 + 4) = 0$$

Graph each of the following. (See Example 12.)

71. $y = \ln(x + 2)$

72. $y = \ln x + 2$

73. $y = \log(x - 3)$

74. $y = \log x - 3$

75. Graph $f(x) = \log x$ and $g(x) = \log(x/4)$ for $-2 \le x \le 8$. How are these graphs related? How does the quotient rule support your answer?

In Exercises 76 and 77 the coordinates of a point on the graph of the indicated function are displayed at the bottom of the screen. Write the logarithmic and exponential equations associated with the display.

76.

77.

78. Match each equation with its graph. Each tick mark represents one unit.
   (a) $y = \log x$
   (b) $y = 10^x$
   (c) $y = \ln x$
   (d) $y = e^x$

   A.   **B.**

   C.   **D.**

79. Management The doubling function

$$D(r) = \dfrac{\ln 2}{\ln(1 + r)}$$

gives the number of years required to double your money when it is invested at interest rate r (expressed as a decimal), compounded annually. How long does it take to double your money at each of the following rates?

(a) 4%
(b) 8%
(c) 18%
(d) 36%
(e) Round each of your answers in parts (a)–(d) to the nearest year and compare them with these numbers: 72/4, 72/8, 72/18, 72/36. Use this evidence to state a "rule of thumb" for determining approximate doubling time without using the function D. This rule, which has long been used by bankers, is called the *rule of 72*.

80. **Management** Suppose the sales of a certain product are approximated by

$$S(t) = 125 + 83 \log(5t + 1),$$

where $S(t)$ is sales in thousands of dollars t years after the product was introduced on the market. Find
(a) $S(0)$;
(b) $S(2)$;
(c) $S(4)$;
(d) $S(31)$.
(e) Graph $y = S(t)$.
(f) Does the graph approach a horizontal asymptote? Explain.

81. **Natural Science** Two people with the flu visited the campus of Big State U. The number of days T that it took for the flu virus to infect n people is given by

$$T = -1.43 \ln\!\left(\frac{10,000 - n}{4998n}\right).$$

How many days will it take for the virus to infect
(a) 500 people?
(b) 5000 people?

82. **Natural Science** (a) Use a graphing calculator to graph the "flu function" in Exercise 81 in a viewing window with $0 \le n \le 11,000$.
(b) How many people are infected on the 16th day?
(c) Explain why this model is not realistic when large numbers of people are involved.

83. **Management** The graph shows the percent increase in commercial rents in California from 1992 to 1999 (1997–1999 figures are estimates). In the two-year period 1990–1992 commercial rents decreased by about 6%. They started to increase in 1992 as the state finally began an economic recovery.

(a) Describe the growth in rents during the period shown in the graph.
(b) The graph can be approximated by the equation $f(x) = -650 + 143 \ln x$, where x is the number of years since 1900 and y is the corresponding percent change in rents. Find $f(92)$ and $f(99)$. Compare your results with the corresponding y-values from the graph.

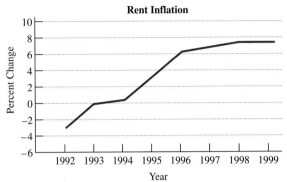

Rent Inflation

Source: CB Commercial/Torto Wheaton Research.

84. **Social Science** The data in the table gives the number of visitors to U.S. national parks from 1950 to 1994 (in millions).*

1950	1960	1970	1980	1990	1993
14	28	46	47	57	60

Suppose x represents the number of years since 1900—thus, 1950 is represented by 50, 1960 by 60, and so on. The logarithmic function defined by $f(x) = -266 + 72 \ln x$ closely approximates the data. Use this function to estimate the number of visitors in the year 2000. What assumption must we make to estimate the number of visitors beyond 1993?

Statistical Abstract of the United States 1995.

5.4 APPLICATIONS OF LOGARITHMIC FUNCTIONS

We begin this section by introducing and illustrating a powerful method of solving exponential and logarithmic equations. We then show some applications that require this method of solution. The method depends on the following fact.

Let x and y be positive numbers. Let a be a positive number, $a \neq 1$.

If $x = y$, then $\log_a x = \log_a y$.
If $\log_a x = \log_a y$, then $x = y$.

For convenience, base e is used in most of these examples.

EXAMPLE 1 Solve $3^x = 5$.

Since 3 and 5 cannot easily be written with the same base, the methods of Section 5.1 cannot be used to solve this equation. Instead, use the result given above and take natural logarithms of both sides.

$$3^x = 5$$
$$\ln 3^x = \ln 5$$
$$x \ln 3 = \ln 5 \qquad \text{Property (c) of logarithms}$$
$$x = \frac{\ln 5}{\ln 3} \approx 1.465$$

To check, evaluate $3^{1.465}$; the answer should be approximately 5, which verifies that the solution of the given equation is 1.465 to the nearest thousandth. ■ ☐ 1

CAUTION Be careful; $\dfrac{\ln 5}{\ln 3}$ is *not* equal to $\ln\left(\dfrac{5}{3}\right)$ or $\ln 5 - \ln 3$. (However, by the change of base theorem, $\dfrac{\ln 5}{\ln 3}$ can be written as $\log_5 3$.) ◆

EXAMPLE 2 Solve $3^{2x-1} = 4^{x+2}$.

Taking natural logarithms on both sides gives

$$\ln 3^{2x-1} = \ln 4^{x+2}.$$

Now use property (c) of logarithms and the fact that ln 3 and ln 4 are constants to rewrite the equation.

$(2x - 1)(\ln 3) = (x + 2)(\ln 4)$	Property (c)
$2x(\ln 3) - 1(\ln 3) = x(\ln 4) + 2(\ln 4)$	Distributive property
$2x(\ln 3) - x(\ln 4) = 2(\ln 4) + 1(\ln 3)$	Collect terms with x on one side.

Factor out x on the left side to get

$$[2(\ln 3) - \ln 4]x = 2(\ln 4) + \ln 3.$$

☐ 1 Solve each equation. Round solutions to the nearest thousandth.

(a) $2^x = 7$

(b) $5^m = 50$

(c) $3^y = 17$

Answers:
(a) 2.807

(b) 2.431

(c) 2.579

2 Solve each equation. Round solutions to the nearest thousandth.

(a) $6^m = 3^{2m-1}$

(b) $5^{6a-3} = 2^{4a+1}$

Answers:
(a) 2.710
(b) .802

Divide both sides by the coefficient of x:

$$x = \frac{2(\ln 4) + \ln 3}{2(\ln 3) - \ln 4}.$$

Using a calculator to evaluate this last expression, we find that

$$x = \frac{2\ln 4 + \ln 3}{2\ln 3 - \ln 4} \approx 4.774. \quad \blacksquare \quad \boxed{2}$$

Recall that $\ln e = 1$ (because 1 is the exponent to which e must be raised to produce e). This fact simplifies the solution of equations involving powers of e.

EXAMPLE 3 Solve $3e^{x^2} = 600$.
First divide each side by 3 to get

$$e^{x^2} = 200.$$

Now take natural logarithms on both sides; then use properties of logarithms.

3 Solve each equation. Round solutions to the nearest thousandth.

(a) $e^{.1x} = 11$

(b) $e^{3+x} = .893$

(c) $e^{2x^2-3} = 9$

Answers:
(a) 23.979
(b) −3.113
(c) ±1.612

$$
\begin{aligned}
e^{x^2} &= 200 \\
\ln e^{x^2} &= \ln 200 \\
x^2 \ln e &= \ln 200 &&\text{Property (c)} \\
x^2 &= \ln 200 &&\ln e = 1 \\
x &= \pm\sqrt{\ln 200} \\
x &\approx \pm 2.302
\end{aligned}
$$

The solutions are ± 2.302, rounding to the nearest thousandth. (The symbol $\pm$ is used as a shortcut for writing the two solutions, 2.302 and −2.302.) $\blacksquare$ $\boxed{3}$

The fact given at the beginning of this section, along with the properties of logarithms from Section 5.3, is useful in solving logarithmic equations, as shown in the next examples.

EXAMPLE 4 Solve $\log(x + 4) - \log(x + 2) = \log x$.
Using property (b) of logarithms, rewrite the equation as

$$\log \frac{x + 4}{x + 2} = \log x.$$

Then

$$\frac{x + 4}{x + 2} = x$$

$$x + 4 = x(x + 2)$$
$$x + 4 = x^2 + 2x$$
$$x^2 + x - 4 = 0.$$

By the quadratic formula,

$$x = \frac{-1 \pm \sqrt{1 + 16}}{2},$$

so that

$$x = \frac{-1 + \sqrt{17}}{2} \quad \text{or} \quad x = \frac{-1 - \sqrt{17}}{2}.$$

Log x cannot be evaluated for $x = (-1 - \sqrt{17})/2$, because this number is negative and not in the domain of log x. By substitution, verify that $(-1 + \sqrt{17})/2$ is a solution. ∎ 4

4 Solve each equation.

(a) $\log_2(p + 9) - \log_2 p$
$= \log_2(p + 1)$

(b) $\log_3(m + 1)$
$- \log_3(m - 1) = \log_3 m$

Answers:

(a) 3

(b) $1 + \sqrt{2} \approx 2.414$

A radioactive substance decays according to a function of the form $y = y_0 e^{-kt}$, where y_0 is the amount present initially (at time $t = 0$) and k is a positive number. The **half-life** of a radioactive substance is the time it takes for exactly half of the substance to decay. We find the half-life by finding the value of t such that $y = (1/2)y_0$. Similarly, we can find the time for any proportion of y_0 to remain.

EXAMPLE 5   Carbon 14, also known as radiocarbon, is a radioactive form of carbon that is found in all living plants and animals. After a plant or animal dies, the radiocarbon disintegrates with a half-life of approximately 5600 years. Scientists can determine the age of the remains by comparing the amount of radiocarbon with the amounts present in living plants and animals. This technique is called *carbon dating*. The amount of radiocarbon present after t years is given by

$$y = y_0 e^{-(\ln 2)(1/5600)t},$$

where y_0 is the amount present in living plants and animals.

A round table hanging in Winchester Castle (England) was alleged to belong to King Arthur, who lived in the 5th century. A recent chemical analysis showed that the table had 91% of the amount of radiocarbon present in living wood. How old is the table?

The amount of radiocarbon present in the round table after y years is $.91y_0$. Therefore, in the equation

$$y = y_0 e^{-(\ln 2)(1/5600)t}$$

replace y with $.91y_0$ and solve for t.

$$.91y_0 = y_0 e^{-(\ln 2)(1/5600)t}$$
$$.91 = e^{-(\ln 2)(1/5600)t} \qquad \text{Divide both sides by } y_0.$$
$$\ln .91 = \ln e^{-(\ln 2)(1/5600)t} \qquad \text{Take logarithms on both sides.}$$
$$\ln .91 = -(\ln 2)(1/5600)t \qquad \text{Property (c) of logarithms and } \ln e = 1$$
$$t = \frac{(5600)\ln .91}{-\ln 2} \approx 761.94$$

5 What is the age of a specimen in which $y = (1/3)y_0$?

Answer:
About 8880 years

The table is about 762 years old and therefore could not have belonged to King Arthur.   ■   **5**

Our next example uses common logarithms (base 10).

EXAMPLE 6   The intensity $R(i)$ of an earthquake, measured on the **Richter scale,** is given by

$$R(i) = \log\left(\frac{i}{i_0}\right),$$

where i is the intensity of the ground motion of the earthquake and i_0 is the intensity of the ground motion of the so-called *zero earthquake* (the smallest detectable earthquake, against which others are measured). The 1989 San Francisco earthquake measured 7.1 on the Richter scale.

(a) How did the ground motion of this earthquake compare with that of the zero earthquake?
In this case $R(i) = 7.1$, that is, $\log(i/i_0) = 7.1$. Thus 7.1 is the exponent to which 10 must be raised to produce i/i_0. In other words,

$$10^{7.1} = \frac{i}{i_0}, \quad \text{or equivalently,} \quad i = 10^{7.1}i_0.$$

So this earthquake had $10^{7.1}$ (approximately 12.6 million) times more ground motion than the zero earthquake.

(b) What is the Richter scale intensity of an earthquake with 10 times as much ground motion as the 1989 San Francisco earthquake?
Using the result from (a), the ground motion of such a quake would be

$$i = 10(10^{7.1}i_0) = 10^1 \cdot 10^{7.1}i_0 = 10^{8.1}i_0$$

so that its Richter scale intensity would be

$$R(i) = \log\left(\frac{i}{i_0}\right) = \log\left(\frac{10^{8.1}i_0}{i_0}\right) = \log 10^{8.1} = 8.1.$$

6 Find the Richter scale intensity of an earthquake whose ground motion is 100 times greater than the ground motion of the 1989 San Francisco earthquake discussed in Example 6.

Answer:
9.1

Therefore, a ten-fold increase in ground motion increases the Richter scale intensity by just 1. ■ 6

Earlier, we saw how rational functions describe cost-benefit functions. The final example illustrates another type of cost-benefit function.

EXAMPLE 7 One action that government could take to reduce carbon emissions into the atmosphere is to place a tax on fossil fuel. This tax would be based on the amount of carbon dioxide that is emitted into the air when the fuel is burned. The *cost-benefit* equation $\ln(1 - P) = -.0034 - .0053T$ describes the approximate relationship between a tax of T dollars per ton of carbon dioxide and the corresponding percent reduction P (in decimals) of emission of carbon dioxide.*

(a) Write P as a function of T.

We begin by writing the cost-benefit equation in exponential form.

$$\ln(1 - P) = -.0034 - .0053T$$
$$1 - P = e^{-.0034 - .0053T}$$
$$P = P(T) = 1 - e^{-.0034 - .0053T}$$

A calculator-generated graph of $P(T)$ is shown in Figure 5.12.

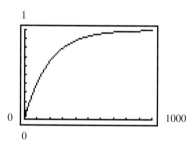

FIGURE 5.12

(b) Discuss the benefit of continuing to raise taxes on carbon dioxide emissions.

From the graph we see that initially there is a rapid reduction of carbon dioxide emissions. However, after a while there is little benefit in raising taxes further. ■

The most important applications of exponential and logarithmic functions for the fields of management and economics are considered in Chapter 6 (Mathematics of Finance).

---

*Nordhause, W., "To Slow or Not to Slow: The Economics of the Greenhouse Effect." Yale University, New Haven, Connecticut.

5.4 EXERCISES

Solve each exponential equation. Round to the nearest thousandth. (See Examples 1–3.)

1. $3^x = 5$

2. $5^x = 4$

3. $2^x = 3^{x-1}$

4. $4^{x+2} = 2^{x-1}$

5. $3^{1-2x} = 5^{x+5}$

6. $4^{3x-1} = 3^{x-2}$

7. $e^{2x} = 5$

8. $e^{-3x} = 2$

9. $2e^{5a+2} = 8$

10. $10e^{3z-7} = 5$

11. $2^{x^2-1} = 12$

12. $3^{2-x^2} = 4$

13. $2(e^x + 1) = 10$

14. $5(e^{2x} - 2) = 15$

Solve each equation for c.

15. $10^{4c-3} = d$

16. $3 \cdot 10^{2c+1} = 4d$

17. $e^{2c-1} = b$

18. $3e^{5c-7} = b$

Solve each logarithmic equation. (See Example 4.)

19. $\ln(3x - 1) - \ln(2 + x) = \ln 2$

20. $\ln(8k - 7) - \ln(3 + 4k) = \ln(9/11)$

21. $\ln x + 1 = \ln(x - 4)$

22. $\ln(4x - 2) = \ln 4 - \ln(x - 2)$

23. $2 \ln(x - 3) = \ln(x + 5) + \ln 4$

24. $\ln(k + 5) + \ln(k + 2) = \ln 14k$

25. $\log_5(r + 2) + \log_5(r - 2) = 1$

26. $\log_4(z + 3) + \log_4(z - 3) = 1$

27. $\log_3(a - 3) = 1 + \log_3(a + 1)$

28. $\log w + \log(3w - 13) = 1$

29. $\log_2 \sqrt{2y^2 - 1} = 1/2$

30. $\log_2(\log_2 x) = 1$

31. $\log z = \sqrt{\log z}$

32. $\log x^2 = (\log x)^2$

Solve each equation for c.

33. $\log(3 + b) = \log(4c - 1)$

34. $\ln(b + 7) = \ln(6c + 8)$

35. $2 - b = \log(6c + 5)$

36. $8b + 6 = \ln(2c) + \ln c$

37. Explain why the equation $3^x = -4$ has no solutions.

38. Explain why the equation $\log(-x) = -4$ does have a solution and find that solution.

Work these exercises. (See Example 5.)

39. Natural Science The amount of cobalt-60 (in grams) in a storage facility at time t is given by

$$C(t) = 25e^{-.14t},$$

where time is measured in years.
(a) How much cobalt-60 was present initially?
(b) What is the half-life of cobalt-60?

40. Natural Science An American Indian mummy was found recently. It had 73.6% of the amount of radiocarbon present in living beings. Approximately how long ago did this person die?

41. Natural Science How old is a piece of ivory that has lost 36% of its radiocarbon?

42. Natural Science A sample from a refuse deposit near the Strait of Magellan had 60% of the carbon 14 of a contemporary living sample. How old was the sample?

43. Natural Science A large cloud of radioactive debris from a nuclear explosion has floated over the Pacific Northwest, contaminating much of the hay supply. Consequently, farmers in the area are concerned that the cows who eat this hay will give contaminated milk. (The tolerance level for radioactive iodine in milk is 0.) The percent of the initial amount of radioactive iodine still present in the hay after t days is approximated by $P(t)$, which is given by

$$P(t) = 100e^{-.1t}.$$

(a) Some scientists feel that the hay is safe after the percent of radioactive iodine has declined to 10% of the original amount. Solve the equation $10 = 100e^{-.1t}$ to find the number of days before the hay may be used.
(b) Other scientists believe that the hay is not safe until the level of radioactive iodine has declined to only 1% of the original level. Find the number of days that this would take.

Natural Science *For Exercises 44–47, refer to Example 6.*

44. Find the Richter scale intensity of earthquakes whose ground motion is
(a) $1000i_0$
(b) $100,000i_0$
(c) $10,000,000i_0$.
(d) Fill the blank in this statement: Increasing the ground motion by a factor of 10^k increases the Richter intensity by _____ units.

45. The great San Francisco earthquake of 1906 measured 8.3 on the Richter scale. How much greater was the ground motion in 1906 than in the 1989 earthquake, which measured 7.1 on the Richter scale?

46. The loudness of sound is measured in units called decibels. The decibel rating of a sound is given by

$$D(i) = 10 \cdot \log\left(\frac{i}{i_0}\right),$$

where i is the intensity of the sound and i_0 is the minimum intensity detectable by the human ear (the so-called *threshold sound*). Find the decibel rating of each of the following sounds whose intensities are given. Round answers to the nearest whole number.

(a) Whisper, $115i_0$
(b) Busy street, $9,500,000i_0$
(c) Rock music, $895,000,000,000i_0$
(d) Jetliner at takeoff, $109,000,000,000,000i_0$

47. (a) How much more intense is a sound that measures 100 decibels than the threshold sound?
(b) How much more intense is a sound that measures 50 decibels than the threshold sound?
(c) How much more intense is a sound measuring 100 decibels than one measuring 50 decibels?

48. **Natural Science** Refer to Example 7.
(a) Determine the percent reduction in carbon dioxide when the tax is $60.
(b) What tax will cause a 50% reduction in carbon dioxide emissions?

Natural Science *To find the maximum permitted levels of certain pollutants in fresh water, the EPA has established the functions defined in Exercises 49–50, where M(h) is the maximum permitted level of pollutant for a water hardness of h milligrams per liter. Find M(h) in each case. (These results give the maximum permitted average concentration in micrograms per liter for a 24-hour period.)*

49. Copper: $M(h) = e^r$, where $r = .65 \ln h - 1.94$ and $h = 9.7$.

50. Lead: $M(h) = e^r$, where $r = 1.51 \ln h - 3.37$ and $h = 8.4$.

51. **Social Science** The number of years, $N(r)$, since two independently evolving languages split off from a common ancestral language is approximated by

$$N(r) = -5000 \ln r,$$

where r is the proportion of the words from the ancestral language that is common to both languages now. Find each of the following.
(a) $N(.9)$　　(b) $N(.5)$　　(c) $N(.3)$
(d) How many years have elapsed since the split if 70% of the words of the ancestral language are common to both languages today?

(e) If two languages split off from a common ancestral language about 1000 years ago, find r.

52. **Natural Science** In the central Sierra Nevada of California, the percent of moisture that falls as snow rather than rain is approximated reasonably well by

$$p = 86.3 \ln h - 680,$$

where p is the percent of moisture as snow at an altitude of h feet (with $3000 \le h < 8500$).
(a) Graph p.
(b) At what altitude is 50 percent of the moisture snow?

53. **Social Science** In the exercises for Section 5.3 we saw that the number of visitors (in millions) to U.S. National Parks from 1950 through 1993 can be approximated by the logarithmic function with $f(x) = -266 + 72 \ln x$.* Here, x represents the number of years since 1900. According to the function, in what year will the number of visitors reach 70 million?

54. **Natural Science** The growth of outpatient surgery as a percent of total surgeries at hospitals is approximated by $f(x) = -1317 + 304 \ln x$, where x represents the number of years since 1900.†
(a) What does this function predict for the percent of outpatient surgeries in 1998?
(b) When did outpatient surgeries reach 50%?

55. **Natural Science** A drug's effectiveness decreases over time. If each hour a drug is only 90% as effective as the previous hour, at some point the patient will not be receiving enough medication and must receive another dose. This situation can be modeled by an exponential function with $y = y_0(.90)^{t-1}$. In this equation, y_0 is the amount of the initial dose and y is the percent of medication still available t hours after the drug was administered. Suppose 200 mg of the drug is administered. How long will it take for this initial dose to reach the dangerously low level of 50 mg?

56. **Physical Science** The table gives some of the planets' average distances D from the sun and their period P of revolution around the sun in years. The distances have been normalized so that Earth is one unit from the sun. Thus, Jupiter's distance of 5.2 means that Jupiter's distance from the sun is 5.2 times farther than Earth's.‡

Planet	D	P
Earth	1	1
Jupiter	5.2	11.9
Saturn	9.54	29.5
Uranus	19.2	84.0

---

*Statistical Abstract of the United States 1994, p. 249.

†American Hospital Association, Chicago.

‡Ronan, C. The Natural History of the Universe, MacMillan Publishing Co., New York, 1991.

(a) Plot the points (D, P) for these planets. Would a straight line or an exponential curve fit these points best?

(b) Plot the points $(\ln D, \ln P)$ for these planets. Do these points appear to lie on a line?

(c) Determine a linear equation that approximates the data points with $x = \ln D$ and $y = \ln P$. Use the data points $(0, 0)$ and $(2.95, 4.43)$. Graph your line and the data on the same coordinate axes.

(d) Use the linear equation to predict the period of the planet Pluto if its distance is 39.5. Compare your answer to the true value of 248.5 years.

57. Social Science   The graph shows that the percent y of American children growing up without a father has increased rapidly since 1950.* If x represents the number of years since 1900, the function defined by

$$y = \frac{25}{1 + 1364.3e^{-x/9.316}}$$

models the data fairly well.

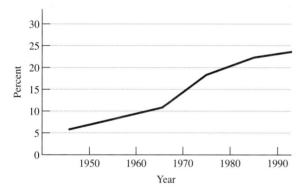

Percent vs. Year

*National Longitudinal Survey of Youth, U.S. Department of Commerce, Bureau of the Census.

(a) In what year were 20% of these children living without a father?

(b) If the percent continues to increase in the same way, in what year will 30% of American children live in a home without a father?

You will need a graphing calculator for Exercise 58.

58. Natural Science   Many environmental situations place effective limits on the size of a population in an area. As mentioned in Section 5.2, many such limited growth situations are described by the *logistic function,* more specifically defined as

$$G(t) = \frac{MG_0}{G_0 + (M - G_0)e^{-kMt}},$$

where G_0 is the initial number present, M is the maximum possible size of the population, and k is a positive constant. (The function defined in Exercise 57 is an example of a logistic function.) Assume $G_0 = 100$, $M = 2500$, $k = .0004$, and t is time in decades (10-year periods).

(a) Use a graphing calculator to graph the function using $0 \le t \le 8$, $0 \le y \le 2500$.

(b) Use the capability of your calculator to find $G(2)$. What does this number represent?

(c) Find the x-coordinate of the intersection of the curve with the horizontal line $y = 1000$. Interpret your answer.

(d) Use logarithms to solve the equation $G(t) = 1000$.

CHAPTER 5   SUMMARY

Key Terms and Symbols

5.1   exponential function
exponential growth and decay
exponential equation

5.2   the number $e \approx 2.71828\ldots$
learning curve
forgetting curve

5.3 $\log x$ common logarithm (base 10 logarithm) of x

$\ln x$ natural logarithm (base e logarithm) of x
$\log_a x$ base a logarithm of x
logarithmic equation
logarithmic function
inverses

5.4   half-life
Richter scale

Key Concepts

An important application of exponents is the **exponential growth function,** defined as $f(t) = y_0 e^{kt}$, where y_0 is the amount of a quantity present at time $t = 0$, $e \approx 2.71828$, and k is a constant.

The **logarithm** of x to the base a is defined as follows. For $a > 0$ and $a \neq 1$, $y = \log_a x$ means $a^y = x$. Thus, $\log_a x$ is an *exponent,* the power to which a must be raised to produce x.

Properties of Logarithms

Let x, y, and a be positive real numbers, $a \neq 1$, and let r be any real number.

$$\log_a xy = \log_a x + \log_a y \qquad \log_a \frac{x}{y} = \log_a x - \log_a y$$

$$\log_a x^r = r \log_a x \qquad\qquad \log_a 1 = 0$$

$$\log_a a = 1 \qquad\qquad\qquad a^{\log_a x} = x$$

$$\log_a a^r = r$$

Solving Exponential and Logarithmic Equations

Let $a > 0$, $a \neq 1$.

If $a^x = a^y$, then $x = y$ and if $x = y$, then $a^x = a^y$.

If $x = y$, then $\log_a x = \log_a y$, $x > 0$, $y > 0$.

If $\log_a x = \log_a y$, then $x = y$, $x > 0$, $y > 0$.

Chapter 5 Review Exercises

Match each equation with the letter of the graph that most closely resembles its graph. Assume that $a > 1$.

1. $y = a^{x+2}$

2. $y = a^x + 2$

3. $y = -a^x + 2$

4. $y = a^{-x} + 2$

(a)   (b)

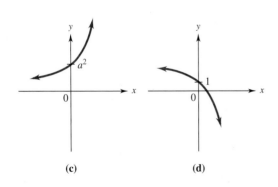

(c)   (d)

Consider the exponential function $y = f(x) = a^x$ graphed here. Answer each question based on the graph.

5. What is true about the value of a in comparison to 1?

6. What is the domain of f?

7. What is the range of f?

8. What is the value of $f(0)$?

Solve each equation.

9. $2^{3x} = \dfrac{1}{8}$

10. $\left(\dfrac{9}{16}\right)^x = \dfrac{3}{4}$

11. $9^{2y-1} = 27^y$

12. $\dfrac{1}{2} = \left(\dfrac{b}{4}\right)^{1/4}$

Graph each function.

13. $f(x) = 4^x$

14. $g(x) = 4^{-x}$

15. $f(x) = \ln x + 5$

16. $g(x) = \log x - 3$

Work these problems.

17. **Management**   A person learning certain skills involving repetition tends to learn quickly at first. Then learning tapers off and approaches some upper limit. Suppose the number of symbols per minute a textbook typesetter can produce is given by $p(t) = 250 - 120(2.8)^{-.5t}$, where t is the number of months the typesetter has been in training. Find each of the following:
    (a) $p(2)$
    (b) $p(4)$
    (c) $p(10)$.
    (d) Graph $y = p(t)$.

18. **Natural Science**   The growth of atmospheric carbon dioxide is modeled by the exponential function $y = 353e^{.00609t}$, where t is the number of years since 1990.* Use this model to estimate the year when the levels of carbon dioxide in the air will double the preindustrial level of 280 ppm, assuming no changes are made.

Translate each exponential statement into an equivalent logarithmic one.

19. $10^{1.6721} = 47$ 20. $5^4 = 625$

21. $e^{3.6636} = 39$ 22. $5^{1/2} = \sqrt{5}$

Translate each logarithmic statement into an equivalent exponential one.

23. $\log 1000 = 3$

24. $\log 16.6 = 1.2201$

25. $\ln 95.4 = 4.5581$

26. $\log_2 64 = 6$

Evaluate each expression without using a calculator.

27. $\ln e^3$ 28. $\log \sqrt{10}$

29. $10^{\log 7.4}$ 30. $\ln e^{4k}$

31. $\log_8 16$ 32. $\log_{25} 5$

Write each expression as a single logarithm. Assume all variables represent positive quantities.

33. $\log 4k + \log 5k^3$ 34. $4 \log x - 2 \log x^3$

35. $2 \log b - 3 \log c$ 36. $4 \ln x - 2(\ln x^3 + 4 \ln x)$

Solve each equation. Round to the nearest thousandth.

37. $8^p = 19$ 38. $3^z = 11$

39. $5 \cdot 2^{-m} = 35$ 40. $2 \cdot 15^{-k} = 18$

41. $e^{-5-2x} = 5$ 42. $e^{3x-1} = 12$

---

*International Panel on Climate Change (IPCC), 1990.

43. $6^{2-m} = 2^{3m+1}$

44. $5^{3r-1} = 6^{2r+5}$

45. $(1 + .003)^k = 1.089$

46. $(1 + .094)^z = 2.387$

47. $\log(m + 2) = 1$

48. $\log x^2 = 2$

49. $\log_2(3k - 2) = 4$

50. $\log_5\left(\dfrac{5z}{z - 2}\right) = 2$

51. $\log x + \log(x + 3) = 1$

52. $\log_2 r + \log_2(r - 2) = 3$

53. $\ln(m + 3) - \ln m = \ln 2$

54. $2 \ln(y + 1) = \ln(y^2 - 1) + \ln 5$

Work these problems.

55. **Management**   Suppose the gross national product (GNP) of a small country (in millions of dollars) is approximated by $G(t) = 15 + 2 \log t$ where t is time in years, for $1 \le t \le 6$. Find the GNP at the following times.
    (a) 1 year
    (b) 2 years
    (c) 5 years

56. **Natural Science**   A population is increasing according to the growth law $y = 2e^{.02t}$, where y is in millions and t is in years. Match each of the questions (a), (b), (c), and (d) with one of the solutions (A), (B), (C), or (D).
    (a) How long will it take for the population to triple?
    (b) When will the population reach 3 million?
    (c) How large will the population be in 3 years?
    (d) How large will the population be in 4 months?

    (A) Evaluate $2e^{.02(1/3)}$.
    (B) Solve $2e^{.02t} = 3 \cdot 2$ for t.
    (C) Evaluate $2e^{.02(3)}$.
    (D) Solve $2e^{.02t} = 3$ for t.

57. **Natural Science**   The amount of polonium (in grams) present after t days is given by

$$A(t) = 10e^{-.00495t}.$$

    (a) How much polonium was present initially?
    (b) What is the half-life of polonium?
    (c) How long will it take for the polonium to decay to 3 grams?

58. **Natural Science**   One earthquake measures 4.6 on the Richter scale. A second earthquake has ground motion 1000 times greater than the first. What does the second one measure on the Richter scale?

Natural Science *Refer to Newton's Law of Cooling, given in Section 5.2 Exercises 17 and 18:* $F(t) = T_0 + Ce^{-kt}$, *where C and k are constants.*

59. A piece of metal is heated to 300° Celsius and then placed in a cooling liquid at 50° Celsius. After 4 minutes the metal has cooled to 175° Celsius. Find its temperature after 12 minutes.

60. A frozen pizza has a temperature of 3.4° Celsius when it is taken from the freezer and left out in a room at 18° Celsius. After half an hour its temperature is 7.2° Celsius. How long will it take for the pizza to thaw to 10° Celsius?

61. Management India has become an important exporter of software to the United States. The chart shows India's software exports, y, (in millions of U.S. dollars) in the years since 1985. The figure for 1997 is an estimate.

Year (x)	1985	1987	1989	1991	1993	1995	1997
$ (y)	6	39	67	128	225	483	1000

Letting x represent the number of years since 1900, the function with

$$f(x) = 6.2(10)^{-12}(1.4)^x$$

approximates the data reasonably well. According to this function, when will software exports double their value in 1997?

62. Management Southwest Airlines has grown exponentially since it began service in 1971. It has been ranked as the nation's best overall carrier by more than one rating system. The number of Southwest Airlines passengers in millions, y, for selected years since 1987 are shown in the table, where x represents the number of years since 1900. The corresponding ordered pairs are plotted in the figure.

x	y
88	14.9
90	19.8
92	27.8
94	42.7
96	49.6

The function defined by $f(x) = 2.757 \cdot 10^{-5}e^{.150x}$, also shown in the figure, fits the data quite well.

(a) Use the function to estimate the number of Southwest passengers in the year 2000.

(b) According to the function, when will the number of passengers reach 100 million?

In Exercises 63 and 64, a graphing calculator will be helpful.

63. Physical Science The atmospheric pressure (in millibars) at a given altitude (in meters) is listed in the table.

Altitude	0	2000	4000	6000	8000	10,000
Pressure	1013	795	617	472	357	265

(a) Plot the data for atmospheric pressure P at altitude x.

(b) Would a linear or exponential function fit the data better?

(c) Graph the function defined by

$$P(x) = 1013(2.78)^{-.0001341x}$$

on the same axes as the data in part (a). Does this graph fit the data points?

(d) Use $P(x)$ to predict the pressure at 1500 m and 11,000 m and compare the results to the actual values of 846 millibars and 227 millibars.

64. Physical Science The power of personal computers has increased dramatically as a result of the ability to place an increasing number of transistors on a single processor chip. The table lists the number of transistors on some popular computer chips by Intel.*

Year	Chip	Transistors
1971	4004	2,300
1986	386DX	275,000
1989	486DX	1,200,000
1993	Pentium	3,300,000
1995	P6	5,500,000

(a) Let x be the year, where $x = 0$ corresponds to 1971, and y be the number of transistors. Plot the data.

(b) Would a linear, exponential, or logarithmic function fit this data best?

(c) Use the ordered pairs $(0, 2300)$ and $(24, 5500000)$ to get an equation in the form $y = ae^{kx}$, by finding values of a and k that fit the data at 1971 and 1995.

(d) Graph the function from part (c) with the data points. Does the function fit the data?

---

*Data provided by Intel.

CASE 5

Characteristics of the Monkeyface Prickleback*

The monkeyface prickleback (*Cebidichthys violaceus*), known to anglers as the monkeyface "eel," is found in rocky intertidal and subtidal habitats ranging from San Quintin Bay, Baja California, to Brookings, Oregon. Pricklebacks are prime targets of the few sports anglers who "poke pole" in the rocky intertidal zone at low tide. Little is known about the life history of this species. The results of a study of the length, weight, and age of this species is discussed in this case.

Data on standard length (*SL*) and total length (*TL*) were collected. Early in the study only *TL* was measured, so a conversion to *SL* was necessary. The equation relating the two lengths, calculated from 177 observations for which both lengths had been measured, is

$$SL = TL(.931) + 1.416.$$

Ages (determined by standard aging techniques) were used to estimate parameters of the von Bertanfany growth model

$$L_t = L_x(1 - e^{-kt}), \qquad (1)$$

where

L_t = length at age t,
L_x = asymptotic age of the species,
k = growth completion rate, and
t_0 = theoretical age at zero length.

The constants a and b in the model

$$W = aL^b, \qquad (2)$$

where

W = weight in g,
L = standard length in cm,

were determined using 139 fish ranging from 27 cm and 145 g to 60 cm and 195 g.

Growth curves giving length as a function of age are shown in Figure 1. For the data marked opercle, the lengths were computed from the ages using Equation (1).

*Characteristics of the Monkeyface Prickleback, by William H. Marshall and Tina Wyllie Echeverria as appeared in *California Fish & Game*, Vol. 78, Spring 1992, Number 2.

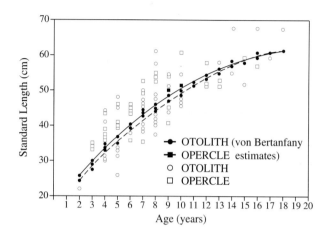

FIGURE 1

Estimated length from Equation (1) at a given age was larger for males than females after age eight. See the table. Weight/length relationships found with Equation (2) are shown in Figure 2, along with data from other studies.

Structure /Sex	Age (yr)	Length (cm)	L_x	k	t_0	n
Otolith						
Est.	2–18	23–67	72	.10	−1.89	91
S.D.			8	.03	1.08	
Opercle						
Est.	2–18	23–67	71	.10	−2.63	91
S.D.			8	.04	1.31	
Opercle-Females						
Est.	0–18	15–62	62	.14	−1.95	115
S.D.			2	.02	.28	
Opercle-Males						
Est.	0–18	13–67	70	.12	−1.91	74
S.D.			5	.02	.29	

FIGURE 2

EXERCISES

1. Use Equation (1) to estimate the lengths at ages 4, 11, and 17. Let $L_x = 71.5$ and $k = .1$. Compare your answers with the results in Figure 1. What do you find?

2. Use Equation (2) with $a = .01289$ and $b = 2.9$ to estimate the weights for lengths of 25 cm, 40 cm, and 60 cm. Compare with the results in Figure 2. Are your answers reasonable compared to the curve?

CHAPTER 6 Mathematics of Finance

Whether you are in a position to invest money or to borrow money, it is important for both business managers and consumers to understand *interest*. The formulas for interest are developed in this chapter.

6.1 SIMPLE INTEREST AND DISCOUNT

Interest is the fee paid for the use of someone else's money. For example, you might pay interest to a bank for money you borrow or the bank might pay you interest on the money in your savings account. The amount of money that is borrowed or deposited is called the **principal.** The fee paid as interest depends on the interest **rate** and the length of **time** for which you have the use of the money. Unless stated otherwise, the time, t, is measured in years and the interest rate, r, is in percent per year, expressed as a decimal; for instance, $8\% = .08$ or $9.5\% = .095$.

There are two common ways of computing interest, the first of which we study in this section. **Simple interest** is interest paid only on the amount deposited and not on past interest.

> The **simple interest,** I, on an amount of P dollars at a rate of interest r per year for t years is
> $$I = Prt.$$

(Interest is rounded to the nearest cent, as is customary in financial problems.)

1 Find the simple interest for the following.

(a) $1000 at 8% for 2 years

(b) $5500 at 10.5% for $1\frac{1}{2}$ years

Answers:
(a) $160
(b) $866.25

EXAMPLE 1 To buy furniture for a new apartment, Sylvia Chang borrowed $5000 at 11% simple interest for 11 months. How much interest will she pay?

From the formula, $I = Prt$, with $P = 5000$, $r = .11$, and $t = 11/12$ (in years). The total interest she will pay is

$$I = 5000(.11)(11/12) = 504.17,$$

or $504.17. ■ **1**

A deposit of P dollars today at a rate of interest r for t years produces interest of $I = Prt$. The interest, added to the original principal P, gives

$$P + Prt = P(1 + rt).$$

This amount is called the **future value** of P dollars at an interest rate r for time t in years. When loans are involved, the future value is often called the **maturity value** of the loan. This idea is summarized as follows.

> The **future value** or **maturity value**, A, of P dollars for t years all at a rate of interest r per year is
>
> $$A = P(1 + rt).$$

EXAMPLE 2 Find the maturity value for each of the following loans at simple interest.

(a) A loan of $2500 to be repaid in 9 months with interest at 12.1%
The loan is for 9 months, or $9/12 = .75$ of a year. The maturity value is

$$A = P(1 + rt)$$
$$A = 2500[1 + .121(.75)]$$
$$\approx 2500(1.09075) = 2726.875,$$

or $2726.88. Because the maturity value is the sum of principal and interest, the interest paid on this loan is

$$\$2726.88 - \$2500 = \$226.88.$$

2 Find the maturity value of each loan.

(a) $10,000 at 10% for 6 months

(b) $8970 at 11% for 9 months

(c) $95,106 at 9.8% for 76 days

Answers:
(a) $10,500
(b) $9710.03
(c) $97,073.64

(b) A loan of $11,280 for 85 days at 11% interest
It is common to assume 360 days in a year when working with simple interest. We shall usually make such an assumption in this book. The maturity value in this example is

$$A = 11,280\left[1 + .11\left(\frac{85}{360}\right)\right] \approx 11,280[1.0259722] \approx 11,572.97,$$

or $11,572.97. ■ **2**

TECHNOLOGY TIP The table feature of a graphing calculator makes it easy to compare the effect on the future value of a loan of a small difference in the interest rate. The screen in Figure 6.1 shows the portion of a year in the X column, and the future value of the loan in Example 2(a) with interest rates of 12.1% in the Y_1 column and 11.6% in the Y_2 column. From the table we see that when X = .75 the future value is $9.40 less at 11.6% than at 12.1%. ✔

FIGURE 6.1

PRESENT VALUE   A sum of money that can be deposited today to yield some larger amount in the future is called the **present value** of that future amount. Present value refers to the principal to be invested or loaned, so we use the same variable P as we did for principal. In interest problems, P always represents the amount at the beginning of the time period, and A always represents the amount at the end of the time period. To find a formula for P, we begin with the future value formula

$$A = P(1 + rt).$$

Dividing each side by $1 + rt$ gives the following formula for present value.

$$P = \frac{A}{1 + rt}$$

> The **present value** P of a future amount of A dollars at a simple interest rate r for t years is
> $$P = \frac{A}{1 + rt}.$$

3 Find the present value of the following future amounts. Assume 12% interest.

(a) $7500 in 1 year

(b) $89,000 in 5 months

(c) $164,200 in 125 days

Answers:

(a) $6696.43

(b) $84,761.90

(c) $157,632.00

EXAMPLE 3   Find the present value of $32,000 in 4 months at 9% interest.

$$P = \frac{32,000}{1 + (.09)\left(\dfrac{4}{12}\right)} = \frac{32,000}{1.03} = 31,067.96$$

A deposit of $31,067.96 today, at 9% interest, would produce $32,000 in 4 months. These two sums, $31,067.96 today and $32,000.00 in 4 months, are equivalent (at 9%) because the first amount becomes the second amount in 4 months.   ■   **3**

EXAMPLE 4   Because of a court settlement, Charlie Dawkins owes $5000 to Joshua Parker. The money must be paid in 10 months, with no interest. Suppose Dawkins wishes to pay the money today. What amount should Parker be willing to accept? Assume an interest rate of 5%.

4 Jerrell Davis is owed $19,500 by Christine O'Brien. The money will be paid in 11 months, with no interest. If the current interest rate is 10%, how much should Davis be willing to accept today in settlement of the debt?

The amount that Parker should be willing to accept is given by the present value:

$$P = \frac{5000}{1 + (.05)\left(\dfrac{10}{12}\right)} = 4800.00.$$

Answer:

$17,862.60

Parker should be willing to accept $4800.00 today in settlement of the obligation.   ■   **4**

EXAMPLE 5   Suppose you borrow $40,000 today and are required to pay $41,400 in 4 months to pay off the loan and interest. What is the simple interest rate?

We can use the future value formula, with $P = 40{,}000$, $A = 41{,}400$, and $t = 4/12 = 1/3$, and solve for r.

$$A = P(1 + rt)$$

$$41{,}400 = 40{,}000\left(1 + r \cdot \frac{1}{3}\right)$$

$$41{,}400 = 40{,}000 + \frac{40{,}000r}{3}$$

$$1400 = \frac{40{,}000r}{3}$$

$$40{,}000r = 3 \cdot 1400 = 4200$$

$$r = \frac{4200}{40{,}000} = .105$$

Therefore, the interest rate is 10.5%. ■

SIMPLE DISCOUNT NOTES   The loans discussed up to this point are called **simple interest notes,** where interest on the face value of the loan is added to the loan and paid at maturity. Another common type of note, called a **simple discount note,** has the interest deducted in advance from the amount of a loan before giving the *balance* to the borrower. The *full* value of the note must be paid at maturity. The money that is deducted is called the **bank discount** or just the **discount,** and the money actually received by the borrower is called the **proceeds.**

For example, consider a loan of $3000 at 6% interest for 9 months. We can compare the two types of loan arrangements as follows.

	Simple Interest Note	*Bank Discount Note*
Interest on the note	3000(.06)(9/12) = $135	3000(.06)(9/12) = $135
Borrower receives	**$3000**	**$2865**
Borrower pays back	**$3135**	**$3000**

EXAMPLE 6   Theresa DePalo needs a loan from her bank and agrees to pay $8500 to her banker in 9 months. The banker subtracts a discount of 12% and gives the balance to DePalo. Find the amount of the discount and the proceeds.

As shown above, the discount is found in the same way that simple interest is found, except that it is based on the amount to be repaid.

$$\textbf{Discount} = 8500(.12)\left(\frac{9}{12}\right) = \textbf{765.00}$$

The proceeds are found by subtracting the discount from the original amount.

$$\textbf{Proceeds} = \$8500 - \$765.00 = \$7735.00 \quad ■ \quad \boxed{5}$$

⑤ Kelly Bell signs an agreement at her bank to pay the bank $25,000 in 5 months. The bank charges a 13% discount rate. Find the amount of the discount and the amount Bell actually receives.

Answer:
$1354.17; $23,645.83

In Example 6, the borrower was charged a discount of 12%. However, 12% is *not* the interest rate paid, since 12% applies to the $8500, while the borrower actually received only $7735. In the next example, we find the rate of interest actually paid by the borrower.

EXAMPLE 7 Find the actual rate of interest paid by DePalo in Example 6.

Use the formula for simple interest, $I = Prt$, with r the unknown. Since the borrower received only $7735 and must repay $8500, $I = 8500 - 7735 = 765$. Here, $P = 7735$ and $t = 9/12 = .75$. Substitute these values into $I = Prt$.

$$I = Prt$$
$$765 = 7735(r)(.75)$$
$$\frac{765}{7735(.75)} = r$$
$$.132 \approx r$$

The actual interest rate paid by the borrower is about 13.2%. ■ 6

[6] Refer to Problem 5 and find the actual rate of interest paid by Bell.

Answer:
13.7% (to the nearest tenth)

Let D represent the amount of discount on a loan. Then $D = Art$, where A is the maturity value of the loan (the amount borrowed plus interest), and r is the stated rate of interest. The amount actually received, the proceeds, can be written as $P = A - D$, or $P = A - Art = A(1 - rt)$.

The formulas for discount are summarized below.

DISCOUNT

If D is the discount on a loan having a maturity value A at a rate of interest r for t years, and if P represents the proceeds, then

$$P = A - D \quad \text{or} \quad P = A(1 - rt).$$

EXAMPLE 8 John Young owes $4250 to Meg Holden. The loan is payable in 1 year at 10% interest. Holden needs cash to buy a new car, so 3 months before the loan is payable she goes to the bank to have the loan discounted. That is, she sells the loan (note) to the bank. The bank charges an 11% discount fee. Find the amount of cash she will receive from the bank.

First find the maturity value of the loan, the amount (with interest) Young must pay Holden. By the formula for maturity value,

$$A = P(1 + rt)$$
$$= 4250[1 + (.10)(1)]$$
$$= 4250(1.10) = 4675$$

or $4675.00.

7 A firm accepts a $21,000 note due in 7 months with interest of 10.5%. Suppose the firm discounts the note at a bank 75 days before it is due. Find the amount the firm would receive if the bank charges a 12.4% discount rate. (Use 360 days in a year.)

Answer:
$21,710.52

The bank applies its discount rate to this total:

Amount of discount $= 4675(.11)(3/12) \approx 128.56$.

(Remember that the loan was discounted 3 months before it was due.) Holden actually receives

$$\$4675 - \$128.56 = \$4546.44$$

in cash from the bank. Three months later, the bank will get $4675.00 from Young. ■ 7

6.1 EXERCISES

1. What factors determine the amount of interest earned on a fixed principal?

Find the simple interest in Exercises 2–5. (See Example 1.)

2. $25,000 at 7% for 9 mo
3. $3850 at 9% for 8 mo
4. $1974 at 6.3% for 7 mo
5. $3724 at 8.4% for 11 mo

Find the simple interest. Assume a 360-day year and a 30-day month. (See Example 2.)

6. $5147.18 at 10.1% for 58 days
7. $2930.42 at 11.9% for 123 days
8. $7980 at 10%; loan made on May 7 and due September 19
9. $5408 at 12%; loan made on August 16 and due December 30

Find the simple interest. Assume 365 days in a year, and use the exact number of days in a month. (Assume 28 days in February.)

10. $7800 at 11%; made on July 7 and due October 25
11. $11,000 at 10%; made on February 19 and due May 31
12. $2579 at 9.6%; made on October 4 and due March 15
13. $37,098 at 11.2%; made on September 12 and due July 30
14. In your own words, describe the *maturity value* of a loan.
15. What is meant by the *present value* of money?

Find the present value of each of the future amounts in Exercises 16–19. Assume 360 days in a year. (See Example 3.)

16. $15,000 for 8 mo; money earns 6%
17. $48,000 for 9 mo; money earns 5%
18. $15,402 for 125 days; money earns 6.3%
19. $29,764 for 310 days; money earns 7.2%

Find the proceeds for the amounts in Exercises 20–23. Assume 360 days in a year. (See Example 6.)

20. $7150; discount rate 12%; length of loan 11 mo

21. $9450; discount rate 10%; length of loan 7 mo
22. $35,800; discount rate 9.1%; length of loan 183 days
23. $50,900; discount rate 8.2%; length of loan 238 days
24. Why is the discount rate charged on a simple discount note different from the actual interest rate paid on the proceeds?

Find the interest rate to the nearest tenth on the proceeds for the following simple discount notes. (See Example 7.)

25. $6200; discount rate 10%; length of loan 8 mo
26. $5000; discount rate 8.1%; length of loan 6 mo
27. $58,000; discount rate 10.8%; length of loan 9 mo
28. $43,000; discount rate 9%; length of loan 4 mo

Management *Work the following applied problems.*

29. Linda Davis borrowed $25,900 from her father to start a flower shop. She repaid him after 11 mo, with interest of 8.4%. Find the total amount she repaid.
30. An accountant for a corporation forgot to pay the firm's income tax of $725,896.15 on time. The government charged a penalty of 12.7% interest for the 34 days the money was late. Find the total amount (tax and penalty) that was paid. (Use a 365-day year.)
31. A $100,000 certificate of deposit held for 60 days is worth $101,133.33. To the nearest tenth of a percent, what interest rate was earned?
32. Tuition of $1769 will be due when the spring term begins in 4 mo. What amount should a student deposit today, at 6.25%, to have enough to pay the tuition?
33. A firm of accountants has ordered 7 new computers at a cost of $5104 each. The machines will not be delivered for 7 mo. What amount could the firm deposit in an account paying 6.42% to have enough to pay for the machines?
34. Sun Kang needs $5196 to pay for remodeling work on his house. He plans to repay the loan in 10 mo. His bank loans money at a discount rate of 13%. Find the amount of his loan.

35. Joan McKee decides to go back to college. To get to school, she buys a small car for $6100. She decides to borrow the money from a bank that charges an 11.8% discount rate. If she will repay the loan in 7 months, find the amount of the loan.

36. John Matthews signs a $4200 note at the bank. The bank charges a 12.2% discount rate. Find the net proceeds if the note is for 10 mo. Find the actual interest rate (to the nearest hundredth) charged by the bank.

37. A stock that sold for $22 at the beginning of the year was selling for $24 at the end of the year. If the stock paid a dividend of $.50 per share, what is the simple interest rate on an investment in this stock? (*Hint:* Consider the interest to be the increase in value plus the dividend.)

38. A bond with a face value of $10,000 in 10 yr can be purchased now for $5988.02. What is the simple interest rate?

39. A building contractor gives a $13,500 note to a plumber. (The plumber loans $13,500 to the contractor.) The note is due in 9 mo, with interest of 9%. Three months after the note is signed, the plumber discounts it at the bank. The bank charges a 10.1% discount rate. How much will the plumber receive? Will it be enough to pay a bill for $13,582?

40. Maria Lopez owes $7000 to the Eastside Music Shop. She has agreed to pay the amount in 7 mo at an interest rate of 10%. Two months before the loan is due, the store needs $7350 to pay a wholesaler's bill. The bank will discount the note at a rate of 10.5%. How much will the store receive? Is it enough to pay the bill?

41. Fay, Inc., received a $30,000, six-month, 12% interest-bearing note from a customer.* The note was discounted the same day by Carr National Bank at 15%. The amount of cash received by Fay from the bank was
    (a) $30,000   (b) $29,550
    (c) $29,415   (d) $27,750.

---

*Uniform CPA Examination, May, 1989, American Institute of Certified Public Accountants.

6.2 COMPOUND INTEREST

Simple interest is normally used for loans or investments of a year or less. For longer periods *compound interest* is used. With compound interest, interest is charged (or paid) on interest as well as on principal. For example, if $1000 is deposited at 5% compounded annually, then the interest for the first year is $1000(.05) = $50, just as with simple interest, so that the account balance is $1050 at the end of the year. During the second year, interest is paid on the entire $1050 (not just on the original $1000 as with simple interest), so the amount in the account at the end of the second year is $1050 + $1050(.05) = $1102.50. This is more than simple interest would produce. ☐1

1 Use the formula
$$A = P(1 + rt)$$
to find the amount in the account after 2 years at 5% simple interest.

Answer:
$1100

To find a formula for compound interest, suppose that P dollars are deposited at interest rate r per year. The amount A on deposit after 1 year is found by the simple interest formula.

$$A = P[1 + r(1)] = P(1 + r)$$

If the deposit earns compound interest, the interest for the second year is paid on the total amount on deposit at the end of the first year, $P(1 + r)$. Using the formula $A = P(1 + rt)$ again, with $P = P(1 + r)$ and $t = 1$ gives the total amount on deposit at the end of the second year.

$$A = [P(1 + r)](1 + r \cdot 1) = P(1 + r)^2$$

In the same way, the total amount on deposit at the end of the third year is

$$P(1 + r)^3.$$

Continuing in this way, the total amount on deposit after t years is

$$A = P(1 + r)^t,$$

called the **compound amount.**

N O T E   Compare this formula for compound interest with the formula for simple interest from the previous section.

$$\text{Compound interest} \quad A = P(1 + r)^t$$
$$\text{Simple interest} \qquad A = P(1 + rt)$$

The important distinction between the two formulas is that in the compound interest formula, the number of years t is an *exponent,* so that money grows much more rapidly when interest is compounded. ◆

Figure 6.2 shows calculator-generated graphs of these two formulas with $P = 1000$ and $r = 10\%$ from 0 to 20 years. The future value after 15 years is shown for each graph. After 15 years at compound interest, $1000 grows to $4177.25, whereas with simple interest, it amounts to $2500.00, a difference of $1677.25.

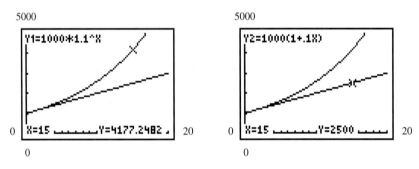

FIGURE 6.2

Interest can be compounded more than once a year. Common **compounding periods** include *semiannually* (two periods per year), *quarterly* (four periods per year), *monthly* (twelve periods), and *daily* (usually 365 periods per year). To find the *interest rate per period, i,* we *divide* the annual interest rate r by the number of compounding periods, m, per year. The total number of compounding periods, n, is found by *multiplying* the number of years t by the number of compounding periods per year, m. Then the general formula can be derived in much the same way as the formula given above.

COMPOUND INTEREST FORMULA

If P dollars are deposited for n compounding periods at a rate of interest i per period, the compound amount (or future value) A is

$$A = P(1 + i)^n.$$

In particular, if the annual interest rate r is compounded m times per year and the number of years is t, then $i = r/m$ and $n = mt$.

EXAMPLE 1   Suppose $1000 is deposited for 6 years in an account paying 8.31% per year compounded annually.

(a) Find the compound amount.

In the formula above, $P = 1000$, $i = .0831$, and $n = 6$. The compound amount is

$$A = P(1 + i)^n$$
$$A = 1000(1.0831)^6$$
$$A = \$1614.40.$$

2 Suppose $17,000 is deposited at 4% compounded semiannually for 11 years.

(a) Find the compound amount.

(b) Find the amount of interest earned.

Answers:
(a) $26,281.65
(b) $9281.65

3 Find the compound amount.
(a) $10,000 at 8% compounded quarterly for 7 years
(b) $36,000 at 6% compounded monthly for 2 years

Answers:
(a) $17,410.24
(b) $40,577.75

(b) Find the amount of interest earned.
Subtract the initial deposit from the compound amount.

$$\text{Amount of interest} = \$1614.40 - \$1000 = \$614.40 \quad \blacksquare \quad \boxed{2}$$

EXAMPLE 2 Find the amount of interest earned by a deposit of $2450 for 6.5 years at 5.25% compounded quarterly.

Interest compounded quarterly is compounded 4 times a year. In 6.5 years, there are $6.5(4) = 26$ periods. Thus, $n = 26$. Interest of 5.25% per year is 5.25/4 per quarter, so $i = .0525/4$. Now use the formula for compound amount.

$$A = P(1 + i)^n$$
$$A = 2450(1 + .0525/4)^{26} = 3438.78$$

Rounded to the nearest cent, the compound amount is $3438.78, so the interest is $3438.78 - \$2450 = \$998.78.$ $\blacksquare$ $\boxed{3}$

CAUTION As shown in Example 2, compound interest problems involve two rates—the nominal or stated annual rate r and the rate per compounding period i. Be sure you understand the distinction between them. When interest is compounded annually, these rates are the same. In all other cases, $i \neq r.$ ◆

The more often interest is compounded within a given time period, the more interest will be earned. Surprisingly, however, there is a limit on the amount of interest, no matter how often it is compounded. To see this, suppose that $1 is invested at 100% interest per year, compounded n times per year. Then the interest rate (in decimal form) is 1.00 and the interest rate per period is $1/n$. According to the formula (with $P = 1$), the compound amount at the end of 1 year will be $A = \left(1 + \dfrac{1}{n}\right)^n.$

A computer gives the following results for various values of n.

Interest Is Compounded	n	$\left(1 + \dfrac{1}{n}\right)^n$
Annually	1	$\left(1 + \dfrac{1}{1}\right)^1 = 2$
Semiannually	2	$\left(1 + \dfrac{1}{2}\right)^2 = 2.25$
Quarterly	4	$\left(1 + \dfrac{1}{4}\right)^4 \approx 2.4414$
Monthly	12	$\left(1 + \dfrac{1}{12}\right)^{12} \approx 2.6130$
Daily	365	$\left(1 + \dfrac{1}{365}\right)^{365} \approx 2.71457$
Hourly	8760	$\left(1 + \dfrac{1}{8760}\right)^{8760} \approx 2.718127$
Every minute	525,600	$\left(1 + \dfrac{1}{525,600}\right)^{525,600} \approx 2.7182792$
Every second	31,536,000	$\left(1 + \dfrac{1}{31,536,000}\right)^{31,536,000} \approx 2.7182818$

Because interest is rounded to the nearest penny, the compound amount never exceeds $2.72, no matter how big n is.

NOTE Try computing the values in the table with your calculator. You may notice your answers do not agree exactly. This is because of round-off error. ◆

The table above suggests that as n takes larger and larger values, then the corresponding values of $\left(1 + \dfrac{1}{n}\right)^n$ get closer and closer to a specific real number, whose decimal expansion begins 2.71828. . . . This is indeed the case, as is shown in calculus, and the number 2.71828 . . . is denoted e.

The preceding example is typical of what happens when interest is compounded n times per year, with larger and larger values of n. It can be shown that no matter what interest rate or principal is used, there is always an upper limit on the compound amount, which is called the compound amount from **continuous compounding.**

CONTINUOUS COMPOUNDING

The compound amount A for a deposit of P dollars at interest rate r per year compounded continuously for t years is given by

$$A = Pe^{rt}.*$$

Many calculators have an e^x key for computing powers of e. See Chapter 5 for more details on using a calculator to evaluate e^x.

| 4 | Find the compound amount, assuming continuous compounding.

(a) $12,000 at 10% for 5 years

(b) $22,867 at 7.2% for 9 years

Answers:
(a) $19,784.66
(b) $43,715.15

EXAMPLE 3 Suppose $5000 is invested at an annual rate of 4% compounded continuously for 5 years. Find the compound amount.

In the formula for continuous compounding, let $P = 5000$, $r = .04$ and $t = 5$. Then a calculator with an e^x key shows that

$$A = 5000e^{(.04)5} = 5000e^{.2} = \$6107.01.$$

You can readily verify that daily compounding would have produced a compound amount about 6¢ less. ■ | 4 |

EFFECTIVE RATE If $1 is deposited at 4% compounded quarterly, a calculator can be used to find that at the end of one year, the compound amount is $1.0406, an increase of 4.06% over the original $1. The actual increase of 4.06% in the money is somewhat higher than the stated increase of 4%. To differentiate between these two numbers, 4% is called the **nominal** or **stated rate** of interest, while 4.06% is called the **effective rate.** To avoid confusion between stated rates and effective rates, we shall continue to use r for the stated rate and we will use r_e for the effective rate.

EXAMPLE 4 Find the effective rate corresponding to a stated rate of 6% compounded semiannually.

---

*Other applications of the exponential function $f(x) = e^x$ are discussed in Chapter 5.

5 Find the effective rate corresponding to a nominal rate of

(a) 12% compounded monthly;

(b) 8% compounded quarterly.

Answers:

(a) 12.68%

(b) 8.24%

A calculator shows that \$100 at 6% compounded semiannually will grow to

$$A = 100\left(1 + \frac{.06}{2}\right)^2 = 100(1.03)^2 = \$106.09.$$

Thus, the actual amount of compound interest is \$106.09 − \$100 = \$6.09. Now if you earn \$6.09 interest on \$100 in 1 year with annual compounding, your rate is 6.09/100 = .0609 = 6.09%. Thus, the effective rate is r_e = 6.09%. ■ **5**

In the preceding example we found the effective rate by dividing compound interest for 1 year by the original principal. The same thing can be done with any principal P and rate r compounded m times per year.

$$\text{Effective rate} = \frac{\text{compound interest}}{\text{principal}}$$

$$r_e = \frac{\text{compound amount} - \text{principal}}{\text{principal}}$$

$$= \frac{P\left(1 + \frac{r}{m}\right)^m - P}{P} = \frac{P\left[\left(1 + \frac{r}{m}\right)^m - 1\right]}{P}$$

$$r_e = \left(1 + \frac{r}{m}\right)^m - 1$$

The **effective rate** corresponding to a stated rate of interest r per year compounded m times per year is

$$r_e = \left(1 + \frac{r}{m}\right)^m - 1.^*$$

Using the effective rate formula, a graphing calculator generated the table in Figure 6.3, in which the interest rate r is 10%, Y_1 denotes the effective rate, and X the number of compoundings.

X	Y1
1	.1
2	.1025
4	.10381
6	.10426
8	.10449
10	.10462
12	.10471

X=4

FIGURE 6.3

*When applied to consumer finance, the effective rate is called the annual percentage rate, APR, or annual percentage yield, APY.

6 Find the effective rate corresponding to a nominal rate of

(a) 15% compounded monthly;

(b) 10% compounded quarterly.

Answers:
(a) 16.08%
(b) 10.38%

EXAMPLE 5 A bank pays interest of 4.9% compounded monthly. Find the effective rate.

Use the formula given above with $r = .049$ and $m = 12$. The effective rate is

$$r_e = \left(1 + \frac{.049}{12}\right)^{12} - 1$$

$$= 1.050115575 - 1 \approx .0501,$$

or 5.01%. ■ 6

EXAMPLE 6 Bank A is now lending money at 13.2% interest compounded annually. The rate at Bank B is 12.6% compounded monthly and the rate at Bank C is 12.7% compounded quarterly. If you need to borrow money, at which bank will you pay the least interest?

Compare the effective rates.

$$\text{Bank A:} \quad \left(1 + \frac{.132}{1}\right)^1 - 1 = .132 = 13.2\%$$

$$\text{Bank B:} \quad \left(1 + \frac{.126}{12}\right)^{12} - 1 \approx .13354 = 13.354\%$$

$$\text{Bank C:} \quad \left(1 + \frac{.127}{4}\right)^4 - 1 \approx .13318 = 13.318\%$$

The lowest effective interest rate is at Bank A, which has the highest nominal rate. ■

PRESENT VALUE WITH COMPOUND INTEREST   The formula for compound interest, $A = P(1 + i)^n$, has four variables, A, P, i, and n. Given the values of any three of these variables, the value of the fourth can be found. In particular, if A (the future amount), i, and n are known, then P can be found. Here P is the amount that should be deposited today to produce A dollars in n periods.

EXAMPLE 7   Joan Ailanjian must pay a lump sum of $6000 in 5 years. What amount deposited today at 6.2% compounded annually will amount to $6000 in 5 years?

Here $A = 6000$, $i = .062$, $n = 5$, and P is unknown. Substituting these values into the formula for the compound amount gives

7 Find P in Example 7 if the interest rate is

(a) 6%;

(b) 10%.

Answers:
(a) $4483.55
(b) $3725.53

$$6000 = P(1.062)^5$$

$$P = \frac{6000}{(1.062)^5} = 4441.49,$$

or $4441.49. If Ailanjian leaves $4441.49 for 5 years in an account paying 6.2% compounded annually, she will have $6000 when she needs it. To check your work, use the compound interest formula with $P = \$4441.49$, $i = .062$, and $n = 5$. You should get $A = \$6000.00$. ■ 7

As Example 7 shows, $6000 in 5 years is the same as $4441.49 today (if money can be deposited at 6.2% annual interest). An amount that can be deposited today to yield a given amount in the future is called the *present value* of the future amount.

Generalizing from Example 7, by solving $A = P(1 + i)^n$ for P, we get the following formula for present value.

> The **present value** of A dollars compounded at an interest rate i per period for n periods is
>
> $$P = \frac{A}{(1 + i)^n} \quad \text{or} \quad P = A(1 + i)^{-n}.$$

Compare this with the present value of an amount at simple interest r for t years, given in the previous section:

$$P = \frac{A}{1 + rt}.$$

NOTE This is just the compound interest formula solved for P. It is not necessary to remember a new formula for present value. You can use the compound interest formula, if you understand what each of the variables represents. ◆

EXAMPLE 8 Find the present value of \$16,000 in 9 years if money can be deposited at 6% compounded semiannually.

In 9 years there are $2 \cdot 9 = 18$ semiannual periods. A rate of 6% per year is 3% in each semiannual period. Apply the formula with $A = 16,000$, $i = .03$, and $n = 18$.

$$P = \frac{A}{(1 + i)^n} = \frac{16,000}{(1 + .03)^{18}} \approx \frac{16,000}{1.702433} \approx 9398.31$$

A deposit of \$9398.31 today, at 6% compounded semiannually, will produce a total of \$16,000 in 9 years. ■ 8

The formula for compound amount also can be solved for n.

EXAMPLE 9 Suppose the general level of inflation in the economy averages 8% per year. Find the number of years it would take for the overall level of prices to double.

To find the number of years it will take for \$1 worth of goods or services to cost \$2, we must solve for n in the equation

$$2 = 1(1 + .08)^n,$$

where $A = 2$, $P = 1$, and $i = .08$. This equation simplifies to

$$2 = (1.08)^n.$$

To solve for n, we will use base 10 logarithms as in Chapter 5.

$\log 2 = \log (1.08)^n$	Take the logarithm of each side.
$\log 2 = n \log 1.08$	Property (c) of logarithms
$n = \dfrac{\log 2}{\log 1.08}$	Divide both sides by log 1.08.
$n \approx 9.01$	

To check, use a calculator to get $1.08^{9.01} = 2.00$ to the nearest hundredth. ■ 9

8 Find the present value in Example 8 if money is deposited at 10% compounded semiannually.

Answer:
\$6648.33

9 Using a calculator, find the number of years it will take for \$500 to increase to \$750 in an account paying 6% interest compounded semiannually.

Answer:
About 7 years

When interest is compounded continuously, the present value can be found by solving the continuous compounding formula $A = Pe^{rt}$ for P.

EXAMPLE 10   How much must be deposited today in an account paying 7.5% interest compounded continuously in order to have $10,000 in 4 years?

Here the future value is $A = 10{,}000$, the interest rate is $r = .075$, and the number of years is $t = 4$. The present value P is found as follows.

$$A = Pe^{rt}$$
$$10{,}000 = Pe^{(.075)4}$$
$$10{,}000 = Pe^{.3}$$
$$P = \frac{10{,}000}{e^{.3}} \approx \$7408.18 \quad \blacksquare$$

At this point, it seems helpful to summarize the notation and the most important formulas for simple and compound interest. We use the following variables.

P = principal or present value

A = future or maturity value

r = annual (stated or nominal) interest rate

t = number of years

m = number of compounding periods per year

i = interest rate per period ($i = r/m$)

r_e = effective rate

n = total number of compounding periods ($n = tm$)

Simple Interest

$$A = P(1 + rt)$$

$$P = \frac{A}{1 + rt}$$

Compound Interest

$$A = P(1 + i)^n$$

$$P = \frac{A}{(1 + i)^n} = A(1 + i)^{-n}$$

$$r_e = \left(1 + \frac{r}{m}\right)^m - 1$$

Continuous Compounding

$$A = Pe^{rt}$$

$$P = \frac{A}{e^{rt}}$$

6.2 EXERCISES

1. Explain the difference between simple interest and compound interest.

2. In the compound interest formula, how do r and i differ? How do t and n differ?

Find the compound amount for each of the following deposits. (See Examples 1–2.)

3. $1000 at 6% compounded annually for 8 yr

4. $1000 at 7% compounded annually for 10 yr

5. $470 at 10% compounded semiannually for 12 yr

6. $15,000 at 6% compounded semiannually for 11 yr

7. $6500 at 12% compounded quarterly for 6 yr

8. $9100 at 8% compounded quarterly for 4 yr

Find the amount of interest earned by each of the following deposits. (See Examples 1–2.)

9. $26,000 at 7% compounded annually for 5 years

10. $32,000 at 5% compounded annually for 10 years

11. $8000 at 4% compounded semiannually for 6.4 years

12. $2500 at 4.5% compounded semiannually for 8 years

13. $5124.98 at 6.3% compounded quarterly for 5.2 years

14. $27,630.35 at 7.1% compounded quarterly for 3.7 years

Find the compound amount if $25,000 is invested at 6% compounded continuously for the following number of years. (See Example 3.)

15. 1 **16.** 5

17. 10 **18.** 15

19. In Figure 6.2, one graph is a straight line and the other is curved. Explain why this is so, and which represents each type of interest (simple or compound).

20. How do the nominal or stated interest rate and the effective interest rate differ?

Find the effective rate corresponding to the following nominal rates. (See Examples 4–6.)

21. 4% compounded semiannually

22. 8% compounded quarterly

23. 8% compounded semiannually

24. 10% compounded semiannually

25. 12% compounded semiannually

26. 12% compounded quarterly

Find the present value of the following future amounts. (See Examples 7 and 8.)

27. $12,000 at 5% compounded annually for 6 years

28. $8500 at 6% compounded annually for 9 years

29. $4253.91 at 6.8% compounded semiannually for 4 years

30. $27,692.53 at 4.6% compounded semiannually for 5 years

31. $17,230 at 4% compounded quarterly for 10 years

32. $5240 at 8% compounded quarterly for 8 years

33. If money can be invested at 8% compounded quarterly, which is larger: $1000 now or $1210 in 5 years? Use present value to decide.

34. If money can be invested at 6% compounded annually, which is larger: $10,000 now or $15,000 in 10 years? Use present value to decide.

Find the present value of $17,200 for the following number of years, if money can be deposited at 11.4% compounded continuously.

35. 2 **36.** 4

37. 7 **38.** 10

Under certain conditions, Swiss banks pay negative interest—they charge you. (You didn't think all that secrecy was free?) Suppose a bank "pays" −2.4% interest compounded annually. Use a calculator and find the compound amount for a deposit of $150,000 after the following.

39. 2 years **40.** 4 years

41. 8 years **42.** 12 years

Management Work the following applied problems.

43. A New York bank offered the following special on C.D. (certificate of deposit) rates. The rates are annual percent-

age yield, or effective rates, which are higher than the corresponding nominal rates. Assume quarterly compounding. Solve for r to approximate the corresponding nominal rates to the nearest hundredth.

Term	6 mo	1 yr	18 mo	2 yr	3 yr
APY (%)	5.00	5.30	5.45	5.68	5.75

44. The pie graph shows the percent of baby boomers ages 46 to 49 who said they had investments with a total value as shown in each category.* Note that 30% say they have saved less than $10,000 and 28% don't know or gave no answer. Assume the money is invested at an average rate of 8% compounded quarterly for 20 years, when this age group will be ready for retirement. Find the range of amounts each group (except the "don't know or no answer" group) in the graph will have saved for retirement if no more is added.

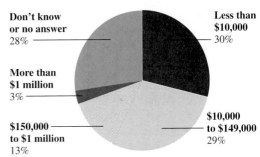

Note: Figures add to more than 100% because of rounding.

Sources: Census Bureau (age distribution); Merrill Lynch Baby Boom Retirement Index (investments); William M. Mercer Inc. (life expectancy).

45. When Lindsay Branson was born, her grandfather made an initial deposit of $3000 in an account for her college education. Assuming an interest rate of 6% compounded quarterly, how much will the account be worth in 18 yr?

46. Frank Capek has $10,000 in an Individual Retirement Account (IRA). Because of new tax laws, he decides to make no further deposits. The account earns 6% interest compounded semiannually. Find the amount on deposit in 15 yr.

47. A small business borrows $50,000 for expansion at 12% compounded monthly. The loan is due in 4 yr. How much interest will the business pay?

48. A developer needs $80,000 to buy land. He is able to borrow the money at 10% per year compounded quarterly. How much will the interest amount to if he pays off the loan in 5 yr?

49. A company has agreed to pay $2.9 million in 5 yr to settle a lawsuit. How much must they invest now in an account paying 8% compounded monthly to have that amount when it is due?

*From The New York Times, December 31, 1995, Section 3, p. 5.

50. Bill Poole wants to have $20,000 available in 5 yr for a down payment on a house. He has inherited $15,000. How much of the inheritance should he invest now to accumulate the $20,000, if he can get an interest rate of 8% compounded quarterly?

51. Two partners agree to invest equal amounts in their business. One will contribute $10,000 immediately. The other plans to contribute an equivalent amount in 3 yr, when she expects to acquire a large sum of money. How much should she contribute at that time to match her partner's investment now, assuming an interest rate of 6% compounded semi-annually?

52. As the prize in a contest, you are offered $1000 now or $1210 in 5 yr. If money can be invested at 6% compounded annually, which is larger?

53. The consumption of electricity has increased historically at 6% per year. If it continues to increase at this rate indefinitely, find the number of years before the electric utilities will need to double their generating capacity.

54. Suppose a conservation campaign coupled with higher rates cause the demand for electricity to increase at only 2% per year, as it has recently. Find the number of years before the utilities will need to double generating capacity.

Use the approach in Example 9 to find the time it would take for the general level of prices in the economy to double at the average annual inflation rates in Exercises 55 and 56.

55. 4%   **56.** 5%

57. In 1995, O. G. McClain of Houston, Texas, mailed a $100 check to a descendant of Texas independence hero Sam Houston to repay a $100 debt of McClain's great-great-grandfather, who died in 1835, to Sam Houston.* A bank

The New York Times, March 30, 1995.

estimated the interest on the loan to be $420 million for the 160 years it was due. Find the interest rate the bank was using, assuming interest is compounded annually.

58. In the New Testament, Jesus commends a widow who contributed 2 mites to the temple treasury (Mark 12:42–44). A mite was worth roughly 1/8 of a cent. Suppose the temple had invested those 2 mites at 4% interest compounded quarterly. How much would the money be worth 2000 years later?

59. On January 1, 1980, Jack deposited $1000 into Bank X to earn interest at the rate of j per annum compounded semi-annually. On January 1, 1985, he transferred his account to Bank Y to earn interest at the rate of k per annum compounded quarterly. On January 1, 1988, the balance at Bank Y was $1990.76. If Jack could have earned interest at the rate of k per annum compounded quarterly from January 1, 1980, through January 1, 1988, his balance would have been $2203.76. Which of the following represents the ratio k/j?*

  (a) 1.25   **(b)** 1.30   **(c)** 1.35   **(d)** 1.40
  (e) 1.45

60. On January 1, 1987, Tone Company exchanged equipment for a $200,000 noninterest bearing note due on January 1, 1990. The prevailing rate of interest for a note of this type at January 1, 1987, was 10%. The present value of $1 at 10% for three periods is 0.75. What amount of interest revenue should be included in Tone's 1988 income statement?†

  (a) $7500   **(b)** $15,000   **(c)** $16,500
  (d) $20,000

---

*Problem from "Course 140 Examination, Mathematics of Compound Interest" of the Education and Examination Committee of The Society of Actuaries. Reprinted by permission of The Society of Actuaries.

†Uniform CPA Examination, May, 1989, American Institute of Certified Public Accountants.

6.3 ANNUITIES

So far in this chapter, only lump sum deposits and payments have been discussed. But many financial situations involve a sequence of equal payments at regular intervals, such as weekly deposits in a savings account or monthly payments on a mortgage or car loan. In order to develop formulas to deal with periodic payments like these, we must first discuss sequences.

GEOMETRIC SEQUENCES   If a and r are fixed nonzero numbers, then the infinite list of numbers, $a,\ ar,\ ar^2,\ ar^3,\ ar^4, \ldots$ is called a **geometric sequence.** For instance, if $a = 5$ and $r = 2$, we have the sequence

$$5, 5 \cdot 2, 5 \cdot 2^2, 5 \cdot 2^3, 5 \cdot 2^4, \ldots,$$

or

$$5, 10, 20, 40, 80, \ldots.$$

In the sequence $a, ar, ar^2, ar^3, ar^4, \ldots$, the number a is the first term of the sequence, ar the second term, ar^2 the third term, ar^3 the fourth term, and so on. Thus, for any $n \geq 1$,

$$ar^{n-1} \text{ is the } n\text{th term of the sequence.}$$

Each term in the sequence is r times the preceding term. The number r is called the **common ratio** of the sequence.

EXAMPLE 1   Find the seventh term of the geometric sequence with $a = 6$ and $r = 4$. Using ar^{n-1}, with $n = 7$, the **seventh** term is

$$ar^6 = 6(4)^6 = 6(4096) = 24,576. \quad \blacksquare$$

TECHNOLOGY TIP Some graphing calculators have the ability to produce a list of the terms of a sequence, given the expression for the nth term. With other calculators you can use the TABLE feature by entering the expression for the nth term as a function. The first four terms of the sequence discussed in Example 1 are shown in Figure 6.4. ✔

FIGURE 6.4

EXAMPLE 2 $100 is deposited in an account that pays interest of 10% compounded annually. How much is in the account at the beginning of each of the first five years?

According to the compound interest formula (with $P = 100$ and $i = .01$), the value of the account at the beginning of each of the first five years is given by the first five terms of the geometric sequence with $a = 100$ and $r = 1.1$. The first five terms are

$$100, \quad 100(1.1), \quad 100(1.1)^2, \quad 100(1.1)^3, \quad 100(1.1)^4,$$

so the value of the account at the beginning of each of the five years is

$$\$100, \quad \$110, \quad \$121, \quad \$133.10, \quad \$146.41. \quad \blacksquare \quad \boxed{1}$$

Next we find the sum S_n of the first n terms of a geometric sequence. That is, we find S_n, where

$$S_n = a + ar + ar^2 + ar^3 + \cdots + ar^{n-1}. \tag{1}$$

If $r = 1$, this is easy because

$$S_n = \underbrace{a + a + a + \cdots + a}_{n \text{ terms}} = na.$$

$\boxed{1}$ Write the first four terms of the geometric sequence with $a = 5$ and $r = -2$. Then find the seventh term.

Answer:
$5, -10, 20, -40;\ 320$

If $r \neq 1$, multiply both sides of equation (1) by r, obtaining

$$rS_n = ar + ar^2 + ar^3 + ar^4 + \cdots + ar^n. \tag{2}$$

Now subtract corresponding sides of equation (1) from equation (2):

$$
\begin{aligned}
rS_n &= ar + ar^2 + ar^3 + \cdots + ar^{n-1} + ar^n \\
-S_n &= -(a + ar + ar^2 + ar^3 + \cdots + ar^{n-1}) \\
\hline
rS_n - S_n &= ar^n - a \\
S_n(r - 1) &= a(r^n - 1) \\
S_n &= \frac{a(r^n - 1)}{r - 1}.
\end{aligned}
$$

Hence, we have this useful formula.

> If a geometric sequence has first term a and common ratio r, then the **sum of the first n terms** is given by
>
> $$S_n = \frac{a(r^n - 1)}{r - 1}, \quad r \neq 1.$$

2 **(a)** Find S_4 and S_7 for the geometric sequence 5, 15, 45, 135, ...

(b) Find S_5 for the geometric sequence having $a = -3$ and $r = -5$.

Answers:
(a) $S_4 = 200$, $S_7 = 5465$
(b) $S_5 = -1563$

EXAMPLE 3   Find the sum of the first six terms of the geometric sequence 3, 12, 48, ...

Here $a = 3$ and $r = 4$. Find S_6 by the result above.

$$
\begin{aligned}
S_6 &= \frac{3(4^6 - 1)}{4 - 1} \qquad \text{Let } n = 6, a = 3, r = 4. \\
&= \frac{3(4096 - 1)}{3} \\
&= 4095 \quad \blacksquare \quad \boxed{2}
\end{aligned}
$$

TECHNOLOGY TIP Graphing calculators with sequence capability can also find the sum of the first n terms of a sequence, given the expression for the nth term. ✔

ORDINARY ANNUITIES   A sequence of equal payments made at equal periods of time is called an **annuity.** If the payments are made at the end of the time period, and if the frequency of payments is the same as the frequency of compounding, the annuity is called an **ordinary annuity.** The time between payments is the **payment period,** and the time from the beginning of the first payment period to the end of the last period is called the **term of the annuity.** The **future value of the annuity,** the final sum on deposit, is defined as the sum of the compound amounts of all the payments, compounded to the end of the term.

Two common uses of annuities are to accumulate funds for some goal or to withdraw funds from an account. For example, an annuity may be used to save money for a large purchase, such as an automobile, an expensive trip, or a down payment on a home. An annuity also may be used to provide monthly payments for retirement. We explore these options in this and the next section.

For example, suppose $1500 is deposited at the end of the year for the next 6 years in an account paying 8% per year compounded annually. Figure 6.5 shows this annuity schematically.

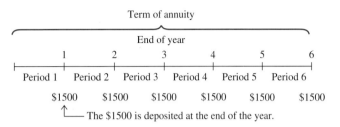

FIGURE 6.5

To find the future value of this annuity, look separately at each of the $1500 payments. The first of these payments will produce a compound amount of

$$1500(1 + .08)^5 = 1500(1.08)^5.$$

Use 5 as the exponent instead of 6 since the money is deposited at the *end* of the first year and earns interest for only 5 years. The second payment of $1500 will produce a compound amount of $1500(1.08)^4$. As shown in Figure 6.6, the future value of the annuity is

$$1500(1.08)^5 + 1500(1.08)^4 + 1500(1.08)^3 + 1500(1.08)^2$$
$$+ 1500(1.08)^1 + 1500.$$

(The last payment earns no interest at all.) Reading this in reverse order, we see that it is just the sum of the first six terms of a geometric sequence with $a = 1500$, $r = 1.08$, and $n = 6$. Therefore, the sum is

$$\frac{a(r^n - 1)}{r - 1} = \frac{1500[(1.08)^6 - 1]}{1.08 - 1} = \$11,003.89. \quad \boxed{3}$$

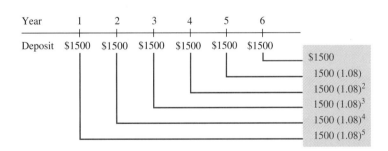

FIGURE 6.6

To generalize this result, suppose that payments of R dollars each are deposited into an account at the *end of each period* for n periods, at a rate of interest i *per period*. The first payment of R dollars will produce a compound amount of $R(1 + i)^{n-1}$ dollars, the second payment will produce $R(1 + i)^{n-2}$ dollars, and so on; the final payment earns no interest and contributes just R dollars to the total. If S represents the future value of the annuity, then (as shown in Figure 6.7),

$$S = R(1 + i)^{n-1} + R(1 + i)^{n-2} + R(1 + i)^{n-3} + \cdots + R(1 + i) + R$$

or, written in reverse order,

$$S = R + R(1 + i)^1 + R(1 + i)^2 + \cdots + R(1 + i)^{n-1}.$$

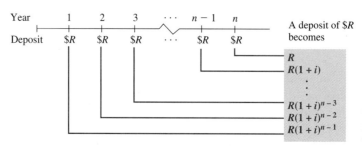

FIGURE 6.7

This result is the sum of the first n terms of the geometric sequence having first term R and common ratio $1 + i$. Using the formula for the sum of the first n terms of a geometric sequence,

$$S = \frac{R[(1 + i)^n - 1]}{(1 + i) - 1} = \frac{R[(1 + i)^n - 1]}{i} = R\left[\frac{(1 + i)^n - 1}{i}\right].$$

The quantity in brackets is commonly written $s_{\overline{n}|i}$ (read "s-angle-n at i"), so that

$$S = R \cdot s_{\overline{n}|i}.$$

We can summarize the preceding work as follows.*

FUTURE VALUE OF AN ORDINARY ANNUITY

$$S = R\left[\frac{(1 + i)^n - 1}{i}\right] \quad \text{or} \quad S = R \cdot s_{\overline{n}|i}$$

where

S is future value,
R is the payment at the end of each period,
i is the interest rate per period,
n is the number of periods.

---

*We use S for the future value here, instead of A as in the compound interest formula, to help avoid confusing the two formulas.

4 Johnson Building Materials deposits $2500 at the end of each year into an account paying 8% per year compounded annually. Find the total amount on deposit after

(a) 6 years;

(b) 10 years.

Answers:
(a) $18,339.82
(b) $36,216.41

EXAMPLE 4 Athan Lee Lewis is an athlete who feels that his playing career will last 7 years. To prepare for his future, he deposits $22,000 at the end of each year for 7 years in an account paying 6% compounded annually. How much will he have on deposit after 7 years?

His payments form an ordinary annuity with $R = 22,000$, $n = 7$, and $i = .06$. The future value of this annuity (by the formula above) is

$$S = 22,000\left[\frac{(1.06)^7 - 1}{.06}\right] = \$184,664.43. \quad \blacksquare \quad \boxed{4}$$

EXAMPLE 5 Experts say that the baby boom generation (born between 1946 and 1960) cannot count on a company pension or Social Security to provide a comfortable retirement, as their parents did. It is recommended that they start to save early and regularly. Michael Karelius, a baby boomer, has decided to deposit $200 at the end of each month in an account that pays interest of 7.2% compounded monthly for retirement in 20 years.

(a) How much will be in the account at that time?

This savings plan is an annuity with $R = 200$, $i = .072/12$, and $n = 12(20)$. The future value is

$$S = 200\left[\frac{\left(1 + (.072/12)^{12(20)}\right) - 1}{.072/12}\right] = 106,752.47,$$

or $106,752.47.

(b) Michael believes he needs to accumulate $130,000 in the 20-year period to have enough for retirement. What interest rate would provide that amount?

Once again we have an annuity with $R = 200$ and $n = 12(20)$. Now the future value $S = 130,000$ is given and we must find the interest rate. If x is the annual interest rate, then the interest rate per month is $i = x/12$ and we have

$$R\left[\frac{(1 + i)^n - 1}{i}\right] = S$$

$$200\left[\frac{(1 + x/12)^{12(20)} - 1}{x/12}\right] = 130,000. \tag{1}$$

5 Helen's Dry Goods deposits $5800 at the end of each quarter for 4 years.

(a) Find the final amount on deposit if the money earns 6.4% compounded quarterly.

(b) Helen wants to accumulate $110,000 in the four-year period. What interest rate (to the nearest tenth) will be required?

Answers:
(a) $104,812.44
(b) 8.9%

This last equation is difficult to solve algebraically. We can get a rough approximation by using a calculator to compute the left side of the equation for various values of x, to see which one comes closest to producing $130,000. For example, $x = .08$ produces 117,804 and $x = .09$ produces 133,577, so the solution is between .08 and .09. Further experimentation of this type leads to the approximate solution $x = .0879$, that is, 8.79%. A more exact approximation can be obtained with a graphing calculator (see the following Technology Tip). $\blacksquare$ $\boxed{5}$

TECHNOLOGY TIP To solve equation (1) in Example 5 with a graphing calculator, graph the equations

$$y = 200\left[\frac{[(1 + x/12)^{12(20)} - 1]}{x/12}\right] \quad \text{and} \quad y = 130,000$$

on the same screen and find the intersection of their graphs, as in Figure 6.8. The same result can be obtained on the TI-83 without graphing, by using the TVM solver in the FINANCE menu. ✔

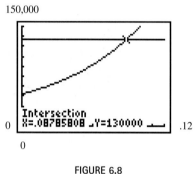

FIGURE 6.8

ANNUITIES DUE   The formula developed above is for *ordinary annuities*—those with payments at the *end* of each time period. These results can be modified slightly to apply to **annuities due**—annuities where payments are made at the *beginning* of each time period.

To find the future value of an annuity due, treat each payment as if it were made at the end of the *preceding* period, that is, find $s_{\overline{n}|\,i}$ for *one additional period*. No payment is made at the end of the last period, so to compensate for this, subtract the amount of one payment. Thus, the **future value of an annuity due** of n payments of R dollars each at the beginning of consecutive interest periods with interest compounded at the rate of i per period is

$$S = R\left[\frac{(1 + i)^{n+1} - 1}{i}\right] - R \quad \text{or} \quad S = R \cdot s_{\overline{n+1}|\,i} - R.$$

EXAMPLE 6   Payments of $500 are made at the beginning of each quarter for 7 years in an account paying 12% compounded quarterly.

(a) Find the future value of this annuity due.

In 7 years there are 28 quarterly periods. Add one period to get $n = 29$, and use the formula with $i = 12\%/4 = 3\%$.

$$S = 500\left[\frac{(1.03)^{29} - 1}{.03}\right] - 500$$

$$S \approx 500(45.21885) - 500 = \$22,109.43$$

The account will contain a total of $22,109.43 after 7 years.

(b) Compare the result from part (a) with the future value of an ordinary annuity. In this case the future value is given by

$$S = 500\left[\frac{1.03^{28} - 1}{.03}\right] = \$21,465.46.$$

The future amount is a little less because the payments are each made one month later than the payments for an annuity due. ■ **6**

SINKING FUNDS A fund set up to receive periodic payments is called a **sinking fund.** The periodic payments, together with the interest earned by the payments, are designed to produce a given sum at some time in the future. For example, a sinking fund might be set up to receive money that will be needed to pay off the principal on a loan at some future time. If the payments are all the same amount and are made at the end of a regular time period, they form an ordinary annuity.

EXAMPLE 7 The Chinns are close to retirement. They agree to sell an antique urn to the local museum for $17,000. Their tax adviser suggests that they defer receipt of this money until they retire, 5 years in the future. (At that time, they might well be in a lower tax bracket.) Find the amount of each payment the museum must make into a sinking fund so that it will have the necessary $17,000 in 5 years. Assume that the museum can earn 6% compounded annually on its money. Also, assume that the payments are made annually.

These payments are the periodic payments into an ordinary annuity. The annuity will amount to $17,000 in 5 years at 6% compounded annually. Using the formula and a calculator,

$$17,000 = R\left[\frac{(1.06)^5 - 1}{.06}\right]$$

$$17,000 \approx R(5.637093)$$

$$R \approx \frac{17,000}{5.637093} = 3015.74$$

or $3015.74. If the museum management deposits $3015.74 at the end of each year for 5 years in an account paying 6% compounded annually, it will have the total amount needed. This result is shown in the following sinking fund table. In these tables the last payment may differ slightly from the others because of rounding in the preceding lines.

Payment Number	Amount of Deposit	Interest Earned	Total in Account
1	$3015.74	$0	$3,015.74
2	3015.74	180.94	6,212.42
3	3015.74	372.75	9,600.91
4	3015.74	576.05	13,192.70
5	3015.74	791.56	17,000.00

To construct the table, notice that the first payment does not earn interest until the second payment is made. Line 2 of the table shows the second payment, 6% interest of $180.94 on the first payment, and the sum of these amounts added to the total in line 1. Line 3 shows the third payment, 6% interest of $372.75 on the total from line 2, and the new total found by adding these amounts to the total in line 2. This procedure is continued to complete the table. ■ **7**

6.3 EXERCISES

Find the fourth term of each of the following geometric sequences. (See Example 1.)

1. $a = 5, r = 3$
2. $a = 20, r = 2$
3. $a = 48, r = .5$
4. $a = 80, r = .1$
5. $a = 2000, r = 1.05$
6. $a = 10,000, r = 1.01$

Find the sum of the first four terms for each of the following geometric sequences. (See Example 3.)

7. $a = 1, r = 2$
8. $a = 3, r = 3$
9. $a = 5, r = .2$
10. $a = 6, r = .5$
11. $a = 128, r = 1.1$
12. $a = 100, r = 1.05$

Find each of the following values.

13. $s_{\overline{12}|.05}$
14. $s_{\overline{20}|.06}$
15. $s_{\overline{16}|.04}$
16. $s_{\overline{40}|.02}$
17. $s_{\overline{20}|.01}$
18. $s_{\overline{18}|.015}$

19. Explain the difference between an ordinary annuity and an annuity due.

Find the future value of the following ordinary annuities. Payments are made and interest is compounded as given. (See Examples 4 and 5.)

20. $R = \$1500$, 4% interest compounded annually for 8 years
21. $R = \$680$, 5% interest compounded annually for 6 years
22. $R = \$12,000$, 6.2% interest compounded annually for 10 years
23. $R = \$20,000$, 5.5% interest compounded annually for 12 years
24. $R = \$865$, 6% interest compounded semiannually for 8 years
25. $R = \$7300$, 9% interest compounded semiannually for 6 years
26. $R = \$1200$, 8% interest compounded quarterly for 10 years
27. $R = \$20,000$, 6% interest compounded quarterly for 12 years

Find the future value of each annuity due. (See Example 6.)

28. Payments of $500 for 10 years at 5% compounded annually
29. Payments of $1050 for 6 years at 3.5% compounded annually
30. Payments of $16,000 for 8 years at 4.7% compounded annually
31. Payments of $25,000 for 12 years at 6% compounded annually
32. Payments of $1000 for 9 years at 8% compounded semiannually

33. Payments of $750 for 15 years at 6% compounded semiannually
34. Payments of $100 for 7 years at 12% compounded quarterly
35. Payments of $1500 for 11 years at 12% compounded quarterly

Find the periodic payment that will amount to the given sums under the given conditions, if payments are made at the end of each period. (See Example 7.)

36. $S = \$14,500$, interest is 5% compounded semiannually for 8 years
37. $S = \$43,000$, interest is 6% compounded semiannually for 5 years
38. $S = \$62,000$, interest is 8% compounded quarterly for 6 years
39. $S = \$12,800$, interest is 6% compounded monthly for 4 years

40. What is meant by a sinking fund? Give an example of a sinking fund.

Find the amount of each payment to be made into a sinking fund to accumulate the following amounts. Payments are made at the end of each period. (See Example 7.)

41. $11,000, money earns 6% compounded semiannually, for 6 years
42. $75,000, money earns 6% compounded semiannually, for $4\frac{1}{2}$ years
43. $50,000, money earns 10% compounded quarterly, for $2\frac{1}{2}$ years
44. $25,000, money earns 12% compounded quarterly, for $3\frac{1}{2}$ years
45. $6000, money earns 8% compounded monthly, for 3 years
46. $9000, money earns 12% compounded monthly, for $2\frac{1}{2}$ years

Management *Work the following applied problems.*

47. A typical pack-a-day smoker spends about $55 per month on cigarettes. Suppose the smoker invests that amount at the end of each month in a savings account at 4.8% compounded monthly. What would the account be worth after 40 years?
48. Pat Quinlan deposits $6000 at the end of each year for 4 years in an account paying 5% interest compounded annually.
    (a) Find the final amount she will have on deposit.
    (b) Pat's brother-in-law works in a bank that pays 4.8% compounded annually. If she deposits her money in this

bank instead of the one above, how much will she have in her account?

(c) How much would Quinlan lose over 5 years by using her brother-in-law's bank?

49. A father opened a savings account for his daughter on the day she was born, depositing $1000. Each year on her birthday he deposits another $1000, making the last deposit on her twenty-first birthday. If the account pays 9.5% interest compounded annually, how much is in the account at the end of the day on the daughter's twenty-first birthday?

50. A 45-year-old man puts $1000 in a retirement account at the end of each quarter until he reaches the age of 60 and makes no further deposits. If the account pays 11% interest compounded quarterly, how much will be in the account when the man retires at age 65?

51. At the end of each quarter a 50-year-old woman puts $1200 in a retirement account that pays 7% interest compounded quarterly. When she reaches age 60, she withdraws the entire amount and places it in a mutual fund that pays 9% interest compounded monthly. From then on she deposits $300 in the mutual fund at the end of each month. How much is in the account when she reaches age 65?

52. Jasspreet Kaur deposits $2435 at the beginning of each semiannual period for 8 years in an account paying 6% compounded semiannually. She then leaves that money alone, with no further deposits, for an additional 5 years. Find the final amount on deposit after the entire 13-year period.

53. Chuck Hickman deposits $10,000 at the beginning of each year for 12 years in an account paying 5% compounded annually. He then puts the total amount on deposit in another account paying 6% compounded semiannually for another 9 years. Find the final amount on deposit after the entire 21-year period.

54. David Horwitz needs $10,000 in 8 years.
(a) What amount should he deposit at the end of each quarter at 8% compounded quarterly so that he will have his $10,000?
(b) Find Horwitz's quarterly deposit if the money is deposited at 6% compounded quarterly.

55. Harv's Meats knows that it must buy a new deboner machine in 4 years. The machine costs $12,000. In order to accumulate enough money to pay for the machine, Harv decides to deposit a sum of money at the end of each 6 months in an account paying 6% compounded semiannually. How much should each payment be?

56. Karin Sandberg wants to buy an $18,000 car in 6 years. How much money must she deposit at the end of each quarter in an account paying 12% compounded quarterly so that she will have enough to pay for her car?

In Exercises 57 and 58, use a graphing calculator to find the value of i that produces the given value of S. (See Example 5(b).)

57. To save for retirement, Karla Harby put $300 each month into an ordinary annuity for 20 years. Interest was compounded monthly. At the end of the 20 years, the annuity was worth $147,126. What annual interest rate did she receive?

58. Jennifer Wall made payments of $250 per month at the end of each month to purchase a piece of property. At the end of 30 years, she completely owned the property, which she sold for $330,000. What annual interest rate would she need to earn on an ordinary annuity for a comparable rate of return?

59. In a 1992 Virginia lottery, the jackpot was $27 million. An Australian investment firm tried to buy all possible combinations of numbers, which would have cost $7 million. In fact, the firm ran out of time and was unable to buy all combinations, but ended up with the only winning ticket anyway. The firm received the jackpot in 20 equal annual payments of $1.35 million.* Assume these payments meet the conditions of an ordinary annuity.
(a) Suppose the firm can invest money at 8% interest compounded annually. How many years would it take until the investors would be further ahead than if they had simply invested the $7 million at the same rate? (*Hint:* Experiment with different values of n, the number of years, or use a graphing calculator to plot the value of both investments as a function of the number of years.)
(b) How many years would it take in part (a) at an interest rate of 12%?

60. Diane Gray sells some land in Nevada. She will be paid a lump sum of $60,000 in 7 yr. Until then, the buyer pays 8% simple interest quarterly.
(a) Find the amount of each quarterly interest payment.
(b) The buyer sets up a sinking fund so that enough money will be present to pay off the $60,000. The buyer wants to make semiannual payments into the sinking fund; the account pays 6% compounded semiannually. Find the amount of each payment into the fund.
(c) Prepare a table showing the amount in the sinking fund after each deposit.

61. Joe Seniw bought a rare stamp for his collection. He agreed to pay a lump sum of $4000 after 5 yr. Until then, he pays 6% simple interest semiannually.
(a) Find the amount of each semiannual interest payment.
(b) Seniw sets up a sinking fund so that enough money will be present to pay off the $4000. He wants to make annual payments into the fund. The account pays 8% compounded annually. Find the amount of each payment.
(c) Prepare a table showing the amount in the sinking fund after each deposit.

*The Washington Post, March 10, 1992, p. A1.

6.4 PRESENT VALUE OF AN ANNUITY; AMORTIZATION

Suppose that at the end of each year, for the next 10 years, $500 is deposited in a savings account paying 7% interest compounded annually. This is an example of an ordinary annuity. The **present value** of this annuity is the amount that would have to be deposited in one lump sum today (at the same compound interest rate) in order to produce exactly the same balance at the end of 10 years. We can find a formula for the present value of an annuity as follows.

Suppose deposits of R dollars are made at the end of each period for n periods at interest rate i per period. Then the amount in the account after n periods is the future value of this annuity:

$$S = R\left[\frac{(1 + i)^n - 1}{i}\right].$$

On the other hand, if P dollars are deposited today at the same compound interest rate i, then at the end of n periods, the amount in the account is $P(1 + i)^n$. This amount must be the same as the amount S in the formula above, that is,

$$P(1 + i)^n = R\left[\frac{(1 + i)^n - 1}{i}\right].$$

To solve this equation for P, multiply both sides by $(1 + i)^{-n}$.

$$P = R\,(1 + i)^{-n}\left[\frac{(1 + i)^n - 1}{i}\right]$$

Use the distributive property; also recall that $(1 + i)^{-n}(1 + i)^n = (1 + i)^0 = 1$.

$$P = R\left[\frac{(1 + i)^{-n}(1 + i)^n - (1 + i)^{-n}}{i}\right]$$

$$P = R\left[\frac{1 - (1 + i)^{-n}}{i}\right]$$

The amount P is called the **present value of the annuity.** The quantity in brackets is abbreviated as $a_{\overline{n}|i}$, (read "a-angle-n at i"), so

$$a_{\overline{n}|i} = \frac{1 - (1 + i)^{-n}}{i}.$$

Compare this quantity with $s_{\overline{n}|i}$ in the previous section.
Here is a summary of what we have done.

PRESENT VALUE OF AN ANNUITY

The present value P of an annuity of n payments of R dollars each at the end of consecutive interest periods with interest compounded at a rate of interest i per period is

$$P = R\left[\frac{1 - (1 + i)^{-n}}{i}\right] \quad \text{or} \quad P = R \cdot a_{\overline{n}|i}.$$

CAUTION Don't confuse the formula for the present value of an annuity with the one for the future value of an annuity. Notice the difference: the numerator of the fraction in the present value formula is $1 - (1 + i)^{-n}$, but in the future value formula, it is $(1 + i)^n - 1$. ◆

EXAMPLE 1 Paul Eldersveld and Maria Gonzalez are both graduates of the Forestvire Institute of Technology. They both agree to contribute to the endowment fund of FIT. Eldersveld says that he will give $500 at the end of each year for 9 years. Gonzalez prefers to give a lump sum today. What lump sum can she give that will equal the present value of Eldersveld's annual gifts, if the endowment fund earns 7.5% compounded annually?

Here $R = 500$, $n = 9$, and $i = .075$ and we have

$$P = 500\left[\frac{1 - (1.075)^{-9}}{.075}\right]$$
$$= 3189.44.$$

Therefore, Gonzalez must donate a lump sum of $3189.44 today.   ■   1

TECHNOLOGY TIP It would be useful to store a program for finding present value in your calculator (if one is not built-in); refer to the Program Appendix. ✔

EXAMPLE 2 A car costs $12,000. After a down payment of $2000, the balance will be paid off in 36 equal monthly payments with interest of 12% per year on the unpaid balance. Find the amount of each payment.

A single lump sum payment of $10,000 today would pay off the loan. So, $10,000 is the present value of an annuity of 36 monthly payments with interest of 12%/12 = 1% per month. Thus, $P = 10,000$, $n = 36$, $i = .01$, and we must find the monthly payment R in the formula

$$P = R\left[\frac{1 - (1 + i)^{-n}}{i}\right]$$
$$10,000 = R\left[\frac{1 - (1.01)^{-36}}{.01}\right]$$
$$R \approx 332.1430981.$$

A monthly payment of $332.14 will be needed.   ■   2

AMORTIZATION A loan is **amortized** if both the principal and interest are paid by a sequence of equal periodic payments. In Example 2 above, a loan of $10,000 at 12% interest compounded monthly could be amortized by paying $332.14 per month for 36 months.

The periodic payment needed to amortize a loan may be found, as in Example 2, by solving the present value equation for R.

---

1 What lump sum deposited today would be equivalent to equal payments of

(a) $650 at the end of each year for 9 years at 4% compounded annually?

(b) $1000 at the end of each quarter for 4 years at 4% compounded quarterly?

Answers:
(a) $4832.97
(b) $14,717.87

2 Kelly Erin buys a small business for $174,000. She agrees to pay off the cost in payments at the end of each semiannual period for 7 years, with interest of 10% compounded semiannually on the unpaid balance. Find the amount of each payment.

Answer:
$17,578.17

AMORTIZATION PAYMENTS

A loan of P dollars at interest rate i per period may be amortized in n equal periodic payments of R dollars made at the end of each period, where

$$R = \frac{P}{a_{\overline{n}|i}} = \frac{P}{\left[\dfrac{1 - (1 + i)^{-n}}{i}\right]} = \frac{Pi}{1 - (1 + i)^{-n}} .$$

EXAMPLE 3 The Beckenstein family buys a house for $94,000 with a downpayment of $16,000. They take out a 30-year mortgage for $78,000 at an annual interest rate of 9.6%.

(a) Find the amount of the monthly payment needed to amortize this loan.

Here $P = 78,000$ and the monthly interest rate is $9.6\%/12 = .096/12 = .008$.* The number of monthly payments is $12 \cdot 30 = 360$. Therefore,

$$R = \frac{78,000}{a_{\overline{360}|.008}} = \frac{78,000}{\left[\dfrac{1 - (1.008)^{-360}}{.008}\right]} \approx \frac{78,000}{117.90229} = 661.56.$$

Monthly payments of $661.56 are required to amortize the loan.

(b) Find the total amount of interest paid when the loan is amortized over 30 years.

The Beckenstein family makes 360 payments of $661.56 each, for a total of $238,161.60. Since the amount of the loan was $78,000, the total interest paid is

$$\$238,161.60 - \$78,000 = \$160,161.60.$$

This large amount of interest is typical of what happens with a long mortgage. A 15-year mortgage would have higher payments but would involve significantly less interest. $\boxed{3}$

(c) Find the part of the first payment that is interest and the part that is applied to reducing the debt.

As we saw in part (a), the monthly interest rate is .008. During the first month, the entire $78,000 is owed. Interest on this amount for 1 month is found by the formula for simple interest.

$$I = Prt = 78,000(.008)(1) = 624$$

At the end of the month, a payment of $661.56 is made; since $624 of this is interest, a total of

$$\$661.56 - \$624 = \$37.56$$

is applied to the reduction of the original debt. ∎

AMORTIZATION SCHEDULES In Example 3, 360 payments are made to amortize a $78,000 loan. The loan balance after the first payment is reduced by only $37.56, which is much less than $(1/360)(78,000) \approx \$216.67$. Therefore, even though equal *payments* are made to amortize a loan, the loan *balance* does not decrease in equal steps. This fact is very important if a loan is paid off early.

$\boxed{3}$ If the mortgage in Example 3 runs for 15 years, find

(a) the monthly payment;

(b) the total amount of interest paid.

Answers:
(a) $819.21
(b) $69,457.80

*Mortgage rates are quoted in terms of annual interest, but it is always understood that the monthly rate is 1/12 of the annual rate and that interest is compounded monthly.

EXAMPLE 4   Jill Stuart borrows $1000 for 1 year at 12% annual interest compounded monthly.

(a) What is her monthly loan payment?
The monthly interest rate is 12%/12 = 1% = .01, so that her payment is

$$R = \frac{1000}{a_{\overline{12}|.01}} = \frac{1000}{\left[\dfrac{1 - (1.01)^{-12}}{.01}\right]} \approx \frac{1000}{11.25508} = \$88.85.$$

(b) After making three payments, she decides to pay off the remaining balance all at once. How much must she pay?
Since nine payments remain to be paid, they can be thought of as an annuity consisting of nine payments of $88.25 at 1% interest per period. The present value of this annuity is

$$88.85\left[\frac{1 - (1.01)^{-9}}{.01}\right] = 761.09.$$

So Jill's remaining balance, computed by this method, is $761.09.
An alternative method of figuring the balance is to consider the payments already made as an annuity of three payments. At the beginning, the present value of this annuity was

$$88.85\left[\frac{1 - (1.01)^{-3}}{.01}\right] = 261.31.$$

That is, if she had borrowed only $261.31, the loan would be repaid after these 3 payments. Considered this way, she has paid nothing toward the rest of her loan, so she still owes the difference $1000 − $261.31 = $738.69. Furthermore, she owes the interest on this amount for 3 months, for a total of

$$(738.69)(1.01)^3 = \$761.07.$$

This balance due differs from the one obtained by the first method by 2¢ because the monthly payment and the other calculations were rounded to the nearest penny. ■   4

4 Find the following for a car loan of $10,000 for 48 months at 9% interest compounded monthly.

(a) the monthly payment

(b) How much is still due after 12 payments have been made?

Answers:
(a) $248.85
(b) $7825.54 or $7825.56

Although most people wouldn't quibble about a 2¢ difference in the balance due in Example 4, the difference in other cases (larger amounts or longer terms) might be more than that. A bank or business must keep its books accurately to the nearest penny, so it must determine the balance due in cases like this unambiguously and exactly. This is done by means of an **amortization schedule,** which lists how much of each payment is interest, how much goes to reduce the balance, and how much is owed after *each* payment.

EXAMPLE 5   Determine the exact amount Jill Stuart in Example 4 owes after three monthly payments.
An amortization table for the loan is shown on the next page. It is obtained as follows. The annual interest rate is 12% compounded monthly, so the interest rate per month is 12%/12 = 1% = .01. When the first payment is made, 1 month's interest, namely .01(1000) = $10, is owed. Subtracting this from the $88.85 payment leaves $78.85 to be applied to repayment. Hence, the principal at the end of the first payment period is 1000 − 78.85 = $921.15, as shown in the "payment 1" line of the table.

When payment 2 is made, 1 month's interest on $921.15 is owed, namely .01(921.15) = $9.21. Subtracting this from the $88.85 payment leaves $79.64 to reduce the principal. Hence, the principal at the end of payment 2 is 921.15 − 79.64 = $841.51. The interest portion of payment 3 is based on this amount and the remaining lines of the table are found in a similar fashion.

Payment Number	Amount of Payment	Interest for Period	Portion to Principal	Principal at End of Period
0	—	—	—	$1000.00
1	$88.85	$10.00	$78.85	921.15
2	88.85	9.21	79.64	841.51
3	88.85	8.42	80.43	761.08
4	88.85	7.61	81.24	679.84
5	88.85	6.80	82.05	597.79
6	88.85	5.98	82.87	514.92
7	88.85	5.15	83.70	431.22
8	88.85	4.31	84.54	346.68
9	88.85	3.47	85.38	261.30
10	88.85	2.61	86.24	175.06
11	88.85	1.75	87.10	87.96
12	88.84	.88	87.96	0

The schedule shows that after three payments, she still owes $761.08, an amount that differs slightly from that obtained by either method in Example 4. ∎

The amortization schedule in Example 5 is typical. In particular, note that all payments are the same except the last one. It is often necessary to adjust the amount of the final payment to account for rounding off earlier and to ensure that the final balance is exactly 0.

An amortization schedule also shows how the periodic payments are applied to interest and principal. The amount going to interest decreases with each payment, while the amount going to reduce the principal owed increases with each payment.

TECHNOLOGY TIP Programs to produce amortization schedules are in the Program Appendix. ✔

6.4 EXERCISES

1. Which of the following is represented by $a_{\overline{n}|i}$?

(a) $\dfrac{(1 + i)^{-n} - 1}{i}$

(b) $\dfrac{(1 + i)^{n} - 1}{i}$

(c) $\dfrac{1 - (1 + i)^{-n}}{i}$

(d) $\dfrac{1 - (1 + i)^{n}}{i}$

2. Which of the choices in Exercise 1 represents $s_{\overline{n}|i}$?

Find each of the following.

3. $a_{\overline{15}|.06}$

4. $a_{\overline{10}|.03}$

5. $a_{\overline{18}|.04}$

6. $a_{\overline{30}|.01}$

7. $a_{\overline{16}|.01}$

8. $a_{\overline{32}|.02}$

9. Explain the difference between the present value of an annuity and the future value of an annuity. For a given annuity, which is larger? Why?

Find the present value of each ordinary annuity. (See Example 1.)

10. Payments of $5200 are made annually for 6 years at 5% compounded annually.

11. Payments of $1250 are made annually for 8 years at 4% compounded annually.

12. Payments of $675 are made semiannually for 10 years at 6% compounded semiannually.

13. Payments of $750 are made semiannually for 8 years at 6% compounded semiannually.

14. Payments of $16,908 are made quarterly for 3 years at 4.4% compounded quarterly.

15. Payments of $12,125 are made quarterly for 5 years at 5.6% compounded quarterly.

Find the lump sum deposited today that will yield the same total amount as payments of $10,000 at the end of each year for 15 yr, at each of the given interest rates. (See Example 1.)

16. 4% compounded annually

17. 5% compounded annually

18. 6% compounded annually

19. What does it mean to amortize a loan?

Find the payment necessary to amortize each of the following loans. (See Example 2.)

20. $800, 10 annual payments at 5%

21. $12,000, 8 quarterly payments at 6%

22. $25,000, 8 quarterly payments at 6%

23. $35,000, 12 quarterly payments at 4%

24. $5000, 36 monthly payments at 12%

25. $472, 48 monthly payments at 12%

Find the monthly house payment necessary to amortize the following loans. Interest is calculated on the unpaid balance. (See Examples 3 and 4.)

26. $49,560 at 10.75% for 25 years

27. $70,892 at 11.11% for 30 years

28. $53,762 at 12.45% for 30 years

29. $96,511 at 10.57% for 25 years

Use the amortization table in Example 5 to answer the questions in Exercises 30–33.

30. How much of the fifth payment is interest?

31. How much of the tenth payment is used to reduce the debt?

32. How much interest is paid in the first 5 months of the loan?

33. How much interest is paid in the last 5 months of the loan?

34. What sum deposited today at 5% compounded annually for 8 yr will provide the same amount as $1000 deposited at the end of each year for 8 yr at 6% compounded annually?

35. What lump sum deposited today at 8% compounded quarterly for 10 yr will yield the same final amount as deposits of $4000 at the end of each six-month period for 10 yr at 6% compounded semiannually?

Management   *Work the following applied problems.*

36. Stereo Shack sells a stereo system for $600 down and monthly payments of $30 for the next 3 yr. If the interest rate is 1.25% per month on the unpaid balance, find
 (a) the cost of the stereo system;
 (b) the total amount of interest paid.

37. Hong Le buys a car costing $6000. He agrees to make payments at the end of each monthly period for 4 years. He pays 12% interest, compounded monthly.
 (a) What is the amount of each payment?
 (b) Find the total amount of interest Le will pay.

38. A speculator agrees to pay $15,000 for a parcel of land; this amount, with interest, will be paid over 4 yr, with semiannual payments, at an interest rate of 10% compounded semiannually. Find the amount of each payment.

39. In the Million-Dollar Lottery, a winner is paid a million dollars at the rate of $50,000 per year for 20 yr. Assume that these payments form an ordinary annuity, and that the lottery managers can invest money at 6% compounded annually. Find the lump sum that the management must put away to pay off the million-dollar winner.

40. The Adams family bought a house for $81,000. They paid $20,000 down and took out a 30-year mortgage for the balance at 11%. Use one of the methods in Example 4 to estimate the mortgage balance after 100 payments.

41. After making 180 payments on their mortgage, the Adams family of Exercise 40 sells their house for $136,000. They must pay closing costs of $3700 plus 2.5% of the sale price of the house. Approximately how much money will they receive at the close of the sale after their current mortgage is paid off and closing costs are deducted?

In Exercises 42–45, prepare an amortization schedule showing the first 4 payments for each loan. (See Example 5.)

42. An insurance firm pays $4000 for a new printer for its computer. It amortizes the loan for the printer in 4 annual payments at 8% compounded annually.

43. Large semitrailer trucks cost $72,000 each. Ace Trucking buys such a truck and agrees to pay for it by a loan that will be amortized with 9 semiannual payments at 6% compounded semiannually.

44. One retailer charges $1048 for a computer monitor. A firm of tax accountants buys 8 of these monitors. They make a down payment of $1200 and agree to amortize the balance with monthly payments at 12% compounded monthly for 4 yr.

45. Joan Varozza plans to borrow $20,000 to stock her small boutique. She will repay the loan with semiannual payments for 5 yr at 7% compounded semiannually.

Student borrowers now have more options to choose from when selecting repayment plans.* The standard plan repays the loan in 10 years with equal monthly payments. The extended plan allows from 12 to 30 years to repay the loan. A student borrows $35,000 at 7.43% compounded monthly.

46. Find the monthly payment and total interest paid under the standard plan.

47. Find the monthly payment and total interest paid under the extended plan with 20 years to pay off the loan.

48. When Teresa Flores opened her law office, she bought $14,000 worth of law books and $7200 worth of office furniture. She paid $1200 down and agreed to amortize the balance with semiannual payments for 5 yr, at 12% compounded semiannually.
   (a) Find the amount of each payment.
   (b) When her loan had been reduced below $5000, Flores received a large tax refund and decided to pay off the loan. How many payments were left at this time?

49. Kareem Adams buys a house for $285,000. He pays $60,000 down and takes out a mortgage at 9.5% on the balance. Find his monthly payment and the total amount of interest he will pay if the length of the mortgage is
   (a) 15 yr;   **(b)** 20 yr;   **(c)** 25 yr.
   (d) When will half the 20-year loan be paid off?

50. Sandi Goldstein has inherited $25,000 from her grandfather's estate. She deposits the money in an account offering 6% interest compounded annually. She wants to make equal annual withdrawals from the account so that the money (principal and interest) lasts exactly 8 years.
   (a) Find the amount of each withdrawal.
   (b) Find the amount of each withdrawal if the money must last 12 years.

**The New York Times*, April 2, 1995, "Money and College," Saul Hansell, p. 28.

51. The Beyes plan to purchase a home for $212,000. They will pay 20% down and finance the remainder for 30 years at 8.9% interest, compounded monthly.*
   (a) How large are their monthly payments?
   (b) What will be their loan balance right after they have made their 96th payment?
   (c) How much interest will they pay during the 7th year of the loan?
   (d) If they were to increase their monthly payments by $150, how long would it take to pay off the loan?

52. Jeni Ramirez plans to retire in 20 years. She will make 240 equal monthly contributions to her retirement account. One month after her last contribution, she will begin the first of 120 monthly withdrawals from the account. She expects to withdraw $3500 per month. How large must her monthly contributions be in order to accomplish her goal if her account is assumed to earn interest of 10.5%, compounded monthly throughout the life of this problem?

53. Ron Okimura also plans to retire in 20 years. Ron will make 120 equal monthly contributions to his account. Ten years after his last contribution (120 months) he will begin the first of 120 monthly withdrawals from his account. Ron also expects to withdraw $3500 per month. His account also earns interest of 10.5%, compounded monthly throughout the life of this problem. How large must Ron's monthly contributions be in order to accomplish his goals?

54. Madcline and Nick Swenson took out a 30-year mortgage for $160,000 at 9.8% interest, compounded monthly. After they had made 12 years of payments (144 payments) they decided to refinance the remaining loan balance for 25 years at 7.2% interest, compounded monthly. What will be the balance on their loan 5 years after they refinance?

*Exercises 51–54 were supplied by Norman Lindquist at Western Washington University.

6.5   APPLYING FINANCIAL FORMULAS

We have presented a lot of new formulas in this chapter. By answering the following questions, you can decide which formula to use for a particular problem.

1. Is simple or compound interest involved?
   Simple interest is normally used for investments or loans of a year or less; compound interest is normally used in all other cases.

2. If simple interest is being used, what is being sought: interest amount, future value, present value, or discount?

3. If compound interest is being used, does it involve a lump sum (single payment) or an annuity (sequence of payments)?
   (a) For a lump sum,
      (i) Is ordinary compound interest or continuous interest involved?
      (ii) What is being sought: present value, future value, number of periods, or effective rate?

(b) For an annuity,
  (i) Is it an ordinary annuity (payment at the end of each period) or an annuity due (payment at the beginning of each period)?
  (ii) What is being sought: present value, future value, or payment amount?

Once you have answered these questions, choose the appropriate formula from the Chapter Summary, as in the following examples.

EXAMPLE 1   Karen La Bonte must pay $27,000 in settlement of an obligation in 3 years. What amount can she deposit today, at 4% compounded monthly, to have enough?

In this problem, the future value of $27,000 is known, and the present value must be found. Because the time is more than one year, we should use the formula for the present value of compound interest. ∎

EXAMPLE 2   A bond sells for $800 and pays interest of 7.5%. How much interest is earned in 6 months?

The time period is for less than a year, so use the simple interest formula. The cost of the bond is the principal P, so $P = 800$, $i = .075$, and $t = 6/12 = 1/2$ year. ∎

EXAMPLE 3   A new car is priced at $11,000. The buyer must pay $3000 down and pay the balance in 48 equal monthly payments at 12% compounded monthly. Find the payment.

The payments form an ordinary annuity with a present value of $11,000 - 3000 = 8000$. Use the formula for the present value of an annuity with $P = 8000$, $i = .01$, and $n = 48$ to find R, the amount of the payment. ∎

EXAMPLE 4   Hassi is paid on the first of each month and $80 is automatically deducted from his pay and deposited in a savings account. If the account pays 4.5% interest compounded monthly, how much will be in the account after 3 years and 9 months?

The time period is for more than a year, so we use compound interest. There will be $3(12) + 9 = 45$ monthly deposits of $80, so this is an annuity due (the payments are at the first of each month). We want the future value of the annuity. Use the formula for an annuity due with $R = 80$, $i = .045$, and $n = 45$. ∎

After you have analyzed the situation and chosen the correct formula, as in the preceding examples, work the problem. As a final step, consider whether the answer you get makes sense. For instance, present value should always be less than future value. Similarly, the future value of an annuity (which includes interest) should be larger than the sum of the payments.

6.5 EXERCISES

For the exercises in this section, round money amounts to the nearest cent, time to the nearest day, and rates to the nearest tenth of a percent.

Find the present value of $82,000 for the following number of years, if the money can be deposited at 12% compounded quarterly.

1. 5 years   **2.** 7 years

Find the amount of the payment necessary to amortize the following amounts.

3. $4250, 13 quarterly payments at 4%

4. $58,000, 23 quarterly payments at 6%

Find the compound amount for each of the following.

5. $4792.35 at 4% compounded semiannually for $5\frac{1}{2}$ years

6. $2500 at 6% compounded quarterly for $3\frac{3}{4}$ years

Find the simple interest for each of the following. Assume 360 days in a year.

7. $42,500 at 5.75% for 10 months

8. $32,662 at 6.882% for 225 days

Find the amount of interest earned by each of the following deposits.

9. $22,500 at 6% compounded quarterly for $5\frac{1}{4}$ years

10. $53,142 at 8% compounded monthly for 32 months

Find the compound amount and the amount of interest earned by a deposit of $32,750 at 5% compounded continuously for the following number of years.

11. $7\frac{1}{2}$ years

12. 9.2 years

Find the present value of the following future amounts. Assume 360 days in a year and use simple interest.

13. $17,320 for 9 months, money earns 3.5%

14. $122,300 for 138 days, money earns 4.75%

Find the proceeds for the following. Assume 360 days in a year and use simple interest.

15. $23,561 for 112 days, discount rate 4.33%

16. $267,100 for 271 days, discount rate 5.72%

Find the future value of each of the following annuities.

17. $2500 is deposited at the end of each semiannual period for $5\frac{1}{2}$ years, money earns 7% compounded semiannually

18. $800 is deposited at the end of each month for $1\frac{1}{4}$ years, money earns 8% compounded monthly

19. $250 is deposited at the end of each quarter for $7\frac{1}{4}$ years, money earns 4% compounded quarterly

20. $100 is deposited at the beginning of each year for 5 years, money earns 8% compounded yearly

Find the amount of each payment to be made to a sinking fund to accumulate the indicated amounts.

21. $10,000, money earns 5% compounded annually, 7 annual payments at the end of each year

22. $42,000, money earns 4% compounded quarterly, 13 quarterly payments at the end of each quarter

23. $100,000, money earns 6% compounded semiannually, 9 semiannual payments at the end of each semiannual period

24. $53,000, money earns 6% compounded monthly, 35 monthly payments at the end of each month

Find the present value of each ordinary annuity.

25. Payments of $1200 are made annually for 7 years at 5% compounded annually

26. Payments of $500 are made semiannually at 4% compounded semiannually for $3\frac{1}{2}$ years

27. Payments of $1500 are made quarterly for $5\frac{1}{4}$ years at 6% compounded quarterly

28. Payments of $905.43 are made monthly for $2\frac{11}{12}$ years at 8% compounded monthly

Prepare an amortization schedule for each of the following loans. Interest is calculated on the unpaid balance.

29. $8500 loan repaid in semiannual payments for $3\frac{1}{2}$ years at 8%

30. $40,000 loan repaid in quarterly payments for $1\frac{3}{4}$ years at 9%

Management   *Solve the following applied problems.*

31. A 10-month loan of $42,000 at simple interest produces $1785 interest. Find the interest rate.

32. Willa Burke deposits $803.47 at the end of each quarter for $3\frac{3}{4}$ years in an account paying 4% compounded quarterly. Find the final amount in the account and the amount of interest earned.

33. Makarim Wibison owes $7850 to a relative. He has agreed to pay the money in 5 months at simple interest of 7%. One month before the loan is due, the relative discounts the loan at the bank. The bank charges a 9.2% discount rate. How much money does the relative receive?

34. A small resort must add a swimming pool to compete with a new resort built nearby. The pool will cost $28,000. The resort borrows the money and agrees to repay it with equal payments at the end of each quarter for $6\frac{1}{2}$ years at an interest rate of 6% compounded quarterly. Find the amount of each payment.

35. According to the terms of a divorce settlement, one spouse must pay the other a lump sum of $2800 in 17 months. What lump sum can be invested today, at 6% compounded monthly, so that enough will be available for the payment?

36. An accountant loans $28,000 at simple interest to her business. The loan is at 9% and earns $3255 interest. Find the time of the loan in months.

37. A firm of attorneys deposits $5000 of profit sharing money at the end of each semiannual period for $7\frac{1}{2}$ years. Find the final amount in the account if the deposit earns 5% compounded semiannually. Find the amount of interest earned.

38. Find the principal that must be invested at 5.25% simple interest to earn $937.50 interest in 8 months.

39. In 3 years Ms. Thompson must pay a pledge of $7500 to her college's building fund. What lump sum can she deposit today, at 4% compounded semiannually, so that she will have enough to pay the pledge?

40. The owner of Eastside Hallmark borrows $48,000 to expand the business. The money will be repaid in equal payments at the end of each year for 7 years. Interest is 8%. Find the amount of each payment.

41. To buy a new computer, Mark Nguyen borrows $3250 from a friend at 9% interest compounded annually for 4 years.

Find the compound amount he must pay back at the end of the 4 years.

42. A small business invests some spare cash for 3 months at 5.2% simple interest and earns $244 in interest. Find the amount of the investment.

43. When the Lee family bought their home, they borrowed $115,700 at 10.5% compounded monthly for 25 years. If they make all 300 payments, repaying the loan on schedule, how much interest will they pay? Assume the last payment is the same as the previous ones.

44. Suppose $84,720 is deposited for 7 months and earns $2372.16 in interest. Find the rate of interest.

CHAPTER 6 SUMMARY

Key Terms and Symbols

6.1 interest
principal
rate
time
simple interest
future value (maturity value)
present value
discount (bank discount)
proceeds

6.2 compound interest
compound amount
compounding period
continuous compounding
nominal rate (stated rate)
effective rate

6.3 geometric sequence
common ratio
annuity
ordinary annuity
payment period
term of an annuity
future value of an annuity
annuity due
sinking fund

6.4 present value of an annuity
amortize a loan
amortization schedule

Key Concepts

Simple Interest

The **simple interest** I on an amount of P dollars for t years at interest rate r per year is $I = Prt$.

The **future value** A of P dollars at simple interest rate r for t years is $A = P(1 + rt)$.

The **present value** P of a future amount of A dollars at simple interest rate r for t years is

$$P = \frac{A}{1 + rt}.$$

If D is the **discount** on a loan having maturity value A at simple interest rate r for t years, then $D = Art$. If D is the discount and P the **proceeds** of a loan having maturity value A at simple interest rate r for t years, then $P = A - D$ or $P = A(1 - rt)$.

Compound Interest

If P dollars is deposited for n time periods at compound interest rate i per period, the **compound amount (future value)** A is $A = P(1 + i)^n$.

The **present value** P of A dollars at compound interest rate i per period for n periods is

$$P = \frac{A}{(1 + i)^n} = A(1 + i)^{-n}.$$

The **effective rate** corresponding to a stated interest rate r per year, compounded m times per year, is

$$r_e = \left(1 + \frac{r}{m}\right)^m - 1.$$

Continuous Compound Interest

If P dollars is deposited for t years at interest rate r per year, compounded continuously, the **compound amount (future value)** A is $A = Pe^{rt}$.

The **present value** P of A dollars at interest rate r per year compounded continuously for t years is

$$P = \frac{A}{e^{rt}}.$$

Annuities

The **future value S of an ordinary annuity** of n payments of R dollars each at the end of consecutive interest periods with interest compounded at rate i per period is

$$S = R\left[\frac{(1 + i)^n - 1}{i}\right] \quad \text{or} \quad S = R \cdot s_{\overline{n}|\,i}.$$

The **present value P of an ordinary annuity** of n payments of R dollars each at the end of consecutive interest periods with interest compounded at rate i per period is

$$P = R\left[\frac{1 - (1 + i)^{-n}}{i}\right] \quad \text{or} \quad P = R \cdot a_{\overline{n}|\,i}.$$

To find the payment, solve the formula for R.

The **future value S of an annuity due** of n payments of R dollars each at the beginning of consecutive interest periods with interest compounded at rate i per period is

$$S = R\left[\frac{(1 + i)^{n+1} - 1}{i}\right] - R \quad \text{or} \quad S = R \cdot s_{\overline{n+1}|\,i} - R.$$

Chapter 6 Review Exercises

Find the simple interest for the following loans.

1. $4902 at 9.5% for 11 months

2. $42,368 at 15.22% for 5 months

3. $3478 at 7.4% for 88 days (assume a 360-day year)

4. $2390 at 18.7% from May 3 to July 28 (assume 365 days in a year)

5. What is meant by the present value of an amount A?

Find the present value of the following future amounts. Assume 360 days in a year; use simple interest.

6. $459.57 in 7 months, money earns 8.5%

7. $80,612 in 128 days, money earns 6.77%

8. Explain what happens when a borrower is charged a discount. What are the proceeds?

Find the proceeds in Exercises 9 and 10. Assume 360 days in a year.

9. $802.34; discount rate 8.6%; length of loan 11 mo

10. $12,000; discount rate 7.09%; length of loan 145 days

11. For a given amount of money at a given interest rate for a given time period greater than 1 year, does simple interest or compound interest produce more interest? Explain.

Find the compound amount and the amount of interest earned in each of the following.

12. $2800 at 6% compounded annually for 10 yr

13. $57,809.34 at 4% compounded quarterly for 5 yr

14. $12,903.45 at 6.37% compounded quarterly for 29 quarters

15. $4677.23 at 4.57% compounded monthly for 32 mo

Find the present value of the following amounts.

16. $42,000 in 7 yr, 12% compounded monthly

17. $17,650 in 4 yr, 8% compounded quarterly

18. $1347.89 in 3.5 yr, 6.77% compounded semiannually

19. $2388.90 in 44 mo, 5.93% compounded monthly

20. Write the first five terms of the geometric sequence with $a = 2$ and $r = 3$.

21. Write the first four terms of the geometric sequence with $a = 4$ and $r = 1/2$.

22. Find the sixth term of the geometric sequence with $a = -3$ and $r = 2$.

23. Find the fifth term of the geometric sequence with $a = -2$ and $r = -2$.

24. Find the sum of the first four terms of the geometric sequence with $a = -3$ and $r = 3$.

25. Find the sum of the first five terms of the geometric sequence with $a = 8000$ and $r = -1/2$.

26. Find $s_{\overline{30}|.01}$.

27. What is meant by the future value of an annuity?

Find the future value of each annuity.

28. $1288 deposited at the end of each year for 14 yr; money earns 8% compounded annually

29. $4000 deposited at the end of each quarter for 7 yr; money earns 6% compounded quarterly

30. $233 deposited at the end of each month for 4 yr; money earns 12% compounded monthly

31. $672 deposited at the beginning of each quarter for 7 yr; money earns 8% compounded quarterly

32. $11,900 deposited at the beginning of each month for 13 mo; money earns 12% compounded monthly

33. What is the purpose of a sinking fund?

Find the amount of each payment that must be made into a sinking fund to accumulate the following amounts. Assume payments are made at the end of each period.

34. $6500; money earns 5% compounded annually; 6 annual payments

35. $57,000; money earns 6% compounded semiannually for $8\frac{1}{2}$ yr

36. $233,188; money earns 5.7% compounded quarterly for $7\frac{3}{4}$ yr

37. $56,788; money earns 6.12% compounded monthly for $4\frac{1}{2}$ yr

Find the present value of each ordinary annuity.

38. Payments of $850 annually for 4 yr at 5% compounded annually

39. Payments of $1500 quarterly for 7 yr at 8% compounded quarterly

40. Payments of $4210 semiannually for 8 yr at 8.6% compounded semiannually

41. Payments of $877.34 monthly for 17 mo at 6.4% compounded monthly

42. Give two examples of the types of loans that are commonly amortized.

Find the amount of the payment necessary to amortize each of the following loans.

43. $32,000 at 9.4% compounded quarterly, 10 quarterly payments

44. $5607 at 7.6% compounded monthly, 32 monthly payments

Find the monthly house payments for the following mortgages.

45. $56,890 at 10.74% for 25 yr

46. $77,110 at 8.45% for 30 yr

A portion of an amortization table is given below for a $127,000 loan at 8.5% interest compounded monthly for 25 yr.

Payment Number	Amount of Payment	Interest for Period	Portion to Principal	Principal at End of Period
0	—	—	—	$127,000.00
1	$1022.64	$899.58	$123.06	126,876.94
2	1022.64	898.71	123.93	126,753.02
3	1022.64	897.83	124.80	126,628.21
4	1022.64	896.95	125.69	126,502.53
5	1022.64	896.06	126.58	126,375.95
6	1022.64	895.16	127.48	126,248.47

Use the table to answer the following questions.

47. How much of the fifth payment is interest?

48. How much of the sixth payment is used to reduce the debt?

49. How much interest is paid in the first 3 months of the loan?

50. How much has the debt been reduced at the end of the first 6 months?

Management *Work the following applied problems.*

51. Larry DiCenso needs $9812 to buy new equipment for his business. The bank charges a discount rate of 12%. Find the amount of DiCenso's loan if he borrows the money for 7 mo.

52. A florist borrows $1400 at simple interest to pay taxes. The loan is at 11.5% and costs $120.75 in interest. Find the time of the loan in months.

53. A developer deposits $84,720 for 7 mo and earns $4055.46 in interest. Find the interest rate.

54. Tom Wilson owes $5800 to his mother. He has agreed to pay back the money in 10 mo, at an interest rate of 10%. Three months before the loan is due, Tom's mother discounts the loan at the bank to get cash for $6000 worth of new furniture. The bank charges a 13.45% discount rate. How much money does Tom's mother receive? Is it enough to buy the furniture?

55. In 3 yr Ms. Flores must pay a pledge of $7500 to her favorite charity. What lump sum can she deposit today, at 10% compounded semiannually, so that she will have enough to pay the pledge?

56. Each year a firm must set aside enough funds to provide employee retirement benefits of $52,000 in 20 years. If the firm can invest money at 7.5% compounded monthly, what amount must be invested at the end of each month for this purpose?

57. Chalon Bridges deposits semiannual payments of $3200, received in payment of a debt, in an ordinary annuity at 6.8% compounded semiannually. Find the final amount in the account and the interest earned at the end of 3.5 years.

58. To finance the $15,000 cost of their kitchen remodeling, the Chews will make equal payments at the end of each month for 36 months. They will pay interest at the rate of 7.2% compounded monthly. Find the amount of each payment.

59. To expand her business, the owner of a small restaurant borrows $40,000. She will repay the money in equal payments at the end of each semiannual period for 8 years at 9% interest compounded semiannually. What payments must she make?

60. Pension experts recommend that you start drawing at least 40% of your full pension as early as possible.* Suppose you

have built up a pension with $12,000 annual payments by working 10 years for a company, when you leave to accept a better job. The company gives you the option of collecting half the full pension when you reach age 55 or the full pension at age 65. Assume an interest rate of 8% compounded annually. By age 75, how much will each plan produce? Which plan would produce the larger amount?

▶ **61.** The Taggart family bought a house for $91,000. They paid $20,000 down and took out a 30-yr mortgage for the balance at 9%.

(a) Find their monthly payment.

(b) How much of the first payment is interest?

After 180 payments, the family sells their house for $136,000. They must pay closing costs of $3700 plus 2.5% of the sale price.

(c) Estimate the current mortgage balance at the time of the sale using one of the methods from Example 4 in Section 6.4.

(d) Find the total closing costs.

(e) Find the amount of money they receive from the sale after paying off the mortgage.

▶ **62.** The proceeds of a $10,000 death benefit are left on deposit with an insurance company for 7 yr at an annual effective interest rate of 5%.* The balance at the end of 7 yr is paid to the beneficiary in 120 equal monthly payments of X, with the first payment made immediately. During the payout period, interest is credited at an annual effective interest rate of 3%. Calculate X.

(a) 117   (b) 118   (c) 129   (d) 135   (e) 158

▶ **63.** Gene deposits $500 per quarter to his nest egg account. The account earns interest of 9%, compounded quarterly. Right after Gene makes his 16th deposit, he loses his job and cannot make any deposits for the next 5 years (20 quarters). Eventually Gene gets another job and again begins making deposits to his account. Since he missed so many deposits while he was out of work, Gene now deposits $750 per quarter. His first $750 deposit comes exactly 20 quarters after his last $500 deposit. What will be Gene's account balance right after he has made his 32nd $750 deposit?†

▶ **64.** Cathy wants to retire on $55,000 per year for her life expectancy of 20 years. She estimates that she will be able to earn interest of 9%, compounded annually, throughout her lifetime. To reach her retirement goal, Cathy will make annual contributions to her account for the next 25 years. One year after making her last contribution, she will take her first retirement check. How large must her yearly contributions be?

---

*Problem from "Course 140 Examination, Mathematics of Compound Interest" of the Education and Examination Committee of The Society of Actuaries. Reprinted by permission of The Society of Actuaries.

†Exercises 63 and 64 supplied by Norman Lindquist, Western Washington University.

*Smart Money, October, 1994, "Pocket That Pension," p. 33.

CASE 6

Time, Money, and Polynomials*

A *time line* is often helpful for evaluating complex investments. For example, suppose you buy a $1000 CD at time t_0. After one year $2500 is added to the CD at t_1. By time t_2, after another year, your money has grown to $3851 with interest. What rate of interest, called *yield to maturity* (*YTM*), did your money earn? A time line for this situation is shown in Figure 1.

FIGURE 1

Assuming interest is compounded annually at a rate i, and using the compound interest formula gives the following description of the YTM.

$$1000(1 + i)^2 + 2500(1 + i) = 3851$$

To determine the yield to maturity, we must solve this equation for i. Since the quantity $1 + i$ is repeated, let $x = 1 + i$ and first solve the second-degree (quadratic) polynomial equation for x.

$$1000x^2 + 2500x - 3851 = 0$$

We can use the quadratic formula with $a = 1000$, $b = 2500$, and $c = -3851$.

$$x = \frac{-2500 \pm \sqrt{2500^2 - 4(1000)(-3851)}}{2(1000)}$$

We get $x = 1.0767$ and $x = -3.5767$. Since $x = 1 + i$, the two values for i are .0767 = 7.67% and −4.5767 = −457.67%. We reject the negative value because the final accumulation is greater than the sum of the deposits. In some applications, however, negative rates may be meaningful. By checking in the first equation, we see that the yield to maturity for the CD is 7.67%.

Now let us consider a more complex but realistic problem. Suppose Curt Reynolds has contributed for four years to a retirement fund. He contributed $6000 at the beginning of the first

year. At the beginning of the next three years, he contributed $5840, $4000, and $5200, respectively. At the end of the fourth year, he had $29,912.38 in his fund. The interest rate earned by the fund varied between 21% and −3%, so Reynolds would like to know the YTM = i for his hard-earned retirement dollars. From a time line (see Figure 2), we set up the following equation in $1 + i$ for Reynolds' savings program.

$$6000(1 + i)^4 + 5840(1 + i)^3 + 4000(1 + i)^2 + 5200(1 + i) = 29{,}912.38$$

Let $x = 1 + i$. We need to solve the fourth-degree polynomial equation

$$f(x) = 6000x^4 + 5840x^3 + 4000x^2 + 5200x - 29{,}912.38 = 0.$$

There is no simple way to solve a fourth-degree polynomial equation, so we will use a guess and check method.

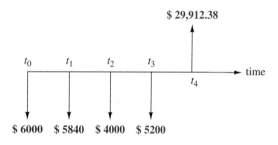

FIGURE 2

We expect that $0 < i < 1$, so that $1 < x < 2$. Let us calculate $f(1)$ and $f(2)$. If there is a change of sign, we will know that there is a solution to $f(x) = 0$ between 1 and 2. We find that

$$f(1) = -8872.38 \text{ and } f(2) = 139{,}207.62.$$

There is a change in sign as expected. Now we find $f(1.1)$, $f(1.2)$, $f(1.3)$, and so on, and look for another change in sign. (A computer or calculator program would help here to try values of x between 1 and 2.) Right away we find

$$f(1.1) = -2794.74 \quad \text{and} \quad f(1.2) = 4620.74.$$

This process can be repeated for values of x between 1.1 and 1.2, to get $f(1.11) = -2116.59$, $f(1.12) = -1424.88$, $f(1.13) = -719.42$, and $f(1.14) = 0$. We were lucky; the solution for $f(x) = 0$ is exactly 1.14, so i = YTM = .14 = 14%.

*From *Time, Money, and Polynomials*, COMAP.

EXERCISES

1. Lucinda Turley received $50 on her 16th birthday, and $70 on her 17th birthday, both of which she immediately invested in the bank with interest compounded annually. On her 18th birthday, she had $127.40 in her account. Draw a time line, set up a polynomial equation, and calculate the YTM.

2. At the beginning of the year, Jay Beckenstein invested $10,000 at 5% for the first year. At the beginning of the second year, he added $12,000 to the account. The total account earned 4.5% for the second year.
   (a) Draw a time line for this investment.
   (b) How much was in the fund at the end of the second year?
   (c) Set up and solve a polynomial equation and determine the YTM. What do you notice about the YTM?

3. On January 2 each year for three years, Earl Karn deposited bonuses of $1025, $2200, and $1850, respectively, in an account. He received no bonus the following year, so he made no deposit. At the end of the fourth year, there was $5864.17 in the account.
   (a) Draw a time line for these investments.
   (b) Write a polynomial equation in x ($x = 1 + i$) and use the guess and check method to find the YTM for these investments.

4. Pat Kelley invested yearly in a fund for his children's college education. At the beginning of the first year, he invested $1000; at the beginning of the second year, $2000; at the third through the sixth, $2500 each year; and at the beginning of the seventh, he invested $5000. At the beginning of the eighth year, there was $21,259 in the fund.
   (a) Draw a time line for this investment program.
   (b) Write a seventh-degree polynomial equation in $1 + i$ that gives the YTM for this investment program.
   (c) Use a graphing calculator to show that the YTM is less than 5.07% and greater than 5.05%.
   (d) Use a graphing calculator to calculate the solution for $1 + i$ and find the YTM.

5. People often lose money on investments. Jim Carlson invested $50 at the beginning of each of two years in a mutual fund, and at the end of two years his investment was worth $90.
   (a) Draw a time line and set up a polynomial equation in $1 + i$. Solve for i.
   (b) Examine each negative solution (rate of return on the investment) to see if it has a reasonable interpretation in the context of the problem. To do this, use the compound interest formula on each value of i to trace each $50 payment to maturity.

CHAPTER 7

Systems of Linear Equations and Matrices

Many applications of mathematics require finding the solution of a *system* of first-degree equations. This chapter presents methods for solving such systems, including matrix methods. Matrix algebra and other applications of matrices are also discussed.

7.1   SYSTEMS OF LINEAR EQUATIONS

This section deals with **linear** (or **first-degree**) **equations** such as

$$2x + 3y = 14 \quad \text{linear equation in two variables,}$$
$$4x - 2y + 5z = 8 \quad \text{linear equation in three variables,}$$

and so on. A **solution** of such an equation is an ordered set of numbers that when substituted for the variables in the order they appear produces a true statement. For instance, $(1, 4)$ is a solution of the equation $2x + 3y = 14$ because substituting $x = 1$ and $y = 4$ produces the true statement $2(1) + 3(4) = 14$. Similarly, $(0, -4, 0)$ is a solution of $4x - 2y + 5z = 8$ because $4(0) - 2(-4) + 5(0) = 8$.

Many applications involve **systems of linear equations,** such as these two:

Two equations in two variables	Three equations in four variables
$5x - 3y = 7$	$2x + y + z = 3$
$2x + 4y = 8$	$x + y + z + w = 5$
	$-4x + z + w = 0$

A **solution of a system** is a solution that satisfies *all* the equations in the system. For instance, in the right-hand system of equations above, $(1, 0, 1, 3)$ is a solution of all three equations (check it) and hence is a solution of the system. On the other hand, $(1, 1, 0, 3)$ is a solution of the first two equations but not the third. Hence $(1, 1, 0, 3)$ is not a solution of the system.   **1**

1 Which of $(-8, 3, 0)$, $(8, -2, -2)$, $(-6, 5, -2)$ are solutions of the system

$$x + 2y + 3z = -2$$
$$2x + 6y + z = 2$$
$$3x + 3y + 10z = -2?$$

Answer:
Only $(8, -2, -2)$

SYSTEMS OF TWO EQUATIONS IN TWO VARIABLES   The graph of a linear equation in two variables is a straight line. Earlier we saw that the coordinates of each point on the graph represent a solution of the equation. Thus, the solution of a system of two such equations is represented by the point or points where the two lines intersect. There are exactly three geometric possibilities for two lines: they intersect at a single point, or they coincide, or they are distinct and parallel. As illustrated in Figure 7.1, each of these geometric possibilities corresponds to an algebraic outcome for which special terminology is used.

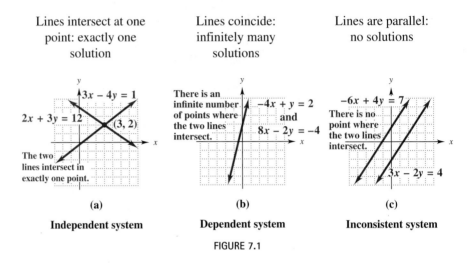

Lines intersect at one point: exactly one solution	Lines coincide: infinitely many solutions	Lines are parallel: no solutions
(a)	(b)	(c)
Independent system	**Dependent system**	**Inconsistent system**

FIGURE 7.1

In theory, every system of two equations in two variables can be solved by graphing the lines and finding their intersection points (if any). In practice, however, algebraic techniques are usually needed to determine the solutions precisely. Such systems can be solved by the **elimination method,** as illustrated in the following examples.

EXAMPLE 1 Solve the system

$$3x - 4y = 1 \qquad (1)$$
$$2x + 3y = 12. \qquad (2)$$

To eliminate one variable by addition of the two equations, the coefficients of either x or y in the two equations must be additive inverses. For example, let us choose to eliminate x. We multiply both sides of equation (1) by 2 and both sides of equation (2) by -3 to get

$$6x - 8y = 2$$
$$-6x - 9y = -36.$$

Adding these equations gives a new equation with just one variable, y. This equation can be solved for y.

$$
\begin{array}{rl}
6x - 8y = & 2 \\
-6x - 9y = & -36 \\
\hline
-17y = & -34 \qquad \text{Variable } x \text{ is eliminated.} \\
y = & 2
\end{array}
$$

2 Solve the system of equations

$$3x + 2y = -1$$
$$5x - 3y = 11.$$

Draw the graph of each equation on the same axes.

Answer:
(1, −2)

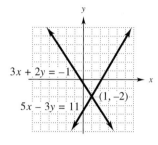

To find the corresponding value of x, substitute 2 for y in either of equations (1) or (2). We choose equation (1).

$$3x - 4(2) = 1$$
$$3x - 8 = 1$$
$$x = 3$$

Therefore, the solution of the system is (3, 2). The graphs of both equations of the system are shown in Figure 7.1(a). They intersect at the point (3, 2), the solution of the system. ■ **2**

TECHNOLOGY TIP To find approximate solutions of a system of two equations in two variables with a graphing calculator, solve each equation for y and graph both of them on the same screen. In Example 1, for instance, we would graph

$$y_1 = \frac{3x - 1}{4} \quad \text{and} \quad y_2 = \frac{12 - 2x}{3}.$$

The point where the graphs intersect (the solution of the system) can then be found by using the calculator's intersection finder, as shown in Figure 7.2. ✔

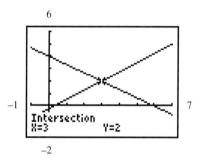

FIGURE 7.2

EXAMPLE 2   Solve the system

$$-4x + y = 2$$
$$8x - 2y = -4.$$

Eliminate x by multiplying both sides of the first equation by 2 and adding the results to the second equation.

$$-8x + 2y = 4$$
$$\underline{8x - 2y = -4}$$
$$0 = 0 \qquad \text{Both variables eliminated.}$$

3 Solve the following system.

$$3x - 4y = 13$$
$$12x - 16y = 52$$

Answer:
All ordered pairs that satisfy the equation $3x - 4y = 13$ (or $12x - 16y = 52$)

Although both variables have been eliminated, the resulting statement "$0 = 0$" is true, which is the algebraic indication that the two equations have the same graph, as shown in Figure 7.1(b). Therefore the system is dependent and has an infinite number of solutions: every ordered pair that is a solution of the equation $-4x + y = 2$ is a solution of the system. ■ **3**

4 Solve the system

$$x - y = 4$$
$$2x - 2y = 3.$$

Draw the graph of each equation on the same axes.

Answer:
No solution

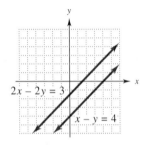

$$2x - 2y = 3$$
$$x - y = 4$$

EXAMPLE 3 Solve the system

$$3x - 2y = 4$$
$$-6x + 4y = 7.$$

The graphs of these equations are parallel lines (each has slope 3/2), as shown in Figure 7.1(c). Therefore the system has no solution. However, you do not need the graphs to discover this fact. If you try to solve the system algebraically by multiplying both sides of the first equation by 2 and adding the results to the second equation, you obtain

$$\begin{array}{r} 6x - 4y = 8 \\ -6x + 4y = 7 \\ \hline 0 = 15. \end{array}$$

The false statement "0 = 15" is the algebraic signal that the system is inconsistent and has no solution. ■ **4**

LARGER SYSTEMS OF LINEAR EQUATIONS Two systems of equations are said to be **equivalent** if they have the same solutions. The basic procedure for solving a large system of equations is to use properties of algebra to transform the system into a simpler, equivalent system and then solve this simpler system.

---

5 Verify that $x = 2$, $y = 1$ is the solution of the system

$$x - 3y = -1$$
$$3x + 2y = 8.$$

(a) Replace the second equation by the sum of itself and -3 times the first equation.

(b) What is the solution of the system in part (a)?

Answers:
(a) The system becomes

$$x - 3y = -1$$
$$11y = 11.$$

(b) $x = 2$, $y = 1$

Performing any one of the following **elementary operations** on a system of linear equations produces an equivalent system.

1. Interchange any two equations.
2. Multiply both sides of an equation by a nonzero constant.
3. Replace an equation by the sum of itself and a constant multiple of another equation in the system.

Performing either of the first two elementary operations produces an equivalent system because rearranging the order of the equations or multiplying both sides of an equation by a constant does not affect the solutions of the individual equations, and hence does not affect the solutions of the system. No formal proof will be given here that performing the third elementary operation produces an equivalent system, but Side Problem 5 illustrates this fact. **5**

Example 4 shows how to use elementary operations on a system to eliminate certain variables and produce an equivalent system that is easily solved. The italicized statements provide an outline of the procedure.

EXAMPLE 4 Use the elimination method to solve the system

$$2x + y - z = 2$$
$$x + 3y + 2z = 1$$
$$x + y + z = 2.$$

First, *use elementary operations to produce an equivalent system in which 1 is the coefficient of x in the first equation.* One way to do this is to interchange the first two equations (another would be to multiply both sides of the first equation by 1/2).

$$\begin{aligned} x + 3y + 2z &= 1 \\ 2x + y - z &= 2 \\ x + y + z &= 2 \end{aligned} \qquad \text{Interchange } R_1, R_2.$$

Here and below we use R_1 to denote the first equation in a system, R_2 the second equation, and so on.

Next, *use elementary operations to produce an equivalent system in which the x-term has been eliminated from the second and third equations.* To eliminate the x term from the second equation above, replace the second equation by the sum of itself and -2 times the first equation.

$$\begin{array}{r|r} -2R_1 & -2x - 6y - 4z = -2 \\ R_2 & 2x + y - z = 2 \\ \hline -2R_1 + R_2 & -5y - 5z = 0 \end{array}$$

$$\begin{aligned} x + 3y + 2z &= 1 \\ -5y - 5z &= 0 \qquad -2R_1 + R_2 \\ x + y + z &= 2 \end{aligned}$$

To eliminate the x-term from the third equation of this last system, replace the third equation by the sum of itself and -1 times the first equation.

$$\begin{array}{r|r} -1R_1 & -x - 3y - 2z = -1 \\ R_3 & x + y + z = 2 \\ \hline -1R_1 + R_3 & -2y - z = 1 \end{array}$$

$$\begin{aligned} x + 3y + 2z &= 1 \\ -5y - 5z &= 0 \\ -2y - z &= 1 \qquad -1R_1 + R_3 \end{aligned}$$

Now that x has been eliminated from all but the first equation, we ignore the first equation and work on the remaining ones. *Use elementary operations to produce an equivalent system in which 1 is the coefficient of y in the second equation.* This can be done by multiplying the second equation in the system above by $-1/5$.

$$\begin{aligned} x + 3y + 2z &= 1 \\ y + z &= 0 \qquad -\frac{1}{5}R_2 \\ -2y - z &= 1 \end{aligned}$$

Then *use elementary operations to obtain an equivalent system in which y has been eliminated from the third equation:* replace the third equation by the sum of itself and 2 times the second equation:

$$\begin{aligned} x + 3y + 2z &= 1 \\ y + z &= 0 \\ z &= 1. \qquad 2R_2 + R_3 \end{aligned}$$

This last system is said to be in **triangular form,** a form in which the solution of the third equation is obvious: $z = 1$. Now work backward in the system. Substitute 1 for

z in the second equation and solve for y, obtaining $y = -1$. Finally, substitute 1 for z and -1 for y in the first equation and solve for x, obtaining $x = 2$. This process is known as **back substitution.** When it is finished we have the solution of the original system, namely, $(2, -1, 1)$. It is always wise to check the solution by substituting the values for x, y, and z in *all* equations of the original system. ■

The procedure used in Example 4 to eliminate variables and produce a system in which back substitution works can be carried out with any system, as summarized below. In this summary, the first variable that appears in an equation with nonzero coefficient is called the **leading variable** of that equation and its nonzero coefficient is called the **leading coefficient.**

THE ELIMINATION METHOD FOR SOLVING LARGE SYSTEMS OF LINEAR EQUATIONS

Use elementary operations to transform the given system into an equivalent one in which the leading coefficient in each equation is 1 and the leading variable in each equation does not appear in any later equations. This can be done systematically as follows:

1. Make the leading coefficient of the first equation 1.
2. Eliminate the leading variable of the first equation from each later equation by replacing the later equation by the sum of itself and a suitable multiple of the first equation.
3. Repeat Steps 1 and 2 for the second equation: make its leading coefficient 1 and eliminate its leading variable from each later equation by replacing the later equation by the sum of itself and a suitable multiple of the second equation.
4. Repeat Steps 1 and 2 for the third equation, fourth equation, and so on, until it is not possible to go further.

Then solve the resulting system by back substitution.

---

[6] Use the elimination method to solve each system.

(a) $2x + y = -1$
$\quad x + 3y = 2$

(b) $2x - y + 3z = 2$
$\quad x + 2y - z = 6$
$\quad -x - y + z = -5$

Answers:

(a) $(-1, 1)$

(b) $(3, 1, -1)$

At various stages in the elimination process, you may have a choice of elementary operations that can be used. As long as the final result is a system in which back substitution can be used, the choice does not matter. To avoid unnecessary errors, choose elementary operations that minimize the amount of computation and, as far as possible, avoid complicated fractions. [6]

MATRIX METHODS You may have noticed that the variables in a system of equations remain unchanged during the solution process. We really need to keep track of just the coefficients and the constants. For instance, consider the system in Example 4:

$$2x + y - z = 2$$
$$x + 3y + 2z = 1$$
$$x + y + z = 2.$$

7 (a) Write the augmented matrix of this system.

$$4x - 2y + 3z = 4$$
$$3x + 5y + z = -7$$
$$5x - y + 4z = 6$$

(b) Write the system of equations associated with this augmented matrix.

$$\begin{bmatrix} 2 & -2 & | & -2 \\ 1 & 1 & | & 4 \\ 3 & 5 & | & 8 \end{bmatrix}$$

Answers:

(a) $\begin{bmatrix} 4 & -2 & 3 & | & 4 \\ 3 & 5 & 1 & | & -7 \\ 5 & -1 & 4 & | & 6 \end{bmatrix}$

(b) $2x - 2y = -2$
$x + y = 4$
$3x + 5y = 8$

8 Perform the following row operations on the matrix

$$\begin{bmatrix} -1 & 5 \\ 3 & -2 \end{bmatrix}.$$

(a) Interchange R_1 and R_2.

(b) $2R_1$

(c) Replace R_2 by $-3R_1 + R_2$.

(d) Replace R_1 by $2R_2 + R_1$.

Answers:

(a) $\begin{bmatrix} 3 & -2 \\ -1 & 5 \end{bmatrix}$

(b) $\begin{bmatrix} -2 & 10 \\ 3 & -2 \end{bmatrix}$

(c) $\begin{bmatrix} -1 & 5 \\ 6 & -17 \end{bmatrix}$

(d) $\begin{bmatrix} 5 & 1 \\ 3 & -2 \end{bmatrix}$

This system can be written in an abbreviated form as

$$\begin{bmatrix} 2 & 1 & -1 & 2 \\ 1 & 3 & 2 & 1 \\ 1 & 1 & 1 & 2 \end{bmatrix} \quad R_1 - R_2 - R_1$$

Such a rectangular array of numbers, consisting of horizontal **rows** and vertical **columns,** is called a **matrix** (plural **matrices**). Each number in the array is an **element** or **entry.** To separate the constants in the last column of the matrix from the coefficients of the variables, we may use a vertical line, producing the following **augmented matrix.**

$$\begin{bmatrix} 2 & 1 & -1 & | & 2 \\ 1 & 3 & 2 & | & 1 \\ 1 & 1 & 1 & | & 2 \end{bmatrix} \quad \boxed{7}$$

The rows of the augmented matrix can be transformed in the same way as the equations of the system, since the matrix is just a shortened form of the system. The following **row operations** on the augmented matrix correspond to the elementary operations used on systems of equations.

Performing any one of the following **row operations** on the augmented matrix of a system of linear equations produces the augmented matrix of an equivalent system.

1. Interchange any two rows.
2. Multiply each element of a row by a nonzero constant.
3. Replace a row by the sum of itself and a constant multiple of another row of the matrix.

Row operations on a matrix are indicated by the same notation we used for elementary operations on a system of equations. For example, $2R_3 + R_1$ indicates the sum of 2 times row 3 and row 1. $\boxed{8}$

EXAMPLE 5 Use matrices to solve the system

$$x - 2y = 6 - 4z$$
$$x + 13z = 6 - y$$
$$-2x + 6y - z = -10.$$

First, put the system in the required form, with the constants on the right side of the equals sign and the terms with variables in the same order in each equation on the left side of the equals sign. Then write the augmented matrix of the system.

$$\begin{array}{c} x - 2y + 4z = 6 \\ x + y + 13z = 6 \\ -2x + 6y - z = -10 \end{array} \qquad \begin{bmatrix} 1 & -2 & 4 & | & 6 \\ 1 & 1 & 13 & | & 6 \\ -2 & 6 & -1 & | & -10 \end{bmatrix}$$

The matrix method is the same as the elimination method, except that row operations are used on the augmented matrix instead of elementary operations on the corresponding system of equations, as shown in this side-by-side comparison.

Equation Method

Replace the second equation by the sum of itself and -1 times the first equation.

$$x - 2y + 4z = 6$$
$$3y + 9z = 0$$
$$-2x + 6y - z = -10$$

Matrix Method

Replace the second row by the sum of itself and -1 times the first row.

$$\begin{bmatrix} 1 & -2 & 4 & | & 6 \\ 0 & 3 & 9 & | & 0 \\ -2 & 6 & -1 & | & -10 \end{bmatrix}$$
$$\leftarrow -1R_1 + R_2 \rightarrow$$

Replace the third equation by the sum of itself and 2 times the first equation.

$$x - 2y + 4z = 6$$
$$3y + 9z = 0$$
$$2y + 7z = 2$$

Replace the third row by the sum of itself and 2 times the first row.

$$\begin{bmatrix} 1 & -2 & 4 & | & 6 \\ 0 & 3 & 9 & | & 0 \\ 0 & 2 & 7 & | & 2 \end{bmatrix}$$
$$\leftarrow 2R_1 + R_3 \rightarrow$$

Multiply both sides of the second equation by 1/3.

$$x - 2y + 4z = 6$$
$$y + 3z = 0$$
$$2y + 7z = 2$$

Multiply each element of row 2 by 1/3.

$$\begin{bmatrix} 1 & -2 & 4 & | & 6 \\ 0 & 1 & 3 & | & 0 \\ 0 & 2 & 7 & | & 2 \end{bmatrix}$$
$$\leftarrow \frac{1}{3}R_2 \rightarrow$$

Replace the third equation by the sum of itself and -2 times the second equation.

$$x - 2y + 4z = 6$$
$$y + 3z = 0$$
$$z = 2$$

Replace the third row by the sum of itself and -2 times the second row.

$$\begin{bmatrix} 1 & -2 & 4 & | & 6 \\ 0 & 1 & 3 & | & 0 \\ 0 & 0 & 1 & | & 2 \end{bmatrix}$$
$$\leftarrow -2R_2 + R_3 \rightarrow$$

The system is now in triangular form. Using back substitution shows that the solution of the system is $(-14, -6, 2)$. ■ 9

9 Complete the matrix solution of the system with this augmented matrix.

$$\begin{bmatrix} 1 & 1 & 1 & | & 2 \\ 1 & -2 & 1 & | & -1 \\ 0 & 3 & 1 & | & 5 \end{bmatrix}$$

Answer:
$(-1, 1, 2)$

TECHNOLOGY TIP Virtually all graphing calculators have matrix capabilities. You should consult your instruction manual to learn how to enter a matrix and perform row operations. With most calculators you will need to enter the matrix dimensions (the number of rows and columns). For example, a matrix with three rows and four columns has dimensions 3×4.

A matrix can be put in triangular form in a single step on TI-83/85/86/92 calculators by using REF (in the MATH or OPS submenu of the MATRIX menu). "REF" stands for "Row Echelon Form," which is the triangular form in which 1 is the first nonzero entry in each row. ✔

DEPENDENT AND INCONSISTENT SYSTEMS The possible number of solutions of a system with more than two variables or equations is the same as for smaller systems. Such a system has exactly one solution (independent system); or infinitely many solutions (dependent system); or no solutions (inconsistent system). Both the equation method and the matrix method always produce the unique solution of an independent system. The matrix method also provides a useful way of describing the infinitely many solutions of a dependent system, as we now see.

EXAMPLE 6   Solve the system

$$2x - 3y + 4z = 6$$
$$x - 2y + z = 9.$$

Use the steps of the matrix method as far as possible. The augmented matrix is

$$\begin{bmatrix} 2 & -3 & 4 & | & 6 \\ 1 & -2 & 1 & | & 9 \end{bmatrix}.$$

$$\begin{bmatrix} 1 & -2 & 1 & | & 9 \\ 2 & -3 & 4 & | & 6 \end{bmatrix} \qquad \text{Interchange } R_1 \text{ and } R_2.$$

$$\begin{bmatrix} 1 & -2 & 1 & | & 9 \\ 0 & 1 & 2 & | & -12 \end{bmatrix} \qquad -2R_1 + R_2 \qquad \begin{array}{l} x - 2y + z = 9 \\ y + 2z = -12 \end{array}$$

The last augmented matrix above represents the system shown to its right. Since there are only two rows in the matrix, it is not possible to continue the process. The fact that the corresponding system has one variable (namely z) that is not the leading variable of an equation indicates a dependent system. Its solutions can be found as follows. Solve the second equation for y.

$$y = -2z - 12$$

Now substitute the result for y in the first equation, and solve for x.

$$x - 2y + z = 9$$
$$x - 2(-2z - 12) + z = 9$$
$$x + 4z + 24 + z = 9$$
$$x + 5z = -15$$
$$x = -5z - 15$$

Each choice of a value for z leads to values for x and y. For example,

$$\text{if } z = 1, \quad \text{then } x = -20 \quad \text{and} \quad y = -14;$$
$$\text{if } z = -6, \text{then } x = 15 \quad \text{and} \quad y = 0;$$
$$\text{if } z = 0, \quad \text{then } x = -15 \quad \text{and} \quad y = -12.$$

There are infinitely many solutions for the original system, since z can take on infinitely many values. The solutions are all ordered triples in the form

$$(-5z - 15, -2z - 12, z),$$

where z is any real number. ∎   **10**

10 Use the following values of z to find additional solutions for the system of Example 6.

(a) $z = 7$

(b) $z = -14$

(c) $z = 5$

Answers:

(a) $(-50, -26, 7)$

(b) $(55, 16, -14)$

(c) $(-40, -22, 5)$

Since both x and y in Example 6 were expressed in terms of z, the variable z is called a **parameter.** If we solved the system in a different way, x or y could be the parameter. The system in Example 6 had one more variable than equations. If there are two more variables than equations, there usually will be two parameters, and so on.

Whenever there are more variables than equations, as in Example 6, then the system cannot have a unique solution. It must be either dependent (infinitely many solutions) or inconsistent (no solutions).

When a system is inconsistent, the matrix method will indicate this fact, too, as in the next example.

EXAMPLE 7   Solve the system

$$4x + 12y + 8z = -4$$
$$2x + 8y + 5z = 0$$
$$3x + 9y + 6z = 2$$
$$3x + 2y - z = 6.$$

Write the augmented matrix and go through the steps of the matrix method.

$$\begin{bmatrix} 4 & 12 & 8 & -4 \\ 2 & 8 & 5 & 0 \\ 3 & 9 & 6 & 2 \\ 3 & 2 & -1 & 6 \end{bmatrix}$$

$$\begin{bmatrix} 1 & 3 & 2 & -1 \\ 2 & 8 & 5 & 0 \\ 3 & 9 & 6 & 2 \\ 3 & 2 & -1 & 6 \end{bmatrix} \quad (1/4)R_1$$

$$\begin{bmatrix} 1 & 3 & 2 & -1 \\ 0 & 2 & 1 & 2 \\ 3 & 9 & 6 & 2 \\ 3 & 2 & -1 & 6 \end{bmatrix} \quad -2R_1 + R_2$$

$$\begin{bmatrix} 1 & 3 & 2 & -1 \\ 0 & 2 & 1 & 2 \\ 0 & 0 & 0 & 5 \\ 3 & 2 & -1 & 6 \end{bmatrix} \quad -3R_1 + R_3$$

11 Complete the matrix solution of the system.

$$\begin{bmatrix} -1 & 3 & -2 & -1 \\ 1 & -2 & 3 & 1 \\ 2 & -4 & 6 & 5 \end{bmatrix}$$

Answer:
No solution

Stop! The third row has all zeros to the left of the vertical bar, so the corresponding equation is $0x + 0y + 0z = 5$. This equation has no solution because its left side is 0 and its right side is 5, resulting in the false statement "$0 = 5$." Therefore the entire system cannot have any solutions and is inconsistent. ■ **11**

CAUTION   It is possible for the elimination method to result in an equation that is 0 on *both* sides, such as $0x + 0y + 0z = 0$. Unlike the situation in Example 7, such an equation has infinitely many solutions, so a system that contains it may also have solutions.

As a general rule, there is no way to determine in advance whether a system is independent, dependent, or inconsistent. So you should carry out the elimination process as far as necessary. If you obtain an equation of the form $0 = c$ for some nonzero c (as in Example 7), then the system is inconsistent and has no solutions. Otherwise, the system will either be independent with a unique solution (Example 4) or dependent with infinitely many solutions (Example 6). ◆

APPLICATIONS   The mathematical techniques in this text will be useful to you only if you are able to apply them to practical problems. To do this, always begin by reading the problem carefully. Next, identify what must be found. Let each unknown quantity be represented by a variable. (It is a good idea to *write down* exactly what each variable represents.) Now reread the problem, looking for all necessary data. Write that down, too. Finally, look for one or more sentences that lead to equations or inequalities.

EXAMPLE 8   Kelly Karpet Kleaners sells rug cleaning machines. The EZ model weighs 10 pounds and comes in a 10 cubic ft box. The compact model weighs 20 pounds and comes in an 8 cubic ft box. The commercial model weighs 60 pounds and comes in a 28 cubic ft box. Each of their delivery vans has 248 cubic feet of space and can hold a maximum of 440 pounds. In order for a van to be fully loaded, how many of each model should it carry?

Let x be the number of EZ, y the number of compact, and z the number of commercial models carried by a van. Then we can summarize the information in this chart.

Model	Number	Weight	Volume
EZ	x	10	10
Compact	y	20	8
Commercial	z	60	28
Total for a load		440	248

Since a fully loaded van can carry 440 pounds and 248 cubic feet, we must solve this system of equations.

$$10x + 20y + 60z = 440 \quad \text{Weight equation}$$
$$10x + 8y + 28z = 248 \quad \text{Volume equation} \quad \boxed{12}$$

As shown in Problem 12 at the side, this system is equivalent to the following one.

$$x + 2y + 6z = 44$$
$$y + \frac{8}{3}z = 16$$

Solving this dependent system by back substitution, we have

$$y = 16 - \frac{8}{3}z$$

$$x = 44 - 2y - 6z = 44 - 2\left(16 - \frac{8}{3}z\right) - 6z = 12 - \frac{2}{3}z,$$

so that all solutions of the system are given by $\left(12 - \frac{2}{3}z, 16 - \frac{8}{3}z, z\right)$. The only

---

$\boxed{12}$ **(a)** Write the augmented matrix for the system of equations in Example 8.

(b) List a sequence of row operations that will transform the matrix in part (a) into triangular form, and give the triangular form.

Answers:

(a) $\begin{bmatrix} 10 & 20 & 60 & | & 440 \\ 10 & 8 & 28 & | & 248 \end{bmatrix}$

(b) Many sequences are possible, including this one:

replace R_1 by $\frac{1}{10}R_1$;

replace R_2 by $\frac{1}{2}R_2$;

replace R_2 by $-5R_1 + R_2$;

replace R_2 by $-\frac{1}{6}R_2$.

$\begin{bmatrix} 1 & 2 & 6 & | & 44 \\ 0 & 1 & \frac{8}{3} & | & 16 \end{bmatrix}$

solutions that apply in this situation, however, are those given by $z = 0$, 3, or 6 because all other values of z lead to fractions or negative numbers (you can't deliver a negative number of boxes, or part of a box). Hence, there are three ways to have a fully loaded van.

Solution	Van Load
$(12, 16, 0)$	12 EZ, 16 compact, 0 commercial
$(10, 8, 3)$	10 EZ, 8 compact, 3 commercial
$(8, 0, 6)$	8 EZ, 0 compact, 6 commercial ■

7.1 EXERCISES

Determine whether the given ordered set of numbers is a solution of the system of equations.

1. $(-1, 3)$

$\quad 2x + y = 1$
$\quad -3x + 2y = 9$

Yes

2. $(2, 1.5, -.5)$

$\quad 3x + 4y - 2z = -.5$
$\quad .5x \quad\quad + 8z = -3$
$\quad x - 3y + 5z = -5$

NO

Solve each of the following systems of two equations in two variables. (See Examples 1–3.)

3. $\quad x - 2y = 5$
$\quad\ 2x + y = 3$

hw

4. $\quad 3x - y = 1$
$\quad -x + 2y = 4$

5. $\quad 2x - 2y = 12$
$\quad -2x + 3y = 10$

6. $3x + 2y = -4$
$\quad 4x - 2y = -10$

7. $\quad x + 3y = -1$
$\quad 2x - y = 5$

8. $4x - 3y = -1$
$\quad x + 2y = 19$

9. $2x + 3y = 15$
$\quad 8x + 12y = 40$

10. $2x + 5y = 8$
$\quad\ 6x + 15y = 18$

11. $2x - 8y = 2$
$\quad 3x - 12y = 3$

12. $3x - 2y = 4$
$\quad\ 6x - 4y = 8$

13. $3x + 2y = 5$
$\quad\ 6x + 4y = 8$

14. $\quad 9x - 5y = 1$
$\quad -18x + 10y = 1$

15. Only one of the following screens gives the correct graphs for the system in Exercise 8. Which is it? (*Hint:* Solve for y first in each equation and use the slope-intercept form to help you answer the question.)

(a)

(b)

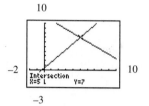

(c)

In Exercises 16–19, multiply both sides of each equation by a common denominator to eliminate the fractions. Then solve the system.

16. $\dfrac{x}{2} + \dfrac{y}{3} = 8$

$\quad \dfrac{2x}{3} + \dfrac{3y}{2} = 17$

17. $\dfrac{x}{5} + 3y = 31$

$\quad 2x - \dfrac{y}{5} = 8$

18. $\dfrac{x}{2} + y = \dfrac{3}{2}$

$\quad \dfrac{x}{3} + y = \dfrac{1}{3}$

19. $x + \dfrac{y}{3} = -6$

$\quad \dfrac{x}{5} + \dfrac{y}{4} = -\dfrac{7}{4}$

In Exercises 20–25, perform row operations on the augmented matrix as far as necessary to determine whether the system is independent, dependent, or inconsistent. (See Examples 6 and 7.)

20. $x + 2y = 0$
$y - z = 2$
$x + y + z = -2$

21. $x + 2y + z = 0$
$y + 2z = 0$
$x + y - z = 0$

22. $x + 2y + 4z = 6$
$y + z = 1$
$x + 3y + 5z = 10$

23. $x + y + 2z + 3w = 1$
$2x + y + 3z + 4w = 1$
$3x + y + 4z + 5w = 2$

24. $a - 3b - 2c = -3$
$3a + 2b - c = 12$
$-a - b + 4c = 3$

25. $2x + 2y + 2z = 6$
$3x - 3y - 4z = -1$
$x + y + 3z = 11$

Write the augmented matrix of the system and use the matrix method to solve the system. If the system is dependent, express the solutions in terms of the parameter z. (See Examples 4–7.)

26. $x + y + z = 2$
$2x + y - z = 5$
$x - y + z = -2$

27. $2x + y + z = 9$
$-x - y + z = 1$
$3x - y + z = 9$

28. $x + 3y + 4z = 14$
$2x - 3y + 2z = 10$
$3x - y + z = 9$

29. $4x - y + 3z = -2$
$3x + 5y - z = 15$
$-2x + y + 4z = 14$

30. $x + 2y + 3z = 8$
$3x - y + 2z = 5$
$-2x - 4y - 6z = 5$

31. $3x - 2y - 8z = 1$
$9x - 6y - 24z = -2$
$x - y + z = 1$

32. $2x - 4y + z = -4$
$x + 2y - z = 0$
$-x + y + z = 6$

33. $4x - 3y + z = 9$
$3x + 2y - 2z = 4$
$x - y + 3z = 5$

34. $5x + 3y + 4z = 19$
$3x - y + z = -4$

35. $3x + y - z = 0$
$2x - y + 3z = -7$

36. $11x + 10y + 9z = 5$
$x + 2y + 3z = 1$
$3x + 2y + z = 1$

37. $x + y = 3$
$5x - y = 3$
$9x - 4y = 1$

38. Find constants a, b, c such that the points $(2, 3)$, $(-1, 0)$, and $(-2, 2)$ lie on the graph of the equation $y = ax^2 + bx + c$. (*Hint:* Since $(2, 3)$ is on the graph, we must have $3 = a(2^2) + b(2) + c$, that is, $4a + 2b + c = 3$. Similarly, the other two points lead to two more equations. Solve the resulting system for a, b, c.)

39. (a) Find the equation of the straight line through $(1, 2)$ and $(3, 4)$.
(b) Find the equation of the line through $(-1, 1)$ with slope 3.
(c) Find a point that lies on both the lines in (a) and (b).

40. Graph the equations in the following system. Then explain why the graphs show that the system is inconsistent.

$$2x + 3y = 8$$
$$x - y = 4$$
$$5x + y = 7$$

41. Explain why a system with more variables than equations cannot have a unique solution (that is, be an independent

system). (*Hint:* When you apply the elimination method to such a system, what must necessarily happen?)

Work the following problems by writing and solving a system of equations. (See Example 8.)

42. Physical Science Linear systems occur in the design of roof trusses for new homes and buildings. The simplest type of roof truss is a triangle. The truss shown in the figure is used to frame roofs of small buildings. If a 100-pound force is applied at the peak of the truss, then the forces or weights W_1 and W_2 exerted parallel to each rafter of the truss are determined by the following linear system of equations.

$$\frac{\sqrt{3}}{2}(W_1 + W_2) = 100$$

$$W_1 - W_2 = 0$$

Solve the system to find W_1 and W_2.*

43. Physical Science (Refer to Exercise 42.) Use the following system of equations to determine the forces or weights W_1 and W_2 exerted on each rafter for the truss shown in the figure.

$$W_1 + \sqrt{2}W_2 = 300$$
$$\sqrt{3}W_1 - \sqrt{2}W_2 = 0$$

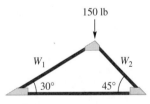

44. Management Shirley Cicero has $16,000 invested in Boeing and GE stock. The Boeing stock currently sells for $30 a share and the GE stock for $70 a share. If GE stock triples in value and Boeing stock goes up 50%, her stock will be worth $34,500. How many shares of each stock does she own?

45. Management A flight leaves New York at 8 P.M. and arrives in Paris at 9 A.M. (Paris time). This 13-hour difference includes the flight time plus the change in time zones. The return flight leaves Paris at 1 P.M. and arrives in New

---

*Hibbeler R., *Structural Analysis*, Prentice-Hall, Englewood Cliffs, 1995.

York at 3 P.M. (New York time). This 2-hour difference includes the flight time *minus* the time zones, plus an extra hour due to the fact that flying westward is against the wind. Find the actual flight time eastward and the difference in time zones.

46. Management The price at which grapefruit is sold affects the consumer demand for grapefruit according to the equation

$$2p = -.2q + 5,$$

where p is the price per pound (in dollars) at which consumers will demand q thousand pounds of grapefruit. The amount q of grapefruit that producers will supply at price p is governed by the equation $5p = .3q + 5.3$. Find the equilibrium quantity and the equilibrium price. (That is, find the quantity and price at which supply equals demand, or the values of p and q that satisfy both equations.)

47. Management If 20 pounds of rice and 10 pounds of potatoes cost $16.20 and 30 pounds of rice and 12 pounds of potatoes cost $23.04, how much will 10 pounds of rice and 50 pounds of potatoes cost?

48. Management An apparel shop sells skirts for $45 and blouses for $35. Its entire stock is worth $51,750. But sales are slow and only half the skirts and two-thirds of the blouses are sold, for a total of $30,600. How many skirts and blouses are left in the store?

49. Management A company produces two models of bicycles, model 201 and model 301. Model 201 requires 2 hours of assembly time and model 301 requires 3 hours of assembly time. The parts for model 201 cost $25 per bike and the parts for model 301 cost $30 per bike. If the company has a total of 34 hours of assembly time and $365 available per day for these two models, how many of each can be made in a day?

50. Social Science The relationship between a professional basketball player's height H (in inches) and weight W (in pounds) was modeled using two different samples of players. The resulting equations that modeled each sample were $W = 7.46H - 374$ and $W = 7.93H - 405$.

(a) Use both equations to predict the weight of a 6'11" professional basketball player.

(b) According to each model, what change in weight is associated with a 1-inch increase in height?

(c) Determine the weight and height where the two models agree.

51. Management Juanita invests $10,000, received from her grandmother, in three ways. With one part, she buys mutual funds which offer a return of 8% per year. The second part, which amounts to twice the first, is used to buy government bonds at 9% per year. She puts the rest in the bank at 5% annual interest. The first year her investments bring a return of $830. How much did she invest in each way?

52. Management To get the necessary funds for a planned expansion, a small company took out three loans totaling $25,000. The company was able to borrow some of the money at 16%. They borrowed $2000 more than one half the amount of the 16% loan at 20%, and the rest at 18%. The total annual interest was $4440. How much did they borrow at each rate?

53. Natural Science Three brands of fertilizer are available that provide nitrogen, phosphoric acid, and soluble potash to the soil. One bag of each brand provides the following units of each nutrient.

Brand	NUTRIENT		
	Nitrogen	*Phosphoric Acid*	*Potash*
A	1	3	2
B	2	1	0
C	3	2	1

For ideal growth, the soil in a certain country needs 18 units of nitrogen, 23 units of phosphoric acid, and 13 units of potash per acre. How many bags of each brand of fertilizer should be used per acre for ideal growth?

54. Management A company produces three color television sets: models X, Y, and Z. Each model X set requires 2 hours of electronics work and 2 hours of assembly time. Each model Y requires 1 hour of electronics work and 3 hours of assembly time. Each model Z requires 3 hours of electronics work and 2 hours of assembly time. There are 100 hours available for electronics and assembly each per week. How many of each model should be produced each week if all available time must be used?

55. Management A restaurant owner orders a replacement set of knives, forks, and spoons. The box arrives containing 40 utensils and weighing 141.3 ounces (ignoring the weight of the box). A knife, fork, and spoon weigh 3.9 ounces, 3.6 ounces, and 3.0 ounces, respectively.

(a) How many solutions are there for the number of knives, forks, and spoons in the box? What are the possible number of spoons?

(b) Which solution has the smallest number of spoons?

56. Management Turley Tailor Inc. makes long-sleeve, short-sleeve, and sleeveless blouses. A sleeveless blouse requires .5 hour of cutting and .6 hour of sewing. A short-sleeve blouse requires 1 hour of cutting and .9 hour of sewing. A long-sleeve blouse requires 1.5 hours of cutting and 1.2 hours of sewing. There are 380 hours of labor available in the cutting department each day and 330 hours in the sewing department. If the plant is to run at full capacity, how many of each type of blouse should be made each day?

7.2 THE GAUSS-JORDAN METHOD

In Example 5 of the previous section, we used matrix methods to rewrite the system

$$x - 2y = 6 - 4z$$
$$x + 13z = 6 - y$$
$$-2x + 6y - z = -10$$

as an augmented matrix. We carried out the steps of the matrix method until the final matrix was

$$\begin{bmatrix} 1 & -2 & 4 & | & 6 \\ 0 & 1 & 3 & | & 0 \\ 0 & 0 & 1 & | & 2 \end{bmatrix}$$

We then used back substitution to solve it. In the **Gauss-Jordan method,** we continue the process with additional elimination of variables replacing back substitution as follows.

$$\begin{bmatrix} 1 & -2 & 4 & | & 6 \\ 0 & 1 & 0 & | & -6 \\ 0 & 0 & 1 & | & 2 \end{bmatrix} \quad -3R_3 + R_2$$

$$\begin{bmatrix} 1 & -2 & 0 & | & -2 \\ 0 & 1 & 0 & | & -6 \\ 0 & 0 & 1 & | & 2 \end{bmatrix} \quad -4R_3 + R_1$$

$$\begin{bmatrix} 1 & 0 & 0 & | & -14 \\ 0 & 1 & 0 & | & -6 \\ 0 & 0 & 1 & | & 2 \end{bmatrix} \quad 2R_2 + R_1$$

The solution of the system is now obvious; it is $(-14, -6, 2)$. Note that this solution is the last column of the final augmented matrix.

In the Gauss-Jordan method row operations can be performed in any order, provided they eventually lead to the augmented matrix of a system in which the leading variable of each equation is not the leading variable in any other equation of the system. When there is a unique solution, as in the example above, this final system will be of the form $x =$ constant, $y =$ constant, $z =$ constant, and so on. ☐1

It is best to transform the matrix systematically. Either follow the procedure in the example (which first puts the system in a form where back substitution could be used and then eliminates additional variable terms) or work column by column from left to right, as in the next example.

EXAMPLE 1 Use the Gauss-Jordan method to solve the system

$$x \quad + 5z = -6 + y$$
$$3x + 3y \quad = 10 + z$$
$$x + 3y + 2z = 5.$$

### Sidebar

☐1 Use the Gauss-Jordan method to solve the system
$$x + 2y = 11$$
$$-4x + y = -8,$$
as follows. Give the shorthand notation and the new matrix in (b)–(d).

(a) Set up the augmented matrix.

(b) Get 0 in row two, column one.

(c) Get 1 in row two, column two.

(d) Finally, get 0 in row one, column two.

(e) The solution for the system is ____.

Answers:

(a) $\begin{bmatrix} 1 & 2 & | & 11 \\ -4 & 1 & | & -8 \end{bmatrix}$

(b) $4R_1 + R_2$; $\begin{bmatrix} 1 & 2 & | & 11 \\ 0 & 9 & | & 36 \end{bmatrix}$

(c) $\frac{1}{9}R_2$; $\begin{bmatrix} 1 & 2 & | & 11 \\ 0 & 1 & | & 4 \end{bmatrix}$

(d) $-2R_2 + R_1$; $\begin{bmatrix} 1 & 0 & | & 3 \\ 0 & 1 & | & 4 \end{bmatrix}$

(e) $(3, 4)$

The system must first be rewritten in proper form as follows.

$$\begin{aligned} x - y + 5z &= -6 \\ 3x + 3y - z &= 10 \\ x + 3y + 2z &= 5 \end{aligned}$$

2 Continue the solution of the system in Example 1 as follows. Give the shorthand notation and the matrix for each step.

(a) Get 1 in row two, column two.

(b) Get 0 in row one, column two.

(c) Now get 0 in row three, column two.

Answers:

(a) $\frac{1}{6}R_2$;

$$\begin{bmatrix} 1 & -1 & 5 & | & -6 \\ 0 & 1 & -\frac{8}{3} & | & \frac{14}{3} \\ 0 & 4 & -3 & | & 11 \end{bmatrix}$$

(b) $R_2 + R_1$;

$$\begin{bmatrix} 1 & 0 & \frac{7}{3} & | & -\frac{4}{3} \\ 0 & 1 & -\frac{8}{3} & | & \frac{14}{3} \\ 0 & 4 & -3 & | & 11 \end{bmatrix}$$

(c) $-4R_2 + R_3$;

$$\begin{bmatrix} 1 & 0 & \frac{7}{3} & | & -\frac{4}{3} \\ 0 & 1 & -\frac{8}{3} & | & \frac{14}{3} \\ 0 & 0 & \frac{23}{3} & | & -\frac{23}{3} \end{bmatrix}$$

(Solution continued in the text.)

Begin the solution by writing the augmented matrix of the linear system.

$$\begin{bmatrix} 1 & -1 & 5 & | & -6 \\ 3 & 3 & -1 & | & 10 \\ 1 & 3 & 2 & | & 5 \end{bmatrix}$$

The first element in column one is already 1. Get 0 for the second element in column one by multiplying each element in the first row by -3 and adding the results to the corresponding elements in row two.

$$\begin{bmatrix} 1 & -1 & 5 & | & -6 \\ 0 & 6 & -16 & | & 28 \\ 1 & 3 & 2 & | & 5 \end{bmatrix} \quad -3R_1 + R_2$$

Now, change the first element in row three to 0 by multiplying each element of the first row by -1 and adding the results to the corresponding elements of the third row.

$$\begin{bmatrix} 1 & -1 & 5 & | & -6 \\ 0 & 6 & -16 & | & 28 \\ 0 & 4 & -3 & | & 11 \end{bmatrix} \quad -1R_1 + R_3$$

This transforms the first column. Transform the second column in a similar manner, as directed in Side Problem 2. **2**

Complete the solution by transforming the third column of the matrix in part (c) of Side Problem 2.

$$\begin{bmatrix} 1 & 0 & \frac{7}{3} & | & -\frac{4}{3} \\ 0 & 1 & -\frac{8}{3} & | & \frac{14}{3} \\ 0 & 0 & 1 & | & -1 \end{bmatrix} \quad \frac{3}{23}R_3$$

$$\begin{bmatrix} 1 & 0 & 0 & | & 1 \\ 0 & 1 & -\frac{8}{3} & | & \frac{14}{3} \\ 0 & 0 & 1 & | & -1 \end{bmatrix} \quad -\frac{7}{3}R_3 + R_1$$

$$\begin{bmatrix} 1 & 0 & 0 & | & 1 \\ 0 & 1 & 0 & | & 2 \\ 0 & 0 & 1 & | & -1 \end{bmatrix} \quad \frac{8}{3}R_3 + R_2$$

The linear system associated with this last augmented matrix is

$$\begin{aligned} x & & &= 1 \\ & y & &= 2 \\ & & z &= -1, \end{aligned}$$

and the solution is $(1, 2, -1)$. ∎

3 Use the Gauss-Jordan method to solve

$$x + y - z = 6$$
$$2x - y + z = 3$$
$$-x + y + z = -4.$$

Answer:
$(3, 1, -2)$

4 Solve each system.

(a) $x - y = 4$
$-2x + 2y = 1$

(b) $3x - 4y = 0$
$2x + y = 0$

Answers:
(a) No solution
(b) $(0, 0)$

5 Use the Gauss-Jordan method to solve the system.

$$x + 3y = 4$$
$$4x + 8y = 4$$
$$6x + 12y = 6$$

Answer:
$(-5, 3)$

N O T E   Notice that the first two row operations are used to get the ones and the third row operation is used to get the zeros. ◆   **3**

EXAMPLE 2   Use the Gauss-Jordan method to solve the system

$$2x + 4y = 4$$
$$3x + 6y = 8$$
$$2x + y = 7.$$

Write the augmented matrix and perform row operations to obtain a first column whose entries (from top to bottom) are 1, 0, 0.

$$\begin{bmatrix} 2 & 4 & | & 4 \\ 3 & 6 & | & 8 \\ 2 & 1 & | & 7 \end{bmatrix}$$

$$\begin{bmatrix} 1 & 2 & | & 2 \\ 3 & 6 & | & 8 \\ 2 & 1 & | & 7 \end{bmatrix} \quad \frac{1}{2}R_1$$

$$\begin{bmatrix} 1 & 2 & | & 2 \\ 0 & 0 & | & 2 \\ 2 & 1 & | & 7 \end{bmatrix} \quad -3R_1 + R_2$$

Stop! The second row of this augmented matrix denotes the equation $0x + 0y = 2$. Since the left side of this equation is always 0 and the right side is 2, it has no solutions. Hence, the original system has no solutions.   ■   **4**

Whenever the Gauss-Jordan method produces a row whose elements are all 0 except the last one, such as $[0 \quad 0 \,|\, 2]$ in Example 2, the system is inconsistent and has no solutions. On the other hand, if a row with *every* element 0 is produced, the system may have solutions. In that case, continue carrying out the Gauss-Jordan method.   **5**

EXAMPLE 3   Use the Gauss-Jordan method to solve the system

$$x + 2y - z = 0$$
$$3x - y + z = 6.$$

Start with the augmented matrix and use row operations to obtain a first column whose entries (from top to bottom) are 1, 0.

$$\begin{bmatrix} 1 & 2 & -1 & | & 0 \\ 3 & -1 & 1 & | & 6 \end{bmatrix}$$

$$\begin{bmatrix} 1 & 2 & -1 & | & 0 \\ 0 & -7 & 4 & | & 6 \end{bmatrix} \quad -3R_1 + R_2$$

Now use row operations to obtain a second column whose entries (from top to bottom) are 0, 1.

$$\begin{bmatrix} 1 & 2 & -1 & | & 0 \\ 0 & 1 & -\frac{4}{7} & | & -\frac{6}{7} \end{bmatrix} \quad -\frac{1}{7}R_2$$

$$\begin{bmatrix} 1 & 0 & \frac{1}{7} & \frac{12}{7} \\ 0 & 1 & -\frac{4}{7} & -\frac{6}{7} \end{bmatrix} \qquad -2R_2 + R_1$$

This last matrix is the augmented matrix of the system

$$x + \frac{1}{7}z = \frac{12}{7}$$

$$y - \frac{4}{7}z = -\frac{6}{7}.$$

[6] Use the Gauss-Jordan method to solve the following.

(a) $3x + 9y = -6$
  $-x - 3y = 2$

(b) $2x + 9y = 12$
  $4x + 18y = 5$

Answers:
(a) y arbitrary,
$x = -3y - 2$
or $(-3y - 2, y)$

(b) No solution

Solving the first equation for x and the second for y gives the solution

z arbitrary

$$y = -\frac{6}{7} + \frac{4}{7}z$$

$$x = \frac{12}{7} - \frac{1}{7}z,$$

or $(12/7 - z/7, -6/7 + 4z/7, z)$. ■ [6]

The techniques used in Examples 1–3 can be summarized as follows.

THE GAUSS–JORDAN METHOD FOR SOLVING A SYSTEM OF LINEAR EQUATIONS

1. Arrange the equations with the variable terms in the same order on the left of the equals sign and the constants on the right.
2. Write the augmented matrix of the system.
3. Use row operations to transform the augmented matrix into this form:
   (a) The rows consisting entirely of zeros are grouped together at the bottom of the matrix.
   (b) In each row that does not consist entirely of zeros, the leftmost nonzero element is 1 (called a *leading* 1).
   (c) Each column that contains a leading 1 has zeros in all other entries.
   (d) The leading 1 in any row is to the left of any leading 1's in the rows below it.
4. Stop the process in Step 3 if you obtain a row whose elements are all zero except the last one. In that case, the system is inconsistent and has no solutions. Otherwise, finish Step 3 and read the solutions of the system from the final matrix.

When doing Step 3, try to choose row operations so that as few fractions as possible are carried through the computation. This makes calculation easier when working by hand and avoids introducing round-off errors when using a calculator or computer.

TECHNOLOGY TIP The Gauss-Jordan method can be carried out in a single step on some calculators by using RREF (in the MATH or OPS submenu of the MATRIX menu of TI-83/85/86/92 and the MATRIX submenu of the MATH menu of HP-38). ✔

EXAMPLE 4 An animal feed is to be made from corn, soybeans, and cottonseed. Determine how many units of each ingredient are needed to make a feed that supplies 1800 units of fiber, 2800 units of fat, and 2200 units of protein, given that one unit of each ingredient provides the numbers of units shown in the table below. The table states, for example, that a unit of corn provides 10 units of fiber, 30 units of fat, and 20 units of protein.

	Corn	Soybeans	Cottonseed	Totals
Units of Fiber	10	20	30	1800
Units of Fat	30	20	40	2800
Units of Protein	20	40	25	2200

[7] List a sequence of row operations that will transform the augmented matrix of system (1) in Example 4 into the augmented matrix of system (2).

Answer:
Many sequences are possible, including this one:

replace R_1 by $\frac{1}{10}R_1$;

replace R_2 by $\frac{1}{10}R_2$;

replace R_3 by $\frac{1}{5}R_3$;

replace R_2 by $-3R_1 + R_2$;
replace R_3 by $-4R_1 + R_3$;

replace R_2 by $-\frac{1}{4}R_2$;

replace R_3 by $-\frac{1}{7}R_3$.

Let x represent the required number of units of corn, y the number of units of soybeans, and z the number of units of cottonseed. Since the total amount of fiber is to be 1800,

$$10x + 20y + 30z = 1800.$$

The feed must supply 2800 units of fat, so

$$30x + 20y + 40z = 2800.$$

Finally, since 2200 units of protein are required,

$$20x + 40y + 25z = 2200.$$

Thus we must solve this system of equations.

$$\begin{aligned} 10x + 20y + 30z &= 1800 \\ 30x + 20y + 40z &= 2800 \\ 20x + 40y + 25z &= 2200 \end{aligned} \tag{1}$$

The elimination method or a matrix method leads to the following equivalent system, as shown in Problem 7 at the side. [7]

$$\begin{aligned} x + 2y + 3z &= 180 \\ y + \frac{5}{4}z &= 65 \\ z &= 40 \end{aligned} \tag{2}$$

Back substitution now shows that $z = 40$,

$$y = 65 - \frac{5}{4}(40) = 15 \quad \text{and} \quad x = 180 - 2(15) - 3(40) = 30.$$

Thus, the feed should contain 30 units of corn, 15 units of soybeans, and 40 units of cottonseed. ∎

EXAMPLE 5 The U-Drive Rent-a-Truck Company plans to spend 3 million dollars on 200 new vehicles. Each van will cost $10,000, each small truck, $15,000, and each large truck, $25,000. Past experience shows that they need twice as many vans as small trucks. How many of each kind of vehicle can they buy?

Let x be the number of vans, y the number of small trucks, and z the number of large trucks. Then $x + y + z = 200$. The cost of x vans at \$10,000 each is $10,000x$, the cost of y small trucks is $15,000y$, and the cost of z large trucks is $25,000z$, so that $10,000x + 15,000y + 25,000z = 3,000,000$. Dividing this equation on each side by 5000 makes it $2x + 3y + 5z = 600$. Finally, the number of vans is twice the number of small trucks: $x = 2y$, or equivalently, $x - 2y = 0$.

To solve the system

$$x + y + z = 200$$
$$2x + 3y + 5z = 600$$
$$x - 2y = 0,$$

we form the augmented matrix and use the indicated row operations.

$$\begin{bmatrix} 1 & 1 & 1 & 200 \\ 2 & 3 & 5 & 600 \\ 1 & -2 & 0 & 0 \end{bmatrix}$$

$$\begin{bmatrix} 1 & 1 & 1 & 200 \\ 0 & 1 & 3 & 200 \\ 0 & -3 & -1 & -200 \end{bmatrix} \quad \begin{matrix} -2R_1 + R_2 \\ -R_1 + R_3 \end{matrix}$$

$$\begin{bmatrix} 1 & 0 & -2 & 0 \\ 0 & 1 & 3 & 200 \\ 0 & 0 & 8 & 400 \end{bmatrix} \quad \begin{matrix} -R_2 + R_1 \\ \\ 3R_2 + R_3 \end{matrix}$$

$$\begin{bmatrix} 1 & 0 & -2 & 0 \\ 0 & 1 & 3 & 200 \\ 0 & 0 & 1 & 50 \end{bmatrix} \quad \frac{1}{8}R_3$$

$$\begin{bmatrix} 1 & 0 & 0 & 100 \\ 0 & 1 & 0 & 50 \\ 0 & 0 & 1 & 50 \end{bmatrix} \quad \begin{matrix} 2R_3 + R_1 \\ -3R_3 + R_2 \end{matrix}$$

The final matrix corresponds to the system

$$x = 100$$
$$y = 50$$
$$z = 50.$$

Therefore, U-Drive should buy 100 vans, 50 small trucks, and 50 large trucks. ■ 8

8 In Example 5, suppose the U-Drive Company can spend only 2 million dollars on 150 new vehicles, and that they need three times as many vans as small trucks. Write a system of equations to express these conditions.

Answer:
$x + y + z = 150$
$2x + 3y + 5z = 400$
$x - 3y = 0$

7.2 EXERCISES

Write the augmented matrix of each of the following systems. Do not solve the systems.

1. $2x + y + z = 3$
$3x - 4y + 2z = -5$
$x + y + z = 2$

2. $3x + 4y - 2z - 3w = 0$
$x - 3y + 7z + 4w = 9$
$2x + 5z - 6w = 0$

Write the system of equations associated with the following augmented matrices. Do not solve the systems.

3. $\begin{bmatrix} 2 & 3 & 8 & 20 \\ 1 & 4 & 6 & 12 \\ 0 & 3 & 5 & 10 \end{bmatrix}$

4. $\begin{bmatrix} 3 & 2 & 6 & 18 \\ 2 & -2 & 5 & 7 \\ 1 & 0 & 5 & 20 \end{bmatrix}$

Use the indicated row operation to transform each matrix.

5. Interchange R_2 and R_3.

$$\begin{bmatrix} 1 & 2 & 3 & | & -1 \\ 6 & 5 & 4 & | & 6 \\ 2 & 0 & 7 & | & -4 \end{bmatrix}$$

6. Replace R_3 by $-3R_1 + R_3$.

$$\begin{bmatrix} 1 & 5 & 2 & 0 & | & -1 \\ 8 & 5 & 4 & 6 & | & 6 \\ 3 & 0 & 7 & 1 & | & -4 \end{bmatrix}$$

7. Replace R_2 by $2R_1 + R_2$.

$$\begin{bmatrix} -4 & -3 & 1 & -1 & | & 2 \\ 8 & 2 & 5 & 0 & | & 6 \\ 0 & -2 & 9 & 4 & | & 5 \end{bmatrix}$$

8. Replace R_3 by $\frac{1}{4}R_3$.

$$\begin{bmatrix} 2 & 5 & 1 & | & -1 \\ -4 & 0 & 4 & | & 6 \\ 6 & 0 & 8 & | & -4 \end{bmatrix}$$

Use the Gauss-Jordan method to solve each of the following systems of equations. (See Examples 1–3.)

9.
$$\begin{aligned} x + 2y + z &= 5 \\ 2x + y - 3z &= -2 \\ 3x + y + 4z &= -5 \end{aligned}$$

10.
$$\begin{aligned} 3x - 2y + z &= 6 \\ 3x + y - z &= -4 \\ -x + 2y - 2z &= -8 \end{aligned}$$

11.
$$\begin{aligned} x + 3y - 6z &= 7 \\ 2x - y + 2z &= 0 \\ x + y + 2z &= -1 \end{aligned}$$

12.
$$\begin{aligned} x &= 1 - y \\ 2x &= z \\ 2z &= -2 - y \end{aligned}$$

13.
$$\begin{aligned} 3x + 5y - z &= 0 \\ 4x - y + 2z &= 1 \\ -6x - 10y + 2z &= 0 \end{aligned}$$

14.
$$\begin{aligned} x + y &= -1 \\ y + z &= 4 \\ x + z &= 1 \end{aligned}$$

15.
$$\begin{aligned} x + y - z &= 6 \\ 2x - y + z &= -9 \\ x - 2y + 3z &= 1 \end{aligned}$$

16.
$$\begin{aligned} y &= x - 1 \\ y &= 6 + z \\ z &= -1 - x \end{aligned}$$

17.
$$\begin{aligned} x - 2y + z &= 5 \\ 2x + y - z &= 2 \\ -2x + 4y - 2z &= 2 \end{aligned}$$

18.
$$\begin{aligned} 2x + 3y + z &= 9 \\ 4x + y - 3z &= -7 \\ 6x + 2y - 4z &= -8 \end{aligned}$$

19.
$$\begin{aligned} -8x - 9y &= 11 \\ 24x + 34y &= 2 \\ 16x + 11y &= -57 \end{aligned}$$

20.
$$\begin{aligned} 2x + y &= 7 \\ x - y &= 3 \\ x + 3y &= 4 \end{aligned}$$

21.
$$\begin{aligned} x + y - z &= -20 \\ 2x - y + z &= 11 \end{aligned}$$

22.
$$\begin{aligned} 4x + 3y + z &= 1 \\ -2x - y + 2z &= 0 \end{aligned}$$

23.
$$\begin{aligned} 2x + y + 3z - 2w &= -6 \\ 4x + 3y + z - w &= -2 \\ x + y + z + w &= -5 \\ -2x - 2y + 2z + 2w &= -10 \end{aligned}$$

24.
$$\begin{aligned} x + y + z + w &= -1 \\ -x + 4y + z - w &= 0 \\ x - 2y + z - 2w &= 11 \\ -x - 2y + z + 2w &= -3 \end{aligned}$$

25.
$$\begin{aligned} x + 2y - z &= 3 \\ 3x + y + w &= 4 \\ 2x - y + z + w &= 2 \end{aligned}$$

26.
$$\begin{aligned} x - 2y - z - 3w &= -3 \\ -x + y + z &= 2 \\ 4y + 3z - 6w &= -2 \end{aligned}$$

Set up a system of equations and use the Gauss-Jordan method to solve it. (See Examples 4 and 5.)

27. Management   McFrugal Snack Shops plan to hire two public relations firms to survey 500 customers by phone, 750 by mail, and 250 by in-person interviews. The Garcia firm has personnel to do 10 phone surveys, 30 mail surveys, and 5 interviews per hour. The Wong firm can handle 20 phone surveys, 10 mail surveys, and 10 interviews per hour. For how many hours should each firm be hired to produce the exact number of surveys needed?

28. Management   A knitting shop ordered yarn from three suppliers, I, II, and III. One month the shop ordered a total of 100 units of yarn from these suppliers. The delivery costs were $80, $50, and $65 per unit for the orders from suppliers I, II, and III, respectively, with total delivery costs of $5990. The shop ordered the same amount from suppliers I and III. How many units were ordered from each supplier?

29. Management   An electronics company produces three models of stereo speakers, models A, B, and C, and can deliver them by truck, van, or station wagon. A truck holds 2 boxes of model A, 1 of model B, and 3 of model C. A van holds 1 box of model A, 3 boxes of model B, and 2 boxes of model C. A station wagon holds 1 box of model A, 3 boxes of model B, and 1 box of model C. If 15 boxes of model A, 20 boxes of model B, and 22 boxes of model C are to be delivered, how many vehicles of each type should be used so that all operate at full capacity?

30. Management   Pretzels cost $3 per pound, dried fruit $4 per pound, and nuts $8 per pound. How many pounds of each should be used to produce 140 pounds of trail mix costing $6 per pound in which there are twice as many pretzels (by weight) as dried fruit?

31. Management   A manufacturer purchases a part for use at both of its two plants—one in Canoga Park, California, the other in Wooster, Ohio. The part is available in limited quantities from two suppliers. Each supplier has 75 units available. The Canoga Park plant needs 40 units and the Wooster plant requires 75 units. The first supplier charges $70 per unit delivered to Canoga Park and $90 per unit delivered to Wooster. Corresponding costs from the second supplier are $80 and $120. The manufacturer wants to order a total of 75 units from the first, less expensive, supplier, with the remaining 40 units to come from the second supplier. If the company spends $10,750 to purchase the required number of units for the two plants, find the number of units that should be purchased from each supplier for each plant as follows.

(a) Assign variables to the four unknowns.

(b) Write a system of five equations with the four variables. (Not all equations will involve all four variables.)

(c) Solve the system of equations.

32. Management   An auto manufacturer sends cars from two plants, I and II, to dealerships A and B located in a midwestern city. Plant I has a total of 28 cars to send, and plant II has 8. Dealer A needs 20 cars, and dealer B needs 16. Transportation costs based on the distance of each dealership from each plant are $220 from I to A, $300 from I to B, $400 from II to A, and $180 from II to B. The manufacturer wants to limit transportation costs to $10,640. How many cars should be sent from each plant to each of the two dealerships?

33. Management The electronics company in Exercise 29 no longer makes model C. Each kind of delivery vehicle can now carry one more box of model B than previously and the same number of boxes of model A. If 16 boxes of model A and 22 boxes of model B are to be delivered, how many vehicles of each type should be used so that all operate at full capacity?

34. Natural Science An animal breeder can buy four types of tiger food. Each case of Brand A contains 25 units of fiber, 30 units of protein, and 30 units of fat. Each case of Brand B contains 50 units of fiber, 30 units of protein, and 20 units of fat. Each case of Brand C contains 75 units of fiber, 30 units of protein, and 20 units of fat. Each case of Brand D contains 100 units of fiber, 60 units of protein, and 30 units of fat. How many cases of each brand should the breeder mix together to obtain a food that provides 1200 units of fiber, 600 units of protein, and 400 units of fat?

35. Management An investment firm recommends that a client invest in AAA, A, and B rated bonds. The average yield on AAA bonds is 6%, on A bonds 7%, and on B bonds 10%. The client wants to invest twice as much in AAA bonds as in B bonds. How much should be invested in each type of bond under the following conditions?
(a) The total investment is $25,000, and the investor wants an annual return of $1810 on the three investments.
(b) The values in part (a) are changed to $30,000 and $2150, respectively.
(c) The values in part (a) are changed to $40,000 and $2900, respectively.

36. Management An electronics company produces transistors, resistors, and computer chips. Each transistor requires 3 units of copper, 1 unit of zinc, and 2 units of glass. Each resistor requires 3, 2, and 1 units of the three materials, and each computer chip requires 2, 1, and 2 units of these materials, respectively. How many of each product can be made with the following amounts of materials?
(a) 810 units of copper, 410 units of zinc, and 490 units of glass
(b) 765 units of copper, 385 units of zinc, and 470 units of glass
(c) 1010 units of copper, 500 units of zinc, and 610 units of glass

37. Management At a pottery factory, fuel consumption for heating the kilns varies with the size of the order being fired. In the past, the company recorded these figures.

x = Number of Platters	y = Fuel Cost per Platter
6	$2.80
8	2.48
10	2.24

(a) Find an equation of the form $y = ax^2 + bx + c$ whose graph contains the three points corresponding to the data in the table.
(b) How many platters should be fired at one time in order to minimize the fuel cost per platter? What is the minimum fuel cost per platter?

38. Management A convenience store sells 23 sodas one summer afternoon in 12, 16, and 20 ounce cups (small, medium, and large). The total volume of soda sold was 376 ounces.
(a) Suppose that the prices for a small, medium, and large soda are $1, $1.25, and $1.40, respectively, and that the total sales were $28.45. How many of each size did the store sell?
(b) Suppose the prices for a small, medium, and large soda are changed to $1, $2, and $3, respectively, but all other information is the same. How many of each size did the store sell?
(c) Suppose the prices are the same as in part (b), but the total revenue is $48. Now how many of each size did the store sell?

39. Management According to data from a 1984 Texas agricultural report, the amount of nitrogen (in lbs/acre), phosphate (in lbs/acre), and labor (in hrs/acre) needed to grow honeydews, yellow onions, and lettuce is given by the following table.*

	Honeydews	Yellow Onions	Lettuce
Nitrogen	120	150	180
Phosphate	180	80	80
Labor	4.97	4.45	4.65

(a) If the farmer has 220 acres, 29,100 lbs of nitrogen, 32,600 lbs of phosphate, and 480 hours of labor, is it possible to use all resources completely? If so, how many acres should he allot for each crop?
(b) Suppose everything is the same as in part (a), except that 1061 hours of labor are available. Is it possible to use all resources completely? If so, how many acres should he allot for each crop?

40. Management The business analyst for Melcher Manufacturing wants to find an equation that can be used to project sales of a relatively new product. For the years 1995, 1996, and 1997, sales were $15,000, $32,000, and $123,000, respectively.
(a) Graph the sales for the years 1995, 1996, and 1997, letting the year 1995 equal 0 on the x-axis. Let the values on the vertical axis be in thousands. (For example, the point (1996, 32,000) will be graphed as (1, 32).)

*Paredes, Miguel, Fatehi, Mohammad, and Hinthorn, Richard, "The Transformation of an Inconsistent Linear System into a Consistent System," *The AMATYC Review*, Vol 13, No. 2, Spring 1992.

(b) Find the equation of the straight line $ax + by = c$ through the points for 1995 and 1997.

(c) Find the equation of the parabola $y = ax^2 + bx + c$ through the three given points.

(d) Find the projected sales for 2000 first by using the equation from part (b) and then by using the equation from part (c). If you were to estimate sales of the product in 2000, which result would you choose? Why?

41. Natural Science Determining the amount of carbon dioxide in the atmosphere is important because carbon dioxide is known to be a greenhouse gas. Carbon dioxide concentrations (in parts per million) have been measured at Mauna Loa, Hawaii, over the past 30 years. The concentrations have increased quadratically.* The table lists readings for three years.

Year	CO_2
1958	315
1973	325
1988	352

(a) If the relationship between the carbon dioxide concentration C and the year t is expressed as $C = at^2 + bt + $

c, where $t = 0$ corresponds to 1958, use a linear system of equations to determine the constants a, b, and c.

(b) Predict the year when the amount of carbon dioxide in the atmosphere will double from its 1958 level.

42. Physical Science For certain aircraft there exists a quadratic relationship between an airplane's maximum speed S (in knots) and its ceiling C, its highest altitude possible (in thousands of feet).* The table lists three airplanes that conform to this relationship.

Airplane	Maximum Speed	Ceiling
Hawkeye	320	33
Corsair	600	40
Tomcat	1283	50

(a) If the relationship between C and S is written as $C = aS^2 + bS + c$, use a linear system of equations to determine the constants a, b, and c.

(b) A new aircraft of this type has a ceiling of 45,000 feet. Predict its top speed.

*Nilsson, A., *Greenhouse Earth*, John Wiley & Sons, New York, 1992.

*Sanders, D., *Statistics: A First Course*, Fifth Edition, McGraw Hill, Inc., 1995.

7.3 BASIC MATRIX OPERATIONS

Until now we have used matrices only as a convenient shorthand to solve systems of equations. However, matrices are also important in the fields of management, natural science, engineering, and social science as a way to organize data, as Example 1 demonstrates.

EXAMPLE 1 The EZ Life Company manufactures sofas and armchairs in three models, A, B, and C. The company has regional warehouses in New York, Chicago, and San Francisco. In its August shipment, the company sends 10 model A sofas, 12 model B sofas, 5 model C sofas, 15 model A chairs, 20 model B chairs, and 8 model C chairs to each warehouse.

This data might be organized by first listing it as follows.

Sofas	10 model A	12 model B	5 model C
Chairs	15 model A	20 model B	8 model C

Alternatively, we might tabulate the data in a chart.

		MODEL		
		A	B	C
FURNITURE	Sofa	10	12	5
	Chair	15	20	8

1 Rewrite this information in a matrix with three rows and two columns.

Answer:
$$\begin{bmatrix} 10 & 15 \\ 12 & 20 \\ 5 & 8 \end{bmatrix}$$

With the understanding that the numbers in each row refer to the furniture type (sofa, chair) and the numbers in each column refer to the model (A, B, C), the same information can be given by a matrix, as follows.

$$M = \begin{bmatrix} 10 & 12 & 5 \\ 15 & 20 & 8 \end{bmatrix}$$ ■ **1**

A matrix with m rows and n columns has dimensions or size $m \times n$. The number of rows is always given first.

2 Give the size of each of the following matrices.

(a) $\begin{bmatrix} 2 & 1 & -5 & 6 \\ 3 & 0 & 7 & -4 \end{bmatrix}$

(b) $\begin{bmatrix} 1 & 2 & 3 \\ 4 & 5 & 6 \\ 9 & 8 & 7 \end{bmatrix}$

Answers:
(a) 2×4

(b) 3×3

EXAMPLE 2   (a) The matrix $\begin{bmatrix} 6 & 5 \\ 3 & 4 \\ 5 & -1 \end{bmatrix}$ is a 3 × 2 matrix.

(b) $\begin{bmatrix} 5 & 8 & 9 \\ 0 & 5 & -3 \\ -4 & 0 & 5 \end{bmatrix}$ is a 3 × 3 matrix.

(c) $[1 \quad 6 \quad 5 \quad -2 \quad 5]$ is a 1 × 5 matrix.

(d) A graphing calculator displays a 4 × 1 matrix like this. ■ **2**

3 Use the numbers 2, 5, −8, 4 to write

(a) a row matrix;

(b) a column matrix;

(c) a square matrix.

Answers:
(a) $[2 \quad 5 \quad -8 \quad 4]$

(b) $\begin{bmatrix} 2 \\ 5 \\ -8 \\ 4 \end{bmatrix}$

(c) $\begin{bmatrix} 2 & 5 \\ -8 & 4 \end{bmatrix}$ or $\begin{bmatrix} 2 & -8 \\ 5 & 4 \end{bmatrix}$

(Other answers are possible.)

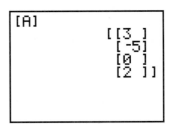

A matrix with only one row, as in Example 2(c), is called a **row matrix** or **row vector.** A matrix with only one column, as in Example 2(d), is called a **column matrix** or **column vector.** A matrix with the same number of rows as columns is called a **square matrix.** The matrix in Example 2(b) above is a square matrix, as are

$$A = \begin{bmatrix} -5 & 6 \\ 8 & 3 \end{bmatrix} \quad \text{and} \quad B = \begin{bmatrix} 0 & 0 & 0 & 0 \\ -2 & 4 & 1 & 3 \\ 0 & 0 & 0 & 0 \\ -5 & -4 & 1 & 8 \end{bmatrix}.$$ **3**

When a matrix is denoted by a single letter, such as the matrix A above, then the element in row i and column j is denoted a_{ij}. For example, $a_{21} = 8$ (the element in row 2, column 1). Similarly, in matrix B above, $b_{42} = -4$ (the element in row 4, column 2).

ADDITION   The matrix given in Example 1,

$$M = \begin{bmatrix} 10 & 12 & 5 \\ 15 & 20 & 8 \end{bmatrix},$$

shows the August shipment from the EZ Life plant to each of its warehouses. If matrix N below gives the September shipment to the New York warehouse, what is the total shipment for each item of furniture to the New York warehouse for these two months?

$$N = \begin{bmatrix} 45 & 35 & 20 \\ 65 & 40 & 35 \end{bmatrix}$$

If 10 model A sofas were shipped in August and 45 in September, then altogether 55 model A sofas were shipped in the 2 months. Adding the other corresponding entries gives a new matrix, Q, that represents the total shipment to the New York warehouse for the 2 months.

$$Q = \begin{bmatrix} 55 & 47 & 25 \\ 80 & 60 & 43 \end{bmatrix}$$

It is convenient to refer to Q as the "sum" of M and N.

The way these two matrices were added illustrates the following definition of addition of matrices.

The **sum** of two $m \times n$ matrices X and Y is the $m \times n$ matrix $X + Y$ in which each element is the sum of the corresponding elements of X and Y.

---

4 Find each sum when possible.

(a) $\begin{bmatrix} 2 & 5 & 7 \\ 3 & -1 & 4 \end{bmatrix}$

$+ \begin{bmatrix} -1 & 2 & 0 \\ 10 & -4 & 5 \end{bmatrix}$

(b) $\begin{bmatrix} 1 \\ 2 \\ 3 \end{bmatrix} + \begin{bmatrix} 2 & -1 \\ 4 & 5 \\ 6 & 0 \end{bmatrix}$

(c) $[5 \quad 4 \quad -1] + [-5 \quad 2 \quad 3]$

Answers:

(a) $\begin{bmatrix} 1 & 7 & 7 \\ 13 & -5 & 9 \end{bmatrix}$

(b) Not possible

(c) $[0 \quad 6 \quad 2]$

It is important to remember that only matrices that are the same size can be added.

TECHNOLOGY TIP Sums (and differences) of matrices that have the same dimensions can be found with a graphing calculator that has matrix capability. See your instruction manual for details. ✔

EXAMPLE 3   Find each sum if possible.

(a) $\begin{bmatrix} 5 & -6 \\ 8 & 9 \end{bmatrix} + \begin{bmatrix} -4 & 6 \\ 8 & -3 \end{bmatrix} = \begin{bmatrix} 5 + (-4) & -6 + 6 \\ 8 + 8 & 9 + (-3) \end{bmatrix} = \begin{bmatrix} 1 & 0 \\ 16 & 6 \end{bmatrix}$

(b)   The matrices

$$A = \begin{bmatrix} 5 & 8 \\ 6 & 2 \end{bmatrix} \quad \text{and} \quad B = \begin{bmatrix} 3 & 9 & 1 \\ 4 & 2 & 5 \end{bmatrix}$$

are different sizes, so it is not possible to find the sum $A + B$. ∎   **4**

EXAMPLE 4   The September shipments of the three models of sofas and chairs from the EZ Life Company to the New York, San Francisco, and Chicago warehouses are given in matrices N, S, and C below.

$$N = \begin{bmatrix} 45 & 35 & 20 \\ 65 & 40 & 35 \end{bmatrix}, \quad S = \begin{bmatrix} 30 & 32 & 28 \\ 43 & 47 & 30 \end{bmatrix}, \quad C = \begin{bmatrix} 22 & 25 & 38 \\ 31 & 34 & 35 \end{bmatrix}$$

What was the total amount shipped to the three warehouses in September?

5 From the result of Example 4, find the total number of the following shipped to the three warehouses.

(a) Model A chairs

(b) Model B sofas

(c) Model C chairs

Answers:
(a) 139
(b) 92
(c) 100

The total of the September shipments is represented by the sum of the three matrices N, S, and C.

$$N + S + C = \begin{bmatrix} 45 & 35 & 20 \\ 65 & 40 & 35 \end{bmatrix} + \begin{bmatrix} 30 & 32 & 28 \\ 43 & 47 & 30 \end{bmatrix} + \begin{bmatrix} 22 & 25 & 38 \\ 31 & 34 & 35 \end{bmatrix}$$

$$= \begin{bmatrix} 97 & 92 & 86 \\ 139 & 121 & 100 \end{bmatrix}$$

For example, from this sum the total number of model C sofas shipped to the three warehouses in September was 86. ■ **5**

As mentioned in Section 1.1, the additive inverse of the real number a is $-a$; a similar definition is given for the additive inverse of a matrix.

The **additive inverse** (or *negative*) of a matrix X is the matrix $-X$ in which each element is the additive inverse of the corresponding element of X.

If

$$A = \begin{bmatrix} 1 & 2 & 3 \\ 0 & -1 & 5 \end{bmatrix} \quad \text{and} \quad B = \begin{bmatrix} -2 & 3 & 0 \\ 1 & -7 & 2 \end{bmatrix},$$

then by the definition of the additive inverse of a matrix,

$$-A = \begin{bmatrix} -1 & -2 & -3 \\ 0 & 1 & -5 \end{bmatrix} \quad \text{and} \quad -B = \begin{bmatrix} 2 & -3 & 0 \\ -1 & 7 & -2 \end{bmatrix}.$$

TECHNOLOGY TIP A graphing calculator with matrix capability gives the additive inverse of a matrix by preceding the matrix with a negative sign. See Figure 7.3. ✔

```
[A]
           [[2 3  ]
            [1 1.5]]
-[A]
           [[-2 -3  ]
            [-1 -1.5]]
```

FIGURE 7.3

By the definition of matrix addition, for each matrix X, the sum $X + (-X)$ is a **zero matrix**, O, whose elements are all zeros. There is an $m \times n$ zero matrix for each pair of values of m and n.

$$\begin{bmatrix} 0 & 0 \\ 0 & 0 \end{bmatrix} \qquad \begin{bmatrix} 0 & 0 & 0 & 0 \\ 0 & 0 & 0 & 0 \end{bmatrix}$$

2×2 zero matrix $\qquad$ 2×4 zero matrix

Zero matrices have the following *identity property.*

> If O is the $m \times n$ zero matrix, and A is any $m \times n$ matrix, then
> $$A + O = O + A = A.$$

Compare this with the identity property for real numbers: for any real number a, $a + 0 = 0 + a = a$.

SUBTRACTION   The **subtraction** of matrices can be defined in a manner comparable to subtraction for real numbers.

> For two $m \times n$ matrices X and Y, the **difference** of X and Y is the $m \times n$ matrix $X - Y$ in which each element is the difference of the corresponding elements of X and Y, or, equivalently,
> $$X - Y = X + (-Y).$$

According to this definition, matrix subtraction can be performed by subtracting corresponding elements. For example, using A and B as defined above,

$$A - B = \begin{bmatrix} 1 & 2 & 3 \\ 0 & -1 & 5 \end{bmatrix} - \begin{bmatrix} -2 & 3 & 0 \\ 1 & -7 & 2 \end{bmatrix}$$
$$= \begin{bmatrix} 1 - (-2) & 2 - 3 & 3 - 0 \\ 0 - 1 & -1 - (-7) & 5 - 2 \end{bmatrix}$$
$$= \begin{bmatrix} 3 & -1 & 3 \\ -1 & 6 & 3 \end{bmatrix}.$$

6 Find each of the following differences when possible.

(a) $\begin{bmatrix} 2 & 5 \\ -1 & 0 \end{bmatrix} - \begin{bmatrix} 6 & 4 \\ 3 & -2 \end{bmatrix}$

(b) $\begin{bmatrix} 1 & 5 & 6 \\ 2 & 4 & 8 \end{bmatrix} - \begin{bmatrix} 2 & 1 \\ 10 & 3 \end{bmatrix}$

(c) $[5 \quad -4 \quad 1] - [6 \quad 0 \quad -3]$

Answers:

(a) $\begin{bmatrix} -4 & 1 \\ -4 & 2 \end{bmatrix}$

(b) Not possible

(c) $[-1 \quad -4 \quad 4]$

EXAMPLE 5   (a) $[8 \quad 6 \quad -4] - [3 \quad 5 \quad -8] = [5 \quad 1 \quad 4]$

(b) The matrices
$$\begin{bmatrix} -2 & 5 \\ 0 & 1 \end{bmatrix} \text{ and } \begin{bmatrix} 3 \\ 5 \end{bmatrix}$$
are different sizes and cannot be subtracted. ■ **6**

EXAMPLE 6   During September the Chicago warehouse of the EZ Life Company shipped out the following numbers of each model.
$$K = \begin{bmatrix} 5 & 10 & 8 \\ 11 & 14 & 15 \end{bmatrix}$$

What was the Chicago warehouse inventory on October 1, taking into account only the number of items received and sent out during the month?

The number of each kind of item received during September is given by matrix C from Example 4; the number of each model sent out during September is given by matrix K above. The October 1 inventory will be represented by the matrix $C - K$ as shown below.

$$\begin{bmatrix} 22 & 25 & 38 \\ 31 & 34 & 35 \end{bmatrix} - \begin{bmatrix} 5 & 10 & 8 \\ 11 & 14 & 15 \end{bmatrix} = \begin{bmatrix} 17 & 15 & 30 \\ 20 & 20 & 20 \end{bmatrix} \quad \blacksquare$$

EXAMPLE 7 A drug company is testing 200 patients to see if Painoff (a new headache medicine) is effective. Half the patients receive Painoff and half receive a placebo. The data on the first 50 patients is summarized in this matrix.

Pain Relief Obtained

Yes   No

Patient took Painoff $\begin{bmatrix} 22 & 3 \\ 8 & 17 \end{bmatrix}$
Patient took placebo

For example, row 2 shows that of the people who took the placebo, 8 got relief, but 17 did not. The test was repeated on three more groups of 50 patients each, with the results summarized by these matrices.

$$\begin{bmatrix} 21 & 4 \\ 6 & 19 \end{bmatrix} \quad \begin{bmatrix} 19 & 6 \\ 10 & 15 \end{bmatrix} \quad \begin{bmatrix} 23 & 2 \\ 3 & 22 \end{bmatrix}$$

7 Later it was discovered that the data in the last group of 50 patients in Example 7 was invalid. Use a matrix to represent the total test results after those data were eliminated.

Answer:

$\begin{bmatrix} 62 & 13 \\ 24 & 51 \end{bmatrix}$

The total results of the test can be obtained by adding these four matrices.

$$\begin{bmatrix} 22 & 3 \\ 8 & 17 \end{bmatrix} + \begin{bmatrix} 21 & 4 \\ 6 & 19 \end{bmatrix} + \begin{bmatrix} 19 & 6 \\ 10 & 15 \end{bmatrix} + \begin{bmatrix} 23 & 2 \\ 3 & 22 \end{bmatrix} = \begin{bmatrix} 85 & 15 \\ 27 & 73 \end{bmatrix}$$

Because 85 of 100 patients got relief with Painoff and only 27 of 100 with the placebo, it appears that Painoff is effective. $\blacksquare$ **7**

Suppose one of the EZ Life Company warehouses receives the following order, written in matrix form, where the entries have the same meaning as given earlier.

$$\begin{bmatrix} 5 & 4 & 1 \\ 3 & 2 & 3 \end{bmatrix}$$

Later, the store that sent the order asks the warehouse to send six more of the same order. The six new orders can be written as one matrix by multiplying each element in the matrix by 6, giving the product

$$6\begin{bmatrix} 5 & 4 & 1 \\ 3 & 2 & 3 \end{bmatrix} = \begin{bmatrix} 30 & 24 & 6 \\ 18 & 12 & 18 \end{bmatrix}.$$

8 Find each product.

(a) $-3\begin{bmatrix} 4 & -2 \\ 1 & 5 \end{bmatrix}$

(b) $4\begin{bmatrix} 2 & 4 & 7 \\ 8 & 2 & 1 \\ 5 & 7 & 3 \end{bmatrix}$

Answers:

(a) $\begin{bmatrix} -12 & 6 \\ -3 & -15 \end{bmatrix}$

(b) $\begin{bmatrix} 8 & 16 & 28 \\ 32 & 8 & 4 \\ 20 & 28 & 12 \end{bmatrix}$

In work with matrices, a real number, like the 6 in the multiplication above, is called a **scalar.**

> The **product** of a scalar k and a matrix X is the matrix kX, each of whose elements is k times the corresponding element of X.

For example,

$$(-3)\begin{bmatrix} 2 & -5 \\ 1 & 7 \end{bmatrix} = \begin{bmatrix} -6 & 15 \\ -3 & -21 \end{bmatrix}. \quad \boxed{8}$$

7.3 EXERCISES

Find the size of each of the following. Identify any square, column, or row matrices. (See Example 2.) Give the additive inverse of each matrix.

1. $\begin{bmatrix} 7 & -8 & 4 \\ 0 & 13 & 9 \end{bmatrix}$

2. $\begin{bmatrix} -7 & 23 \\ 5 & -6 \end{bmatrix}$

3. $\begin{bmatrix} -3 & 0 & 11 \\ 1 & \frac{1}{4} & -7 \\ 5 & -3 & 9 \end{bmatrix}$

4. $\begin{bmatrix} 6 & -4 & \frac{2}{3} & 12 & 2 \end{bmatrix}$

5. $\begin{bmatrix} 7 \\ 11 \end{bmatrix}$

6. $[-5]$

7. If A is a 5×3 matrix and $A + B = A$, what do you know about B?

8. If C is a 3×3 matrix and D is a 3×4 matrix, then $C + D$ is _____.

Perform the indicated operations where possible. (See Examples 3–6.)

9. $\begin{bmatrix} 1 & 2 & 5 & -1 \\ 3 & 0 & 2 & -4 \end{bmatrix} + \begin{bmatrix} 8 & 10 & -5 & 3 \\ -2 & -1 & 0 & 0 \end{bmatrix}$

10. $\begin{bmatrix} 1 & 5 \\ 2 & -3 \\ 3 & 7 \end{bmatrix} + \begin{bmatrix} 2 & 3 \\ 8 & 5 \\ -1 & 9 \end{bmatrix}$

11. $\begin{bmatrix} 1 & 5 & 7 \\ 2 & 2 & 3 \end{bmatrix} + \begin{bmatrix} 4 & 8 & -7 \\ 1 & -1 & 5 \end{bmatrix}$

12. $\begin{bmatrix} 2 & 4 \\ -8 & 1 \end{bmatrix} + \begin{bmatrix} 9 & -3 \\ 8 & 5 \end{bmatrix}$

13. $\begin{bmatrix} 4 & -2 & 5 \\ 3 & 7 & 0 \end{bmatrix} - \begin{bmatrix} 1 & 5 & -2 \\ -1 & 3 & 8 \end{bmatrix}$

14. $\begin{bmatrix} 9 & 1 \\ 0 & -3 \\ 4 & 10 \end{bmatrix} - \begin{bmatrix} 1 & 9 & -4 \\ -1 & 1 & 0 \end{bmatrix}$

Let $A = \begin{bmatrix} -2 & 4 \\ 0 & 3 \end{bmatrix}$ and $B = \begin{bmatrix} -6 & 2 \\ 4 & 0 \end{bmatrix}$. Find each of the following.

15. $2A$

16. $-3B$

17. $-4B$

18. $5A$

19. $-4A + 5B$

20. $3A - 10B$

Let $A = \begin{bmatrix} 1 & -2 \\ 4 & 3 \end{bmatrix}$ and $B = \begin{bmatrix} 2 & -1 \\ 0 & 5 \end{bmatrix}$. Find a matrix X satisfying the given equation.

21. $2X = 2A + 3B$

22. $3X = A - 3B$

Using matrices

$$O = \begin{bmatrix} 0 & 0 \\ 0 & 0 \end{bmatrix}, P = \begin{bmatrix} m & n \\ p & q \end{bmatrix}, T = \begin{bmatrix} r & s \\ t & u \end{bmatrix}, \text{ and}$$

$$X = \begin{bmatrix} x & y \\ z & w \end{bmatrix},$$

verify that the statements in Exercises 23–28 are true.

23. $X + T$ is a 2×2 matrix.

24. $X + T = T + X$ (Commutative property of addition of matrices)

25. $X + (T + P) = (X + T) + P$ (Associative property of addition of matrices)

26. $X + (-X) = O$ (Inverse property of addition of matrices)

27. $P + O = P$ (Identity property of addition of matrices)

28. Which of the above properties are valid for matrices that are not square?

29. **Management** An investment group planning a shopping center decided to include a market, a barber shop, a variety store, a drug store, and a bakery. They estimated the initial cost and the guaranteed rent (both in dollars per square foot) for each type of store, respectively, as follows. Initial cost: 18, 10, 8, 10, and 10; guaranteed rent: 2.7, 1.5, 1.0, 2.0, and 1.7. Write this information first as a 5×2 matrix and then as a 2×5 matrix. (See Example 1.)

30. Natural Science A dietician prepares a diet specifying the amounts a patient should eat of four basic food groups: group I, meats; group II, fruits and vegetables; group III, breads and starches; group IV, milk products. Amounts are given in "exchanges" which represent 1 ounce (meat), 1/2 cup (fruits and vegetables), 1 slice (bread), 8 ounces (milk), or other suitable measurements.

(a) The number of "exchanges" for breakfast for each of the four food groups, respectively, are 2, 1, 2, and 1; for lunch, 3, 2, 2, and 1; and for dinner, 4, 3, 2, and 1. Write a 3×4 matrix using this information.

(b) The amounts of fat, carbohydrates, and protein in each food group respectively are as follows.

Fat: 5, 0, 0, 10
Carbohydrates: 0, 10, 15, 12
Protein: 7, 1, 2, 8

Use this information to write a 4×3 matrix.

(c) There are 8 calories per unit of fat, 4 calories per unit of carbohydrates, and 5 calories per unit of protein; summarize this data in a 3×1 matrix.

31. Natural Science At the beginning of a laboratory experiment, five baby rats measured 5.6, 6.4, 6.9, 7.6, and 6.1 centimeters in length, and weighed 144, 138, 149, 152, and 146 grams, respectively.

(a) Write a 2×5 matrix using this information.

(b) At the end of 2 weeks, their lengths were 10.2, 11.4, 11.4, 12.7, and 10.8 centimeters, and they weighed 196, 196, 225, 250, and 230 grams. Write a 2×5 matrix with this information.

(c) Use matrix subtraction and the matrices found in (a) and (b) to write a matrix that gives the amount of change in length and weight for each rat. (See Examples 5, 6, and 7.)

(d) The following week the rats grew 1.8, 1.5, 2.3, 1.8, and 2.0 centimeters, respectively, and gained 25, 22, 29, 33, and 20 grams, respectively. Set up a matrix with these increases and use matrix addition to find their lengths and weights at the end of this week.

32. Management There are three convenience stores in Gambier. This week, Store I sold 88 loaves of bread, 48 quarts of milk, 16 jars of peanut butter, and 112 pounds of cold cuts. Store II sold 105 loaves of bread, 72 quarts of milk, 21 jars of peanut butter, and 147 pounds of cold cuts. Store III sold 60 loaves of bread, 40 quarts of milk, no peanut butter, and 50 pounds of cold cuts.

(a) Use a 3×4 matrix to express the sales information for the three stores.

(b) During the following week, sales on these products at Store I increased by 25%; sales at Store II increased by 1/3; and sales at Store III increased by 10%. Write the sales matrix for that week.

(c) Write a matrix that represents total sales over the two-week period.

33. Management A toy company has plants in Boston, Chicago, and Seattle that manufacture toy rockets and robots. The matrix below gives the production costs (in dollars) for each item at the Boston plant:

$$\begin{array}{cc} & \text{Rockets} \quad \text{Robots} \\ \begin{array}{c} \text{Material} \\ \text{Labor} \end{array} & \begin{bmatrix} 4.27 & 6.94 \\ 3.45 & 3.65 \end{bmatrix} \end{array}$$

(a) In Chicago, a rocket costs $4.05 for materials and $3.27 for labor; a robot costs $7.01 for material and $3.51 for labor. In Seattle, material costs are $4.40 for rockets and $6.90 for robots; labor costs are $3.54 for rockets and $3.76 for robots. Write the production cost matrices for Chicago and Seattle.

(b) Assume each plant makes the same number of each item. Write a matrix that expresses the average production costs for all three plants.

(c) Suppose labor costs increase by $.11 per item in Chicago and material costs there increase by $.37 for a rocket and $.42 for a robot. What is the new production cost matrix for Chicago?

(d) After the Chicago cost increases, the Boston plant is closed and production divided evenly among the other two plants. What is the matrix that now expresses the average production costs for the entire country?

34. Social Sciences The following table gives the educational attainment of the U.S. population 25 years and older.*

	MALE		FEMALE	
	Four Years of High School or More	*Four Years of College or More*	*Four Years of High School or More*	*Four Years of College or More*
1940	22.7%	5.5%	26.3%	3.8%
1950	32.6	7.3	36.0	5.2
1959	42.2	10.3	45.2	6.0
1970	55.0	14.1	55.4	8.2
1980	69.1	20.8	68.1	13.5
1987	76.0	23.6	75.3	16.5
1991	78.5	24.3	78.3	18.8

(a) Write a matrix for the educational attainment of males.

(b) Write a matrix for the educational attainment of females.

(c) Use the matrices from parts (a) and (b) to write a matrix showing how much more (or less) education males have attained than females.

*"Educational Attainment by Percentage of Population 25+ Years, 1940–91" from "The Universal Almanac, 1993," John W. Wright, General Editor, Kansas City, New York: Andrews and McMeel.

7.4   MATRIX PRODUCTS AND INVERSES

In the previous section we showed how to multiply a matrix by a scalar. Finding the product of two matrices is more involved, but is important in solving practical problems. To understand the reasoning behind the definition of matrix multiplication, look again at the EZ Life Company. Suppose sofas and chairs of the same model are often sold as sets with matrix W showing the number of each model set in each warehouse.

$$
\begin{array}{c}
\text{New York} \\
\text{Chicago} \\
\text{San Francisco}
\end{array}
\begin{array}{ccc}
A & B & C \\
\end{array}
\left[\begin{array}{ccc}
10 & 7 & 3 \\
5 & 9 & 6 \\
4 & 8 & 2
\end{array}\right] = W
$$

If the selling price of a model A set is $800, of a model B set $1000, and of a model C set $1200, find the total value of the sets in the New York warehouse as follows.

Type	Number of Sets		Price of Set		Total
A	10	×	$ 800	=	$ 8,000
B	7	×	1000	=	7,000
C	3	×	1200	=	3,600
			Total for New York		$18,600

1 In this example of the EZ Life Company, find the total value of the New York sets if model A sets sell for $1200, model B for $1600, and model C for $1300.

Answer:
$27,100

The total value of the three kinds of sets in New York is $18,600.   **1**

The work done in the table above is summarized as follows:

$$10(\$800) + 7(\$1000) + 3(\$1200) = \$18,600.$$

In the same way, the Chicago sets have a total value of

$$5(\$800) + 9(\$1000) + 6(\$1200) = \$20,200,$$

and in San Francisco, the total value of the sets is

$$4(\$800) + 8(\$1000) + 2(\$1200) = \$13,600.$$

The selling prices can be written as a column matrix, P, and the total value in each location as a column matrix, V.

$$
\left[\begin{array}{c} 800 \\ 1000 \\ 1200 \end{array}\right] = P \quad \text{and} \quad \left[\begin{array}{c} 18,600 \\ 20,200 \\ 13,600 \end{array}\right] = V
$$

Consider how the first row of the matrix W and the single column P lead to the first entry of V.

Similarly, adding the products of corresponding entries in the second row of W and the column P produces the second entry in V. The third entry in V is obtained in the same way by using the third row of W and column P. This suggests that it is reasonable to *define* the product WP to be V.

$$WP = \begin{bmatrix} 10 & 7 & 3 \\ 5 & 9 & 6 \\ 4 & 8 & 2 \end{bmatrix} \begin{bmatrix} 800 \\ 1000 \\ 1200 \end{bmatrix} = \begin{bmatrix} 18{,}600 \\ 20{,}200 \\ 13{,}600 \end{bmatrix} = V$$

Note the sizes of the matrices here: the product of a 3×3 matrix and a 3×1 matrix is a 3×1 matrix.

MULTIPLYING MATRICES   We first define the **product of a row of a matrix and a column of a matrix** (with the same number of entries in each) to be the *number* obtained by multiplying the corresponding entries (first by first, second by second, and so on) and adding the results. For instance,

$$[3 \quad -2 \quad 1] \cdot \begin{bmatrix} 4 \\ 5 \\ 0 \end{bmatrix} = 3 \cdot 4 + (-2) \cdot 5 + 1 \cdot 0 = 12 - 10 + 0 = 2.$$

Now **matrix multiplication** is defined as follows.

> Let A be an $m \times n$ matrix and let B be an $n \times k$ matrix. The **product matrix** AB is the $m \times k$ matrix whose entry in the ith row and jth column is
>
> the product of the ith row of A and the jth column of B.

CAUTION   Be careful when multiplying matrices. Remember that the number of *columns* of A must equal the number of *rows* of B in order to get the product matrix AB. The final product will have as many rows as A and as many columns as B. ◆

EXAMPLE 1   Suppose matrix A is 2×2 and matrix B is 2×4. Can the product AB be calculated? What is the size of the product?

The following diagram helps decide the answers to these questions.

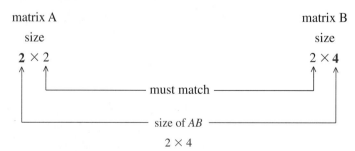

|2| Matrix A is 4×6 and matrix B is 2×4.

(a) Can AB be found? If so, give its size.

(b) Can BA be found? If so, give its size.

Answers:

(a) No

(b) Yes; 2×6

The product AB can be calculated because A has two columns and B has two rows. The product will be a 2×4 matrix. ■ |2|

EXAMPLE 2   Find the product CD given

$$C = \begin{bmatrix} -3 & 4 & 2 \\ 5 & 0 & 4 \end{bmatrix} \quad \text{and} \quad D = \begin{bmatrix} -6 & 4 \\ 2 & 3 \\ 3 & -2 \end{bmatrix}.$$

Here matrix C is 2×3 and matrix D is 3×2, so matrix CD can be found and will be 2×2.

Step 1

$$\begin{bmatrix} -3 & 4 & 2 \\ 5 & 0 & 4 \end{bmatrix}\begin{bmatrix} -6 & 4 \\ 2 & 3 \\ 3 & -2 \end{bmatrix} \qquad (-3) \cdot (-6) + 4 \cdot 2 + 2 \cdot 3 = 32$$

Step 2

$$\begin{bmatrix} -3 & 4 & 2 \\ 5 & 0 & 4 \end{bmatrix}\begin{bmatrix} -6 & 4 \\ 2 & 3 \\ 3 & -2 \end{bmatrix} \qquad (-3) \cdot 4 + 4 \cdot 3 + 2 \cdot (-2) = -4$$

Step 3

$$\begin{bmatrix} -3 & 4 & 2 \\ 5 & 0 & 4 \end{bmatrix}\begin{bmatrix} -6 & 4 \\ 2 & 3 \\ 3 & -2 \end{bmatrix} \qquad 5 \cdot (-6) + 0 \cdot 2 + 4 \cdot 3 = -18$$

Step 4

$$\begin{bmatrix} -3 & 4 & 2 \\ 5 & 0 & 4 \end{bmatrix}\begin{bmatrix} -6 & 4 \\ 2 & 3 \\ 3 & -2 \end{bmatrix} \qquad 5 \cdot 4 + 0 \cdot 3 + 4 \cdot (-2) = 12$$

Step 5   The product is

$$CD = \begin{bmatrix} -3 & 4 & 2 \\ 5 & 0 & 4 \end{bmatrix}\begin{bmatrix} -6 & 4 \\ 2 & 3 \\ 3 & -2 \end{bmatrix} = \begin{bmatrix} 32 & -4 \\ -18 & 12 \end{bmatrix}. \quad \blacksquare \quad \boxed{3}$$

EXAMPLE 3   Find BA given

$$A = \begin{bmatrix} 1 & -3 \\ 7 & 2 \end{bmatrix} \quad \text{and} \quad B = \begin{bmatrix} 1 & 0 & -1 \\ 3 & 1 & 4 \end{bmatrix}.$$

Since B is a 2×3 matrix and A is a 2×2 matrix, the product BA cannot be found. $\blacksquare$ $\boxed{4}$

TECHNOLOGY TIP Graphing calculators can find matrix products. However, if you use a graphing calculator to try to find the product in Example 3, the calculator will display an error message.   ✔

Matrix multiplication has some similarities with the multiplication of numbers.

$\boxed{3}$ Find the product CD given

$$C = \begin{bmatrix} 1 & 3 & 5 \\ 2 & -4 & -1 \end{bmatrix}$$

and

$$D = \begin{bmatrix} 2 & -1 \\ 4 & 3 \\ 1 & -2 \end{bmatrix}.$$

Answer:

$$CD = \begin{bmatrix} 19 & -2 \\ -13 & -12 \end{bmatrix}$$

$\boxed{4}$ Give the size of each of the following products that can be found.

(a) $\begin{bmatrix} 2 & 4 \\ 6 & 8 \end{bmatrix}\begin{bmatrix} 1 & 2 & 3 \\ 0 & -1 & 2 \end{bmatrix}$

(b) $\begin{bmatrix} 1 & 2 \\ 5 & 10 \\ 12 & 7 \end{bmatrix}\begin{bmatrix} 2 & 4 \\ 3 & 6 \\ 9 & 1 \end{bmatrix}$

(c) $\begin{bmatrix} 5 \\ 2 \\ 4 \end{bmatrix}\begin{bmatrix} 1 & 0 & 6 \end{bmatrix}$

Answers:

(a) 2×3

(b) Not possible

(c) 3×3

> For any matrices A, B, C, such that all the indicated sums and products exist, matrix multiplication is associative and distributive.
>
> $$A(BC) = (AB)C \quad A(B + C) = AB + AC \quad (B + C)A = BA + CA$$

However, there are important differences between matrix multiplication and multiplication of numbers. (See Exercises 19–22 at the end of this section.) In particular, matrix multiplication is *not* commutative.

> If A and B are matrices such that the products AB and BA exist,
>
> $$AB \text{ may not equal } BA.$$

The next example illustrates matrix multiplication used in an application.

EXAMPLE 4 A contractor builds three kinds of houses, models A, B, and C, with a choice of two styles, Spanish or contemporary. Matrix P shows the number of each kind of house planned for a new 100-home subdivision.

$$\begin{array}{c} \\ \text{Model A} \\ \text{Model B} \\ \text{Model C} \end{array} \begin{array}{c} \text{Spanish} \quad \text{Contemporary} \\ \begin{bmatrix} 0 & 30 \\ 10 & 20 \\ 20 & 20 \end{bmatrix} \end{array} = P$$

The amounts for each of the exterior materials used depend primarily on the style of the house. These amounts are shown in matrix Q. (Concrete is in cubic yards, lumber in units of 1000 board feet, brick in 1000s, and shingles in units of 100 square feet.)

$$\begin{array}{c} \\ \text{Spanish} \\ \text{Contemporary} \end{array} \begin{array}{c} \text{Concrete} \quad \text{Lumber} \quad \text{Brick} \quad \text{Shingles} \\ \begin{bmatrix} 10 & 2 & 0 & 2 \\ 50 & 1 & 20 & 2 \end{bmatrix} \end{array} = Q$$

Matrix R gives the cost for each kind of material.

$$\begin{array}{c} \\ \text{Concrete} \\ \text{Lumber} \\ \text{Brick} \\ \text{Shingles} \end{array} \begin{array}{c} \text{Cost per Unit} \\ \begin{bmatrix} 20 \\ 180 \\ 60 \\ 25 \end{bmatrix} \end{array} = R$$

(a) What is the total cost for each model house?

First find PQ. The product PQ shows the amount of each material needed for each model house.

$$PQ = \begin{bmatrix} 0 & 30 \\ 10 & 20 \\ 20 & 20 \end{bmatrix} \begin{bmatrix} 10 & 2 & 0 & 2 \\ 50 & 1 & 20 & 2 \end{bmatrix}$$

Concrete Lumber Brick Shingles

$$PQ = \begin{bmatrix} 1500 & 30 & 600 & 60 \\ 1100 & 40 & 400 & 60 \\ 1200 & 60 & 400 & 80 \end{bmatrix} \begin{matrix} \text{Model A} \\ \text{Model B} \\ \text{Model C} \end{matrix}$$

Now multiply PQ and R, the cost matrix, to get the total cost for each model house.

Cost

$$\begin{bmatrix} 1500 & 30 & 600 & 60 \\ 1100 & 40 & 400 & 60 \\ 1200 & 60 & 400 & 80 \end{bmatrix} \begin{bmatrix} 20 \\ 180 \\ 60 \\ 25 \end{bmatrix} = \begin{bmatrix} 72{,}900 \\ 54{,}700 \\ 60{,}800 \end{bmatrix} \begin{matrix} \text{Model A} \\ \text{Model B} \\ \text{Model C} \end{matrix}$$

(b) How much of each of the four kinds of material must be ordered?

The totals of the columns of matrix PQ will give a matrix whose elements represent the total amounts of each material needed for the subdivision. Call this matrix T, and write it as a row matrix.

$$T = [3800 \quad 130 \quad 1400 \quad 200]$$

(c) What is the total cost for material?

Find the total cost of all the materials by taking the product of matrix T, the matrix showing the total amounts of each material, and matrix R, the cost matrix. (To multiply these and get a 1×1 matrix, representing total cost, we must multiply a 1×4 matrix by a 4×1 matrix. This is why T was written as a row matrix in (b) above.)

$$TR = [3800 \quad 130 \quad 1400 \quad 200] \begin{bmatrix} 20 \\ 180 \\ 60 \\ 25 \end{bmatrix} = [188{,}400]$$

(d) Suppose the contractor builds the same number of homes in five subdivisions. Calculate the total amount of each material for each model for all five subdivisions.

Multiply PQ by the scalar 5, as follows.

$$5 \begin{bmatrix} 1500 & 30 & 600 & 60 \\ 1100 & 40 & 400 & 60 \\ 1200 & 60 & 400 & 80 \end{bmatrix} = \begin{bmatrix} 7500 & 150 & 3000 & 300 \\ 5500 & 200 & 2000 & 300 \\ 6000 & 300 & 2000 & 400 \end{bmatrix} \blacksquare$$

We can introduce a notation to help keep track of the quantities a matrix represents. For example, we can say that matrix P, from Example 4, represents models/styles, matrix Q represents styles/materials, and matrix R represents materials/cost. In each case, the meaning of the rows is written first and the columns second. When we found the product PQ in Example 4, the rows of the matrix represented models and the columns represented materials. Therefore, we can say the matrix product PQ represents models/materials. The common quantity, styles, in both P and Q was eliminated in the product PQ. Do you see that the product $(PQ)R$ represents models/cost?

In practical problems this notation helps decide in what order to multiply two matrices so that the results are meaningful. In Example 4(c) we could have found either product RT or product TR. However, since T represents subdivisions/materials and R represents materials/cost, the product TR gives subdivisions/cost. [5]

5 Let matrix A be

Vitamin

$$\begin{matrix} & & \text{C} & \text{E} & \text{K} \\ \textit{Brand} & \begin{matrix} \text{X} \\ \text{Y} \end{matrix} & \begin{bmatrix} 2 & 7 & 5 \\ 4 & 6 & 9 \end{bmatrix} \end{matrix}$$

and matrix B be

Cost

$$\begin{matrix} & & \text{X} & \text{Y} \\ \textit{Vitamin} & \begin{matrix} \text{C} \\ \text{E} \\ \text{K} \end{matrix} & \begin{bmatrix} 12 & 14 \\ 18 & 15 \\ 9 & 10 \end{bmatrix} \end{matrix}.$$

(a) What quantities do matrices A and B represent?

(b) What quantities does the product AB represent?

(c) What quantities does the product BA represent?

Answers:

(a) A = brand/vitamin, B = vitamin/cost

(b) AB = brand/cost

(c) Not meaningful, although the product BA can be found

6 Let $A = \begin{bmatrix} 3 & -2 \\ 4 & -1 \end{bmatrix}$

and $I = \begin{bmatrix} 1 & 0 \\ 0 & 1 \end{bmatrix}$.

Find IA and AI.

Answer:

$IA = \begin{bmatrix} 3 & -2 \\ 4 & -1 \end{bmatrix} = A$ and

$AI = \begin{bmatrix} 3 & -2 \\ 4 & -1 \end{bmatrix} = A$

IDENTITY AND INVERSE MATRICES   Recall from Section 1.1 that the real number 1 is the identity element for multiplication of real numbers: for any real number a, $a \cdot 1 = 1 \cdot a = a$. In this section, an **identity matrix I** is defined that has properties similar to those of the number 1.

If I is to be the identity matrix, the products AI and IA must both equal A. The 2×2 identity matrix that satisfies these conditions is

$$I = \begin{bmatrix} 1 & 0 \\ 0 & 1 \end{bmatrix}. \quad \boxed{6}$$

To check that I is really the 2×2 identity matrix, let

$$A = \begin{bmatrix} a & b \\ c & d \end{bmatrix}.$$

Then AI and IA should both equal A.

$$AI = \begin{bmatrix} a & b \\ c & d \end{bmatrix}\begin{bmatrix} 1 & 0 \\ 0 & 1 \end{bmatrix} = \begin{bmatrix} a(1)+b(0) & a(0)+b(1) \\ c(1)+d(0) & c(0)+d(1) \end{bmatrix} = \begin{bmatrix} a & b \\ c & d \end{bmatrix} = A$$

$$IA = \begin{bmatrix} 1 & 0 \\ 0 & 1 \end{bmatrix}\begin{bmatrix} a & b \\ c & d \end{bmatrix} = \begin{bmatrix} 1(a)+0(c) & 1(b)+0(d) \\ 0(a)+1(c) & 0(b)+1(d) \end{bmatrix} = \begin{bmatrix} a & b \\ c & d \end{bmatrix} = A$$

This verifies that I has been defined correctly. (It can also be shown that I is the only 2×2 identity matrix.)

The identity matrices for 3×3 matrices and 4×4 matrices, respectively, are

$$I = \begin{bmatrix} 1 & 0 & 0 \\ 0 & 1 & 0 \\ 0 & 0 & 1 \end{bmatrix} \quad \text{and} \quad I = \begin{bmatrix} 1 & 0 & 0 & 0 \\ 0 & 1 & 0 & 0 \\ 0 & 0 & 1 & 0 \\ 0 & 0 & 0 & 1 \end{bmatrix}.$$

By generalizing, an identity matrix can be found for any n by n matrix: this identity matrix will have 1's on the main diagonal from upper left to lower right, with all other entries equal to 0.

TECHNOLOGY TIP On Casio 9850, HP-38, and most TI calculators you can display an $n \times n$ identity matrix by using IDENTITY n or IDENT n or IDENMAT(n). Look in the MATH or OPS submenu of the TI MATRIX menu, or the OPTN MAT menu of Casio 9850, or the MATRIX submenu of the HP-38 MATH menu. ✔

Recall that for every nonzero real number a, the equation $ax = 1$ has a solution, namely, $x = 1/a = a^{-1}$. Similarly, for a square matrix A it is natural to consider the matrix equation $AX = I$. This equation does not always have a solution, but when it does, we use special terminology. If there is a matrix A^{-1} satisfying

$$AA^{-1} = I,$$

(that is A^{-1} is a solution of $AX = I$), then A^{-1} is called the **inverse matrix** of A. In this case it can be proved that $A^{-1}A = I$ and that A^{-1} is unique (that is, a square matrix has no more than one inverse). When a matrix has an inverse, it can be found by using the row operations given in Section 7.2, as we shall see below.

CAUTION Only square matrices have inverses, but not every square matrix has one. A matrix that does not have an inverse is called a **singular matrix.** Note that the symbol A^{-1} (read *A*-inverse) does *not* mean $1/A$ or I/A; the symbol A^{-1} is just the notation for the inverse of matrix *A*. There is no such thing as matrix division. ◆

7 Given

$$A = \begin{bmatrix} 1 & 2 \\ 4 & 6 \end{bmatrix}$$

and

$$B = \begin{bmatrix} -3 & 1 \\ 2 & -\frac{1}{2} \end{bmatrix},$$

decide if they are inverses.

Answer:
Yes because $AB = BA = I$.

EXAMPLE 5 Given matrices *A* and *B* below, decide if they are inverses.

$$A = \begin{bmatrix} 2 & 3 \\ 1 & 8 \end{bmatrix} \qquad B = \begin{bmatrix} -1 & 3 \\ 1 & -2 \end{bmatrix}$$

The matrices are inverses if *AB* and *BA* both equal *I*.

$$AB = \begin{bmatrix} 2 & 3 \\ 1 & 8 \end{bmatrix} \begin{bmatrix} -1 & 3 \\ 1 & -2 \end{bmatrix} = \begin{bmatrix} 1 & 0 \\ 7 & -13 \end{bmatrix} \neq I$$

Since $AB \neq I$, the two matrices are not inverses of each other. ■ **7**

To see how to find the multiplicative inverse of a matrix, let us look for the inverse of

$$A = \begin{bmatrix} 2 & 4 \\ 1 & -1 \end{bmatrix}.$$

Let the unknown inverse matrix be

$$A^{-1} = \begin{bmatrix} x & y \\ z & w \end{bmatrix}.$$

By the definition of matrix inverse, $AA^{-1} = I$, or

$$AA^{-1} = \begin{bmatrix} 2 & 4 \\ 1 & -1 \end{bmatrix} \begin{bmatrix} x & y \\ z & w \end{bmatrix} = \begin{bmatrix} 1 & 0 \\ 0 & 1 \end{bmatrix}.$$

Use matrix multiplication to get

$$\begin{bmatrix} 2x + 4z & 2y + 4w \\ x - z & y - w \end{bmatrix} = \begin{bmatrix} 1 & 0 \\ 0 & 1 \end{bmatrix}.$$

Setting corresponding elements equal gives the system of equations

$$2x + 4z = 1 \qquad (1)$$
$$2y + 4w = 0 \qquad (2)$$
$$x - z = 0 \qquad (3)$$
$$y - w = 1. \qquad (4)$$

Since equations (1) and (3) involve only *x* and *z*, while equations (2) and (4) involve only *y* and *w*, these four equations lead to two systems of equations,

$$\begin{array}{cc} 2x + 4z = 1 \\ x - z = 0 \end{array} \quad \text{and} \quad \begin{array}{cc} 2y + 4w = 0 \\ y - w = 1. \end{array}$$

Writing the two systems as augmented matrices gives

$$\begin{bmatrix} 2 & 4 & | & 1 \\ 1 & -1 & | & 0 \end{bmatrix} \quad \text{and} \quad \begin{bmatrix} 2 & 4 & | & 0 \\ 1 & -1 & | & 1 \end{bmatrix}.$$

Each of these systems can be solved by the Gauss-Jordan method. However, since the elements to the left of the vertical bar are identical, the two systems can be combined into one matrix

$$\left[\begin{array}{cc|cc} 2 & 4 & 1 & 0 \\ 1 & -1 & 0 & 1 \end{array}\right] \tag{5}$$

and solved simultaneously as follows.

$$\left[\begin{array}{cc|cc} 1 & -1 & 0 & 1 \\ 2 & 4 & 1 & 0 \end{array}\right] \qquad \text{Interchange } R_1, R_2$$

$$\left[\begin{array}{cc|cc} 1 & -1 & 0 & 1 \\ 0 & 6 & 1 & -2 \end{array}\right] \qquad -2R_1 + R_2$$

$$\left[\begin{array}{cc|cc} 1 & -1 & 0 & 1 \\ 0 & 1 & \frac{1}{6} & -\frac{1}{3} \end{array}\right] \qquad \frac{1}{6}R_2$$

$$\left[\begin{array}{cc|cc} 1 & 0 & \frac{1}{6} & \frac{2}{3} \\ 0 & 1 & \frac{1}{6} & -\frac{1}{3} \end{array}\right] \qquad R_2 + R_1 \tag{6}$$

The left half of the augmented matrix (6) is the identity matrix, so the Gauss-Jordan process is finished and the solutions can be read from the right half of the augmented matrix. The numbers in the first column to the right of the vertical bar give the values of x and z. The second column to the right of the bar gives the values of y and w. That is,

$$\left[\begin{array}{cc|cc} 1 & 0 & x & y \\ 0 & 1 & z & w \end{array}\right] = \left[\begin{array}{cc|cc} 1 & 0 & \frac{1}{6} & \frac{2}{3} \\ 0 & 1 & \frac{1}{6} & -\frac{1}{3} \end{array}\right]$$

so that

$$A^{-1} = \left[\begin{array}{cc} x & y \\ z & w \end{array}\right] = \left[\begin{array}{cc} \frac{1}{6} & \frac{2}{3} \\ \frac{1}{6} & -\frac{1}{3} \end{array}\right].$$

Thus the original augmented matrix (5) has A as its left half and the identity matrix as its right half, while the final augmented matrix (6), at the end of the Gauss-Jordan process, has the identity matrix as its left half and the inverse matrix A^{-1} as its right half.

$$[A \mid I] \rightarrow [I \mid A^{-1}]$$

Check by multiplying A and A^{-1}. The result should be I.

$$AA^{-1} = \left[\begin{array}{cc} 2 & 4 \\ 1 & -1 \end{array}\right]\left[\begin{array}{cc} \frac{1}{6} & \frac{2}{3} \\ \frac{1}{6} & -\frac{1}{3} \end{array}\right] = \left[\begin{array}{cc} \frac{1}{3} + \frac{2}{3} & \frac{4}{3} - \frac{4}{3} \\ \frac{1}{6} - \frac{1}{6} & \frac{2}{3} + \frac{1}{3} \end{array}\right] = \left[\begin{array}{cc} 1 & 0 \\ 0 & 1 \end{array}\right] = I \quad \boxed{8}$$

This procedure for finding the inverse of a matrix can be generalized as follows.

> To obtain an **inverse matrix** A^{-1} for any $n \times n$ matrix A for which A^{-1} exists, follow these steps.
>
> 1. Form the augmented matrix $[A \mid I]$ where I is the $n \times n$ identity matrix.
> 2. Perform row operations on $[A \mid I]$ to get a matrix of the form $[I \mid B]$.
> 3. Matrix B is A^{-1}.

$\boxed{8}$ **(a)** Find A^{-1} if

$$A = \left[\begin{array}{cc} 2 & 2 \\ 4 & 1 \end{array}\right].$$

(b) Check your answer by finding AA^{-1}.

Answers:

(a) $\left[\begin{array}{cc} -\frac{1}{6} & \frac{1}{3} \\ \frac{2}{3} & -\frac{1}{3} \end{array}\right]$

(b) $\left[\begin{array}{cc} 1 & 0 \\ 0 & 1 \end{array}\right]$

EXAMPLE 6   Find A^{-1} if $A = \begin{bmatrix} 1 & 0 & 1 \\ 2 & -2 & -1 \\ 3 & 0 & 0 \end{bmatrix}$.

First write the augmented matrix $[A \mid I]$.

$$[A \mid I] = \begin{bmatrix} 1 & 0 & 1 & \mid & 1 & 0 & 0 \\ 2 & -2 & -1 & \mid & 0 & 1 & 0 \\ 3 & 0 & 0 & \mid & 0 & 0 & 1 \end{bmatrix}$$

The augmented matrix already has 1 in the upper left-hand corner as needed, so begin by selecting the row operation which will result in a 0 for the first element in row two. Multiply row one by -2 and add the result to row two. This gives

$$\begin{bmatrix} 1 & 0 & 1 & \mid & 1 & 0 & 0 \\ 0 & -2 & -3 & \mid & -2 & 1 & 0 \\ 3 & 0 & 0 & \mid & 0 & 0 & 1 \end{bmatrix}. \qquad -2R_1 + R_2$$

Get 0 for the first element in row three by multiplying row one by -3 and adding to row three as directed in Side Problem 9. **9**

Get 1 for the second element in row two by multiplying row two of the matrix found in Problem 9 at the side by $-1/2$, obtaining the new matrix

$$\begin{bmatrix} 1 & 0 & 1 & \mid & 1 & 0 & 0 \\ 0 & 1 & \frac{3}{2} & \mid & 1 & -\frac{1}{2} & 0 \\ 0 & 0 & -3 & \mid & -3 & 0 & 1 \end{bmatrix}. \qquad -\frac{1}{2}R_2$$

Get 1 for the third element in row three by multiplying row three by $-1/3$, with the result

$$\begin{bmatrix} 1 & 0 & 1 & \mid & 1 & 0 & 0 \\ 0 & 1 & \frac{3}{2} & \mid & 1 & -\frac{1}{2} & 0 \\ 0 & 0 & 1 & \mid & 1 & 0 & -\frac{1}{3} \end{bmatrix}. \qquad -\frac{1}{3}R_3$$

Now do Side Problem 10 to get 0s for the third elements in rows one and two. **10**

The answer for Problem 10(a) at the side gives the desired inverse:

$$A^{-1} = \begin{bmatrix} 0 & 0 & \frac{1}{3} \\ -\frac{1}{2} & -\frac{1}{2} & \frac{1}{2} \\ 1 & 0 & -\frac{1}{3} \end{bmatrix}.$$

Verify that AA^{-1} is I. ∎

9 (a) Complete this step.
(b) Write this row transformation as _____.

Answers:
(a)
$$\begin{bmatrix} 1 & 0 & 1 & \mid & 1 & 0 & 0 \\ 0 & -2 & -3 & \mid & -2 & 1 & 0 \\ 0 & 0 & -3 & \mid & -3 & 0 & 1 \end{bmatrix}$$
(b) $-3R_1 + R_3$

10 (a) Complete these steps.
(b) Write these row transformations as _____.

Answers:
(a)
$$\begin{bmatrix} 1 & 0 & 0 & \mid & 0 & 0 & \frac{1}{3} \\ 0 & 1 & 0 & \mid & -\frac{1}{2} & -\frac{1}{2} & \frac{1}{2} \\ 0 & 0 & 1 & \mid & 1 & 0 & -\frac{1}{3} \end{bmatrix}$$
(b) $-1R_3 + R_1;\ -\frac{3}{2}R_3 + R_2$

TECHNOLOGY TIPS **1.** With 3×3 or larger matrices, a graphing calculator is a much easier way to find an inverse. For a matrix A, A^{-1} is found by defining A and using the x^{-1} key. The calculator-generated inverse matrix for Example 6 is shown in Figure 7.4. The inverse is given with decimal approximations for 1/3 and 1/2. The three dots at the end of each row indicate that more of the matrix can be seen by moving the cursor to the right.

FIGURE 7.4

2. On TI calculators, you can use FRAC (in the MATH menu) to display the matrix A^{-1} shown in Figure 7.4 in the more conventional fractional form shown in Figure 7.5. On HP calculators you can do the same thing by changing the "number format" (in the MODES menu) to "fraction."

FIGURE 7.5

3. If you attempt to find the inverse of a singular matrix (one without an inverse) with a calculator, it will produce an error message. Sometimes, however, because of round-off error, the calculator will display a matrix that it erroneously says is A^{-1}. To verify, you should multiply A by A^{-1} to see if the product is the identity matrix. If it is not, then A does not have an inverse. ✔

7.4 EXERCISES

In Exercises 1–6, the sizes of two matrices A and B are given. Find the sizes of the product AB and the product BA whenever these products exist. (See Example 1.)

1. A is 2×2 and B is 2×2.

2. A is 3×3 and B is 3×3.

3. A is 3×5 and B is 5×2.

4. A is 4×3 and B is 3×6.

5. A is 4×2 and B is 3×4.

6. A is 7×3 and B is 2×7.

7. To find the product matrix AB, the number of _____ of A must be the same as the number of _____ of B.

8. The product matrix AB has the same number of _____ as A and the same number of _____ as B.

Find each of the following matrix products. (See Examples 2–4.)

9. $\begin{bmatrix} 1 & 2 \\ 3 & 4 \end{bmatrix} \begin{bmatrix} -1 \\ 7 \end{bmatrix}$

10. $\begin{bmatrix} -1 & 5 \\ 7 & 0 \end{bmatrix} \begin{bmatrix} 6 \\ 2 \end{bmatrix}$

11. $\begin{bmatrix} 2 & 2 & -1 \\ 3 & 0 & 1 \end{bmatrix} \begin{bmatrix} 0 & 2 \\ -1 & 4 \\ 0 & 2 \end{bmatrix}$

12. $\begin{bmatrix} -9 & 2 & 1 \\ 3 & 0 & 0 \end{bmatrix} \begin{bmatrix} 2 \\ -1 \\ 4 \end{bmatrix}$

13. $\begin{bmatrix} -4 & 1 \\ 2 & -3 \end{bmatrix} \begin{bmatrix} 1 & 0 \\ 0 & 1 \end{bmatrix}$

14. $\begin{bmatrix} 1 & 0 \\ 0 & 1 \end{bmatrix} \begin{bmatrix} 3 & -2 \\ 1 & -5 \end{bmatrix}$

15. $\begin{bmatrix} 1 & 0 & 0 \\ 0 & 1 & 0 \\ 0 & 0 & 1 \end{bmatrix} \begin{bmatrix} 3 & -5 & 7 \\ -2 & 1 & 6 \\ 0 & -3 & 4 \end{bmatrix}$

16. $\begin{bmatrix} -8 & 9 \\ 3 & -4 \\ -1 & 6 \end{bmatrix} \begin{bmatrix} 1 & 0 & 0 \\ 0 & 1 & 0 \end{bmatrix}$

17. $\begin{bmatrix} 1 & 2 & 3 \\ 4 & 5 & 6 \\ 7 & 8 & 9 \end{bmatrix} \begin{bmatrix} -1 & 5 \\ 7 & 0 \\ 1 & 2 \end{bmatrix}$

18. $\begin{bmatrix} -2 & 0 & 3 \\ 5 & -3 & -1 \end{bmatrix} \begin{bmatrix} 2 & 0 & -1 & 3 \\ 0 & 1 & 0 & -1 \\ 4 & 2 & 5 & -4 \end{bmatrix}$

In Exercises 19–21, use the matrices

$$A = \begin{bmatrix} -3 & -9 \\ 2 & 6 \end{bmatrix} \quad and \quad B = \begin{bmatrix} 4 & 6 \\ 2 & 3 \end{bmatrix}.$$

19. Show that $AB \neq BA$. Hence, matrix multiplication is not commutative.

20. Show that $(A + B)^2 \neq A^2 + 2AB + B^2$.

21. Show that $(A + B)(A - B) \neq A^2 - B^2$.

22. Show that $D^2 = D$, where

$$D = \begin{bmatrix} 1 & 0 & 0 \\ \frac{1}{2} & 0 & \frac{1}{2} \\ 0 & 0 & 1 \end{bmatrix}.$$

Given matrices

$$P = \begin{bmatrix} m & n \\ p & q \end{bmatrix}, \quad X = \begin{bmatrix} x & y \\ z & w \end{bmatrix}, \quad T = \begin{bmatrix} r & s \\ t & u \end{bmatrix},$$

verify that the statements in Exercises 23–26 are true.

23. $(PX)T = P(XT)$ (Associative property)

24. $P(X + T) = PX + PT$ (Distributive property)

25. $k(X + T) = kX + kT$ for any real number k

26. $(k + h)P = kP + hP$ for any real numbers k and h

Determine whether the given matrices are inverses of each other by computing their product. (See Example 5.)

27. $\begin{bmatrix} 5 & 2 \\ 3 & -1 \end{bmatrix}$ and $\begin{bmatrix} -1 & 2 \\ 3 & -4 \end{bmatrix}$

28. $\begin{bmatrix} 0 & 1 \\ 1 & 0 \end{bmatrix}$ and $\begin{bmatrix} 3 & 5 \\ 7 & 9 \end{bmatrix}$

29. $\begin{bmatrix} 3 & -1 \\ -4 & 2 \end{bmatrix}$ and $\begin{bmatrix} 1 & \frac{1}{2} \\ 2 & \frac{3}{2} \end{bmatrix}$

30. $\begin{bmatrix} 1 & 1 \\ .1 & .2 \end{bmatrix}$ and $\begin{bmatrix} 2 & -10 \\ -1 & 10 \end{bmatrix}$

31. $\begin{bmatrix} 1 & 1 & 1 \\ 2 & 3 & 0 \\ 1 & 2 & 1 \end{bmatrix}$ and $\begin{bmatrix} 1.5 & .5 & -1.5 \\ -1 & 0 & 1 \\ .5 & -2 & 2 \end{bmatrix}$

32. $\begin{bmatrix} 2 & 5 & 4 \\ 1 & 4 & 3 \\ 1 & 3 & 2 \end{bmatrix}$ and $\begin{bmatrix} 1 & 2 & 1 \\ -5 & 8 & 2 \\ 7 & -11 & -3 \end{bmatrix}$

Find the inverse, if it exists, for each of the following matrices. (See Example 6.)

33. $\begin{bmatrix} 2 & 3 \\ 1 & 2 \end{bmatrix}$

34. $\begin{bmatrix} -1 & 2 \\ 1 & -1 \end{bmatrix}$

35. $\begin{bmatrix} 2 & 4 \\ 3 & 6 \end{bmatrix}$

36. $\begin{bmatrix} -3 & -5 \\ 6 & 10 \end{bmatrix}$

37. $\begin{bmatrix} 2 & 6 \\ 1 & 4 \end{bmatrix}$

38. $\begin{bmatrix} 1 & 2 \\ 3 & 4 \end{bmatrix}$

39. $\begin{bmatrix} 1 & -1 & 1 \\ 0 & 2 & -1 \\ 2 & 3 & 0 \end{bmatrix}$

40. $\begin{bmatrix} 1 & 2 & 3 \\ 1 & 1 & 2 \\ 0 & 1 & 2 \end{bmatrix}$

41. $\begin{bmatrix} 1 & 4 & 3 \\ 1 & -3 & -2 \\ 2 & 5 & 4 \end{bmatrix}$

42. $\begin{bmatrix} 1 & 2 & 0 \\ 3 & -1 & 2 \\ -2 & 3 & -2 \end{bmatrix}$

43. $\begin{bmatrix} 1 & -3 & 4 \\ 2 & -5 & 7 \\ 0 & -1 & 1 \end{bmatrix}$

44. $\begin{bmatrix} 5 & 0 & 2 \\ 2 & 2 & 1 \\ -3 & 1 & -1 \end{bmatrix}$

Use a graphing calculator to find the inverse of each matrix.

45. $\begin{bmatrix} 2 & 4 & 6 \\ -1 & -4 & -3 \\ 0 & 1 & -1 \end{bmatrix}$

46. $\begin{bmatrix} 2 & 2 & -4 \\ 2 & 6 & 0 \\ -3 & -3 & 5 \end{bmatrix}$

47. $\begin{bmatrix} 1 & -2 & 3 & 0 \\ 0 & 1 & -1 & 1 \\ -2 & 2 & -2 & 4 \\ 0 & 2 & -3 & 1 \end{bmatrix}$

48. $\begin{bmatrix} 1 & 1 & 0 & 2 \\ 2 & -1 & 1 & -1 \\ 3 & 3 & 2 & -2 \\ 1 & 2 & 1 & 0 \end{bmatrix}$

49. Management Burger Barn's three locations sell hamburgers, fries, and soft drinks. Barn I sells 900 burgers, 600 orders of fries, and 750 soft drinks each day. Barn II sells 1500 burgers a day and Barn III sells 1150. Soft drink sales number 900 a day at Barn II and 825 a day at Barn III. Barn II sells 950 and Barn III sells 800 orders of fries per day.

(a) Write a 3×3 matrix S that displays daily sales figures for all locations.

(b) Burgers cost $1.50 each, fries $.90 an order, and soft drinks $.60 each. Write a 1×3 matrix P that displays the prices.

(c) What matrix product displays the daily revenue at each of the three locations?

(d) What is the total daily revenue from all locations?

50. Management The four departments of Stagg Enterprises need to order the following amounts of the same products.

	Paper	Tape	Printer Ribbon	Memo Pads	Pens
Department 1	10	4	3	5	6
Department 2	7	2	2	3	8
Department 3	4	5	1	0	10
Department 4	0	3	4	5	5

The unit price (in dollars) of each product is given below for two suppliers.

	Supplier A	*Supplier B*
Paper	2	3
Tape	1	1
Printer Ribbon	4	3
Memo Pads	3	3
Pens	1	2

(a) Use matrix multiplication to get a matrix showing the comparative costs for each department for the products from the two suppliers.

(b) Find the total cost to buy products from each supplier. From which supplier should the company make the purchase?

51. **Management** The Perulli Candy Company makes three types of chocolate candy: Cheery Cherry, Mucho Mocha, and Almond Delight. The company produces its products in San Diego, Mexico City, and Managua using two main ingredients: chocolate and sugar.

(a) Each kilogram of Cheery Cherry requires .5 kg of sugar and .2 kg of chocolate; each kilogram of Mucho Mocha requires .4 kg of sugar and .3 kg of chocolate; and each kilogram of Almond Delight requires .3 kg of sugar and .3 kg of chocolate. Put this information into a 2 × 3 matrix, labeling the rows and columns.

(b) The cost of 1 kg of sugar is $3 in San Diego, $2 in Mexico City, and $1 in Managua. The cost of 1 kg of chocolate is $3 in San Diego, $3 in Mexico City, and $4 in Managua. Put this information into a matrix in such a way that when you multiply it with your matrix from part (a), you get a matrix representing the ingredient cost of producing each type of candy in each city.

(c) Multiply the matrices in parts (a) and (b), labeling the product matrix.

(d) From part (c) what is the combined sugar-and-chocolate cost to produce 1 kg of Mucho Mocha in Managua?

(e) Perulli Candy needs to quickly produce a special shipment of 100 kg of Cheery Cherry, 200 kg of Mucho Mocha, and 500 kg of Almond Delight, and it decides to select one factory to fill the entire order. Use matrix multiplication to determine in which city the total sugar-and-chocolate cost to produce the order is the smallest.

52. **Natural Science** Label the matrices $\begin{bmatrix} 2 & 1 & 2 & 1 \\ 3 & 2 & 2 & 1 \\ 4 & 3 & 2 & 1 \end{bmatrix}$,

$\begin{bmatrix} 5 & 0 & 7 \\ 0 & 10 & 1 \\ 0 & 15 & 2 \\ 10 & 12 & 8 \end{bmatrix}$, and $\begin{bmatrix} 8 \\ 4 \\ 5 \end{bmatrix}$ found in parts (a), (b), and (c) of

Section 7.3 Exercise 30, respectively, X, Y, and Z.

(a) Find the product matrix XY. What do the entries of this matrix represent?

(b) Find the product matrix YZ. What do the entries represent?

(c) Find the products (XY)Z and X(YZ) and verify that they are equal. What do the entries represent?

53. **Social Sciences** The average birth and death rates per million for several regions and the world population (in millions) by region are given below.*

	Births	*Deaths*
Asia	.027	.009
Latin America	.030	.007
North America	.015	.009
Europe	.013	.011
Soviet Union	.019	.011

	Asia	*Latin America*	*North America*	*Europe*	*USSR*
1960	1596	218	199	425	214
1970	1996	286	226	460	243
1980	2440	365	252	484	266
1990	2906	455	277	499	291

(a) Write the information in each table as a matrix.

(b) Use the matrices from part (a) to find the total number (in millions) of births and deaths in each year.

(c) Using the results of part (b), compare the number of births in 1960 and in 1990. Also compare the birth rates from part (a). Which gives better information?

(d) Using the results of part (b), compare the number of deaths in 1980 and in 1990. Discuss how this comparison differs from comparing death rates from part (a).

54. Explain why the system of equations

$$x - 3y = 4$$
$$2x + y = 1$$

is equivalent to the matrix equation $AX = B$, where

$A = \begin{bmatrix} 1 & -3 \\ 2 & 1 \end{bmatrix}$, $X = \begin{bmatrix} x \\ y \end{bmatrix}$, and $B = \begin{bmatrix} 4 \\ 1 \end{bmatrix}$. (*Hint:* What is the product AX?)

Solve the matrix equation AX = B. (See Exercise 54.)

55. $A = \begin{bmatrix} 1 & 2 \\ 5 & -4 \end{bmatrix}$, $X = \begin{bmatrix} x \\ y \end{bmatrix}$, $B = \begin{bmatrix} 3 \\ -6 \end{bmatrix}$.

56. $A = \begin{bmatrix} 1 & 2 & 3 \\ 2 & 6 & 1 \\ 3 & 3 & 10 \end{bmatrix}$, $X = \begin{bmatrix} x \\ y \\ z \end{bmatrix}$, $B = \begin{bmatrix} -2 \\ 2 \\ -2 \end{bmatrix}$.

*"Vital Events and Rates by Region and Development Category, 1987" and "World Population by Region and Development Category, 1950–2025" from U.S. Bureau of the Census, *World Population Profile: 1987.*

7.5   APPLICATIONS OF MATRICES

This section gives a variety of applications of matrices.

SOLVING SYSTEMS WITH MATRICES   Consider this system of linear equations.

$$2x - 3y = 4$$
$$x + 5y = 2$$

Let

$$A = \begin{bmatrix} 2 & -3 \\ 1 & 5 \end{bmatrix}, \quad X = \begin{bmatrix} x \\ y \end{bmatrix}, \quad B = \begin{bmatrix} 4 \\ 2 \end{bmatrix}.$$

Since

$$AX = \begin{bmatrix} 2 & -3 \\ 1 & 5 \end{bmatrix}\begin{bmatrix} x \\ y \end{bmatrix} = \begin{bmatrix} 2x - 3y \\ x + 5y \end{bmatrix} \quad \text{and} \quad B = \begin{bmatrix} 4 \\ 2 \end{bmatrix},$$

the original system is equivalent to the single matrix equation $AX = B$. Similarly, any system of linear equations can be written as a matrix equation $AX = B$. The matrix A is called the **coefficient matrix.**   **1**

A matrix equation $AX = B$ can be solved if A^{-1} exists. Assuming A^{-1} exists and using the facts that $A^{-1}A = I$ and $IX = X$ along with the associative property of multiplication of matrices gives

$$AX = B$$
$$A^{-1}(AX) = A^{-1}B \qquad \text{Multiply both sides by } A^{-1}.$$
$$(A^{-1}A)X = A^{-1}B \qquad \text{Associative property}$$
$$IX = A^{-1}B \qquad \text{Inverse property}$$
$$X = A^{-1}B. \qquad \text{Identity property}$$

When multiplying by matrices on both sides of a matrix equation, be careful to multiply in the same order on both sides of the equation, since multiplication of matrices is not commutative (unlike multiplication of real numbers). This discussion is summarized below.

> A system of equations $AX = B$, where A is the matrix of coefficients, X is the matrix of variables, and B is the matrix of constants, is solved by first finding A^{-1}. Then, if A^{-1} exists, $X = A^{-1}B$.

This method is very efficient when a graphing calculator with matrix capability is available. The elimination method (the matrix version) or the Gauss-Jordan method is the best choice when solving a system by hand.

EXAMPLE 1   Use the inverse of the coefficient matrix to solve the system

$$-x - 2y + 2z = 9$$
$$2x + y - z = -3$$
$$3x - 2y + z = -6.$$

We use the fact that $X = A^{-1}B$, with

$$A = \begin{bmatrix} -1 & -2 & 2 \\ 2 & 1 & -1 \\ 3 & -2 & 1 \end{bmatrix} \quad \text{and} \quad B = \begin{bmatrix} 9 \\ -3 \\ -6 \end{bmatrix}.$$

Using either the method explained in Section 7.4 or a graphing calculator,

$$A^{-1} = \begin{bmatrix} \frac{1}{3} & \frac{2}{3} & 0 \\ \frac{5}{3} & \frac{7}{3} & -1 \\ \frac{7}{3} & \frac{8}{3} & -1 \end{bmatrix}$$

2 Use the inverse matrix to solve the system in Example 1, if the constants for the three equations are 12, 0, and 8, respectively.

and

$$X = A^{-1}B = \begin{bmatrix} \frac{1}{3} & \frac{2}{3} & 0 \\ \frac{5}{3} & \frac{7}{3} & -1 \\ \frac{7}{3} & \frac{8}{3} & -1 \end{bmatrix} \begin{bmatrix} 9 \\ -3 \\ -6 \end{bmatrix} = \begin{bmatrix} 1 \\ 14 \\ 19 \end{bmatrix}.$$

Answer:
(4, 12, 20)

Thus, the solution is (1, 14, 19). ■ **2**

EXAMPLE 2 Use the inverse of the coefficient matrix to solve the system

$$x + 1.5y = 8$$
$$2x + 3y = 10.$$

The coefficient matrix is $A = \begin{bmatrix} 1 & 1.5 \\ 2 & 3 \end{bmatrix}$. A calculator will indicate that A^{-1} does not exist. If we try to carry out the row operations, we see why.

$$\begin{bmatrix} 1 & 1.5 & | & 1 & 0 \\ 2 & 3 & | & 0 & 1 \end{bmatrix}$$

3 Solve the system in Example 2 if the constants are, respectively, 3 and 6.

$$\begin{bmatrix} 1 & 1.5 & | & 1 & 0 \\ 0 & 0 & | & -2 & 1 \end{bmatrix}$$

Answer:
$(3 - 1.5y, y)$ for all real numbers y

The next step cannot be performed because of the zero in the second row second column. Verify that the original system has no solution. ■ **3**

INPUT-OUTPUT ANALYSIS An interesting application of matrix theory to economics was developed by Nobel Prize winner Wassily Leontief. His application of matrices to the interdependencies in an economy is called **input-output** analysis. In practice, input-output analysis is very complicated with many variables. We shall discuss only simple examples with a few variables.

Input-output models are concerned with the production and flow of goods (and perhaps services). In an economy with n basic commodities (or sectors), the production of each commodity uses some (perhaps all) of the commodities in the economy as inputs. The amounts of each commodity used in the production of 1 unit of each commodity can be written as an $n \times n$ matrix A, called the **technological** or **input-output matrix** of the economy.

EXAMPLE 3 Suppose a simplified economy involves just three commodity categories: agriculture, manufacturing, and transportation, all in appropriate units.

Production of 1 unit of agriculture requires 1/2 unit of manufacturing and 1/4 unit of transportation. Production of 1 unit of manufacturing requires 1/4 unit of agriculture and 1/4 unit of transportation; while production of 1 unit of transportation requires 1/3 unit of agriculture and 1/4 unit of manufacturing. Write the input-output matrix of this economy.

The matrix is shown below.

$$\text{Input}\begin{array}{c}\text{Agriculture}\\\text{Manufacturing}\\\text{Transportation}\end{array}\begin{bmatrix}0 & \frac{1}{4} & \frac{1}{3}\\\frac{1}{2} & 0 & \frac{1}{4}\\\frac{1}{4} & \frac{1}{4} & 0\end{bmatrix}=A$$

Output: Agriculture, Manufacturing, Transportation

The first column of the input-output matrix represents the amount of each of the three commodities consumed in the production of 1 unit of agriculture. The second column gives the corresponding amounts required to produce 1 unit of manufacturing, and the last column gives the amounts needed to produce 1 unit of transportation. (Although it is unrealistic, perhaps, that production of 1 unit of a commodity requires none of that commodity, the simpler matrix involved is useful for our purposes.) ■ **4**

[4] Write a 2 × 2 technological matrix in which 1 unit of electricity requires 1/2 unit of water and 1/3 unit of electricity, while 1 unit of water requires no water but 1/4 unit of electricity.

Answer:

$$\begin{array}{c}\text{Elec.}\\\text{Water}\end{array}\begin{array}{cc}\text{Elec.} & \text{Water}\\\begin{bmatrix}\frac{1}{3} & \frac{1}{4}\\\frac{1}{2} & 0\end{bmatrix}\end{array}$$

Another matrix used with the input-output matrix is a matrix giving the amount of each commodity produced, called the **production matrix,** or the **vector of gross output.** In an economy producing n commodities, the production matrix can be represented by a column matrix x with entries $x_1, x_2, x_3, \ldots, x_n$.

EXAMPLE 4 In Example 3, suppose the production matrix is

$$X = \begin{bmatrix}60\\52\\48\end{bmatrix}.$$

Then 60 units of agriculture, 52 units of manufacturing, and 48 units of transportation are produced. As 1/4 unit of agriculture is used for each unit of manufacturing produced, $1/4 \times 52 = 13$ units of agriculture must be used up in the "production" of manufacturing. Similarly, $1/3 \times 48 = 16$ units of agriculture will be used up in the "production" of transportation. Thus $13 + 16 = 29$ units of agriculture are used for production in the economy. Look again at the matrices A and X. Since X gives the number of units of each commodity produced and A gives the amount (in units) of each commodity used to produce 1 unit of the various commodities, the matrix product AX gives the amount of each commodity used up in production.

$$AX = \begin{bmatrix}0 & \frac{1}{4} & \frac{1}{3}\\\frac{1}{2} & 0 & \frac{1}{4}\\\frac{1}{4} & \frac{1}{4} & 0\end{bmatrix}\begin{bmatrix}60\\52\\48\end{bmatrix}=\begin{bmatrix}29\\42\\28\end{bmatrix}$$

This product shows that 29 units of agriculture, 42 units of manufacturing, and 28 units of transportation are used to produce 60 units of agriculture, 52 units of manufacturing, and 48 units of transportation. ■

5 **(a)** Write a 2 × 1 matrix X to represent gross production of 9000 units of electricity and 12,000 units of water.

(b) Find AX using A from the last side problem.

(c) Find D using $D = X - AX$.

Answers:

(a) $\begin{bmatrix} 9000 \\ 12{,}000 \end{bmatrix}$

(b) $\begin{bmatrix} 6000 \\ 4500 \end{bmatrix}$

(c) $\begin{bmatrix} 3000 \\ 7500 \end{bmatrix}$

We have seen that the matrix product AX represents the amount of each commodity used in the production process. The remainder (if any) must be enough to satisfy the demand for the various commodities from outside the production system. In an n-commodity economy, this demand can be represented by a **demand matrix** D with entries $d_1, d_2, \ldots, d_n$. The difference between the production matrix, X, and the amount, AX, used in the production process must equal the demand, D, or

$$D = X - AX.$$

In Example 4,

$$D = \begin{bmatrix} 60 \\ 52 \\ 48 \end{bmatrix} - \begin{bmatrix} 29 \\ 42 \\ 28 \end{bmatrix} = \begin{bmatrix} 31 \\ 10 \\ 20 \end{bmatrix}.$$

This result shows that production of 60 units of agriculture, 52 units of manufacturing, and 48 units of transportation would satisfy a demand of 31, 10, and 20 units of each, respectively. **5**

In practice, A and D usually are known and X must be found. That is, we need to decide what amounts of production are necessary to satisfy the required demands. Matrix algebra can be used to solve the equation $D = X - AX$ for X.

$$D = X - AX$$
$$D = IX - AX \qquad \text{Identity property}$$
$$D = (I - A)X \qquad \text{Distributive property}$$

If the matrix $I - A$ has an inverse, then

$$X = (I - A)^{-1}D.$$

EXAMPLE 5 Suppose, in the 3-commodity economy of Examples 3 and 4, there is a demand for 516 units of agriculture, 258 units of manufacturing, and 129 units of transportation. What should production of each commodity be?

The demand matrix is

$$D = \begin{bmatrix} 516 \\ 258 \\ 129 \end{bmatrix}.$$

Find the production matrix by first calculating $I - A$.

$$I - A = \begin{bmatrix} 1 & 0 & 0 \\ 0 & 1 & 0 \\ 0 & 0 & 1 \end{bmatrix} - \begin{bmatrix} 0 & \frac{1}{4} & \frac{1}{3} \\ \frac{1}{2} & 0 & \frac{1}{4} \\ \frac{1}{4} & \frac{1}{4} & 0 \end{bmatrix} = \begin{bmatrix} 1 & -\frac{1}{4} & -\frac{1}{3} \\ -\frac{1}{2} & 1 & -\frac{1}{4} \\ -\frac{1}{4} & -\frac{1}{4} & 1 \end{bmatrix}$$

Using a calculator with matrix capability or row operations, find the inverse of $I - A$.

$$(I - A)^{-1} = \begin{bmatrix} 1.3953 & .4961 & .5891 \\ .8372 & 1.3643 & .6202 \\ .5581 & .4651 & 1.3023 \end{bmatrix}$$

(The entries are rounded to four decimal places.)* Since $X = (I - A)^{-1}D$,

$$X = \begin{bmatrix} 1.3953 & .4961 & .5891 \\ .8372 & 1.3643 & .6202 \\ .5581 & .4651 & 1.3023 \end{bmatrix} \begin{bmatrix} 516 \\ 258 \\ 129 \end{bmatrix} = \begin{bmatrix} 924 \\ 864 \\ 576 \end{bmatrix},$$

(rounded to the nearest whole numbers).

From the last result, we see that production of 924 units of agriculture, 864 units of manufacturing, and 576 units of transportation is required to satisfy demands of 516, 258, and 129 units, respectively. ■

TECHNOLOGY TIP If you are using a graphing calculator to determine X, you can calculate $(I - A)^{-1}D$ in one step without finding the intermediate matrices $I - A$ and $(I - A)^{-1}$. ✔

6 A simple economy depends on just two products, beer and pretzels.

(a) Suppose 1/2 unit of beer and 1/2 unit of pretzels are needed to make 1 unit of beer, and 3/4 unit of beer is needed to make 1 unit of pretzels. Write the technological matrix A for the economy.

(b) Find $I - A$.

(c) Find $(I - A)^{-1}$.

(d) Find the gross production X that will be needed to get a net production of

$$D = \begin{bmatrix} 100 \\ 1000 \end{bmatrix}.$$

Answers:

(a) $\begin{bmatrix} \frac{1}{2} & \frac{3}{4} \\ \frac{1}{2} & 0 \end{bmatrix}$

(b) $\begin{bmatrix} \frac{1}{2} & -\frac{3}{4} \\ -\frac{1}{2} & 1 \end{bmatrix}$

(c) $\begin{bmatrix} 8 & 6 \\ 4 & 4 \end{bmatrix}$

(d) $\begin{bmatrix} 6800 \\ 4400 \end{bmatrix}$

EXAMPLE 6   An economy depends on two basic products, wheat and oil. To produce 1 metric ton of wheat requires .25 metric tons of wheat and .33 metric tons of oil. Production of 1 metric ton of oil consumes .08 metric tons of wheat and .11 metric tons of oil. Find the production that will satisfy a demand of 500 metric tons of wheat and 1000 metric tons of oil.

The input-output matrix, A, and the matrix $I - A$, are

$$A = \begin{bmatrix} .25 & .08 \\ .33 & .11 \end{bmatrix} \quad \text{and} \quad I - A = \begin{bmatrix} .75 & -.08 \\ -.33 & .89 \end{bmatrix}.$$

Next, calculate $(I - A)^{-1}$.

$$(I - A)^{-1} = \begin{bmatrix} 1.3882 & .1248 \\ .5147 & 1.1699 \end{bmatrix} \quad \text{(rounded)}$$

To find the production matrix X, use the equation $X = (I - A)^{-1}D$, with

$$D = \begin{bmatrix} 500 \\ 1000 \end{bmatrix}.$$

The production matrix is

$$X = \begin{bmatrix} 1.3882 & .1248 \\ .5147 & 1.1699 \end{bmatrix} \begin{bmatrix} 500 \\ 1000 \end{bmatrix} = \begin{bmatrix} 819 \\ 1427 \end{bmatrix}.$$

The production numbers were rounded to the nearest whole number. Production of 819 metric tons of wheat and 1427 metric tons of oil is required to satisfy the indicated demand. ■ **6**

CODE THEORY   Governments need sophisticated methods of coding and decoding messages. One example of such an advanced code uses matrix theory. Such a code takes the letters in the words and divides them into groups. (Each space between words is treated as a letter; punctuation is disregarded.) Then, numbers are assigned to the letters of the alphabet. For our purposes, let the letter a correspond to 1, b to 2, and so on. Let the number 27 correspond to a space between words.

*Although we show the matrix $(I - A)^{-1}$ with entries rounded to four decimal places, we did not round off in calculating $(I - A)^{-1}D$. If the rounded figures are used, the numbers in the product may vary slightly in the last digit.

For example, the message

mathematics is for the birds

can be divided into groups of three letters each.

mat   hem   ati   cs–   is–   for   –th   e–b   ird   s––

(We used – to represent a space between words.) We now write a column matrix for each group of three symbols using the corresponding numbers, as determined above, instead of letters. For example, the letters *mat* can be encoded as

$$\begin{bmatrix} 13 \\ 1 \\ 20 \end{bmatrix}.$$

7 Write the message "*when*" using 2×1 matrices.

Answer:
$$\begin{bmatrix} 23 \\ 8 \end{bmatrix}, \begin{bmatrix} 5 \\ 14 \end{bmatrix}$$

The coded message then consists of the 3×1 column matrices:

$$\begin{bmatrix} 13 \\ 1 \\ 20 \end{bmatrix}, \begin{bmatrix} 8 \\ 5 \\ 13 \end{bmatrix}, \begin{bmatrix} 1 \\ 20 \\ 9 \end{bmatrix}, \begin{bmatrix} 3 \\ 19 \\ 27 \end{bmatrix}, \begin{bmatrix} 9 \\ 19 \\ 27 \end{bmatrix}, \begin{bmatrix} 6 \\ 15 \\ 18 \end{bmatrix}, \begin{bmatrix} 27 \\ 20 \\ 8 \end{bmatrix}, \begin{bmatrix} 5 \\ 27 \\ 2 \end{bmatrix}, \begin{bmatrix} 9 \\ 18 \\ 4 \end{bmatrix}, \begin{bmatrix} 19 \\ 27 \\ 27 \end{bmatrix}.$$ **7**

We can further complicate the code by choosing a matrix that has an inverse (in this case a 3×3 matrix, call it *M*) and finding the products of this matrix and each of the above column matrices. The size of each group, the assignment of numbers to letters, and the choice of matrix *M* must all be predetermined.

Suppose we choose

$$M = \begin{bmatrix} 1 & 3 & 3 \\ 1 & 4 & 3 \\ 1 & 3 & 4 \end{bmatrix}.$$

8 Use the matrix given below to find the 2×1 matrices to be transmitted for the message you encoded in Problem 7 at the side.

$$\begin{bmatrix} 2 & 1 \\ 5 & 0 \end{bmatrix}$$

Answer:
$$\begin{bmatrix} 54 \\ 115 \end{bmatrix}, \begin{bmatrix} 24 \\ 25 \end{bmatrix}$$

If we find the products of *M* and the column matrices above, we have a new set of column matrices,

$$\begin{bmatrix} 1 & 3 & 3 \\ 1 & 4 & 3 \\ 1 & 3 & 4 \end{bmatrix}\begin{bmatrix} 13 \\ 1 \\ 20 \end{bmatrix} = \begin{bmatrix} 76 \\ 77 \\ 96 \end{bmatrix}, \text{ and so on.}$$

The entries of these matrices can then be transmitted to an agent as the message 76, 77, 96, and so on.   **8**

When the agent receives the message, it is divided into groups of three numbers with each group formed into a column matrix. After multiplying each column matrix by the matrix M^{-1}, the message can be read. For example,

$$M^{-1} \cdot \begin{bmatrix} 76 \\ 77 \\ 96 \end{bmatrix} = \begin{bmatrix} 13 \\ 1 \\ 20 \end{bmatrix}.$$

Although this type of code is relatively simple, it is actually difficult to break. Many complications are possible. For example, a long message might be placed in groups of 20, thus requiring a 20×20 matrix for coding and decoding. Finding the inverse of such a matrix would require an impractical amount of time if calculated by hand. For this reason some of the largest computers are used by government agencies involved in coding.

ROUTING   The diagram in Figure 7.6 shows the roads connecting four cities. Another way of representing this information is shown in matrix A, where the entries represent the number of roads connecting two cities without passing through another city.* For example, from the diagram we see that there are two roads connecting city 1 to city 4 without passing through either city 2 or 3. This information is entered in row one, column four and again in row four, column one of matrix A.

$$A = \begin{bmatrix} 0 & 1 & 2 & 2 \\ 1 & 0 & 1 & 0 \\ 2 & 1 & 0 & 1 \\ 2 & 0 & 1 & 0 \end{bmatrix}$$

Note that there are zero roads connecting each city to itself. Also, there is one road connecting cities 3 and 2.

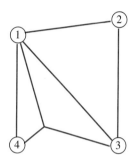

FIGURE 7.6

How many ways are there to go from city 1 to city 2, for example, by going through exactly one other city? Because we must go through one other city, we must go through either city 3 or city 4. On the diagram in Figure 7.6, we see that we can go from city 1 to city 2 through city 3 in two ways. We can go from city 1 to city 3 in two ways and then from city 3 to city 2 in one way, so there are $2 \cdot 1 = 2$ ways to get from city 1 to city 2 through city 3. It is not possible to go from city 1 to city 2 through city 4, because there is no direct route between cities 4 and 2.

The matrix A^2 gives the number of ways to travel between any two cities by passing through exactly one other city. Multiply matrix A by itself, to get A^2. Let the first row, second column entry of A^2 be b_{12}. (We use a_{ij} to denote the entry in the ith row and jth column of matrix A.) The entry b_{12} is found as follows.

$$b_{12} = a_{11}a_{12} + a_{12}a_{22} + a_{13}a_{32} + a_{14}a_{42}$$
$$= 0 \cdot 1 + 1 \cdot 0 + 2 \cdot 1 + 2 \cdot 0$$
$$= 2$$

The first product $0 \cdot 1$ in the calculations above represents the number of ways to go from city 1 to city 1, (that is, 0), and then from city 1 to city 2, (that is, 1). The 0 result indicates that such a trip does not involve a third city. The only nonzero product $(2 \cdot 1)$

*From *Matrices with Applications*, section 3.2, example 5, by Hugh G. Campbell, Copyright © 1968, pp. 50–51. Adapted by permission of Prentice-Hall, Englewood Cliffs, New Jersey.

represents the two routes from city 1 to city 3 and the one route from city 3 to city 2 which result in the $2 \cdot 1$ or 2 routes from city 1 to city 2 by going through city 3.

Similarly, A^3 gives the number of ways to travel between any two cities by passing through exactly two cities. Also, $A + A^2$ represents the total number of ways to travel between two cities with at most one intermediate city.

The diagram can be given many other interpretations. For example, the lines could represent lines of mutual influence between people or nations, or they could represent communication lines such as telephone lines.

7.5 EXERCISES

Solve the matrix equation $AX = B$ for X. (See Example 1.)

1. $A = \begin{bmatrix} 1 & -1 \\ 5 & -6 \end{bmatrix}$, $B = \begin{bmatrix} 2 \\ 4 \end{bmatrix}$

2. $A = \begin{bmatrix} 3 & -2 \\ -1 & 1 \end{bmatrix}$, $B = \begin{bmatrix} -3 \\ 5 \end{bmatrix}$

3. $A = \begin{bmatrix} 3 & 1 \\ 4 & 2 \end{bmatrix}$, $B = \begin{bmatrix} 3 & 4 \\ 5 & 6 \end{bmatrix}$

4. $A = \begin{bmatrix} 7 & -3 \\ -2 & 1 \end{bmatrix}$, $B = \begin{bmatrix} 0 & 8 \\ 4 & 1 \end{bmatrix}$

5. $A = \begin{bmatrix} 1 & -2 & -3 \\ -1 & 4 & 6 \\ 1 & -1 & -2 \end{bmatrix}$, $B = \begin{bmatrix} 2 \\ 7 \\ 4 \end{bmatrix}$

6. $A = \begin{bmatrix} 3 & -1 & 0 \\ 0 & 1 & 2 \\ 6 & 0 & 5 \end{bmatrix}$, $B = \begin{bmatrix} -6 \\ 12 \\ 15 \end{bmatrix}$

Use the inverse of the coefficient matrix to solve each system of equations. (The inverses for Exercises 9–14 were found in Exercises 41 and 44–48 of Section 7.4.) (See Example 1.)

7.
$\begin{aligned} x + 2y + 3z &= 5 \\ 2x + 3y + 2z &= 2 \\ -x - 2y - 4z &= -1 \end{aligned}$

8.
$\begin{aligned} x + y - 3z &= 4 \\ 2x + 4y - 4z &= 8 \\ -x + y + 4z &= -3 \end{aligned}$

9.
$\begin{aligned} x + 4y + 3z &= -12 \\ x - 3y - 2z &= 0 \\ 2x + 5y + 4z &= 7 \end{aligned}$

10.
$\begin{aligned} 5x + 2z &= 3 \\ 2x + 2y + z &= 4 \\ -3x + y - z &= 5 \end{aligned}$

11.
$\begin{aligned} 2x + 4y + 6z &= 4 \\ -x - 4y - 3z &= 8 \\ y - z &= -4 \end{aligned}$

12.
$\begin{aligned} 2x + 2y - 4z &= 12 \\ 2x + 6y &= 16 \\ -3x - 3y + 5z &= -20 \end{aligned}$

13.
$\begin{aligned} x - 2y + 3z &= 4 \\ y - z + w &= -8 \\ -2x + 2y - 2z + 4w &= 12 \\ 2y - 3z + w &= -4 \end{aligned}$

14.
$\begin{aligned} x + y + 2w &= 3 \\ 2x - y + z - w &= 3 \\ 3x + 3y + 2z - 2w &= 5 \\ x + 2y + z &= 3 \end{aligned}$

Use matrix algebra to solve the following matrix equations for X. Then use the given matrices to find X and check your work.

15. $N = X - MX$, $N = \begin{bmatrix} 8 \\ -12 \end{bmatrix}$, $M = \begin{bmatrix} 0 & 1 \\ -2 & 1 \end{bmatrix}$

16. $A = BX + X$, $A = \begin{bmatrix} 4 & 6 \\ -2 & 2 \end{bmatrix}$, $B = \begin{bmatrix} -2 & -2 \\ 3 & 3 \end{bmatrix}$

Find the production matrix given the following input-output and demand matrices. (See Examples 3–6.)

17. $A = \begin{bmatrix} \frac{1}{2} & \frac{2}{5} \\ \frac{1}{4} & \frac{1}{5} \end{bmatrix}$, $D = \begin{bmatrix} 2 \\ 4 \end{bmatrix}$

18. $A = \begin{bmatrix} \frac{1}{5} & \frac{1}{25} \\ \frac{3}{5} & \frac{1}{20} \end{bmatrix}$, $D = \begin{bmatrix} 3 \\ 10 \end{bmatrix}$

19. $A = \begin{bmatrix} .1 & .03 \\ .07 & .6 \end{bmatrix}$, $D = \begin{bmatrix} 5 \\ 10 \end{bmatrix}$

20. $A = \begin{bmatrix} .01 & .03 \\ .05 & .05 \end{bmatrix}$, $D = \begin{bmatrix} 100 \\ 200 \end{bmatrix}$

21. $A = \begin{bmatrix} .4 & 0 & .3 \\ 0 & .8 & .1 \\ 0 & .2 & .4 \end{bmatrix}$, $D = \begin{bmatrix} 1 \\ 3 \\ 2 \end{bmatrix}$

22. $A = \begin{bmatrix} .1 & .5 & 0 \\ 0 & .3 & .4 \\ .1 & .2 & .1 \end{bmatrix}$, $D = \begin{bmatrix} 10 \\ 4 \\ 2 \end{bmatrix}$

Write a system of equations and use the inverse of the coefficient matrix to solve the system.

23. **Management** Felsted Furniture makes dining room furniture. A buffet requires 30 hours for construction and 10 hours for finishing. A chair requires 10 hours for construction and 10 hours for finishing. A table requires 10 hours for construction and 30 hours for finishing. The construction department has 350 hours of labor and the finishing department 150 hours of labor available each week. How many pieces of each type of furniture should be produced each week if the factory is to run at full capacity?

24. **Natural Science** (a) A hospital dietician is planning a special diet for a certain patient. The total amount per meal of food groups A, B, and C must equal 400 grams. The diet should include one-third as much of group A as of group B, and the sum of the amounts of group A and group C should equal twice the amount of group B. How many grams of each food group should be included?
    (b) Suppose we drop the requirement that the diet include one-third as much of group A as of group B. Describe the set of all possible solutions.
    (c) Suppose that, in addition to the conditions given in part (a), foods A and B cost 2 cents per gram and food C costs 3 cents per gram, and that a meal must cost $8. Is a solution possible?

25. **Natural Science** Three species of bacteria are fed three foods, I, II, and III. A bacterium of the first species consumes 1.3 units each of foods I and II and 2.3 units of food III each day. A bacterium of the second species consumes 1.1 units of food I, 2.4 units of food II, and 3.7 units of food III each day. A bacterium of the third species consumes 8.1 units of I, 2.9 units of II, and 5.1 units of III each day. If 16,000 units of I, 28,000 units of II, and 44,000 units of III are supplied each day, how many of each species can be maintained in this environment?

26. **Management** A company produces three combinations of mixed vegetables which sell in 1 kilogram packages. Italian style combines .3 kilogram of zucchini, .3 of broccoli, and .4 of carrots. French style combines .6 kilogram of broccoli and .4 of carrots. Oriental style combines .2 kilogram of zucchini, .5 of broccoli, and .3 of carrots. The company has a stock of 16,200 kilograms of zucchini, 41,400 kilograms of broccoli, and 29,400 kilograms of carrots. How many packages of each style should they prepare to use up their supplies?

Exercises 27 and 28 refer to Example 6.

27. **Management** If the demand is changed to 690 metric tons of wheat and 920 metric tons of oil, how many units of each commodity should be produced?

28. **Management** Change the technological matrix so that production of 1 metric ton of wheat requires 1/5 metric ton of oil (and no wheat), and the production of 1 metric ton of oil requires 1/3 metric ton of wheat (and no oil). To satisfy the same demand matrix, how many units of each commodity should be produced?

29. **Management** A simplified economy has only two industries, the electric company and the gas company. Each dollar's worth of the electric company's output requires $.40 of its own output and $.50 of the gas company's output. Each dollar's worth of the gas company's output requires $.25 of its own output and $.60 of the electric company's output. What should the production of electricity and gas be (in dollars) if there is a $12 million demand for gas and a $15 million demand for electricity?

30. **Management** A two-segment economy consists of manufacturing and agriculture. To produce one unit of manufacturing output requires .40 unit of its own output and .20 unit of agricultural output. To produce one unit of agricultural output requires .30 unit of its own output and .40 unit of manufacturing output. If there is a demand of 240 units of manufacturing and 90 units of agriculture, what should be the output of each segment?

31. **Management** A primitive economy depends on two basic goods, yams and pork. Production of 1 bushel of yams requires 1/4 bushel of yams and 1/2 of a pig. To produce 1 pig requires 1/6 bushel of yams. Find the amount of each commodity that should be produced to get
    (a) 1 bushel of yams and 1 pig;
    (b) 100 bushels of yams and 70 pigs.

32. **Management** A simplified economy is based on agriculture, manufacturing, and transportation. Each unit of agricultural output requires .4 unit of its own output, .3 unit of manufacturing, and .2 unit of transportation output. One unit of manufacturing output requires .4 unit of its own output, .2 unit of agricultural, and .3 unit of transportation output. One unit of transportation output requires .4 unit of its own output, .1 unit of agricultural, and .2 unit of manufacturing output. There is demand for 35 units of agricultural, 90 units of manufacturing, and 20 units of transportation output. How many units should each segment of the economy produce?

33. **Management** How many units of each segment should the economy in Exercise 32 produce, if the demand is for 55 units of agricultural, 20 units of manufacturing, and 10 units of transportation output?

34. **Social Science** Use the method discussed in the text to encode the message

Anne is home.

Break the message into groups of two letters and use the matrix

$$M = \begin{bmatrix} 1 & 3 \\ 2 & 7 \end{bmatrix}.$$

35. **Social Science** Use the matrix of Exercise 34 to encode the message

Head for the hills!

36. Social Science Decode the following message, which was encoded by using the matrix M of Exercise 34.

$$\begin{bmatrix} 90 \\ 207 \end{bmatrix}, \begin{bmatrix} 39 \\ 87 \end{bmatrix}, \begin{bmatrix} 26 \\ 57 \end{bmatrix}, \begin{bmatrix} 66 \\ 145 \end{bmatrix}, \begin{bmatrix} 61 \\ 142 \end{bmatrix}, \begin{bmatrix} 89 \\ 205 \end{bmatrix}.$$

37. Social Science Use matrix A in the discussion on routing in the text to find A^2. Then answer the following questions. How many ways are there to travel from
   (a) City 1 to city 3 by passing through exactly one city?
   (b) City 2 to city 4 by passing through exactly one city?
   (c) City 1 to city 3 by passing through at most one city?
   (d) City 2 to city 4 by passing through at most one city?

38. Social Science Find A^3. (See Exercise 37.) Then answer the following questions.
   (a) How many ways are there to travel between cities 1 and 4 by passing through exactly two cities?
   (b) How many ways are there to travel between cities 1 and 4 by passing through at most two cities?

39. Management A small telephone system connects three cities. There are four lines between cities 3 and 2, three lines connecting city 3 with city 1, and two lines between cities 1 and 2.
   (a) Write a matrix B to represent this information.
   (b) Find B^2.
   (c) How many lines which connect cities 1 and 2 go through exactly one other city (city 3)?
   (d) How many lines which connect cities 1 and 2 go through at most one other city?

40. Management The figure shows four southern cities served by Supersouth Airlines.

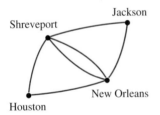

(a) Write a matrix to represent the number of nonstop routes between cities.
(b) Find the number of one-stop flights between Houston and Jackson.
(c) Find the number of flights between Houston and Shreveport which require at most one stop.
(d) Find the number of one-stop flights between New Orleans and Houston.

41. Natural Science The figure shows a food web. The arrows indicate the food sources of each population. For example, cats feed on rats and on mice.

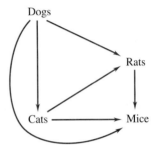

(a) Write a matrix C in which each row and corresponding column represents a population in the food chain. Enter a 1 when the population in a given row feeds on the population in the given column.
(b) Calculate and interpret C^2.

CHAPTER 7 SUMMARY

Key Terms and Symbols

7.1 linear equation
system of linear equations
solution of a system
independent system
dependent system
inconsistent system
elimination method
equivalent systems
elementary operations

triangular form
row
column
matrix (matrices)
element (entry)
augmented matrix
row operations
parameter

7.2 Gauss-Jordan method

Key Concepts

Solving Systems of Equations

The following **elementary operations** are used to transform a system of equations into a simpler equivalent system.

1. Interchange any two equations.
2. Multiply both sides of an equation by a nonzero constant.
3. Replace an equation by the sum of itself and a constant multiple of another equation in the system.

The **elimination method** is a systematic way of using elementary operations to transform a system into an equivalent system that can be solved by **back substitution.** See Section 7.1 for details.

The matrix version of the elimination method uses the following **matrix row operations** that correspond to using elementary row operations with back substitution on a system of equations.

1. Interchange any two rows.
2. Multiply each element of a row by a nonzero constant.
3. Replace a row by the sum of itself and a constant multiple of another row in the matrix.

The **Gauss-Jordan method** is an extension of the elimination method for solving a system of linear equations. It uses row operations on the augmented matrix of the system. See Section 7.2 for details.

Operations on Matrices

The **sum** of two $m \times n$ matrices X and Y is the $m \times n$ matrix $X + Y$ in which each element is the sum of the corresponding elements of X and Y. The **difference** of two $m \times n$ matrices X and Y is the $m \times n$ matrix $X - Y$ in which each element is the difference of the corresponding elements of X and Y.

The **product** of a scalar k and a matrix X is the matrix kX, with each element k times the corresponding element of X.

The **product matrix** AB of an $m \times n$ matrix A and an $n \times k$ matrix B is the $m \times k$ matrix whose entry in the ith row and jth column is the product of the ith row of A and the jth column of B.

The **inverse matrix** A^{-1} for any $n \times n$ matrix A for which A^{-1} exists is found as follows. Form the augmented matrix $[A \mid I]$; perform row operations on $[A \mid I]$ to get the matrix $[I \mid A^{-1}]$.

Chapter 7 Review Exercises

Use the elimination or matrix method to solve each of the following systems. Identify any dependent or inconsistent systems.

1. $-5x - 3y = 4$
$2x + y = -3$

2. $3x - y = 6$
$2x + 3y = 7$

3. $3x - 5y = 10$
$4x - 3y = 6$

4. $\dfrac{1}{4}x - \dfrac{1}{3}y = -\dfrac{1}{4}$

$\dfrac{1}{10}x + \dfrac{2}{5}y = \dfrac{2}{5}$

5. $x - 2y = 1$
$4x + 4y = 2$
$10x + 8y = 4$

6. $x + y - 4z = 0$
$2x + y - 3z = 2$

7. $3x + y - z = 13$
$x + 2z = 9$
$-3x - y + 2z = 9$

8. $4x - y - 2z = 4$
$x - y - \dfrac{1}{2}z = 1$
$2x - y - z = 8$

9. Management An office supply manufacturer makes two kinds of paper clips, standard and extra large. To make 1000 standard paper clips requires 1/4 hr on a cutting machine and 1/2 hr on a machine that shapes the clips. One thousand extra large paper clips require 1/3 hr on each machine. The manager of paper clip production has 4 hr per day available on the cutting machine and 6 hr per day on the shaping machine. How many of each kind of clip can he make?

10. Management Gretchen Schmidt plans to buy shares of two stocks. One costs $32 per share and pays dividends of $1.20 per share. The other costs $23 per share and pays dividends of $1.40 per share. She has $10,100 to spend and wants to earn dividends of $540. How many shares of each stock should she buy?

11. Management Joyce Pluth has money in two investment funds. Last year the first fund paid a dividend of 8% and the second a dividend of 2% and Joyce received a total of $780. This year the first fund paid a 10% dividend and the second only 1% and Joyce received $810. How much does she have invested in each fund?

12. You are given $144 in one, five, and ten dollar bills. There are 35 bills. There are two more ten dollar bills than five dollar bills. How many bills of each type are there?

13. Social Science A social service agency provides counseling, meals, and shelter to clients referred by sources I, II, and III. Clients from source I require an average of $100 for food, $250 for shelter, and no counseling. Source II clients require an average of $100 for counseling, $200 for food, and nothing for shelter. Source III clients require an average of $100 for counseling, $150 for food, and $200 for shelter. The agency has funding of $25,000 for counseling, $50,000 for food, and $32,500 for shelter. How many clients from each source can be served?

14. Management The Waputi Indians make woven blankets, rugs, and skirts. Each blanket requires 24 hr for spinning the yarn, 4 hr for dying the yarn, and 15 hr for weaving. Rugs require 30, 5, and 18 hr and skirts 12, 3, and 9 hr, respectively. If there are 306, 59, and 201 hr available for spinning, dying, and weaving, respectively, how many of each item can be made? (*Hint:* Simplify the equations you write, if possible, before solving the system.)

Use the Gauss-Jordan method to solve the following systems.

15. $x - z = -3$
$y + z = 6$
$2x - 3z = -9$

16. $2x - y + 4z = -1$
$-3x + 5y - z = 5$
$2x + 3y + 2z = 3$

17. $5x - 8y + z = 1$
$3x - 2y + 4z = 3$
$10x - 16y + 2z = 3$

18. $x - 2y + 3z = 4$
$2x + y - 4z = 3$
$-3x + 4y - z = -2$

19. $3x + 2y - 6z = 9$
$x + y + 2z = 4$
$2x + 2y + 5z = 0$

20. Management Each week at a furniture factory, there are 2000 work hours available in the construction department, 1400 work hours in the painting department, and 1300 work hours in the packing department. Producing a chair requires 2 hours of construction, 1 hour of painting, and 2 hours for packing. Producing a table requires 4 hours of construction, 3 hours of painting, and 3 hours for packing. Producing a chest requires 8 hours of construction, 6 hours of painting, and 4 hours for packing. If all available time is used in every department, how many of each item are produced each week?

For each of the following, find the dimensions of the matrix and identify any square, row, or column matrices.

21. $\begin{bmatrix} 2 & 3 \\ 5 & 9 \end{bmatrix}$

22. $\begin{bmatrix} 2 & -1 \\ 4 & 6 \\ 5 & 7 \end{bmatrix}$

23. $[12 \quad 4 \quad -8 \quad -1]$

24. $\begin{bmatrix} -7 & 5 & 6 \\ 3 & 2 & -1 \\ -1 & 12 & 8 \end{bmatrix}$

25. $\begin{bmatrix} 6 & 8 & 10 \\ 5 & 3 & -2 \end{bmatrix}$

26. $\begin{bmatrix} -9 \\ 15 \\ 4 \end{bmatrix}$

27. Natural Science The activities of a grazing animal can be classified roughly into three categories: grazing, moving, and resting. Suppose horses spend 8 hours grazing, 8 moving, and 8 resting; cattle spend 10 grazing, 5 moving, and 9 resting; sheep spend 7 grazing, 10 moving, and 7 resting; and goats spend 8 grazing, 9 moving, and 7 resting. Write this information as a 4 × 3 matrix.

28. Management The New York Stock Exchange reports in the daily newspapers give the dividend, price-to-earnings ratio, sales (in hundreds of shares), last price, and change in price for each company. Write the following stock reports as a 4×5 matrix. American Telephone & Telegraph: 5, 7, 2532, $52\frac{3}{8}$, $-\frac{1}{4}$. General Electric: 3, 9, 1464, 56, $+\frac{1}{8}$. Gulf Oil: 2.50, 5, 4974, 41, $-1\frac{1}{2}$. Sears: 1.36, 10, 1754, 18, $+\frac{1}{2}$.

Given the matrices

$$A = \begin{bmatrix} 4 & 10 \\ -2 & -3 \\ 6 & 9 \end{bmatrix}, \quad B = \begin{bmatrix} 2 & 3 & -2 \\ 2 & 4 & 0 \\ 0 & 1 & 2 \end{bmatrix}, \quad C = \begin{bmatrix} 5 & 0 \\ -1 & 3 \\ 4 & 7 \end{bmatrix},$$

$$D = \begin{bmatrix} 6 \\ 1 \\ 0 \end{bmatrix}, \quad E = [1 \quad 3 \quad -4], \quad F = \begin{bmatrix} -1 & 4 \\ 3 & 7 \end{bmatrix},$$

$$G = \begin{bmatrix} 2 & 5 \\ 1 & 6 \end{bmatrix},$$

find each of the following (if possible).

29. $-B$ **30.** $-D$ **31.** $3A - 2C$

32. $F + 3G$ **33.** $2B - 5C$ **34.** $G - 2F$

35. Management Refer to Exercise 28. Write a 4×2 matrix using the sales and price changes for the four companies. The next day's sales and price changes for the same four companies were 2310, 1258, 5061, 1812 and $-1/4$, $-1/4$, $+1/2$, $+1/2$, respectively. Write a 4×2 matrix using these new sales and price change figures. Use matrix addition to find the total sales and price changes for the two days.

36. Management An oil refinery in Tulsa sent 110,000 gallons of oil to a Chicago distributor, 73,000 to a Dallas distributor, and 95,000 to an Atlanta distributor. Another refinery in New Orleans sent the following amounts to the same three distributors: 85,000, 108,000, 69,000. The next month the two refineries sent the same distributors new shipments of oil as follows: from Tulsa, 58,000 to Chicago, 33,000 to Dallas, and 80,000 to Atlanta; from New Orleans, 40,000, 52,000, and 30,000, respectively.
(a) Write the monthly shipments from the two distributors to the three refineries as 3×2 matrices.
(b) Use matrix addition to find the total amounts sent to the refineries from each distributor.

Use the matrices given above Exercise 29 to find each of the following (if possible).

37. AG **38.** EB **39.** GF

40. CA **41.** AGF **42.** B^2D

43. Management An office supply manufacturer makes two kinds of paper clips, standard and extra large. To make a unit of standard paper clips requires 1/4 hour on a cutting machine and 1/2 hour on a machine that shapes the clips. A

unit of extra large paper clips requires 1/3 hour on each machine.
(a) Write this information as a 2×2 matrix (size/machine).
(b) If 48 units of standard and 66 units of extra large clips are to be produced, use matrix multiplication to find out how many hours each machine will operate. (*Hint:* Write the units as a 1×2 matrix.)

44. Management Theresa DePalo buys shares of three stocks. Their cost per share and dividend earnings per share are $32, $23, and $54, and $1.20, $1.49, and $2.10, respectively. She buys 50 shares of the first stock, 20 shares of the second, and 15 shares of the third.
(a) Write the cost per share and earnings per share of the stocks as a 3×2 matrix.
(b) Write the number of shares of each stock as a 1×3 matrix.
(c) Use matrix multiplication to find the total cost and total dividend earnings of these stocks.

45. If $A = \begin{bmatrix} 3 & 0 \\ 2 & 1 \end{bmatrix}$, find a matrix B such that both AB and BA are defined and $AB \neq BA$.

46. Is it possible to do Exercise 45 if $A = \begin{bmatrix} 4 & 0 \\ 0 & 4 \end{bmatrix}$? Explain why.

Find the inverse of each of the following matrices that has an inverse.

47. $\begin{bmatrix} -4 & 2 \\ 0 & 3 \end{bmatrix}$ **48.** $\begin{bmatrix} 2 & 1 \\ 5 & 3 \end{bmatrix}$

49. $\begin{bmatrix} 6 & 4 \\ 3 & 2 \end{bmatrix}$ **50.** $\begin{bmatrix} 2 & 0 \\ -1 & 5 \end{bmatrix}$

51. $\begin{bmatrix} 2 & 0 & 4 \\ 1 & -1 & 0 \\ 0 & 1 & -2 \end{bmatrix}$ **52.** $\begin{bmatrix} 2 & -1 & 0 \\ 1 & 0 & 1 \\ 1 & -2 & 0 \end{bmatrix}$

53. $\begin{bmatrix} 2 & 3 & 5 \\ -2 & -3 & -5 \\ 1 & 4 & 2 \end{bmatrix}$ **54.** $\begin{bmatrix} 1 & 3 & 6 \\ 4 & 0 & 9 \\ 5 & 15 & 30 \end{bmatrix}$

55. $\begin{bmatrix} 1 & 3 & -2 & -1 \\ 0 & 1 & 1 & 2 \\ -1 & -1 & 1 & -1 \\ 1 & -1 & -3 & -2 \end{bmatrix}$ **56.** $\begin{bmatrix} 3 & 2 & 0 & -1 \\ 2 & 0 & 1 & 2 \\ 1 & 2 & -1 & 0 \\ 2 & -1 & 1 & 1 \end{bmatrix}$

Refer again to the matrices given above Exercise 29 to find each of the following (if possible).

57. F^{-1} **58.** G^{-1} **59.** $(G - F)^{-1}$

60. $(F + G)^{-1}$ **61.** B^{-1}

62. Explain why the matrix $\begin{bmatrix} a & 0 \\ c & 0 \end{bmatrix}$, where a and c are nonzero constants, cannot possibly have an inverse.

Solve each of the following matrix equations AX = B for X.

63. $A = \begin{bmatrix} 2 & 4 \\ -1 & -3 \end{bmatrix}$, $B = \begin{bmatrix} 8 \\ 3 \end{bmatrix}$

64. $A = \begin{bmatrix} 1 & 3 \\ -2 & 4 \end{bmatrix}$, $B = \begin{bmatrix} 15 \\ 10 \end{bmatrix}$

65. $A = \begin{bmatrix} 1 & 0 & 2 \\ -1 & 1 & 0 \\ 3 & 0 & 4 \end{bmatrix}$, $B = \begin{bmatrix} 8 \\ 4 \\ -6 \end{bmatrix}$

66. $A = \begin{bmatrix} 2 & 4 & 0 \\ 1 & -2 & 0 \\ 0 & 0 & 3 \end{bmatrix}$, $B = \begin{bmatrix} 72 \\ -24 \\ 48 \end{bmatrix}$

Use the method of matrix inverses to solve each of the following systems.

67. $\begin{aligned} x + y &= 4 \\ 2x + 3y &= 10 \end{aligned}$

68. $\begin{aligned} 5x - 3y &= -2 \\ 2x + 7y &= -9 \end{aligned}$

69. $\begin{aligned} 2x + y &= 5 \\ 3x - 2y &= 4 \end{aligned}$

70. $\begin{aligned} x - 2y &= 7 \\ 3x + y &= 7 \end{aligned}$

71. $\begin{aligned} x + y + z &= 1 \\ 2x - y &= -2 \\ 3y + z &= 2 \end{aligned}$

72. $\begin{aligned} x &= -3 \\ y + z &= 6 \\ 2x - 3z &= -9 \end{aligned}$

73. $\begin{aligned} 3x - 2y + 4z &= 4 \\ 4x + y - 5z &= 2 \\ -6x + 4y - 8z &= -2 \end{aligned}$

74. $\begin{aligned} x + 2y &= -1 \\ 3y - z &= -5 \\ x + 2y - z &= -3 \end{aligned}$

Solve each of the following problems by any method.

75. Management A wine maker has two large casks of wine. One is 8% alcohol and the other is 18% alcohol. How many liters of each wine should be mixed to produce 30 liters of wine that is 12% alcohol?

76. Management A gold merchant has some 12 carat gold (12/24 pure gold), and some 22 carat gold (22/24 pure). How many grams of each could be mixed to get 25 grams of 15 carat gold?

77. Natural Science A chemist has some 40% acid solution and some 60% solution. How many liters of each should be used to get 40 liters of a 45% solution?

78. Management How many pounds of tea worth $4.60 a pound should be mixed with tea worth $6.50 a pound to get 10 pounds of a mixture worth $5.74 a pound?

79. Management A machine in a pottery factory takes 3 minutes to form a bowl and 2 minutes to form a plate. The material for a bowl costs $.25 and the material for a plate costs $.20. If the machine runs for 8 hours and exactly $44 is spent for material, how many bowls and plates can be produced?

80. A boat travels at a constant speed a distance of 57 km downstream in 3 hours, then turns around and travels 55 km upstream in 5 hours. What is the speed of the boat and of the current?

81. Management Ms. Tham invests $50,000 three ways—at 8%, $8\frac{1}{2}$%, and 11%. In total, she receives $4436.25 per year in interest. The interest from the 11% investment is $80 more than the interest on the 8% investment. Find the amount she has invested at each rate.

82. Tickets to a band concert cost $2 for children, $3 for teenagers, and $5 for adults. 570 people attended the concert and total ticket receipts were $1950. Three fourths as many teenagers as children attended. How many children, teenagers, and adults were at the concert?

Find the production matrix given the following input-output and demand matrices.

83. $A = \begin{bmatrix} .01 & .05 \\ .04 & .03 \end{bmatrix}$, $D = \begin{bmatrix} 200 \\ 300 \end{bmatrix}$

84. $A = \begin{bmatrix} .2 & .1 & .3 \\ .1 & 0 & .2 \\ 0 & 0 & .4 \end{bmatrix}$, $D = \begin{bmatrix} 500 \\ 200 \\ 100 \end{bmatrix}$

85. Given the input-output matrix $A = \begin{bmatrix} 0 & \frac{1}{4} \\ \frac{1}{2} & 0 \end{bmatrix}$ and the demand matrix $D = \begin{bmatrix} 2100 \\ 1400 \end{bmatrix}$, find each of the following.
(a) $I - A$ **(b)** $(I - A)^{-1}$
(c) the production matrix X

86. Management An economy depends on two commodities, goats and cheese. It takes 2/3 of a unit of goats to produce 1 unit of cheese and 1/2 unit of cheese to produce 1 unit of goats.
(a) Write the input-output matrix for this economy.
(b) Find the production required to satisfy a demand of 400 units of cheese and 800 units of goats.

87. Management In a simple economic model, a country has two industries: agriculture and manufacturing. To produce $1 of agricultural output requires $.10 of agricultural output and $.40 of manufacturing output. To produce $1 of manufacturing output requires $.70 of agricultural output and $.20 of manufacturing output. If agricultural demand is $60,000 and manufacturing demand is $20,000, what must each industry produce? (Round answers to the nearest whole number.)

88. Management The matrix below represents the number of direct flights between four cities.

$$\begin{array}{cc} & \begin{array}{cccc} \text{A} & \text{B} & \text{C} & \text{D} \end{array} \\ \begin{array}{c} \text{A} \\ \text{B} \\ \text{C} \\ \text{D} \end{array} & \begin{bmatrix} 0 & 1 & 0 & 1 \\ 1 & 0 & 0 & 1 \\ 0 & 0 & 0 & 1 \\ 1 & 1 & 1 & 0 \end{bmatrix} \end{array}$$

(a) Find the number of one-stop flights between cities A and C.

(b) Find the total number of flights between cities B and C that are either direct or one-stop.

(c) Find the matrix that gives the number of two-stop flights between these cities.

89. **Social Science**   (a) Use the matrix $M = \begin{bmatrix} 2 & 6 \\ 1 & 4 \end{bmatrix}$ to encode the message "leave now."

(b) What matrix should be used to decode this message?

CASE 7

Leontief's Model of the American Economy

In the April 1965 issue of *Scientific American,* Wassily Leontief explained his input-output system using the 1958 American economy as an example.* He divided the economy into 81 sectors, grouped into six families of related sectors. In order to keep the discussion reasonably simple, we will treat each family of sectors as a single sector and so, in effect, work with a six-sector model. The sectors are listed in Table 1.

Table 1

Sector	Examples
Final Nonmetal (FN)	Furniture, processed food
Final Metal (FM)	Household appliances, motor vehicles
Basic Metal (BM)	Machine-shop products, mining
Basic Nonmetal (BN)	Agriculture, printing
Energy (E)	Petroleum, coal
Services (S)	Amusements, real estate

The workings of the American economy in 1958 are described in the input-output table (Table 2) based on Leontief's figures. We will demonstrate the meaning of Table 2 by considering the first left-hand column of numbers. The numbers in this column mean that 1 unit of final nonmetal production requires the consumption of .170 unit of (other) final nonmetal production, .003 unit of final metal output, .025 unit of basic metal products, and so on down the column. Since the unit of measurement that Leontief used for this table is millions of dollars, we conclude that the production of $1 million worth of final nonmetal production consumes $.170 million, or $170,000, worth of other final nonmetal products, $3000 of final metal products, $25,000 of basic metal products, and so on. Similarly, the entry in the column headed FM and opposite S of .074 means that $74,000 worth of input from the service industries goes into the production of $1 million worth of final metal products, and

Table 2

	FN	FM	BM	BN	E	S
FN	.170	.004	0	.029	0	.008
FM	.003	.295	.018	.002	.004	.016
BM	.025	.173	.460	.007	.011	.007
BN	.348	.037	.021	.403	.011	.048
E	.007	.001	.039	.025	.358	.025
S	.120	.074	.104	.123	.173	.234

the number .358 in the column headed E and opposite E means that $358,000 worth of energy must be consumed to produce $1 million worth of energy.

By the underlying assumption of Leontief's model, the production of n units (n = any number) of final nonmetal production consumes $.170n$ units of final nonmetal output, $.003n$ units of final metal output, $.025n$ units of basic metal production, and so on. Thus, production of $50 million worth of products from the final nonmetal section of the 1958 American economy required $(.170)(50) = 8.5$ units ($8.5 million) worth of final nonmetal input, $(.003)(50) = .15$ units of final metal input, $(.025)(50) = 1.25$ units of basic metal production, and so on.

EXAMPLE 1   According to the simplified input-output table for the 1958 American economy, how many dollars worth of final metal products, basic nonmetal products, and services are required to produce $120 million worth of basic metal products?

Each unit ($1 million worth) of basic metal products requires .018 units of final metal products because the number in the BM column of the table opposite FM is .018. Thus, $120 million, or 120 units, requires $(.018)(120) = 2.16$ units, or $2.16 million worth of final metal products. Similarly, 120 units of basic metal production uses $(.021)(120) = 2.52$ units of basic nonmetal production and $(.104)(120) = 12.48$ units of services, or $2.52 million and $12.48 million worth of basic nonmetal output and services, respectively.   ∎

The Leontief model also involves a *bill of demands,* that is, a list of requirements for units of output beyond that required for

*Adapted from *Applied Finite Mathematics* by Robert F. Brown and Brenda W. Brown. Copyright © 1977 by Robert F. Brown and Brenda W. Brown. Reprinted by permission.

its inner workings as described in the input-output table. These demands represent exports, surpluses, government and individual consumption, and the like. The bill of demands (in millions) for the simplified version of the 1958 American economy we have been using is shown below.

FN	$99,640
FM	$75,548
BM	$14,444
BN	$33,501
E	$23,527
S	$263,985

We can now use the methods developed in Section 7.5 to answer this question: How many units of output from each sector are needed in order to run the economy and fill the bill of demands? The units of output from each sector required to run the economy and fill the bill of demands is unknown, so we denote them by variables. In our example, there are six quantities which are, at the moment, unknown. The number of units of final nonmetal production required to solve the problem will be our first unknown, because this sector is represented by the first row of the input-output matrix. The unknown quantity of final nonmetal units will be represented by the symbol x_1. Following the same pattern, we represent the unknown quantities in the following manner.

x_1 = units of final nonmetal production required

x_2 = units of final metal production required

x_3 = units of basic metal production required

x_4 = units of basic nonmetal production required

x_5 = units of energy required

x_6 = units of services required

These six numbers are the quantities we are attempting to calculate, and the variables will make up the entries in our production matrix.

To find these numbers, first let A be the 6×6 matrix corresponding to the input-output table.

$$A = \begin{bmatrix} .170 & .004 & 0 & .029 & 0 & .008 \\ .003 & .295 & .018 & .002 & .004 & .016 \\ .025 & .173 & .460 & .007 & .011 & .007 \\ .348 & .037 & .021 & .403 & .011 & .048 \\ .007 & .001 & .039 & .025 & .358 & .025 \\ .120 & .074 & .104 & .123 & .173 & .234 \end{bmatrix}$$

A is the input-output matrix. The bill of demands leads to a 6×1 demand matrix D, and X is the matrix of unknowns.

$$D = \begin{bmatrix} 99,640 \\ 75,548 \\ 14,444 \\ 33,501 \\ 23,527 \\ 263,985 \end{bmatrix} \quad \text{and} \quad X = \begin{bmatrix} x_1 \\ x_2 \\ x_3 \\ x_4 \\ x_5 \\ x_6 \end{bmatrix}$$

To use the formula $D = (I - A)X$, or $X = (I - A)^{-1}D$, we need to find $I - A$.

$$I - A = \begin{bmatrix} 1 & 0 & 0 & 0 & 0 & 0 \\ 0 & 1 & 0 & 0 & 0 & 0 \\ 0 & 0 & 1 & 0 & 0 & 0 \\ 0 & 0 & 0 & 1 & 0 & 0 \\ 0 & 0 & 0 & 0 & 1 & 0 \\ 0 & 0 & 0 & 0 & 0 & 1 \end{bmatrix}$$

$$- \begin{bmatrix} .170 & .004 & 0 & .029 & 0 & .008 \\ .003 & .295 & .018 & .002 & .004 & .016 \\ .025 & .173 & .460 & .007 & .011 & .007 \\ .348 & .037 & .021 & .403 & .011 & .048 \\ .007 & .001 & .039 & .025 & .358 & .025 \\ .120 & .074 & .104 & .123 & .173 & .234 \end{bmatrix}$$

$$= \begin{bmatrix} .830 & -.004 & 0 & -.029 & 0 & -.008 \\ -.003 & .705 & -.018 & -.002 & -.004 & -.016 \\ -.025 & -.173 & .540 & -.007 & -.011 & -.007 \\ -.348 & -.037 & -.021 & .597 & -.011 & -.048 \\ -.007 & -.001 & -.039 & -.025 & .642 & -.025 \\ -.120 & -.074 & -.104 & -.123 & -.173 & .766 \end{bmatrix}$$

By the method given in Section 7.5, we find the inverse (actually an approximation).

$$(I - A)^{-1} = \begin{bmatrix} 1.234 & .014 & .007 & .064 & .006 & .017 \\ .017 & 1.436 & .056 & .014 & .019 & .032 \\ .078 & .467 & 1.878 & .036 & .044 & .031 \\ .752 & .133 & .101 & 1.741 & .065 & .123 \\ .061 & .045 & .130 & .083 & 1.578 & .059 \\ .340 & .236 & .307 & .315 & .376 & 1.349 \end{bmatrix}$$

Therefore,

$$X = (I - A)^{-1}D =$$

$$\begin{bmatrix} 1.234 & .014 & .007 & .064 & .006 & .017 \\ .017 & 1.436 & .056 & .014 & .019 & .032 \\ .078 & .467 & 1.878 & .036 & .044 & .031 \\ .752 & .133 & .101 & 1.741 & .065 & .123 \\ .061 & .045 & .130 & .083 & 1.578 & .059 \\ .340 & .236 & .307 & .315 & .376 & 1.349 \end{bmatrix} \begin{bmatrix} 99,640 \\ 75,548 \\ 14,444 \\ 33,501 \\ 23,527 \\ 263,985 \end{bmatrix}$$

$$= \begin{bmatrix} 131,033 \\ 120,459 \\ 80,681 \\ 178,732 \\ 66,929 \\ 431,562 \end{bmatrix}.$$

From this result,

$$x_1 = 131,033$$
$$x_2 = 120,459$$
$$x_3 = 80,681$$
$$x_4 = 178,732$$
$$x_5 = 66,929$$
$$x_6 = 431,562.$$

In other words, it would require 131,033 units ($131,033 million worth) of final nonmetal production, 120,459 units of final metal output, 80,681 units of basic metal products, and so on to run the 1958 American economy and completely fill the stated bill of demands.

EXERCISES

Use a graphing calculator to work the following problems.

1. A much simplified version of Leontief's 42-sector analysis of the 1947 American economy divides the economy into just 3 sectors: agriculture, manufacturing, and the household (i.e., the sector of the economy that produces labor). It is summarized in the following input-output table.

	Agriculture	*Manufacturing*	*Household*
Agriculture	.245	.102	.051
Manufacturing	.099	.291	.279
Household	.433	.372	.011

The bill of demands (in billions of dollars) is shown below.

Agriculture	2.88
Manufacturing	31.45
Household	30.91

(a) Write the input-output matrix A, the demand matrix D, and the matrix X.
(b) Compute $I - A$.

(c) Check that $(I - A)^{-1} = \begin{bmatrix} 1.453 & .291 & .157 \\ .532 & 1.762 & .525 \\ .836 & .790 & 1.277 \end{bmatrix}$ is an approximation of the inverse of $I - A$ by calculating $(I - A)^{-1}(I - A)$.
(d) Use the matrix of part (c) to compute X.
(e) Explain the meaning of the numbers in X in dollars.

2. An analysis of the 1958 Israeli economy* is here simplified by grouping the economy into three sectors: agriculture, manufacturing, and energy. The input-output table is given below.

	Agriculture	*Manufacturing*	*Energy*
Agriculture	.293	0	0
Manufacturing	.014	.207	.017
Energy	.044	.010	.216

Exports (in thousands of Israeli pounds) were as follows.

Agriculture	138,213
Manufacturing	17,597
Energy	1,786

(a) Write the input-output matrix A and the demand (export) matrix D.
(b) Compute $I - A$.
(c) Check that $(I - A)^{-1} = \begin{bmatrix} 1.414 & 0 & 0 \\ .027 & 1.261 & .027 \\ .080 & .016 & 1.276 \end{bmatrix}$ is an approximation of the inverse of $I - A$ by calculating $(I - A)^{-1}(I - A)$.
(d) Use the matrix of part (c) to determine the number of Israeli pounds worth of agricultural products, manufactured goods, and energy required to run this model of the Israeli economy and export the stated value of products.

---

*Leontief, Wassily, *Input-Output Economics* (New York: Oxford University Press, 1966), pp. 54–57.

Linear Programming

Many realistic problems involve inequalities. For example, a factory may have no more than 200 workers on a shift and must manufacture at least 3000 units at a cost of no more than $35 each. How many workers should it have per shift in order to produce the required units at minimal cost? *Linear programming* is a method for finding the optimal (best possible) solution for such problems, if there is one.

In this chapter we shall study two methods of solving linear programming problems: the graphical method and the simplex method. The graphical method requires a knowledge of **linear inequalities,** those involving only first-degree polynomials in *x* and *y*. So we begin with a study of such inequalities.

8.1 GRAPHING LINEAR INEQUALITIES IN TWO VARIABLES

Examples of linear inequalities in two variables include

$$x + 2y < 4, \quad 3x + 2y > 6, \quad \text{and} \quad 2x - 5y \geq 10.$$

A solution of a linear inequality is an ordered pair that satisfies the inequality. For example (4, 4) is a solution of

$$3x - 2y \leq 6.$$

(Check by substituting 4 for *x* and 4 for *y*.) A linear inequality has an infinite number of solutions, one for every choice of a value for *x*. The best way to show these solutions is with a graph that consists of all the points in the plane whose coordinates satisfy the inequality.

EXAMPLE 1   Graph the linear inequality $3x - 2y \leq 6$.

Because of the "=" portion of $\leq$, the points of the line $3x - 2y = 6$ satisfy the linear inequality $3x - 2y \leq 6$ and are part of its graph. As in Chapter 2, we find the intercepts by first letting $x = 0$ and then letting $y = 0$; we use these points to get the graph of $3x - 2y = 6$ shown in Figure 8.1.

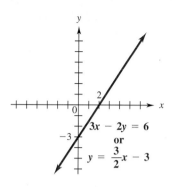

FIGURE 8.1

1 Graph.

(a) $2x + 5y \le 10$

(b) $x - y \ge 4$

Answers:

(a)

(b)

The points on the line satisfy "$3x - 2y$ *equals* 6." The points satisfying "$3x - 2y$ *is less than* 6" can be found by first solving $3x - 2y \le 6$ for y.

$$3x - 2y \le 6$$

$$-2y \le -3x + 6$$

$$y \ge \frac{3}{2}x - 3 \qquad \text{Multiply by } -1/2;$$
$$\text{reverse the inequality.}$$

As shown in Figure 8.2, the points *above* the line $3x - 2y = 6$ satisfy

$$y > \frac{3}{2}x - 3,$$

while those below the line satisfy

$$y < \frac{3}{2}x - 3.$$

The line itself is the **boundary.** In summary, the inequality $3x - 2y \le 6$ is satisfied by all points *on or above* the line $3x - 2y = 6$. Indicate the points above the line by shading, as in Figure 8.3. The line and shaded region of Figure 8.3 make up the graph of the linear inequality $3x - 2y \le 6$. ■   **1**

FIGURE 8.2

FIGURE 8.3

Graph.

(a) $2x + 3y > 12$

(b) $3x - 2y < 6$

Answers:

(a)

(b)

EXAMPLE 2 Graph $x + 4y < 4$.

The boundary here is the line $x + 4y = 4$. Since the points on this line do not satisfy $x + 4y < 4$, when drawing the graph by hand we use a dashed line, as in Figure 8.4. Solving the inequality for y gives $y < -(1/4)x + 1$. The $<$ symbol tells us we should shade the region below the line, as in Figure 8.4. ■

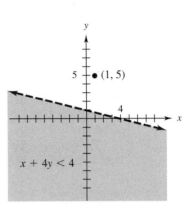

FIGURE 8.4

In Example 2, another way to decide which region to shade is to use a test point. For example, test the coordinates of the point $(1, 5)$ in the inequality.

$$1 + 4(5) = 21 \not< 4$$

Since the coordinates do not satisfy the inequality, shade the region on the other side of the boundary line, as shown in Figure 8.4.

As these examples suggest, the graph of a linear inequality is a region in the plane, perhaps including the line that is the boundary of the region. Each of the shaded regions is an example of a **half-plane,** a region on one side of a line. For example, in Figure 8.5 line r divides the plane into half-planes P and Q. The points of line r belong to neither P nor Q. Line r is the boundary of each half-plane.

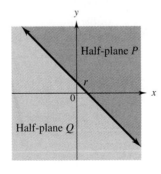

FIGURE 8.5

EXAMPLE 3 Graph each of the following.

(a) $x \le -1$

Recall that the graph of $x = -1$ is the vertical line through $(-1, 0)$. The points where $x < -1$ lie in the half-plane to the left of the boundary line, as shown in Figure 8.6(a).

Graph each of the following.

(a) $x \geq 3$

(b) $y - 3 \leq 0$

Answers:

(a)

(b)

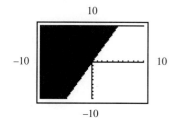 Use your calculator to graph $2x < y$.

Answer:

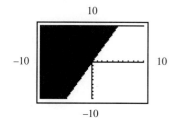

(b) $y \geq 2$

The graph of $y = 2$ is the horizontal line through $(0, 2)$. The points where $y \geq 2$ lie in the half-plane above this boundary line or on the line, as shown in Figure 8.6(b). ■

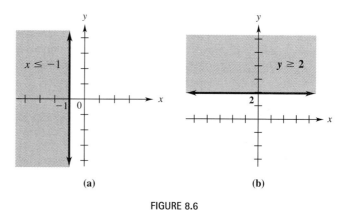

(a)   (b)

FIGURE 8.6

A graphing calculator can be used to shade regions in the plane as shown in the next example.

EXAMPLE 4* Use a graphing calculator to graph $x \geq 3y$.

Begin by solving the inequality for y to get $y \leq (1/3)x$. Use your calculator to graph the corresponding line $y = (1/3)x$. Because of the $\leq$ sign, we want to shade below the line. Use your calculator and the Technology Tip on the next page to shade this area, as in Figure 8.7. Note that you cannot tell from the graph if the boundary line is solid or dashed (it should be solid here). In fact, because of the way the calculator graphs, the boundary looks like a staircase (with very tiny stairs) rather than a straight line. ■

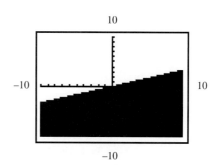

FIGURE 8.7

*If you don't have a graphing calculator, skip this example.

TECHNOLOGY TIP Most calculators can shade the region under or over a graph in the plane. On TI calculators, use SHADE in the DRAW menu. On Sharp 9300, use FILL BELOW or FILL ABOVE in the EQTN menu. On HP-38, use AREA in the FCN menu that appears after graphing. On some Casio calculators, use inequality graphing mode. Check your instruction manual for the correct syntax and procedure. On TI calculators and HP-38, you must shade the area *between* two graphs. So to shade the area below the graph of $y = f(x)$, as in Example 4, have the calculator shade the area between the graph of y and the horizontal line at the bottom of the screen. In Figure 8.7, for example, the calculator shaded the area between $y = (1/3)x$ and the horizontal line $y = -10$. ✔

The steps used to graph a linear inequality are summarized below.

GRAPHING A LINEAR INEQUALITY

1. Graph the boundary line. Decide whether the line is part of the solution. (If graphing by hand, make the line solid if the inequality involves $\leq$ or $\geq$; make it dashed if the inequality involves $<$ or $>$.)
2. Solve the inequality for y: Shade the region above the line if $y > mx + b$; shade the region below the line if $y < mx + b$.

SYSTEMS OF INEQUALITIES Realistic problems often involve many inequalities. For example, a manufacturing problem might produce inequalities resulting from production requirements, as well as inequalities about cost requirements. A set of at least two inequalities is called a **system of inequalities.** The graph of a system of inequalities is made up of all those points that satisfy all the inequalities of the system at the same time. The next example shows how to graph a system of linear inequalities.

EXAMPLE 5 Graph the system

$$y < -3x + 12$$
$$x < 2y.$$

First, graph the solution of $y < -3x + 12$. The boundary, the line with equation $y = -3x + 12$, is dashed. The points below the boundary satisfy $y < -3x + 12$. Now graph the solution of $x < 2y$ on the same axes. Again, the boundary line $x = 2y$ is dashed. Use a test point to see that the region above the boundary satisfies $x < 2y$. The heavily shaded region in Figure 8.8 shows all the points that satisfy both inequalities of the system. ■ 5

5 Graph the system

$x + y \leq 6$, $2x + y \geq 4$.

Answer:

FIGURE 8.8

The heavily shaded region of Figure 8.8 is sometimes called the **region of feasible solutions,** or just the **feasible region,** since it is made up of all the points that satisfy (are feasible for) each inequality of the system.

6 Graph the feasible region of the system

$$x + 4y \le 8$$
$$x - y \ge 3$$
$$x \ge 0, y \ge 0.$$

Answer:

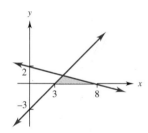

EXAMPLE 6   Graph the feasible region for the system

$$2x - 5y \le 10$$
$$x + 2y \le 8$$
$$x \ge 0, y \ge 0.$$

On the same axes, graph each inequality by graphing the boundary and choosing the appropriate half-plane. Graph the solid boundary line $2x - 5y = 10$ by first locating the intercepts that give the points $(5, 0)$ and $(0, -2)$. Use a test point to see that $2x - 5y < 10$ is satisfied by the points above the boundary. In the same way, the solid boundary line $x + 2y = 8$ goes through $(8, 0)$ and $(0, 4)$. A test point will show that the graph of $x + 2y < 8$ includes all points below the boundary. The inequalities $x \ge 0$ and $y \ge 0$ restrict the feasible region to the first quadrant. Find the feasible region by locating the intersection (overlap) of *all* the half-planes. This feasible region is shaded in Figure 8.9(a).   ■   **6**

(a)

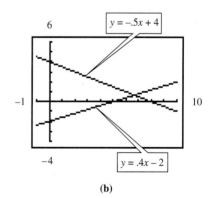

(b)

FIGURE 8.9

TECHNOLOGY TIP   To use a graphing calculator to solve the system in Example 6, solve the first two inequalities for y to get

$$y \ge .4x - 2 \quad \text{and} \quad y \le -.5x + 4.$$

Graph the boundary lines $y = .4x - 2$ and $y = -.5x + 4$ in the same window. See Figure 8.9(b). Because $x \geq 0$ and $y \geq 0$, the feasible region consists of all points that are in the first quadrant, above $y = .4x - 2$ and below $y = -.5x + 4$, as shown in Figure 8.9(a). ✔

APPLICATIONS   As we shall see in the rest of this chapter, many realistic problems lead to systems of linear inequalities. The next example is typical of such problems.

EXAMPLE 7   Midtown Manufacturing Company makes plastic plates and cups, both of which require time on two machines. A unit of plates requires 1 hour on machine A and 2 on machine B, while a unit of cups requires 3 hours on machine A and 1 on machine B. Each machine is operated for at most 15 hours per day. Write a system of inequalities expressing these conditions and graph the feasible region.

Let x represent the number of units of plates to be made, and y represent the number of units of cups. Then make a chart that summarizes the given information.

	Number Made	TIME ON MACHINE	
		A	B
Plates	x	1	2
Cups	y	3	1
Maximum time available		15	15

On machine A, x units of plates require a total of $1 \cdot x = x$ hours while y units of cups require $3 \cdot y = 3y$ hours. Since machine A is available no more than 15 hours a day,

$$x + 3y \leq 15. \qquad \text{Machine A}$$

The requirement that machine B be used no more than 15 hours a day gives

$$2x + y \leq 15. \qquad \text{Machine B}$$

It is not possible to produce a negative number of cups or plates, so that

$$x \geq 0 \quad \text{and} \quad y \geq 0.$$

The feasible region for this system of inequalities is shown in Figure 8.10. ■

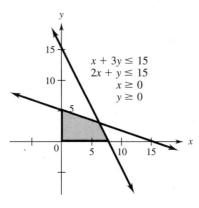

FIGURE 8.10

8.1 EXERCISES

Graph each of the following linear inequalities. (See Examples 1–4.)

1. $y < 5 - 2x$
2. $y > x + 3$
3. $3x - 2y \geq 18$
4. $2x + 5y \leq 10$
5. $2x - y \leq 4$
6. $4x - 3y \geq 24$
7. $y \leq -4$
8. $x \geq -2$
9. $x + 4y \leq 2$
10. $3x + 2y \geq 6$
11. $4x + 3y > -3$
12. $5x + 3y > 15$
13. $2x - 4y < 3$
14. $4x - 3y < 12$
15. $x \leq 5y$
16. $2x \geq y$
17. $-3x < y$
18. $-x > 6y$
19. $y < x$
20. $y > -2x$

21. In your own words, explain how to determine whether the boundary of an inequality is solid or dashed.

22. When graphing $y \leq 3x - 6$, would you shade above or below the line $y = 3x - 6$? Explain your answer.

Graph the feasible region for the following systems of inequalities. (See Examples 5 and 6.)

23. $x - y \geq 1$
    $x \leq 3$
24. $2x + y \leq 5$
    $x + 2y \leq 5$
25. $4x + y \geq 9$
    $2x + 3y \leq 7$
26. $2x + y > 8$
    $4x - y < 3$
27. $x + y > 5$
    $x - 2y < 2$
28. $3x - 4y < 6$
    $2x + 5y > 15$
29. $2x - y < 1$
    $3x + y < 6$
30. $x + 3y \leq 6$
    $2x + 4y \geq 7$
31. $-x - y < 5$
    $2x - y < 4$
32. $6x - 4y > 8$
    $3x + 2y > 4$
33. $3x + y \geq 6$
    $x + 2y \geq 7$
    $x \geq 0$
    $y \geq 0$
34. $2x + 3y \geq 12$
    $x + y \geq 4$
    $x \geq 0$
    $y \geq 0$
35. $-2 < x < 3$
    $-1 \leq y \leq 5$
    $2x + y < 6$
36. $-2 < x < 2$
    $y > 1$
    $x - y > 0$
37. $2y - x \geq -5$
    $y \leq 3 + x$
    $x \geq 0$
    $y \geq 0$
38. $2x + 3y \leq 12$
    $2x + 3y > -6$
    $3x + y < 4$
    $x \geq 0$
    $y \geq 0$
39. $3x + 4y > 12$
    $2x - 3y < 6$
    $0 \leq y \leq 2$
    $x \geq 0$
40. $0 \leq x \leq 9$
    $x - 2y \geq 4$
    $3x + 5y \leq 30$
    $y \geq 0$

In Exercises 41 and 42, find a system of inequalities whose feasible region is the interior of the given polygon.

41. Rectangle with vertices $(2, 3)$, $(2, -1)$, $(7, 3)$, $(7, -1)$
42. Triangle with vertices $(2, 4)$, $(-4, 0)$, $(2, -1)$

43. **Management** Cindi Herring and Kent Merrill produce handmade shawls and afghans. They spin the yarn, dye it, and then weave it. A shawl requires 1 hour of spinning, 1 hour of dyeing, and 1 hour of weaving. An afghan needs 2 hours of spinning, 1 of dyeing, and 4 of weaving. Together, they spend at most 8 hours spinning, 6 hours dyeing, and 14 hours weaving.
    (a) Complete the following chart.

	Number	Hours Spinning	Hours Dyeing	Hours Weaving
Shawls	x	1	1	1
Afghans	y	2	1	4
Maximum number of hours available		8	6	14

   (b) Use the chart to write a system of inequalities that describe the situation.
   (c) Graph the feasible region of this system of inequalities.

44. **Management** An electric shaver manufacturer makes two models, the regular and the flex. Because of demand, the number of regular shavers made is never more than half the number of flex shavers. The factory's production cannot exceed 1200 shavers per week.
    (a) Write a system of inequalities that describe the possibilities for making x regular and y flex shavers per week.
    (b) Graph the feasible region of this system of inequalities.

In each of the following, write a system of inequalities that describes all the conditions and graph the feasible region of the system. (See Example 7.)

45. **Management** Southwestern Oil supplies two distributors located in the Northwest. One distributor needs at least 3000 barrels of oil, and the other needs at least 5000 barrels. Southwestern can send out at most 10,000 barrels. Let x = the number of barrels of oil sent to distributor 1 and y = the number sent to distributor 2.

46. **Management** The California Almond Growers have 2400 boxes of almonds to be shipped from their plant in Sacramento to Des Moines and San Antonio. The Des Moines market needs at least 1000 boxes, while the San Antonio market must have at least 800 boxes. Let x = the

number of boxes to be shipped to Des Moines and $y =$ the number of boxes to be shipped to San Antonio.

47. Management A cement manufacturer produces at least 3.2 million barrels of cement annually. He is told by the Environmental Protection Agency that his operation emits 2.5 pounds of dust for each barrel produced. The EPA has ruled that annual emissions must be reduced to 1.8 million pounds. To do this, the manufacturer plans to replace the present dust collectors with two types of electronic precipitators. One type would reduce emissions to .5 pound per barrel and would cost 16¢ per barrel. The other would reduce the dust to .3 pound per barrel and would cost 20¢ per barrel. The manufacturer does not want to spend more than .8 million dollars on the precipitators. He needs to know how many barrels he should produce with each type. Let $x =$ the number of barrels in millions produced with the first type and $y =$ the number of barrels in millions produced with the second type.

48. Natural Science A dietician is planning a snack package of fruit and nuts. Each ounce of fruit will supply 1 unit of protein, 2 units of carbohydrates, and 1 unit of fat. Each ounce of nuts will supply 1 unit of protein, 1 unit of carbohydrates, and 1 unit of fat. Every package must provide at least 7 units of protein, at least 10 units of carbohydrates, and no more than 9 units of fat. Let x be the ounces of fruit and y the ounces of nuts to be used in each package.

8.2 LINEAR PROGRAMMING: THE GRAPHICAL METHOD

Many problems in business, science, and economics involve finding the optimal value of a function (for instance, the maximum value of the profit function or the minimum value of the cost function) subject to various **constraints** (such as transportation costs, environmental protection laws, availability of parts, interest rates, etc.). **Linear programming** deals with such situations in which the function to be optimized, called the **objective function,** is linear and the constraints are given by linear inequalities. Linear programming problems that involve only two variables can be solved by the graphical method, explained in Example 1.

EXAMPLE 1 Find the maximum and minimum values of the objective function $z = 2x + 5y$, subject to the following constraints.

$$3x + 2y \leq 6$$
$$-2x + 4y \leq 8$$
$$x + y \geq 1$$
$$x \geq 0, y \geq 0$$

First, graph the feasible region of the system of inequalities (Figure 8.11). The points in this region or on its boundaries are the only ones that satisfy all the constraints. However, each such point may produce a different value of the objective function. For instance, the points (.5, 1) and (1, 0) in the feasible region lead to the values

$$z = 2(.5) + 5(1) = 6 \quad \text{and} \quad z = 2(1) + 5(0) = 2.$$

We must find the points that produce the maximum and minimum values of z.

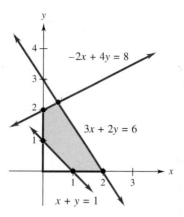

FIGURE 8.11

To find the maximum value, consider various possible values for z. For instance, when $z = 0$, then the objective function is $0 = 2x + 5y$, whose graph is a straight line. Similarly, when z is 5, 10, and 15, the objective function becomes (in turn)

$$5 = 2x + 5y, \qquad 10 = 2x + 5y, \qquad 15 = 2x + 5y.$$

These four lines are graphed in Figure 8.12. (All the lines are parallel because they have the same slope.) The figure shows that z cannot take on the value 15 because the graph for $z = 15$ is entirely outside the feasible region. The maximum possible value of z will be obtained from a line parallel to the others and between the lines representing the objective function when $z = 10$ and $z = 15$. The value of z will be as large as possible and all constraints will be satisfied if this line just touches the feasible region. This occurs at point A.

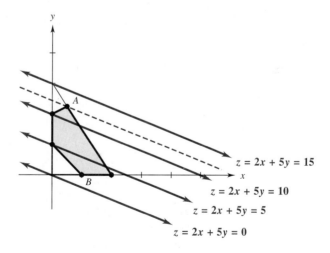

FIGURE 8.12

The point A is the intersection of the graphs of $3x + 2y = 6$ and $-2x + 4y = 8$. Its coordinates can be found either algebraically or graphically (using a graphing calculator).

1 Suppose the objective function in Example 1 is changed to $z = 5x + 2y$.

(a) Sketch the graphs of the objective function when $z = 0$, $z = 5$, and $z = 10$ on the region of feasible solutions given in Figure 8.11.

(b) From the graph, decide what values of x and y will maximize the objective function.

Answers:

(a)

(b) $(2, 0)$

Algebraic Method

Solve the system

$$3x + 2y = 6$$
$$-2x + 4y = 8,$$

as in Section 7.1, to get $x = 1/2$ and $y = 9/4$. Hence A has coordinates $(1/2, 9/4) = (.5, 2.25)$.

The value of z at the point A is

$$z = 2x + 5y = 2(.5) + 5(2.25) = 12.25.$$

Thus, the maximum possible value of z is 12.25. Similarly, the minimum value of z occurs at point B, which has coordinates $(1, 0)$. The minimum value of z is $2(1) + 5(0) = 2$. ■ **1**

Graphical Method

Solve the two equations for y.

$$y = -1.5x + 3$$
$$y = .5x + 2$$

Graph both equations on the same screen and use the intersection finder to find that the coordinates of the intersection point A are $(.5, 2.25)$.

Points such as A and B in Example 1 are called corner points. A **corner point** is a point in the feasible region where the boundary lines of two constraints cross. The feasible region in Figure 8.11 is **bounded,** because the region is enclosed by boundary lines on all sides. Linear programming problems with bounded regions always have solutions. However, if Example 1 did not include the constraint $3x + 2y \leq 6$, the feasible region would be **unbounded,** and there would be no way to *maximize* the value of the objective function.

Some general conclusions can be drawn from the method of solution used in Example 1. Figure 8.13 shows various feasible regions and the lines that result from various values of z. (Figure 8.13 shows the situation in which the lines are in order from left to right as z increases.) In part (a) of the figure, the objective function takes on its minimum value at corner point Q and its maximum value at P. The minimum is again at Q in part (b), but the maximum occurs at P_1 or P_2, or any point on the line segment connecting them. Finally, in part (c), the minimum value occurs at Q, but the objective function has no maximum value because the feasible region is unbounded.

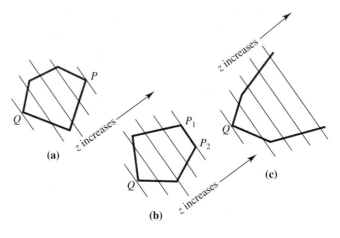

FIGURE 8.13

The preceding discussion suggests the truth of the **corner point theorem.**

> **CORNER POINT THEOREM**
>
> If the feasible region is bounded, then the objective function has both a maximum and a minimum value and each occurs at one or more corner points.
>
> If the feasible region is unbounded, the objective function may not have a maximum or minimum. But if a maximum or minimum value exists, it will occur at one or more corner points.

2 **(a)** Identify the corner points in the graph.

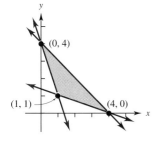

(b) Which corner point would minimize $z = 2x + 3y$?

Answers:
(a) $(0, 4)$, $(1, 1)$, $(4, 0)$

(b) $(1, 1)$

This theorem simplifies the job of finding an optimum value: First, graph the feasible region and find all corner points. Then test each point in the objective function. Finally, identify the corner point producing the optimum solution.

With the theorem, the problem in Example 1 could have been solved by identifying the five corner points of Figure 8.11: $(0, 1)$, $(0, 2)$, $(.5, 2.25)$, $(2, 0)$, and $(1, 0)$. Then, substituting each of these points into the objective function $z = 2x + 5y$ would identify the corner points that produce the maximum and the minimum values of z.

Corner Point	Value of $z = 2x + 5y$
$(0, 1)$	$2(0) + 5(1) = 5$
$(0, 2)$	$2(0) + 5(2) = 10$
$(.5, 2.25)$	$2(.5) + 5(2.25) = 12.25$ **(maximum)**
$(2, 0)$	$2(2) + 5(0) = 4$
$(1, 0)$	$2(1) + 5(0) = 2$ **(minimum)**

From these results, the corner point $(.5, 2.25)$ yields the maximum value of 12.25 and the corner point $(1, 0)$ gives the minimum value of 2. These are the same values found earlier. **2**

A summary of the steps in solving a linear programming problem by the graphical method is given here.

> **SOLVING A LINEAR PROGRAMMING PROBLEM GRAPHICALLY**
>
> **1.** Write the objective function and all necessary constraints.
> **2.** Graph the feasible region.
> **3.** Determine the coordinates of each of the corner points.
> **4.** Find the value of the objective function at each corner point.
> **5.** If the feasible region is bounded, the solution is given by the corner point producing the optimum value of the objective function.
> **6.** If the feasible region is an unbounded region in the first quadrant and both coefficients of the objective function are positive,* then the minimum value of the objective function occurs at a corner point and there is no maximum value.

---

*This is the only case of an unbounded region that occurs in the applications considered here.

EXAMPLE 2 Sketch the feasible region for the following set of constraints:

$$3y - 2x \geq 0$$
$$y + 8x \leq 52$$
$$y - 2x \leq 2$$
$$x \geq 3.$$

Then find the maximum and minimum values of the objective function $z = 5x + 2y$.

The graph in Figure 8.14(a) shows that the feasible region is bounded. The corner points are found by solving systems of two equations either algebraically using the methods of Chapter 7 or graphically using the intersection finder on a graphing calculator. Figure 8.14(b) shows the calculator graphs of the first three inequalities. With the graphical method, corner points on the line $x = 3$ are found by observation.

3 Use the region of feasible solutions in the sketch to find the following.

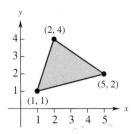

(a) The values of x and y that maximize $z = 2x - y$.

(b) The maximum value of $z = 2x - y$.

(c) The values of x and y that minimize $z = 4x + 3y$.

(d) The minimum value of $z = 4x + 3y$.

Answers:
(a) $(5, 2)$

(b) 8

(c) $(1, 1)$

(d) 7

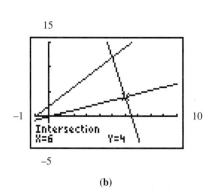

(a)   (b)

FIGURE 8.14

Use the corner points from the graph to find the maximum and minimum values of the objective function.

Corner Point	Value of $z = 5x + 2y$
$(3, 2)$	$5(3) + 2(2) = 19$ (minimum)
$(6, 4)$	$5(6) + 2(4) = 38$
$(5, 12)$	$5(5) + 2(12) = 49$ (maximum)
$(3, 8)$	$5(3) + 2(8) = 31$

The minimum value of $z = 5x + 2y$ is 19 at the corner point $(3, 2)$. The maximum value is 49 at $(5, 12)$. ■ **3**

EXAMPLE 3 Solve the following linear programming problem.

$$\text{Minimize} \quad z = x + 2y$$
$$\text{subject to:} \quad x + y \leq 10$$
$$3x + 2y \geq 6$$
$$x \geq 0, y \geq 0.$$

The feasible region is shown in Figure 8.15. From the figure, the corner points are $(0, 3)$, $(0, 10)$, $(10, 0)$, and $(2, 0)$. These corner points give the following values of z.

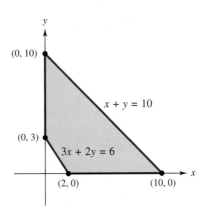

FIGURE 8.15

4 The sketch shows a feasible region. Let $z = x + 3y$. Use the sketch to find the values of x and y that

(a) minimize z;

(b) maximize z.

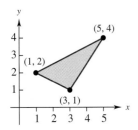

Answers:
(a) $(3, 1)$
(b) $(5, 4)$

Corner Point	Value of $z = x + 2y$
$(0, 3)$	$0 + 2(3)\ = 6$
$(0, 10)$	$0 + 2(10) = 20$
$(10, 0)$	$10 + 2(0)\ = 10$
$(2, 0)$	$2 + 2(0)\ = 2$ (minimum)

The minimum value of z is 2; it occurs at $(2, 0)$. ■ **4**

EXAMPLE 4   Solve the following linear programming problem.

$$\text{Minimize} \quad z = 2x + 4y$$
$$\text{subject to:} \quad x + 2y \geq 10$$
$$3x + \ y \geq 10$$
$$x \geq 0, y \geq 0.$$

Figure 8.16 shows the hand-drawn graph with corner points $(0, 10)$, $(2, 4)$, and $(10, 0)$, as well as the calculator graph with the corner point $(2, 4)$. Find the value of z for each point.

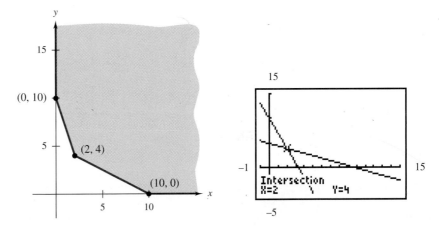

FIGURE 8.16

5 The sketch below shows a region of feasible solutions. From the sketch decide what ordered pair would minimize $z = 2x + 4y$.

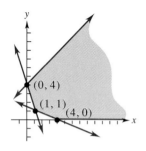

Answer:
$(1, 1)$

Corner Point	Value of $z = 2x + 4y$
$(0, 10)$	$2(0) + 4(10) = 40$
$(2, 4)$	$2(2) + 4(4) = 20$ (minimum)
$(10, 0)$	$2(10) + 4(0) = 20$ (minimum)

In this case, both $(2, 4)$ and $(10, 0)$, as well as all the points on the boundary line between them, give the same optimum value of z. There is an infinite number of equally "good" values of x and y which give the same minimum value of the objective function $z = 2x + 4y$. The minimum value is 20. ■ **5**

8.2 EXERCISES

Exercises 1–6 show regions of feasible solutions. Use these regions to find maximum and minimum values of each given objective function. (See Examples 1–2.)

1. $z = 3x + 5y$

2. $z = 6x + y$

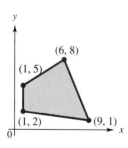

3. $z = .40x + .75y$

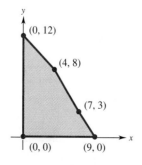

4. $z = .35x + 1.25y$

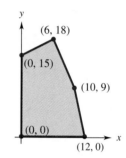

5. (a) $z = 4x + 2y$
(b) $z = 2x + yy$
(c) $z = 2x + 4y$
(d) $z = x + 4y$

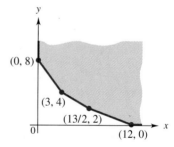

6. (a) $z = 4x + y$
(b) $z = 5x + 6y$
(c) $z = x + 2y$
(d) $z = x + 6y$

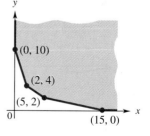

Use graphical methods to solve Exercises 7–12. (See Examples 2–4.)

7. Maximize $z = 5x + 2y$
subject to: $2x + 3y \leq 6$
$4x + y \leq 6$
$x \geq 0, y \geq 0.$

8. Minimize $z = x + 3y$
subject to: $2x + y \leq 10$
$5x + 2y \geq 20$
$-x + 2y \geq 0$
$x \geq 0, y \geq 0.$

9. Minimize $z = 2x + y$
subject to: $3x - y \geq 12$
$x + y \leq 15$
$x \geq 2, y \geq 3.$

10. Maximize $z = x + 3y$
subject to: $2x + 3y \leq 100$
$5x + 4y \leq 200$
$x \geq 10, y \geq 20.$

11. Maximize $z = 4x + 2y$
subject to: $x - y \leq 10$
$5x + 3y \leq 75$
$x \geq 0, y \geq 0.$

12. Maximize $z = 4x + 5y$
subject to: $10x - 5y \leq 100$
$20x + 10y \geq 150$
$x \geq 0, y \geq 0.$

Find the minimum and maximum values of $z = 3x + 4y$ (if possible) for each of the following sets of constraints. (See Examples 2–4.)

13. $3x + 2y \geq 6$
$x + 2y \geq 4$
$x \geq 0, y \geq 0$

14. $2x + y \leq 20$
$10x + y \geq 36$
$2x + 5y \geq 36$

15. $x + y \leq 6$
$-x + y \leq 2$
$2x - y \leq 8$

16. $-x + 2y \leq 6$
$3x + y \geq 3$
$x \geq 0, y \geq 0$

17. Find values of $x \geq 0$ and $y \geq 0$ that maximize $z = 10x + 12y$ subject to each of the following sets of constraints.

(a) $x + y \leq 20$
$x + 3y \leq 24$

(b) $3x + y \leq 15$
$x + 2y \leq 18$

(c) $x + 2y \geq 10$
$2x + y \geq 12$
$x - y \leq 8$

18. Find values of $x \geq 0$ and $y \geq 0$ that minimize $z = 3x + 2y$ subject to each of the following sets of constraints.

(a) $10x + 7y \leq 42$
$4x + 10y \geq 35$

(b) $6x + 5y \geq 25$
$2x + 6y \geq 15$

(c) $2x + 5y \geq 22$
$4x + 3y \leq 28$
$2x + 2y \leq 17$

19. Explain why it is impossible to maximize the function $z = 3x + 4y$ subject to the constraints:

$x + y \geq 8$
$2x + y \leq 10$
$x + 2y \leq 8$
$x \geq 0, \ y \geq 0$

20. You are given the following linear programming problem:*

Maximize $z = c_1x_1 + c_2x_2$
subject to: $2x_1 + x_2 \leq 11$
$-x_1 + 2x_2 \leq 2$
$x_1 \geq 0, x_2 \geq 0.$

If $c_2 > 0$, determine the range of c_1/c_2 for which $(x_1, x_2) = (4, 3)$ is an optimal solution.

(a) $[-2, 1/2]$

(b) $[-1/2, 2]$

(c) $[-11, -1]$

(d) $[1, 11]$

(e) $[-11, 11]$

---

*Problem from "Course 130 Examination Operations Research" of the *Education and Examination Committee of The Society of Actuaries*. Reprinted by permission of The Society of Actuaries.

8.3 APPLICATIONS OF LINEAR PROGRAMMING

In this section we show several applications of linear programming with two variables.

EXAMPLE 1 A 4-H Club member raises only geese and pigs. She wants to raise no more than 16 animals including no more than 10 geese. She spends $15 to raise a goose and $45 to raise a pig, and has $540 available for this project. Find the maximum profit she can make if each goose produces a profit of $7 and each pig a profit of $20.

The total profit is determined by the number of geese and pigs. So let x be the number of geese to be produced, and let y be the number of pigs. Then summarize the information of the problem in a table.

	Number	Cost to Raise	Profit Each
Geese	x	$ 15	$ 7
Pigs	y	45	20
Maximum available	16	$540	

Use this table to write the necessary constraints. Since the total number of animals cannot exceed 16, the first constraint is

$$x + y \le 16.$$

"No more than 10 geese" leads to

$$x \le 10.$$

The cost to raise x geese at $15 per goose is $15x$ dollars, while the cost for y pigs at $45 each is $45y$ dollars. Only $540 is available, so

$$15x + 45y \le 540.$$

Dividing both sides by 15 gives the equivalent inequality

$$x + 3y \le 36.$$

The number of geese and pigs cannot be negative, so

$$x \ge 0, \quad y \ge 0.$$

The 4-H Club member wants to know the number of geese and the number of pigs that should be raised for maximum profit. Each goose produces a profit of $7 and each pig, $20. If z represents total profit, then

$$z = 7x + 20y$$

is the objective function, which is to be maximized.

We must solve the following linear programming problem.

Maximize $z = 7x + 20y$ Objective function

subject to: $\begin{aligned} x + y &\le 16 \\ x &\le 10 \\ x + 3y &\le 36 \\ x &\ge 0, y \ge 0. \end{aligned}$ Constraints

Using the methods of the previous section, graph the feasible region for the system of inequalities given by the constraints as in Figure 8.17.

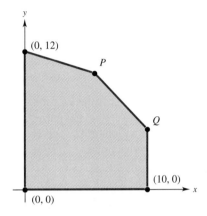

FIGURE 8.17

1 Find the corner points P and Q in Figure 8.17.

Answer:
$P = (6, 10)$
$Q = (10, 6)$

The corner points $(0, 12)$, $(0, 0)$, and $(10, 0)$ can be read directly from the graph. Find the coordinates of the other corner points by solving a system of equations or with a graphing calculator.   **1**

Test each corner point in the objective function to find the maximum profit.

Corner Point	$z = 7x + 20y$
(0, 12)	$7(0) + 20(12) = 240$
(6, 10)	$7(6) + 20(10) = 242$ (**maximum**)
(10, 6)	$7(10) + 20(6) = 190$
(10, 0)	$7(10) + 20(0) = 70$
(0, 0)	$7(0) + 20(0) = 0$

The maximum value for z of 242 occurs at $(6, 10)$. Thus 6 geese and 10 pigs will produce a maximum profit of $242.   ■

EXAMPLE 2   An office manager needs to purchase new filing cabinets. He knows that Ace cabinets cost $40 each, require 6 square feet of floor space, and hold 8 cubic feet of files. On the other hand, each Excello cabinet costs $80, requires 8 square feet of floor space, and holds 12 cubic feet. His budget permits him to spend no more than

$560 on files, while the office has room for no more than 72 square feet of cabinets. The manager desires the greatest storage capacity within the limitations imposed by funds and space. How many of each type of cabinet should he buy?

Let x represent the number of Ace cabinets to be bought and let y represent the number of Excello cabinets. The information given in the problem can be summarized as follows.

	Number	Cost of Each	Space Required	Storage Capacity
Ace	x	$ 40	6 sq ft	8 cu ft
Excello	y	$ 80	8 sq ft	12 cu ft
Maximum available		$560	72 sq ft	

The constraints imposed by cost and space are

$$40x + 80y \le 560 \quad \text{Cost}$$
$$6x + 8y \le 72. \quad \text{Floor space}$$

The number of cabinets cannot be negative, so $x \ge 0$ and $y \ge 0$. The objective function to be maximized gives the amount of storage capacity provided by some combination of Ace and Excello cabinets. From the information in the chart, the objective function is

$$\text{Storage space} = z = 8x + 12y.$$

In summary, the given problem has produced the following linear programming problem.

$$\text{Maximize} \quad z = 8x + 12y$$
$$\text{subject to:} \quad 40x + 80y \le 560$$
$$6x + 8y \le 72$$
$$x \ge 0, y \ge 0.$$

2 Find the corner point labeled Q on the region of feasible solutions given below.

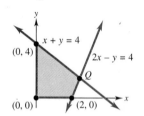

Answer:
(8/3, 4/3)

A graph of the feasible region is shown in Figure 8.18. Three of the corner points can be identified from the graph as (0, 0), (0, 7), and (12, 0). The fourth corner point, labeled Q in the figure, can be found algebraically or with a graphing calculator to be (8, 3). **2**

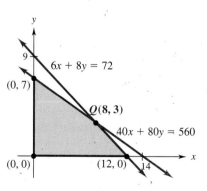

FIGURE 8.18

Use the corner point theorem to find the maximum value of z.

3 A popular cereal combines oats and corn. At least 27 tons of the cereal is to be made. For the best flavor, the amount of corn should be no more than twice the amount of oats. Corn costs $200 per ton and oats cost $300 per ton. How much of each grain should be used to minimize the cost?

(a) Make a chart to organize the information given in the problem.

(b) Write an equation for the objective function.

(c) Write four inequalities for the constraints.

Answers:

(a)

	Number of Tons	Cost/Ton
Oats	x	$300
Corn	y	200
	27	

(b) $z = 300x + 200y$

(c) $x + y \geq 27$
$y \leq 2x$
$x \geq 0$
$y \geq 0$

Corner Point	Value of $z = 8x + 12y$
(0, 0)	0
(0, 7)	84
(12, 0)	96
(8, 3)	100 (maximum)

The objective function, which represents storage space, is maximized when $x = 8$ and $y = 3$. The manager should buy 8 Ace cabinets and 3 Excello cabinets. ■ **3**

EXAMPLE 3 Certain laboratory animals must have at least 30 grams of protein and at least 20 grams of fat per feeding period. These nutrients come from food A, which costs 18¢ per unit and supplies 2 grams of protein and 4 of fat, and food B, with 6 grams of protein and 2 of fat, costing 12¢ per unit. Food B is bought under a long-term contract requiring that at least 2 units of B be used per serving. How much of each food must be bought to produce minimum cost per serving?

Let x represent the amount of food A needed, and y the amount of food B. Use the given information to produce the following table.

Food	Number of Units	Grams of Protein	Grams of Fat	Cost
A	x	2	4	18¢
B	y	6	2	12¢
Minimum required		30	20	

The linear programming problem can be stated as follows.

$$\text{Minimize} \quad z = .18x + .12y$$
$$\text{subject to:} \quad 2x + 6y \geq 30$$
$$4x + 2y \geq 20$$
$$y \geq 2$$
$$x \geq 0, y \geq 0.$$

(The constraint $y \geq 0$ is redundant because of the constraint $y \geq 2$.) A graph of the feasible region with the corner points identified is shown in Figure 8.19.

FIGURE 8.19

[4] Use the information in Side Problem 3 to do the following.

(a) Graph the feasible region and find the corner points.

(b) Determine the minimum value of the objective function and the point where it occurs.

(c) Is there a maximum cost?

Answers:

(a)

Corner points: (27, 0), (9, 18)

(b) $6300 at (9, 18)

(c) No

Use the corner point theorem to find the minimum value of z as shown in the chart below.

Corner Points	$z = .18x + .12y$
(0, 10)	$.18(0) + .12(10) = 1.20$
(3, 4)	$\mathbf{.18(3) + .12(4) = 1.02}$ **(minimum)**
(9, 2)	$.18(9) + .12(2) = 1.86$

The minimum value of 1.02 occurs at (3, 4). Thus, 3 units of A and 4 units of B will produce a minimum cost of $1.02 per serving. ■ [4]

The feasible region in Figure 8.19 is an unbounded feasible region—the region extends indefinitely to the upper right. With this region it would not be possible to *maximize* the objective function, because the total cost of the food could always be increased by encouraging the animals to eat more.

8.3 EXERCISES

Write the constraints in Exercises 1–5 as linear inequalities and identify all variables used. In some instances, not all of the information is needed to write the constraints. (See Examples 1–3.)

1. A canoe requires 6 hours of fabrication and a rowboat 4 hours. The fabrication department has at most 90 hours of labor available each week.

2. Doug Gilbert needs at least 2400 mg of Vitamin C per day. Each Supervite pill provides 250 mg and each Vitahealth pill provides 350 mg.

3. A candidate can afford to spend no more than $8500 on radio and TV advertising. Each radio spot costs $150 and each TV ad costs $750.

4. A hospital dietician has two meal choices, one for patients on solid food that costs $2.25, and one for patients on liquids that costs $3.75. There are a maximum of 400 patients in the hospital.

5. Cashews costing $8 per pound are to be mixed with peanuts costing $3 per pound to obtain at least 30 pounds of mixed nuts.

6. An agricultural advisor looks at the results of Example 1 and claims that it cannot possibly be correct. He says that since the 4-H Club member could raise 16 animals, but is only raising 12, she could earn more profit by raising all 16 animals. How would you respond?

Solve the following linear programming problems. (See Examples 1–3.)

7. Management   Audio City Corporation has warehouses in Meadville and Cambridge. It has 80 stereo systems stored in Meadville and 70 in Cambridge. Superstore orders 35 systems and ValueHouse orders 60. It costs $8 to ship a system from Meadville to Superstore and $12 to ship one to ValueHouse. It costs $10 to ship a system from Cambridge to Superstore and $13 to ship one to Value-House. How should the orders be filled to keep shipping costs as low as possible? What is the minimum shipping cost? (*Hint:* If x systems are shipped from Meadville to Superstore, then $35 - x$ systems are shipped from Cambridge to Superstore.)

8. Management   A manufacturer of refrigerators must ship at least 100 refrigerators to its two West Coast warehouses. Each warehouse holds a maximum of 100 refrigerators. Warehouse A holds 25 refrigerators already, and warehouse B has 20 on hand. It costs $12 to ship a refrigerator to warehouse A and $10 to ship one to warehouse B. Union rules require that at least 300 workers be hired. Shipping a refrigerator to Warehouse A requires 4 workers, while shipping a refrigerator to Warehouse B requires 2 workers. How many refrigerators should be shipped to each warehouse to minimize costs? What is the minimum cost?

9. **Management** A company is considering two insurance plans with the types of coverage and premiums shown in the table below.

	Policy A	Policy B
Fire/Theft	$10,000	$15,000
Liability	$180,000	$120,000
Premium	$50	$40

(For example, this means that $50 buys one unit of plan A, consisting of $10,000 fire and theft insurance and $180,000 of liability insurance.)

(a) The company wants at least $300,000 fire/theft insurance and at least $3,000,000 liability insurance from these plans. How many units should be purchased from each plan to minimize the cost of the premiums? What is the minimum premium?

(b) Suppose the premium for policy A is reduced to $25. Now how many units should be purchased from each plan to minimize the cost of the premiums? What is the minimum premium?

10. **Management** Hotnews Magazine publishes a U.S. and a Canadian edition each week. There are 30,000 subscribers in the United States and 20,000 in Canada. Other copies are sold at newsstands. Postage and shipping costs average $80 per thousand copies for the U.S. and $60 per thousand copies for Canada. Surveys show that no more than 120,000 copies of each issue can be sold (including subscriptions) and that the number of copies of the Canadian edition should not exceed twice the number of copies of the U.S. edition. The publisher can spend at most $8400 a month on postage and shipping. If the profit is $200 for each thousand copies of the U.S. edition and $150 for each thousand copies of the Canadian edition, how many copies of each version should be printed to earn as large a profit as possible? What is that profit?

11. **Management** The manufacturing process requires that oil refineries must manufacture at least 2 gallons of gasoline for every gallon of fuel oil. To meet the winter demand for fuel oil, at least 3 million gallons a day must be produced. The demand for gasoline is no more than 6.4 million gallons per day. It takes .25 hour to ship each million gallons of gasoline and 1 hour to ship each million gallons of fuel oil out of the warehouse. No more than 4.65 hours are available for shipping. If the refinery sells gasoline for $1.25 per gallon and fuel oil for $1 per gallon, how much of each should be produced to maximize revenue? Find the maximum revenue.

12. **Natural Science** Kim Walrath has a nutritional deficiency and is told to take at least 2400 mg of iron, 2100 mg of Vitamin B-1, and 1500 mg of Vitamin B-2. One Maxivite pill contains 40 mg of iron, 10 mg of B-1, and 5 mg of B-2,

and costs 6¢. One Healthovite pill provides 10 mg of iron, 15 mg of B-1, and 15 mg of B-2, and costs 8¢. What combination of Maxivite and Healthovite pills will meet the requirement at lowest cost? What is the minimum cost?

13. **Management** A machine shop manufactures two types of bolts. The bolts require time on each of three groups of machines, but the time required on each group differs, as shown in the table below.

		MACHINE GROUP		
		I	II	III
BOLTS	Type 1	.1 min	.1 min	.1 min
	Type 2	.1 min	.4 min	.02 min

Production schedules are made up one day at a time. In a day there are 240, 720, and 160 min available, respectively, on these machines. Type 1 bolts sell for 10¢ and type 2 bolts for 12¢. How many of each type of bolt should be manufactured per day to maximize revenue? What is the maximum revenue?

14. **Management** The Miers Company produces small engines for several manufacturers. The company receives orders from two assembly plants for their Topflight engine. Plant I needs at least 50 engines, and plant II needs at least 27 engines. The company can send at most 85 engines to these two assembly plants. It costs $20 per engine to ship to plant I and $35 per engine to ship to plant II. Plant I gives Miers $15 in rebates towards its products for each engine they buy, while plant II gives similar $10 rebates. Miers estimates that they need at least $1110 in rebates to cover products they plan to buy from the two plants. How many engines should be shipped to each plant to minimize shipping costs? What is the minimum cost?

15. **Management** A small country can grow only two crops for export, coffee and cocoa. The country has 500,000 hectares of land available for the crops. Long-term contracts require that at least 100,000 hectares be devoted to coffee and at least 200,000 hectares to cocoa. Cocoa must be processed locally, and production bottlenecks limit cocoa to 270,000 hectares. Coffee requires two workers per hectare, with cocoa requiring five. No more than 1,750,000 people are available for working with these crops. Coffee produces a profit of $220 per hectare and cocoa a profit of $310 per hectare. How many hectares should the country devote to each crop in order to maximize profit? Find the maximum profit.

16. **Management** 60 pounds of chocolates and 100 pounds of mints are available to make up 5-pound boxes of candy. A regular box has 4 pounds of chocolates and 1 pound of mints and sells for $10. A deluxe box has 2 pounds of chocolates and 3 pounds of mints and sells for $16. How

Test 3/22 thursday chapt. 7 B 8

many boxes of each kind should be made to maximize revenue?

17. **Management**   A greeting card manufacturer has 370 boxes of a particular card in warehouse I and 290 boxes of the same card in warehouse II. A greeting card shop in San Jose orders 350 boxes of the card, and another shop in Memphis orders 300 boxes. The shipping costs per box to these shops from the two warehouses are shown in the following table.

		DESTINATION	
		San Jose	Memphis
WAREHOUSE	I	$.25	$.22
	II	$.23	$.21

How many boxes should be shipped to each city from each warehouse to minimize shipping costs? What is the minimum cost? (*Hint:* Use x, $350 - x$, y, and $300 - y$ as the variables.)

18. **Management**   A pension fund manager decides to invest at most $40 million in U.S. Treasury Bonds paying 8% annual interest and in mutual funds paying 12% annual interest. He plans to invest at least $20 million in bonds and at least $15 million in mutual funds. Bonds have an initial fee of $300 per million dollars, while the fee for mutual funds is $100 per million. The fund manager is allowed to spend no more than $8400 on fees. How much should be invested in each to maximize annual interest? What is the maximum annual interest?

19. **Natural Science**   A certain predator requires at least 10 units of protein and 8 units of fat per day. One prey of Species I provides 5 units of protein and 2 units of fat; one prey of Species II provides 3 units of protein and 4 units of fat. Capturing and digesting each Species II prey requires 3 units of energy, and capturing and digesting each Species I prey requires 2 units of energy. How many of each prey would meet the predator's daily food requirements with the least expenditure of energy? Are the answers reasonable? How could they be interpreted?

20. **Management**   In a small town in South Carolina, zoning rules require that the window space (in square feet) in a house be at least one-sixth of the space used up by solid walls. The cost to build windows is $10 per square foot, while the cost to build solid walls is $20 per square foot. The total amount available for building walls and windows is no more than $12,000. The estimated monthly cost to heat the house is $.32 for each square foot of windows and $.20 for each square foot of solid walls. Find the maximum total area (windows plus walls) if no more than $160 per month is available to pay for heat.

21. **Social Science**   Students at Upscale U are required to take at least 3 humanities and 4 science courses. The maximum allowable number of science courses is 12. Each humanities course carries 4 credits and each science course 5 credits. The total number of credits in science and humanities cannot exceed 80. Quality points for each course are assigned in the usual way: the number of credit hours times 4 for an A grade; times 3 for a B grade; times 2 for a C grade. Susan Katz expects to get B's in all her science courses. She expects to get C's in half her humanities courses, B's in one-fourth of them, and A's in the rest. Under these assumptions, how many courses of each kind should she take in order to earn the maximum possible number of quality points?

22. **Social Science**   In Exercise 21, find Susan's grade point average (the total number of quality points divided by the total number of credit hours) at each corner point of the feasible region. Does the distribution of courses that produces the highest number of quality points also yield the highest grade point average? Is this a contradiction?

The importance of linear programming is shown by the inclusion of linear programming problems on most qualification examinations for Certified Public Accountants. Exercises 23–25 are reprinted from one such examination.*

The Random Company manufactures two products, Zeta and Beta. Each product must pass through two processing operations. All materials are introduced at the start of Process No. 1. There are no work-in-process inventories. Random may produce either one product exclusively or various combinations of both products subject to the following constraints.

	Process No. 1	Process No. 2	Contribution Margin Per Unit
Hours required to produce 1 unit of:			
Zeta	1 hour	1 hour	$4.00
Beta	2 hours	3 hours	5.25
Total capacity in hours per day	1000 hours	1275 hours	

A shortage of technical labor has limited Beta production to 400 units per day. There are no constraints on the production of Zeta other than the hour constraints in the above schedule. Assume that all the relationships between capacity and production are linear.

---

*Material from *Uniform CPA Examinations and Unofficial Answers*, copyright © 1973, 1974, 1975 by the American Institute of Certified Public Accountants, Inc., is reprinted with permission.

23. Given the objective to maximize total contribution margin, what is the production constraint for Process No. 1?
(a) Zeta + Beta ≤ 1000
(b) Zeta + 2 Beta ≤ 1000
(c) Zeta + Beta ≥ 1000
(d) Zeta + 2 Beta ≥ 1000

24. Given the objective to maximize total contribution margin, what is the labor constraint for production of Beta?

(a) Beta ≤ 400   (b) Beta ≥ 400
(c) Beta ≤ 425   (d) Beta ≥ 425

25. What is the objective function of the data presented?
(a) Zeta + 2 Beta = $9.25
(b) $4.00 Zeta + 3($5.25) Beta = total contribution margin
(c) $4.00 Zeta + $5.25 Beta = total contribution margin
(d) 2($4.00) Zeta + 3($5.25) Beta = total contribution margin

8.4   THE SIMPLEX METHOD: MAXIMIZATION

For linear programming problems with more than two variables or with two variables and many constraints, the graphical method is usually too inefficient, so the **simplex method** is used. The simplex method, which is introduced here, was developed for the U.S. Air Force by George B. Danzig in 1947. It was used successfully during the Berlin airlift in 1948–49 to maximize the amount of cargo delivered under very severe constraints and is widely used today in a variety of industries.

Because the simplex method is used for problems with many variables, it usually is not convenient to use letters such as x, y, z, or w as variable names. Instead, the symbols x_1 (read "x-sub-one"), x_2, x_3, and so on, are used. In the simplex method, all constraints must be expressed in the linear form

$$a_1x_1 + a_2x_2 + a_3x_3 + \cdots \leq b,$$

where x_1, x_2, x_3, . . . are variables, a_1, a_2, a_3, . . . are coefficients, and b is a constant.

We first discuss the simplex method for linear programming problems in *standard maximum form*.

STANDARD MAXIMUM FORM

A linear programming problem is in **standard maximum form** if

1. the objective function is to be maximized;
2. all variables are nonnegative ($x_i \geq 0$, $i = 1, 2, 3, \ldots$);
3. all constraints involve ≤;
4. the constants on the right side in the constraints are all nonnegative ($b \geq 0$).

Problems that do not meet all of these conditions are considered in Sections 8.6 and 8.7.

The "mechanics" of the simplex method are demonstrated in Examples 1–5. Although the procedures to be followed will be made clear, as will the fact that they result in an optimal solution, the reasons why these procedures are used may not be immediately apparent. Examples 6 and 7 will supply these reasons and explain the connection between the simplex method and the graphical method used in Section 8.3.

SETTING UP THE PROBLEM   The first step is to convert each constraint, a linear inequality, into a linear equation. This is done by adding a nonnegative variable, called a **slack variable,** to each constraint. For example, convert the inequality $x_1 + x_2 \leq 10$ into an equation by adding the slack variable x_3, to get

$$x_1 + x_2 + x_3 = 10, \quad \text{where } x_3 \geq 0.$$

The inequality $x_1 + x_2 \leq 10$ says that the sum $x_1 + x_2$ is less than or perhaps equal to 10. The variable x_3 "takes up any slack" and represents the amount by which $x_1 + x_2$ fails to equal 10. For example, if $x_1 + x_2$ equals 8, then x_3 is 2. If $x_1 + x_2 = 10$, the value of x_3 is 0.

CAUTION A different slack variable must be used for each constraint. ◆

EXAMPLE 1 Restate the following linear programming problem by introducing slack variables.

$$\text{Maximize} \quad z = 2x_1 + 3x_2 + x_3$$
$$\text{subject to:} \quad x_1 + x_2 + 4x_3 \leq 100$$
$$x_1 + 2x_2 + x_3 \leq 150$$
$$3x_1 + 2x_2 + x_3 \leq 320$$
$$\text{with} \quad x_1 \geq 0, x_2 \geq 0, x_3 \geq 0.$$

1 Rewrite the following set of constraints as equations by adding nonnegative slack variables.

$$x_1 + x_2 + x_3 \leq 12$$
$$2x_1 + 4x_2 \leq 15$$
$$x_2 + 3x_3 \leq 10$$

Answer:
$$x_1 + x_2 + x_3 + x_4 = 12$$
$$2x_1 + 4x_2 + x_5 = 15$$
$$x_2 + 3x_3 + x_6 = 10$$

Rewrite the three constraints as equations by introducing nonnegative slack variables x_4, x_5, and x_6, one for each constraint. Then the problem can be restated as

$$\text{Maximize} \quad z = 2x_1 + 3x_2 + x_3$$
$$\text{subject to:} \quad x_1 + x_2 + 4x_3 + x_4 = 100$$
$$x_1 + 2x_2 + x_3 + x_5 = 150$$
$$3x_1 + 2x_2 + x_3 + x_6 = 320$$
$$\text{with} \quad x_1 \geq 0, x_2 \geq 0, x_3 \geq 0, x_4 \geq 0, x_5 \geq 0, x_6 \geq 0. \quad ■ \quad \boxed{1}$$

Adding slack variables to the constraints converts a linear programming problem into a system of linear equations. These equations should have all variables on the left of the equals sign and all constants on the right. All the equations of Example 1 satisfy this condition except for the objective function, $z = 2x_1 + 3x_2 + x_3$, which may be written with all variables on the left as

$$-2x_1 - 3x_2 - x_3 + z = 0.$$

Now the equations of Example 1 (with the constraints listed first) can be written as the following augmented matrix.

x_1	x_2	x_3	x_4	x_5	x_6	z	
1	1	4	1	0	0	0	100
1	2	1	0	1	0	0	150
3	2	1	0	0	1	0	320
−2	−3	−1	0	0	0	1	0

Indicators

2 Set up the initial simplex tableau for the following linear programming problem:

Maximize $z = 2x_1 + 3x_2$

subject to: $x_1 + 2x_2 \leq 85$

$2x_1 + x_2 \leq 92$

$x_1 + 4x_2 \leq 104$

with $x_1 \geq 0, x_2 \geq 0$.

Answer:

$$
\begin{array}{ccccccc}
x_1 & x_2 & x_3 & x_4 & x_5 & z & \\
\left[\begin{array}{cccccc|c}
1 & 2 & 1 & 0 & 0 & 0 & 85 \\
2 & 1 & 0 & 1 & 0 & 0 & 92 \\
1 & 4 & 0 & 0 & 1 & 0 & 104 \\
\hline
-2 & -3 & 0 & 0 & 0 & 1 & 0
\end{array}\right]
\end{array}
$$

This matrix is the initial **simplex tableau.** Except for the last entries, the 1 and 0 on the right end, the numbers in the bottom row of a simplex tableau are called **indicators.** **2**

This simplex tableau represents a system of 4 linear equations in 7 variables. Since there are more variables than equations, the system is dependent and has infinitely many solutions. Our goal is to find a solution in which all the variables are non-negative and z is as large as possible. This will be done by using row operations to replace the given system by an equivalent one in which certain variables are eliminated from some of the equations. The process will be repeated until the optimum solution can be read from the matrix, as explained below.

SELECTING THE PIVOT   Recall how row operations are used to eliminate variables in the Gauss-Jordan method. A particular nonzero entry in the matrix is chosen and changed to a 1; then all other entries in that column are changed to zeros. A similar process is used in the simplex method. The chosen entry is called the **pivot.** The procedure for selecting the appropriate pivot in the simplex method is explained in the next example. The reason why this procedure is used is discussed in Example 7.

TECHNOLOGY TIP A graphing calculator provides an efficient method for doing the row operations in the discussion below, so you may want to enter the initial simplex tableau of Example 1 on your calculator and perform the various operations on it as you read the following examples. Because of its size this simplex tableau will not fit on a calculator screen. Instead, you will see something like Figure 8.20(a), which shows only the left half of the tableau. Use the left and right arrow keys to scroll through the entire matrix and obtain its other half, as in Figure 8.20(b). Note that the calculator does not put a vertical line before the last column, as we do when working by hand. ✔

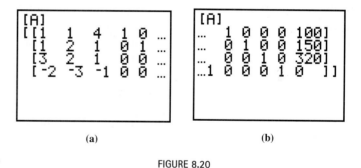

(a)   (b)

FIGURE 8.20

EXAMPLE 2   Determine the pivot in the simplex tableau for the problem in Example 1.

Look at the indicators (the last row of the tableau) and choose the most negative one.

$$
\begin{array}{ccccccc}
x_1 & x_2 & x_3 & x_4 & x_5 & x_6 & z \\
\left[\begin{array}{ccccccc|c}
1 & 1 & 4 & 1 & 0 & 0 & 0 & 100 \\
1 & 2 & 1 & 0 & 1 & 0 & 0 & 150 \\
3 & 2 & 1 & 0 & 0 & 1 & 0 & 320 \\
\hline
-2 & \mathbf{-3} & -1 & 0 & 0 & 0 & 1 & 0
\end{array}\right]
\end{array}
$$

Most negative indicator

The most negative indicator identifies the variable that is to be eliminated from all but one of the equations (rows), in this case x_2. The column containing the most negative indicator is called the **pivot column.** Now for each *positive* entry in the pivot column, divide the number in the far right column of the same row by the positive number in the pivot column.

x_1	x_2	x_3	x_4	x_5	x_6	z		Quotients
1	**1**	4	1	0	0	0	**100**	$100/1 = 100$
1	**2**	1	0	1	0	0	**150**	$150/2 = 75 \leftarrow$ Smallest
3	**2**	1	0	0	1	0	**320**	$320/2 = 160$
−2	−3	−1	0	0	0	1	0	

The row with the smallest quotient (in this case, the second row) is called the **pivot row.** The entry in the pivot row and pivot column is the pivot.

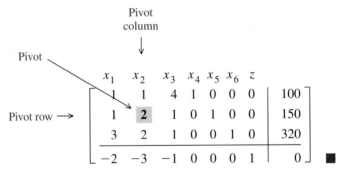

3 Find the pivot for the following tableau.

x_1	x_2	x_3	x_4	x_5	z	
0	1	1	0	0	0	50
−2	3	0	1	0	0	78
2	4	0	0	1	0	65
−5	−3	0	0	0	1	0

Answer:
2 (in first column)

CAUTION In some simplex tableaus the pivot column may contain zeros or negative entries. Only the positive entries in the pivot column should be used to form the quotients and determine the pivot row. If there are no positive entries in the pivot column (so that a pivot row cannot be chosen), then no maximum solution exists. ◆ **3**

PIVOTING Once the pivot has been selected, row operations are used to replace the initial simplex tableau by another simplex tableau in which the pivot column variable is eliminated from all but one of the equations. Since this new tableau is obtained by row operations, it represents an equivalent system of equations (that is, a system with the same solutions as the original system). This process, which is called **pivoting,** is explained in the next example.

EXAMPLE 3 Use the indicated pivot, 2, to perform the pivoting on the simplex tableau of Example 2:

x_1	x_2	x_3	x_4	x_5	x_6	z	
1	1	4	1	0	0	0	100
1	**2**	1	0	1	0	0	150
3	2	1	0	0	1	0	320
−2	−3	−1	0	0	0	1	0

Start by multiplying each entry of row two by 1/2 in order to change the pivot to 1.

$$\begin{array}{ccccccc}
x_1 & x_2 & x_3 & x_4 & x_5 & x_6 & z \\
\end{array}$$

$$\left[\begin{array}{ccccccc|c}
1 & 1 & 4 & 1 & 0 & 0 & 0 & 100 \\
\frac{1}{2} & 1 & \frac{1}{2} & 0 & \frac{1}{2} & 0 & 0 & 75 \\
3 & 2 & 1 & 0 & 0 & 1 & 0 & 320 \\
\hline
-2 & -3 & -1 & 0 & 0 & 0 & 1 & 0
\end{array}\right] \quad \frac{1}{2}R_2$$

Now use row operations to make the entry in row one, column two a 0.

$$\left[\begin{array}{ccccccc|c}
\frac{1}{2} & 0 & \frac{7}{2} & 1 & -\frac{1}{2} & 0 & 0 & 25 \\
\frac{1}{2} & 1 & \frac{1}{2} & 0 & \frac{1}{2} & 0 & 0 & 75 \\
3 & 2 & 1 & 0 & 0 & 1 & 0 & 320 \\
\hline
-2 & -3 & -1 & 0 & 0 & 0 & 1 & 0
\end{array}\right] \quad -R_2 + R_1$$

Change the 2 in row three, column two to a 0 by a similar process.

$$\left[\begin{array}{ccccccc|c}
\frac{1}{2} & 0 & \frac{7}{2} & 1 & -\frac{1}{2} & 0 & 0 & 25 \\
\frac{1}{2} & 1 & \frac{1}{2} & 0 & \frac{1}{2} & 0 & 0 & 75 \\
2 & 0 & 0 & 0 & -1 & 1 & 0 & 170 \\
\hline
-2 & -3 & -1 & 0 & 0 & 0 & 1 & 0
\end{array}\right] \quad -2R_2 + R_3$$

Finally, add 3 times row 2 to the last row in order to change the indicator -3 to 0.

$$\left[\begin{array}{ccccccc|c}
\frac{1}{2} & 0 & \frac{7}{2} & 1 & -\frac{1}{2} & 0 & 0 & 25 \\
\frac{1}{2} & 1 & \frac{1}{2} & 0 & \frac{1}{2} & 0 & 0 & 75 \\
2 & 0 & 0 & 0 & -1 & 1 & 0 & 170 \\
\hline
-\frac{1}{2} & 0 & \frac{1}{2} & 0 & \frac{3}{2} & 0 & 1 & 225
\end{array}\right] \quad 3R_2 + R_4$$

The pivoting is now complete because the pivot column variable x_2 has been eliminated from all equations except the one represented by the pivot row. The initial simplex tableau has been replaced by a new simplex tableau, which represents an equivalent system of equations. ■

CAUTION During pivoting, do not interchange rows of the matrix. Make the pivot entry 1 by multiplying the pivot row by an appropriate constant, as in Example 3. ◆ 4

When at least one of the indicators in the last row of a simplex tableau is negative (as is the case with the tableau obtained in Example 3), the simplex method requires that a new pivot be selected and the pivoting be performed again. This procedure is repeated until a simplex tableau with no negative indicators in the last row is obtained or a tableau is reached in which no pivot row can be chosen.

EXAMPLE 4 In the simplex tableau obtained in Example 3, select a new pivot and perform the pivoting.

4 For the simplex tableau below,

(a) find the pivot.

(b) Perform the pivoting and write the new tableau.

$$\left[\begin{array}{ccccc|c}
1 & 2 & 6 & 1 & 0 & 0 & 16 \\
1 & 3 & 0 & 0 & 1 & 0 & 25 \\
\hline
-1 & -4 & -3 & 0 & 0 & 1 & 0
\end{array}\right]$$

Answers:
(a) 2
(b)

$$\left[\begin{array}{ccccc|c}
\frac{1}{2} & 1 & 3 & \frac{1}{2} & 0 & 0 & 8 \\
-\frac{1}{2} & 0 & -9 & -\frac{3}{2} & 1 & 0 & 1 \\
\hline
1 & 0 & 9 & 2 & 0 & 1 & 32
\end{array}\right]$$

First, locate the pivot column by finding the most negative indicator in the last row. Then locate the pivot row by computing the necessary quotients and finding the smallest one, as shown here.

$$
\begin{array}{c}
\begin{array}{ccccccc}
x_1 & x_2 & x_3 & x_4 & x_5 & x_6 & z
\end{array}\\
\text{Pivot row} \longrightarrow \left[\begin{array}{ccccccc|c}
\frac{1}{2} & 0 & \frac{7}{2} & 1 & -\frac{1}{2} & 0 & 0 & 25\\[4pt]
\frac{1}{2} & 1 & \frac{1}{2} & 0 & \frac{1}{2} & 0 & 0 & 75\\[4pt]
2 & 0 & 0 & 0 & -1 & 1 & 0 & 170\\[4pt]
\hline
-\frac{1}{2} & 0 & \frac{1}{2} & 0 & \frac{3}{2} & 0 & 1 & 225
\end{array}\right]
\end{array}
\qquad
\begin{array}{l}
\text{Quotients}\\[6pt]
\dfrac{25}{1/2}=50 \quad \text{Smallest}\\[8pt]
\dfrac{75}{1/2}=150\\[8pt]
170/2 = 85
\end{array}
$$

Pivot column

So the pivot is the number 1/2 in row one, column one. Begin the pivoting by multiplying every entry in row one by 2. Then continue as indicated below, to obtain the following simplex tableau.

$$
\begin{array}{c}
\begin{array}{ccccccc}
x_1 & x_2 & x_3 & x_4 & x_5 & x_6 & z
\end{array}\\
\left[\begin{array}{ccccccc|c}
1 & 0 & 7 & 2 & -1 & 0 & 0 & 50\\
0 & 1 & -3 & -1 & 1 & 0 & 0 & 50\\
0 & 0 & -14 & -4 & 1 & 1 & 0 & 70\\
\hline
0 & 0 & 4 & 1 & 1 & 0 & 1 & 250
\end{array}\right]
\end{array}
\qquad
\begin{array}{l}
2R_1\\[6pt]
-\dfrac{1}{2}R_1 + R_2\\[6pt]
-2R_1 + R_3\\[6pt]
\dfrac{1}{2}R_1 + R_4
\end{array}
$$

Since there are no negative indicators in the last row, no further pivoting is necessary and we call this the **final simplex tableau.** ∎

READING THE SOLUTION The next example shows how to read an optimal solution of the original linear programming problem from the final simplex tableau.

EXAMPLE 5 Solve the linear programming problem introduced in Example 1.
    Look at the final simplex tableau for this problem, which was obtained in Example 4.

$$
\begin{array}{c}
\begin{array}{ccccccc}
x_1 & x_2 & x_3 & x_4 & x_5 & x_6 & z
\end{array}\\
\left[\begin{array}{ccccccc|c}
1 & 0 & 7 & 2 & -1 & 0 & 0 & 50\\
0 & 1 & -3 & -1 & 1 & 0 & 0 & 50\\
0 & 0 & -14 & -4 & 1 & 1 & 0 & 70\\
\hline
0 & 0 & 4 & 1 & 1 & 0 & 1 & 250
\end{array}\right]
\end{array}
$$

The last row of this matrix represents the equation

$$4x_3 + x_4 + x_5 + z = 250, \quad \text{or equivalently,} \quad z = 250 - 4x_3 - x_4 - x_5.$$

If $x_3, x_4,$ and x_5 are all 0, then the value of z is 250. If any one of $x_3, x_4,$ or x_5 is positive, then z will have a smaller value than 250 (why?). Consequently, since we want a solution for this system in which all the variables are nonnegative and z is as large as possible, we must have

$$x_3 = 0, \quad x_4 = 0, \quad x_5 = 0.$$

When these values are substituted in the first equation (represented by the first row of the final simplex tableau), the result is

$$x_1 + 7 \cdot 0 + 2 \cdot 0 - 1 \cdot 0 = 50, \quad \text{that is,} \quad x_1 = 50.$$

Similarly, substituting 0 for x_3, x_4, and x_5 in the last three equations represented by the final simplex tableau shows that

$$x_2 = 50, \quad x_6 = 70, \quad z = 250.$$

Therefore, the maximum value of $z = 2x_1 + 3x_2 + x_3$ occurs when

$$x_1 = 50, \quad x_2 = 50, \quad x_3 = 0,$$

in which case $z = 2 \cdot 50 + 3 \cdot 50 + 0 = 250$. (The values of the slack variables are irrelevant in stating the solution of the original problem.) ■

In any simplex tableau, some columns look like columns of an identity matrix (one entry is 1, the rest are 0). The variables corresponding to these columns are called **basic variables** and the variables corresponding to the other columns **nonbasic variables.** In the tableau of Example 5, for instance, the basic variables are x_1, x_2, x_6, and z (shown in color below), and the nonbasic variables are x_3, x_4, and x_5.

5 A linear programming problem with slack variables x_4 and x_5 has the final simplex tableau shown below. What is the optimal solution?

$$
\begin{array}{c}
\begin{array}{cccccc} x_1 & x_2 & x_3 & x_4 & x_5 & z \end{array} \\
\left[\begin{array}{cccccc|c}
0 & 3 & 1 & 5 & 2 & 0 & 9 \\
1 & -2 & 0 & 4 & 1 & 0 & 6 \\
\hline
0 & 5 & 0 & 1 & 0 & 1 & 21
\end{array}\right]
\end{array}
$$

Answer:
$z = 21$ when $x_1 = 6$, $x_2 = 0$, $x_3 = 9$.

$$
\begin{array}{c}
\begin{array}{ccccccc} x_1 & x_2 & x_3 & x_4 & x_5 & x_6 & z \end{array} \\
\left[\begin{array}{ccccccc|c}
1 & 0 & 7 & 2 & -1 & 0 & 0 & 50 \\
0 & 1 & -3 & -1 & 1 & 0 & 0 & 50 \\
0 & 0 & -14 & -4 & 1 & 1 & 0 & 70 \\
\hline
0 & 0 & 4 & 1 & 1 & 0 & 1 & 250
\end{array}\right]
\end{array}
$$

The optimal solution in Example 5 was obtained from the final simplex tableau by setting the nonbasic variables equal to 0 and solving for the basic variables. Furthermore, the values of the basic variables are easy to read off the matrix: find the 1 in the column representing a basic variable; the last entry in that row is the value of that basic variable in the optimal solution. In particular, *the entry in the lower right-hand corner of the final simplex tableau is the maximum value of z.* **5**

CAUTION If there are two identical columns in a tableau, each of which is a column in an identity matrix, only one of the variables corresponding to these columns can be a basic variable. The other is treated as a nonbasic variable. You may choose either one to be the basic variable, unless one of them is z, in which case z must be the basic variable. ◆

The steps involved in solving a standard maximum linear programming problem by the simplex method have been illustrated in Examples 1–5 and are summarized here.

SIMPLEX METHOD

1. Determine the objective function.
2. Write all necessary constraints.
3. Convert each constraint into an equation by adding slack variables.
4. Set up the initial simplex tableau.
5. Locate the most negative indicator. If there are two such indicators, choose one. This indicator determines the pivot column.
6. Use the positive entries in the pivot column to form the quotients necessary for determining the pivot. If there are no positive entries in the pivot column,

continued

no maximum solution exists. If two quotients are equally the smallest, let either determine the pivot.*

7. Multiply every entry in the pivot row by the reciprocal of the pivot to change the pivot to 1. Then use row operations to change all other entries in the pivot column to 0 by adding suitable multiples of the pivot row to the other rows. (These steps can be done with a graphing calculator.)

8. If the indicators are all positive or 0, this is the final tableau. If not, go back to Step 5 above and repeat the process until a tableau with no negative indicators is obtained.†

9. Determine the basic and nonbasic variables and read the solution from the final tableau. The maximum value of the objective function is the number in the lower right-hand corner of the final tableau.

6 A linear programming problem has the initial tableau given below. Use the simplex method to solve the problem.

$$\begin{array}{ccccc} x_1 & x_2 & x_3 & x_4 & z \\ \begin{bmatrix} 1 & 1 & 1 & 0 & 0 \\ 2 & 1 & 0 & 1 & 0 \\ -300 & -200 & 0 & 0 & 1 \end{bmatrix} & \begin{array}{c} 40 \\ 24 \\ 0 \end{array} \end{array}$$

Answer:
$x_1 = 0, x_2 = 24, x_3 = 16,$
$x_4 = 0, z = 4800$

The solution found by the simplex method may not be unique, especially when choices are possible in steps 5, 6, or 9. There may be other solutions that produce the same maximum value of the objective function. (See Exercises 37 and 38.) **6**

GEOMETRIC INTERPRETATION OF THE SIMPLEX METHOD Although it may not be immediately apparent, the simplex method is based on the same geometrical considerations as the graphical method. This can be seen by looking at a problem that can be readily solved by both methods.

EXAMPLE 6 In Example 2 of Section 8.3 the following problem was solved graphically (using x, y instead of x_1, x_2):

$$\text{Maximize} \quad z = 8x_1 + 12x_2$$
$$\text{subject to:} \quad 40x_1 + 80x_2 \le 560$$
$$6x_1 + 8x_2 \le 72$$
$$x_1 \ge 0, x_2 \ge 0.$$

Graphing the feasible region (Figure 8.21) and evaluating z at each corner point shows that the maximum value of z occurs at (8, 3).

Corner Point	Value of $z = 8x_1 + 12x_2$
(0, 0)	0
(0, 7)	84
(12, 0)	96
(8, 3)	**100** **(maximum)**

FIGURE 8.21

---

*It may be that the first choice of a pivot does not produce a solution. In that case try the other choice.

†Some special circumstances are noted at the end of Section 8.7.

To solve the same problem by the simplex method, add a slack variable to each constraint.

$$40x_1 + 80x_2 + x_3 \qquad = 560$$
$$6x_1 + 8x_2 \qquad + x_4 = 72$$

Then write the initial simplex tableau.

$$
\begin{array}{ccccc}
x_1 & x_2 & x_3 & x_4 & z \\
\end{array}
$$

$$
\left[
\begin{array}{ccccc|c}
40 & 80 & 1 & 0 & 0 & 560 \\
6 & 8 & 0 & 1 & 0 & 72 \\
\hline
-8 & -12 & 0 & 0 & 1 & 0 \\
\end{array}
\right]
$$

In this tableau the basic variables are x_3, x_4, and z (why?). By setting the nonbasic variables (namely, x_1 and x_2) equal to 0 and solving for the basic variables, we obtain the following solution (which will be called a **basic feasible solution**):

$$x_1 = 0, \quad x_2 = 0, \quad x_3 = 560, \quad x_4 = 72, \quad z = 0.$$

Since $x_1 = 0$ and $x_2 = 0$, this solution corresponds to the corner point at the origin in the graphical solution (Figure 8.21). The value $z = 0$ at the origin is obviously not maximal and pivoting in the simplex method is designed to improve it.

The most negative indicator in the initial tableau is -12 and the necessary quotients are

$$\frac{560}{80} = 7 \quad \text{and} \quad \frac{72}{8} = 9.$$

The smaller quotient is 7, giving 80 as the pivot. Performing the pivoting leads to this tableau.

$$
\begin{array}{ccccc}
x_1 & x_2 & x_3 & x_4 & z \\
\end{array}
$$

$$
\left[
\begin{array}{ccccc|c}
\frac{1}{2} & 1 & \frac{1}{80} & 0 & 0 & 7 \\
2 & 0 & -\frac{1}{10} & 1 & 0 & 16 \\
\hline
-2 & 0 & \frac{3}{20} & 0 & 1 & 84 \\
\end{array}
\right]
\begin{array}{l}
\frac{1}{80}R_1 \\
-8R_1 + R_2 \\
12R_1 + R_3 \\
\end{array}
$$

The basic variables here are x_2, x_4, and z and the basic feasible solution (found by setting the nonbasic variables equal to 0 and solving for the basic variables) is

$$x_1 = 0, \quad x_2 = 7, \quad x_3 = 0, \quad x_4 = 16, \quad z = 84,$$

which corresponds to the corner point $(0, 7)$ in Figure 8.21. Note that the new value of the pivot variable x_2 is precisely the smallest quotient, 7, that was used to select the pivot row. Although this value of z is better, further improvement is possible.

Now the most negative indicator is -2 and the quotients are

$$\frac{7}{1/2} = 14 \quad \text{and} \quad \frac{16}{2} = 8.$$

Since 8 is smaller, the pivot is the number 2 in row two, column one. Pivoting produces the final tableau.

$$
\begin{array}{ccccc|c}
x_1 & x_2 & x_3 & x_4 & z & \\
0 & 1 & \frac{3}{80} & -\frac{1}{4} & 0 & 3 \\
1 & 0 & -\frac{1}{20} & \frac{1}{2} & 0 & 8 \\
\hline
0 & 0 & \frac{1}{20} & 1 & 1 & 100
\end{array}
\qquad
\begin{array}{l}
-\frac{1}{2}R_2 + R_1 \\[4pt]
\frac{1}{2}R_2 \\[4pt]
2R_2 + R_3
\end{array}
$$

Here the basic feasible solution is

$$x_1 = 8, \quad x_2 = 3, \quad x_3 = 0, \quad x_4 = 0, \quad z = 100,$$

which corresponds to the corner point $(8, 3)$ in Figure 8.21. Once again, the new value of the pivot variable x_1 is the smallest quotient, 8, that was used to select the pivot. From the graphical method we know that this solution provides the maximum value of the objective function. This fact can also be seen algebraically by using an algebraic argument similar to the one used in Example 5. Thus there is no need to move to another corner point and the simplex method ends. ■

As illustrated in Example 6, the basic feasible solution obtained from a simplex tableau corresponds to a corner point of the feasible region. Pivoting, which replaces one tableau with another, is a systematic way of moving from one corner point to another, each time improving the value of the objective function. The simplex method ends when a corner point that produces the maximum value of the objective function is reached (or when it becomes clear that the problem has no maximum solution).

When there are three or more variables in a linear programming problem, it may be difficult or impossible to draw a picture, but it can be proved that the optimal value of the objective function occurs at a basic feasible solution (corresponding to a corner point in the two variable case). The simplex method provides a means of moving from one basic feasible solution to another until one that produces the optimal value of the objective function is reached.

EXPLANATION OF PIVOTING The rules for selecting the pivot in the simplex method can be understood by examining how the first pivot was chosen in Example 6.

EXAMPLE 7 The initial simplex tableau of Example 6 provides a basic feasible solution with $x_1 = 0, x_2 = 0$.

$$
\begin{array}{ccccc|c}
x_1 & x_2 & x_3 & x_4 & z & \\
40 & 80 & 1 & 0 & 0 & 560 \\
6 & 8 & 0 & 1 & 0 & 72 \\
\hline
-8 & -12 & 0 & 0 & 1 & 0
\end{array}
$$

This solution certainly does not give a maximum value for the objective function $z = 8x_1 + 12x_2$. Since x_2 has the largest coefficient, z will be increased most if x_2 is increased. In other words, the most negative indicator in the tableau (which corresponds to the largest coefficient in the objective function) identifies the variable that will provide the greatest change in the value of z.

To determine how much x_2 can be increased without leaving the feasible region, look at the first two equations

$$40x_1 + 80x_2 + x_3 \qquad = 560$$
$$6x_1 + 8x_2 \qquad + x_4 = 72$$

and solve for the basic variables x_3 and x_4.

$$x_3 = 560 - 40x_1 - 80x_2$$
$$x_4 = 72 - 6x_1 - 8x_2$$

Now x_2 is to be increased while x_1 is to keep the value 0. Hence

$$x_3 = 560 - 80x_2$$
$$x_4 = 72 - 8x_2.$$

Since $x_3 \geq 0$ and $x_4 \geq 0$, we must have:

$$0 \leq 560 - 80x_2 \quad \text{and} \quad 0 \leq 72 - 8x_2$$
$$80x_2 \leq 560 \qquad\qquad 8x_2 \leq 72$$
$$x_2 \leq \frac{560}{80} = 7 \qquad\qquad x_2 \leq \frac{72}{8} = 9.$$

The right sides of these last inequalities are the quotients used to select the pivot row. Since x_2 must satisfy both inequalities, x_2 can be at most 7. In other words, the smallest quotient formed from positive entries in the pivot column identifies the value of x_2 that produces the largest change in z while remaining in the feasible region. By pivoting with the pivot determined in this way, we obtain the second tableau and a basic feasible solution in which $x_2 = 7$, as was shown in Example 6. ■

An analysis similar to that in Example 7 applies to each occurrence of pivoting in the simplex method. The idea is to improve the value of the objective function by adjusting one variable at a time. The most negative indicator identifies the variable that will account for the largest increase in z. The smallest quotient determines the largest value of that variable that will produce a feasible solution. Pivoting leads to a solution in which the selected variable has this largest value.

The simplex method is easily implemented on a computer and some graphing calculators. A computer is essential for the simplex method in any situation where there are a large number of variables and constraints (and hence an enormous number of corner points to check).

$\mathcal{T}$ECHNOLOGY $\mathcal{T}$IP Simplex method programs for several calculators are in the Program Appendix. ✔

8.4 EXERCISES

In Exercises 1–4, (a) determine the number of slack variables needed; (b) name them; (c) use the slack variables to convert each constraint into a linear equation. (See Example 1.)

1. Maximize $z = 32x_1 + 9x_2$
   subject to: $4x_1 + 2x_2 \leq 20$
   $5x_1 + x_2 \leq 50$
   $2x_1 + 3x_2 \leq 25$
   $x_1 \geq 0, x_2 \geq 0.$

2. Maximize $z = 3.7x_1 + 4.3x_2$
   subject to: $2.4x_1 + 1.5x_2 \leq 10$
   $1.7x_1 + 1.9x_2 \leq 15$
   $x_1 \geq 0, x_2 \geq 0.$

3. Maximize $z = 8x_1 + 3x_2 + x_3$
subject to: $3x_1 - x_2 + 4x_3 \le 95$
$7x_1 + 6x_2 + 8x_3 \le 118$
$4x_1 + 5x_2 + 10x_3 \le 220$
$x_1 \ge 0, x_2 \ge 0, x_3 \ge 0.$

4. Maximize $z = 12x_1 + 15x_2 + 10x_3$
subject to: $2x_1 + 2x_2 + x_3 \le 8$
$x_1 + 4x_2 + 3x_3 \le 12$
$x_1 \ge 0, x_2 \ge 0, x_3 \ge 0.$

Introduce slack variables as necessary and then write the initial simplex tableau for each of these linear programming problems.

5. Maximize $z = 5x_1 + x_2$
subject to: $2x_1 + 3x_2 \le 6$
$4x_1 + x_2 \le 6$
$5x_1 + 2x_2 \le 15$
$x_1 \ge 0, x_2 \ge 0.$

6. Maximize $z = 5x_1 + 3x_2 + 4x_3$
subject to: $4x_1 + 3x_2 + 2x_3 \le 60$
$3x_1 + 4x_2 + x_3 \le 24$
$x_1 \ge 0, x_2 \ge 0, x_3 \ge 0.$

7. Maximize $z = x_1 + 5x_2 + 10x_3$
subject to: $x_1 + 2x_2 + 3x_3 \le 10$
$2x_1 + x_2 + x_3 \le 8$
$3x_1 + 2x_3 \le 6$
$x_1 \ge 0, x_2 \ge 0, x_3 \ge 0.$

8. Maximize $z = 5x_1 - x_2 + 3x_3$
subject to: $3x_1 + 2x_2 + x_3 \le 36$
$x_1 + 4x_2 + x_3 \le 24$
$x_1 - x_2 - x_3 \le 32$
$x_1 \ge 0, x_2 \ge 0, x_3 \ge 0.$

Find the pivot in each of the following simplex tableaus. (See Example 2.)

9.
x_1	x_2	x_3	x_4	x_5	z	
2	2	0	3	1	0	15
3	4	1	6	0	0	20
-2	-1	0	1	0	1	10

10.
x_1	x_2	x_3	x_4	x_5	z	
0	2	1	1	3	0	5
1	-5	0	1	2	0	8
0	-2	0	-1	1	1	10

11.
x_1	x_2	x_3	x_4	x_5	x_6	z	
6	2	1	3	0	0	0	8
0	2	0	1	0	1	0	7
2	1	0	3	1	0	0	6
-3	-2	0	2	0	0	1	12

12.
x_1	x_2	x_3	x_4	x_5	x_6	z	
0	2	0	1	2	2	0	3
0	3	1	0	1	2	0	2
1	4	0	0	3	5	0	5
0	-4	0	0	4	-3	1	20

In Exercises 13–16, use the indicated entry as the pivot and perform the pivoting. (See Examples 3 and 4.)

13.
x_1	x_2	x_3	x_4	x_5	z	
1	2	4	1	0	0	56
2	**2**	1	0	1	0	40
-1	-3	-2	0	0	1	0

14.
x_1	x_2	x_3	x_4	x_5	x_6	z	
2	2	**1**	1	0	0	0	12
1	2	3	0	1	0	0	45
3	1	1	0	0	1	0	20
-2	-1	-3	0	0	0	1	0

15.
x_1	x_2	x_3	x_4	x_5	x_6	z	
1	1	1	1	0	0	0	60
3	1	**2**	0	1	0	0	100
1	2	3	0	0	1	0	200
-1	-1	-2	0	0	0	1	0

16.
x_1	x_2	x_3	x_4	x_5	x_6	z	
4	2	3	1	0	0	0	22
2	2	**5**	0	1	0	0	28
1	3	2	0	0	1	0	45
-3	-2	-4	0	0	0	1	0

For each simplex tableau in Exercises 17–20, (a) list the basic and the nonbasic variables; (b) find the basic feasible solution determined by setting the nonbasic variables equal to 0; (c) decide whether this is a maximum solution. (See Examples 5 and 6.)

17.
x_1	x_2	x_3	x_4	x_5	z	
3	2	0	-3	1	0	29
4	0	1	-2	0	0	16
-5	0	0	-1	0	1	11

18.
x_1	x_2	x_3	x_4	x_5	x_6	z	
-3	0	$\frac{1}{2}$	1	-2	0	0	22
2	0	-3	0	1	1	0	10
4	1	4	0	$\frac{3}{4}$	0	0	17
-1	0	0	0	1	0	1	120

19.

$$
\begin{array}{ccccccc}
x_1 & x_2 & x_3 & x_4 & x_5 & x_6 & z \\
\end{array}
$$

$$
\left[
\begin{array}{ccccccc|c}
1 & 0 & 2 & \frac{1}{2} & 0 & \frac{1}{3} & 0 & 6 \\
0 & 1 & -1 & 5 & 0 & -1 & 0 & 13 \\
0 & 0 & 1 & \frac{3}{2} & 1 & -\frac{1}{3} & 0 & 21 \\
\hline
0 & 0 & 2 & \frac{1}{2} & 0 & 3 & 1 & 18 \\
\end{array}
\right]
$$

20.

$$
\begin{array}{cccccccc}
x_1 & x_2 & x_3 & x_4 & x_5 & x_6 & x_7 & z \\
\end{array}
$$

$$
\left[
\begin{array}{cccccccc|c}
-1 & 0 & 0 & 1 & 0 & 3 & -2 & 0 & 47 \\
2 & 0 & 1 & 0 & 0 & 2 & -\frac{1}{2} & 0 & 37 \\
3 & 0 & 0 & 0 & 1 & -1 & 6 & 0 & 43 \\
\hline
4 & 1 & 0 & 0 & 0 & 6 & 0 & 1 & 86 \\
\end{array}
\right]
$$

Use the simplex method to solve Exercises 21–36.

21. Maximize $z = x_1 + 3x_2$
subject to: $x_1 + x_2 \le 10$
$5x_1 + 2x_2 \le 20$
$x_1 + 2x_2 \le 36$
$x_1 \ge 0, x_2 \ge 0.$

22. Maximize $z = 5x_1 + x_2$
subject to: $2x_1 + 3x_2 \le 8$
$4x_1 + 8x_2 \le 12$
$5x_1 + 2x_2 \le 30$
$x_1 \ge 0, x_2 \ge 0.$

23. Maximize $z = 2x_1 + x_2$
subject to: $x_1 + 3x_2 \le 12$
$2x_1 + x_2 \le 10$
$x_1 + x_2 \le 4$
$x_1 \ge 0, x_2 \ge 0.$

24. Maximize $z = 4x_1 + 2x_2$
subject to: $-x_1 - x_2 \le 12$
$3x_1 - x_2 \le 15$
$x_1 \ge 0, x_2 \ge 0.$

25. Maximize $z = 5x_1 + 4x_2 + x_3$
subject to: $-2x_1 + x_2 + 2x_3 \le 3$
$x_1 - x_2 + x_3 \le 1$
$x_1 \ge 0, x_2 \ge 0, x_3 \ge 0.$

26. Maximize $z = 3x_1 + 2x_2 + x_3$
subject to: $2x_1 + 2x_2 + x_3 \le 10$
$x_1 + 2x_2 + 3x_3 \le 15$
$x_1 \ge 0, x_2 \ge 0, x_3 \ge 0.$

27. Maximize $z = 2x_1 + x_2 + x_3$
subject to: $x_1 - 3x_2 + x_3 \le 3$
$x_1 - 2x_2 + 2x_3 \le 12$
$x_1 \ge 0, x_2 \ge 0, x_3 \ge 0.$

28. Maximize $z = 4x_1 + 5x_2 + x_3$
subject to: $x_1 + 2x_2 + 4x_3 \le 10$
$2x_1 + 2x_2 + x_3 \le 10$
$x_1 \ge 0, x_2 \ge 0, x_3 \ge 0.$

29. Maximize $z = 2x_1 + 2x_2 - 4x_3$
subject to: $3x_1 + 3x_2 - 6x_3 \le 51$
$5x_1 + 5x_2 + 10x_3 \le 99$
$x_1 \ge 0, x_2 \ge 0, x_3 \ge 0.$

30. Maximize $z = 4x_1 + x_2 + 3x_3$
subject to: $x_1 + 3x_3 \le 6$
$6x_1 + 3x_2 + 12x_3 \le 40$
$x_1 \ge 0, x_2 \ge 0, x_3 \ge 0.$

31. Maximize $z = 300x_1 + 200x_2 + 100x_3$
subject to: $x_1 + x_2 + x_3 \le 100$
$2x_1 + 3x_2 + 4x_3 \le 320$
$2x_1 + x_2 + x_3 \le 160$
$x_1 \ge 0, x_2 \ge 0, x_3 \ge 0.$

32. Maximize $z = x_1 + 5x_2 - 10x_3$
subject to: $8x_1 + 4x_2 + 12x_3 \le 18$
$x_1 + 6x_2 + 2x_3 \le 45$
$5x_1 + 7x_2 + 3x_3 \le 60$
$x_1 \ge 0, x_2 \ge 0, x_3 \ge 0.$

33. Maximize $z = 4x_1 - 3x_2 + 2x_3$
subject to: $2x_1 - x_2 + 8x_3 \le 40$
$4x_1 - 5x_2 + 6x_3 \le 60$
$2x_1 - 2x_2 + 6x_3 \le 24$
$x_1 \ge 0, x_2 \ge 0, x_3 \ge 0.$

34. Maximize $z = 3x_1 + 2x_2 - 4x_3$
subject to: $x_1 - x_2 + x_3 \le 10$
$2x_1 - x_2 + 2x_3 \le 30$
$-3x_1 + x_2 + 3x_3 \le 40$
$x_1 \ge 0, x_2 \ge 0, x_3 \ge 0.$

35. Maximize $z = x_1 + 2x_2 + x_3 + 5x_4$
subject to: $x_1 + 2x_2 + x_3 + x_4 \le 50$
$3x_1 + x_2 + 2x_3 + x_4 \le 100$
$x_1 \ge 0, x_2 \ge 0, x_3 \ge 0, x_4 \ge 0.$

36. Maximize $z = x_1 + x_2 + 4x_3 + 5x_4$
subject to: $x_1 + 2x_2 + 3x_3 + x_4 \le 115$
$2x_1 + x_2 + 8x_3 + 5x_4 \le 200$
$x_1 + x_3 \le 50$
$x_1 \ge 0, x_2 \ge 0, x_3 \ge 0, x_4 \ge 0.$

37. The initial simplex tableau of a linear programming problem is given below.

$$
\begin{array}{cccccc}
x_1 & x_2 & x_3 & x_4 & x_5 & z \\
\end{array}
$$

$$
\left[
\begin{array}{cccccc|c}
1 & 1 & 1 & 1 & 0 & 0 & 12 \\
2 & 1 & 2 & 0 & 1 & 0 & 30 \\
\hline
-2 & -2 & -1 & 0 & 0 & 1 & 0 \\
\end{array}
\right]
$$

(a) Use the simplex method to solve the problem, with column one as the first pivot column.

(b) Now use the simplex method to solve the problem, with column two as the first pivot column.

(c) Does this problem have a unique maximum solution? Why?

38. The final simplex tableau of a linear programming problem is given here.

$$\begin{array}{ccccc} x_1 & x_2 & x_3 & x_4 & z \end{array}$$
$$\left[\begin{array}{ccccc|c} 1 & 1 & 2 & 0 & 0 & 24 \\ 2 & 0 & 2 & 1 & 0 & 8 \\ \hline 4 & 0 & 0 & 0 & 1 & 40 \end{array}\right]$$

(a) What is the solution given by this tableau?

(b) Even though all the indicators are nonnegative, perform one more round of pivoting on this tableau, using col-umn three as the pivot column and choosing the pivot row by forming quotients in the usual way.

(c) Show that there is more than one solution to the linear programming problem by comparing your answer in part (a) to the basic feasible solution given by the tableau found in part (b). Does it give the same value of z as the solution in part (a)?

8.5 MAXIMIZATION APPLICATIONS

Applications of linear programming that use the simplex method are considered in this section. First, however, we make a slight change in notation. You have probably noticed that the column representing z in a simplex tableau never changes during pivoting. Furthermore, the value of z in the basic feasible solution associated with the tableau is the number in the lower right-hand corner. Consequently, the z column is unnecessary and will be omitted from all simplex tableaus hereafter.

EXAMPLE 1 A farmer has 100 acres of available land he wishes to plant with a mixture of potatoes, corn, and cabbage. It costs him $400 to produce an acre of pota-toes, $160 to produce an acre of corn, and $280 to produce an acre of cabbage. He has a maximum of $20,000 to spend. He makes a profit of $120 per acre of potatoes, $40 per acre of corn, and $60 per acre of cabbage. How many acres of each crop should he plant to maximize his profit?

Begin by summarizing the given information as follows.

Crop	Number of Acres	Cost per Acre	Profit per Acre
Potatoes	x_1	$400	$120
Corn	x_2	160	40
Cabbage	x_3	280	60
Maximum available	100	$20,000	

If the number of acres allotted to each of the three crops is represented by x_1, x_2, and x_3, respectively, then the constraints can be expressed as

$$x_1 + x_2 + x_3 \le 100 \qquad \text{Number of acres}$$
$$400x_1 + 160x_2 + 280x_3 \le 20{,}000 \qquad \text{Production costs}$$

where x_1, x_2, and x_3 are all nonnegative. The first of these constraints says that $x_1 + x_2 + x_3$ is less than or perhaps equal to 100. Use x_4 as the slack variable, giving the equation

$$x_1 + x_2 + x_3 + x_4 = 100.$$

Here x_4 represents the amount of the farmer's 100 acres that will not be used. (x_4 may be 0 or any value up to 100.)

In the same way, the constraint $400x_1 + 160x_2 + 280x_3 \leq 20,000$ can be converted into an equation by adding a slack variable, x_5:

$$400x_1 + 160x_2 + 280x_3 + x_5 = 20,000.$$

The slack variable x_5 represents any unused portion of the farmer's $20,000 capital. (Again, x_5 may have any value from 0 to 20,000.)

The farmer's profit on potatoes is the product of the profit per acre ($120) and the number x_1 of acres, that is, $120x_1$. His profit on corn and cabbage is computed similarly. Hence his total profit is given by

$$z = \text{profit on potatoes} + \text{profit on corn} + \text{profit on cabbage}$$
$$z = 120x_1 + 40x_2 + 60x_3.$$

The linear programming problem can now be stated as follows:

$$\text{Maximize} \quad z = 120x_1 + 40x_2 + 60x_3$$
$$\text{subject to:} \quad x_1 + x_2 + x_3 + x_4 \qquad\quad = 100$$
$$400x_1 + 160x_2 + 280x_3 + \qquad x_5 = 20,000$$
$$\text{with} \quad x_1 \geq 0, x_2 \geq 0, x_3 \geq 0, x_4 \geq 0, x_5 \geq 0.$$

The initial simplex tableau (without the z column) is

$$
\begin{array}{ccccc}
x_1 & x_2 & x_3 & x_4 & x_5 \\
\end{array}
$$
$$
\left[
\begin{array}{ccccc|c}
1 & 1 & 1 & 1 & 0 & 100 \\
400 & 160 & 280 & 0 & 1 & 20,000 \\
\hline
-120 & -40 & -60 & 0 & 0 & 0 \\
\end{array}
\right].
$$

The most negative indicator is -120; column one is the pivot column. The quotients needed to determine the pivot row are $100/1 = 100$ and $20,000/400 = 50$. So the pivot is 400 in row two, column one. Multiplying row two by 1/400 and completing the pivoting leads to the final simplex tableau.

$$
\begin{array}{ccccc}
x_1 & x_2 & x_3 & x_4 & x_5 \\
\end{array}
$$
$$
\left[
\begin{array}{ccccc|c}
0 & .6 & .3 & 1 & -.0025 & 50 \\
1 & .4 & .7 & 0 & .0025 & 50 \\
\hline
0 & 8 & 24 & 0 & .3 & 6000 \\
\end{array}
\right]
\quad
\begin{array}{l}
-1R_2 + R_1 \\[4pt]
\dfrac{1}{400}R_2 \\[4pt]
120R_2 + R_3 \\
\end{array}
$$

Setting the nonbasic variables x_2, x_3, and x_5 equal to 0, solving for the basic variables x_1 and x_4, and remembering that the value of z is in the lower right-hand corner leads to this maximum solution:

$$x_1 = 50, \quad x_2 = 0, \quad x_3 = 0, \quad x_4 = 50, \quad x_5 = 0, \quad z = 6000.$$

Therefore, the farmer will make a maximum profit of $6000 by planting 50 acres of potatoes, and no corn or cabbage. Thus 50 acres are left unplanted (represented by x_4, the slack variable for potatoes). The farmer has spent his $20,000 most effectively in this way and has no more money to plant the remaining 50 acres. If he had more cash, he would plant more crops. ∎

EXAMPLE 2 Set up the initial simplex tableau for the following problem.

Ana Pott, who is a candidate for the state legislature, has $96,000 to buy TV advertising time. Ads cost $400 per minute on a local cable channel, $4000 per minute on a regional independent channel, and $12,000 per minute on a national network channel. Because of existing contracts the TV stations can provide at most 30 minutes of advertising time, with a maximum of 6 minutes on the national network channel. At any given time during the evening, approximately 100,000 people watch the cable channel, 200,000 the independent channel, and 600,000 the network channel. To get maximum exposure, how much time should Ana buy from each station?

Let x_1 be the number of minutes of ads on the cable channel, x_2 the number of minutes on the independent channel, and x_3 the number of minutes on the network channel. Exposure is measured in viewer-minutes. For instance, 100,000 people watching x_1 minutes of ads on the cable channel produces $100,000x_1$ viewer-minutes. The amount of exposure is given by the total number of viewer-minutes for all three channels, namely,

$$100,000x_1 + 200,000x_2 + 600,000x_3.$$

Since 30 minutes are available,

$$x_1 + x_2 + x_3 \leq 30.$$

The fact that only 6 minutes can be used on the network channel means that

$$x_3 \leq 6.$$

Expenditures are limited to $96,000, so

Cable cost + independent cost + network cost $\leq 96,000$

$$400x_1 + 4000x_2 + 12,000x_3 \leq 96,000.$$

Therefore, Ana must solve the following linear programming problem:

Maximize $\quad z = 100,000x_1 + 200,000x_2 + 600,000x_3$

subject to: $\qquad\qquad\qquad x_1 + x_2 + x_3 \leq 30$

$$x_3 \leq 6$$

$$400x_1 + 4000x_2 + 12,000x_3 \leq 96,000$$

with $\quad x_1 \geq 0, x_2 \geq 0, x_3 \geq 0.$

Introducing slack variables x_4, x_5, and x_6 (one for each constraint), rewriting the constraints as equations, and expressing the objective function as

$$-100,000x_1 - 200,000x_2 - 600,000x_3 + z = 0,$$

leads to the initial simplex tableau:

x_1	x_2	x_3	x_4	x_5	x_6	
1	1	1	1	0	0	30
0	0	1	0	1	0	6
400	4000	12,000	0	0	1	96,000
−100,000	−200,000	−600,000	0	0	0	0

1 What is the optimal solution in Example 2?

Answer:
Buying 20 minutes of time on the cable channel, 4 minutes on the independent channel, and 6 minutes on the network channel produces the maximum of 6,400,000 viewer-minutes.

Working by hand or using a computer or graphing calculator simplex method program leads to the final simplex tableau:

$$
\begin{array}{ccccccc}
x_1 & x_2 & x_3 & x_4 & x_5 & x_6 \\
\left[\begin{array}{cccccc|c}
1 & 0 & 0 & \frac{10}{9} & \frac{20}{9} & \frac{-25}{90,000} & 20 \\
0 & 0 & 1 & 0 & 1 & 0 & 6 \\
0 & 1 & 0 & -\frac{1}{9} & -\frac{29}{9} & \frac{25}{90,000} & 4 \\
\hline
0 & 0 & 0 & \frac{800,000}{9} & \frac{1,600,000}{9} & \frac{250}{9} & 6,400,000
\end{array}\right]
\end{array}
$$

TECHNOLOGY TIP When using a graphing calculator, you can minimize scrolling and make the final simplex tableau in Example 2 easier to read by using the FRAC key (on most TI calculators) or by setting the number format mode to "fraction" (on HP-38). Doing so will convert most (but not necessarily all) the entries in the tableau into fractional form. ✔

8.5 EXERCISES

Set up the initial simplex tableau for each of the following problems.

1. **Management** A cat breeder has the following amounts of cat food: 90 units of tuna, 80 units of liver, and 50 units of chicken. To raise a Siamese cat, the breeder must use 2 units of tuna, 1 of liver, and 1 of chicken per day, while raising a Persian cat requires 1, 2, and 1 units, respectively, per day. If a Siamese cat sells for $12 while a Persian cat sells for $10, how many of each should be raised in order to obtain maximum gross income? What is the maximum gross income?

2. **Management** Banal, Inc., produces art for motel rooms. Its painters can turn out mountain scenes, seascapes, and pictures of clowns. Each painting is worked on by three different artists, T, D, and H. Artist T works only 25 hours per week, while D and H work 45 and 40 hours per week, respectively. Artist T spends 1 hour on a mountain scene, 2 hours on a seascape, and 1 hour on a clown. Corresponding times for D and H are 3, 2, and 2 hours, and 2, 1, and 4 hours respectively. Banal makes $20 on a mountain scene, $18 on a seascape, and $22 on a clown. The head painting packer can't stand clowns, so that no more than 4 clown paintings may be done in a week. Find the number of each type of painting that should be made weekly in order to maximize profit. Find the maximum possible profit.

3. **Management** A manufacturer makes two products, toy trucks and toy fire engines. Both are processed in four different departments, each of which has a limited capacity. The sheet metal department can handle at least $1\frac{1}{2}$ times as many trucks as fire engines. The truck assembly department can handle at most 6700 trucks per week, while the fire engine assembly department assembles at most 5500 fire engines weekly. The painting department, which finishes both toys, has a maximum capacity of 12,000 per week. If the profit is $8.50 for a toy truck and $12.10 for a toy fire engine, how many of each item should the company produce to maximize profit?

4. **Natural Science** A lake is stocked each spring with three species of fish, A, B, and C. The average weights of the fish are 1.62, 2.12, and 3.01 kilograms for species A, B, and C, respectively. Three foods, I, II, and III, are available in the lake. Each fish of species A requires 1.32 units of food I, 2.9 units of food II, and 1.75 units of food III on the average each day. Species B fish each require 2.1 units of food I, .95 units of food II, and .6 units of food III daily. Species C fish require .86, 1.52, and 2.01 units of I, II, and III per day, respectively. If 490 units of food I, 897 units of food II, and 653 units of food III are available daily, how should the lake be stocked to maximize the weight of the fish supported by the lake?

Use the simplex method to solve the following problems.

5. **Management** A manufacturer of bicycles builds one-, three-, and ten-speed models. The bicycles need both aluminum and steel. The company has available 91,800 units of steel and 42,000 units of aluminum. The one-, three-, and ten-speed models need, respectively, 20, 30, and 40 units of steel and 12, 21, and 16 units of aluminum. How many of each type of bicycle should be made in order to maximize profit if the company makes $8 per one-speed bike, $12 per three-speed, and $24 per ten-speed? What is the maximum possible profit?

6. **Social Science** Jayanta is working to raise money for the homeless by sending information letters and making

follow-up calls to local labor organizations and church groups. She discovered that each church group requires 2 hours of letter writing and 1 hour of follow-up, while for each labor union she needs 2 hours of letter writing and 3 hours of follow-up. Jayanta can raise $100 from each church group and $200 from each union local, and she has a maximum of 16 hours of letter-writing time and a maximum of 12 hours of follow-up time available per month. Determine the most profitable mixture of groups she should contact and the most money she can raise in a month.

7. **Management** A baker has 60 units of flour, 132 units of sugar, and 102 units of raisins. A loaf of raisin bread requires 1 unit of flour, 1 unit of sugar, and 2 units of raisins, while a raisin cake needs 2, 4, and 1 units respectively. If raisin bread sells for $3 a loaf and a raisin cake for $4, how many of each should be baked so that the gross income is maximized? What is the maximum gross income?

8. **Management** Mellow Sounds Inc. produces three types of compact discs: easy listening, jazz, and rock. Each easy listening disc requires 6 hours of recording, 12 hours of mixing, and 2 hours of editing. Each jazz disc requires 8 hours of recording, 8 hours of mixing, and 4 hours of editing. Each rock disc requires 3 hours of recording, 6 hours of mixing, and 1 hour of editing. Each week 288 hours of recording studio time are available, 312 hours of mixing board time are available, and editors are available for at most 124 hours. Mellow Sounds receives $6 for each easy listening and rock disc and $8 for each jazz disc. How many of each type of disc should the company produce each week to maximize their income? What is the maximum income?

9. **Management** The Cut-Right Company sells sets of kitchen knives. The Basic Set consists of 2 utility knives and 1 chef's knife. The Regular Set consists of 2 utility knives, 1 chef's knife, and 1 slicer. The Deluxe Set consists of 3 utility knives, 1 chef's knife, and 1 slicer. Their profit is $30 on a Basic Set, $40 on a Regular Set, and $60 on a Deluxe Set. The factory has on hand 800 utility knives, 400 chef's knives, and 200 slicers. Assuming that all sets will be sold, how many of each type should be made up in order to maximize profit? What is the maximum profit?

10. **Management** Super Souvenir Company makes paper weights, plaques, and ornaments. Each paperweight requires 8 units of plastic, 3 units of metal, and 2 units of paint. Each plaque requires 4 units of plastic, and 1 unit each of metal and paint. Each ornament requires 2 units each of plastic and metal, and 1 unit of paint. They make a profit of $3 on each paperweight and each ornament, and $4 on each plaque. If 36 units of plastic, 24 units of metal, and 30 units of paint are available today, how many of each kind of souvenir should be made in order to maximize profit?

11. **Management** The Fancy Fashions Store has $8000 available each month for advertising. Newspaper ads cost $400 each and no more than 20 can be run per month. Radio ads

cost $200 each and no more than 30 can run per month. TV ads cost $1200 each, with a maximum of 6 available each month. Approximately 2000 women will see each newspaper ad, 1200 will hear each radio commercial, and 10,000 will see each TV ad. How much of each type of advertising should be used if the store wants to maximize its ad exposure?

12. **Management** Caroline's Quality Candy Confectionery is famous for fudge, chocolate cremes, and pralines. Its candy-making equipment is set up to make 100-pound batches at a time. Currently there is a chocolate shortage and the company can get only 120 pounds of chocolate in the next shipment. On a week's run, the confectionery's cooking and processing equipment is available for a total of 42 machine hours. During the same period the employees have a total of 56 work hours available for packaging. A batch of fudge requires 20 pounds of chocolate while a batch of cremes uses 25 pounds of chocolate. The cooking and processing take 120 minutes for fudge, 150 minutes for chocolate cremes, and 200 minutes for pralines. The packaging times measured in minutes per 1-pound box are 1, 2, 3, respectively, for fudge, cremes, and pralines. Determine how many batches of each type of candy the confectionery should make, assuming that the profit per pound box is 50¢ on fudge, 40¢ on chocolate cremes, and 45¢ on pralines. Also, find the maximum profit for the week.

Management *The next two problems come from past CPA examinations.* Select the appropriate answer for each question.*

13. The Ball Company manufactures three types of lamps, labeled A, B, and C. Each lamp is processed in two departments, I and II. Total available man-hours per day for departments I and II are 400 and 600, respectively. No additional labor is available. Time requirements and profit per unit for each lamp type are as follows.

	A	B	C
Man-hours in I	2	3	1
Man-hours in II	4	2	3
Profit per unit	$5	$4	$3

The company has assigned you as the accounting member of its profit planning committee to determine the numbers of types of A, B, and C lamps that it should produce in order to maximize its total profit from the sale of lamps. The following questions relate to a linear programming model that your group has developed.

---

*Material from *Uniform CPA Examination Questions and Unofficial Answers,* copyright © 1973, 1974, 1975 by the American Institute of Certified Public Accountants, Inc., is reprinted with permission.

(a) The coefficients of the objective function would be
  (1) 4, 2, 3;
  (2) 2, 3, 1;
  (3) 5, 4, 3;
  (4) 400,600.
(b) The constraints in the model would be
  (1) 2, 3, 1;
  (2) 5, 4, 3;
  (3) 4, 2, 3;
  (4) 400,600.
(c) The constraint imposed by the available man-hours in department I could be expressed as
  (1) $4X_1 + 2X_2 + 3X_3 \leq 400$;
  (2) $4X_1 + 2X_2 + 3X_3 \geq 400$;
  (3) $2X_1 + 3X_2 + 1X_3 \leq 400$;
  (4) $2X_1 + 3X_2 + 1X_3 \geq 400$.

14. The Golden Hawk Manufacturing Company wants to maximize the profits on products A, B, and C. The contribution margin for each product follows.

Product	Contribution Margin
A	$2
B	5
C	4

The production requirements and departmental capacities, by departments, are as follows.

	PRODUCTION REQUIREMENTS BY PRODUCT (HOURS)			DEPARTMENTAL CAPACITY (TOTAL HOURS)
Department	A	B	C	
Assembling	2	3	2	30,000
Painting	1	2	2	38,000
Finishing	2	3	1	28,000

(a) What is the profit-maximization formula for the Golden Hawk Company?
  (1) $2A + $5B + $4C = X$ (where X = profit)
  (2) $5A + 8B + 5C \leq 96,000$
  (3) $2A + $5B + $4C \leq X$
  (4) $2A + $5B + $4C = 96,000$
(b) What is the constraint for the Painting Department of the Golden Hawk Company?
  (1) $1A + 2B + 2C \geq 38,000$
  (2) $2A + $5B + $4C \geq 38,000$
  (3) $1A + 2B + 2C \leq 38,000$
  (4) $2A + 3B + 2C \leq 30,000$

15. Solve the problem in Exercise 1.

Use a graphing calculator or a computer program for the simplex method to solve the following linear programming problems.

16. Exercise 2. Your final answer should consist of whole numbers (Banal can't sell half a painting).

17. Exercise 3

18. Exercise 4

8.6 THE SIMPLEX METHOD: DUALITY AND MINIMIZATION

In this section the simplex method is extended to linear programming problems satisfying the following conditions.

1. The objective function is to be *minimized.*
2. All the coefficients of the objective function are nonnegative.
3. All constraints involve $\geq$.
4. All variables are nonnegative.

The method of solving minimization problems presented here is based on an interesting connection between maximizing and minimizing problems: any solution of a maximizing problem produces the solution of an associated minimizing problem, or vice-versa. Each of the associated problems is called the **dual** of the other. Thus, duals enable us to solve minimization problems of the type described above by the simplex method introduced in Section 8.4. (An alternative approach for solving minimization problems is given in the next section.)

When dealing with minimization problems, we use y_1, y_2, y_3, etc. as variables and denote the objective function by w. An example will explain the idea of a dual.

EXAMPLE 1 Minimize $w = 8y_1 + 16y_2$

subject to: $y_1 + 5y_2 \geq 9$

$2y_1 + 2y_2 \geq 10$

$y_1 \geq 0, y_2 \geq 0.$

Before considering slack variables, write the augmented matrix of the system of inequalities, and include the coefficients of the objective function (not their negatives) as the last row in the matrix.

$$
\begin{array}{c}
\overset{\displaystyle\downarrow}{}\text{Constants} \\
\begin{bmatrix}
1 & 5 & 9 \\
2 & 2 & 10 \\
\hline
8 & 16 & 0
\end{bmatrix}
\end{array}
$$

Objective function $\longrightarrow$

Look now at the following new matrix, obtained from the one above by interchanging rows and columns.

$$
\begin{array}{c}
\overset{\displaystyle\downarrow}{}\text{Constants} \\
\begin{bmatrix}
1 & 2 & 8 \\
5 & 2 & 16 \\
\hline
9 & 10 & 0
\end{bmatrix}
\end{array}
$$

Objective function $\longrightarrow$

1 Use the corner points in Figure 8.22(a) to find the minimum value of $w = 8y_1 + 16y_2$ and where it occurs.

Answer:
48 when $y_1 = 4, y_2 = 1$

2 Use Figure 8.22(b) to find the maximum value of $z = 9x_1 + 10x_2$ and where it occurs.

Answer:
48 when $x_1 = 2, x_2 = 3$

The *rows* of the first matrix (for the minimizing problem) are the *columns* of the second matrix.

The entries in this second matrix could be used to write the following maximizing problem in standard form (again ignoring the fact that the numbers in the last row are not negative).

Maximize $z = 9x_1 + 10x_2$

subject to: $x_1 + 2x_2 \leq 8$

$5x_1 + 2x_2 \leq 16$

$x_1 \geq 0, x_2 \geq 0.$

Figure 8.22(a) shows the region of feasible solutions for the minimization problem given above, while Figure 8.22(b) shows the region of feasible solutions for the maximization problem produced by exchanging rows and columns. ■ **1** **2**

(a)

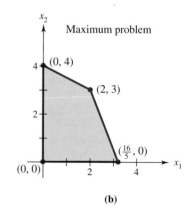

(b)

FIGURE 8.22

The two feasible regions in Figure 8.22 are different and the corner points are different, but the values of the objective functions found in Problems 1 and 2 at the side are equal—both are 48. An even closer connection between the two problems is shown by using the simplex method to solve the maximization problem given above.

Maximization problem

$$\begin{array}{cccc} x_1 & x_2 & x_3 & x_4 \end{array}$$
$$\left[\begin{array}{cccc|c} 1 & 2 & 1 & 0 & 8 \\ 5 & 2 & 0 & 1 & 16 \\ \hline -9 & -10 & 0 & 0 & 0 \end{array}\right]$$

$$\begin{array}{cccc} x_1 & x_2 & x_3 & x_4 \end{array}$$
$$\left[\begin{array}{cccc|c} \frac{1}{2} & 1 & \frac{1}{2} & 0 & 4 \\ 4 & 0 & -1 & 1 & 8 \\ \hline -4 & 0 & 5 & 0 & 40 \end{array}\right] \quad \begin{array}{l} \frac{1}{2}R_1 \\ -2R_1 + R_2 \\ 10R_1 + R_3 \end{array}$$

$$\begin{array}{cccc} x_1 & x_2 & x_3 & x_4 \end{array}$$
$$\left[\begin{array}{cccc|c} 0 & 1 & \frac{5}{8} & -\frac{1}{8} & 3 \\ 1 & 0 & -\frac{1}{4} & \frac{1}{4} & 2 \\ \hline 0 & 0 & 4 & 1 & 48 \end{array}\right] \quad \begin{array}{l} -\frac{1}{2}R_2 + R_1 \\ \frac{1}{4}R_2 \\ 4R_2 + R_3 \end{array}$$

The maximum is 48 when $x_1 = 2$, $x_2 = 3$.

Notice that the solution to the *minimization problem* (namely, $y_1 = 4$, $y_2 = 1$) is found in the bottom row and slack variable columns of the final simplex tableau for the maximization problem. This result suggests that a minimization problem can be solved by forming the dual maximization problem, solving it by the simplex method, and then reading the solution for the minimization problem from the bottom row of the final simplex tableau.

Before using this method to solve a minimization problem, let us find the duals of some typical linear programming problems. The process of exchanging the rows and columns of a matrix, which is used to find the dual, is called **transposing** the matrix, and each of the two matrices is the **transpose** of the other.

EXAMPLE 2 Find the transpose of each matrix.

(a) $A = \begin{bmatrix} 2 & -1 & 5 \\ 6 & 8 & 0 \\ -3 & 7 & -1 \end{bmatrix}$

Write the rows of matrix A as the columns of the transpose.

$$\text{Transpose of } A = \begin{bmatrix} 2 & 6 & -3 \\ -1 & 8 & 7 \\ 5 & 0 & -1 \end{bmatrix}$$

(b) The transpose of $\begin{bmatrix} 1 & 2 & 4 & 0 \\ 2 & 1 & 7 & 6 \end{bmatrix}$ is $\begin{bmatrix} 1 & 2 \\ 2 & 1 \\ 4 & 7 \\ 0 & 6 \end{bmatrix}$. ■ ③

③ Give the transpose of each matrix.

(a) $\begin{bmatrix} 2 & 4 \\ 6 & 3 \\ 1 & 5 \end{bmatrix}$

(b) $\begin{bmatrix} 4 & 7 & 10 \\ 3 & 2 & 6 \\ 5 & 8 & 12 \end{bmatrix}$

Answers:

(a) $\begin{bmatrix} 2 & 6 & 1 \\ 4 & 3 & 5 \end{bmatrix}$

(b) $\begin{bmatrix} 4 & 3 & 5 \\ 7 & 2 & 8 \\ 10 & 6 & 12 \end{bmatrix}$

TECHNOLOGY TIP Most graphing calculators can find the transpose of a matrix. Look for this feature in the MATRIX MATH menu (TI) or the OPTN MAT menu (Casio 9850) or the MATH MATRIX menu (HP-38). The transpose of matrix *A* from Example 2(a) is shown in Figure 8.23. ✔

```
[ [2   -1   5 ]
  [6    8   0 ]
  [-3   7  -1]]
Ans T
  [ [2    6  -3]
    [-1   8   7 ]
    [5    0  -1]]
```

FIGURE 8.23

EXAMPLE 3 Write the duals of the following minimization linear programming problems.

(a) Minimize $w = 10y_1 + 8y_2$

subject to: $y_1 + 2y_2 \geq 2$

$y_1 + y_2 \geq 5$

$y_1 \geq 0, y_2 \geq 0.$

Begin by writing the augmented matrix for the given problem.

$$\left[\begin{array}{cc|c} 1 & 2 & 2 \\ 1 & 1 & 5 \\ \hline 10 & 8 & 0 \end{array} \right]$$

Form the transpose of this matrix to get

$$\left[\begin{array}{cc|c} 1 & 1 & 10 \\ 2 & 1 & 8 \\ \hline 2 & 5 & 0 \end{array} \right].$$

The dual problem is stated from this second matrix as follows (using *x* instead of *y*).

Maximize $z = 2x_1 + 5x_2$

subject to: $x_1 + x_2 \leq 10$

$2x_1 + x_2 \leq 8$

$x_1 \geq 0, x_2 \geq 0.$

(b) Minimize $w = 7y_1 + 5y_2 + 8y_3$

subject to: $3y_1 + 2y_2 + y_3 \geq 10$

$y_1 + y_2 + y_3 \geq 8$

$4y_1 + 5y_2 \geq 25$

$y_1 \geq 0, y_2 \geq 0, y_3 \geq 0.$

4 Write the dual of the following linear programming problem.
Minimize $w = 2y_1 + 5y_2 + 6y_3$
subject to:

$$2y_1 + 3y_2 + y_3 \geq 15$$
$$y_1 + y_2 + 2y_3 \geq 12$$
$$5y_1 + 3y_2 \geq 10$$
$$y_1 \geq 0, y_2 \geq 0, y_3 \geq 0.$$

Answer:
Maximize $z = 15x_1 + 12x_2 + 10x_3$
subject to:

$$2x_1 + x_2 + 5x_3 \leq 2$$
$$3x_1 + x_2 + 3x_3 \leq 5$$
$$x_1 + 2x_2 \leq 6$$
$$x_1 \geq 0, x_2 \geq 0, x_3 \geq 0.$$

The dual problem is stated as follows.

$$\text{Maximize} \quad z = 10x_1 + 8x_2 + 25x_3$$
$$\text{subject to:} \quad 3x_1 + x_2 + 4x_3 \leq 7$$
$$2x_1 + x_2 + 5x_3 \leq 5$$
$$x_1 + x_2 \leq 8$$
$$x_1 \geq 0, x_2 \geq 0, x_3 \geq 0. \quad \blacksquare \quad \boxed{4}$$

In Example 3, all the constraints of the minimization problems were $\geq$ inequalities, while all those in the dual maximization problems were $\leq$ inequalities. This is generally the case; inequalities are reversed when the dual problem is stated.

The following table shows the close connection between a problem and its dual.

Given Problem	Dual Problem
m variables	n variables
n constraints	m constraints
Coefficients from objective function	Constants
Constants	Coefficients from objective function

The next theorem, whose proof requires advanced methods, guarantees that a minimization problem can be solved by forming a dual maximization problem.

THEOREM OF DUALITY

The objective function w of a minimizing linear programming problem takes on a minimum value if and only if the objective function z of the corresponding dual maximizing problem takes on a maximum value. The maximum value of z equals the minimum value of w.

This method is illustrated in the following example.

EXAMPLE 4 Minimize $w = 3y_1 + 2y_2$
subject to: $y_1 + 3y_2 \geq 6$
$2y_1 + y_2 \geq 3$
$y_1 \geq 0, y_2 \geq 0.$

Use the given information to write the matrix.

$$\begin{bmatrix} 1 & 3 & | & 6 \\ 2 & 1 & | & 3 \\ \hline 3 & 2 & | & 0 \end{bmatrix}$$

Transpose to get the following matrix for the dual problem.

$$\begin{bmatrix} 1 & 2 & | & 3 \\ 3 & 1 & | & 2 \\ \hline 6 & 3 & | & 0 \end{bmatrix}$$

Write the dual problem from this matrix, as follows.

$$\text{Maximize} \quad z = 6x_1 + 3x_2$$
$$\text{subject to:} \quad x_1 + 2x_2 \le 3$$
$$3x_1 + x_2 \le 2$$
$$x_1 \ge 0, x_2 \ge 0.$$

Solve this standard maximization problem using the simplex method. Start by introducing slack variables to give the system

$$x_1 + 2x_2 + x_3 \qquad = 3$$
$$3x_1 + x_2 \qquad + x_4 = 2$$
$$-6x_1 - 3x_2 - 0x_3 - 0x_4 + z = 0$$

with $x_1 \ge 0, x_2 \ge 0, x_3 \ge 0, x_4 \ge 0.$

The initial tableau for this system is given below with the pivot as indicated.

x_1	x_2	x_3	x_4		Quotients
1	2	1	0	3	3/1 = 3
3	1	0	1	2	2/3
-6	-3	0	0	0	

The simplex method gives the following final tableau.

x_1	x_2	x_3	x_4	
0	1	$\frac{3}{5}$	$-\frac{1}{5}$	$\frac{7}{5}$
1	0	$-\frac{1}{5}$	$\frac{2}{5}$	$\frac{1}{5}$
0	0	$\frac{3}{5}$	$\frac{9}{5}$	$\frac{27}{5}$

The last row of this final tableau shows that the solution of the given minimization problem is as follows:

The minimum value of $w = 3y_1 + 2y_2$, subject to the given constraints, is $27/5$ and occurs when $y_1 = 3/5$ and $y_2 = 9/5$.

The minimum value of w, $27/5$, is the same as the maximum value of z. ■ 5

A minimizing problem that meets the conditions listed at the beginning of the section can be solved by the method of duals, as illustrated in Examples 1 and 4 and summarized here.

SOLVING MINIMUM PROBLEMS WITH DUALS

1. Find the dual standard maximum problem.*
2. Solve the maximum problem using the simplex method.
3. The minimum value of the objective function w is the maximum value of the objective function z.
4. The optimum solution is given by the entries in the bottom row of the columns corresponding to the slack variables.

---

*The coefficients of the objective function in the minimization problem are the constants on the right side of the constraints in the dual maximization problem. So when all these coefficients are nonnegative (condition 2), the dual problem is in standard maximum form.

5 Minimize $w = 10y_1 + 8y_2$
subject to:

$$y_1 + 2y_2 \ge 2$$
$$y_1 + y_2 \ge 5$$
$$y_1 \ge 0, y_2 \ge 0.$$

Answer:
$y_1 = 0, y_2 = 5$, for a minimum of 40

FURTHER USES OF THE DUAL   The dual is useful not only in solving minimization problems, but also in seeing how small changes in one variable will affect the value of the objective function. For example, suppose an animal breeder needs at least 6 units per day of nutrient A and at least 3 units of nutrient B and that the breeder can choose between two different feeds, feed 1 and feed 2. Find the minimum cost for the breeder if each bag of feed 1 costs $3 and provides 1 unit of nutrient A and 2 units of B, while each bag of feed 2 costs $2 and provides 3 units of nutrient A and 1 of B.

   If y_1 represents the number of bags of feed 1 and y_2 represents the number of bags of feed 2, the given information leads to

$$\text{Minimize} \quad w = 3y_1 + 2y_2$$
$$\text{subject to:} \quad y_1 + 3y_2 \geq 6$$
$$2y_1 + y_2 \geq 3$$
$$y_1 \geq 0, y_2 \geq 0.$$

This minimization linear programming problem is the one we solved in Example 4 of this section. In that example, we formed the dual and reached the following final tableau.

$$
\begin{array}{cccc}
x_1 & x_2 & x_3 & x_4 \\
\left[\begin{array}{cccc|c}
0 & 1 & \frac{3}{5} & -\frac{1}{5} & \frac{7}{5} \\
1 & 0 & -\frac{1}{5} & \frac{2}{5} & \frac{1}{5} \\
\hline
0 & 0 & \frac{3}{5} & \frac{9}{5} & \frac{27}{5}
\end{array}\right]
\end{array}
$$

   This final tableau shows that the breeder will obtain minimum feed costs by using 3/5 bag of feed 1 and 9/5 bags of feed 2 per day, for a daily cost of 27/5 = 5.40 dollars.

   Now look at the data from the feed problem shown in the table below.

	UNITS OF NUTRIENT (PER BAG)		Cost per Bag
	A	B	
Feed 1	1	2	$3
Feed 2	3	1	$2
Minimum nutrient needed	6	3	

If x_1 and x_2 are the cost per unit of nutrients A and B, the constraints of the dual problem can be stated as follows.

$$\text{Cost of feed 1:} \quad x_1 + 2x_2 \leq 3$$
$$\text{Cost of feed 2:} \quad 3x_1 + x_2 \leq 2$$

The solution of the dual problem, which maximizes nutrients, also can be read from the final tableau:

$$x_1 = \frac{1}{5} = .20 \quad \text{and} \quad x_2 = \frac{7}{5} = 1.40,$$

6 The final tableau of the dual of the problem about filing cabinets in Example 2, Section 8.3 and Example 6, Section 8.4 is given below.

$$
\begin{array}{cccc}
x_1 & x_2 & x_3 & x_4 \\
\end{array}
$$

$$
\left[
\begin{array}{cccc|c}
0 & 1 & \frac{1}{2} & -\frac{1}{4} & 1 \\
1 & 0 & -\frac{1}{20} & -\frac{1}{80} & \frac{1}{20} \\
\hline
0 & 0 & 8 & 3 & 100 \\
\end{array}
\right]
$$

(a) What are the imputed amounts of storage for each unit of cost and floor space?

(b) What are the shadow values of the cost and the floor space?

Answers:
(a) Cost: 28 sq ft
floor space: 72 sq ft

(b) $\dfrac{1}{20}$, 1

which means that a unit of nutrient A costs 1/5 of a dollar = $.20, while a unit of nutrient B costs 7/5 dollars = $1.40. The minimum daily cost, $5.40, is found by the following procedure.

$$
\begin{aligned}
(\$.20 \text{ per unit of A}) \times (6 \text{ units of A}) &= \$1.20 \\
+ (\$1.40 \text{ per unit of B}) \times (3 \text{ units of B}) &= \$4.20 \\
\hline
\text{Minimum daily cost} &= \$5.40
\end{aligned}
$$

The numbers .20 and 1.40 are called the **shadow costs** of the nutrients. These two numbers from the dual, $.20 and $1.40, also allow the breeder to estimate feed costs for "small" changes in nutrient requirements. For example, an increase of 1 unit in the requirement for each nutrient would produce total cost as follows.

$5.40	6 units of A, 3 of B
.20	1 extra unit of A
1.40	1 extra unit of B
$7.00	Total cost per day **6**

8.6 EXERCISES

Find the transpose of each matrix. (See Example 2.)

1.
$$
\begin{bmatrix}
3 & -4 & 5 \\
1 & 10 & 7 \\
0 & 3 & 6 \\
\end{bmatrix}
$$

2.
$$
\begin{bmatrix}
3 & -5 & 9 & 4 \\
1 & 6 & -7 & 0 \\
4 & 18 & 11 & 9 \\
\end{bmatrix}
$$

3.
$$
\begin{bmatrix}
3 & 0 & 14 & -5 & 3 \\
4 & 17 & 8 & -6 & 1 \\
\end{bmatrix}
$$

4.
$$
\begin{bmatrix}
15 & -6 & -2 \\
13 & -1 & 11 \\
10 & 12 & -3 \\
24 & 1 & 0 \\
\end{bmatrix}
$$

State the dual problem for each of the following, but do not solve it. (See Example 3.)

5. Minimize $w = 3y_1 + 5y_2$
subject to: $3y_1 + y_2 \geq 4$
$-y_1 + 2y_2 \geq 6$
$y_1 \geq 0, y_2 \geq 0.$

6. Minimize $w = 4y_1 + 7y_2$
subject to: $y_1 + y_2 \geq 17$
$3y_1 + 6y_2 \geq 21$
$2y_1 + 4y_2 \geq 19$
$y_1 \geq 0, y_2 \geq 0.$

7. Minimize $w = 2y_1 + 8y_2$
subject to: $y_1 + 7y_2 \geq 18$
$4y_1 + y_2 \geq 15$
$5y_1 + 3y_2 \geq 20$
$y_1 \geq 0, y_2 \geq 0.$

8. Minimize $w = 5y_1 + y_2 + 3y_3$
subject to: $7y_1 + 6y_2 + 8y_3 \geq 18$
$4y_1 + 5y_2 + 10y_3 \geq 20$
$y_1 \geq 0, y_2 \geq 0, y_3 \geq 0.$

9. Minimize $w = y_1 + 2y_2 + 6y_3$
subject to: $3y_1 + 4y_2 + 6y_3 \geq -8$
$y_1 + 5y_2 + 2y_3 \geq 12$
$y_1 \geq 0, y_2 \geq 0, y_3 \geq 0.$

10. Minimize $w = 4y_1 + 3y_2 + y_3$
subject to: $y_1 + 2y_2 + 3y_3 \geq 115$
$2y_1 + y_2 + 8y_3 \geq 200$
$y_1 - y_3 \geq 50$
$y_1 \geq 0, y_2 \geq 0, y_3 \geq 0.$

11. Minimize $w = 8y_1 + 9y_2 + 3y_3$
subject to: $y_1 + y_2 + y_3 \geq 5$
$y_1 + y_2 \geq 4$
$2y_1 + y_2 + 3y_3 \geq 15$
$y_1 \geq 0, y_2 \geq 0, y_3 \geq 0.$

12. Minimize $w = y_1 + 2y_2 + y_3 + 5y_4$
subject to: $y_1 + y_2 + y_3 + y_4 \geq 50$
$3y_1 + y_2 + 2y_3 + y_4 \geq 100$
$y_1 \geq 0, y_2 \geq 0, y_3 \geq 0, y_4 \geq 0.$

Use duality to solve the following problems. (See Example 4.)

13. Minimize $w = 2y_1 + y_2 + 3y_3$
   subject to: $y_1 + y_2 + y_3 \geq 100$
                           $2y_1 + y_2 \qquad \geq 50$
   $y_1 \geq 0, y_2 \geq 0, y_3 \geq 0.$

14. Minimize $w = 2y_1 + 4y_2$
   subject to: $4y_1 + 2y_2 \geq 10$
                        $4y_1 + y_2 \geq 8$
                        $2y_1 + y_2 \geq 12$
   $y_1 \geq 0, y_2 \geq 0.$

15. Minimize $w = 3y_1 + y_2 + 4y_3$
   subject to: $2y_1 + y_2 + y_3 \geq 6$
                        $y_1 + 2y_2 + y_3 \geq 8$
                        $2y_1 + y_2 + 2y_3 \geq 12$
   $y_1 \geq 0, y_2 \geq 0, y_3 \geq 0.$

16. Minimize $w = y_1 + y_2 + 3y_3$
   subject to: $2y_1 + 6y_2 + y_3 \geq 8$
                        $y_1 + 2y_2 + 4y_3 \geq 12$
   $y_1 \geq 0, y_2 \geq 0, y_3 \geq 0.$

17. Minimize $w = 6y_1 + 4y_2 + 2y_3$
   subject to: $2y_1 + 2y_2 + y_3 \geq 2$
                        $y_1 + 3y_2 + 2y_3 \geq 3$
                        $y_1 + y_2 + 2y_3 \geq 4$
   $y_1 \geq 0, y_2 \geq 0, y_3 \geq 0.$

18. Minimize $w = 12y_1 + 10y_2 + 7y_3$
   subject to: $2y_1 + y_2 + y_3 \geq 7$
                        $y_1 + 2y_2 + y_3 \geq 4$
   $y_1 \geq 0, y_2 \geq 0, y_3 \geq 0.$

19. Minimize $w = 20y_1 + 12y_2 + 40y_3$
   subject to: $y_1 + y_2 + 5y_3 \geq 20$
                           $2y_1 + y_2 + y_3 \geq 30$
   $y_1 \geq 0, y_2 \geq 0, y_3 \geq 0.$

20. Minimize $w = 4y_1 + 5y_2$
   subject to: $10y_1 + 5y_2 \geq 100$
                        $20y_1 + 10y_2 \geq 150$
   $y_1 \geq 0, y_2 \geq 0.$

21. Minimize $w = 4y_1 + 2y_2 + y_3$
   subject to: $y_1 + y_2 + y_3 \geq 4$
                           $3y_1 + y_2 + 3y_3 \geq 6$
                           $y_1 + y_2 + 3y_3 \geq 5$
   $y_1 \geq 0, y_2 \geq 0, y_3 \geq 0.$

22. Minimize $w = 3y_1 + 2y_2$
   subject to: $2y_1 + 3y_2 \geq 60$
                        $y_1 + 4y_2 \geq 40$
   $y_1 \geq 0, y_2 \geq 0.$

23. Natural Science   Glenn Russell, who is dieting, requires two food supplements, I and II. He can get these supplements from two different products, A and B, as shown in the following table.

PRODUCT	SUPPLEMENT (GRAMS PER SERVING)	
	I	II
A	4	2
B	2	5

Glenn's physician has recommended that he include at least 20 grams of supplement I and 18 grams of supplement II in his diet. If product A costs 24¢ per serving and product B costs 40¢ per serving, how can he satisfy these requirements most economically?

24. Management   An animal food must provide at least 54 units of vitamins and 60 calories per serving. One gram of soybean meal provides at least 2.5 units of vitamins and 5 calories. One gram of meat byproducts provides at least 4.5 units of vitamins and 3 calories. One gram of grain provides at least 5 units of vitamins and 10 calories. If a gram of soybean meal costs 8¢, a gram of meat byproducts 9¢, and a gram of grain 10¢, what mixture of these three ingredients will provide the required vitamins and calories at minimum cost?

25. Management   A furniture company makes tables and chairs. Union contracts require that the total number of tables and chairs produced must be at least 60 per week. Sales experience has shown that at least 1 table must be made for every 3 chairs that are made. If it costs $152 to make a table and $40 to make a chair, how many of each should be produced each week to minimize the cost? What is the minimum cost?

26. Management   Brand X canners produce canned corn, beans, and carrots. Labor contracts require them to produce at least 1000 cases per month. Based on past sales, they should produce at least twice as many cases of corn as of beans. At least 340 cases of carrots must be produced to fulfill commitments to a major distributor. It costs $10 to produce a case of beans, $15 to produce a case of corn, and $25 to produce a case of carrots. How many cases of each vegetable should be produced to minimize costs?

27. Management   Refer to the end of this section, to the text on minimizing the daily cost of feeds.
  (a) Find a combination of feeds that will cost $7.00 and give 7 units of A and 4 units of B.
  (b) Use the dual variables to predict the daily cost of feed if the requirements change to 5 units of A and 4 units of B. Find a combination of feeds to meet these requirements at the predicted price.

28. Management   A small toy manufacturing firm has 200 squares of felt, 600 ounces of stuffing, and 90 feet of trim available to make two types of toys, a small bear and a monkey. The bear requires 1 square of felt and 4 ounces of stuffing. The monkey requires 2 squares of felt, 3 ounces of stuffing, and 1 foot of trim. The firm makes $1 profit on each bear and $1.50 profit on each monkey. The linear program to maximize profit is

$$\text{Maximize} \quad x_1 + 1.5x_2 = z$$
$$\text{subject to:} \quad x_1 + 2x_2 \le 200$$
$$4x_1 + 3x_2 \le 600$$
$$x_2 \le 90$$
$$x_1 \ge 0, x_2 \ge 0.$$

The final simplex tableau is

$$\begin{bmatrix} 0 & 1 & .8 & -.2 & 0 & | & 40 \\ 1 & 0 & -.6 & .4 & 0 & | & 120 \\ 0 & 0 & -.8 & .2 & 1 & | & 50 \\ \hline 0 & 0 & .6 & .1 & 0 & | & 180 \end{bmatrix}.$$

(a) What is the corresponding dual problem?
(b) What is the optimal solution to the dual problem?
(c) Use the shadow values to estimate the profit the firm will make if their supply of felt increases to 210 squares.

(d) How much profit will the firm make if their supply of stuffing is cut to 590 ounces and their supply of trim is cut to 80 feet?

29. Refer to Example 1 in Section 8.5.
(a) Give the dual problem.
(b) Use the shadow values to estimate the farmer's profit if land is cut to 90 acres but capital increases to $21,000.
(c) Suppose the farmer has 110 acres but only $19,000. Find the optimum profit and the planting strategy that will produce this profit.

Use duality and a graphing calculator or a computer program for doing the simplex method to solve this problem.

30. Management Natural Brand plant food is made from three chemicals. In a batch of the plant food there must be at least 81 kilograms of the first chemical and the ratio of the second chemical to the third chemical must be at least 4 to 3. If the three chemicals cost $1.09, $.87, and $.65 per kilogram, respectively, how much of each should be used to minimize the cost of producing at least 750 kilograms of the plant food?

8.7 THE SIMPLEX METHOD: NONSTANDARD PROBLEMS

So far we have used the simplex method to solve linear programming problems in standard maximum or minimum form only. Now this work is extended to include linear programming problems with mixed $\le$ and $\ge$ constraints.

The first step is to write each constraint so that the constant on the right side is nonnegative. For instance, the inequality

$$4x_1 + 5x_2 - 12x_3 \le -30$$

can be replaced by the equivalent one obtained by multiplying both sides by -1 and reversing the direction of the inequality sign:

$$-4x_1 - 5x_2 + 12x_3 \ge 30.$$

The next step is to write each constraint as an equation. Recall that constraints involving $\le$ are converted to equations by adding a nonnegative slack variable. Similarly, constraints involving $\ge$ are converted to equations by *subtracting* a nonnegative **surplus variable.** For example, the inequality $2x_1 - x_2 + 5x_3 \ge 12$ means that

$$2x_1 - x_2 + 5x_3 - x_4 = 12$$

for some nonnegative x_4. The surplus variable x_4 represents the amount by which $2x_1 - x_2 + 5x_3$ exceeds 12.

1 **(a)** Restate this problem in terms of equations:

Maximize $z = 3x_1 - 2x_2$

subject to: $2x_1 + 3x_2 \leq 8$

$6x_1 - 2x_2 \geq 3$

$x_1 + 4x_2 \geq 1$

$x_1 \geq 0, x_2 \geq 0.$

(b) Write the initial simplex tableau.

Answers:

(a) Maximize $z = 3x_1 - 2x_2$

subject to:

$2x_1 + 3x_2 + x_3 \qquad = 8$

$6x_1 - 2x_2 \qquad - x_4 \qquad = 3$

$x_1 + 4x_2 \qquad - x_5 = 1$

$x_1 \geq 0, x_2 \geq 0, x_3 \geq 0, x_4 \geq 0,$

$x_5 \geq 0.$

(b)

$$
\begin{array}{ccccc}
x_1 & x_2 & x_3 & x_4 & x_5 \\
\end{array}
$$

$$
\left[
\begin{array}{ccccc|c}
2 & 3 & 1 & 0 & 0 & 8 \\
6 & -2 & 0 & -1 & 0 & 3 \\
1 & 4 & 0 & 0 & -1 & 1 \\
\hline
-3 & 2 & 0 & 0 & 0 & 0 \\
\end{array}
\right]
$$

2 State the basic solution given by each tableau. Is it feasible?

(a)

$$
\begin{array}{ccccc}
x_1 & x_2 & x_3 & x_4 & x_5 \\
\end{array}
$$

$$
\left[
\begin{array}{ccccc|c}
3 & -5 & 1 & 0 & 0 & 12 \\
4 & 7 & 0 & 1 & 0 & 6 \\
1 & 3 & 0 & 0 & -1 & 5 \\
\hline
-7 & 4 & 0 & 0 & 0 & 0 \\
\end{array}
\right]
$$

(b)

$$
\begin{array}{ccccc}
x_1 & x_2 & x_3 & x_4 & x_5 \\
\end{array}
$$

$$
\left[
\begin{array}{ccccc|c}
9 & 8 & -1 & 1 & 0 & 12 \\
-5 & 3 & 0 & 0 & 1 & 7 \\
\hline
4 & 2 & 3 & 0 & 0 & 0 \\
\end{array}
\right]
$$

Answers:

(a) $x_1 = 0, x_2 = 0, x_3 = 12,$ $x_4 = 6, x_5 = -5$; no.

(b) $x_1 = 0, x_2 = 0, x_3 = 0,$ $x_4 = 12, x_5 = 7$; yes.

EXAMPLE 1 Restate the following problem in terms of equations and write its initial simplex tableau.

Maximize $z = 4x_1 + 10x_2 + 6x_3$

subject to: $x_1 + 4x_2 + 4x_3 \geq 8$

$x_1 + 3x_2 + 2x_3 \leq 6$

$3x_1 + 4x_2 + 8x_3 \leq 22$

$x_1 \geq 0, x_2 \geq 0, x_3 \geq 0.$

In order to write the constraints as equations, subtract a surplus variable from the $\geq$ constraint and add a slack variable to each $\leq$ constraint. So the problem becomes

Maximize $z = 4x_1 + 10x_2 + 6x_3$

subject to: $x_1 + 4x_2 + 4x_3 - x_4 \qquad = 8$

$x_1 + 3x_2 + 2x_3 \qquad + x_5 \qquad = 6$

$3x_1 + 4x_2 + 8x_3 \qquad + x_6 = 22$

$x_1 \geq 0, x_2 \geq 0, x_3 \geq 0, x_4 \geq 0, x_5 \geq 0, x_6 \geq 0.$

Write the objective function as $z - 4x_1 - 10x_2 - 6x_3 = 0$ and use the coefficients of the four equations to write the initial simplex tableau (omitting the z column):

$$
\begin{array}{cccccc}
x_1 & x_2 & x_3 & x_4 & x_5 & x_6 \\
\end{array}
$$

$$
\left[
\begin{array}{cccccc|c}
1 & 4 & 4 & -1 & 0 & 0 & 8 \\
1 & 3 & 2 & 0 & 1 & 0 & 6 \\
3 & 4 & 8 & 0 & 0 & 1 & 22 \\
\hline
-4 & -10 & -6 & 0 & 0 & 0 & 0 \\
\end{array}
\right]. \quad \blacksquare \quad \boxed{1}
$$

The tableau in Example 1 resembles those that have appeared previously, and similar terminology is used. The variables whose columns have one entry ± 1 and the rest 0 will be called **basic variables;** the other variables are nonbasic. A solution obtained by setting the nonbasic variables equal to 0 and solving for the basic variables (by looking at the constants in the right-hand column) will be called a **basic solution.** A basic solution that is feasible is called a **basic feasible solution.** In the tableau of Example 1, for instance, the basic variables are $x_4, x_5,$ and $x_6,$ and the basic solution is:

$$x_1 = 0, \quad x_2 = 0, \quad x_3 = 0, \quad x_4 = -8, \quad x_5 = 6, \quad x_6 = 22.$$

However, because one variable is negative, this solution is not feasible. $\quad \boxed{2}$

Stage I of the two-stage method for nonstandard maximization problems consists of finding a basic *feasible* solution that can be used as the starting point for the simplex method. (This stage is unnecessary in a standard maximization problem because the solution given by the initial tableau is always feasible.) There are many systematic ways of finding a feasible solution, all of which depend on the fact that row operations (such as pivoting) produce a tableau that represents a system with the same solutions as the original one. One such technique is explained in the next example. Since the immediate goal is to find a feasible solution, not necessarily an optimal one, the procedures for choosing pivots differ from those in the ordinary simplex method.

EXAMPLE 2 Find a basic feasible solution for the problem in Example 1, whose initial tableau is

$$
\begin{array}{cccccc}
x_1 & x_2 & x_3 & x_4 & x_5 & x_6 \\
\end{array}
$$
$$
\left[
\begin{array}{cccccc|c}
1 & 4 & 4 & -1 & 0 & 0 & 8 \\
1 & 3 & 2 & 0 & 1 & 0 & 6 \\
3 & 4 & 8 & 0 & 0 & 1 & 22 \\
\hline
-4 & -10 & -6 & 0 & 0 & 0 & 0
\end{array}
\right].
$$

In the basic solution given by this tableau, x_4 has a negative value. The only nonzero entry in its column is the -1 in row one. Choose any *positive* entry in row one except the entry on the far right. The column that the chosen entry is in will be the pivot column. We choose the first positive entry in row one, the 1 in column one. The pivot row is determined in the usual way by considering quotients (constant at the right end of the row divided by the positive entry in the pivot column) in each row except the objective row:

$$
8/1 = 8, \quad 6/1 = 6, \quad 22/3 = 7\frac{1}{3}.
$$

The smallest quotient is 6, so the pivot is the 1 in row two, column one. Pivoting in the usual way leads to this tableau

$$
\begin{array}{cccccc}
x_1 & x_2 & x_3 & x_4 & x_5 & x_6 \\
\end{array}
$$
$$
\left[
\begin{array}{cccccc|c}
0 & 1 & 2 & -1 & -1 & 0 & 2 \\
1 & 3 & 2 & 0 & 1 & 0 & 6 \\
0 & -5 & 2 & 0 & -3 & 1 & 4 \\
\hline
0 & 2 & 2 & 0 & 4 & 0 & 24
\end{array}
\right]
\begin{array}{l}
-R_2 + R_1 \\
\\
-3R_2 + R_3 \\
4R_2 + R_4
\end{array}
$$

and the basic solution

$$
x_1 = 6, \quad x_2 = 0, \quad x_3 = 0, \quad x_4 = -2, \quad x_5 = 0, \quad x_6 = 4.
$$

Since the basic variable x_4 is negative, this solution is not feasible. So we repeat the pivoting process described above. The x_4 column has a -1 in row one, so we choose a positive entry in that row, namely, the 1 in row one, column two. This choice makes column two the pivot column. The pivot row is determined by the quotients $2/1 = 2$ and $6/3 = 2$ (negative entries in the pivot column and the entry in the objective row are not used). Since there is a tie, we can choose either row one or row two. We choose row one and use the 1 in row one, column 2 as the pivot. Pivoting produces this tableau

$$
\begin{array}{cccccc}
x_1 & x_2 & x_3 & x_4 & x_5 & x_6 \\
\end{array}
$$
$$
\left[
\begin{array}{cccccc|c}
0 & 1 & 2 & -1 & -1 & 0 & 2 \\
1 & 0 & -4 & 3 & 4 & 0 & 0 \\
0 & 0 & 12 & -5 & -8 & 1 & 14 \\
\hline
0 & 0 & -2 & 2 & 6 & 0 & 20
\end{array}
\right]
\begin{array}{l}
\\
-3R_1 + R_2 \\
5R_1 + R_3 \\
-2R_1 + R_4
\end{array}
$$

and the basic *feasible* solution

$$
x_1 = 0, \quad x_2 = 2, \quad x_3 = 0, \quad x_4 = 0, \quad x_5 = 0, \quad x_6 = 14. \quad \blacksquare
$$

Once a basic feasible solution has been found, Stage I is ended. The procedures used in Stage I are summarized below.*

3 The initial tableau of a maximization problem is given below. Use column one as the pivot column for carrying out Stage I and state the basic feasible solution that results.

$$
\begin{array}{cccc|c}
x_1 & x_2 & x_3 & x_4 & \\
1 & 3 & 1 & 0 & 70 \\
2 & 4 & 0 & -1 & 50 \\
\hline
-8 & -10 & 0 & 0 & 0
\end{array}
$$

Answer:

$$
\begin{array}{cccc|c}
x_1 & x_2 & x_3 & x_4 & \\
0 & 1 & 1 & \frac{1}{2} & 45 \\
1 & 2 & 0 & -\frac{1}{2} & 25 \\
\hline
0 & 6 & 0 & -4 & 200
\end{array}
$$

$x_1 = 25, x_2 = 0, x_3 = 45, x_4 = 0.$

One way to make the required choices systematically is to choose the first possibility in each case (going from the top for rows or from the left for columns). However, any choice meeting the required conditions may be used. For maximum efficiency, it is usually best to choose the pivot column in Step 2 so that the pivot is in the same row chosen in Step 1, if this is possible. **3**

In **Stage II,** the simplex method is applied as usual to the tableau that produced the basic feasible solution in Stage I. Just as in Section 8.4, each round of pivoting replaces the basic feasible solution of one tableau with the basic feasible solution of a new tableau in such a way that the value of the objective function is increased, until an optimal value is obtained (or it becomes clear that no optimal solution exists).

EXAMPLE 3 Solve the linear programming problem in Example 1.

A basic feasible solution for this problem was found in Example 2 by using the tableau shown below. However, this solution is not maximal because there is a negative indicator in the objective row. So we use the simplex method: the most negative indicator determines the pivot column and the usual quotients determine that the number 2 in row one, column three is the pivot.

$$
\begin{array}{cccccc|c}
x_1 & x_2 & x_3 & x_4 & x_5 & x_6 & \\
0 & 1 & \mathbf{2} & -1 & -1 & 0 & 2 \\
1 & 0 & -4 & 3 & 4 & 0 & 0 \\
0 & 0 & 12 & -5 & -8 & 1 & 14 \\
\hline
0 & 0 & -2 & 2 & 6 & 0 & 20
\end{array}
$$

Quotients

$2/2 \leftarrow$ Smallest

$14/12$

↑ Most negative indicator

---

*Except in rare cases that do not occur in this book, this method eventually produces a basic feasible solution or shows that one does not exist. The *two-phase method* using artificial variables, which is discussed in more advanced texts, works in all cases and often is more efficient.

Pivoting leads to the final tableau.

$$
\begin{array}{c}
\begin{array}{cccccc} x_1 & x_2 & x_3 & x_4 & x_5 & x_6 \end{array} \\
\left[\begin{array}{cccccc|c}
0 & \frac{1}{2} & 1 & -\frac{1}{2} & -\frac{1}{2} & 0 & 1 \\
1 & 0 & -4 & 3 & 4 & 0 & 0 \\
0 & 0 & 12 & -5 & -8 & 1 & 14 \\
\hline
0 & 0 & -2 & 2 & 6 & 0 & 20
\end{array}\right]
\end{array}
\quad \frac{1}{2}R_1
$$

4 Complete Stage II and find an optimal solution for Side Problem 3 above. What is the optimal value of the objective function z?

Answer:
The optimal value $z = 560$ occurs when $x_1 = 70$, $x_2 = 0$, $x_3 = 0$, $x_4 = 90$.

$$
\begin{array}{c}
\begin{array}{cccccc} x_1 & x_2 & x_3 & x_4 & x_5 & x_6 \end{array} \\
\left[\begin{array}{cccccc|c}
0 & \frac{1}{2} & 1 & -\frac{1}{2} & -\frac{1}{2} & 0 & 1 \\
1 & 2 & 0 & 1 & 2 & 0 & 4 \\
0 & -6 & 0 & 1 & -2 & 1 & 2 \\
\hline
0 & 1 & 0 & 1 & 5 & 0 & 22
\end{array}\right]
\end{array}
\quad
\begin{array}{l}
\\ 4R_1 + R_2 \\ -12R_1 + R_3 \\ 2R_1 + R_4 \end{array}
$$

Therefore, the maximum value of z occurs when $x_1 = 4$, $x_2 = 0$, and $x_3 = 1$, in which case $z = 22$. ■ **4**

The two-stage method for maximization problems illustrated in Examples 1–3 also provides a means of solving minimization problems. To see why, consider this simple fact: when a number t gets smaller, then $-t$ gets larger, and vice-versa. For instance, if t goes from 6 to 1 to 0 to -8, then $-t$ goes from -6 to -1 to 0 to 8. Thus, if w is the objective function of a linear programming problem, the feasible solution that produces the minimum value of w also produces the maximum value of $-w$, and vice-versa. Therefore, to solve a minimization problem with objective function w, we need only solve the maximization problem with the same constraints and objective function $z = -w$.

EXAMPLE 4   Minimize $w = 2y_1 + y_2 - y_3$
subject to: $-y_1 - y_2 + y_3 \leq -4$
$y_1 + 3y_2 + 3y_3 \geq 6$
$y_1 \geq 0, y_2 \geq 0, y_3 \geq 0.$

Make the constant in the first constraint positive by multiplying both sides by -1. Then solve this maximization problem:

Maximize $z = -w = -2y_1 - y_2 + y_3$
subject to: $y_1 + y_2 - y_3 \geq 4$
$y_1 + 3y_2 + 3y_3 \geq 6$
$y_1 \geq 0, y_2 \geq 0, y_3 \geq 0.$

Convert the constraints to equations by subtracting surplus variables, and set up the first tableau.

$$
\begin{array}{c}
\begin{array}{ccccc} y_1 & y_2 & y_3 & y_4 & y_5 \end{array} \\
\left[\begin{array}{ccccc|c}
1 & 1 & -1 & -1 & 0 & 4 \\
1 & 3 & 3 & 0 & -1 & 6 \\
\hline
2 & 1 & -1 & 0 & 0 & 0
\end{array}\right]
\end{array}
$$

The basic solution given by this tableau, $y_1 = 0$, $y_2 = 0$, $y_3 = 0$, $y_4 = -4$, $y_5 = -6$ is not feasible, so the procedures of Stage I must be used to find a basic feasible solution. In the column of the negative basic variable y_4, there is a -1 in row one; we choose the first positive entry in that row, so that column one will be the pivot column. The quotients $4/1 = 4$ and $6/1 = 6$ show that the pivot is the 1 in row one, column one. Pivoting produces this tableau:

$$
\begin{array}{ccccc}
y_1 & y_2 & y_3 & y_4 & y_5
\end{array}
$$
$$
\left[
\begin{array}{ccccc|c}
1 & 1 & -1 & -1 & 0 & 4 \\
0 & 2 & 4 & 1 & -1 & 2 \\
\hline
0 & -1 & 1 & 2 & 0 & -8
\end{array}
\right].
\quad
\begin{array}{l}
\\
-R_1 + R_2 \\
\\
-2R_1 + R_3
\end{array}
$$

The basic solution $y_1 = 4$, $y_2 = 0$, $y_3 = 0$, $y_4 = 0$, $y_5 = -2$ is not feasible because y_5 is negative, so we repeat the process. We choose the first positive entry in row two (the row containing the -1 in the y_5 column), which is in column two, so that column two is the pivot column. The relevant quotients are $4/1 = 4$ and $2/2 = 1$, so the pivot is the 2 in row two, column two. Pivoting produces a new tableau.

$$
\begin{array}{ccccc}
y_1 & y_2 & y_3 & y_4 & y_5
\end{array}
$$
$$
\left[
\begin{array}{ccccc|c}
1 & 1 & -1 & -1 & 0 & 4 \\
0 & 1 & 2 & \frac{1}{2} & -\frac{1}{2} & 1 \\
\hline
0 & -1 & 1 & 2 & 0 & -8
\end{array}
\right]
\quad
\begin{array}{l}
\\
\frac{1}{2}R_2 \\
\\
\end{array}
$$

$$
\begin{array}{ccccc}
y_1 & y_2 & y_3 & y_4 & y_5
\end{array}
$$
$$
\left[
\begin{array}{ccccc|c}
1 & 0 & -3 & -\frac{3}{2} & \frac{1}{2} & 3 \\
0 & 1 & 2 & \frac{1}{2} & -\frac{1}{2} & 1 \\
\hline
0 & 0 & 3 & \frac{5}{2} & -\frac{1}{2} & -7
\end{array}
\right]
\quad
\begin{array}{l}
-R_2 + R_1 \\
\\
\\
R_2 + R_3
\end{array}
$$

The basic solution $y_1 = 3$, $y_2 = 1$, $y_3 = 0$, $y_4 = 0$, $y_5 = 0$, is feasible, so Stage I is complete. However, this solution is not optimal because the objective row contains the negative indicator $-1/2$ in column five. According to the simplex method, column five is the next pivot column. The only positive ratio $3/\frac{1}{2} = 6$ is in row one, so the pivot is $1/2$ in row one, column five. Pivoting produces the final tableau.

$$
\begin{array}{ccccc}
y_1 & y_2 & y_3 & y_4 & y_5
\end{array}
$$
$$
\left[
\begin{array}{ccccc|c}
2 & 0 & -6 & -3 & 1 & 6 \\
0 & 1 & 2 & \frac{1}{2} & -\frac{1}{2} & 1 \\
\hline
0 & 0 & 3 & \frac{5}{2} & -\frac{1}{2} & -7
\end{array}
\right]
\quad
\begin{array}{l}
2R_1 \\
\\
\\
\end{array}
$$

$$
\begin{array}{ccccc}
y_1 & y_2 & y_3 & y_4 & y_5
\end{array}
$$
$$
\left[
\begin{array}{ccccc|c}
2 & 0 & -6 & -3 & 1 & 6 \\
1 & 1 & -1 & -1 & 0 & 4 \\
\hline
1 & 0 & 0 & 1 & 0 & -4
\end{array}
\right]
\quad
\begin{array}{l}
\\
\frac{1}{2}R_1 + R_2 \\
\\
\frac{1}{2}R_1 + R_3
\end{array}
$$

Since there are no negative indicators, the solution given by this tableau ($y_1 = 0$, $y_2 = 4$, $y_3 = 0$, $y_4 = 0$, $y_5 = 6$) is optimal. The maximum value of $z = -w$ is -4.

[5] Minimize $w = 2y_1 + 3y_2$
subject to:

$$y_1 + y_2 \geq 10$$
$$2y_1 + y_2 \geq 16$$
$$y_1 \geq 0, y_2 \geq 0.$$

Answer:
$y_1 = 10, y_2 = 0; w = 20$

Therefore, the minimum value of the original objective function w is $-(-4) = 4$, which occurs when $y_1 = 0, y_2 = 4, y_3 = 0$. ■ [5]

Here is a summary of the two-stage method that was illustrated in Examples 1–4.

SOLVING NONSTANDARD PROBLEMS

1. If necessary, write each constraint with a positive constant and convert the problem to a maximum problem by letting $z = -w$.
2. Add slack variables and subtract surplus variables as needed to convert the constraints into equations.
3. Write the initial simplex tableau.
4. Find a basic feasible solution for the problem, if one exists (Stage I).
5. When a basic feasible solution is found, use the simplex method to solve the problem (Stage II).

N O T E It may happen that the tableau that gives the basic feasible solution in Stage I has no negative indicators in its last row. In this case, the solution found is already optimal and Stage II is not necessary. ◆

EXAMPLE 5 A college textbook publisher has received orders from two colleges, C_1 and C_2. C_1 needs at least 500 books, and C_2 needs at least 1000. The publisher can supply the books from either of two warehouses. Warehouse W_1 has 900 books available and warehouse W_2 has 700. The costs to ship a book from each warehouse to each college are given below.

		TO	
		C_1	C_2
FROM	W_1	$1.20	$1.80
	W_2	$2.10	$1.50

How many books should be sent from each warehouse to each college to minimize the shipping costs?
To begin, let

$$y_1 = \text{the number of books shipped from } W_1 \text{ to } C_1;$$
$$y_2 = \text{the number of books shipped from } W_2 \text{ to } C_1;$$
$$y_3 = \text{the number of books shipped from } W_1 \text{ to } C_2;$$
$$y_4 = \text{the number of books shipped from } W_2 \text{ to } C_2.$$

C_1 needs at least 500 books, so

$$y_1 + y_2 \geq 500.$$

Similarly,

$$y_3 + y_4 \geq 1000.$$

Since W_1 has 900 books available and W_2 has 700 available,

$$y_1 + y_3 \le 900 \quad \text{and} \quad y_2 + y_4 \le 700.$$

The company wants to minimize shipping costs, so the objective function is

$$w = 1.20y_1 + 2.10y_2 + 1.80y_3 + 1.50y_4.$$

Now write the problem as a system of linear equations, adding slack or surplus variables as needed, and let $z = -w$.

$$
\begin{aligned}
y_1 + y_2 -y_5 &= 500 \\
y_3 + y_4 -y_6 &= 1000 \\
y_1 + y_3 + y_7 &= 900 \\
y_2 + y_4 + y_8 &= 700 \\
1.20y_1 + 2.10y_2 + 1.80y_3 + 1.50y_4 + z &= 0
\end{aligned}
$$

Set up the initial simplex tableau.

y_1	y_2	y_3	y_4	y_5	y_6	y_7	y_8	
1	1	0	0	−1	0	0	0	500
0	0	1	1	0	−1	0	0	1000
1	0	1	0	0	0	1	0	900
0	1	0	1	0	0	0	1	700
1.20	2.10	1.80	1.50	0	0	0	0	0

The indicated solution is

$$y_5 = -500, \quad y_6 = -1000, \quad y_7 = 900, \quad y_8 = 700,$$

which is not feasible since y_5 and y_6 are negative.

There is a -1 in row one of the y_5 column. We choose the first positive entry in row one, which makes column one the pivot column. The quotients $500/1 = 500$ and $900/1 = 900$ show that the pivot is the 1 in row one, column one. Pivoting produces this tableau.

y_1	y_2	y_3	y_4	y_5	y_6	y_7	y_8	
1	1	0	0	−1	0	0	0	500
0	0	1	1	0	−1	0	0	1000
0	−1	1	0	1	0	1	0	400
0	1	0	1	0	0	0	1	700
0	.9	1.8	1.5	1.2	0	0	0	−600

The basic variable y_6 is negative and there is a -1 in row two of its column. We choose the first positive entry in that row, which makes column three the pivot column. The quotients $1000/1 = 1000$ and $400/1 = 400$ show that the pivot is the 1 in row three, column three. Pivoting leads to the next tableau.

$$
\begin{array}{cccccccc}
y_1 & y_2 & y_3 & y_4 & y_5 & y_6 & y_7 & y_8 \\
\end{array}
$$

$$
\left[\begin{array}{cccccccc|c}
1 & 1 & 0 & 0 & -1 & 0 & 0 & 0 & 500 \\
0 & 1 & 0 & 1 & -1 & -1 & -1 & 0 & 600 \\
0 & -1 & 1 & 0 & 1 & 0 & 1 & 0 & 400 \\
0 & 1 & 0 & 1 & 0 & 0 & 0 & 1 & 700 \\
\hline
0 & 2.7 & 0 & 1.5 & -.6 & 0 & -1.8 & 0 & -1320
\end{array}\right]
$$

The basic variable y_6 is still negative, so we must choose a positive entry in row two (the row containing the -1 in the y_6 column). If we choose the first positive entry, as we usually have done, then the quotients for determining the pivot will be 500/1 and 700/1 so that the pivot will be in row one. However, it is usually more efficient to have the pivot in the same row as the -1 of the basic variable (in this case row two). So we choose the 1 in row two, column four. Then the quotients for determining the pivot are 600/1 and 700/1 and the pivot is the 1 in row two, column four. Pivoting produces this tableau

$$
\begin{array}{cccccccc}
y_1 & y_2 & y_3 & y_4 & y_5 & y_6 & y_7 & y_8 \\
\end{array}
$$

$$
\left[\begin{array}{cccccccc|c}
1 & 1 & 0 & 0 & -1 & 0 & 0 & 0 & 500 \\
0 & 1 & 0 & 1 & -1 & -1 & -1 & 0 & 600 \\
0 & -1 & 1 & 0 & 1 & 0 & 1 & 0 & 400 \\
0 & 0 & 0 & 0 & 1 & 1 & 1 & 1 & 100 \\
\hline
0 & 1.2 & 0 & 0 & .9 & 1.5 & -.3 & 0 & -2220
\end{array}\right]
$$

and the basic feasible solution $y_1 = 500$, $y_2 = 0$, $y_3 = 400$, $y_4 = 600$, $y_5 = 0$, $y_6 = 0$, $y_7 = 0$, $y_8 = 100$. Hence Stage I is ended. This solution is not optimal because there is a negative indicator in the objective row, so we proceed to Stage II and the usual simplex method. Column seven has the only negative indicator, $-.3$, and row four the smallest quotient, $100/1 = 100$. Pivoting on the 1 in row four, column seven produces the final simplex tableau.

$$
\begin{array}{cccccccc}
y_1 & y_2 & y_3 & y_4 & y_5 & y_6 & y_7 & y_8 \\
\end{array}
$$

$$
\left[\begin{array}{cccccccc|c}
1 & 1 & 0 & 0 & -1 & 0 & 0 & 0 & 500 \\
0 & 1 & 0 & 1 & 0 & 0 & 0 & 1 & 700 \\
0 & -1 & 1 & 0 & 0 & -1 & 0 & -1 & 300 \\
0 & 0 & 0 & 0 & 1 & 1 & 1 & 1 & 100 \\
\hline
0 & 1.2 & 0 & 0 & 1.2 & 1.8 & 0 & .3 & -2190
\end{array}\right]
$$

Since there are no negative indicators, the solution given by this tableau ($y_1 = 500$, $y_3 = 300$, $y_4 = 700$) is optimal. The publisher should ship 500 books from W_1 to C_1, 300 books from W_1 to C_2, and 700 books from W_2 to C_2 for a minimum shipping cost of $2190 (remember that the optimal value for the original minimization problem is the negative of the optimal value for the associated maximization problem). ■

Although they will not occur in this book, various complications can arise in using the simplex method. Some of the possible difficulties, which are treated in more advanced texts, include the following:

1. Some of the constraints may be *equations* instead of inequalities. In this case, *artificial variables* must be used.
2. Occasionally, a transformation will cycle—that is, produce a "new" solution which was an earlier solution in the process. These situations are known as *degeneracies* and special methods are available for handling them.
3. It may not be possible to convert a nonfeasible basic solution to a feasible basic solution. In that case, no solution can satisfy all the constraints. Graphically, this means there is no region of feasible solutions.

A linear programming model on preparing a fertilizer mix to certain specifications is presented at the end of this chapter. The model illustrates the usefulness of linear programming. In most real applications, the number of variables is so large that these problems could not be solved without the use of a method, like the simplex method, that can be adapted to a computer.

8.7 EXERCISES

In Exercises 1–4, (a) *restate the problem in terms of equations by introducing slack and surplus variables;* (b) *write the initial simplex tableau. (See Example 1.)*

1. Maximize $z = 5x_1 + 2x_2 - x_3$
   subject to: $2x_1 + 3x_2 + 5x_3 \geq 8$
   $4x_1 - x_2 + 3x_3 \leq 7$
   $x_1 \geq 0, x_2 \geq 0, x_3 \geq 0.$

2. Maximize $z = x_1 + 4x_2 + 6x_3$
   subject to: $5x_1 + 8x_2 - 5x_3 \leq 10$
   $6x_1 + 2x_2 + 3x_3 \geq 7$
   $x_1 \geq 0, x_2 \geq 0, x_3 \geq 0.$

3. Maximize $z = 2x_1 - 3x_2 + 4x_3$
   subject to: $x_1 + x_2 + x_3 \leq 100$
   $x_1 + x_2 + x_3 \geq 75$
   $x_1 + x_2 \geq 27$
   $x_1 \geq 0, x_2 \geq 0, x_3 \geq 0.$

4. Maximize $z = -x_1 + 5x_2 + x_3$
   subject to: $2x_1 + x_3 \leq 40$
   $x_1 + x_2 \geq 18$
   $x_1 + x_3 \geq 20$
   $x_1 \geq 0, x_2 \geq 0, x_3 \geq 0.$

Convert Exercises 5–8 into maximization problems with positive constants on the right side of each constraint and write the initial simplex tableau. (See Example 4.)

5. Minimize $w = 2y_1 + 5y_2 - 3y_3$
   subject to: $y_1 + 2y_2 + 3y_3 \geq 115$
   $2y_1 + y_2 + y_3 \leq 200$
   $y_1 + y_3 \geq 50$
   $y_1 \geq 0, y_2 \geq 0, y_3 \geq 0.$

6. Minimize $w = 7y_1 + 6y_2 + y_3$
   subject to: $y_1 + y_2 + y_3 \geq 5$
   $-y_1 + y_2 \leq -4$
   $2y_1 + y_2 + 3y_3 \geq 15$
   $y_1 \geq 0, y_2 \geq 0, y_3 \geq 0.$

7. Minimize $w = y_1 - 4y_2 + 2y_3$
   subject to: $-7y_1 + 6y_2 - 8y_3 \leq -18$
   $4y_1 + 5y_2 + 10y_3 \geq 20$
   $y_1 \geq 0, y_2 \geq 0, y_3 \geq 0.$

8. Minimize $w = y_1 + 2y_2 + y_3 + 5y_4$
   subject to: $-y_1 + y_2 + y_3 + y_4 \leq -50$
   $3y_1 + y_2 + 2y_3 + y_4 \geq 100$
   $y_1 \geq 0, y_2 \geq 0, y_3 \geq 0, y_4 \geq 0.$

Use the two-stage method to solve Exercises 9–18. (See Examples 1–4.)

9. Maximize $z = 12x_1 + 10x_2$
   subject to: $x_1 + 2x_2 \geq 24$
   $x_1 + x_2 \leq 40$
   $x_1 \geq 0, x_2 \geq 0.$

10. Find $x_1 \geq 0, x_2 \geq 0,$ and $x_3 \geq 0$ such that
    $x_1 + x_2 + x_3 \leq 150$
    $x_1 + x_2 + x_3 \geq 100$
    and $z = 2x_1 + 5x_2 + 3x_3$ is maximized.

11. Find $x_1 \geq 0, x_2 \geq 0,$ and $x_3 \geq 0$ such that
    $x_1 + x_2 + 2x_3 \leq 38$
    $2x_1 + x_2 + x_3 \geq 24$
    and $z = 3x_1 + 2x_2 + 2x_3$ is maximized.

12. Maximize $z = 6x_1 + 8x_2$
subject to: $3x_1 + 12x_2 \geq 48$
$2x_1 + 4x_2 \leq 60$
$x_1 \geq 0, x_2 \geq 0.$

13. Find $x_1 \geq 0$ and $x_2 \geq 0$ such that

$$x_1 + 2x_2 \leq 18$$
$$x_1 + 3x_2 \geq 12$$
$$2x_1 + 2x_2 \leq 30$$

and $z = 5x_1 + 10x_2$ is maximized.

14. Find $x_1 \geq 0$ and $x_2 \geq 0$ such that

$$x_1 + x_2 \leq 100$$
$$x_1 + x_2 \geq 50$$
$$2x_1 + x_2 \leq 110$$

and $z = 2x_1 + 3x_2$ is maximized.

15. Find $y_1 \geq 0$, $y_2 \geq 0$ such that

$$10y_1 + 5y_2 \geq 100$$
$$20y_1 + 10y_2 \geq 160$$

and $w = 4y_1 + 5y_2$ is minimized.

16. Minimize $w = 3y_1 + 2y_2$
subject to: $2y_1 + 3y_2 \geq 60$
$y_1 + 4y_2 \geq 40$
$y_1 \geq 0, y_2 \geq 0.$

17. Minimize $w = 3y_1 + 4y_2$
subject to: $y_1 + 2y_2 \geq 10$
$y_1 + y_2 \geq 8$
$2y_1 + y_2 \leq 22$
$y_1 \geq 0, y_2 \geq 0.$

18. Minimize $w = 4y_1 + 2y_2$
subject to: $y_1 + y_2 \geq 20$
$y_1 + 2y_2 \geq 25$
$-5y_1 + y_2 \leq 4$
$y_1 \geq 0, y_2 \geq 0.$

In Exercises 19–22, set up the initial simplex tableau, but do not solve the problem. (See Example 5.)

19. Management The manufacturer of a popular personal computer has orders from two dealers. Dealer D_1 wants at least 32 computers, and dealer D_2 wants at least 20 computers. The manufacturer can fill the orders from either of two warehouses, W_1 or W_2. W_1 has 25 of the computers on hand, and W_2 has 30. The costs (in dollars) to ship one computer to each dealer from each warehouse are given below.

		TO	
		D_1	D_2
FROM	W_1	$14	$22
	W_2	$12	$10

How should the orders be filled to minimize shipping costs?

20. Natural Science Mark, who is ill, takes vitamin pills. Each day he must have at least 16 units of vitamin A, 5 units of vitamin B_1, and 20 units of vitamin C. He can choose between pill #1 which costs 10¢ and contains 8 units of A, 1 of B_1, and 2 of C, and pill #2 which costs 20¢ and contains 2 units of A, 1 of B_1, and 7 of C. How many of each pill should he buy in order to minimize his cost?

21. Management A company is developing a new additive for gasoline. The additive is a mixture of three liquid ingredients, I, II, and III. For proper performance, the total amount of additive must be at least 10 ounces per gallon of gasoline. However, for safety reasons, the amount of additive should not exceed 15 ounces per gallon of gasoline. At least 1/4 ounce of ingredient I must be used for every ounce of ingredient II and at least 1 ounce of ingredient III must be used for every ounce of ingredient I. If the costs of I, II, and III are $.30, $.09, and $.27 per ounce, respectively, find the mixture of the three ingredients that produces the minimum cost of the additive. How much of the additive should be used per gallon of gasoline?

22. Management A popular soft drink called Sugarlo, which is advertised as having a sugar content of no more than 10%, is blended from five ingredients, each of which has some sugar content. Water may also be added to dilute the mixture. The sugar content of the ingredients and their costs per gallon are given below.

	INGREDIENT					
	1	2	3	4	5	Water
Sugar content (%)	.28	.19	.43	.57	.22	0
Cost ($/gal.)	.48	.32	.53	.28	.43	.04

At least .01 of the content of Sugarlo must come from ingredients 3 or 4, .01 must come from ingredients 2 or 5, and .01 from ingredients 1 or 4. How much of each ingredient should be used in preparing at least 15,000 gallons of Sugarlo to minimize the cost?

Use the two-stage method to solve Exercises 23–30. (See Example 5.)

23. Management Southwestern Oil supplies two distributors in the Northwest from two outlets. Distributor D_1 needs at least 3000 barrels of oil, and distributor D_2 needs at least 5000 barrels. The two outlets can each furnish up to 5000 barrels of oil. The costs per barrel to send the oil are given below.

		TO	
		D_1	D_2
FROM	S_1	$30	$20
	S_2	$25	$22

How should the oil be supplied to minimize shipping costs?

24. Management Change Exercise 23 so that there is also a shipping tax per barrel as given in the table below. Southwestern Oil is determined to spend no more than $40,000 on shipping tax.

	D_1	D_2
S_1	\$2	\$6
S_2	\$5	\$4

How should the oil be supplied to minimize shipping costs?

25. Management A bank has set aside a maximum of $25 million for commercial and home loans. Every million dollars in commercial loans requires 2 lengthy application forms, while every million dollars in home loans requires 3 lengthy application forms. The bank cannot process more than 72 application forms at this time. The bank's policy is to loan at least four times as much for home loans as for commercial loans. Because of prior commitments, at least $10 million will be used for these two types of loans. The bank earns 12% on home loans and 10% on commercial loans. What amount of money should be allotted for each type of loan to maximize the interest income?

26. Management Virginia Keleske has decided to invest a $100,000 inheritance in government securities that earn 7% per year, municipal bonds that earn 6% per year, and mutual funds that earn an average of 10% per year. She will spend at least $40,000 on government securities, and she wants at least half the inheritance to go to bonds and mutual funds. Government securities have an initial fee of 2%, municipal bonds have an initial fee of 1%, and mutual funds have an initial fee of 3%. Virginia has $2400 available to pay initial fees. How much should be invested in each way to maximize the interest yet meet the constraints? What is the maximum interest she can earn?

27. Management Brand X Canners produce canned whole tomatoes and tomato sauce. This season, they have available 3,000,000 kilograms of tomatoes for these two products. To meet the demands of regular customers, they must produce at least 80,000 kilograms of sauce and 800,000 kilograms of whole tomatoes. The cost per kilogram is $4 to produce canned whole tomatoes and $3.25 to produce tomato sauce. Labor agreements require that at least 110,000 person hours be used. Each can of sauce requires 3 minutes for one worker, and each can of whole tomatoes requires 6 minutes for one worker. How many kilograms of tomatoes should Brand X use for each product to minimize cost? (For simplicity, assume production of y_1 kg of canned whole tomatoes and y_2 kilograms of tomato sauce requires $y_1 + y_2$ kg of tomatoes.)

28. Management A brewery produces regular beer and a lower-carbohydrate "light" beer. Steady customers of the brewery buy 12 units of regular beer and 10 units of light beer. While setting up the brewery to produce the beers, the management decides to produce extra beer, beyond that needed to satisfy the steady customers. The cost per unit of regular beer is $36,000 and the cost per unit of light beer is $48,000. The number of units of light beer should not exceed twice the number of units of regular beer. At least 20 additional units of beer can be sold. How much of each type beer should be made so as to minimize total production costs?

29. Management The chemistry department at a local college decides to stock at least 800 small test tubes and 500 large test tubes. It wants to buy at least 1500 test tubes to take advantage of a special price. Since the small tubes are broken twice as often as the larger, the department will order at least twice as many small tubes as large. If the small test tubes cost 15¢ each and the large ones, made of a cheaper glass, cost 12¢ each, how many of each size should they order to minimize cost?

30. Management Topgrade Turf lawn seed mixture contains three types of seeds: bluegrass, rye, and bermuda. The costs per pound of the three types of seed are 12¢, 15¢ and 5¢. In each mixture there must be at least 20% bluegrass seed, and the amount of bermuda must be no more than 2/3 the amount of rye. To fill current orders, the company must make at least 5000 pounds of the mixture. How much of each kind of seed should be used to minimize cost?

CHAPTER 8 SUMMARY

Key Terms and Symbols

8.1
linear inequality
boundary
half-plane
system of inequalities
region of feasible solutions
 (feasible region)

8.2
linear programming
objective function

constraints
corner point
bounded feasible region
unbounded feasible region
corner point theorem

8.4
standard maximum form
slack variable
simplex tableau

indicator
pivot and pivoting
final simplex tableau
basic variables
nonbasic variables
basic feasible solution
8.6 dual
transpose of a matrix

theorem of duality
shadow costs
8.7 surplus variable
basic variables
basic solution
basic feasible solution
Stage I
Stage II

Key Concepts

Graphing a Linear Inequality

Graph the boundary line as a solid line if the inequality includes "or equal," a dashed line otherwise. Shade the half-plane that includes a test point that makes the inequality true. The graph of a system of inequalities, called the **region of feasible solutions,** includes all points that satisfy all the inequalities of the system at the same time.

Solving Linear Programming Problems

Graphically: Determine the objective function and all necessary constraints. Graph the region of feasible solutions. The maximum or minimum value will occur at one or more of the corner points of this region.

Simplex Method: Determine the objective function and all necessary constraints. Convert each constraint into an equation by adding slack variables. Set up the initial simplex tableau. Locate the most negative indicator. Form the quotients to determine the pivot. Use row operations to change the pivot to 1 and all other numbers in that column to 0. If the indicators are all positive or 0, this is the final tableau. If not, choose a new pivot and repeat the process until no indicators are negative. Read the solution from the final tableau. The optimum value of the objective function is the number in the lower right corner of the final tableau. For problems with **mixed constraints,** add surplus variables as well as slack variables. In Stage I, use row operations to transform the matrix until the solution is feasible. In Stage II, use the simplex method as described above. For **minimum** problems, let the objective function be w and set $-w = z$. Then proceed as with mixed constraints.

Solving Minimum Problems with Duals

Find the dual maximum problem. Solve the dual using the simplex method. The minimum value of the objective function w is the maximum value of the dual objective function z. The optimal solution is found in the entries in the bottom row of the columns corresponding to the slack variables.

Chapter 8 Review Exercises

Graph each of the following linear inequalities.

1. $y \le 3x + 2$

2. $2x - y \ge 6$

3. $3x + 4y \ge 12$

4. $y \le 4$

Graph the solution of each of the following systems of inequalities.

5. $x + y \le 6$
$2x - y \ge 3$

6. $4x + y \ge 8$
$2x - 3y \le 6$

7. $2 \le x \le 5$
$1 \le y \le 7$
$x - y \le 3$

8. $x + 2y \le 4$
$2x - 3y \le 6$
$x \ge 0$
$y \ge 0$

Set up a system of inequalities for each of the following problems; then graph the region of feasible solutions.

9. Management A bakery makes both cakes and cookies. Each batch of cakes requires 2 hours in the oven and 3 hours in the decorating room. Each batch of cookies needs $1\frac{1}{2}$ hours in the oven and 2/3 of an hour in the decorating room. The oven is available no more than 15 hours a day, while the decorating room can be used no more than 13 hours a day.

10. Management A company makes two kinds of pizza, special and basic. The special has toppings of cheese, tomatoes, and vegetables. Basic has just cheese and tomatoes.

The company sells at least 6 units a day of the special pizza and 4 units a day of the basic. The cost of the vegetables (including tomatoes) is $2 per unit for special and $1 per unit for basic. No more than $32 per day can be spent on vegetables (including tomatoes). The cheese used for the special is $5 per unit, and the cheese for the basic is $4 per unit. The company can spend no more than $100 per day on cheese.

Use the given regions to find the maximum and minimum values of the objective function z = 2x + 4y.

11.

12.

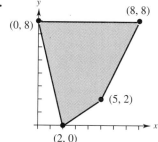

Use the graphical method to solve Exercises 13–16.

13. Maximize $z = 3x + 2y$
   subject to: $2x + 7y \le 14$
       $2x + 3y \le 10$
       $x \ge 0, y \ge 0.$

14. Find $x \ge 0$ and $y \ge 0$ such that
       $8x + 9y \ge 72$
       $6x + 8y \ge 72$
   and $w = 4x + 12y$ is minimized.

15. Find $x \ge 0$ and $y \ge 0$ such that
       $x + y \le 50$
       $2x + y \ge 20$
       $x + 2y \ge 30$
   and $w = 8x + 3y$ is minimized.

16. Maximize $z = 2x - 5y$
   subject to: $3x + 2y \le 12$
       $5x + y \ge 5$
       $x \ge 0, y \ge 0.$

17. Management How many batches of cakes and cookies should the bakery of Exercise 9 make in order to maximize

profits if cookies produce a profit of $20 per batch and cakes produce a profit of $30 per batch?

18. Management How many units of each kind of pizza should the company of Exercise 10 make in order to maximize profits if special sells for $20 per unit and basic for $15 per unit?

For Exercises 19–22, (a) *select appropriate variables,* (b) *write the objective function,* (c) *write the constraints as inequalities.*

19. Management Roberta Hernandez sells three items, A, B, and C, in her gift shop. Each unit of A costs her $2 to buy, $1 to sell, and $2 to deliver. For each unit of B, the costs are $3, $2, and $2, respectively, and for each unit of C the costs are $6, $2, and $4, respectively. The profit on A is $4, on B it is $3, and on C, $3. How many of each should she order to maximize her profit if she can spend $1200 to buy, $800 on selling costs, and $500 on delivery costs?

20. Management An investor is considering three types of investment: a high-risk venture into oil leases with a potential return of 15%, a medium-risk investment in bonds with a 9% return, and a relatively safe stock investment with a 5% return. He has $50,000 to invest. Because of the risk, he will limit his investment in oil leases and bonds to 30% and his investment in oil leases and stock to 50%. How much should he invest in each to maximize his return, assuming investment returns are as expected?

21. Management The Aged Wood Winery makes two white wines, Fruity and Crystal, from two kinds of grapes and sugar. The wines require the following amounts of each ingredient per gallon and produce a profit per gallon as shown below.

	Grape A (bushels)	Grape B (bushels)	Sugar (pounds)	Profit (dollars)
Fruity	2	2	2	12
Crystal	1	3	1	15

The winery has available 110 bushels of grape A, 125 bushels of grape B, and 90 pounds of sugar. How much of each wine should be made to maximize profit?

22. Management A company makes three sizes of plastic bags: 5 gallon, 10 gallon and 20 gallon. The production time in hours for cutting, sealing, and packaging a unit of each size is shown below.

Size	Cutting	Sealing	Packaging
5 gallon	1	1	2
10 gallon	1.1	1.2	3
20 gallon	1.5	1.3	4

There are at most 8 hours available each day for each of the three operations. If the profit per unit is $1 for 5-gallon bags, $.90 for 10-gallon bags, and $.95 for 20-gallon bags, how many of each size should be made per day to maximize the profit?

23. When is it necessary to use the simplex method rather than the graphical method?

24. What types of problems can be solved using slack, surplus, and artificial variables?

25. What kind of problem can be solved using the method of duals?

▷ 26. In solving a linear programming problem, you are given the following initial tableau.

$$\begin{bmatrix} 4 & 2 & 3 & 1 & 0 & 0 & | & 9 \\ 5 & 4 & 1 & 0 & 1 & 0 & | & 10 \\ -6 & -7 & -5 & 0 & 0 & 1 & | & 0 \end{bmatrix}$$

(a) What is the problem being solved?
(b) If the 1 in row 1, column 4 were a -1 rather than a 1, how would it change your answer to part (a)?
(c) After several steps of the simplex algorithm, the following tableau results.

$$\begin{bmatrix} 3 & 0 & 5 & 2 & -1 & 0 & | & 8 \\ 11 & 10 & 0 & -1 & 3 & 0 & | & 21 \\ 47 & 0 & 0 & 13 & 11 & 10 & | & 227 \end{bmatrix}$$

What is the solution? (List only the values of the original variables and the objective function. Do not include slack or surplus variables.)
(d) What is the dual of the problem you found in part (a)?
(e) What is the solution of the dual you found in part (d)? (Do not perform any steps of the simplex algorithm; just examine the tableau given in part (c).)

For each of the following problems, (a) add slack variables and (b) set up the initial simplex tableau.

27. Maximize $z = 2x_1 + 7x_2$
    subject to: $3x_1 + 5x_2 \le 47$
    $x_1 + x_2 \le 25$
    $5x_1 + 2x_2 \le 35$
    $2x_1 + x_2 \le 30$
    $x_1 \ge 0, x_2 \ge 0.$

28. Maximize $z = 15x_1 + 10x_2$
    subject to: $2x_1 + 5x_2 \le 50$
    $x_1 + 3x_2 \le 25$
    $4x_1 + x_2 \le 18$
    $x_1 + x_2 \le 12$
    $x_1 \ge 0, x_2 \ge 0.$

29. Maximize $z = 4x_1 + 6x_2 + 3x_3$
    subject to: $x_1 + x_2 + x_3 \le 100$
    $2x_1 + 3x_2 \le 500$
    $x_1 + 2x_3 \le 350$
    $x_1 \ge 0, x_2 \ge 0, x_3 \ge 0.$

30. Maximize $z = x_1 + 4x_2 + 2x_3$
    subject to: $x_1 + x_2 + x_3 \le 90$
    $2x_1 + 5x_2 + x_3 \le 120$
    $x_1 + 3x_2 \le 80$
    $x_1 \ge 0, x_2 \ge 0, x_3 \ge 0.$

For each of the following, use the simplex method to solve the maximization linear programming problems with initial tableaus as given.

31.

x_1	x_2	x_3	x_4	x_5		
1	2	3	1	0		28
2	4	8	0	1		32
-5	-2	-3	0	0		0

32.

x_1	x_2	x_3	x_4		
2	1	1	0		10
9	3	0	1		15
-2	-3	0	0		0

33.

x_1	x_2	x_3	x_4	x_5	x_6		
1	2	2	1	0	0		50
4	24	0	0	1	0		20
1	0	2	0	0	1		15
-5	-3	-2	0	0	0		0

34.

x_1	x_2	x_3	x_4	x_5		
1	-2	1	0	0		38
1	-1	0	1	0		12
2	1	0	0	1		30
-1	-2	0	0	0		0

Convert the following problems into maximization problems without using duals.

35. Minimize $w = 18y_1 + 10y_2$
    subject to: $y_1 + y_2 \ge 17$
    $5y_1 + 8y_2 \ge 42$
    $y_1 \ge 0, y_2 \ge 0.$

36. Minimize $w = 12y_1 + 20y_2 - 8y_3$
    subject to: $y_1 + y_2 + 2y_3 \ge 48$
    $y_1 + y_2 \ge 12$
    $y_3 \ge 10$
    $3y_1 + y_3 \ge 30$
    $y_1 \ge 0, y_2 \ge 0, y_3 \ge 0.$

37. Minimize $w = 6y_1 - 3y_2 + 4y_3$
subject to: $2y_1 + y_2 + y_3 \geq 112$
$y_1 + y_2 + y_3 \geq 80$
$y_1 + y_2 \quad\;\; \geq 45$
$y_1 \geq 0, y_2 \geq 0, y_3 \geq 0.$

Use the simplex method to solve the following mixed constraint problems.

38. Maximize $z = 2x_1 + 4x_2$
subject to: $3x_1 + 2x_2 \leq 12$
$5x_1 + \;\; x_2 \geq 5$
$x_1 \geq 0, x_2 \geq 0.$

39. Minimize $w = 4y_1 - 8y_2$
subject to: $y_1 + \;\; y_2 \leq 50$
$2y_1 - 4y_2 \geq 20$
$y_1 - \;\; y_2 \leq 22$
$y_1 \geq 0, y_2 \geq 0.$

The following tableaus are the final tableaus of minimizing problems solved by letting $w = -z$. Give the solution and the minimum value of the objective function for each problem.

40.
$$\begin{bmatrix} 0 & 1 & 0 & 2 & 5 & 0 & | & 17 \\ 0 & 0 & 1 & 3 & 1 & 1 & | & 25 \\ 1 & 0 & 0 & 4 & 2 & \frac{1}{2} & | & 8 \\ \hline 0 & 0 & 0 & 2 & 5 & 0 & | & -427 \end{bmatrix}$$

41.
$$\begin{bmatrix} 0 & 0 & 2 & 1 & 0 & 6 & 6 & | & 92 \\ 1 & 0 & 3 & 0 & 0 & 0 & 2 & | & 47 \\ 0 & 1 & 0 & 0 & 0 & 1 & 0 & | & 68 \\ 0 & 0 & 4 & 0 & 1 & 0 & 3 & | & 35 \\ \hline 0 & 0 & 5 & 0 & 0 & 2 & 9 & | & -1957 \end{bmatrix}$$

The tableaus in Exercises 42–44 are the final tableaus of minimizing problems solved by the method of duals. State the solution and the minimum value of the objective function for each problem.

42.
$$\begin{bmatrix} 1 & 0 & 0 & 3 & 1 & 2 & | & 12 \\ 0 & 0 & 1 & 4 & 5 & 3 & | & 5 \\ 0 & 1 & 0 & -2 & 7 & -6 & | & 8 \\ \hline 0 & 0 & 0 & 5 & 7 & 3 & | & 172 \end{bmatrix}$$

43.
$$\begin{bmatrix} 0 & 0 & 1 & 6 & 3 & 1 & | & 2 \\ 1 & 0 & 0 & 4 & -2 & 2 & | & 8 \\ 0 & 1 & 0 & 10 & 7 & 0 & | & 12 \\ \hline 0 & 0 & 0 & 9 & 5 & 8 & | & 62 \end{bmatrix}$$

44.
$$\begin{bmatrix} 1 & 0 & 7 & -1 & | & 100 \\ 0 & 1 & 1 & 3 & | & 27 \\ \hline 0 & 0 & 7 & 2 & | & 640 \end{bmatrix}$$

45. Solve Exercise 19.

46. Solve Exercise 20.

47. Solve Exercise 21.

48. Solve Exercise 22.

Management   *Solve the following minimization problems.*

49. A contractor builds boathouses in two basic models, the atlantic and pacific. Each atlantic model requires 1000 feet of framing lumber, 3000 cubic feet of concrete, and $2000 for advertising. Each pacific model requires 2000 feet of framing lumber, 3000 cubic feet of concrete, and $3000 for advertising. Contracts call for using at least 8000 feet of framing lumber, 18,000 cubic feet of concrete, and $15,000 worth of advertising. If the total spent on each atlantic model is $3000 and the total spent on each pacific model is $4000, how many of each model should be built to minimize costs?

50. A steel company produces two types of alloys. A run of type I requires 3000 pounds of molybdenum and 2000 tons of iron ore pellets as well as $2000 in advertising. A run of type II requires 3000 pounds of molybdenum and 1000 tons of iron ore pellets as well as $3000 in advertising. Total costs are $15,000 on a run of type I and $6000 on a run of type II. Because of various contracts, the company must use at least 18,000 pounds of molybdenum and 7000 tons of iron ore pellets and spend at least $14,000 on advertising. How much of each type should be produced to minimize costs?

CASE 8

A Minimum Cost Balanced Organic Fertilizer Mix*

Organic fertilizers build on three key elements that are essential to plant growth and health. Nitrogen (N) promotes plant growth and resistance to insects. Phosphorus (P) is essential for proper root growth. Potassium (K) aids in plant growth and resistance to disease. The percent of these three elements in a fertilizer is always listed in the order N-P-K. Thus, a 3-2-2 fertilizer is 3% nitrogen, 2% phosphorus, and 2% potassium.

Many garden centers and seed catalogs sell pre-mixed organic fertilizers. However, to get a mixture with a specific N-P-K combination at a minimum cost, a gardener must solve an optimization problem. A fertilizer with a b-b-b mix for some number b, is called a balanced fertilizer. To get a balanced organic fertilizer with a b-b-b mix, a gardener must use a combination of the following choices.

> Blood Meal (11-0-0) @ $6.00 per five-pound bag
> Bone Meal (6-12-0) @ $5.00 per five-pound bag
> Greensand (0-1-7) @ $3.00 per five-pound bag
> Sul-Po-Mag (0-0-22) @ $2.80 per five-pound bag
> Sustane (5-2-4) @ $4.25 per five-pound bag

Linear programming can be used to model this problem. Let x_1 represent the number of pounds of blood meal to be used, x_2 represent the number of pounds of bone meal to be used, and so on. The gardener should minimize the objective function (using costs per pound)

$$z = 1.20x_1 + 1.00x_2 + .60x_3 + .56x_4 + .85x_5,$$

subject to the constraints

$$.11x_1 + .06x_2 \qquad\qquad + .05x_5 = 1$$
$$.12x_2 + .01x_3 \qquad + .02x_5 = 1$$
$$.07x_3 + .22x_4 + .04x_5 = 1,$$

where the element proportions are expressed as percents in decimal form. By setting all three sums equal to the same number, we ensure that the percents will be equal to provide a balanced fertilizer, as required. We use 1 on the right hand side to indicate 1 unit of mixed fertilizer. This makes the calculations easier. The nonnegativity constraints are

$$x_1 \geq 0, \quad x_2 \geq 0, \quad x_3 \geq 0, \quad x_4 \geq 0, \quad x_5 \geq 0.$$

Using a graphing calculator or a computer with linear programming software, we get the following solution.

$$x_1 = 4.5454, \quad x_2 = 8.3333, \quad x_3 = 0, \quad x_4 = 4.5454, \quad x_5 = 0$$

This mixture would cost $16.33 (rounded), and since the mixture weighs 17.4242 pounds, the minimum cost per pound is $.94 (rounded). Verify that a minimum cost organic fertilizer can be obtained by mixing 6 parts blood meal with 11 parts bone meal and 6 parts sul-po-mag.

EXERCISES

1. Show that the minimum cost mixture has proportions 6-11-6.

2. To determine the number b, the percent of each nutrient present in the fertilizer, we must divide the sum on the right of each constraint by the weight of the total mixture. (The results should be equal.) Do this and express the results in the form N-P-K. Is the result a balanced fertilizer?

---

*Turner, Steven J., "Using Linear Programming to Obtain a Minimum Cost Balanced Organic Fertilizer Mix," *The AMATYC Review,* Vol. 14, No. 2, Spring 1993, pp. 21–25.

CHAPTER 9

Sets and Probability

Federal officials cannot predict exactly how traffic deaths are affected by the trends toward fewer drunken drivers and increased use of seat belts. Economists cannot tell exactly how stricter federal regulations on bank loans affect the U.S. economy. The number of traffic deaths and the growth of the economy are subject to many factors that cannot be predicted precisely.

Probability theory enables us to deal with uncertainty. The basic concepts of probability are discussed in this chapter and applications of probability are discussed in the next chapter. Sets and set operations are the basic tools for the study of probability, so we begin with them.

9.1 SETS

Think of a **set** as a collection of objects. For example, a set of coins might contain one of each type of coin now put out by the United States government. Another set might include all the students in a class. The set containing the numbers 3, 4, and 5 is written

$$\{3, 4, 5\},$$

where **set braces**, { }, are used to enclose the numbers belonging to the set. The numbers, 3, 4, and 5 are called the **elements** or **members** of this set. To show that 4 is an element of the set {3, 4, 5}, we use the symbol $\in$ and write

$$4 \in \{3, 4, 5\},$$

read "4 is an element of the set containing 3, 4, and 5."

Also, $5 \in \{3, 4, 5\}$. Place a slash through the symbol $\in$ to show that 8 is *not* an element of this set.

$$8 \notin \{3, 4, 5\}$$

This is read "8 is not an element of the set {3, 4, 5}."

1 Write *true* or *false*.

(a) $9 \in \{8, 4, -3, -9, 6\}$ ~~False~~

(b) $4 \notin \{3, 9, 7\}$ ~~True~~

(c) If $M = \{0, 1, 2, 3, 4\}$, then $0 \in M.$ ~~TRue~~

Answers:
(a) False
(b) True
(c) True

Sets are often named with capital letters, so that if

$$B = \{5, 6, 7\},$$

then, for example, $6 \in B$ and $10 \notin B$. **1**

Sometimes a set has no elements. Some examples are the set of woman presidents of the United States in the period 1788–2000, the set of counting numbers less than 1, and the set of men more than 10 feet tall. A set with no elements is called the **empty set.** The symbol $\emptyset$ is used to represent the empty set.

CAUTION Be careful to distinguish between the symbols 0, $\emptyset$ and $\{0\}$. The symbol 0 represents a *number;* $\emptyset$ represents a *set* with no elements; and $\{0\}$ represents a *set* with one element, the number 0. Do not confuse the empty set symbol $\emptyset$ with the zero $\emptyset$ on a computer screen or printout. ◆

Two sets are **equal** if they contain exactly the same elements. The sets $\{5, 6, 7\}$, $\{7, 6, 5\}$, and $\{6, 5, 7\}$ all contain exactly the same elements and are equal. In symbols,

$$\{5, 6, 7\} = \{7, 6, 5\} = \{6, 5, 7\}.$$

This means that the ordering of the elements in a set is unimportant. Sets that do not contain exactly the same elements are *not equal.* For example, the sets $\{5, 6, 7\}$ and $\{5, 6, 7, 8\}$ do not contain exactly the same elements and are not equal. We show this by writing

$$\{5, 6, 7\} \neq \{5, 6, 7, 8\}.$$

Sometimes we describe a set by a common property of its elements rather than by a list of elements. This common property can be expressed with **set-builder notation;** for example,

2 List the elements in the following sets.

(a) $\{x \mid x$ is a counting number more than 5 and less than 8$\}$

(b) $\{x \mid x$ is an integer, $-3 < x \le 1\}$

Answers:
(a) $\{6, 7\}$
(b) $\{-2, -1, 0, 1\}$

$$\{x \mid x \text{ has property } P\}$$

(read "the set of all elements x such that x has property P") represents the set of all elements x having some property P.

EXAMPLE 1 List the elements belonging to each of the following sets.

(a) $\{x \mid x$ is a natural number less than 5$\}$
The natural numbers less than 5 make up the set $\{1, 2, 3, 4\}$.

(b) $\{x \mid x$ is a state that borders Florida$\} = \{$Alabama, Georgia$\}$ ■ **2**

The **universal set** in a particular discussion is a set that contains all of the objects being discussed. In grade-school arithmetic, for example, the set of whole numbers might be the universal set, whereas in a college calculus class the universal set might be the set of all real numbers. When it is necessary to consider the universal set being used, it will be clearly specified or easily understood from the context of the problem.

Sometimes every element of one set also belongs to another set. For example, if

$$A = \{3, 4, 5, 6\}$$

and

$$B = \{2, 3, 4, 5, 6, 7, 8\},$$

Symbols
$\in$ = Element of
$\notin$ = not an element of
$\subseteq$ = Subset
$\nsubseteq$ = not subset
$\cap$ = intersect—formed from Sets Common in both
$\cup$ = Union — all in common

then every element of A is also an element of B. This is an example of the following definition.

> A set A is a **subset** of a set B (written $A \subseteq B$) provided that every element of A is also an element of B.

3 Write *true* or *false*.

(a) $\{3, 4, 5\} \subseteq \{2, 3, 4, 6\}$ *False*

(b) $\{x \mid x$ is an automobile$\}$ $\subseteq \{x \mid x$ is a motor vehicle$\}$ *True*

(c) $\{3, 6, 9, 10\}$ $\subseteq \{3, 9, 11, 13\}$ *False*

Answers:
(a) False
(b) True
(c) False

EXAMPLE 2 Decide whether or not $M \subseteq N$.

(a) M is the set of all small businesses with less than 20 employees. N is the set of all businesses.

Each business with less than 20 employees is also a business, so $M \subseteq N$.

(b) M is the set of all fourth-grade students in a school at the end of the school year, and N is the set of all nine-year-old students in the school at the end of the school year.

By the end of the school year, some fourth-grade students are ten years old, so there are elements in M that are not in N. Thus, M is not a subset of N, written $M \not\subseteq N$. ■ **3**

Every set A is a subset of itself because the statement "every element of A is also an element of A" is always true. It is also true that the empty set is a subset of every set.*

> For any set A,
>
> $$\emptyset \subseteq A \quad \text{and} \quad A \subseteq A.$$

EXAMPLE 3 List all possible subsets for each of the following sets.

(a) $\{7, 8\}$
There are 4 subsets of $\{7, 8\}$: $\quad n^2 = 2^2 = 4$

$$\emptyset, \quad \{7\}, \quad \{8\}, \quad \{7, 8\}.$$

4 List all subsets of $\{w, x, y, z\}$.

Answer:
$\emptyset, \{w\}, \{x\}, \{y\}, \{z\}, \{w, x\},$
$\{w, y\}, \{w, z\}, \{x, y\}, \{x, z\},$
$\{y, z\}, \{w, x, y\}, \{w, x, z\},$
$\{w, y, z\}, \{x, y, z\}, \{w, x, y, z\}$

(b) $\{a, b, c\}$
There are 8 subsets of $\{a, b, c\}$: $\quad n^3$

$$\emptyset, \quad \{a\}, \quad \{b\}, \quad \{c\}, \quad \{a, b\}, \quad \{a, c\}, \quad \{b, c\}, \quad \{a, b, c\}. \quad ■$$

In Example 3, the subsets of $\{7, 8\}$ and the subsets of $\{a, b, c\}$ were found by trial and error. An alternative method uses a **tree diagram**, a systematic way of listing all the subsets of a given set. Figures 9.1(a) and (b) on the next page show tree diagrams for finding the subsets of $\{7, 8\}$ and $\{a, b, c\}$. **4**

2^4

---

*This fact is not intuitively obvious to most people. If you wish, you can think of it as a convention that we agree to adopt in order to simplify the statements of several results later.

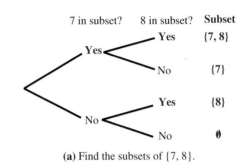

(a) Find the subsets of $\{7, 8\}$.

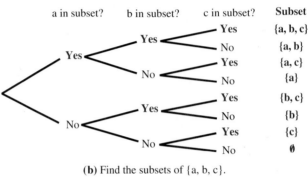

(b) Find the subsets of $\{a, b, c\}$.

FIGURE 9.1

5 Find the number of subsets for each of the following sets.

(a) $\{x \mid x \text{ is a season of the year}\}$

(b) $\{-6, -5, -4, -3, -2, -1, 0\}$

(c) $\{6\}$

Answers:
(a) 16
(b) 128
(c) 2

6 Refer to sets A, B, C, and U in the diagram.

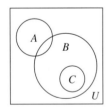

(a) Is $A \subseteq B$?
(b) Is $C \subseteq B$?
(c) Is $C \subseteq U$?
(d) Is $\emptyset \subseteq A$?

Answers:
(a) No
(b) Yes
(c) Yes
(d) Yes

By using the fact that there are two possibilities for each element (either it is in the subset or it is not) we have found that a set with 2 elements has 4 ($= 2^2$) subsets and a set with 3 elements has 8 ($= 2^3$) subsets. Similar arguments work for any finite set and lead to this conclusion.

> A set of n distinct elements has 2^n subsets.

EXAMPLE 4 Find the number of subsets for each of the following sets.

(a) $\{3, 4, 5, 6, 7\}$

Since this set has 5 elements, it has 2^5 or 32 subsets.

(b) $\{x \mid x \text{ is a day of the week}\}$

This set has 7 elements and therefore has $2^7 = 128$ subsets.

(c) $\emptyset$

Since the empty set has 0 elements, it has $2^0 = 1$ subset, $\emptyset$ itself. ■ **5**

Figure 9.2 shows a set A, which is a subset of a set B, because A is entirely in B. (The areas of the regions are not meant to be proportional to the size of the corresponding sets.) The rectangle represents the universal set, U. Such diagrams, called **Venn diagrams,** are used to illustrate relationships among sets. **6**

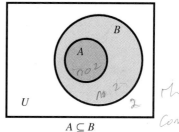

A ⊆ B

FIGURE 9.2

[handwritten: then 2 is a Complement of A + B]

We can operate on sets to get other sets, just as operations on numbers (such as finding the negative, addition, and multiplication) produce other numbers. For example, given a set A and a universal set U, the set of all elements of U which do *not* belong to A is called the **complement** of set A. For example, if set A is the set of all the female students in your class and U is the set of all students in the class, then the complement of A would be the set of all male students in the class. The complement of set A is written A' (read "A-prime"). The Venn diagram of Figure 9.3 shows a set B. Its complement, B', is shown in color.

[handwritten: A' This symbol means Complement it is called prime]

[handwritten: A = {1 2 3}, A∩B = {2 3}, B = {2 3 4}]

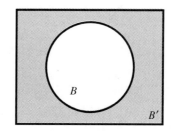

B

B'

FIGURE 9.3

[7] Let $U = \{a, b, c, d, e, f, g\}$, with $K = \{c, d, f, g\}$ and $R = \{a, c, d, e, g\}$. Find

(a) K'; *[handwritten: {a, b, e}]*

(b) R'. *[handwritten: {b f}]*

Answers:

(a) $\{a, b, e\}$

(b) $\{b, f\}$

EXAMPLE 5 Let $U = \{1, 2, 3, 4, 5, 6, 7\}$, $A = \{1, 3, 5, 7\}$, and $B = \{3, 4, 6\}$. Find the following sets.

(a) A'

Set A' contains the elements of U that are not in A.

$$A' = \{2, 4, 6\}$$

(b) $B' = \{1, 2, 5, 7\}$

(c) $\emptyset' = U$ and $U' = \emptyset$ ■ **[7]**

Given two sets A and B, the set of all elements belonging to *both* set A and set B is called the **intersection** of the two sets, written $A \cap B$. For example, the elements that belong to both $A = \{1, 2, 4, 5, 7\}$ and $B = \{2, 4, 5, 7, 9, 11\}$ are 2, 4, 5, and 7, so that

$$A \cap B = \{1, 2, 4, 5, 7\} \cap (2, 4, 5, 7, 9, 11\} = \{2, 4, 5, 7\}.$$

The Venn diagram of Figure 9.4 shows two sets A and B with their intersection, $A \cap B$, shown in color.

8 Find the following.

(a) $\{1, 2, 3, 4\} \cap \{3, 5, 7, 9\}$

(b) Suppose set K is the set of all blue-eyed blondes in a class and J is the set of all blue-eyed brunettes in the class. If the class has only blondes or brunettes, describe $K \cap J$ in words.

(c) Let $P = \{x \mid x$ is a brown-eyed redhead$\}$. Describe $K \cap P$.

Answers:

(a) $\{3\}$

(b) All members of the class with blue eyes

(c) $\emptyset$

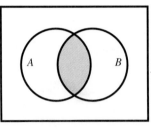

$A \cap B$

FIGURE 9.4

EXAMPLE 6 **(a)** $\{9, 15, 25, 36\} \cap \{15, 20, 25, 30, 35\} = \{15, 25\}$

The elements 15 and 25 are the only ones belonging to both sets.

(b) $\{x \mid x$ is a teen-ager$\} \cap \{x \mid x$ is a senior citizen$\}$ is an empty set. ■ **8**

Two sets that have no elements in common are called **disjoint sets.** For example, there are no elements common to both $\{50, 51, 54\}$ and $\{52, 53, 55, 56\}$, so that these two sets are disjoint, and

$$\{50, 51, 54\} \cap \{52, 53, 55, 56\} = \emptyset.$$

The result of this example can be generalized:

For any sets A and B,

if A and B are disjoint sets then $A \cap B = \emptyset$.

Figure 9.5 is a Venn diagram of disjoint sets.

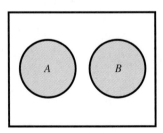

A and B are disjoint sets.

FIGURE 9.5

The set of all elements belonging to set A, to set B, or to both sets is called the **union** of the two sets, written $A \cup B$. For example,

$$\{1, 3, 5\} \cup \{3, 5, 7, 9\} = \{1, 3, 5, 7, 9\}.$$

The Venn diagram of Figure 9.6 shows two sets A and B, with their union $A \cup B$ shown in color.

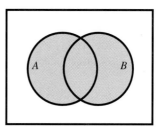

$A \cup B$

FIGURE 9.6

$\{a, b, c, e\}$

$\{1, 2, 3, 4, 5, 8, 9, 14\}$

9 Find the following.

(a) $\{a, b, c\} \cup \{a, c, e\}$

(b) Describe $K \cup J$ in words for the sets given in Side Problem 8(b).

Answers:

(a) $\{a, b, c, e\}$

(b) All members of the class

EXAMPLE 7 **(a)** Find the union of $\{1, 2, 5, 9, 14\}$ and $\{1, 3, 4, 8\}$.

Begin by listing the elements of the first set, $\{1, 2, 5, 9, 14\}$. Then include any elements from the second set *that are not already listed.* Doing this gives

$$\{1, 2, 5, 9, 14\} \cup \{1, 3, 4, 8\} = \{1, 2, 3, 4, 5, 8, 9, 14\}.$$

(b) {terriers, spaniels, chows, dalmatians} $\cup$ {spaniels, collies, bulldogs} = {terriers, spaniels, chows, dalmatians, collies, bulldogs} ■ **9**

Finding the complement of a set, the intersection of two sets, or the union of two sets are examples of *set operations.*

The set operations are summarized below.

OPERATIONS ON SETS

Let A and B be any sets with U the universal set. Then

the **complement** of A, written A', is

$$A' = \{x \mid x \notin A \text{ and } x \in U\};$$

the **intersection** of A and B is

$$A \cap B = \{x \mid x \in A \text{ and } x \in B\};$$

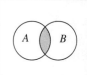

the **union** of A and B is

$$A \cup B = \{x \mid x \in A \text{ or } x \in B \text{ or both}\}.$$

CAUTION As shown in the definitions given above, an element is in the *intersection* of sets A and B if it is in *both A and B*. On the other hand, an element is in the *union* of sets A and B if it is in *A or B or both*. ◆

EXAMPLE 8 The table gives the annual high and low prices, the last price, and the weekly change for six stocks, as listed on the New York Stock Exchange at the end of a recent week.*

Stock	High	Low	Last	Change
ATT	68 7/8	49 1/4	51 7/8	−1 1/2
GnMill	60 1/2	50 3/8	53 1/2	−1 1/4
Hershey	81 1/4	56 1/2	78 3/8	+5 1/2
IBM	128 7/8	83 1/8	103 7/8	+10 1/8
Mobil	120 1/8	93 1/2	109 5/8	−5 1/2
PepsiCo	35 7/8	22 1/8	31 7/8	−5/8

Let set A include all stocks with a high price greater than \$100; B all stocks with a last price between \$75 and \$100; C all stocks with a positive price change. Find the following.

(a) A'

Set A' contains all the listed stocks outside set A, those with a high price less than or equal to \$100, so

$$A' = \{\text{ATT, GnMill, Hershey, PepsiCo}\}.$$

(b) $A \cap C$

The intersection of A and C will contain those stocks with a high price greater than \$100 and a positive price change.

$$A \cap C = \{\text{IBM}\}$$

(c) $A \cup B$

The union of A and B contains all stocks with a high price greater than \$100 or a last price between \$75 and \$100.

$$A \cup C = \{\text{Hershey, IBM, Mobil}\} \quad ■ \quad \boxed{10}$$

*The Sacramento Bee, July 27, 1996, pp. F4 and F7.

10. If $U = \{-3, -2, -1, 0, 1, 2, 3\}$, $A = \{-1, -2, -3\}$, and $B = \{-2, 0, 2\}$, find

(a) $A \cup B$;

(b) B';

(c) $A \cap B$.

Answers:

(a) $\{-3, -2, -1, 0, 2\}$

(b) $\{-3, -1, 1, 3\}$

(c) $\{-2\}$

9.1 EXERCISES

Write true or false for each statement.

1. $3 \in \{2, 5, 7, 9, 10\}$

2. $6 \in \{-2, 6, 9, 5\}$

3. $9 \notin \{2, 1, 5, 8\}$

4. $3 \notin \{7, 6, 5, 4\}$

5. $\{2, 5, 8, 9\} = \{2, 5, 9, 8\}$

6. $\{3, 7, 12, 14\} = \{3, 7, 12, 14, 0\}$

7. $\{\text{all whole numbers greater than 7 and less than 10}\} = \{8, 9\}$

8. $\{\text{all counting numbers not greater than 3}\} = \{0, 1, 2\}$

9. $\{x \mid x \text{ is an odd integer}, 6 \leq x \leq 18\} = \{7, 9, 11, 15, 17\}$

10. $\{x \mid x \text{ is a vowel}\} = \{a, e, i, o, u\}$

11. The elements of a set may be sets themselves, as in $\{1, \{1, 3\}, \{2\}, 4\}$. Explain why the set $\{\emptyset\}$ is not the same set as $\{0\}$.

12. What is set-builder notation? Give an example.

Let $A = \{-3, 0, 3\}$, $B = \{-2, -1, 0, 1, 2\}$, $C = \{-3, -1\}$, $D = \{0\}$, $E = \{-2\}$, and $U = \{-3, -2, -1, 0, 1, 2, 3\}$. Insert $\subseteq$ or $\not\subseteq$ to make the following statements true. (See Example 2.)

13. $A \underline{\subseteq} U$

14. $E \underline{\subseteq} A$

15. $A \underline{\not\subseteq} E$

16. $B \underline{\not\subseteq} C$

17. $\emptyset \underline{\subseteq} A$

18. $\{0, 2\} \underline{\quad} D$

19. $D \underline{\quad} B$

20. $A \underline{\quad} C$

Find the number of subsets for each set. (See Example 4.)

21. $\{A, B, C\}$

22. {red, yellow, blue, black, white}

23. $\{x \mid x$ is an integer between 0 and 7$\}$

24. $\{x \mid x$ is a whole number less than 4$\}$

25. Describe the intersection and union of sets. How do they differ?

Insert $\cap$ or $\cup$ to make each statement true. (See Examples 6 and 7.)

26. $\{5, 7, 9, 19\} \underline{\quad} \{7, 9, 11, 15\} = \{7, 9\}$

27. $\{8, 11, 15\} \underline{\quad} \{8, 11, 19, 20\} = \{8, 11\}$

28. $\{2, 1, 7\} \underline{\quad} \{1, 5, 9\} = \{1\}$

29. $\{6, 12, 14, 16\} \underline{\quad} \{6, 14, 19\} = \{6, 14\}$

30. $\{3, 5, 9, 10\} \underline{\quad} \emptyset = \emptyset$

31. $\{3, 5, 9, 10\} \underline{\quad} \emptyset = \{3, 5, 9, 10\}$

32. $\{1, 2, 4\} \underline{\quad} \{1, 2, 4\} = \{1, 2, 4\}$

33. $\{1, 2, 4\} \underline{\quad} \{1, 2\} = \{1, 2, 4\}$

34. Is it possible for two nonempty sets to have the same intersection and union? If so, give an example.

Let $U = \{2, 3, 4, 5, 7, 9\}$; $X = \{2, 3, 4, 5\}$; $Y = \{3, 5, 7, 9\}$; and $Z = \{2, 4, 5, 7, 9\}$.

List the members of each of the following sets, using set braces. (See Examples 5–7.)

35. $X \cap Y$

36. $X \cup Y$

37. X'

38. Y'

39. $X' \cap Y'$

40. $X' \cap Z$

41. $X \cup (Y \cap Z)$

42. $Y \cap (X \cup Z)$

Let $U = \{$all students in this school$\}$; $M = \{$all students taking this course$\}$; $N = \{$all students taking accounting$\}$; $P = \{$all students taking zoology$\}$.

Describe each of the following sets in words.

43. M'

44. $M \cup N$

45. $N \cap P$

46. $N' \cap P'$

47. Refer to the sets listed in the directions for Exercises 13–20. Which pairs of sets are disjoint?

48. Refer to the sets listed in the directions for Exercises 35–42. Which pairs are disjoint?

Refer to Example 8 in the text. Describe each of the sets in Exercises 49–52 in words; then list the elements of each set.

49. B'

50. $A \cap B$

51. $(A \cup B)'$

52. $(A \cap C)'$

Management   A department store classifies credit applicants by sex, marital status, and employment status. Let the universal set be the set of all applicants, M be the set of male applicants, S be the set of all single applicants, and E be the set of employed applicants. Describe the following sets in words.

53. $M \cap E$

54. $M' \cap S$

55. $M' \cup S'$

Management   Computing power of personal computers has increased dramatically as a result of the ability to place an increasing number of transistors on a single processor chip. The table lists the number of transistors on some popular computer chips made by Intel.*

Year	Chip	Transistors
1971	4004	2,300
1986	386DX	275,000
1989	486DX	1,200,000
1993	Pentium	3,300,000
1995	P6	5,500,000

List the elements of each set.

56. The set of chips produced in the 1980s.

57. The years listed when there were more than 1 million transistors on a chip.

58. The chips with names consisting of both letters and numerals.

Social Science   The top five pay-cable services in 1994 are listed below.† Use this information for Exercises 59–64.

Network	Subscribers (in thousands)	Content
Home Box Office (HBO)	19,200	Movies, variety sports, documentaries
The Disney Channel	12,600	Movies, cartoons
Showtime/The Movie Channel	11,900	Movies, variety, comedy, specials
Spice	11,000	Adult movies
Cinemax	7,800	Movies, comedy, specials

---

*Intel.

†"Top Five Pay-Cable Services, 1994" from *The Universal Almanac 1996*, edited by John W. Wright (Kansas City and New York: Andrews and McMeel, 1996).

List the elements of the following sets.

59. *F*, the set of networks with more than 12,000 thousand subscribers

60. *G*, the set of networks that show cartoons

61. *H*, the set of networks that show adult movies

62. $F \cap G$ **63.** $H \cup G$ **64.** F'

Natural Science   *The table below shows some symptoms of an overactive thyroid and an underactive thyroid.*

Underactive Thyroid	Overactive Thyroid
Sleepiness, s	Insomnia, i
Dry hands, d	Moist hands, m
Intolerance of cold, c	Intolerance of heat, h
Goiter, g	Goiter, g

Let U be the smallest possible set that includes all the symptoms listed, N be the set of symptoms for an underactive thyroid, and O be the set of symptoms for an overactive thyroid. Use the italic letters to list the elements of each set.

65. O'

66. N'

67. $N \cap O$

68. $N \cup O$

69. $N \cap O'$

9.2 APPLICATIONS OF VENN DIAGRAMS

Venn diagrams were used in the last section to illustrate set union and intersection. The rectangular region in a Venn diagram represents the universal set, *U*. Including only a single set, *A,* inside the universal set, as in Figure 9.7, divides *U* into two nonoverlapping regions. Region 1 represents A', those elements outside set *A*, while region 2 represents those elements belonging to set *A*. (The numbering of these regions is arbitrary.)

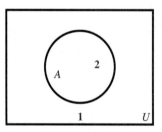

One set leads to 2 regions (numbering is arbitrary).

FIGURE 9.7

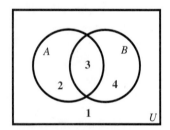

Two sets lead to 4 regions (numbering is arbitrary).

FIGURE 9.8

The Venn diagram of Figure 9.8 shows two sets inside *U*. These two sets divide the universal set into four nonoverlapping regions. As labeled in Figure 9.8, region 1 includes those elements outside both set *A* and set *B*. Region 2 includes those elements belonging to *A* and not to *B*. Region 3 includes those elements belonging to both *A* and *B*. Which elements belong to region 4? (Again, the numbering is arbitrary.)

1 Draw Venn diagrams for the following.

(a) $A \cup B'$

(b) $A' \cap B'$

Answers:

(a)

$A \cup B'$

(b)

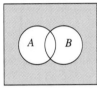

$A' \cap B'$

2 Draw a Venn diagram for the following.

(a) $B' \cap A$

(b) $(A \cup B)' \cap C$

Answers:

(a)

$B' \cap A$

(b)

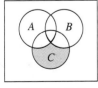

$(A \cup B)' \cap C$

EXAMPLE 1 Draw Venn diagrams similar to Figure 9.8 and shade the regions representing the following sets.

(a) $A' \cap B$

Set A' contains all the elements outside set A. As labeled in Figure 9.8, A' is represented by regions 1 and 4. Set B is represented by the elements in regions 3 and 4. The intersection of sets A' and B, the set $A' \cap B$, is given by the region common to regions 1 and 4 and regions 3 and 4. The result, region 4, is shaded in Figure 9.9.

(b) $A' \cup B'$

Again, set A' is represented by regions 1 and 4, and set B' by regions 1 and 2. To find $A' \cup B'$, identify the region that represents the set of all elements in A', B', or both. The result, which is shaded in Figure 9.10, includes regions 1, 2, and 4. ■ **1**

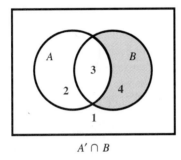

$A' \cap B$

FIGURE 9.9

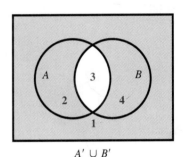

$A' \cup B'$

FIGURE 9.10

Venn diagrams also can be drawn with three sets inside U. These three sets divide the universal set into eight nonoverlapping regions which can be numbered (arbitrarily) as in Figure 9.11.

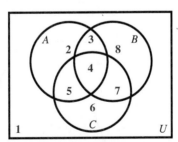

Three sets lead to 8 regions.

FIGURE 9.11

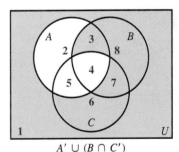

$A' \cup (B \cap C')$

FIGURE 9.12

EXAMPLE 2 Shade $A' \cup (B \cap C')$ on a Venn diagram.

First find $B \cap C'$. See Figure 9.12. Set B is represented by regions 3, 4, 7, and 8, and C' by regions 1, 2, 3, and 8. The overlap of these regions, regions 3 and 8, represents the set $B \cap C'$. Set A' is represented by regions 1, 6, 7, and 8. The union of regions 3 and 8 and regions 1, 6, 7, and 8 includes regions 1, 3, 6, 7, and 8, which are shaded in Figure 9.12. ■ **2**

Venn diagrams can be used to solve problems that result from surveying groups of people. As an example, suppose a researcher collecting data on 100 households finds that

21 have a home computer;

56 have a videocassette recorder (VCR); and

12 have both.

The researcher wants to answer the following questions.

(a) How many do not have a VCR?

(b) How many have neither a computer nor a VCR?

(c) How many have a computer but not a VCR?

A Venn diagram like the one in Figure 9.13 will help sort out the information. In Figure 9.13(a), we put the number 12 in the region common to both a VCR and a computer, because 12 households have both. Of the 21 with a home computer, $21 - 12 = 9$ have no VCR, so in Figure 9.13(b) we put 9 in the region for a computer but no VCR. Similarly, $56 - 12 = 44$ households have a VCR but not a computer, so we put 44 in that region. Finally, the diagram shows that $100 - 44 - 12 - 9 = 35$ households have neither a VCR nor a computer. Now we can answer the questions:

(a) $35 + 9 = 44$ do not have a VCR;

(b) 35 have neither;

(c) 9 have a computer but not a VCR. 3

(a) Place numbers in the regions on a Venn diagram if the data on the 100 households showed

    29 home computers;

    63 VCRs;

    20 with both.

(b) How many have a VCR but not a computer?

Answers:

(a)

(b) 43

(a)

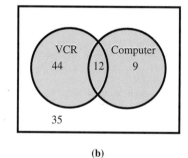

(b)

FIGURE 9.13

EXAMPLE 3 A group of 60 freshman business students at a large university was surveyed, with the following results.

    19 of the students read *Business Week;*

    18 read *The Wall Street Journal;*

    50 read *Fortune;*

    13 read *Business Week* and *The Journal;*

    11 read *The Journal* and *Fortune;*

    13 read *Business Week* and *Fortune;*

     9 read all three.

Use these data to answer the following questions.

    (a) How many students read none of the publications?
    (b) How many read only *Fortune?*
    (c) How many read *Business Week* and *The Journal,* but not *Fortune?*

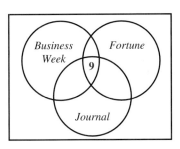

FIGURE 9.14(a)

Once again, use a Venn diagram to represent the data. Since 9 students read all three publications, begin by placing 9 in the area in Figure 9.14(a) that belongs to all three regions.

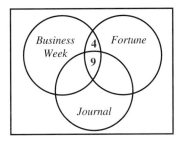

FIGURE 9.14(b)

Of the 13 students who read *Business Week* and *Fortune,* 9 also read *The Journal.* Therefore only $13 - 9 = 4$ students read just *Business Week* and *Fortune.* So place a 4 in the region common only to *Business Week* and *Fortune* readers, as in Figure 9.14(b).

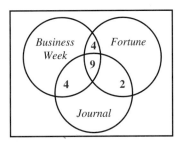

FIGURE 9.14(c)

In the same way, place a 4 in the region of Figure 9.14(c) common only to *Business Week* and *The Journal,* and 2 in the region common only to *Fortune* and *The Journal.*

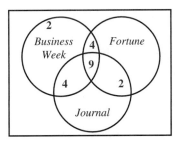

FIGURE 9.14(d)

    The data show that 19 students read *Business Week.* However, $4 + 9 + 4 = 17$ readers have already been placed in the *Business Week* region. The balance of this region in Figure 9.14(d) will contain only $19 - 17 = 2$ students. These 2 students read *Business Week* only—not *Fortune* and not *The Journal.*

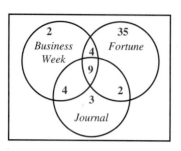

FIGURE 9.14(e)

In the same way, 3 students read only *The Journal* and 35 read only *Fortune*, as shown in Figure 9.14(e).

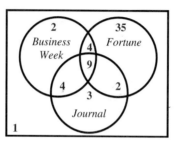

FIGURE 9.14(f)

A total of 2 + 4 + 3 + 4 + 9 + 2 + 35 = 59 students are placed in the various regions of Figure 9.14(f). Since 60 students were surveyed, 60 − 59 = 1 student reads none of the three publications and 1 is placed outside the other regions in Figure 9.14(f).

Figure 9.14(f) can now be used to answer the questions asked above.

(a) Only 1 student reads none of the publications.
(b) There are 35 students who read only *Fortune*.
(c) The overlap of the regions representing *Business Week* and *The Journal* shows that 4 students read *Business Week* and *The Journal* but not *Fortune*. ■ [4]

[4] In the example about the three publications, how many students read exactly

(a) 1 of the publications;

(b) 2 of the publications?

Answers:
(a) 40
(b) 10

EXAMPLE 4   Jeff Friedman is a section chief for an electric utility company. The employees in his section cut down tall trees, climb poles, and splice wire. Friedman reported the following information to the management of the utility.

Of the 100 employees in my section,

45 can cut tall trees;

50 can climb poles;

57 can splice wire;

28 can cut trees and climb poles;

20 can climb poles and splice wire;

25 can cut trees and splice wire;

11 can do all three;

 9 can't do any of the three (management trainees).

The data supplied by Friedman lead to the numbers shown in Figure 9.15. Add the numbers from all the regions to get the total number of Friedman's employees.

$$9 + 3 + 14 + 23 + 11 + 9 + 17 + 13 = 99$$

5 In Example 4, suppose 46 employees can cut tall trees. Then how many

(a) can only cut tall trees?

(b) can cut trees or climb poles?

(c) can cut trees or climb poles or splice wire?

Answers:
(a) 4
(b) 68
(c) 91

Friedman claimed to have 100 employees, but his data indicate only 99. The management decided that Friedman didn't qualify as a section chief, and reassigned him as a nightshift meter reader in Guam. (Moral: He should have taken this course.) ▪ **5**

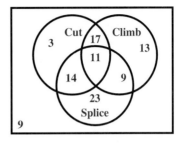

FIGURE 9.15

CAUTION Note that in all the examples above, we started in the innermost region with the intersection of the categories. This is usually the best way to begin solving problems of this type. ◆

We use the symbol $n(A)$ to denote the *number* of elements in A. For instance, if $A = \{w, x, y, z\}$, then $n(A) = 4$. The following useful fact is proved below.

Note: ✱

UNION RULE FOR COUNTING

$$n(A \cup B) = n(A) + n(B) - n(A \cap B)$$

For example, if $A = \{r, s, t, u, v\}$ and $B = \{r, t, w\}$, then $A \cap B = \{r, t\}$, so that $n(A) = 5$, $n(B) = 3$, and $n(A \cap B) = 2$. By the formula in the box, $n(A \cup B) = 5 + 3 - 2 = 6$, which is certainly true since $A \cup B = \{r, s, t, u, v, w\}$.

Here is a proof of the statement in the box: let x be the number of elements in A that are not in B, y be the number of elements in $A \cap B$, and z be the number of elements in B that are not in A, as indicated in Figure 9.16. That diagram shows that $n(A \cup B) = x + y + z$. It also shows that $n(A) = x + y$ and $n(B) = y + z$, so that

$$n(A) + n(B) - n(A \cap B) = (x + y) + (z + y) - y$$
$$= x + y + z$$
$$= n(A \cup B).$$

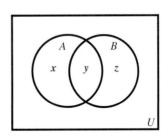

FIGURE 9.16

EXAMPLE 5 A group of 10 students meets to plan a school function. All are majors in accounting or economics or both. Five of the students are economics majors and 7 are majors in accounting. How many major in both subjects?

Use the union rule, with $n(A) = 5$, $n(B) = 7$, and $n(A \cup B) = 10$. We must find $n(A \cap B)$.

$$n(A \cup B) = n(A) + n(B) - n(A \cap B)$$
$$10 = 5 + 7 - n(A \cap B),$$

so

$$n(A \cap B) = 5 + 7 - 10 = 2. \quad \blacksquare \quad \boxed{6}$$

6 If $n(A) = 10$, $n(B) = 7$, and $n(A \cap B) = 3$, find $n(A \cup B)$.

Answer:
14

EXAMPLE 6 Suppose a random sample of 200 voters was selected in the 1994 elections for the U.S. House of Representatives. The approximate numbers from the East, Midwest, South, or West who voted Democrat or Republican is given by the following table.*

	East (E)	Midwest (M)	South (S)	West (W)	Total
Democrat (D)	24	22	27	26	99
Republican (R)	22	28	33	18	101
Total	46	50	60	44	200

Using the letters given in the table, find the number of people in each of the following sets.

(a) $D \cap S$

The set $D \cap S$ consists of all those who voted Democrat *and* were from the South. From the table, we see that there were 27 such people.

(b) $D \cup S$

The set $D \cup S$ consists of all those who voted Democrat *or* were from the South. We include all 99 who voted Democrat, plus the 33 who were from the South and did not vote Democrat, for a total of 132. Alternatively, we could use the formula $n(D \cup S) = n(D) + n(S) - n(D \cap S) = 99 + 60 - 27 = 132$.

7 Find the number of people in each set.

(a) $R \cup M$

(b) $(R \cap W) \cup E'$

Answers:
(a) 123
(b) 154

(c) $(E \cup W) \cap R'$

Begin with the set $E \cup W$, which is everyone from the East or the West. This consists of the four categories with 24, 22, 26, and 18 people. Of this set, take those who did *not* vote Republican, for a total of $24 + 26 = 50$ people. This is the number of people from the East or the West who did not vote Republican. $\quad \blacksquare \quad \boxed{7}$

---

*The New York Times, Nov. 13, 1994, p. 24.

9.2 EXERCISES

Sketch a Venn diagram like the one at the right and use shading to show each of the following sets. (See Example 1.)

1. $B \cap A'$

2. $A \cup B'$

3. $A' \cup B$

4. $A' \cap B'$

5. $B' \cup (A' \cap B')$

6. $(A \cap B) \cup B'$

7. U'

8. $\emptyset'$

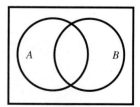

Sketch a Venn diagram like the one shown, and use shading to show each of the following sets. (See Example 2.)

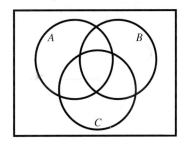

9. $(A \cap B) \cap C$ **10.** $(A \cap C') \cup B$ **11.** $A \cap (B \cup C')$

12. $A' \cap (B \cap C)$ **13.** $(A' \cap B') \cap C$ **14.** $(A \cap B') \cup C$

15. $(A \cap B') \cap C$ **16.** $A' \cap (B' \cup C)$

Use Venn diagrams to answer the following questions. (See Examples 3 and 4.)

17. Social Science A survey of people attending a Lunar New Year celebration in Chinatown yielded the following results:

    120 were women;
    150 spoke Cantonese;
    170 lit firecrackers;
    108 of the men spoke Cantonese;
    100 of the men did not light firecrackers;
    18 of the non-Cantonese-speaking women lit firecrackers;
    78 non-Cantonese-speaking men did not light firecrackers;
    30 of the women who spoke Cantonese lit firecrackers.

(a) How many attended? 342
(b) How many of those who attended did not speak Cantonese? 192
(c) How many women did not light firecrackers? 72
(d) How many of those who lit firecrackers were Cantonese-speaking men? 86

18. Management Jeff Friedman, of Example 4 in the text, was again reassigned, this time to the home economics department of the electric utility. He interviewed 140 people in a suburban shopping center to find out some of their cooking habits. He obtained the following results. Should he be reassigned yet one more time?

    58 use microwave ovens;
    63 use electric ranges;
    58 use gas ranges;
    19 use microwave ovens and electric ranges;
    17 use microwave ovens and gas ranges;
    4 use both gas and electric ranges;
    1 uses all three;
    2 cook only with solar energy.

19. Social Science A recent nationwide survey revealed the following data concerning unemployment in the United States. Out of the total population of people who are working or seeking work,

    9% are unemployed;
    11% are black;
    30% are youths;
    1% are unemployed black youths;
    5% are unemployed youths;
    7% are employed black adults;
    23% are nonblack employed youths.

What percent are
(a) not black?
(b) employed adults?
(c) unemployed adults?
(d) unemployed black adults?
(e) unemployed nonblack youths?

20. Country-western songs emphasize three basic themes: love, prison, and trucks. A survey of the local country-western radio station produced the following data.

    12 songs were about a truck driver who was in love while in prison;
    13 about a prisoner in love;
    28 about a person in love;
    18 about a truck driver in love;
    3 about a truck driver in prison who was not in love;
    2 about a prisoner who was not in love and did not drive a truck;
    8 about a person out of jail who was not in love, and did not drive a truck;
    16 about truck drivers who were not in prison.

(a) How many songs were surveyed?
Find the number of songs about
(b) truck drivers;
(c) prisoners;
(d) truck drivers in prison;
(e) people not in prison;
(f) people not in love.

21. Natural Science After a genetics experiment, the number of pea plants having certain characteristics was tallied, with the results as follows.

    22 were tall;
    25 had green peas;
    39 had smooth peas;
    9 were tall and had green peas;
    17 were tall and had smooth peas;
    20 had green peas and smooth peas;
    6 had all three characteristics;
    4 had none of the characteristics.

(a) Find the total number of plants counted.
(b) How many plants were tall and had peas that were neither smooth nor green?
(c) How many plants were not tall but had peas that were smooth and green?

22. Natural Science Human blood can contain either no antigens, the A antigen, the B antigen, or both the A and B antigens. A third antigen, called the Rh antigen, is important in human reproduction, and again may or may not be present in an individual. Blood is called type A-positive if the individual has the A and Rh antigens, but not the B antigen. A person having only the A and B antigens is said to have type AB-negative blood. A person having only the Rh antigen has type O-positive blood. Other blood types are defined in a similar manner. Identify the blood type of the individuals in regions (a)–(g) of the Venn diagram.

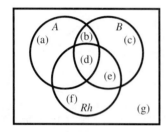

23. Natural Science Use the diagram from Exercise 22. In a certain hospital, the following data were recorded.

> 25 patients had the A antigen;
> 17 had the A and B antigens;
> 27 had the B antigen;
> 22 had the B and Rh antigens;
> 30 had the Rh antigen;
> 12 had none of the antigens;
> 16 had the A and Rh antigens;
> 15 had all three antigens.

How many patients
(a) were represented?
(b) had exactly one antigen?
(c) had exactly two antigens?
(d) had O-positive blood?
(e) had AB-positive blood?
(f) had B-negative blood?
(g) had O-negative blood?
(h) had A-positive blood?

24. Social Science At a pow-wow in Arizona, Native Americans from all over the Southwest came to participate in the ceremonies. A coordinator of the pow-wow took a survey and found that

> 15 families brought food, costumes, and crafts;
> 25 families brought food and crafts;
> 42 families brought food;
> 20 families brought costumes and food;
> 6 families brought costumes and crafts, but not food;
> 4 families brought crafts, but neither food nor costumes;
> 10 families brought none of the three items;
> 18 families brought costumes but not crafts.

(a) How many families were surveyed?
(b) How many families brought costumes?
(c) How many families brought crafts, but not costumes?
(d) How many families did not bring crafts?
(e) How many families brought food or costumes?

25. Social Science The U.S. population by age and race in 1995 (in millions) is given in the table.*

	Non-Hispanic White (A)	Hispanic (B)	African-American (C)	Asian-American (D)	Total
Under 45 (E)	124.5	21.6	23.9	7.0	177.0
45–64 (F)	40.8	3.7	5.0	1.6	51.1
65 and over (G)	28.6	1.5	2.7	.6	33.4
Total	193.9	26.8	31.6	9.2	261.5

Using the letters given in the table, find the number of people in each set.
(a) $A \cap E$ **(b)** $F \cup B$ **(c)** $F \cup (C \cap G)$
(d) $E' \cap (B \cup G)$ **(e)** $G' \cup D$ **(f)** $F' \cap (A' \cap C')$

26. Social Science The projected U.S. population by age and race in 2025 (in millions) is given in the table.*

	Non-Hispanic White (A)	Hispanic (B)	African-American (C)	Asian-American (D)	Total
Under 45 (E)	110.5	33.9	30.0	15.9	190.3
45–64 (F)	52.2	11.0	9.2	5.2	77.6
65 and over (G)	47.2	6.0	5.5	2.9	61.6
Total	209.9	50.9	44.7	24.0	329.5

Using the letters given in the table, find the number of people in each set.
(a) $D \cap E$ **(b)** $F \cup B$ **(c)** $G \cap (C \cup B)$
(d) $E' \cup (A \cup D)$ **(e)** $(F' \cup G') \cap B$

27. Restate the union rule in words.

Use Venn diagrams to answer the following questions. (See Example 5.)

28. If $n(A) = 5$, $n(B) = 8$, and $n(A \cap B) = 4$, what is $n(A \cup B)$?

Projections of Hispanic and Non-Hispanic Populations by Age and Sex: 1995 to 2025, U.S. Bureau of the Census.

29. If $n(A) = 12$, $n(B) = 27$, and $n(A \cup B) = 30$, what is $n(A \cap B)$?

30. Suppose $n(B) = 7$, $n(A \cap B) = 3$, and $n(A \cup B) = 20$. What is $n(A)$?

31. Suppose $n(A \cap B) = 5$, $n(A \cup B) = 35$, and $n(A) = 13$. What is $n(B)$?

Draw a Venn diagram and use the given information to fill in the number of elements for each region.

32. $n(U) = 38$, $n(A) = 16$, $n(A \cap B) = 12$, $n(B') = 20$

33. $n(A) = 26$, $n(B) = 10$, $n(A \cup B) = 30$, $n(A') = 17$

34. $n(A \cup B) = 17$, $n(A \cap B) = 3$, $n(A) = 8$, $n(A' \cup B') = 21$

35. $n(A') = 28$, $n(B) = 25$, $n(A' \cup B') = 45$, $n(A \cap B) = 12$

36. $n(A) = 28$, $n(B) = 34$, $n(C) = 25$, $n(A \cap B) = 14$, $n(B \cap C) = 15$, $n(A \cap C) = 11$, $n(A \cap B \cap C) = 9$, $n(U) = 59$

37. $n(A) = 54$, $n(A \cap B) = 22$, $n(A \cup B) = 85$, $n(A \cap B \cap C) = 4$, $n(A \cap C) = 15$, $n(B \cap C) = 16$, $n(C) = 44$, $n(B') = 63$

38. $n(A \cap B) = 6$, $n(A \cap B \cap C) = 4$, $n(A \cap C) = 7$, $n(B \cap C) = 4$, $n(A \cap C') = 11$, $n(B \cap C') = 8$, $n(C) = 15$, $n(A' \cap B' \cap C') = 5$

39. $n(A) = 13$, $n(A \cap B \cap C) = 4$, $n(A \cap C) = 6$, $n(A \cap B') = 6$, $n(B \cap C) = 6$, $n(B \cap C') = 11$, $n(B \cup C) = 22$, $n(A' \cap B' \cap C') = 5$

*In Exercises 40–43, show that the statements are true by drawing Venn diagrams and shading the regions representing the sets on each side of the equals signs.**

40. $(A \cup B)' = A' \cap B'$

41. $(A \cap B)' = A' \cup B'$

42. $A \cap (B \cup C) = (A \cap B) \cup (A \cap C)$

43. $A \cup (B \cap C) = (A \cup B) \cap (A \cup C)$

---

*The statements in Exercises 40 and 41 are known as De Morgan's laws. They are named for the English mathematician Augustus De Morgan (1806–71).

9.3 PROBABILITY

If you go to a supermarket and buy five pounds of peaches at $1.26 per pound, you can easily find the *exact* price of your purchase: $6.30. On the other hand, the produce manager of the market is faced with the problem of ordering peaches. The manager may have a good estimate of the number of pounds of peaches that will be sold during the day, but there is no way to know *exactly*. The number of pounds that customers will purchase during a day cannot be predicted exactly.

Many problems that come up in applications of mathematics involve phenomena for which exact prediction is impossible. The best that we can do is to determine the *probability* of the possible outcomes. In this section and the next we introduce some of the terminology and basic premises of probability theory.

In probability, an **experiment** is an activity or occurrence with an observable result. Each repetition of an experiment is called a **trial.** The possible results of each trial are called **outcomes.** An example of a probability experiment is the tossing of a coin. Each trial of the experiment (each toss) has two possible outcomes: heads, *h*, and tails, *t*. If the outcomes *h* and *t* are equally likely to occur, then the coin is not "loaded" to favor one side over the other. Such a coin is called **fair.** For a coin that is fair, this "equally likely" assumption is made for each trial.

Because *two* equally likely outcomes are possible, *h* and *t*, theoretically a coin tossed many, many times should come up heads approximately 1/2 of the time. Also, the more times the coin is tossed, the closer the occurrence of heads should be to 1/2. Therefore, it is reasonable to define the probability of a head coming up to be 1/2. Similarly, we define the probability of tails coming up to be 1/2.

For now we shall concentrate on experiments, like coin tossing, in which each outcome is equally likely and use the probabilities suggested by this fact.

> If there are n equally likely outcomes of an experiment, then the probability of any one outcome is $1/n$.

EXAMPLE 1 Suppose a spinner like the one shown in Figure 9.17 is spun.

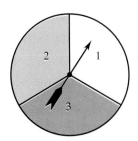

FIGURE 9.17

1 Find the probability that the spinner of Figure 9.17

(a) points to 3;

(b) does not point to 1.

Answers:
(a) 1/3
(b) 2/3

(a) Find the probability that the spinner will point to 1.
It is natural to assume that if this spinner were spun many, many times, it would point to 1 about 1/3 of the time, so that the required probability is 1/3.

(b) Find the probability that the spinner will point to 2.
Since 2 is one of three possible outcomes, the probability is 1/3. ■ ☐1

An ordinary die is a cube whose six faces show the following numbers of dots: 1, 2, 3, 4, 5, and 6. If the die is not "loaded" to favor certain faces over others (a fair die), then any one of the six faces is equally likely to come up when the die is rolled.

2 A jar holds a red marble, a blue marble, and a green marble. If one marble is drawn at random, what is the probability it is

(a) green?

(b) red?

Answers:
(a) 1/3
(b) 1/3

EXAMPLE 2 **(a)** If a single fair die is rolled, find the probability of rolling the number 4.
One out of six faces is a 4, so the required probability is 1/6.

(b) Using the same die, find the probability of rolling a 6.
One out of the six faces shows a 6, so the probability is 1/6. ■ ☐2

SAMPLE SPACES The set of all possible outcomes for an experiment is called the **sample space** for the experiment. A sample space for the experiment of tossing a coin has two outcomes: heads, h, and tails, t. If S represents this sample space, then

$$S = \{h, t\}.$$

EXAMPLE 3 Give a sample space for each experiment.

(a) A spinner like the one in Figure 9.17 is spun.
The three outcomes are 1, 2, or 3, so one sample space is

$$\{1, 2, 3\}.$$

(b) For the purposes of a public opinion poll, people are classified as young, middle-aged, or older, and as male or female.

A sample space for this poll could be written as a set of ordered pairs of these types.

{(young, male), (young, female), (middle-aged, male),
(middle-aged, female), (older, male), (older, female)}

(c) An experiment consists of studying the number of boys and girls in families with exactly 3 children. Let *b* represent *boy* and *g* represent *girl*.

3 **(a)** Write an equally likely sample space for the experiment of rolling a single fair die.

(b) Write an equally likely sample space for the experiment of tossing 2 fair coins.

(c) Write a sample space for the experiment of tossing 2 fair coins, if we are interested only in the number of heads. Are the outcomes in this sample space equally likely?

Answers:
(a) {1, 2, 3, 4, 5, 6}
(b) {hh, ht, th, tt}
(c) {0, 1, 2}; no

A 3-child family can have 3 boys, written *bbb*, 3 girls, *ggg*, or various combinations, such as *bgg*. A sample space with four outcomes (not equally likely) is

$$S_1 = \{3 \text{ boys}, 2 \text{ boys and } 1 \text{ girl}, 1 \text{ boy and } 2 \text{ girls}, 3 \text{ girls}\}.$$

Another sample space that considers the ordering of the births of the 3 children with, for example, *bgg* different from *gbg* or *ggb*, is

$$S_2 = \{bbb, bbg, bgb, gbb, bgg, gbg, ggb, ggg\}.$$

The second sample space, S_2, has equally likely outcomes; S_1 does not, since there is more than one way to get a family with 2 boys and 1 girl or a family of 2 girls and 1 boy, but only one way to get 3 boys or 3 girls. ■

CAUTION An experiment may have more than one sample space, as shown in Example 3(c). The most useful sample spaces have equally likely outcomes, but it is not always possible to choose such a sample space. ◆ **3**

EVENTS An **event** is a subset of outcomes from a sample space. If the sample space for tossing a coin is $S = \{h, t\}$, then one event is the subset $\{h\}$, which represents the outcome "heads." For the experiment of rolling a single fair die, with $S = \{1, 2, 3, 4, 5, 6\}$, some possible events are listed below.

The die shows an even number: $E_1 = \{2, 4, 6\}$.
The die shows a 1: $E_2 = \{1\}$.
The die shows a number less than 5: $E_3 = \{1, 2, 3, 4\}$.
The die shows a multiple of 3: $E_4 = \{3, 6\}$.

EXAMPLE 4 For the sample space S_2 in Example 3(c), write the following events.
(a) Event *H*: the family has exactly 2 girls.
Families can have exactly 2 girls with either *bgg*, *gbg*, or *ggb*, so that event *H* is

$$H = \{bgg, gbg, ggb\}.$$

4 Suppose a die is tossed. Write the following events.

(a) The number showing is less than 3.

(b) The number showing is 5.

(c) The number showing is 8.

Answers:
(a) {1, 2}
(b) {5}
(c) ∅

(b) Event *K*: the 3 children are the same sex.
Two outcomes satisfy this condition, all boys or all girls.

$$K = \{bbb, ggg\}$$

(c) Event *J*: the family has 3 girls.
Only *ggg* satisfies this condition, so

$$J = \{ggg\}. \quad ■ \quad \boxed{4}$$

In Example 4(c), event J had only one possible outcome, ggg. Such an event, with only one possible outcome, is a **simple event.** If an event E equals the sample space S, then E is a **certain event.** If event $E = \emptyset$, then E is an **impossible event.**

5 Which of the events listed in Side Problem 4 is

(a) simple?

(b) certain?

(c) impossible?

Answers:

(a) Part (b)

(b) None

(c) Part (c)

EXAMPLE 5 Suppose a die is rolled. As shown in Side Problem 3(a), the sample space is $\{1, 2, 3, 4, 5, 6\}$.

(a) The event "the die shows a 4," $\{4\}$, has only one possible outcome. It is a simple event.

(b) The event "the number showing is less than 10" equals the sample space $S = \{1, 2, 3, 4, 5, 6\}$. This event is a certain event; if a die is rolled the number showing (either 1, 2, 3, 4, 5, or 6) must be less than 10.

(c) The event "the die shows a 7" is the empty set, $\emptyset$; this is an impossible event. ■ **5**

PROBABILITY For sample spaces with *equally likely* outcomes, the *probability of an event* is defined as follows.

BASIC PROBABILITY PRINCIPLE

Suppose event E is a subset of a sample space S. Then the **probability that event E occurs,** written $P(E)$, is

$$P(E) = \frac{n(E)}{n(S)}.$$

This definition means that the probability of an event is a number that indicates the relative likelihood of the event.

EXAMPLE 6 Suppose a single fair die is rolled. Use the sample space $S = \{1, 2, 3, 4, 5, 6\}$ and give the probability of each of the following events.

(a) E: the die shows an even number.

Here, $E = \{2, 4, 6\}$, a set with three elements. Because S contains six elements,

$$P(E) = \frac{3}{6} = \frac{1}{2}.$$

(b) F: the die shows a number greater than 4.

Since F contains two elements, 5 and 6,

$$P(F) = \frac{2}{6} = \frac{1}{3}.$$

6 A fair die is rolled. Find the probability of rolling

(a) an odd number;

(b) 2, 4, 5, or 6;

(c) a number greater than 5;

(d) the number 7.

Answers:

(a) 1/2

(b) 2/3

(c) 1/6

(d) 0

(c) G: the die shows a number less than 10.

Event G is a certain event, with

$$G = \{1, 2, 3, 4, 5, 6\}, \quad \text{so} \quad P(G) = \frac{6}{6} = 1.$$

(d) H: the die shows an 8.

This event is impossible, so

$$P(H) = \frac{0}{6} = 0. \quad ■ \quad \boxed{6}$$

EXAMPLE 7 If a single card is drawn at random from an ordinary well-shuffled, 52-card deck (shown in Figure 9.18), find the probability of each of the following events.

(a) Drawing an ace

There are 4 aces in the deck. The event "drawing an ace" is

{heart ace, diamond ace, club ace, spade ace}.

Therefore,

$$P(\text{ace}) = \frac{4}{52} = \frac{1}{13}.$$

(b) Drawing a face card (K, Q, J of any suit)

Since there are 12 face cards,

$$P(\text{face card}) = \frac{12}{52} = \frac{3}{13}.$$

(c) Drawing a spade

The deck contains 13 spades, so

$$P(\text{spade}) = \frac{13}{52} = \frac{1}{4}.$$

(d) Drawing a spade or a heart

Besides the 13 spades, the deck contains 13 hearts, so

$$P(\text{spade or heart}) = \frac{26}{52} = \frac{1}{2}. \ \blacksquare \ \boxed{7}$$

7 A single playing card is drawn at random from an ordinary 52-card deck. Find the probability of drawing

(a) a queen;

(b) a diamond;

(c) a red card.

Answers:
(a) 1/13
(b) 1/4
(c) 1/2

FIGURE 9.18

The probability of each event in the preceding examples was a number between 0 and 1, inclusive. The same thing is true in the general case. If an event E consists of m of the possible n outcomes, then $0 \le m \le n$ and hence, $0 \le m/n \le 1$. Since $P(E) = m/n$, $P(E)$ is between 0 and 1, inclusive.

For any event E, $\quad 0 \le P(E) \le 1.$

For any sample space S, $\quad P(S) = 1$ and $P(\emptyset) = 0.$

EXAMPLE 8 The table gives the number of bachelor's and graduate degrees granted in the U.S. in millions.*

Year	Degrees
1989–90	1.49
1990–91	1.54
1991–92	1.60
1992–93	1.62
1993–94	1.65
1994–95	1.67

Find the probability that a randomly selected person with a degree granted during these years received the degree in the 1993–94 year.

We must first find the total number of degrees granted during 1989–95. Verify that the amounts in the table add up to 9.57. Then,

$$P(1993\text{–}94) = \frac{1.65}{9.57} = .172. \quad \blacksquare \quad \boxed{8}$$

8 Find the probability that a randomly selected person with a degree from the years shown in Example 8 received the degree in 1990–91.

Answer:
.161

*The Universal Almanac, 1996, John W. Wright, General Editor, Andrews and McMeel, Kansas City, p. 244.

9.3 EXERCISES

1. What is meant by a "fair" coin or die?

2. What is meant by the sample space for an experiment?

Write sample spaces for the experiments in Exercises 3–8. (See Example 3.)

3. A day in June is chosen for a picnic.

4. At a sleep center, a person is asked the number of hours (to the nearest hour) he slept yesterday.

5. The manager of a small company must decide whether to buy office space or lease it.

6. A student must decide which one of three courses that interest her to take.

7. A CPA chooses a month as the end of the fiscal year for a new corporation.

8. A coin is tossed and a die is rolled.

9. Define an event.

10. What is a simple event?

For the experiments in Exercises 11–16, write out an equally likely sample space, and then write the indicated events in set notation. (See Examples 3–5.)

11. A marble is drawn at random from a bowl containing 3 yellow, 4 white, and 8 blue marbles.
 (a) A yellow marble is drawn.
 (b) A blue marble is drawn.
 (c) A white marble is drawn.
 (d) A black marble is drawn.

12. Slips of paper marked with the numbers 1, 2, 3, 4, and 5 are placed in a box. After being mixed, two slips are drawn.
   (a) Both slips are marked with even numbers.
   (b) One slip is marked with an odd number and the other is marked with an even number.
   (c) Both slips are marked with the same number.

13. An unprepared student takes a three-question true/false quiz in which he guesses the answers to all three questions.
   (a) The student gets three answers wrong.
   (b) The student gets exactly two answers correct.
   (c) The student gets only the first answer correct.

14. A die is tossed twice, with the tosses recorded as ordered pairs.
   (a) The first die shows a 3.
   (b) The sum of the numbers showing is 8.
   (c) The sum of the numbers showing is 13.

15. One urn contains four balls, labeled 1, 2, 3, and 4. A second urn contains five balls, labeled 1, 2, 3, 4, and 5. An experiment consists of taking one ball from the first urn, and then taking a ball from the second urn.
   (a) The number on the first ball is even.
   (b) The number on the second ball is even.
   (c) The sum of the numbers on the two balls is 5.
   (d) The sum of the numbers on the two balls is 1.

16. From 5 employees, Strutz, Martin, Hampton, Williams, and Ewing, 2 are selected to attend a conference.
   (a) Hampton is selected.
   (b) Strutz and Martin are not both selected.
   (c) Both Williams and Ewing are selected.

A single fair die is rolled. Find the probabilities of the following events. (See Example 6.)

17. Getting a 5

18. Getting a number less than 4

19. Getting a number greater than 4

20. Getting a 2 or a 5

A card is drawn from a well-shuffled deck of 52 cards. Find the probability of drawing each of the following. (See Example 7.)

21. A 9
22. A black card
23. A black 9
24. A heart
25. The 9 of hearts
26. A 2 or a queen
27. A black 7 or a red 8
28. A red face card

A jar contains 5 red, 4 black, 7 purple, and 9 green marbles. If a marble is drawn at random, what is the probability that the marble is

29. red?
30. black?
31. green?
32. purple?

33. not black?
34. not purple?
35. red or black?
36. black or purple?

Solve each problem. (See Example 8.)

37. Management   The table gives U.S. advertising volume in millions of dollars by medium in 1994.*

Medium	Expenditures
Newspapers	34.4
Magazines	7.9
Television	34.2
Radio	10.5
Direct Mail	29.6
Yellow Pages	9.8
Other	23.6
Total	150.0

Find the probability that a company randomly decides to spend its advertising dollars on
(a) direct mail,
(b) television,
(c) newspapers.

38. Social Science   The population of the United States by race in 1995, and the projected population by race for the year 2025 are given below.†

U.S. Population by Race (in thousands)

Race	1995	2025
White	193,900	209,900
Hispanic	26,800	50,900
African-Amer.	31,600	44,700
Asian-Amer.	9,200	24,000
Other	1,900	2,800

Find the probability of a randomly selected person being each of the following.
(a) Hispanic in 1995
(b) Hispanic in 2025
(c) African-American in 1995
(d) African-American in 2025

The Universal Almanac, 1996, John W. Wright, General Editor, Andrews and McMeel, Kansas City, p. 270.

†Projections of Hispanic and Non-Hispanic Populations, by Age and Sex: 1995 to 2025, from U.S. Bureau of the Census, *Current Population Reports.*

9.4 BASIC CONCEPTS OF PROBABILITY

We discuss the probability of more complex events in this section. Since events are sets, we can use set operations to find unions, intersections, and complements of events.

EXAMPLE 1   Suppose a die is tossed. Let E be the event "the die shows a number greater than 3," and let F be the event "the die shows an even number." Then

$$E = \{4, 5, 6\} \quad \text{and} \quad F = \{2, 4, 6\}.$$

Find each of the following events (sets).

(a) The die shows an even number greater than 3.

We are looking for an even number *and* a number that is greater than 3. The event $E \cap F$ includes the outcomes common to *both* E and F. Here,

$$E \cap F = \{4, 6\}.$$

(b) The die shows an even number or a number greater than 3.

We want numbers that are even, or greater than 3, *or* both even and greater than 3, which is the event $E \cup F$.

$$E \cup F = \{2, 4, 5, 6\}$$

(c) The die shows a number less than or equal to 3.

This set is the complement of E,

$$E' = \{1, 2, 3\}.$$

The Venn diagrams of Figure 9.19 show the events $E \cap F$, $E \cup F$, and E'. ■ 1

1 Give the following events for the experiment of Example 1 if $E = \{1, 3\}$ and $F = \{2, 3, 4, 5\}$.

(a) $E \cap F$

(b) $E \cup F$

(c) E'

Answers:

(a) $\{3\}$

(b) $\{1, 2, 3, 4, 5\}$

(c) $\{2, 4, 5, 6\}$

$E \cap F$

$E \cup F$

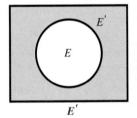

E'

FIGURE 9.19

Two events that cannot both occur at the same time, such as getting both a head and a tail on the same toss of a coin, are called **mutually exclusive events.**

Events A and B are mutually exclusive events if $A \cap B = \emptyset$.

For any event E, the events E and E' are mutually exclusive. See Figure 9.19. By definition, mutually exclusive events are disjoint sets.

2 In Example 2, let $F = \{2, 4, 6\}$, $K = \{1, 3, 5\}$, and G remain the same. Are the following events mutually exclusive?

(a) F and K

(b) F and G

Answers:
(a) Yes
(b) No

EXAMPLE 2   Let $S = \{1, 2, 3, 4, 5, 6\}$, the sample space for tossing a die. Let $E = \{4, 5, 6\}$, and let $G = \{1, 2\}$. Then E and G are mutually exclusive events because they have no outcomes in common; $E \cap G = \emptyset$. See Figure 9.20.   ■   2

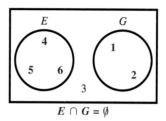

$$E \cap G = \emptyset$$

FIGURE 9.20

A summary of the set operations for events is given below.

Let E and F be events for a sample space S.

$E \cap F$ occurs when both E and F occur.

$E \cup F$ occurs when either E or F or both occur.

E' occurs when E does not.

E and F are mutually exclusive if $E \cap F = \emptyset$.

To determine the probability of the union of the two events E and F in a sample space S, we use the union rule for counting from Section 9.2:

$$n(E \cup F) = n(E) + n(F) - n(E \cap F).$$

Dividing both sides by $n(S)$ shows that

$$\frac{n(E \cup F)}{n(S)} = \frac{n(E)}{n(S)} + \frac{n(F)}{n(S)} - \frac{n(E \cap F)}{n(S)}$$

$$P(E \cup F) = P(E) + P(F) - P(E \cap F).$$

This discussion is summarized below.

UNION RULE FOR PROBABILITY

For any events E and F from a sample space S,

$$P(E \cup F) = P(E) + P(F) - P(E \cap F).$$

EXAMPLE 3   If a single card is drawn from an ordinary deck of cards, find the probability that it will be red or a face card.

Let R represent the event "red card" and F the event "face card." There are 26 red cards in the deck, so $P(R) = 26/52$. There are 12 face cards in the deck, so

[3] A single card is drawn from an ordinary deck. Find the probability that it is black or a 9.

Answer:
7/13

$P(F) = 12/52$. Since there are 6 red face cards in the deck, $P(R \cap F) = 6/52$. By the union rule, the probability that the card will be red or a face card is

$$P(R \cup F) = P(R) + P(F) - P(R \cap F)$$

$$= \frac{26}{52} + \frac{12}{52} - \frac{6}{52} = \frac{32}{52} = \frac{8}{13}. \quad \blacksquare \quad \boxed{3}$$

CAUTION Recall from Section 9.1, the word "or" always indicates a *union*. ◆

EXAMPLE 4 Suppose two fair dice (plural of *die*) are rolled. Find each of the following probabilities.

(a) The first die shows a 2 or the sum of the results is 6 or 7.

The sample space for the throw of two dice is shown in Figure 9.21, where 1-1 represents the event "the first die shows a 1 and the second die shows a 1," 1-2 represents "the first die shows a 1 and the second die shows a 2," and so on. Let *A* represent the event "the first die shows a 2" and *B* represent the event "the sum of the results is 6 or 7." These events are indicated in Figure 9.21. From the diagram, event *A* has 6 elements, *B* has 11 elements, and the sample space has 36 elements. Thus,

$$P(A) = \frac{6}{36}, \quad P(B) = \frac{11}{36}, \quad \text{and} \quad P(A \cap B) = \frac{2}{36}.$$

By the union rule,

$$P(A \cup B) = P(A) + P(B) - P(A \cap B),$$

$$P(A \cup B) = \frac{6}{36} + \frac{11}{36} - \frac{2}{36} = \frac{15}{36} = \frac{5}{12}.$$

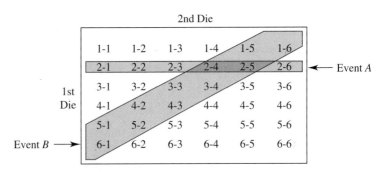

FIGURE 9.21

[4] In the experiment of Example 4, find the following probabilities.

(a) The sum is 5 or the second die shows a 3.

(b) Both dice show the same number or the sum is at least 11.

Answers:
(a) 1/4
(b) 2/9

(b) The sum is 11 or the second die is 5.

$P(\text{sum is } 11) = 2/36$, $P(\text{second die is } 5) = 6/36$, and $P(\text{sum is } 11 \text{ and second die is } 5) = 1/36$, so

$$P(\text{sum is } 11 \text{ or second die is } 5) = \frac{2}{36} + \frac{6}{36} - \frac{1}{36} = \frac{7}{36}. \quad \blacksquare \quad \boxed{4}$$

When events E and F are mutually exclusive, then $E \cap F = \emptyset$ by definition; hence, $P(E \cap F) = 0$. The converse of this fact can be used to determine that two events are mutually exclusive.

**For any events E and F, if $P(E \cap F) = 0$,
then E and F are mutually exclusive.**

Applying the union rule to mutually exclusive events yields a special case.

For mutually exclusive events E and F,
$$P(E \cup F) = P(E) + P(F).$$

EXAMPLE 5 Assume that the probability of a couple having a boy is the same as the probability of their having a girl. If the couple has 3 children, find the probability that at least 2 of them are girls.

The event of having at least 2 girls is the union of the mutually exclusive events $E =$ "the family has exactly 2 girls" and $F =$ "the family has exactly 3 girls." Using the equally likely sample space

$$\{ggg, ggb, gbg, bgg, gbb, bgb, bbg, bbb\},$$

we see that $P(2 \text{ girls}) = 3/8$ and $P(3 \text{ girls}) = 1/8$. Therefore,

$$P(\text{at least 2 girls}) = P(2 \text{ girls}) + P(3 \text{ girls})$$
$$= \frac{3}{8} + \frac{1}{8} = \frac{1}{2}. \ \blacksquare \ \boxed{5}$$

By definition of E', for any event E from a sample space S,

$$E \cup E' = S \quad \text{and} \quad E \cap E' = \emptyset.$$

Because $E \cap E' = \emptyset$, events E and E' are mutually exclusive, so that

$$P(E \cup E') = P(E) + P(E').$$

However, $E \cup E' = S$, the sample space, and $P(S) = 1$. Thus

$$P(E \cup E') = P(E) + P(E') = 1.$$

Rearranging these terms gives the following useful rule.

COMPLEMENT RULE
For any event E,
$$P(E') = 1 - P(E) \quad \text{and} \quad P(E) = 1 - P(E').$$

EXAMPLE 6 If a fair die is rolled, what is the probability that any number but 5 will come up?

If E is the event that 5 comes up, then E' is the event that any number but 5 comes up. $P(E) = 1/6$, so we have $P(E') = 1 - 1/6 = 5/6$. $\ \blacksquare \ \boxed{6}$

**Sidebar:**

$\boxed{5}$ In Example 5, find the probability of no more than 2 girls.

Answer:
7/8

$\boxed{6}$ **(a)** Let $P(K) = 2/3$. Find $P(K')$.

(b) If $P(X') = 3/4$, find $P(X)$.

Answers:
(a) 1/3
(b) 1/4

EXAMPLE 7 If two fair dice are rolled, find the probability that the sum of the numbers showing is greater than 3.

To calculate this probability directly, we must find the probabilities that the sum is 4, 5, 6, 7, 8, 9, 10, 11, or 12 and then add them. It is much simpler to first find the probability of the complement, the event that the sum is less than or equal to 3.

$$P(\text{sum} \le 3) = P(\text{sum is } 2) + P(\text{sum is } 3)$$

$$= \frac{1}{36} + \frac{2}{36} = \frac{3}{36} = \frac{1}{12}$$

Now use the fact the $P(E) = 1 - P(E')$ to get

$$P(\text{sum} > 3) = 1 - P(\text{sum} \le 3) = 1 - \frac{1}{12} = \frac{11}{12}. \quad ■ \quad \boxed{7}$$

|7| In Example 7, find the probability that the sum of the numbers rolled is at least 5.

Answer:
5/6

Venn diagrams are useful in finding probabilities, as shown in the next example.

EXAMPLE 8 Susan is a college student who receives heavy sweaters from her aunt at the first sign of cold weather. She estimates that the probability that a sweater is the wrong size is .47, the probability that it is a loud color is .59, and the probability that it is both the wrong size and a loud color is .31.

(a) Find the probability that the sweater is the correct size and not a loud color.

Let W be the event "wrong size" and L be the event "loud color." Draw a Venn diagram and determine the probabilities of the mutually exclusive events I–IV in Figure 9.22 as follows. We are given $P(\text{II}) = P(W \cap L) = .31$. Because regions I and II are mutually exclusive, $P(W) = P(\text{I}) + P(\text{II})$, so that $P(\text{I}) = P(W) - P(\text{II}) = .47 - .31 = .16$. Similarly, $P(\text{III}) = P(L) - P(\text{II}) = .59 - .31 = .28$. Finally, since IV is the complement of $W \cup L$, we have

$$P(\text{IV}) = 1 - P(W \cup L) = 1 - (.16 + .31 + .28) = .25.$$

The event "correct size *and* not a loud color" is region IV, the area outside of both W and L. Therefore, its probability is .25.

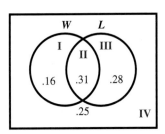

FIGURE 9.22

|8| Find the probability that the sweater in Example 8 is

(a) not a loud color;

(b) the correct size *and* a loud color.

Answers:
(a) .41
(b) .28

(b) Find the probability that the sweater is the correct size *or* not a loud color.

The event "correct size *or* not a loud color" is the region outside of region II (in II every sweater is both the wrong size and a loud color), that is, the complement of II. Hence, its probability is $1 - .31 = .69.$ ■ |8|

In many realistic problems, it is not possible to establish exact probabilities for events. Useful approximations, however, often can be found by using past experience as a guide to the future, as in the next example.

EXAMPLE 9 The manager of a store has decided to make a study of the amounts of money spent by people coming into the store. To begin, he chooses a day that seems fairly typical and gathers the following data. (Purchases have been rounded to the nearest dollar, with sales tax ignored.)

Amount Spent	Number of Customers
$0	158
$1–10	94
$11–20	203
$21–50	126
$51–100	47
$101–200	38
$201 and over	53

First, the manager might add the numbers of customers to find that 719 people came into the store that day. Of these 719 people, $126/719 \approx .175$ made a purchase of at least $21 but no more than $50. Also, $53/719 \approx .074$ of the customers spent $201 or more. For this day, the probability that a customer entering the store will spend from $21 to $50 is .175, and the probability that the customer will spend $201 or more is .074. Probabilities for the other purchase amounts can be assigned in the same way, giving the results below.

Amount Spent	Probability
$0	.220
$1–10	.131
$11–20	.282
$21–50	.175
$51–100	.065
$101–200	.053
$201 and over	.074
Total	1.000

The categories in the table are mutually exclusive simple events. Thus, by the union rule,

$$P(\text{spending} < \$21) = .220 + .131 + .282 = .633.$$

From the table, .282 of the customers spend from $11 to $20, inclusive. Since this price range attracts more than a quarter of the customers, perhaps the store's advertising should emphasize items in, or near, this price range.

The manager should use this table of probabilities to help in predicting the results on other days only if the manager is reasonably sure that the day when the measurements were made is fairly typical of the other days the store is open. For example, on the last few days before Christmas the probabilities might be quite different. ■ 9

9 A traffic engineer gathered the following data on the number of accidents at four busy intersections.

Intersection	Number of Accidents
1	25
2	32
3	28
4	35

If an accident occurs at one of these four intersections, what is the probability that it occurs at intersections 1 or 3?

Answer:
.44

In this section and the previous one, we have used more than one way to assign probabilities of events. There are three types of probability: *theoretical* (or *classical*); *empirical;* and *subjective.* We use theoretical probability when we assign probabilities for experiments with a coin or with dice, as in Examples 1–7. Probabilities determined by the data from experiments, using the frequency definition, $P(E) = n(E)/n(S)$, are empirical probabilities. See Example 9. Subjective probabilities are assigned when no experimental results or long-run frequencies are available. Since these probabilities are based on individual experience, different people may assign different probabilities to the same event. Example 8 gives an example of subjective probability. In the next chapter the use of subjective probability is discussed in more detail. No matter how probabilities are assigned, they must satisfy the following properties.

PROPERTIES OF PROBABILITY

Let S be a sample space consisting of n distinct outcomes $s_1, s_2, \ldots, s_n$. An acceptable probability assignment consists of assigning to each outcome s_i a number p_i (the probability of s_i) according to these rules.

1. The probability of each outcome is a number between 0 and 1.

$$0 \leq p_1 \leq 1, \quad 0 \leq p_2 \leq 1, \ldots, \quad 0 \leq p_n \leq 1$$

2. The sum of the probabilities of all possible outcomes is 1.

$$p_1 + p_2 + p_3 + \cdots + p_n = 1.$$

ODDS Sometimes probability statements are given in terms of **odds,** a comparison of $P(E)$ with $P(E')$. For example, suppose $P(E) = \frac{4}{5}$. Then $P(E') = 1 - \frac{4}{5} = \frac{1}{5}$. These probabilities predict that E will occur 4 out of 5 times and E' will occur 1 out of 5 times. Then we say the **odds in favor** of E are 4 to 1 or 4:1. This ratio may be found from the fraction $P(E)/P(E') = \frac{4}{5}/\frac{1}{5} = 4/1$.

ODDS

The **odds in favor** of an event E are defined as the ratio of $P(E)$ to $P(E')$, or

$$\frac{P(E)}{P(E')}, \quad P(E') \neq 0.$$

$$\frac{P(s)}{P(P)} \quad or$$

EXAMPLE 10 Suppose the weather forecaster says that the probability of rain tomorrow is 1/3. Find the odds in favor of rain tomorrow.

Let E be the event "rain tomorrow." Then E' is the event "no rain tomorrow." Since $P(E) = 1/3$, $P(E') = 2/3$. By the definition of odds, the odds in favor of rain are

$$\frac{1/3}{2/3} = \frac{1}{2}, \quad \text{written 1 to 2} \quad \text{or} \quad \text{1:2}.$$

[10] Suppose $P(E) = 9/10$. Find the odds

(a) in favor of E;

(b) against E.

Answers:
(a) 9 to 1
(b) 1 to 9

On the other hand, the odds that it will *not* rain, or the odds *against* rain, are

$$\frac{2/3}{1/3} = \frac{2}{1}, \quad \text{written 2 to 1.} \quad \blacksquare \quad \boxed{10}$$

If the odds in favor of an event are, say, 3 to 5, then the probability of the event is 3/8, while the probability of the complement of the event is 5/8. (Odds of 3 to 5 indicate 3 outcomes in favor of the event out of a total of 8 outcomes.) This example suggests the following generalization.

> If the odds favoring event E are m to n, then
> $$P(E) = \frac{m}{m + n} \quad \text{and} \quad P(E') = \frac{n}{m + n}.$$

EXAMPLE 11 The odds that a particular bid will be the low bid are 4 to 5.

(a) Find the probability that the bid will be the low bid.

Odds of 4 to 5 show 4 favorable chances out of $4 + 5 = 9$ chances altogether.

$$P(\text{bid will be low bid}) = \frac{4}{4 + 5} = \frac{4}{9}$$

[11] If the odds in favor of event E are 1 to 5, find

(a) $P(E)$;

(b) $P(E')$.

Answers:
(a) 1/6
(b) 5/6

(b) Find the odds against that bid being the low bid.

There is a 5/9 chance that the bid will not be the low bid, so the odds against a low bid are

$$\frac{P(\text{bid will not be low})}{P(\text{bid will be low})} = \frac{5/9}{4/9} = \frac{5}{4},$$

or 5:4. $\quad \blacksquare \quad \boxed{11}$

9.4 EXERCISES

Decide whether the events in Exercises 1–6 are mutually exclusive. (See Example 2.)

1. Being 15 and being a teenager

2. Wearing jogging shoes and wearing sandals

3. Being male and being a dancer

4. Owning a bicycle and owning a car

5. Being a U.S. Senator and being a U.S. Congressman concurrently

6. Being female and owning a truck

7. If the probability of an event is .857, what is the probability that the event will not occur?

8. Given $P(A) = .5$, $P(B) = .35$, $P(A \cup B) = .85$. Can this be correct? Explain under what conditions it is correct and under what conditions it is not correct.

Find the probabilities in Exercises 9–16. (See Examples 3–7.)

9. If a marble is drawn from a bag containing 2 yellow, 5 red, and 3 blue marbles, what are the probabilities of the following?
   (a) The marble is red.
   (b) The marble is either yellow or blue.
   (c) The marble is yellow or red.
   (d) The marble is green.

10. The law firm of Alam, Bartolini, Chinn, Dickinson, and Ellsberg has two senior partners, Alam and Bartolini. Two of the attorneys are to be selected to attend a conference. Assuming that all are equally likely to be selected, find the following probabilities.

(exercise continues)

(a) Chinn is selected.

(b) Ellsberg is not selected.

(c) Alam and Dickinson are selected.

(d) At least one senior partner is selected.

11. Ms. Bezzone invites 10 relatives to a party: her mother, two uncles, three brothers, and four cousins. If the chances of any one guest arriving first are equally likely, find the following probabilities.

(a) The first guest is an uncle or a cousin.

(b) The first guest is a brother or a cousin.

(c) The first guest is an uncle or her mother.

12. A card is drawn from a well-shuffled deck of 52 cards. Find the probability that the card is the following.

(a) a queen

(b) red

(c) a black 3

(d) a club or red

13. In Exercise 12, find the probability of the following.

(a) a face card (K, Q, J of any suit)

(b) red or a 3

(c) less than a four (consider aces as 1s)

14. Two dice are rolled. Find the probability of the following.

(a) The sum of the points is at least 10.

(b) The sum of the points is either 7 or at least 10.

(c) The sum of the points is 3 or the dice both show the same number.

15. **Management** The management of a firm wants to survey its workers, who are classified as follows for the purpose of an interview: 30% have worked for the company more than 5 years; 28% are female; 65% contribute to a voluntary retirement plan; 1/2 of the female workers contribute to the retirement plan. Find the following probabilities.

(a) A male worker is selected.

(b) A worker with less than 5 years in the company is selected.

(c) A worker who contributes to the retirement plan or a female worker is selected.

16. The numbers 1, 2, 3, 4, and 5 are written on five slips of paper, and two slips are drawn at random without replacement. Find the probability of each of the following.

(a) Both numbers are even.

(b) One of the numbers is even or greater than 3.

(c) The sum of the two numbers is 5 or the second number is 2.

Use Venn diagrams to work Exercises 17–22. (See Example 8.)

17. **Social Science** In a refugee camp in southern Mexico, it was found that 90% of the refugees came to escape political oppression, 80% came to escape abject poverty, and 70% came to escape both. What is the probability that a refugee in the camp was not poor nor seeking political asylum?

18. **Social Science** A study on body types gave the following results: 45% were short, 25% were short and overweight, and 24% were tall and not overweight. Find the probability that a person is

(a) overweight;

(b) short, but not overweight;

(c) tall and overweight.

19. **Social Science** A teacher found that 85% of the students in her math class had passed a course in algebra, 60% had passed a course in geometry, and 55% had passed both courses. Find the probability that a student selected randomly from the math class has passed at least one of the two courses.

20. **Social Science** The following data were gathered for 130 adult U.S. workers: 55 were women, 3 women earned more than $40,000, 62 men earned less than $40,000. Find the probability that an individual is

(a) a woman earning less than $40,000;

(b) a man earning more than $40,000;

(c) a man or is earning more than $40,000;

(d) a woman or is earning less than $40,000.

21. **Management** Suppose that 8% of a certain batch of calculators have a defective case, and that 11% have defective batteries. Also, 3% have both a defective case and defective batteries. A calculator is selected from the batch at random. Find the probability that the calculator has a good case and good batteries.

22. **Social Science** Fifty students in a Texas school were interviewed with the following results: 45 spoke Spanish, 10 spoke Vietnamese, and 8 spoke both languages. Find the probability that a randomly selected student from this group

(a) speaks both languages;

(b) speaks neither language;

(c) speaks only one of the two languages.

Work Exercises 23–29 on odds. (See Examples 10–11.)

23. A single die is rolled. Find the odds in favor of getting the following results.

(a) 3

(b) 5 or 6

(c) a number greater than 3

(d) a number less than 2

24. A marble is drawn from a box containing 3 yellow, 4 white, and 8 blue marbles. Find the odds in favor of drawing the following.

(a) A yellow marble

(b) A blue marble

(c) A white marble

25. The probability that a company will make a profit this year is .74. Find the odds against the company making a profit.

26. If the odds that a given candidate will win an election are 3 to 2, what is the probability that the candidate will lose?

27. Social Science A nationwide survey showed that the odds that an individual uses a seat belt when in a car's front seat are 17:8.* What is the probability that such an individual does not use a seat belt?

28. Social Science The survey mentioned in Exercise 27 also showed that the odds that a driver does not drink or does not drive after drinking are 4:1.* What is the probability that a driver was drinking?

29. On page 134 of Roger Staubach's autobiography, *First Down, Lifetime to Go,* Staubach makes the following statement regarding his experience in Vietnam: "Odds against a direct hit are very low but when your life is in danger, you don't worry too much about the odds." Is this wording consistent with our definition of odds for and against? How could it have been said so as to be technically correct?

An experiment is conducted for which the sample space is $S = \{s_1, s_2, s_3, s_4, s_5\}$. *Which of the probability assignments in Exercises 30–35 is possible for this experiment? If an assignment is not possible, tell why.*

30.

Outcomes	s_1	s_2	s_3	s_4	s_5
Probabilities	.02	.27	.35	.22	.14

31.

Outcomes	s_1	s_2	s_3	s_4	s_5
Probabilities	.50	.30	.10	.08	.02

32.

Outcomes	s_1	s_2	s_3	s_4	s_5
Probabilities	1/8	1/6	1/5	1/3	1/2

33.

Outcomes	s_1	s_2	s_3	s_4	s_5
Probabilities	1/10	1/8	1/5	1/5	1/4

34.

Outcomes	s_1	s_2	s_3	s_4	s_5
Probabilities	.23	.17	.32	.38	$-.10$

35.

Outcomes	s_1	s_2	s_3	s_4	s_5
Probabilities	.3	.4	$-.4$	.4	.3

Work the following problems. (See Example 9.)

36. Social Science A consumer survey randomly selects 2000 people and asks them about their income and their TV-watching habits. The results are shown in this table.

	HOURS OF TV WATCHED PER WEEK				
Annual Income	*0–8*	*9–15*	*16–22*	*23–30*	*More than 30*
Less than $12,000	11	15	8	14	120
$12,000–$24,999	10	19	28	96	232
$25,000–$39,999	18	32	88	327	189
$40,000–$59,999	31	85	160	165	100
$60,000 or more	73	60	52	55	12

As a reward for participation, each person questioned is given a small prize. Then the names of all participants are placed in a box and one name is randomly drawn to receive the grand prize of a free vacation trip. What is the probability that the winner of the grand prize
(a) watches TV at least 16 hours per week?
(b) has an income of at least $25,000?
(c) has an income of at least $40,000 and watches TV at least 16 hours per week?
(d) has an income of $0–$24,999 and watches TV more than 30 hours per week?

37. Management GT Global Theme Funds has world-wide assets distributed as follows.*

Geographic Location	*Percent*
Africa & Middle East	4.2
Latin America	6.0
Asia-Pacific	18.7
Europe	23.1
U.S. & Canada	48.0

If an asset is chosen randomly, what is the probability that it is from the following regions?
(a) Asia-Pacific
(b) North or South America
(c) Not from Europe

38. Management GT Global Health Care Fund has invested in the following sectors.*

Sector	*Percent*
Biotechnology	30.7
Pharmaceuticals	25.4
Medical Technology & Supplies	22.3
Health Care Services	8.2
Short-term & Other	13.4

Find the probability that an investment selected at random from this fund is as follows.
(a) In biotechnology or health care services
(b) In pharmaceuticals or medical technology & supplies
(c) Not in biotechnology

39. Natural Science The results of a study relating blood cholesterol level to coronary disease are given in the following table.

(exercise continues)

---

*Exercises 27 and 28 based on an article in *The Sacramento Bee,* June 4, 1992.

*Exercises 37 and 38 from GT Global Theme Funds Annual Report, October 31, 1996.

Cholesterol Level	Probability of Coronary Disease
Under 200	.10
200–219	.15
220–239	.20
240–259	.26
Over 259	.29

Find the probability of coronary disease if the cholesterol level is
(a) less than 240;
(b) 220 or more;
(c) from 200 to 239;
(d) from 220 to 259.

Natural Science   *Color blindness is an inherited characteristic which is sex-linked, so that it is more common in males than in females. If M represents male and C represents red-green color blindness, using the relative frequencies of the incidence of males and red-green color blindness as probabilities, P(C) = .049, P(M ∩ C) = .042, P(M ∪ C) = .534. Find the following. (Hint: Use a Venn diagram with two circles labeled M and C.)*

40. $P(C')$
41. $P(M)$
42. $P(M')$
43. $P(M' \cap C')$
44. $P(C \cap M')$
45. $P(C \cup M')$

Natural Science   *Gregor Mendel, an Austrian monk, was the first to use probability in the study of genetics. In an effort to understand the mechanisms of character transmittal from one generation to the next in plants, he counted the number of occur-rences of various characteristics. Mendel found that the flower color in certain pea plants obeyed this scheme:*

Pure red crossed with pure white produces red.

The red offspring received from its parents genes for both red (R) and white (W), but in this case red is dominant and white reces-sive, so the offspring exhibits the color red. However, the offspring still carries both genes, and when two such offspring are crossed, several things can happen in the third generation, as shown in the table below, which is called a Punnet square.

		2ND PARENT	
		R	W
1ST PARENT	R	RR	RW
	W	WR	WW

This table shows the possible combinations of genes. Use the fact that red is dominant over white to find

46. $P(\text{red})$;
47. $P(\text{white})$.

Natural Science   *Mendel found no dominance in snapdragons, with one red gene and one white gene producing pink-flowering offspring. These second-generation pinks, however, still carry one red and one white gene, and when they are crossed, the next gen-eration still yields the Punnet square above. Find*

48. $P(\text{red})$;
49. $P(\text{pink})$;
50. $P(\text{white})$.

(Mendel verified these probability ratios experimentally with large numbers of observations, and did the same for many character units other than flower color. The importance of his work, pub-lished in 1866, was not recognized until 1900.)

Natural Science   *In most animals and plants, it is very unusual for the number of main parts of the organism (arms, legs, toes, flower petals, etc.) to vary from generation to generation. Some species, however, have meristic variability, in which the number of certain body parts varies from generation from generation. One researcher* studied the front feet of certain guinea pigs and pro-duced the probabilities shown below.*

$$P(\text{only four toes, all perfect}) = .77$$
$$P(\text{one imperfect toe and four good ones}) = .13$$
$$P(\text{exactly five good toes}) = .10$$

Find the probability of having the following.

51. No more than four good toes
52. Five toes, whether perfect or not

*One way to solve a probability problem is to repeat the experiment (or a simulation of the experiment) many times, keeping track of the results. Then the probability can be approximated using the basic definition of the probability of an event E: P(E) = m/n, where m favorable outcomes occur in n trials of an experiment. This is called the **Monte Carlo method** of approximating probabilities.*
  Use a calculator with a random number generator or appro-priate computer software and the Monte Carlo method to simulate the experiments in Exercises 53–58.

Approximate the probabilities in Exercises 53 and 54 if five coins are tossed. Then calculate the theoretical probabilities using the methods of the text and compare the results.

53. $P(4 \text{ heads})$
54. $P(2 \text{ heads}, 1 \text{ tail}, 2 \text{ heads})$ (in the order given)

*From "An Analysis of Variability in Guinea Pigs" by J. R. Wright, from *Genetics*, Vol. 19, 1934, pp. 506–36. Reprinted by permission.

Approximate the following probabilities if 4 cards are drawn from a 52-card deck.

55. P(any 2 cards and then 2 kings)

56. P(exactly 2 kings)

Approximate the probabilities in Exercises 57 and 58.

57. A jeweler received 8 identical watches each in a box marked with the series number of the watch. An assistant, who does not know that the boxes are marked, is told to polish the watches and then put them back in the boxes. She puts them in the boxes at random. What is the probability that she gets at least one watch in the right box?

58. A check room attendant has 10 hats but has lost the numbers identifying them. If he gives them back randomly, what is the probability that at least 2 of the hats are given back correctly?

9.5 CONDITIONAL PROBABILITY; INDEPENDENT EVENTS

The training manager for a large brokerage firm has noticed that some of the firm's brokers use the firm's research advice, while other brokers tend to go with their own beliefs about which stocks will go up. To see whether the research department performs better than the beliefs of the brokers, the manager conducted a survey of 100 brokers, with results as shown in the following table.

	Picked a Stock that Went Up	Didn't Pick a Stock that Went Up	Totals
Used Research	30	15	45
Didn't Use Research	30	25	55
Totals	60	40	100

1 Use the data in the table to find

(a) $P(B)$,

(b) $P(A')$,

(c) $P(B')$.

Answers:
(a) .45
(b) .4
(c) .55

Letting A be the event "picked a stock that went up" and letting B be the event "used research," $P(A)$, $P(A')$, $P(B)$, and $P(B')$ can be found. For example, the chart shows that a total of 60 brokers picked stocks that went up, so $P(A) = 60/100 = .6$. **1**

Suppose we want to find the probability that a broker using research will pick a stock that goes up. From the table above, of the 45 brokers who use research, there are 30 who picked stocks that went up, so

$$P(\text{broker who uses research picks stocks that go up}) = \frac{30}{45} \approx .667.$$

This is a different number than the probability that a broker picks a stock that goes up, .6, because *we have additional information* (the broker uses research) *that has reduced the sample space.* In other words, we found the probability that a broker picks a stock that goes up, A, given the additional information that the broker uses research, B. This is called the *conditional probability* of event A, given that event B has occurred, written $P(A \mid B)$. ($P(A \mid B)$ may also be read as "the probability of A given B.")

In the example above,

$$P(A \mid B) = \frac{30}{45}.$$

If we divide the numerator and denominator by 100 (the size of the sample space), this can be written as

$$P(A \mid B) = \frac{\dfrac{30}{100}}{\dfrac{45}{100}} = \frac{P(A \cap B)}{P(B)},$$

where $P(A \cap B)$ represents, as usual, the probability that both A and B will occur.

To generalize this result, assume that E and F are two events for a particular experiment. Assume that the sample space S for this experiment has n possible equally likely outcomes. Suppose event F has m elements, and $E \cap F$ has k elements $(k \le m)$. Using the fundamental principle of probability,

$$P(F) = \frac{m}{n} \quad \text{and} \quad P(E \cap F) = \frac{k}{n}.$$

We now want $P(E \mid F)$, the probability that E occurs given that F has occurred. Since we assume F has occurred, reduce the sample space to F: look only at the m elements inside F. (See Figure 9.23.) Of these m elements, there are k elements where E also occurs, because $E \cap F$ has k elements. This makes

$$P(E \mid F) = \frac{k}{m}.$$

Divide numerator and denominator by n to get

$$P(E \mid F) = \frac{k/n}{m/n} = \frac{P(E \cap F)}{P(F)}.$$

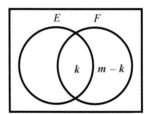

Event F has a total of m elements.

FIGURE 9.23

The last result gives the definition of conditional probability.

The **conditional probability** of an event E given event F, written $P(E \mid F)$, is

$$P(E \mid F) = \frac{P(E \cap F)}{P(F)}, \quad P(F) \neq 0.$$

This definition tells us that, for equally likely outcomes, conditional probability is found by *reducing the sample space to event F,* and then finding the number of outcomes in *F* that are also in event *E.* Thus,

$$P(E \mid F) = \frac{n(E \cap F)}{n(F)}.$$

2 The table shows the results of a survey of a buffalo herd.

	Males	Females	Totals
Adults	500	1300	1800
Calves	520	500	1020
Totals	1020	1800	2820

Let *M* represent "male" and *A* represent "adult." Find each of the following.

(a) $P(M \mid A)$

(b) $P(M' \mid A)$

(c) $P(A \mid M')$

(d) $P(A' \mid M)$

(e) State the probability in part (d) in words.

Answers:
(a) 5/18

(b) 13/18

(c) 13/18

(d) 26/51

(e) The probability that a buffalo is a calf given that it is a male

EXAMPLE 1 Use the chart in the stockbroker's problem at the beginning of this section to find the following probabilities, where *A* is the event "picked a stock that went up" and *B* is the event "used research."

(a) $P(B \mid A)$

This represents the probability that the broker used research, given that the broker picked a stock that went up. Reduce the sample space to *A.* Then find $n(A \cap B)$ and $n(A)$.

$$P(B \mid A) = \frac{P(B \cap A)}{P(A)} = \frac{n(A \cap B)}{n(A)} = \frac{30}{60} = \frac{1}{2}$$

If a broker picked a stock that went up, then the probability is 1/2 that the broker used research.

(b) $P(A' \mid B)$

In words, this is the probability that a broker picks a stock that does not go up, even though he used research.

$$P(A' \mid B) = \frac{n(A' \cap B)}{n(B)} = \frac{15}{45} = \frac{1}{3}$$

(c) $P(B' \mid A')$

Here, we want the probability that a broker who picked a stock that did not go up did not use research.

$$P(B' \mid A') = \frac{n(B' \cap A')}{n(A')} = \frac{25}{40} = \frac{5}{8}$$ ■ **2**

Venn diagrams can be used to illustrate problems in conditional probability. A Venn diagram for Example 1, in which the probabilities are used to indicate the number in the set defined by each region, is shown in Figure 9.24. In the diagram, $P(B \mid A)$ is found by *reducing the sample space to just set A.* Then $P(B \mid A)$ is the ratio of the number in that part of set *B* which is also in *A* to the number in set *A,* or .3/.6 = .5.

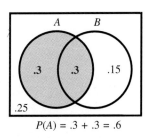

$P(A) = .3 + .3 = .6$

FIGURE 9.24

EXAMPLE 2 Given $P(E) = .4$, $P(F) = .5$, and $P(E \cup F) = .7$, find $P(E \mid F)$.

Find $P(E \cap F)$ first. Then use a Venn diagram to find $P(E \mid F)$. By the union rule,

$$P(E \cup F) = P(E) + P(F) - P(E \cap F)$$
$$.7 = .4 + .5 - P(E \cap F)$$
$$P(E \cap F) = .2.$$

3 Find $P(F \mid E)$ if $P(E) = .3$, $P(F) = .4$, and $P(E \cup F) = .6$.

Answer:
1/3

Now use the probabilities to indicate the number in each region of the Venn diagram in Figure 9.25. $P(E \mid F)$ is the ratio of the probability of that part of E which is in F to the probability of F or

$$P(E \mid F) = \frac{P(E \cap F)}{P(F)} = \frac{.2}{.5} = \frac{2}{5} = .4. \quad \blacksquare \quad \boxed{3}$$

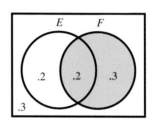

FIGURE 9.25

EXAMPLE 3 Two fair coins were tossed, and it is known that at least one was a head. Find the probability that both were heads.

The sample space has four equally likely outcomes, $S = \{hh, ht, th, tt\}$. Define two events:

$$E_1 = \text{at least 1 head} = \{hh, ht, th\}$$

and

$$E_2 = \text{2 heads} = \{hh\}.$$

Because there are four equally likely outcomes, $P(E_1) = 3/4$. Also, $P(E_1 \cap E_2) = 1/4$. We want the probability that both were heads, given that at least one was a head: that is, we want to find $P(E_2 \mid E_1)$. Because of the condition that at least one coin was a head, the reduced sample space is

$$\{hh, ht, th\}.$$

Since only one outcome in this reduced sample space is 2 heads,

4 In Example 3, find the probability that exactly one coin showed a head, given that at least one was a head.

$$P(E_2 \mid E_1) = \frac{1}{3}.$$

Answer:
2/3

Alternatively, use the definition given above.

$$P(E_2 \mid E_1) = \frac{P(E_2 \cap E_1)}{P(E_1)} = \frac{1/4}{3/4} = \frac{1}{3} \quad \blacksquare \quad \boxed{4}$$

PRODUCT RULE If $P(E) \neq 0$ and $P(F) \neq 0$, then the definition of conditional probability shows that

$$P(E \mid F) = \frac{P(E \cap F)}{P(F)} \quad \text{and} \quad P(F \mid E) = \frac{P(F \cap E)}{P(E)}.$$

Using the fact that $P(E \cap F) = P(F \cap E)$, and solving each of these equations for $P(E \cap F)$, we obtain the following rule.

PRODUCT RULE OF PROBABILITY

If E and F are events, then $P(E \cap F)$ may be found by either of these formulas.

$$P(E \cap F) = P(F) \cdot P(E \mid F) \quad \text{or} \quad P(E \cap F) = P(E) \cdot P(F \mid E).$$

The **product rule** gives a method for finding the probability that events E and F both occur, as illustrated by the next few examples.

EXAMPLE 4   In a class with 2/5 women and 3/5 men, 25% of the women are business majors. Find the probability that a student chosen at random from the class is a female business major.

⎡5⎤ In a litter of puppies, 3 were female and 4 were male. Half the males were black. Find the probability that a puppy chosen at random from the litter would be a black male.

Answer:
2/7

Let B and W represent the events "business major" and "women," respectively. We want to find $P(B \cap W)$. By the product rule,

$$P(B \cap W) = P(W) \cdot P(B \mid W).$$

From the given information, $P(W) = 2/5 = .4$ and the probability that a woman is a business major is $P(B \mid W) = .25$. Then

$$P(B \cap W) = .4(.25) = .10. \quad \blacksquare \quad \boxed{5}$$

In Section 9.1 we used a tree diagram to find the number of subsets of a given set. By including the probabilities for each branch of a tree diagram, we convert it to a **probability tree.** The next examples show how conditional probability is used with probability trees.

EXAMPLE 5   A company needs to hire a new director of advertising. It has decided to try to hire either person A or person B, who are assistant advertising directors for its major competitor. To decide between A and B, the company does research on the campaigns managed by A or B (none are managed by both), and finds that A is in charge of twice as many advertising campaigns as B. Also, A's campaigns have satisfactory results three out of four times, while B's campaigns have satisfactory results only two out of five times. Suppose one of the competitor's advertising campaigns (managed by A or B) is selected randomly.

We can represent this situation schematically as follows. Let A denote the event "Person A does the job" and B the event "Person B does the job." Let S be the event "satisfactory results" and U the event "unsatisfactory results." Then the given information can be summarized in the probability tree in Figure 9.26 on the next page. Since A does twice as many jobs as B, $P(A) = 2/3$ and $P(B) = 1/3$, as noted on the first-stage branches of the tree. When A does a job, the probability of satisfactory results is 3/4 and of unsatisfactory results 1/4, as noted on the second-stage branches. Similarly, the probabilities when B does the job are noted on the remaining second-stage branches. The composite branches labeled 1–4 represent the four mutually exclusive possibilities for the running and outcome of the campaign.

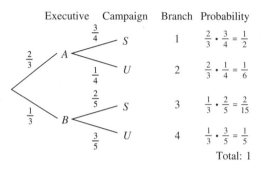

FIGURE 9.26

(a) Find the probability that A is in charge of a campaign that produces satisfactory results.

We are asked to find $P(A \cap S)$. We know that when A does the job, the probability of success is 3/4, that is, $P(S \mid A) = 3/4$. Hence, by the product rule,

$$P(A \cap S) = P(A) \cdot P(S \mid A) = \frac{2}{3} \cdot \frac{3}{4} = \frac{1}{2}.$$

The event $A \cap S$ is represented by branch 1 of the tree, and as we have just seen, its probability is the product of the probabilities that make up that branch.

(b) Find the probability that B runs a campaign that produces satisfactory results.

We must find $P(B \cap S)$. The event is represented by branch 3 of the tree and, as before, its probability is the product of the probabilities of the pieces of that branch:

$$P(B \cap S) = P(B) \cdot P(S \mid B) = \frac{1}{3} \cdot \frac{2}{5} = \frac{2}{15}.$$

(c) What is the probability that the campaign is satisfactory?

The event S is the union of the mutually exclusive events $A \cap S$ and $B \cap S$, which are represented by branches 1 and 3 of the tree diagram. By the union rule,

$$P(S) = P(A \cap S) + P(B \cap S) = \frac{1}{2} + \frac{2}{15} = \frac{19}{30}.$$

Thus, the probability of an event that appears on several branches is the sum of the probabilities of each of these branches.

(d) What is the probability that the campaign is unsatisfactory?

$P(U)$ can be read from branches 2 and 4 of the tree.

$$P(U) = \frac{1}{6} + \frac{1}{5} = \frac{11}{30}.$$

Alternatively, because U is the complement of S,

$$P(U) = 1 - P(S) = 1 - \frac{19}{30} = \frac{11}{30}.$$

6 Find each of the following probabilities for Example 5.

(a) $P(U \mid A)$

(b) $P(U \mid B)$

Answers:

(a) 1/4

(b) 3/5

7 Find the probability of drawing a green marble and then a white marble.

Answer:

1/5

8 In Example 6, find the probability of drawing 1 white and 1 red marble.

Answer:

1/5

(e) Find the probability that either A runs the campaign or the results are satisfactory (or possibly both).

Event A combines branches 1 and 2, while event S combines branches 1 and 3, so use branches 1, 2, and 3.

$$P(A \cup S) = \frac{1}{2} + \frac{1}{6} + \frac{2}{15} = \frac{4}{5} \quad \blacksquare \quad \boxed{6}$$

EXAMPLE 6   From a box containing 1 red, 3 white, and 2 green marbles, two marbles are drawn one at a time without replacing the first before the second is drawn. Find the probability that one white and one green marble are drawn.

A probability tree showing the various possible outcomes is given in Figure 9.27. In this diagram, W represents the event "drawing a white marble" and G represents "drawing a green marble." On the first draw, $P(W$ on the 1st$) = 3/6 = 1/2$ because three of the six marbles in the box are white. On the second draw, $P(G$ on the 2nd $\mid W$ on the 1st$) = 2/5$. One white marble has been removed, leaving 5, of which 2 are green.

We want to find the probability of drawing exactly one white marble and exactly one green marble. Two events satisfy this condition: drawing a white marble first and then a green one (branch 2 of the tree), or drawing a green marble first and then a white one (branch 4). For branch 2,

$$P(W \text{ on 1st}) \cdot P(G \text{ on 2nd} \mid W \text{ on 1st}) = \frac{1}{2} \cdot \frac{2}{5} = \frac{1}{5}. \quad \boxed{7}$$

For branch 4, where the green marble is drawn first,

$$P(G \text{ first}) \cdot P(W \text{ second} \mid G \text{ first}) = \frac{1}{3} \cdot \frac{3}{5} = \frac{1}{5}.$$

Since these two events are mutually exclusive, the final probability is the sum of the two probabilities.

$$P(\text{one } W, \text{ one } G) = P(W \text{ on 1st}) \cdot P(G \text{ on 2nd} \mid W \text{ on 1st})$$

$$+ P(G \text{ on 1st}) \cdot P(W \text{ on 2nd} \mid G \text{ on 1st}) = \frac{2}{5} \quad \blacksquare \quad \boxed{8}$$

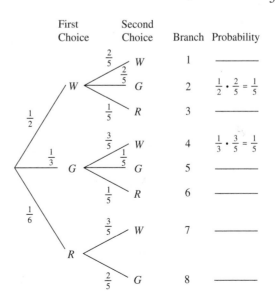

FIGURE 9.27

The product rule is often used when dealing with *stochastic processes,* which are mathematical models that evolve over time in a probabilistic manner. For example, drawing different colored marbles from a box (without replacing them) is such a process, in which the probabilities change with each successive draw. (Particular stochastic processes are studied further in Section 10.4.)

EXAMPLE 7   Two cards are drawn without replacement from an ordinary deck (52 cards). Find the probability that the first card is a heart and the second card is red.

Start with the probability tree of Figure 9.28. (You may wish to refer to the deck of cards shown in Figure 9.18.) On the first draw, since there are 13 hearts in the 52 cards, the probability of drawing a heart first is $13/52 = 1/4$. On the second draw, since a (red) heart has been drawn already, there are 25 red cards in the remaining 51 cards. Thus the probability of drawing a red card on the second draw, given that the first is a heart, is 25/51. By the product rule of probability,

$$P(\text{heart on 1st and red on 2nd})$$
$$= P(\text{heart on 1st}) \cdot P(\text{red on 2nd} \mid \text{heart on 1st})$$
$$= \frac{1}{4} \cdot \frac{25}{51} = \frac{25}{204} \approx .1225. \quad \blacksquare \quad \boxed{9}$$

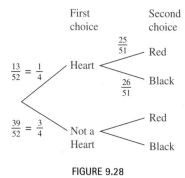

FIGURE 9.28

EXAMPLE 8   Three cards are drawn, without replacement, from an ordinary deck. Find the probability that exactly 2 of the cards are red.

Here we need a probability tree with three stages, as shown in Figure 9.29. The three branches indicated with arrows produce exactly 2 red cards from the draws. Multiply the probabilities along each of these branches and then add.

$$P(\text{exactly 2 red cards}) = \frac{26}{52} \cdot \frac{25}{51} \cdot \frac{26}{50} + \frac{26}{52} \cdot \frac{26}{51} \cdot \frac{25}{50} + \frac{26}{52} \cdot \frac{26}{51} \cdot \frac{25}{50}$$
$$= \frac{50,700}{132,600} = \frac{13}{34} \approx .382 \quad \blacksquare \quad \boxed{10}$$

INDEPENDENT EVENTS   Suppose a fair coin is tossed and shows heads. The probability of heads on the next toss is still 1/2; the fact that heads was the result on a given toss has no effect on the outcome of the next toss. Coin tosses are **independent events,** since the outcome of one toss does not help decide the outcome of the next toss. Rolls of a fair die are independent events; the fact that a 2 came up on one

9 Find the probability of drawing a heart on the first draw and a black card on the second, if two cards are drawn without replacement.

Answer:
13/102 or .1275

10 Use the tree in Example 8 to find the probability that exactly one of the cards is red.

Answer:
13/34 ≈ .382

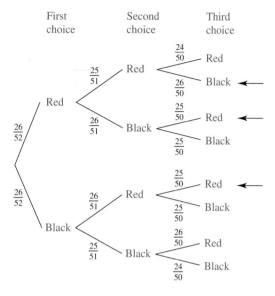

FIGURE 9.29

roll does not increase our knowledge of the outcome of the next roll. On the other hand, the events "today is cloudy" and "today is rainy" are **dependent events;** if it is cloudy, then there is an increased chance of rain. Similarly, in the example at the beginning of this section, the events A (broker picked a stock that went up) and B (broker used research) are dependent events, because information about the use of research affects the probability of picking a stock that goes up. That is, $P(A \mid B)$ is different from $P(A)$.

   If events E and F are independent, then the knowledge that E has occurred gives no additional (probability) information about the occurrence or nonoccurrence of event F. That is, $P(F)$ is exactly the same as $P(F \mid E)$, or

$$P(F \mid E) = P(F).$$

In fact, this is the formal definition of independent events.

> E and F are **independent events** if
>
> $$P(F \mid E) = P(F) \quad \text{or} \quad P(E \mid F) = P(E).$$

When E and F are independent events, then $P(F \mid E) = P(F)$ and the product rule becomes

$$P(E \cap F) = P(E) \cdot P(F \mid E) = P(E) \cdot P(F).$$

Conversely, if this equation holds, then it follows that $P(F) = P(F \mid E)$. Consequently, we have this useful fact.

> **PRODUCT RULE FOR INDEPENDENT EVENTS**
>
> E and F are independent events if and only if
>
> $$P(E \cap F) = P(E) \cdot P(F).$$

EXAMPLE 9 A calculator requires a key-stroke assembly and a logic circuit. Assume that 99% of the key-stroke assemblies are satisfactory and 97% of the logic circuits are satisfactory. Find the probability that a finished calculator will be satisfactory.

If the failure of a key-stroke assembly and the failure of a logic circuit are independent events, then

P(satisfactory calculator)

$= P$(satisfactory key-stroke assembly) $\cdot P$(satisfactory logic circuit)

$= (.99)(.97) \approx .96.$ ■ 11

11 Find the probability of getting 4 successive heads on 4 tosses of a fair coin.

Answer:
1/16

CAUTION It is common for students to confuse the ideas of *mutually exclusive* events and *independent* events. Events E and F are mutually exclusive if $E \cap F = \emptyset$. For example, if a family has exactly one child, the only possible outcomes are $B = \{boy\}$ and $G = \{girl\}$. These two events are mutually exclusive. However, the events are *not* independent, since $P(G \mid B) = 0$ (if a family with only one child has a boy, the probability it has a girl is then 0). Since $P(G \mid B) \neq P(G)$, the events are not independent. Of all the families with exactly *two* children, the events $G_1 = \{first \ child \ is \ a \ girl\}$ and $G_2 = \{second \ child \ is \ a \ girl\}$ are independent, because $P(G_2 \mid G_1)$ equals $P(G_2)$. However, G_1 and G_2 are not mutually exclusive, since $G_1 \cap G_2 = \{both \ children \ are \ girls\} \neq \emptyset.$ ◆

To show that two events E and F are independent, we can show that $P(F \mid E) = P(F)$ or that $P(E \mid F) = P(E)$ or that $P(E \cap F) = P(E) \cdot P(F)$. Another way is to observe that knowledge of one outcome does not influence the probability of the other outcome, as we did for coin tosses.

EXAMPLE 10 On a typical January day in Manhattan the probability of snow is .10, the probability of a traffic jam is .80, and the probability of snow or a traffic jam (or both) is .82. Are the event "it snows" and the event "a traffic jam occurs" independent?

Let S represent the event "it snows" and T represent the event "a traffic jam occurs." We must determine whether

$$P(T \mid S) = P(T) \quad \text{or} \quad P(S \mid T) = P(S).$$

We know $P(S) = .10$, $P(T) = .8$, and $P(S \cup T) = .82$. We can use the union rule (or a Venn diagram) to find $P(S \cap T) = .08$, $P(T \mid S) = .8$, and $P(S \mid T) = .1$. Since

$$P(T \mid S) = P(T) = .8 \quad \text{and} \quad P(S \mid T) = P(S) = .1,$$

the events "it snows" and "a traffic jam occurs" are independent. ■ 12

12 In the U.S. population, the probability of being Hispanic is .11, the probability of living in California is .12, and the probability of being a Hispanic living in California is .04. Are the events being Hispanic and living in California independent?

Answer:
No

Although we showed $P(T \mid S) = P(T)$ and $P(S \mid T) = P(S)$ in Example 10, only one of these results is needed to establish independence.

It is important not to confuse $P(A \mid B)$ with $P(B \mid A)$. For example, in a criminal trial, a prosecutor may point out to the jury that the probability of the defendant's DNA profile matching that of a sample taken at the scene of the crime, given that the defendant is innocent, is very small. What the jury must decide, however, is the probability that the defendant is innocent, given that the defendant's DNA profile matches the sample. Confusing the two is an error sometimes called "the prosecutor's fallacy," and the 1990 conviction of a rape suspect in England was overturned by a panel of judges, who ordered a retrial, because the fallacy made the original trial unfair.*

In the next section, we will see how to compute $P(A \mid B)$ when we know $P(B \mid A)$.

---

*"Who's the DNA fingerprinting pointing at?", David Pringle, *New Scientist,* Jan. 29, 1994, pp. 51–52.

9.5 EXERCISES

If a single fair die is rolled, find the probability of rolling the following. (See Examples 1–3.)

1. 3, given that the number rolled was odd

2. 5, given that the number rolled was even

3. An odd number, given that the number rolled was 3

If two fair dice are rolled (recall the 36-outcome sample space), find the probability of rolling the following.

4. A sum of 8, given the sum was greater than 7

5. A sum of 6, given the roll was a "double" (two identical numbers)

6. A double, given that the sum was 9

If two cards are drawn without replacement from an ordinary deck (see Example 7), find the following probabilities.

7. The second is a heart, given that the first is a heart.

8. The second is black, given that the first is a spade.

9. The second is a face card, given that the first is a jack.

10. The second is an ace, given that the first is not an ace.

11. A jack and a 10 are drawn.

12. An ace and a 4 are drawn.

13. Two black cards are drawn.

14. Two hearts are drawn.

15. In your own words, explain how to find the conditional probability $P(E \mid F)$.

16. Your friend asks you to explain how the product rule for independent events differs from the product rule for dependent events. How would you respond?

17. Another friend asks you to explain how to tell whether two events are dependent or independent. How would you reply? (Use your own words.)

18. A student reasons that the probability in Example 3 of both coins being heads is just the probability that the other coin is a head, that is, 1/2. Explain why this reasoning is wrong.

Use a probability tree or Venn diagram in Exercises 19–30. (See Examples 2 and 5–8.)

Social Science *Marriages between cousins are very common in some countries, where about 50% of marriages are consanguineous*

(between first cousins or people even more closely related). A recent study in Pakistan has shown that 16% of children from unrelated marriages died by age 10, while 21% of children from consanguineous marriages died by age 10. Find the following probabilities.

19. a child survives

20. a child from a consanguineous marriage survives

Management *Among users of automated teller machines (ATMs), 92% use ATMs to withdraw cash, and 32% use them to check their account balance.* Suppose that 96% use ATMs to either withdraw cash or check their account balance (or both). Find the probability that

21. An ATM user withdraws funds if that person uses the ATM to check an account balance.

22. An ATM user checks an account balance given that the ATM is used to withdraw funds.

Natural Science *The following table based on data from the World Health Organization, gives the number of people (in thousands) infected with the AIDS virus by geographical location and method of transmission.†*

	United States	*Rest of World*	*Totals*
Homosexual transmission	1410	1090	2,500
Heterosexual transmission	90	7410	7,500
Totals	1500	8500	10,000

23. (a) Find the probability that a resident of the United States with AIDS received it via homosexual transmission.

(b) Find the probability that a person with AIDS who is not a resident of the United States received it via homosexual transmission.

---

*Chicago Tribune, Sec. 4, p. 1, Dec. 18, 1995.

†Jane M. Watson, "Conditional Probability: Its Place in the Mathematics Curriculum," *The Mathematics Teacher,* Jan. 1995, Vol. 88, No. 1, pp. 12–17.

24. (a) Find the probability a person with AIDS is a resident of the United States.
(b) Find the probability a person who received AIDS via homosexual transmission is a resident of the United States.
(c) Are the events that a person with AIDS is a United States resident and that the person received AIDS via homosexual transmission independent? Explain your answer.

Social Science *A survey has shown that 52% of the women in a certain community work outside the home. Of these women, 64% are married, while 86% of the women who do not work outside the home are married. Find the probability that a woman in that community is*

25. married;

26. a single woman working outside the home.

Social Science *A study showed that, in 1991, 31.6% of men and 35.0% of women were obese.* Given that 48.7% of Americans are men and 51.3% are women, find the probability that a randomly selected adult fits the following description.*

27. An obese man

28. Obese

Management *A shop that produces custom kitchen cabinets has two employees, Sitlington and Capek. 95% of Capek's work is satisfactory and 10% of Sitlington's work is unsatisfactory. 60% of the shop's cabinets are made by Capek (the rest by Sitlington). Find the following probabilities.*

29. An unsatisfactory cabinet was made by Capek.

30. A finished cabinet is unsatisfactory.

Management *The table below shows employment figures for managerial/professional occupations in 1991 for U.S. civilians with four or more years of college.†*

	White	Black	Totals
Women	6,813	617	7,430
Men	9,453	435	9,888
Totals	16,266	1052	17,318

Letting A represent white, B represent black, C represent women, and D represent men, express each of the following probabilities in words and find its value. (See Example 1.)

31. $P(A \mid D)$
32. $P(C \mid A)$
33. $P(B \mid C)$
34. $P(D \cap A)$

*The New York Times, July 17, 1994, p. 18.

†U.S. Bureau of the Census, Statistical Abstract of the United States: 1992 (112th edition), Washington, D.C., 1992.

Natural Science *The following table shows frequencies for red-green color blindness, where M represents male and C represents color-blind.*

	M	M'	Totals
C	.042	.007	.049
C'	.485	.466	.951
Totals	.527	.473	1.000

Use this table to find the following probabilities.

35. $P(M)$
36. $P(C)$
37. $P(M \cap C)$
38. $P(M \cup C)$
39. $P(M \mid C)$
40. $P(M' \mid C)$

41. Are the events C and M described above dependent? (Recall that two events E and F are dependent if $P(E \mid F) \neq P(E)$. See Example 10.)

42. Natural Science A scientist wishes to determine if there is any dependence between color blindness (C) and deafness (D). Given the probabilities listed in the table below, what should his findings be? (See Example 10.)

	D	D'	Totals
C	.0004	.0796	.0800
C'	.0046	.9154	.9200
Totals	.0050	.9950	1.0000

Social Science *The Motor Vehicle Department has found that the probability of a person passing the test for a driver's license on the first try is .75. The probability that an individual who fails on the first test will pass on the second try is .80, and the probability that an individual who fails the first and second tests will pass the third time is .70. Find the probability that an individual*

43. fails both the first and second tests;

44. will fail three times in a row;

45. will require at least two tries to pass the test.

Natural Science *Four different medications, C, D, E, and F, may be used to control high blood pressure. A physician usually prescribes C first because it is least likely to cause side effects. If blood pressure remains high, the patient is switched to D. If this fails to work, the patient is switched to E, and if necessary to F. The probability that C will work is .7. If C fails, the probability that D will work is .8. If D fails, the probability that E will work is .62. If E fails, the probability that F will work is .45. Find the probability that*

46. A patient's blood pressure will not be reduced by any of the medications.

47. A patient will have to take at least two medications and will have his or her blood pressure reduced.

48. If medications C and D fail, what is the probability that a patient's blood pressure will be reduced by medication E or F?

Management *The number of vehicles (in thousands) on the road from the major worldwide producers in selected years is shown in the following table.*

	U.S.	Europe	Japan	Canada
1975	4495	6,737	3,471	712
1980	6008	11,584	8,282	1031
1985	9322	12,767	9,817	1546
1990	9291	17,683	12,812	1801

Find the following probabilities for a vehicle selected at random.

49. It was made in the U.S.

50. It was made in 1990.

51. It was made in Japan in 1985.

52. It was made in 1980, given that it was made in Europe.

The probability that the first record by a singing group will be a hit is .32. If their first record is a hit, so are all their subsequent records. If their first record is not a hit, the probability of their second record and all subsequent ones being hits is .16. If the first two records are not hits, the probability that the third is a hit is .08. The probability of a hit continues to decrease by half with each successive nonhit record. Find the probability that

53. the group will have at least one hit in their first four records.

54. the group will have exactly one hit in their first three records.

55. the group will have a hit in their first six records if the first three are not hits.

Work the following problems on independent events. (See Examples 9 and 10.)

56. Management Corporations such as banks, where a computer is essential to day-to-day operations, often have a second, backup computer in case of failure by the main

computer. Suppose that there is a .003 chance that the main computer will fail in a given time period and a .005 chance that the backup computer will fail while the main computer is being repaired. Assume these failures represent independent events, and find the fraction of the time that the corporation can assume it will have computer service. How realistic is our assumption of independence?

57. Management According to a booklet put out by Eastwest Airlines, 98% of all scheduled Eastwest flights actually take place. (The other flights are canceled due to weather, equipment problems, and so on.) Assume that the event that a given flight takes place is independent of the event that another flight takes place.

(a) Elisabeta Guervara plans to visit her company's branch offices; her journey requires 3 separate flights on Eastwest Airlines. What is the probability that all of these flights will take place?

(b) Based on the reasons we gave for a flight to be canceled, how realistic is the assumption of independence that we made?

58. Social Science In one area, 4% of the population drives luxury cars. However, 17% of the CPAs drive luxury cars. Are the events "person drives a luxury car" and "person is a CPA" independent?

59. Social Science The probability that a key component of a space rocket will fail is .03.

(a) How many such components must be used as backups to ensure that the probability of at least one of the components' working is .999999?

(b) Is it reasonable to assume independence here?

60. Natural Science A medical experiment showed that the probability that a new medicine is effective is .75, the probability that a patient will have a certain side effect is .4, and the probability that both events occur is .3. Decide whether these events are dependent or independent.

61. Social Science A teacher has found that the probability that a student studies for a test is .6, the probability that a student gets a good grade on a test is .7, and the probability that both occur is .52. Are these events independent?

9.6 BAYES' FORMULA

Suppose the probability that a person gets lung cancer, given that the person smokes a pack or more of cigarettes daily, is known. For a research project, it might be necessary to know the probability that a person smokes a pack or more of cigarettes daily, given that the person has lung cancer. More generally, if $P(E \mid F)$ is known for two events E and F, can $P(F \mid E)$ be found? The answer is yes, we can find $P(F \mid E)$ using the formula to be developed in this section. To develop this formula, we can use a probability tree to find $P(F \mid E)$. Since $P(E \mid F)$ is known, the first outcome is either F or F'. Then for each of these outcomes, either E or E' occurs, as shown in Figure 9.30.

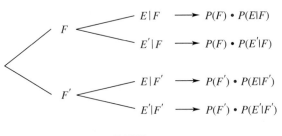

FIGURE 9.30

The four cases have the probabilities shown on the right. Notice that $P(E)$ is the sum of the first and third cases. By the definition of conditional probability and the product rule

$$P(F \mid E) = \frac{P(F \cap E)}{P(E)} = \frac{P(F) \cdot P(E \mid F)}{P(F) \cdot P(E \mid F) + P(F') \cdot P(E \mid F')}.$$

We have proved a special case of Bayes' formula, which is generalized later in this section.

1 Use the special case of Bayes' formula to find $P(F \mid E)$ if $P(F) = .2$, $P(E \mid F) = .1$, and $P(E \mid F') = .3$. (*Hint*: $P(F') = 1 - P(F)$.)

Answer:
$1/13 \approx .077$

BAYES' FORMULA (SPECIAL CASE)

$$P(F \mid E) = \frac{P(F) \cdot P(E \mid F)}{P(F) \cdot P(E \mid F) + P(F') \cdot P(E \mid F')}. \quad \boxed{1}$$

EXAMPLE 1 For a fixed length of time, the probability of worker error on a certain production line is .1, the probability that an accident will occur when there is a worker error is .3, and the probability that an accident will occur when there is no worker error is .2. Find the probability of a worker error if there is an accident.

Let E represent the event of an accident, and let F represent the event of worker error. From the information above,

$$P(F) = .1, \quad P(E \mid F) = .3, \quad \text{and} \quad P(E \mid F') = .2.$$

These probabilities are shown on the probability tree in Figure 9.31.

		Branch	Probability
$P(E \mid F) = .3$	E	1	$P(F) \cdot P(E \mid F)$
$P(E' \mid F) = .7$	E'	2	$P(F) \cdot P(E' \mid F)$
$P(E \mid F') = .2$	E	3	$P(F') \cdot P(E \mid F')$
$P(E' \mid F') = .8$	E'	4	$P(F') \cdot P(E' \mid F')$

$P(F) = .1$ F
$P(F') = .9$ F'

FIGURE 9.31

Find $P(F \mid E)$ using the tree or Bayes' formula.

2 In Example 1, find $P(F' \mid E)$.

Answer:
$6/7 \approx .857$

$$P(F \mid E) = \frac{P(F) \cdot P(E \mid F)}{P(F) \cdot P(E \mid F) + P(F') \cdot P(E \mid F')}$$

$$= \frac{(.1)(.3)}{(.1)(.3) + (.9)(.2)} \approx .143 \quad \blacksquare \quad \boxed{2}$$

Bayes' formula can be generalized to more than two possibilities with the probability tree of Figure 9.32. This diagram shows the paths that can produce an event E. We assume that events $F_1, F_2, \ldots, F_n$ are pairwise mutually exclusive events (that is, events which, taken two at a time, are disjoint), whose union is the sample space, and E is an event that has occurred. See Figure 9.33.

		Branch	Probability
F_1	$\xrightarrow{P(E\mid F_1)} E$	1	$P(F_1) \cdot P(E\mid F_1)$
F_2	$\xrightarrow{P(E\mid F_2)} E$	2	$P(F_2) \cdot P(E\mid F_2)$
$\vdots$	$\vdots$	$\vdots$	$\vdots$
F_n	$\xrightarrow{P(E\mid F_n)} E$	n	$P(F_n) \cdot P(E\mid F_n)$

with $P(F_1), P(F_2), \ldots, P(F_n)$.

FIGURE 9.32

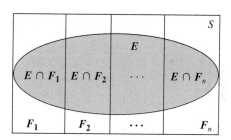

FIGURE 9.33

The probability $P(F_i \mid E)$, where $1 \le i \le n$, can be found by dividing the probability for the branch containing $P(E \mid F_i)$ by the sum of the probabilities of all the branches producing event E.

BAYES' FORMULA

$$P(F_i \mid E) = \frac{P(F_i) \cdot P(E \mid F_i)}{P(F_1) \cdot P(E \mid F_1) + \cdots + P(F_n) \cdot P(E \mid F_n)}.$$

This result is known as **Bayes' formula,** after the Reverend Thomas Bayes (1702–61), whose paper on probability was published about two hundred years ago.

The statement of Bayes' formula can be daunting. Actually, it is easier to remember the formula by thinking of the probability tree that produced it. Go through the following steps.

USING BAYES' FORMULA

Step 1 Start a probability tree with branches representing events F_1, $F_2, \ldots, F_n$. Label each branch with its corresponding probability.

Step 2 From the end of each of these branches, draw a branch for event E. Label this branch with the probability of getting to it, or $P(E \mid F_i)$.

Step 3 There are now n different paths that result in event E. Next to each path, put its probability—the product of the probabilities that the first branch occurs, $P(F_i)$, and that the second branch occurs, $P(E \mid F_i)$: that is, $P(F_i) \cdot P(E \mid F_i)$.

Step 4 $P(F_i \mid E)$ is found by dividing the probability of the branch for F_i by the sum of the probabilities of all the branches producing event E.

EXAMPLE 2 Based on past experience, a company knows that an experienced machine operator (one or more years of experience) will produce a defective item 1% of the time. Operators with some experience (up to one year) have a 2.5% defect rate, while new operators have a 6% defect rate. At any one time, the company has 60% experienced employees, 30% with some experience, and 10% new employees. Find the probability that a particular defective item was produced by a new operator.

Let E represent the event "item is defective," with F_1 representing "item was made by an experienced operator," F_2 "item was made by an operator with some experience," and F_3 "item was made by a new operator." Then

$$P(F_1) = .60 \qquad P(E \mid F_1) = .01$$
$$P(F_2) = .30 \qquad P(E \mid F_2) = .025$$
$$P(F_3) = .10 \qquad P(E \mid F_3) = .06.$$

We need to find $P(F_3 \mid E)$, the probability that an item was produced by a new operator, given that it is defective. First, draw a probability tree using the given information, as in Figure 9.34 on the next page. The steps leading to event E are shown.

Find $P(F_3 \mid E)$ using the bottom branch of the tree in Figure 9.34: divide the probability for this branch by the sum of the probabilities of all the branches leading to E.

$$P(F_3 \mid E) = \frac{.10(.06)}{.60(.01) + .30(.025) + .10(.06)} = \frac{.006}{.0195} \approx .3077 \quad \blacksquare \quad \boxed{3}$$

After working Problem 3 at the side, check that $P(F_1 \mid E) + P(F_2 \mid E) + P(F_3 \mid E) = 1$. (That is, the defective item was made by *someone*.)

$\boxed{3}$ In Example 2, find
(a) $P(F_1 \mid E)$;
(b) $P(F_2 \mid E)$.

Answers:
(a) $4/13 \approx .3077$
(b) $5/13 \approx .3846$

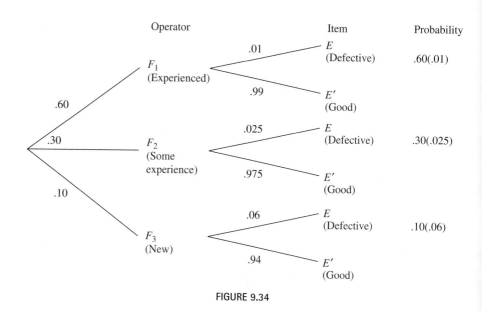

FIGURE 9.34

EXAMPLE 3 A manufacturer buys items from six different suppliers. The fraction of the total number of items obtained from each supplier, along with the probability that an item purchased from that supplier is defective, is shown in the following table.

Supplier	Fraction of Total Supplied	Probability of Defective
1	.05	.04
2	.12	.02
3	.16	.07
4	.23	.01
5	.35	.03
6	.09	.05

Find the probability that a defective item came from supplier 5.

Let F_1 be the event that an item came from supplier 1, with F_2, F_3, F_4, F_5, and F_6 defined in a similar manner. Let E be the event that an item is defective. We want to find $P(F_5 \mid E)$. Use the probabilities in the table above to prepare a probability tree. By Bayes' formula,

$$P(F_5 \mid E)$$

$$= \frac{(.35)(.03)}{(.05)(.04) + (.12)(.02) + (.16)(.07) + (.23)(.01) + (.35)(.03) + (.09)(.05)}$$

$$= \frac{.0105}{.0329} \approx .319.$$

There is about a 32% chance that a defective item came from supplier 5. ∎ 4

4 In Example 3, find the probability that the defective item came from

(a) supplier 3;

(b) supplier 6.

Answers:
(a) .340

(b) .137

9.6 EXERCISES

For two events, M and N, $P(M) = .4$, $P(N \mid M) = .3$, and $P(N \mid M') = .4$. Find each of the following. (See Example 1.)

1. $P(M \mid N)$ **2.** $P(M' \mid N)$

For mutually exclusive events R_1, R_2, R_3, $P(R_1) = .05$, $P(R_2) = .6$, and $P(R_3) = .35$. In addition, $P(Q \mid R_1) = .40$, $P(Q \mid R_2) = .30$, and $P(Q \mid R_3) = .60$. Find each of the following. (See Examples 2 and 3.)

3. $P(R_1 \mid Q)$ **4.** $P(R_2 \mid Q)$

5. $P(R_3 \mid Q)$ **6.** $P(R_1' \mid Q)$

Suppose three jars have the following contents: 2 black balls and 1 white ball in the first; 1 black ball and 2 white balls in the second; 1 black ball and 1 white ball in the third. If the probability of selecting one of the three jars is 1/2, 1/3, and 1/6, respectively, find the probability that if a white ball is drawn, it came from the

7. second jar;   **8.** third jar.

Social Science *In 1990, 1% of the United States population was Native American. The probability that a Native American lives in the West or South is .76. The probability is .56 that a person who is not Native American lives in the West or South. Find the following probabilities.*

9. A person living in the West or South is Native American.

10. A person who does not live in the West or South is not Native American.

Social Science *A federal study showed that in 1990, 49% of all those involved in a fatal car crash wore seat belts. Of those in a fatal crash who wore seat belts, 44% were injured and 27% were killed. For those not wearing seat belts, the comparable figures were 41% and 50%, respectively.**

11. Find the probability that a randomly selected person who was killed in a car crash was wearing a seat belt.

12. Find the probability that a randomly selected person who was unharmed in a fatal crash was not wearing a seat belt.

Management *The probability that a customer of a local department store will be a "slow pay" is .02. The probability that a "slow pay" will make a large down payment when buying a refrigerator is .14. The probability that a person who is not a "slow pay" will make a large down payment when buying a refrigerator is .50. Suppose a customer makes a large down payment on a refrigerator. Find the probability that the customer is*

13. a "slow pay";   **14.** not a "slow pay."

Management *Companies A, B, and C produce 15%, 40%, and 45%, respectively, of the major appliances sold in a certain area. In that area, 1% of the Company A appliances, $1\frac{1}{2}$% of the Company B appliances, and 2% of the Company C appliances need service within the first year. Suppose an appliance that needs service within the first year is chosen at random; find the probability that it was manufactured by Company*

15. A;   **16.** B.

Management *On a given weekend in the fall, a tire company can buy television advertising time for a college football game, a baseball game, or a professional football game. If the company sponsors the college game, there is a 70% chance of a high rating, a 50% chance if they sponsor a baseball game, and a 60% chance if they sponsor a professional football game. The probability of the company sponsoring these various games is .5, .2, and .3, respectively. Suppose the company does get a high rating; find the probability that it sponsored*

17. a college game;   **18.** a professional football game.

19. Social Science A 1995 survey showed that of Baby Boomers (Americans born between 1946 and 1960), 78% own a home and 56% have a family income of more than $40,000.* The probability that a Baby Boomer's family own their own home if their income is over $40,000 is .51, while the probability that they own their home if the family income is less than or equal to $40,000 is .27. Find the probability that a Baby Boomer has family income over $40,000, if it is known that the family owns its home.

20. Social Science During the murder trial of O.J. Simpson, Alan Dershowitz, an advisor to the defense team, stated on television that only about .1% of men who batter their wives actually murder them. Statistician I. J. Good observed that even if, given that a husband is a batterer, the probability he is guilty of murdering his wife is .001, what we really want to know is the probability that the husband is guilty, given that the wife was murdered.† Good estimates the probability of a battered wife being murdered, given that her husband is not guilty, as .001. The probability that she is murdered if her husband is guilty is 1, of course. Using these numbers and Dershowitz' .001 probability of the husband being guilty, find the probability that the husband is guilty, given that the wife was murdered.

21. Management A manufacturing firm finds that 70% of its new hires turn out to be good workers and 30% poor

---

*National Highway Traffic Safety Administration, Office of Driver and Pedestrian Research: "Occupant Protection Trends in 19 Cities" (November 1989) and "Use of Automatic Safety Belt Systems in 19 Cities" (February 1991).

**The New York Times*, Sec. 3, p. 1, Dec. 31, 1995.

†I. J. Good, "When Batterer Turns Murderer," *Nature*, Vol. 375, No. 15, June 15, 1995, p. 541.

workers. All current workers are given a reasoning test. Of the good workers, 80% pass it; 40% of the poor workers pass it. Assume that these figures will hold true in the future. If the company makes the test part of its hiring procedure and only hires people who meet the previous requirements and pass the test, what percent of the new hires will turn out to be good workers?

22. **Management**   A bank finds that the relationship between mortgage defaults and the size of the down payment is given by this table.

Down Payment (%)	5%	10%	20%	25%
Number of mortgages with this down payment	1260	700	560	280
Probability of default	.05	.03	.02	.01

If a default occurs, what is the probability that it is on a mortgage with a 5% down payment? (See Examples 2 and 3.)

23. **Management**   The following information pertains to three shipping terminals operated by Krag Corp.:*

Terminal	Percentage of Cargo Handled	Percentage of Error
Land	50	2
Air	40	4
Sea	10	14

Krag's internal auditor randomly selects one set of shipping documents, ascertaining that the set selected contains an error. Which of the following gives the probability that the error occurred in the Land Terminal?

(a) .02   (b) .10   (c) .25   (d) .50

Natural Science   *In a test for toxemia, a disease that affects pregnant women, the woman lies on her left side and then rolls over on her back. The test is considered positive if there is a 20 mm rise in her blood pressure within one minute. The results have produced the following probabilities, where T represents having toxemia at some time during the pregnancy, and N represents a negative test.*

$$P(T' \mid N) = .90 \quad and \quad P(T \mid N') = .75$$

Assume that $P(N') = .11$, and find each of the following.

24. $P(N \mid T)$

25. $P(N' \mid T)$

26. **Natural Science**   The probability that a person with certain symptoms has hepatitis is .8. The blood test used to confirm this diagnosis gives positive results for 90% of those who have the disease and 5% of those without the disease. What is the probability that an individual with the symptoms who reacts positively to the test has hepatitis?

27. **Natural Science**   Suppose the probability that an individual has AIDS is .01, the probability of a person testing positive if he or she has AIDS is .95, and the probability of a person testing positive if he or she does not have AIDS is .05.

(a) Find the probability that a person who tests positive has AIDS.

(b) It has been argued that everyone should be tested for AIDS. Based on the results of part (a), how useful would the results of such testing be?

Social Science   *The following table gives the proportions of adult men and women in the U.S. population, and the proportions of adult men and women who have never married, in 1987.*

	MEN	
Age	Proportion of Men	Proportion Never Married
18–24	.151	.875
25–29	.126	.433
30–34	.126	.250
35–39	.110	.140
40 or over	.487	.054

	WOMEN	
Age	Proportion of Women	Proportion Never Married
18–24	.142	.752
25–29	.117	.295
30–34	.116	.161
35–39	.103	.090
40 or over	.522	.033

28. Find the probability that a randomly selected man who has never married is between 30 and 34 years old (inclusive).

29. Find the probability that a randomly selected woman who has been married is between 18 and 24 (inclusive).

*Uniform CPA Examination, November 1989.

*From *The Sacramento Bee*, September 10, 1987.

30. Find the probability that a randomly selected woman who has never been married is between 35 and 39 (inclusive).

31. Natural Science A recent study by the Harvard School of Public Health reported that 86% of male students who live in a fraternity house are binge drinkers. The figure for fraternity members who are not residents of a fraternity house is 71%, while the figure for men who do not belong to a fraternity is 45%.* Suppose that 10% of U.S. students live in a fraternity house, 15% belong to a fraternity, but do not live in a fraternity house, and 75% do not belong to a fraternity.

(a) What is the probability that a randomly selected male student is a binge drinker?

(b) If a randomly selected male student is a binge drinker, what is the probability that he lives in a fraternity house?

*New York Times, December 6, 1995, p. B16.

CHAPTER 9 SUMMARY

Key Terms and Symbols

{ }	set braces
$\in$	is an element of
$\notin$	is not an element of
$\emptyset$	empty set
$\subseteq$	is a subset of
$\nsubseteq$	is not a subset of
A'	complement of set A
$\cap$	set intersection
$\cup$	set union

9.1 set
element (member) of a set
empty set
equal sets
set-builder notation
universal set
subset
tree diagram
Venn diagram
complement
intersection
disjoint sets
union

9.2 Union Rule for Counting

9.3 $P(E)$ probability of event E

experiment
trial
outcome
fair coin
sample space
event
simple event
certain event
impossible event
basic probability principle
probability of an event

9.4 mutually exclusive events
Union Rule for Probability
Complement Rule
odds in favor

9.5 $P(F \mid E)$ probability of F, given that E has occurred
conditional probability
Product Rule of Probability
probability tree
independent events
dependent events
Product Rule for Independent Events

9.6 Bayes' formula

Key Concepts

Sets   Set A is a **subset** of set B if every element of A is also an element of B.

A set of n elements has 2^n subsets.

Let A and B be any sets with universal set U.
The **complement** of A is $A' = \{x \mid x \notin A$ and $x \in U\}$.
The **intersection** of A and B is $A \cap B = \{x \mid x \in A$ and $x \in B\}$.
The **union** of A and B is $A \cup B = \{x \mid x \in A$ or $x \in B$ or both$\}$.

$$n(A \cup B) = n(A) + n(B) - n(A \cap B)$$

Probability   If $n(S) = n$ and $n(E) = m$, where S is the sample space, then $P(E) = \dfrac{m}{n}$.

The probability of any outcome is a number between 0 and 1, inclusive.

The sum of the probabilities of all possible distinct outcomes in a sample space is 1.

Let E and F be events from a sample space S.

$$P(E') = 1 - P(E) \text{ and } P(E) = 1 - P(E')$$

$$P(E \cup F) = P(E) + P(F) - P(E \cap F)$$

$$P(E \mid F) = \frac{P(E \cap F)}{P(F)}, \quad P(F) \neq 0$$

$$P(E \cap F) = P(F) \cdot P(E \mid F) \quad \text{or} \quad P(E \cap F) = P(E) \cdot P(F \mid E)$$

Odds: The odds in favor of event E are given by the ratio of $P(E)$ to $P(E')$.

Events E and F are **mutually exclusive** if $E \cap F = \emptyset$. In that case, $P(E \cup F) = P(E) + P(F)$.

Events E and F are **independent events** if $P(F \mid E) = P(F)$ or $P(E \mid F) = P(E)$. In that case, $P(E \cap F) = P(E) \cdot P(F)$.

Bayes' Formula: $P(F_i \mid E) = \dfrac{P(F_i) \cdot P(E \mid F_i)}{P(F_1) \cdot P(E \mid F_1) + \cdots + P(F_n) \cdot P(E \mid F_n)}$

Chapter 9 Review Exercises

Write true *or* false *for each of the following.*

1. $9 \in \{8, 4, -3, -9, 6\}$

2. $4 \in \{3, 9, 7\}$

3. $2 \notin \{0, 1, 2, 3, 4\}$

4. $0 \notin \{0, 1, 2, 3, 4\}$

5. $\{3, 4, 5\} \subseteq \{2, 3, 4, 5, 6\}$

6. $\{1, 2, 5, 8\} \subseteq \{1, 2, 5, 10, 11\}$

7. $\emptyset \subseteq \{1\}$

8. $0 \subseteq \emptyset$

List the elements in the following sets.

9. $\{x \mid x \text{ is a national holiday}\}$

10. $\{x \mid x \text{ is an integer, } -3 \leq x < 1\}$

11. $\{$all counting numbers less than 5$\}$

12. $\{x \mid x \text{ is a leap year between 1989 and 1999}\}$

Let $U = \{Vitamins\ A, B_1, B_2, B_3, B_6, B_{12}, C, D, E\}$, $M = \{A, C, D, E\}$, *and* $N = \{A, B_1, B_2, C, E\}$. *Find the following.*

13. M'

14. N'

15. $M \cap N$

16. $M \cup N$

17. $M \cup N'$

18. $M' \cap N$

Let $U = \{all\ students\ in\ a\ class\}$, $A = \{all\ male\ students\}$, $B = \{all\ A\ students\}$, $C = \{all\ students\ with\ red\ hair\}$, and $D = \{all\ students\ younger\ than\ 21\}$. Describe the following sets in words.

19. $A \cap C$

20. $B \cap D$

21. $A \cup D$

22. $A' \cap D$

23. $B' \cap C'$

Draw a Venn diagram and shade the given set.

24. $B \cup A'$

25. $A' \cap B$

26. $A' \cap (B' \cap C)$

27. $(A \cup B)' \cap C$

Social Science *A survey of a group of military personnel revealed the following information.*

> 20 officers
> 27 minorities
> 19 women
> 5 women officers
> 8 minority women
> 10 minority officers
> 3 women minority officers
> 6 Caucasian male enlisted personnel

28. How many were interviewed?

29. How many were enlisted minority women?

30. How many were male minority officers?

Write sample spaces for the following.

31. A die is rolled and the number of points showing is noted.

32. A card is drawn from a deck containing only 4 aces.

33. A color is selected from the set {red, blue, green}, and then a number is chosen from the set {10, 20, 30}.

A jar contains 5 discs labeled 2, 4, 6, 8, 10, and another jar contains 2 blue and 3 yellow balls. One disc is drawn and then a ball is drawn. Give the following.

34. The sample space

35. Event F, the ball is blue.

36. Event E, the disc shows a number greater than 5.

37. Are the outcomes in this sample space equally likely?

Management *A company sells typewriters and copiers. Let E be the event "a customer buys a typewriter," and let F be the event "a customer buys a copier." In Exercises 38 and 39, write each of the following using ∩, ∪, or ', as necessary.*

38. A customer buys neither.

39. A customer buys at least one.

40. A student gives the answer to a probability problem as 6/5. Explain why this answer must be incorrect.

41. Describe what is meant by disjoint sets and give an example.

42. Describe what is meant by mutually exclusive events and give an example.

43. How are disjoint sets and mutually exclusive events related?

A single card is drawn from an ordinary deck. Find the probability of drawing each of the following.

44. A black king

45. A face card

46. A red card or a face card

47. A black card, given it is a 2

48. A jack, given it is a face card

49. A face card, given it is a jack

Find the odds in favor of drawing the following.

50. A spade

51. A red queen

52. A black face card or a 7

Management *A sample shipment of five electric motors is chosen at random. The probability of exactly 0, 1, 2, 3, 4, or 5 motors being defective is given in the following table.*

Number defective	0	1	2	3	4	5
Probability	.31	.25	.18	.12	.08	.06

Find the probability that

53. no more than 3 are defective.

54. at least 3 are defective.

Natural Science *The square shows the four possible (equally likely) combinations when both parents are carriers of the sickle cell anemia trait. Each carrier parent has normal cells (N) and trait cells (T).*

	2ND PARENT	
	N_2	T_2
1ST PARENT N_1 T_1		$N_1 T_2$

55. Complete the table.

56. If the disease occurs only when two trait cells combine, find the probability that a child born to these parents will have sickle cell anemia.

57. The child will carry the trait but not have the disease if a normal cell combines with a trait cell. Find this probability.

58. Find the probability that the child is neither a carrier nor has the disease.

Find the probabilities for the following sums when two fair dice are rolled.

59. 8

60. At least 10

61. No more than 5

62. Odd and greater than 8

63. 12, given the sum is greater than 10

64. 7, given that at least one die shows a 4

Suppose $P(E) = .51$, $P(F) = .37$, and $P(E \cap F) = .22$. Find each of the following probabilities.

65. $P(E \cup F)$

66. $P(E \cap F')$

67. $P(E' \cup F)$

68. $P(E' \cap F')$

69. For the events E and F, $P(E) = .2$, $P(E \mid F) = .3$, and $P(F \mid E') = .2$. Find each of the following.
   (a) $P(E \mid F)$ **(b)** $P(E \mid F')$

70. Define independent events and give an example.

71. Are independent events always mutually exclusive? Are they ever mutually exclusive? Give examples.

Management   *Of the appliance repair shops listed in the phone book, 80% are competent and 20% are not. A competent shop can repair an appliance correctly 95% of the time; an incompetent shop can repair an appliance correctly 60% of the time. Suppose an appliance was repaired correctly. Find the probability that it was repaired by*

72. a competent shop;

73. an incompetent shop.

Suppose an appliance was repaired incorrectly. Find the probability that it was repaired by

74. a competent shop;

75. an incompetent shop.

76. Four red and one orange slips of paper are placed in a box. Two red and three orange slips are placed in a second box. A box is chosen at random, and a slip of paper is selected from it. The probability of choosing the first box is 3/8. If the selected slip of paper is orange, what is the probability that it came from the first box?

77. Find the probability that the slip of paper in Exercise 76 came from the second box, given that it is red.

78. Management   A manufacturer buys items from four different suppliers. The fraction of the total number of items that is obtained from each supplier, along with the probability that an item purchased from that supplier is defective, is shown in the following table.

Supplier	Fraction of Total Supplied	Probability of Defective
1	.17	.04
2	.39	.02
3	.35	.07
4	.09	.03

(a) Find the probability that a defective item came from supplier 4.
(b) Find the probability that a defective item came from supplier 2.

79. Management   The table below shows the results of a survey of buyers of a certain model of car.

Car Type	Satisfied	Not Satisfied	Totals
New	300	100	
Used	450		600
Totals		250	

(a) Complete the table.
(b) How many buyers were surveyed?
(c) How many bought a new car and were satisfied?
(d) How many were not satisfied?
(e) How many bought used cars?
(f) How many of those who were not satisfied had bought a used car?
(g) Rewrite the event stated in part (f) using the expression "given that."
(h) Find the probability of the outcome in parts (f) and (g).
(i) Find the probability that a used-car buyer is not satisfied.
(j) You should have different answers in parts (h) and (i). Explain why.

CASE 9

Medical Diagnosis

When a patient is examined, information, typically incomplete, is obtained about his or her state of health. Probability theory provides a mathematical model appropriate for this situation, as well as a procedure for quantitatively interpreting such partial information to arrive at a reasonable diagnosis.*

To do this, list the states of health that can be distinguished in such a way that the patient can be in one and only one state at the time of the examination. Each state of health H is associated with a number $P(H)$ between 0 and 1 such that the sum of all these numbers is 1. This number $P(H)$ represents the probability, before examination, that a patient is in the state of health H, and $P(H)$ may be chosen subjectively from medical experience, using any information available prior to the examination. The probability may be most conveniently established from clinical records; that is, a mean probability is established for patients in general, although the number would vary from patient to patient. Of course, the more information that is brought to bear in establishing $P(H)$, the better the diagnosis.

For example, limiting the discussion to the condition of a patient's heart, suppose there are exactly 3 states of health, with probabilities as follows.

	State of Health H	P(H)
H_1	patient has a normal heart	.8
H_2	patient has minor heart irregularities	.15
H_3	patient has a severe heart condition	.05

Having selected $P(H)$, the information from the examination is processed. First, the results of the examination must be classified. The examination itself consists of observing the state of a number of characteristics of the patient. Let us assume that the examination for a heart condition consists of a stethoscope examination and a cardiogram. The outcome of such an examination, C, might be one of the following:

C_1 = stethoscope shows normal heart and cardiogram shows normal heart;

C_2 = stethoscope shows normal heart and cardiogram shows minor irregularities;

and so on.

*From "Probabilistic Medical Diagnosis," Roger Wright, *Some Mathematical Models in Biology,* Robert M. Thrall, ed., (The University of Michigan, 1967), by permission of Robert M. Thrall.

It remains to assess for each state of health H the conditional probability $P(C \mid H)$ of each examination outcome C using only the knowledge that a patient is in a given state of health. (This may be based on the medical knowledge and clinical experience of the doctor.) The conditional probabilities $P(C \mid H)$ will not vary from patient to patient, so that they may be built into a diagnostic system, although they should be reviewed periodically.

Suppose the result of the examination is C_1. Let us assume the following probabilities.

$$P(C_1 \mid H_1) = .9$$
$$P(C_1 \mid H_2) = .4$$
$$P(C_1 \mid H_3) = .1$$

Now, for a given patient, the appropriate probability associated with each state of health H, after examination, is $P(H \mid C)$ where C is the outcome of the examination. This can be calculated by using Bayes' formula. For example, to find $P(H_1 \mid C_1)$—that is, the probability that the patient has a normal heart given that the examination showed a normal stethoscope examination and a normal cardiogram—we use Bayes' formula as follows.

$$P(H_1 \mid C_1)$$
$$= \frac{P(C_1 \mid H_1)P(H_1)}{P(C_1 \mid H_1)P(H_1) + P(C_1 \mid H_2)P(H_2) + P(C_1 \mid H_3)P(H_3)}$$
$$= \frac{(.9)(.8)}{(.9)(.8) + (.4)(.15) + (.1)(.05)} \approx .92$$

Hence, the probability is about .92 that the patient has a normal heart on the basis of the examination results. This means that in 8 out of 100 patients, some abnormality will be present and not be detected by the stethoscope or the cardiogram.

EXERCISES

1. Find $P(H_2 \mid C_1)$.
2. Assuming the following probabilities, find $P(H_1 \mid C_2)$:
   $$P(C_2 \mid H_1) = .2, \quad P(C_2 \mid H_2) = .8, \quad P(C_2 \mid H_3) = .3.$$
3. Assuming the probabilities of Exercise 2, find $P(H_3 \mid C_2)$.

Further Topics in Probability

A survey by *Money* magazine found that supermarket scanners are overcharging customers at 30% of stores. If you shop at 3 supermarkets that use scanners, what is the probability that you will be overcharged? Techniques for finding this probability and solving other special kinds of probability problems are introduced in this chapter.

10.1 PERMUTATIONS AND COMBINATIONS

Up to this point we have simply listed the outcomes in a sample space S and an event E in order to find $P(E)$. However, when S has many outcomes, listing them becomes very tedious. In this section we discuss counting methods that do not require listing.

Let us begin with a simple example. If there are 3 roads from town A to town B and 2 roads from town B to town C, in how many ways can someone travel from A to C by way of B? For each of the 3 roads from A there are 2 different routes leading from B to C, making $3 \cdot 2 = 6$ different trips, as shown in Figure 10.1.

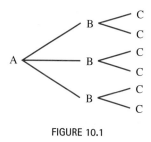

FIGURE 10.1

This example illustrates a general principle of counting called the *multiplication principle*.

> ### MULTIPLICATION PRINCIPLE
>
> Suppose n choices must be made, with
>
> $$m_1 \text{ ways to make choice 1,}$$
>
> and for each of these,
>
> $$m_2 \text{ ways to make choice 2,}$$
>
> and so on, with
>
> $$m_n \text{ ways to make choice } n.$$
>
> Then there are
>
> $$m_1 \cdot m_2 \cdot \cdots \cdot m_n$$
>
> different ways to make the entire sequence of choices.

EXAMPLE 1   A combination lock can be set to open to any 3-letter sequence. How many such sequences are possible?

Since there are 26 letters in the alphabet, there are 26 choices for each of the 3 letters, and, by the multiplication principle, $26 \cdot 26 \cdot 26 = 17{,}576$ different sequences. ■

EXAMPLE 2   Morse code uses a sequence of dots and dashes to represent letters and words. How many sequences are possible with at most 3 symbols?

"At most 3" means "1 or 2 or 3." Each symbol may be either a dot or a dash. Thus, the following number of sequences are possible in each case.

Number of Symbols	Number of Sequences
1	2
2	$2 \cdot 2 = 4$
3	$2 \cdot 2 \cdot 2 = 8$

Altogether, $2 + 4 + 8 = 14$ different sequences of at most 3 symbols are possible. Because there are 26 letters in the alphabet, some letters must be represented by 4 symbols in Morse code. ■

1 **(a)** In how many ways can 6 business tycoons line up their golf carts at the country club?

(b) How many ways can 4 pupils be seated in a row with 4 seats?

Answers:
(a) $6 \cdot 5 \cdot 4 \cdot 3 \cdot 2 \cdot 1 = 720$
(b) $4 \cdot 3 \cdot 2 \cdot 1 = 24$

EXAMPLE 3   A teacher has 5 different books to be arranged side by side. How many different arrangements are possible?

Five choices will be made, 1 for each space that will hold a book. Any of the 5 possible books could be chosen for the first space. There are 4 possible choices for the second space, since 1 book has already been placed in the first space, 3 possible choices for the third space, and so on. By the multiplication principle, the number of different possible arrangements (sequence of choices) is $5 \cdot 4 \cdot 3 \cdot 2 \cdot 1 = 120$. ■   1

The use of the multiplication principle often leads to products such as $5 \cdot 4 \cdot 3 \cdot 2 \cdot 1$, the product of all the natural numbers from 5 down to 1. If n is a natural number, the symbol $n!$ (read "n factorial") denotes the product of all the natural numbers from n down to 1. If $n = 1$, this formula is understood to give $1! = 1$.

2 Evaluate:

(a) 4!

(b) 6!

(c) 1!

(d) 6!/4!

Answers:

(a) 24

(b) 720

(c) 1

(d) 30

> ### n-FACTORIAL
> For any natural number n,
> $$n! = n(n - 1)(n - 2) \cdots (3)(2)(1).$$
> Also, 0! is defined to be the number 1.

With this symbol, the product $5 \cdot 4 \cdot 3 \cdot 2 \cdot 1$ can be written as 5!. Also, $3! = 3 \cdot 2 \cdot 1 = 6$. The definition of $n!$ could be used to show that $n! = n \cdot (n - 1)!$ for all natural numbers $n \geq 2$. It is helpful if this result also holds for $n = 1$. This can only happen if 0! equals 1, as defined above. **2**

Many scientific calculators and all graphing calculators can find the exact value of $n!$ for small values of n (typically, $n \leq 13$) and the approximate value of $n!$ for larger values of n (typically, $14 \leq n \leq 69$). Some graphing calculators can approximate $n!$ for much larger values of n.

TECHNOLOGY TIP The factorial key on a graphing calculator is usually located in the PROB submenu of the MATH menu. ✔

3 In how many ways can 3 of 7 items be arranged?

Answer:
$7 \cdot 6 \cdot 5 = 210$

EXAMPLE 4 Suppose the teacher in Example 3 wishes to place only 3 of the 5 books on his desk. How many arrangements of 3 books are possible?

The teacher again has 5 ways to fill the first space, 4 ways to fill the second space, and 3 ways to fill the third. Because he wants to use only 3 books, there are only 3 spaces to be filled giving $5 \cdot 4 \cdot 3 = 60$ arrangements. ■ **3**

PERMUTATIONS The answer 60 in Example 4 is called the number of *permutations* of 5 things taken 3 at a time. A **permutation** of r elements (where $r \geq 1$) from a set of n elements is any arrangement, *without repetition*, of the r elements. The number of permutations of n things taken r at a time (with $r \leq n$) is written $_nP_r$.* Based on the work in Example 4,

$$_5P_3 = 5 \cdot 4 \cdot 3 = 60.$$

Factorial notation can be used to express this product as follows.

$$5 \cdot 4 \cdot 3 = 5 \cdot 4 \cdot 3 \cdot \frac{2 \cdot 1}{2 \cdot 1} = \frac{5 \cdot 4 \cdot 3 \cdot 2 \cdot 1}{2 \cdot 1} = \frac{5!}{2!} = \frac{5!}{(5 - 3)!}$$

This example illustrates the general rule of permutations, which is stated below.

*Another notation that is sometimes used is $P(n, r)$.

PERMUTATIONS

If $_nP_r$ (where $r \leq n$) is the number of permutations of n elements taken r at a time, then

$$_nP_r = \frac{n!}{(n-r)!}.$$

---

4 Find the number of permutations of

(a) 5 things taken 2 at a time;

(b) 9 things taken 3 at a time.

Find each of the following.

(c) $_3P_1$

(d) $_7P_3$

(e) $_{12}P_2$

Answers:
(a) 20

(b) 504

(c) 3

(d) 210

(e) 132

TECHNOLOGY TIP The permutation function on a graphing calculator is also in the PROB submenu of the MATH menu. As with $n!$, for large values of n and r, the calculator display for $_nP_r$ may be an approximation. ✔

To find $_nP_r$, we can use either the rule above or direct application of the multiplication principle, as the following example shows.

EXAMPLE 5 Early in 1996, eight candidates sought the Republican nomination for president. In how many ways could voters rank their first, second, and third choices?

This is the same as finding the number of permutations of 8 elements taken 3 at a time. Since there are 3 choices to be made, the multiplication principle gives $_8P_3 = 8 \cdot 7 \cdot 6 = 336$. Alternatively, by the formula for $_nP_r$,

$$_8P_3 = \frac{8!}{(8-3)!} = \frac{8!}{5!} = \frac{8 \cdot 7 \cdot 6 \cdot 5 \cdot 4 \cdot 3 \cdot 2 \cdot 1}{5 \cdot 4 \cdot 3 \cdot 2 \cdot 1} = 8 \cdot 7 \cdot 6 = 336. \quad ■ \quad \boxed{4}$$

EXAMPLE 6 Find each of the following.

(a) The number of permutations of the letters A, B, and C
By the formula for $_nP_r$ with both n and r equal to 3,

$$_3P_3 = \frac{3!}{(3-3)!} = \frac{3!}{0!} = \frac{3!}{1} = 3! = 6.$$

The 6 permutations (or arrangements) are

ABC, ACB, BAC, BCA, CAB, CBA.

(b) The number of permutations possible using just 2 of the letters A, B, and C
Find $_3P_2$:

$$_3P_2 = \frac{3!}{(3-2)!} = \frac{3!}{1!} = 3! = 6.$$

This result is exactly the same answer as in part (a). This is because, in the case of $_3P_3$, after the first 2 choices are made, the third is already determined, as shown in the table below.

5 Find the number of permutations of the letters C, O, D, and E

(a) using all the letters;

(b) using 2 of the 4 letters;

(c) using 3 of the 4 letters. (Can you find this without calculating?)

Answers:
(a) 24

(b) 12

(c) 24 (yes)

First two letters	AB	AC	BA	BC	CA	CB
Third letter	C	B	C	A	B	A

■ $\boxed{5}$

EXAMPLE 7 A televised talk show will include 4 women and 3 men as panelists.

(a) In how many ways can the panelists be seated in a row of 7 chairs?
Find $_7P_7$, the total number of ways to seat 7 panelists in 7 chairs.

$$_7P_7 = \frac{7!}{(7-7)!} = \frac{7!}{0!} = \frac{7!}{1} = 7 \cdot 6 \cdot 5 \cdot 4 \cdot 3 \cdot 2 \cdot 1 = 5040$$

There are 5040 ways to seat the 7 panelists.

(b) In how many ways can the panelists be seated if the men and women are to be alternated?

In order to alternate men and women, a woman must be seated in the first chair (since there are 4 women and only 3 men), any of the men next, and so on. Thus, there are 4 ways to fill the first seat, 3 ways to fill the second seat, 3 ways to fill the third seat (with any of the 3 remaining women), and so on. Use the multiplication principle. There are

$$4 \cdot 3 \cdot 3 \cdot 2 \cdot 2 \cdot 1 \cdot 1 = 144$$

ways to seat the panelists. ■ 6

6 A collection of 3 paintings by one artist and 2 by another is to be displayed. In how many ways can the paintings be shown

(a) in a row?

(b) if the works of the artists are to be alternated?

Answers:
(a) 120
(b) 12

COMBINATIONS In Example 4, we found that there are 60 ways that a teacher can arrange 3 of 5 different books on a desk. That is, there are 60 permutations of 5 things taken 3 at a time. Suppose now that the teacher does not wish to arrange the books on his desk, but rather wishes to choose, at random, any 3 of the 5 books to give to a book sale to raise money for his school. In how many ways can he do this?

At first glance, we might say 60 again, but this is incorrect. The number 60 counts all possible *arrangements* of 3 books chosen from 5. However, the following arrangements would all lead to the same set of 3 books being given to the book sale.

mystery-biography-textbook	biography-textbook-mystery
mystery-textbook-biography	textbook-biography-mystery
biography-mystery-textbook	textbook-mystery-biography

The list shows 6 different *arrangements* of 3 books, but only one subset of 3 books selected from the 5 books for the book sale. A subset of items selected *without regard to order* is called a **combination**. The number of combinations of 5 things taken 3 at a time is written $\binom{5}{3}$ or $_5C_3$. Since they are subsets, combinations are *not ordered*.

To evaluate $\binom{5}{3}$, start with the $5 \cdot 4 \cdot 3$ *permutations* of 5 things taken 3 at a time. Combinations are unordered, therefore, find the number of combinations by dividing the number of permutations by the number of ways each group of 3 can be ordered—that is, by 3!.

$$\binom{5}{3} = \frac{5 \cdot 4 \cdot 3}{3!} = \frac{5 \cdot 4 \cdot 3}{3 \cdot 2 \cdot 1} = 10$$

There are 10 ways that the teacher can choose 3 books at random for the book sale.

Generalizing this discussion gives the formula for the number of combinations of n elements taken r at a time, written $\binom{n}{r}$ or $_nC_r$.

7 Evaluate $\dfrac{{}_nP_r}{r!}$ for the following values.

(a) $n = 6, r = 2$

(b) $n = 8, r = 4$

(c) $n = 7, r = 0$

Answers:
(a) 15
(b) 70
(c) 1

$$\binom{n}{r} = {}_nC_r = \frac{{}_nP_r}{r!}$$

$$= \frac{n!}{(n-r)!} \cdot \frac{1}{r!} \qquad \text{Definition of } {}_nP_r$$

$$= \frac{n!}{(n-r)!r!}$$

This last form is the most useful for setting up the calculation.   7

COMBINATIONS

The number of combinations of n elements taken r at a time, where $r \le n$, is

$$\binom{n}{r} = {}_nC_r = \frac{n!}{(n-r)!r!}.$$

TECHNOLOGY TIP The combination function on a graphing calculator is in the PROB submenu of the MATH menu. ✔

Replacing r by $n - r$ in the combinations formula gives

$$\binom{n}{n-r} = \frac{n!}{(n - [n-r])! \, (n-r)!}$$

$$= \frac{n!}{r! \, (n - r)!} = \frac{n!}{(n - r)! \, r!}.$$

Therefore,

8 Use $\dfrac{n!}{(n-r)!r!}$ to evaluate $\binom{n}{r}$.

(a) $\binom{6}{2}$

(b) $\binom{8}{4}$

(c) $\binom{7}{0}$

Compare your answers with the answers for Side Problem 7.

Answers:
(a) 15
(b) 70
(c) 1

$$\binom{n}{r} = \binom{n}{n-r}, \quad \text{or equivalently,} \quad {}_nC_r = {}_nC_{n-r}.$$

For example, by this result,

$$\binom{5}{3} = \binom{5}{2} \quad \text{and} \quad \binom{10}{4} = \binom{10}{6}.$$

EXAMPLE 8   How many committees of 3 people can be formed from a group of 8 people?

A committee is an unordered group, so find $\binom{8}{3}$. By the formula for combinations,

$$\binom{8}{3} = \frac{8!}{5! \, 3!} = \frac{8 \cdot 7 \cdot 6 \cdot 5 \cdot 4 \cdot 3 \cdot 2 \cdot 1}{5 \cdot 4 \cdot 3 \cdot 2 \cdot 1 \cdot 3 \cdot 2 \cdot 1} = \frac{8 \cdot 7 \cdot 6}{3 \cdot 2 \cdot 1} = 56. \ \blacksquare \quad 8$$

EXAMPLE 9   Three managers are to be selected from a group of 30 to work on a special project.

(a) In how many different ways can the managers be selected?

Here we wish to know the number of 3-element combinations that can be formed from a set of 30 elements. (We want combinations and not permutations, since order within the group of 3 does not matter.)

$$\binom{30}{3} = 4060$$

There are 4060 ways to select the project group.

(b) In how many ways can the group of 3 be selected if a certain manager must work on the project?

Since 1 manager has already been selected for the project, the problem is reduced to selecting 2 more from the remaining 29 managers.

$$\binom{29}{2} = 406$$

In this case, the project group can be selected in 406 ways.

(c) In how many ways can a nonempty group of at most 3 managers be selected from these 30 managers?

The group is to be nonempty; therefore, "at most 3" means "1 or 2 or 3." Find the number of ways for each case.

Case	*Number of Ways*
1	$\binom{30}{1} = \dfrac{30!}{29!\,1!} = \dfrac{30 \cdot 29!}{29!\,(1)!} = 30$
2	$\binom{30}{2} = \dfrac{30!}{28!\,2!} = \dfrac{30 \cdot 29 \cdot 28!}{28! \cdot 2 \cdot 1} = 435$
3	$\binom{30}{3} = \dfrac{30!}{27!\,3!} = \dfrac{30 \cdot 29 \cdot 28 \cdot 27!}{27! \cdot 3 \cdot 2 \cdot 1} = 4060$

The total number of ways to select at most 3 managers will be the sum

$$30 + 435 + 4060 = 4525. \quad \blacksquare \quad \boxed{9}$$

The formulas for permutations and combinations given in this section will be very useful in solving probability problems in later sections. Any difficulty in using these formulas usually comes from being unable to differentiate between them. Both permutations and combinations give the number of ways to choose r objects from a set of n objects. The differences between permutations and combinations are outlined below.

PERMUTATIONS	**COMBINATIONS**
Different orderings or arrangements of the r objects are different permutations.	Each choice or subset of r objects gives 1 combination. Order within the r objects does not matter.
$${}_nP_r = \dfrac{n!}{(n-r)!}$$	$$\binom{n}{r} = {}_nC_r = \dfrac{n!}{(n-r)!\,r!}$$
Clue words: Arrangement, Schedule, Order	Clue words: Group, Committee, Sample

---

Margin notes:

9 Five orchids from a collection of 20 are to be selected for a flower show.

(a) In how many ways can this be done? *Comb.*

(b) In how many different ways can the group of 5 be selected if 2 particular orchids must be included?

(c) In how many ways can at least 1 and at most 5 orchids be selected? (*Hint:* Use a calculator or refer to Table 1 in Appendix B.)

Answers:

(a) $\binom{20}{5} = 15{,}504$

(b) $\binom{18}{3} = 816$

(c) 21,699

In the next examples, concentrate on recognizing which of the formulas should be applied.

EXAMPLE 10 For each of the following problems, tell whether permutations or combinations should be used to solve the problem.

(a) How many 4-digit code numbers are possible if no digits are repeated?

Since changing the order of the 4 digits results in a different code, we use permutations.

(b) A sample of 3 light bulbs is randomly selected from a batch of 15 bulbs. How many different samples are possible?

The order in which the 3 light bulbs are selected is not important. The sample is unchanged if the bulbs are rearranged, so combinations should be used.

(c) In a basketball tournament with 8 teams, how many games must be played so that each team plays every other team exactly once?

Selection of 2 teams for a game is an *unordered* subset of 2 from the set of 8 teams. Use combinations again.

(d) In how many ways can 4 patients be assigned to 6 hospital rooms so that each patient has a private room?

The room assignments are an *ordered* selection of 4 rooms from the 6 rooms. Exchanging the rooms of any 2 patients within a selection of 4 rooms gives a different assignment, so permutations should be used. ■ 10

EXAMPLE 11 A manager must select 4 employees for promotion. 12 employees are eligible.

(a) In how many ways can the 4 be chosen?

Because there is no reason to differentiate among the 4 who are selected, we use combinations.

$$\binom{12}{4} = \frac{12!}{4!\,8!} = 495$$

(b) In how many ways can 4 employees be chosen (from 12) to be placed in 4 different jobs?

In this case, once a group of 4 is selected, they can be assigned in many different ways (or arrangements) to the 4 jobs. Therefore, this problem requires permutations.

$$_{12}P_4 = \frac{12!}{8!} = 11,880 \quad ■ \quad \boxed{11}$$

The following problems involve a standard deck of 52 playing cards, shown in Figure 9.18 on page 413.

EXAMPLE 12 In how many ways can a full house of aces and eights (3 aces and 2 eights) be dealt in 5-card poker?

The arrangement of the 3 aces or the 2 eights does not matter, so use combinations and the multiplication principle. There are $\binom{4}{3}$ ways to get 3 aces from the 4

Comb.

10 Solve the problems in Example 10.

Answers:
(a) 5040
(b) 455
(c) 28
(d) 360

11 A salesman has the names of 6 prospects.

(a) In how many ways can he arrange his schedule if he calls on all 6?

(b) In how many ways can he do it if he decides to call on only 4 of the 6?

Answers:
(a) 720
(b) 360

12 In how many ways can 4 aces and any other card be dealt?

Answer:
48

aces in the deck, and $\binom{4}{2}$ ways to get 2 eights. By the multiplication principle, the number of ways to get 3 aces *and* 2 eights is

$$\binom{4}{3} \cdot \binom{4}{2} = 4 \cdot 6 = 24. \quad \blacksquare \quad \boxed{12}$$

EXAMPLE 13 Five cards are dealt from a standard 52-card deck.

(a) How many such hands have all face cards?

The face cards are the king, queen, and jack of each suit. There are 4 suits, so there are 12 face cards. The arrangement of the 5 cards is not important, so use combinations to get

$$\binom{12}{5} = \frac{12!}{5! \, 7!} = 792.$$

(b) How many 5-card hands have all cards of the same suit?

The arrangement of the 5 cards is not important, so use combinations. The total number of ways that 5 cards of a particular suit of 13 cards can be dealt is $\binom{13}{5}$.

Since there are 4 different suits, by the multiplication principle, there are

$$4 \cdot \binom{13}{5} = 4 \cdot 1287 = 5148$$

13 In how many ways can 5 red cards be dealt?

Answer:
65,780

ways to deal 5 cards of the same suit. $\quad \blacksquare \quad \boxed{13}$

As Examples 12 and 13 show, often both combinations and the multiplication principle must be used in the same problem.

EXAMPLE 14 To illustrate the differences between permutations and combinations in another way, suppose 2 cans of soup are to be selected from 4 cans on a shelf: noodle (N), bean (B), mushroom (M), and tomato (T). As shown in Figure 10.2(a), there are 12 ways to select 2 cans from the 4 cans if the order matters (if noodle first and bean second is considered different from bean, then noodle, for example). On the other hand, if order is unimportant, then there are 6 ways to choose 2 cans of soup from the 4, as illustrated in Figure 10.2(b). $\quad \blacksquare$

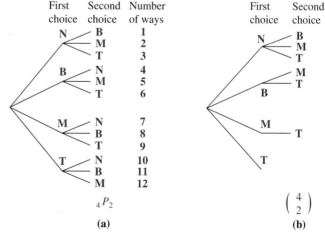

FIGURE 10.2

CAUTION It should be stressed that not all counting problems lend themselves to either permutations or combinations. Whenever a tree diagram or the multiplication principle can be used directly, as in the example at the beginning of this section, then use it. ◆

10.1 EXERCISES

Evaluate the following factorials, permutations, and combinations.

1. $_4P_2$

2. $3!$

3. $\binom{8}{3}$

4. $7!$

5. $_8P_1$

6. $\binom{8}{1}$

7. $4!$

8. $_4P_4$

9. $_{12}C_5$

10. $\binom{10}{8}$

11. $_{13}P_2$

12. $_{12}P_3$

Use a calculator to find values for Exercises 13–20.

13. $_{25}P_5$

14. $_{38}P_4$

15. $_{14}P_5$

16. $_{17}P_8$

17. $\binom{21}{10}$

18. $\binom{34}{25}$

19. $_{25}C_{16}$

20. $\binom{30}{15}$

Use the multiplication principle to solve the following problems. (See Examples 1–4.)

21. **Social Science** A social security number has 9 digits. How many social security numbers are there? The United States population in 1989 was about 248 million. Was it possible for every U.S. resident to have a unique social security number? (Assume no restrictions.)

22. **Social Science** The United States Postal Service uses 5-digit zip codes in most areas. How many zip codes are possible if there are no restrictions on the digits used? How many would be possible if the first number could not be 0?

23. **Social Science** The Postal Service is encouraging the use of 9-digit zip codes in some areas, adding 4 digits after the usual 5-digit code. How many such zip codes are possible with no restrictions?

24. **Management** How many different types of homes are available if a builder offers a choice of 5 basic plans, 3 roof styles, and 2 exterior finishes?

25. **Management** An auto manufacturer produces 7 models, each available in 6 different colors, with 4 different upholstery fabrics, and 5 interior colors. How many varieties of the auto are available?

26. **Management** How many different 4-letter radio station call letters can be made
    (a) if the first letter must be K or W and no letter may be repeated?
    (b) if repeats are allowed (but the first letter is K or W)?
    (c) How many of the 4-letter call letters (starting with K or W) with no repeats end in R?

27. **Social Science** For many years, the state of California used 3 letters followed by 3 digits on its automobile license plates.
    (a) How many different license plates are possible with this arrangement?
    (b) When the state ran out of new numbers, the order was reversed to 3 digits followed by 3 letters. How many new license plate numbers were then possible?
    (c) Several years ago, the numbers described in part (b) were also used up. The state then issued plates with 1 letter followed by 3 digits and then 3 letters. How many new license plate numbers will this provide?

Social Science *The United States is rapidly running out of telephone numbers. In large cities, telephone companies have introduced new area codes as numbers are used up.*

28. (a) Until recently, all area codes had a 0 or a 1 as the middle digit and the first digit could not be 0 or 1. How many area codes were possible with this arrangement? How many telephone numbers does the current 7-digit sequence permit per area code? (The 3-digit sequence that follows the area code cannot start with 0 or 1. Assume there are no other restrictions.)
    (b) The actual number of area codes under the previous system was 152. Explain the discrepancy between this number and your answer to part (a).

29. The shortage of area codes under the previous system was avoided by removing the restriction on the second digit. How many area codes are available under this new system?

30. A problem with the plan in Exercise 29 was that the second digit in the area code had been used to tell phone company equipment that a long-distance call was being made. To avoid changing all equipment, an alternative plan proposed a 4-digit area code and restricted the first and second digits as before. How many area codes would this plan have provided?

31. Still another alternative solution is to increase the local dialing sequence to 8 digits instead of 7. How many additional numbers would this plan create? (Assume the same restrictions.)

32. Define permutation in your own words.

Use permutations to solve each of the following problems. (See Examples 5–7.)

33. A baseball team has 20 players. How many 9-player batting orders are possible?

34. In a game of musical chairs, 12 children will sit in 11 chairs arranged in a row (one will be left out). In how many ways can the 11 children find seats?

35. Management From a carton of 12 cans of a soft drink, 2 are selected for 2 people. In how many ways can this be done?

36. Social Science In an election with 3 candidates for one office and 5 candidates for another office, how many different ballot orders may be used?

37. Management From a pool of 7 secretaries, 3 are selected to be assigned to 3 managers. In how many ways can they be selected?

38. A chapter of union Local 715 has 35 members. In how many different ways can the chapter select a president, a vice-president, a treasurer, and a secretary?

39. The television schedule for a certain evening shows 8 choices from 8 to 9 P.M., 5 choices from 9 to 10 P.M., and 6 choices from 10 to 11 P.M. In how many different ways could a person schedule that evening of television viewing from 8 to 11 P.M.? (Assume each program that is selected is watched for an entire hour.)

40. In a club with 15 members, how many ways can a slate of 3 officers consisting of president, vice-president, and secretary/treasurer be chosen?

Use combinations to solve each of the following problems. (See Examples 8–9 and 12–13.)

41. Management Five items are to be randomly selected from the first 50 items on an assembly line to determine the defect rate. How many different samples of 5 items can be chosen?

42. Social Science A group of 3 students is to be selected from a group of 12 students to take part in a class in cell biology.
(a) In how many ways can this be done?
(b) In how many ways can the group which will *not* take part be chosen?

43. Natural Science From a group of 16 smokers and 20 nonsmokers, a researcher wants to randomly select 8 smokers and 8 nonsmokers for a study. In how many ways can the study group be selected?

44. Five cards are drawn from an ordinary deck. In how many ways is it possible to draw

(a) all queens;
(b) all face cards (face cards are the Jack, Queen, and King);
(c) no face card;
(d) exactly 2 face cards;
(e) 1 heart, 2 diamonds, and 2 clubs.

Exercises 45–64 are mixed problems that may require permutations, combinations, or the multiplication principle. (See Examples 10, 11, and 14.)

45. Use a tree diagram to find the number of ways 2 letters can be chosen from the set {L, M, N} if order is important and
(a) if repetition is allowed;
(b) if no repeats are allowed.
(c) Find the number of combinations of 3 elements taken 2 at a time. Does this answer differ from parts (a) or (b)?

46. Repeat Exercise 45 using the set {L, M, N, P}.

47. Explain the difference between a permutation and a combination.

48. Padlocks with digit dials are often referred to as "combination locks." According to the mathematical definition of combination, is this an accurate description? Explain.

49. Social Science A legislative committee consists of 5 Democrats and 4 Republicans. A delegation of 3 is to be selected to visit a small Pacific Island republic.
(a) How many different delegations are possible?
(b) How many delegations would have all Democrats?
(c) How many delegations would have 2 Democrats and 1 Republican?
(d) How many delegations would include at least 1 Republican?

50. Natural Science In an experiment on plant hardiness, a researcher gathers 6 wheat plants, 3 barley plants, and 2 rye plants. She wishes to select 4 plants at random.
(a) In how many ways can this be done?
(b) In how many ways can this be done if 2 wheat plants must be included?

51. Baskin-Robbins advertises that it has 31 flavors of ice cream.
(a) How many different double-scoop cones can be made?
(b) How many different triple-scoop cones can be made?

52. A concert to raise money for an economics prize is to consist of 5 works: 2 overtures, 2 sonatas, and a piano concerto.
(a) In how many ways can the program be arranged?
(b) In how many ways can the program be arranged if an overture must come first?

53. A state lottery game requires that you pick 6 different numbers from 1 to 99. If you pick all 6 winning numbers, you win $1 million.
(a) How many ways are there to choose 6 numbers if order is not important?
(b) How many ways are there to choose 6 numbers if order matters?

54. In Exercise 53, if you pick 5 of the 6 numbers correctly, you win $250,000. In how many ways can you pick exactly 5 of the 6 winning numbers without regard to order?

55. The game of Sets* consists of a special deck of cards. Each card has on it either one, two, or three shapes. The shapes on each card are all the same color, either green, purple, or red. The shapes on each card are the same style, either solid, shaded, or outline. There are three possible shapes: squiggle, diamond, and oval, and only one type of shape appears on a card. The deck consists of all possible combinations of shape, color, style, and number. How many cards are in a deck?

56. A bag contains 5 black, 1 red, and 3 yellow jelly beans; you take 3 at random. How many samples are possible in which the jelly beans are
(a) all black;   (b) all red;
(c) all yellow;   (d) 2 black, 1 red;
(e) 2 black, 1 yellow;   (f) 2 yellow, 1 black;
(g) 2 red, 1 yellow.

57. In Example 6, there are six 3-letter permutations of the letters A, B, and C. How many 3-letter subsets (unordered groups of letters) are there?

58. In Example 6, how many unordered 2-letter subsets of the letters A, B, and C are there?

59. Natural Science Eleven drugs have been found to be effective in the treatment of a disease. It is believed that the sequence in which the drugs are administered is important in the effectiveness of the treatment. In how many orders can 5 of the 11 drugs be administered?

60. Natural Science A biologist is attempting to classify 52,000 species of insects by assigning 3 initials to each species. Is it possible to classify all the species in this way? If not, how many initials should be used?

61. One play in a state lottery consists of choosing 6 numbers from 1 to 44. If your 6 numbers are drawn (in any order), you win the jackpot.
(a) How many possible ways are there to draw the 6 numbers?
(b) If you get 2 plays for a dollar, how much would it cost to guarantee that 1 of your choices would be drawn?
(c) Assuming that you work alone and can fill out a betting ticket (for 2 plays) every second and the lotto drawing will take place 3 days from now, can you place enough bets to guarantee that 1 of your choices will be drawn?

62. Powerball is a lottery game played in 15 states across the United States. For $1 a ticket, a player selects five numbers from 1 to 45 and one Powerball number from 1 to 45. A match of all six numbers wins the jackpot. How many different selections are possible?

63. In the game of Bingo, each card has five columns. Column 1 has spaces for 5 numbers, chosen from 1 to 15. Column 2 similarly has 5 numbers, chosen from 16 to 30. Column 3 has a free space in the middle, plus 4 numbers chosen from 31 to 45. The 5 numbers in columns 4 and 5 are chosen from 46 to 60 and from 61 to 75, respectively. The numbers in each card can be in any order. How many different Bingo cards are there?

64. A television commercial for Little Caesars pizza announced that with the purchase of two pizzas, one could receive free any combination of up to five toppings on each pizza. The commercial shows a young child waiting in line at Little Caesars who calculates that there are 1,048,576 possibilities for the toppings on the two pizzas.* Verify the child's calculation. Use the fact that Little Caesars has 11 toppings to choose from. Assume that the order of the two pizzas matters; that is, if the first pizza has combination 1 and the second pizza has combination 2, that is different from combination 2 on the first pizza and combination 1 on the second.

If the n objects in a permutations problem are not all distinguishable, that is, there are n_1 of type 1, n_2 of type 2, and so on for r different types, then the number of **distinguishable permutations** *is*

$$\frac{n!}{n_1!\, n_2! \ldots n_r!}$$

Example *In how many ways can you arrange the letters in the word* Mississippi?
This word contains 1 m, 4 i's, 4 s's, and 2 p's. To use the formula, let n = 11, $n_1 = 1$, $n_2 = 4$, $n_3 = 4$, and $n_4 = 2$ to get

$$\frac{11!}{1!\,4!\,4!\,2!} = 34,650$$

arrangements. The letters in a word with 11 different letters can be arranged in 11! = 39,916,800 ways.

65. Find the number of distinguishable permutations of the letters in each of the following words.
(a) initial   (b) little   (c) decreed

66. A printer has 5 A's, 4 B's, 2 C's, and 2 D's. How many different "words" are possible which use all these letters? (A "word" does not have to have any meaning here.)

67. Mike has 4 blue, 3 green, and 2 red books to arrange on a shelf. In how many distinguishable ways can this be done
(a) if they can be arranged in any order?
(b) if books of the same color are identical and must be grouped together?
(c) if books of the same color are identical but need not be grouped together?

*Joseph F. Heiser, "Pascal and Gauss meet Little Caesars," *The Mathematics Teacher,* Vol. 87, Sept. 1994, p. 389. In a letter to *The Mathematics Teacher,* Heiser argued that the two combinations should be counted as the same, so the child has actually overcounted. In that case there would be 524,800 possibilities.

68. A child has a set of different shaped plastic objects. There are 3 pyramids, 4 cubes, and 7 spheres. In how many ways can she arrange them in a row
  (a) if they are all different colors?

  (b) if the same shapes must be grouped?
  (c) In how many distinguishable ways can they be arranged in a row if objects of the same shape are also the same color, but need not be grouped?

10.2   APPLICATIONS OF COUNTING

Many of the probability problems involving *dependent* events that were solved with probability trees in Chapter 9 can also be solved by using combinations. Combinations are especially helpful when the numbers involved would require a tree with a large number of branches.

   The use of combinations to solve probability problems depends on the basic probability principle introduced in Section 9.3 and repeated here.

   If event E is a subset of sample space S, then the probability that event E occurs, written $P(E)$, is

$$P(E) = \frac{n(E)}{n(S)}$$

It is also helpful to keep in mind that in probability statements

"and" corresponds to multiplication,

"or" corresponds to addition.

   To compare the method of using combinations with the method of probability trees used in Section 9.5, the first example repeats Example 6 from that section.

EXAMPLE 1   From a box containing 3 white, 2 green, and 1 red marble, 2 marbles are drawn one at a time without replacement. Find the probability that 1 white and 1 green marble are drawn.

   In Example 6 of Section 9.5, it was necessary to consider the order in which the marbles were drawn. With combinations, it is not necessary. Simply count the number of ways in which 1 white and 1 green marble can be drawn from the given selection. The white marble can be drawn from the 3 white marbles in $\binom{3}{1}$ ways, and the green marble can be drawn from the 2 green marbles in $\binom{2}{1}$ ways. By the multiplication principle, both results can occur in

$$\binom{3}{1} \cdot \binom{2}{1} \text{ ways,}$$

giving the numerator of the probability fraction, $P(E) = m/n$. For the denominator, 2 marbles are to be drawn from a total of 6 marbles. This can occur in $\binom{6}{2}$ ways. The required probability is

$$P(\textbf{1 white and 1 green}) = \frac{\binom{3}{1}\binom{2}{1}}{\binom{6}{2}} = \frac{\dfrac{3!}{2!\,1!} \cdot \dfrac{2!}{1!\,1!}}{\dfrac{6!}{4!\,2!}} = \frac{6}{15} = \frac{2}{5}.$$

This agrees with the answer found earlier.   ∎

EXAMPLE 2   From a group of 22 nurses, 4 are to be selected to present a list of grievances to management.

(a) In how many ways can this be done?

Four nurses from a group of 22 can be selected in $\binom{22}{4}$ ways. (Use combinations, since the group of 4 is an unordered set.)

$$\binom{22}{4} = \frac{22!}{4!\ 18!} = \frac{22(21)(20)(19)}{4(3)(2)(1)} = 7315$$

There are 7315 ways to choose 4 people from 22.

(b) One of the nurses is Michael Branson. Find the probability that Branson will be among the 4 selected.

The probability that Branson will be selected is given by m/n, where m is the number of ways the chosen group includes him, and n is the total number of ways the group of 4 can be chosen. If Branson must be one of the 4 selected, the problem reduces to finding the number of ways that the 3 additional nurses can be chosen. The 3 are chosen from 21 nurses; this can be done in

$$\binom{21}{3} = \frac{21!}{3!\ 18!} = 1330$$

ways, so $m = 1330$. Since n is the number of ways 4 nurses can be selected from 22,

$$n = \binom{22}{4} = 7315.$$

The probability that Branson will be one of the 4 chosen is

$$P(\text{Branson is chosen}) = \frac{1330}{7315} \approx .182.$$

(c) Find the probability that Branson will not be selected.

The probability that he will not be chosen is $1 - .182 = .818$.   ■   1

> **1** A jar contains 1 white and 4 red jelly beans.
>
> **(a)** What is the probability that out of 2 jelly beans selected at random from the jar, 1 will be white?
>
> **(b)** What is the probability of choosing 3 red jelly beans?
>
> **Answers:**
> **(a)** 2/5
> **(b)** 2/5

EXAMPLE 3   When shipping diesel engines abroad, it is common to pack 12 engines in 1 container which is then loaded on a railcar and sent to a port. Suppose that a company has received complaints from its customers that many of the engines arrive in nonworking condition. To help solve this problem, the company decides to make a spot check of containers after loading—the company will test 3 engines from a container at random; if any of the 3 is nonworking, the container will not be shipped until each engine in it is checked. Suppose a given container has 2 nonworking engines. Find the probability that the container will not be shipped.

The container will not be shipped if the sample of 3 engines contains 1 or 2 defective engines. Thus, letting $P(1 \text{ defective})$ represent the probability of exactly 1 defective engine in the sample,

$$P(\text{not shipping}) = P(1 \text{ defective}) + P(2 \text{ defectives}).$$

There are $\binom{12}{3}$ ways to choose the 3 engines for testing:

$$\binom{12}{3} = \frac{12!}{3!\ 9!} = \frac{12(11)(10)}{3(2)(1)} = 220.$$

There are $\binom{2}{1}$ ways of choosing 1 defective engine from the 2 in the container, and for each of these ways, there are $\binom{10}{2}$ ways of choosing 2 good engines from among the 10 in the container. By the multiplication principle, there are

$$\binom{2}{1}\binom{10}{2}$$

ways of choosing a sample of 3 engines containing 1 defective. **2**

Using the result from Problem 2 at the side,

$$P(1 \text{ defective}) = \frac{90}{220}.$$

There are $\binom{2}{2}$ ways of choosing 2 defective engines from the 2 defective engines in the container, and $\binom{10}{1}$ ways of choosing 1 good engine from among the 10 good engines, giving

$$\binom{2}{2}\binom{10}{1}$$

ways of choosing a sample of 3 engines containing 2 defectives. **3**

Then, using the result from Problem 3 at the side,

$$P(2 \text{ defectives}) = \frac{10}{220}$$

and

$$P(\text{not shipping}) = P(1 \text{ defective}) + P(2 \text{ defectives})$$
$$= \frac{90}{220} + \frac{10}{220} = \frac{100}{220} = \frac{5}{11} \approx .455.$$

The probability is $1 - .455 = .545$ that the container *will* be shipped, even though it has 2 defective engines. The management must decide if this probability is acceptable; if not, it may be necessary to test more than 3 engines from a container. ■

Instead of finding the sum $P(1 \text{ defective}) + P(2 \text{ defectives})$, the result in Example 3 could be found by calculating $1 - P(\text{no defectives})$.

$$P(\text{not shipping}) = 1 - P(\text{no defectives in sample})$$

$$= 1 - \frac{\binom{2}{0}\binom{10}{3}}{\binom{12}{3}} = 1 - \frac{1(120)}{220}$$

$$= 1 - \frac{120}{220} = \frac{100}{220} \approx .455 \quad \boxed{4}$$

---

2 Calculate $\binom{2}{1}\binom{10}{2}$.

Answer:
90

3 Calculate $\binom{2}{2}\binom{10}{1}$.

Answer:
10

4 In Example 3, if a sample of 2 engines is tested, what is the probability that the container will not be shipped?

Answer:
.318

EXAMPLE 4 In a common form of the card game *poker,* a hand of 5 cards is dealt to each player from a deck of 52 cards. There are a total of

$$\binom{52}{5} = \frac{52!}{5!47!} = 2{,}598{,}960$$

such hands possible. Find each of the following probabilities.

(a) A hand containing only hearts, called a *heart flush*

There are 13 hearts in a deck; there are

$$\binom{13}{5} = \frac{13!}{5!8!} = \frac{13(12)(11)(10)(9)}{5(4)(3)(2)(1)} = 1287$$

different hands containing only hearts. The probability of a heart flush is

$$P(\text{heart flush}) = \frac{1287}{2{,}598{,}960} = \frac{33}{66{,}640} \approx .000495.$$

(b) A flush of any suit (5 cards, all from 1 suit)

There are 4 suits to a deck, so

$$P(\text{flush}) = 4 \cdot P(\text{heart flush}) = 4(.000495) \approx .00198.$$

(c) A full house of aces and eights (3 aces and 2 eights)

There are $\binom{4}{3}$ ways to choose 3 aces from among the 4 in the deck, and $\binom{4}{2}$ ways to choose 2 eights.

$$P(\text{3 aces, 2 eights}) = \frac{\binom{4}{3} \cdot \binom{4}{2}}{2{,}598{,}960} = \frac{1}{108{,}290} \approx .00000923$$

(d) Any full house (3 cards of one value, 2 of another)

There are 13 values in a deck, so there are 13 choices for the first value mentioned, leaving 12 choices for the second value (order *is* important here, since a full house of aces and eights, for example, is not the same as a full house of eights and aces). Because, from part (c), the probability of any particular full house is .00000923,

$$P(\text{full house}) = 13 \cdot 12(.00000923) \approx .00144. \quad \blacksquare \quad \boxed{5}$$

EXAMPLE 5 A music teacher has 3 violin pupils, Fred, Carl, and Helen. For a recital, the teacher selects a first violinist and a second violinist. The third pupil will play with the others, but not solo. If the teacher selects randomly, what is the probability that Helen is first violinist, Carl is second violinist, and Fred does not solo?

Use *permutations* to find the number of arrangements in the sample space.

$$_3P_3 = 3! = 6$$

The 6 arrangements are equally likely, since the teacher will select randomly. Thus, the required probability is 1/6. $\quad \blacksquare \quad \boxed{6}$

EXAMPLE 6 Suppose a group of 5 people is in a room. Find the probability that at least 2 of the people have the same birthday.

---

5 Find the probability of being dealt a poker hand (5 cards) with 4 kings.

Answer:
.00001847

6 An office manager must select 4 employees, A, B, C, and D, to work on a project. Each employee will be responsible for a specific task and 1 will be the coordinator.

(a) If the manager assigns the tasks randomly, what is the probability that B is the coordinator?

(b) What is the probability of one specific assignment of tasks?

Answers:
(a) 1/4
(b) 1/24

"Same birthday" refers to the month and the day, not necessarily the same year. Also, ignore leap years, and assume that each day in the year is equally likely as a birthday. First find the probability that *no 2 people* among 5 people have the same birthday. There are 365 different birthdays possible for the first of the 5 people, 364 for the second (so that the people have different birthdays), 363 for the third, and so on. The number of ways the 5 people can have different birthdays is thus the number of permutations of 365 things (days) taken 5 at a time or

$$_{365}P_5 = 365 \cdot 364 \cdot 363 \cdot 362 \cdot 361.$$

The number of ways that the 5 people can have the same or different birthdays is

$$365 \cdot 365 \cdot 365 \cdot 365 \cdot 365 = (365)^5.$$

Finally, the *probability* that none of the 5 people have the same birthday is

$$\frac{_{365}P_5}{(365)^5} = \frac{365 \cdot 364 \cdot 363 \cdot 362 \cdot 361}{365 \cdot 365 \cdot 365 \cdot 365 \cdot 365} \approx .973.$$

The probability that at least 2 of the 5 people *do* have the same birthday is $1 - .973 = .027.$ ■

Example 6 can be extended for more than 5 people. In general, the probability that no 2 people among n people have the same birthday is

$$\frac{_{365}P_n}{(365)^n}.$$

The probability that at least 2 of the n people *do* have the same birthday is

$$1 - \frac{_{365}P_n}{(365)^n}. \quad \boxed{7}$$

The following table shows this probability for various values of n.

Number of People, n	Probability that 2 Have the Same Birthday
5	.027
10	.117
15	.253
20	.411
22	.476
23	.507
25	.569
30	.706
35	.814
40	.891
50	.970
365	1

The probability that 2 people among 23 have the same birthday is .507, a little more than half. Many people are surprised at this result—somehow it seems that a larger number of people should be required. $\boxed{8}$

$\boxed{7}$ Evaluate $1 - \dfrac{_{365}P_n}{(365)^n}$ for

(a) $n = 3$;

(b) $n = 6$.

Answers:
(a) .008
(b) .040

$\boxed{8}$ Set up (do not calculate) the probability that at least 2 of the 7 astronauts in the Mercury project have the same birthday.

Answer:
$1 - {_{365}P_7}/365^7$

10.2 EXERCISES

Management *Refer to Example 3. The management feels that the probability of .545 that a container will be shipped even though it contains 2 defectives is too high. They decide to increase the sample size chosen. Find the probability that a container will be shipped even though it contains 2 defectives if the sample size is increased to*

1. 4.

2. 5.

Management *A shipment of 9 computers contains 2 with defects. Find the probability that a sample of the following size, drawn from the 9, will not contain a defective. (See Example 3.)*

3. 1   **4.** 2   **5.** 3   **6.** 4

A grab-bag contains 10 $1 prizes, 6 $5 prizes, and 2 $20 prizes. Three prizes are randomly drawn. Find the following probabilities. (See Examples 1 and 2.)

7. All $1 prizes

8. All $5 prizes

9. Exactly two $20 prizes

10. One prize of each kind

11. At least one $20 prize

12. No $1 prize

Two cards are drawn at random from an ordinary deck of 52 cards. (See Example 4.)

13. How many 2-card hands are possible?

Find the probability that the two-card hand in Exercise 13 contains the following.

14. 2 queens

15. No aces

16. 2 face cards

17. Different suits

18. At least 1 black card

19. No more than 1 heart

20. Discuss the relative merits of using probability trees versus combinations to solve probability problems. When would each approach be most appropriate?

21. Several examples in this section used the rule $P(E') = 1 - P(E)$. Explain the advantage (especially in Example 6) of using this rule.

A bridge hand consists of 13 cards from a deck of 52. Set up the probabilities that a bridge hand includes each of the following.

22. 6 face cards

23. 2 aces and 3 kings

24. 7 cards of one suit and 6 of another

25. In Exercise 53 in the previous section, we found the number of ways to pick 6 different numbers from 1 to 99 in a state lottery.
    (a) Assuming order is unimportant, what is the probability of picking all 6 numbers correctly to win the big prize?
    (b) What is the probability, if order matters?

26. In Exercise 25, what is the probability of picking exactly 5 of the 6 numbers correctly?

27. In December 1993, Percy Ray Pridgen, a District of Columbia resident, and Charles Gill of Richmond, Virginia shared a $90 million Powerball jackpot. The winning numbers were 1, 3, 13, 15, and 29. The Powerball number was 12. The answer to Exercise 62 in Section 10.1 gives the number of ways to win a Powerball jackpot as about 55 million. What is the probability of two individuals independently selecting the winning numbers?

28. Management A cellular phone manufacturer randomly selects 5 of every 50 phones from the assembly line and tests them. If at least 4 of the 5 pass inspection, the batch of 50 is considered acceptable. Find the probability that a batch is considered acceptable if it contains the following.
    (a) 2 defective phones
    (b) No defective phones
    (c) 3 defective phones

29. Social Science Of the 16 members of President Clinton's first cabinet, 4 were women. Suppose the president randomly selected 4 advisors from the cabinet for a meeting. Find the probability that the group of 4 would be composed as follows.
    (a) 2 women and 2 men
    (b) All men
    (c) At least 1 woman

30. Management A car dealer has 8 red, 11 gray, and 9 blue cars in stock. Ten cars are randomly chosen to be displayed in front of the dealership. Find the probability that
    (a) 4 are red and the others are blue.
    (b) 3 are red, 3 are blue, and 4 are gray.
    (c) exactly 5 are gray and none are blue.
    (d) all 10 are gray.

31. Social Science The teacher of a multicultural first-grade class on the first day of school found that in her class of 25 students, 10 spoke only English, 6 spoke only Spanish, 4 spoke only Russian, 3 spoke only Vietnamese, and 2 spoke only Hmong. The children were assigned randomly to groups of 5. Find the probability that a group included the following.
    (a) 2 English-speaking and 3 Russian-speaking children
    (b) All English-speaking children
    (c) No English-speaking children
    (d) At least two children who spoke Vietnamese or Hmong

For Exercises 32–34 refer to Example 6 in this section.

32. Set up the probability that at least 2 of the 41 men who have served as president of the United States have had the same birthday.*

33. Set up the probability that at least 2 of the 100 U.S. Senators have the same birthday.

---

*In fact, James Polk and Warren Harding were both born on November 2.

34. What is the probability that 2 of the 435 members of the House of Representatives have the same birthday?

35. An elevator has 4 passengers and stops at 7 floors. It is equally likely that a person will get off at any one of the 7 floors. Find the probability that no 2 passengers leave at the same floor.

36. Show that the probability that in a group of n people *exactly one pair* have the same birthday is

$$\binom{n}{2} \cdot \frac{_{365}P_{n-1}}{(365)^n}.$$

37. On National Public Radio, the "Weekend Edition" program on Sunday, September 7, 1991, posed the following probability problem: Given a certain number of balls, of which some are blue, pick 5 at random. The probability that all 5 are blue is 1/2. Determine the original number of balls and decide how many were blue.

38. The Big Eight Conference during the 1988 college football season ended the season in a "perfect progression," as shown in the table below.*

Won	Lost	Team
7	0	Nebraska (NU)
6	1	Oklahoma (OU)
5	2	Oklahoma State (OSU)
4	3	Colorado (CU)
3	4	Iowa State (ISU)
2	5	Missouri (MU)
1	6	Kansas (KU)
0	7	Kansas State (KSU)

*From Richard Madsen. "On the Probability of a Perfect Progression," *The American Statistician,* August 1991, vol. 45, no. 3, p. 214.

Someone wondered what the probability of such an outcome might be.

(a) Assuming no ties and that each team had an equally likely probability of winning each game, find the probability of the perfect progression shown above.

(b) Find a general expression for the probability of a perfect progression in an n-team league with the same assumptions.

39. Use a computer or a graphing calculator and the Monte Carlo method with $n = 50$ to estimate the probabilities of the following hands at poker. (See the directions for Exercises 53–58 on page 426.) Assume aces are either high or low. Since each hand has 5 cards, you will need $50 \cdot 5 = 250$ random numbers to "look at" 50 hands. Compare these experimental results with the theoretical results.

(a) A pair of aces

(b) Any two cards of the same value

(c) Three of a kind

40. Use a computer or a graphing calculator and the Monte Carlo method with $n = 20$ to estimate the probabilities of the following 13-card bridge hands. Since each hand has 13 cards, you will need $20 \cdot 13 = 260$ random numbers to "look at" 20 hands.

(a) No aces

(b) 2 kings and 2 aces

(c) No cards of any one suit—that is, only 3 suits represented

10.3 BINOMIAL PROBABILITY

Many probability problems are concerned with experiments in which an event is repeated many times. Some examples include finding the probability of getting 7 heads in 8 tosses of a coin, of hitting a target 6 times out of 6, and of finding 1 defective item in a sample of 15 items. Probability problems of this kind are called **Bernoulli trials** problems, or **Bernoulli processes,** named after the Swiss mathematician Jakob Bernoulli (1654–1705). In each case, some outcome is designated a success, and any other outcome is considered a failure. Thus, if the probability of a success in a single trial is p, the probability of failure will be $1 - p$. A Bernoulli trials problem, or **binomial experiment,** must satisfy the following conditions.

> **BINOMIAL EXPERIMENT**
> 1. The same experiment is repeated several times.
> 2. There are only two possible outcomes, success or failure.
> 3. The repeated trials are independent so that the probability of each outcome remains the same for each trial.

Consider a typical problem of this type: find the probability of getting 5 ones on 5 rolls of a die. Here, the experiment, rolling a die, is repeated 5 times. If getting a one is designated as a success, then getting any other outcome is a failure. The 5 trials are independent; the probability of success (getting a one) is 1/6 on each trial, while the probability of a failure (any other result) is 5/6. Thus, the required probability is

$$P(5 \text{ ones in 5 rolls}) = P(1) \cdot P(1) \cdot P(1) \cdot P(1) \cdot P(1) = \left(\frac{1}{6}\right)^5$$

$$\approx .00013.$$

Now suppose the problem is changed to that of finding the probability of getting a one exactly 4 times in 5 rolls of the die. Again, a success on any roll of the die is defined as getting a one, and the trials are independent, with the same probability of success each time. The desired outcome for this experiment can occur in more than one way, as shown below, where s represents a success (getting a one), and f represents a failure (getting any other result), and each row represents a different outcome.

$$
\begin{array}{ccccc}
s & s & s & s & f \\
s & s & s & f & s \\
s & s & f & s & s \\
s & f & s & s & s \\
f & s & s & s & s \\
\end{array}
$$

Using combinations, the total number of ways in which 4 successes (and 1 failure) can occur is $\binom{5}{4} = 5$. The probability of each of these 5 outcomes is

$$\left(\frac{1}{6}\right)^4 \left(\frac{5}{6}\right).$$

Since the 5 outcomes represent mutually exclusive alternative events, add the 5 probabilities, or multiply by $\binom{5}{4} = 5$.

$$P(4 \text{ ones in 5 rolls}) = \binom{5}{4}\left(\frac{1}{6}\right)^4\left(\frac{5}{6}\right)^{5-4} = 5\left(\frac{1}{6}\right)^4\left(\frac{5}{6}\right)^1 = \frac{5^2}{6^5} \approx .0032$$

The probability of rolling a one exactly 3 times in 5 rolls of a die can be computed in the same way. The probability of any one way of achieving 3 successes and 2 failures will be

$$\left(\frac{1}{6}\right)^3\left(\frac{5}{6}\right)^2.$$

Again the desired outcome can occur in more than one way. The number of ways in which 3 successes (and 2 failures) can occur, using combinations, is $\binom{5}{3} = 10$.

$$P(3 \text{ ones in 5 rolls}) = \binom{5}{3}\left(\frac{1}{6}\right)^3\left(\frac{5}{6}\right)^{5-3} = 10\left(\frac{1}{6}\right)^3\left(\frac{5}{6}\right)^2 = \frac{250}{6^5} \approx .032 \quad \boxed{1}$$

A similar argument works in the general case.

BINOMIAL PROBABILITY

If p is the probability of success in a single trial of a binomial experiment, the probability of x successes and $n - x$ failures in n independent repeated trials of the experiment is

$$\binom{n}{x}p^x(1-p)^{n-x}.$$

This formula plays an important role in the study of *binomial distributions*, which are discussed in Section 11.4.

TECHNOLOGY TIP On the TI-83 calculator, use "binompdf(n,p,x)" in the DISTR menu to compute the probability of exactly x successes in n trials (where p is the probability of success in a single trial). Use "binomcdf(n,p,x)" to compute the probability of at most x successes in n trials. ✔

EXAMPLE 1 The advertising agency that handles the Diet Supercola account believes that 40% of all consumers prefer this product over its competitors. Suppose a random sample of 6 people is chosen. Assume that all responses are independent of each other. Find the probability of the following.

(a) Exactly 4 of the 6 people prefer Diet Supercola.

We can think of the 6 responses as 6 independent trials. A success occurs if a person prefers Diet Supercola. Then this is a binomial experiment with $p = P(\text{success}) = P(\text{prefer Diet Supercola}) = .4$. The sample is made up of 6 people, so $n = 6$. To find the probability that exactly 4 people prefer this drink, we let $x = 4$ and use the result in the box.

$$P(\text{exactly 4}) = \binom{6}{4}(.4)^4(1 - .4)^{6-4}$$
$$= 15(.4)^4(.6)^2$$
$$= 15(.0256)(.36) \quad \text{Use a calculator.}$$
$$= .13824$$

(b) None of the 6 people prefers Diet Supercola.
Let $x = 0$.

$$P(\text{exactly 0}) = \binom{6}{0}(.4)^0(1 - .4)^6$$
$$= 1(1)(.6)^6$$
$$\approx .0467 \quad \blacksquare \quad \boxed{2}$$

**Sidebar:**

$\boxed{1}$ Find the probability of rolling a one
(a) exactly twice in 5 rolls of a die;
(b) exactly once in 5 rolls of a die.
(c) Do the probabilities found in the text, plus the two probabilities found here cover all possible outcomes for this experiment? If not, what other outcomes are possible?

Answers:
(a) .161
(b) .402
(c) No; 0 ones

$\boxed{2}$ Eighty percent of all students at a certain school ski. If a sample of 5 students at this school is selected, and if their responses are independent, find the probability that exactly
(a) 1 of the 5 students skis;
(b) 4 of the 5 students ski.

Answers:
(a) $\binom{5}{1}(.8)^1(.2)^4 = .0064$
(b) $\binom{5}{4}(.8)^4(.2)^1 = .4096$

3 Find the probability of getting 2 fours in 8 tosses of a die.

Answer:
$$\binom{8}{2}\left(\frac{1}{6}\right)^2\left(\frac{5}{6}\right)^6 \approx .2605$$

EXAMPLE 2 Find the probability of getting exactly 7 heads in 8 tosses of a fair coin.

The probability of success, getting a head in a single toss, is 1/2, so the probability of failure, getting a tail, is 1/2.

$$P(7 \text{ heads in 8 tosses}) = \binom{8}{7}\left(\frac{1}{2}\right)^7\left(\frac{1}{2}\right)^1 = 8\left(\frac{1}{2}\right)^8 = .03125 \quad \blacksquare \quad \boxed{3}$$

EXAMPLE 3 Assuming that selection of items for a sample can be treated as independent trials, and that the probability that any 1 item is defective is .01, find the following.

(a) The probability of 1 defective item in a random sample of 15 items from a production line

Here, a "success" is a defective item. Selecting each item for the sample is assumed to be an independent trial, so the binomial probability formula applies. The probability of success (a defective item) is .01, while the probability of failure (an acceptable item) is .99. This makes

$$P(1 \text{ defective in 15 items}) = \binom{15}{1}(.01)^1(.99)^{14}$$
$$= 15(.01)(.99)^{14}$$
$$\approx .130.$$

(b) The probability of at most 1 defective item in a random sample of 15 items from a production line

"At most 1" means 0 defective items or 1 defective item. Since 0 defective items is equivalent to 15 acceptable items,

$$P(0 \text{ defective}) = (.99)^{15} \approx .860.$$

Use the union rule, noting that 0 defective and 1 defective are mutually exclusive events, to get

$$P(\text{at most 1 defective}) = P(0 \text{ defective}) + P(1 \text{ defective})$$
$$\approx .860 + .130$$
$$= .990. \quad \blacksquare \quad \boxed{4}$$

4 Five percent of the clay pots fired in a certain way are defective. Find the probability of getting exactly 2 defective pots in a sample of 12.

Answer:
$$\binom{12}{2}(.05)^2(.95)^{10} \approx .0988$$

In the next example we return to the problem posed at the beginning of this chapter.

EXAMPLE 4 A survey by *Money* magazine found that supermarket scanners are overcharging customers at 30% of stores.* Assume all such scanners consistently overcharge. If you shop at 3 supermarkets that use the scanners, what is the probability that you will be overcharged in at least one store?

We can treat this as a binomial experiment, letting $n = 3$ and $p = .3$. At least 1 of 3 means 1 or 2 or 3. It will be simpler here to find the probability of being

*"Don't Get Cheated by Supermarket Scanners," pp. 132–138, by Vanessa O'Connell reprinted from the April 1993 issue of *Money* by special permission (reprinted in the *Chicago Tribune* as "Beware, When Scanners Err, You Pay Price"). Copyright © 1993, Time Inc.

[5] In Example 4, find the probability that

(a) you are overcharged in more than 1 store.

(b) you are overcharged in less than 1 store.

Answers:
(a) .216
(b) .343

overcharged in none of the 3 stores, that is, $P(0 \text{ overcharges})$, and then find the probability of at least $1 = 1 - P(0 \text{ overcharges})$.

$$P(0 \text{ overcharges}) = \binom{3}{0}(.3)^0(.7)^3$$
$$= 1(1)(.343)$$
$$= .343$$
$$P(\text{at least } 1) = 1 - P(0 \text{ overcharges})$$
$$= 1 - .343$$
$$= .657 \quad \blacksquare \quad \boxed{5}$$

10.3 EXERCISES

Social Science *In 1995 12.7% of the U.S. population was 65 or older, an increase from 4.1% in 1900.* Find the probabilities that the following number of persons selected at random from 20 U.S. residents in 1995 were 65 or older. (See Examples 1–3.)*

1. Exactly 5

2. Exactly 1

3. None

4. All

5. At least 1

6. At most 3

A die is rolled 12 times. Find the probability of rolling

7. Exactly 12 ones;

8. Exactly 6 ones;

9. Exactly 1 one;

10. Exactly 2 ones;

11. No more than 3 ones;

12. No more than 1 one.

A coin is tossed 5 times. Find the probability of getting

13. All heads;

14. Exactly 3 heads;

15. No more than 3 heads;

16. At least 3 heads.

17. How do you identify a probability problem that involves a binomial experiment?

18. Why do combinations occur in the binomial probability formula?

Management *The survey discussed in Example 4 also found the following:*

Customers overpay for 1 out of every 10 items, on average. Scanner errors account for more than half of supermarket profits. In 7% of the stores customers were undercharged.

Suppose a customer purchases 15 items. Find the following probabilities.

19. A customer overpays on 3 items.

20. A customer is not overcharged for any item.

21. A customer overpays on no more than one item.

Suppose a customer shops in 15 stores that use scanners. Find the following probabilities.

22. A customer underpays in 3 stores.

23. A customer underpays in no stores.

24. A customer underpays in at least one store.

Management *The insurance industry has found that the probability is .9 that a life insurance applicant will qualify at the regular rates. Find the probabilities that of the next 10 applicants for life insurance, the following numbers will qualify at the regular rates.*

25. Exactly 10

26. Exactly 9

27. At least 9

28. Less than 8

Natural Science *The probability that a birth will result in twins is .012. Assuming independence (perhaps not a valid assumption), what are the probabilities that out of 100 births in a hospital, there will be the following numbers of sets of twins?*

29. Exactly 2 sets of twins

30. At most 2 sets of twins

Natural Science *Six mice from the same litter, all suffering from a vitamin A deficiency, are fed a certain dose of carrots. If the probability of recovery under such treatment is .70, find the probabilities of the following results.*

31. None of the mice recover.

32. Exactly 3 of the 6 mice recover.

33. All of the mice recover.

34. No more than 3 mice recover.

35. Natural Science In an experiment on the effects of a radiation dose on cells, a beam of radioactive particles is aimed at a group of 10 cells. Find the probability that 8 of the cells will be hit by the beam if the probability that any single cell will be hit is .6. (Assume independence.)

36. Natural Science The probability of a mutation of a given gene under a dose of 1 roentgen of radiation is approximately 2.5×10^{-7}. What is the probability that in 10,000 genes, at least 1 mutation occurs?

37. Social Science The probability that an American falls asleep with the TV on at least three nights a week is 1/4.*

(exercise continues)

Suppose a researcher selects 5 Americans at random and is interested in the probability that one or more is a "TV sleeper." Find the following probabilities.

(a) All 5 are TV sleepers.

(b) Three of the 5 are TV sleepers.

38. Natural Science Children in a certain school are monitored for a year and it is found that 30% of those who do not brush their teeth regularly don't have a cavity during the year. A child who brushes regularly has a 70% chance of having no cavities during the year.

(a) If 10 children brush regularly, what is the probability that at least 9 will have no cavities during the year?

(b) 150 children who brush regularly are divided into 15 groups of 10 each. What is the probability that in at least one group of 10, at least 9 will have no cavities during the year?

39. Social Science A recent study found that 33% of women would prefer to work part time rather than full time if money were not a concern.* Find the probability that if 10 women are selected at random, at least 3 of them would prefer to work part time.

40. Social Science A study several years ago by the College Board and the Western Interstate Commission for Higher Education predicted that by 1995, one-third of U.S. public elementary and secondary school students would belong to ethnic minorities. Assuming independence, find the probability that of 5 students selected at random in 1995 from schools throughout the nation, no more than 3 belonged to ethnic minorities.

41. Social Science According to the state of California, 33% of all state community college students belong to ethnic

minorities. Find the probabilities of the following results in a random sample of 10 California community college students.

(a) Exactly 2 belong to an ethnic minority.

(b) Three or fewer belong to an ethnic minority.

(c) Exactly 5 do not belong to an ethnic minority.

(d) Six or more do not belong to an ethnic minority.

Use a calculator or computer to calculate each of the following probabilities.

42. Natural Science A flu vaccine has a probability of 80% of preventing a person who is inoculated from getting flu. A county health office inoculates 134 people. What is the probability that

(a) exactly 10 of them get the flu?

(b) no more than 10 get the flu?

(c) none of them get the flu?

43. Natural Science The probability that a male will be color-blind is .042. What is the probability that in a group of 53 men

(a) exactly 5 are color-blind?

(b) no more than 5 are color-blind?

(c) at least 1 is color-blind?

44. Management The probability that a certain machine turns out a defective item is .05. What is the probability that in a run of 75 items

(a) exactly 5 defectives are produced?

(b) no defectives are produced?

(c) at least 1 defective is produced?

45. Management A company is taking a survey to find out if people like their product. Their last survey indicated that 70% of the population likes their product. Based on that, of a sample of 58 people, what is the probability that

(a) all 58 like the product?

(b) from 28 to 30 (inclusive) like the product?

---

*Cathleen Ferraro, "Feelings of the Working Women," *The Sacramento Bee,* May 11, 1995, p. A1, A22.

10.4 MARKOV CHAINS

In Section 9.5, we touched on **stochastic processes,** mathematical models that evolve over time in a probabilistic manner. In this section we study a special kind of stochastic process called a **Markov chain,** where the outcome of an experiment depends only on the outcome of the previous experiment. In other words, the next **state** of the system depends only on the present state, not on preceding states. Such experiments are common enough in applications to make their study worthwhile. Markov chains are named after the Russian mathematician A. A. Markov, 1856–1922, who started the theory of stochastic processes. To see how Markov chains work, we look at an example.

EXAMPLE 1 A small town has only two dry cleaners, Johnson and NorthClean. Johnson's manager hopes to increase the firm's market share by an extensive advertising campaign. After the campaign, a market research firm finds that there is a probability of .8 that a Johnson customer will bring his next batch of dirty items to Johnson, and a .35 chance that a NorthClean customer will switch to Johnson for his next batch. Assume that the probability that a customer comes to a given cleaners depends only on where the last load of clothes was taken. If there is an .8 chance that a Johnson customer will return to Johnson, then there must be a $1 - .8 = .2$ chance that the customer will switch to NorthClean. In the same way, there is a $1 - .35 = .65$ chance that a NorthClean customer will return to NorthClean. If an individual bringing a load to Johnson is said to be in state 1 and an individual bringing a load to NorthClean is said to be in state 2, then these probabilities of change from one cleaner to the other are as shown in the following table.

		SECOND LOAD	
	State	1	2
FIRST LOAD	1	.8	.2
	2	.35	.65

1 Given the transition matrix

$$\begin{array}{c} \\ State \end{array} \begin{array}{c} \\ \begin{array}{c} 1 \\ 2 \end{array} \end{array} \begin{array}{c} State \\ \begin{array}{cc} 1 & 2 \end{array} \\ \begin{bmatrix} .3 & .7 \\ .1 & .9 \end{bmatrix} \end{array}.$$

(a) What is the probability of changing from state 1 to state 2?

(b) What does the number .1 represent?

(c) Draw a transition diagram for this information.

Answers:

(a) .7

(b) The probability of changing from state 2 to state 1

(c)

The information from the table can be written in other forms. Figure 10.3 is a **transition diagram** that shows the two states and the probabilities of going from one state to another.

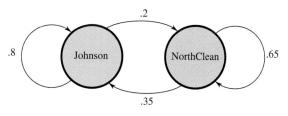

FIGURE 10.3

In a **transition matrix,** the states are indicated at the side and top, as follows.

$$\begin{array}{c} \\ First\ Load \end{array} \begin{array}{c} \\ \begin{array}{c} Johnson \\ NorthClean \end{array} \end{array} \begin{array}{c} Second\ Load \\ \begin{array}{cc} Johnson & NorthClean \end{array} \\ \begin{bmatrix} .8 & .2 \\ .35 & .65 \end{bmatrix} \end{array} \quad \boxed{1}$$

A **transition matrix** has the following features.

1. It is square, since all possible states must be used both as rows and as columns.
2. All entries are between 0 and 1, inclusive, because all entries represent probabilities.
3. The sum of the entries in any row must be 1, because the numbers in the row give the probability of changing from the state at the left to one of the states indicated across the top.

EXAMPLE 2 Suppose that when the new promotional campaign began, Johnson had 40% of the market and NorthClean had 60%. Use the probability tree in Figure 10.4 to find how these proportions would change after another week of advertising.

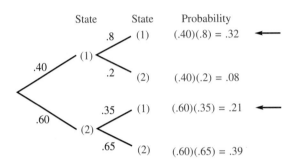

State 1: taking cleaning to Johnson
State 2: taking cleaning to NorthClean

FIGURE 10.4

Add the numbers indicated with arrows to find the proportion of people taking their cleaning to Johnson after 1 week.

$$.32 + .21 = .53$$

Similarly, the proportion taking their cleaning to NorthClean is

$$.08 + .39 = .47.$$

The initial distribution of 40% and 60% becomes, after 1 week, 53% and 47%. ■

These distributions can be written as the *probability vectors*

$$[.40 \quad .60] \quad \text{and} \quad [.53 \quad .47].$$

A **probability vector** is a matrix of only one row, having nonnegative entries with the sum of the entries equal to 1.

The results from the probability tree above are exactly the same as the result of multiplying the initial probability vector by the transition matrix. (Multiplication of matrices was discussed in Section 7.3.)

$$[.4 \quad .6]\begin{bmatrix} .8 & .2 \\ .35 & .65 \end{bmatrix} = [.53 \quad .47]$$

Find the market share after 2 weeks by multiplying the vector [.53   .47] and the transition matrix. **2**

The product from Problem 2 at the side (with the numbers rounded to the nearest hundredth) shows that after 2 weeks of advertising, Johnson's share of the market has increased to 59%. To get the share after 3 weeks, multiply the probability vector [.59   .41] and the transition matrix. Continuing this process gives each cleaner's share of the market after any number of weeks. **3**

2 Find the product

$$[.53 \quad .47]\begin{bmatrix} .8 & .2 \\ .35 & .65 \end{bmatrix}.$$

Answer:
[.59   .41] (rounded)

3 Find each cleaner's market share after 3 weeks.

Answer:
[.62   .38] (rounded)

The following table gives the market share (rounded) for each cleaner at the end of week n for several values of n.

Week	Johnson	NorthClean
Start	.4	.6
1	.53	.47
2	.59	.41
3	.62	.38
4	.63	.37
5	.63	.37
12	.64	.36

The results seem to approach the probability vector [.64   .36]. This vector is called the **equilibrium vector** or **fixed vector** for the given transition matrix. The equilibrium vector gives a long-range prediction—the shares of the market will stabilize (under the same conditions) at 64% for Johnson and 36% for NorthClean.

Starting with some other initial probability vector and going through the steps above would give the same equilibrium vector. In fact, the long-range trend is the same no matter what the initial vector is. The long-range trend depends only on the transition matrix, not on the initial distribution.

One of the many applications of Markov chains is in finding these long-range predictions. It is not possible to make long-range predictions with all transition matrices, but there is a large set of transition matrices for which long-range predictions *are* possible. Such predictions are always possible with *regular transition matrices*. A transition matrix is **regular** if some power of the matrix contains all positive entries. A Markov chain is a **regular Markov chain** if its transition matrix is regular. A regular Markov chain always has an equilibrium vector.

EXAMPLE 3   Decide whether the following transition matrices are regular.

(a) $A = \begin{bmatrix} .75 & .25 & 0 \\ 0 & .5 & .5 \\ .6 & .4 & 0 \end{bmatrix}$

Square the matrix A by multiplying it by itself.

$$A^2 = \begin{bmatrix} .5625 & .3125 & .125 \\ .3 & .45 & .25 \\ .45 & .35 & .2 \end{bmatrix}$$

All entries in A^2 are positive, so that matrix A is regular.

(b) $B = \begin{bmatrix} .5 & 0 & .5 \\ 0 & 1 & 0 \\ 0 & 0 & 1 \end{bmatrix}$

Find various powers of B.

$$B^2 = \begin{bmatrix} .25 & 0 & .75 \\ 0 & 1 & 0 \\ 0 & 0 & 1 \end{bmatrix} \quad B^3 = \begin{bmatrix} .125 & 0 & .875 \\ 0 & 1 & 0 \\ 0 & 0 & 1 \end{bmatrix} \quad B^4 = \begin{bmatrix} .0625 & 0 & .9375 \\ 0 & 1 & 0 \\ 0 & 0 & 1 \end{bmatrix}$$

4 Decide if the following transition matrices are regular.

(a) $\begin{bmatrix} 0 & 1 \\ 1 & 0 \end{bmatrix}$

(b) $\begin{bmatrix} .3 & .7 \\ 1 & 0 \end{bmatrix}$

Answers:
(a) No
(b) Yes

Notice that all of the powers of B shown here have zeros in the same locations. Thus, further powers of B will still give the same zero entries, so that no power of matrix B contains all positive entries. For this reason, B is not regular. ■ **4**

The equilibrium vector can be found without doing all the work shown above. Let $V = \begin{bmatrix} v_1 & v_2 \end{bmatrix}$ represent the desired vector. Let P represent the transition matrix. By definition, V is the fixed probability vector if $VP = V$. In our example,

$$\begin{bmatrix} v_1 & v_2 \end{bmatrix}\begin{bmatrix} .8 & .2 \\ .35 & .65 \end{bmatrix} = \begin{bmatrix} v_1 & v_2 \end{bmatrix}.$$

Multiply on the left to get

$$\begin{bmatrix} .8v_1 + .35v_2 & .2v_1 + .65v_2 \end{bmatrix} = \begin{bmatrix} v_1 & v_2 \end{bmatrix}.$$

Set corresponding entries from the two matrices equal to get

$$.8v_1 + .35v_2 = v_1 \qquad .2v_1 + .65v_2 = v_2.$$

Simplify each of these equations.

$$-.2v_1 + .35v_2 = 0 \qquad .2v_1 - .35v_2 = 0$$

These last two equations are really the same. (The equations in the system obtained from $VP = V$ are always dependent.) To find the values of v_1 and v_2, recall that $V = \begin{bmatrix} v_1 & v_2 \end{bmatrix}$ is a probability vector, so that

$$v_1 + v_2 = 1.$$

Find v_1 and v_2 by solving the system

$$-.2v_1 + .35v_2 = 0$$
$$v_1 + v_2 = 1.$$

5 Solve the equation for v_2.
Round to the nearest thousandth.

Answer:
.364

From the second equation, $v_1 = 1 - v_2$. Substitute for v_1 in the first equation:

$$-.2(1 - v_2) + .35v_2 = 0. \quad \boxed{5}$$

Since $v_2 = .364$ (from Problem 5 at the side) and $v_1 = 1 - v_2$, then $v_1 = 1 - .364 = .636$ and the equilibrium vector is $[.636 \quad .364]$.

EXAMPLE 4 Find the equilibrium vector for the transition matrix

$$P = \begin{bmatrix} \frac{1}{3} & \frac{2}{3} \\ \frac{3}{4} & \frac{1}{4} \end{bmatrix}.$$

Look for a vector $\begin{bmatrix} v_1 & v_2 \end{bmatrix}$ such that

$$\begin{bmatrix} v_1 & v_2 \end{bmatrix}\begin{bmatrix} \frac{1}{3} & \frac{2}{3} \\ \frac{3}{4} & \frac{1}{4} \end{bmatrix} = \begin{bmatrix} v_1 & v_2 \end{bmatrix}.$$

Multiply on the left, obtaining the two equations

$$\frac{1}{3}v_1 + \frac{3}{4}v_2 = v_1 \qquad \frac{2}{3}v_1 + \frac{1}{4}v_2 = v_2.$$

Simplify each of these equations.

$$-\frac{2}{3}v_1 + \frac{3}{4}v_2 = 0 \qquad \frac{2}{3}v_1 - \frac{3}{4}v_2 = 0$$

We know that $v_1 + v_2 = 1$, so $v_1 = 1 - v_2$. Substitute this into either of the two given equations. Using the first equation gives

$$-\frac{2}{3}(1 - v_2) + \frac{3}{4}v_2 = 0$$

$$-\frac{2}{3} + \frac{2}{3}v_2 + \frac{3}{4}v_2 = 0$$

$$-\frac{2}{3} + \frac{17}{12}v_2 = 0$$

$$\frac{17}{12}v_2 = \frac{2}{3}$$

$$v_2 = \frac{8}{17},$$

6 Find the equilibrium vector for the transition matrix

$$P = \begin{bmatrix} .3 & .7 \\ .5 & .5 \end{bmatrix}.$$

Answer:
[5/12   7/12]

and $v_1 = 1 - 8/17 = 9/17$. The equilibrium vector is [9/17   8/17]. ∎ **6**

EXAMPLE 5 The probability that a complex assembly line works correctly depends on whether or not the line worked correctly the last time it was used. The various probabilities are as given in the following transition matrix.

	Works Properly Now	Does Not
Worked Properly Before	.9	.1
Did Not	.7	.3

Find the long-range probability that the assembly line will work properly.
Begin by finding the equilibrium vector $\begin{bmatrix} v_1 & v_2 \end{bmatrix}$, where

$$\begin{bmatrix} v_1 & v_2 \end{bmatrix} \begin{bmatrix} .9 & .1 \\ .7 & .3 \end{bmatrix} = \begin{bmatrix} v_1 & v_2 \end{bmatrix}.$$

Multiplying on the left and setting corresponding entries equal gives the equations

$$.9v_1 + .7v_2 = v_1 \quad \text{and} \quad .1v_1 + .3v_2 = v_2$$

or

$$-.1v_1 + .7v_2 = 0 \quad \text{and} \quad .1v_1 - .7v_2 = 0.$$

Substitute $v_1 = 1 - v_2$ in the first of these equations to get

7 In Example 5, suppose the company modifies the line so that the transition matrix becomes

$$\begin{bmatrix} .95 & .05 \\ .8 & .2 \end{bmatrix}.$$

Find the long-range probability that the assembly line will work properly.

Answer:
16/17

$$-.1(1 - v_2) + .7v_2 = 0$$

$$-.1 + .1v_2 + .7v_2 = 0$$

$$-.1 + .8v_2 = 0$$

$$.8v_2 = .1$$

$$v_2 = \frac{1}{8},$$

and $v_1 = 1 - 1/8 = 7/8$. The equilibrium vector is [7/8   1/8]. In the long run, the company can expect the assembly line to run properly 7/8 of the time. ∎ **7**

The examples suggest the following generalization. If a regular Markov chain has transition matrix P and initial probability vector v, the products

$$vP$$
$$vPP = vP^2$$
$$vP^2P = vP^3$$
$$\cdot$$
$$\cdot$$
$$\cdot$$
$$vP^n$$

lead to the equilibrium vector V. The results of this section can be summarized as follows.

Suppose a regular Markov chain has a transition matrix P.

1. For any initial probability vector v, the products vP^n approach a unique vector V as n gets larger and larger. Vector V is called the equilibrium or fixed vector.
2. Vector V has the property that $VP = V$.
3. Vector V is found by solving a system of equations obtained from the matrix equation $VP = V$ and from the fact that the sum of the entries of V is 1.

10.4 EXERCISES

Decide which of the following could be a probability vector.

1. $\begin{bmatrix} \frac{1}{2} & \frac{2}{3} \end{bmatrix}$

2. $\begin{bmatrix} \frac{3}{4} & \frac{1}{2} \end{bmatrix}$

3. $[0 \quad 1]$

4. $[.4 \quad .2 \quad 0]$

5. $[.3 \quad -.1 \quad .6]$

6. $\begin{bmatrix} \frac{1}{4} & \frac{1}{8} & \frac{5}{8} \end{bmatrix}$

Decide which of the following could be a transition matrix. Sketch a transition diagram for any transition matrices.

7. $\begin{bmatrix} .7 & .1 \\ .5 & .5 \end{bmatrix}$

8. $\begin{bmatrix} \frac{1}{4} & \frac{3}{4} \\ 0 & 1 \end{bmatrix}$

9. $\begin{bmatrix} \frac{2}{3} & \frac{1}{3} \\ \frac{1}{5} & \frac{4}{5} \end{bmatrix}$

10. $\begin{bmatrix} 0 & 1 & 0 \\ \frac{1}{3} & \frac{1}{3} & \frac{1}{3} \\ 1 & 0 & 0 \end{bmatrix}$

11. $\begin{bmatrix} \frac{1}{2} & \frac{1}{4} & 1 \\ \frac{2}{3} & 0 & \frac{1}{3} \\ \frac{1}{3} & 1 & 0 \end{bmatrix}$

12. $\begin{bmatrix} .2 & .3 & .5 \\ 0 & 0 & 1 \\ .1 & .9 & 0 \end{bmatrix}$

In Exercises 13–15, write any transition diagrams as transition matrices.

13.

14.

15.

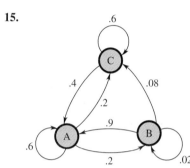

Decide whether the following transition matrices are regular. (See Example 3.)

16. $\begin{bmatrix} 1 & 0 \\ .6 & .4 \end{bmatrix}$

17. $\begin{bmatrix} .2 & .8 \\ .9 & .1 \end{bmatrix}$

18. $\begin{bmatrix} .3 & .5 & .2 \\ 1 & 0 & 0 \\ .5 & .1 & .4 \end{bmatrix}$

19. $\begin{bmatrix} 0 & 1 & 0 \\ .4 & .2 & .4 \\ 1 & 0 & 0 \end{bmatrix}$

20. $\begin{bmatrix} .25 & .40 & .30 & .05 \\ .18 & .23 & .59 & 0 \\ 0 & .15 & .36 & .49 \\ .28 & .32 & .24 & .16 \end{bmatrix}$

21. $\begin{bmatrix} .12 & .68 & 0 & .20 \\ 0 & .33 & .40 & .27 \\ 0 & 0 & 1 & 0 \\ .52 & 0 & .43 & .05 \end{bmatrix}$

Find the equilibrium vector for each of the following transition matrices. (See Examples 4 and 5.)

22. $\begin{bmatrix} .3 & .7 \\ .4 & .6 \end{bmatrix}$

23. $\begin{bmatrix} .8 & .2 \\ .1 & .9 \end{bmatrix}$

24. $\begin{bmatrix} \frac{1}{4} & \frac{3}{4} \\ \frac{1}{2} & \frac{1}{2} \end{bmatrix}$

25. $\begin{bmatrix} \frac{2}{3} & \frac{1}{3} \\ \frac{1}{8} & \frac{7}{8} \end{bmatrix}$

26. $\begin{bmatrix} .25 & .35 & .4 \\ .1 & .3 & .6 \\ .55 & .4 & .05 \end{bmatrix}$

27. $\begin{bmatrix} .16 & .28 & .56 \\ .43 & .12 & .45 \\ .86 & .05 & .09 \end{bmatrix}$

28. $\begin{bmatrix} .1 & .1 & .8 \\ .4 & .4 & .2 \\ .1 & .2 & .7 \end{bmatrix}$

29. $\begin{bmatrix} .5 & .2 & .3 \\ .1 & .4 & .5 \\ .2 & .2 & .6 \end{bmatrix}$

For each of the following transition matrices, use a graphing calculator or computer to find the first five powers of the matrix. Then find the probability that state 2 changes to state 4 after 5 repetitions of the experiment.

30. $\begin{bmatrix} .1 & .2 & .2 & .3 & .2 \\ .2 & .1 & .1 & .2 & .4 \\ .2 & .1 & .4 & .2 & .1 \\ .3 & .1 & .1 & .2 & .3 \\ .1 & .3 & .1 & .1 & .4 \end{bmatrix}$

31. $\begin{bmatrix} .3 & .2 & .3 & .1 & .1 \\ .4 & .2 & .1 & .2 & .1 \\ .1 & .3 & .2 & .2 & .2 \\ .2 & .1 & .3 & .2 & .2 \\ .1 & .1 & .4 & .2 & .2 \end{bmatrix}$

32. Social Science The following chart shows the percent of the poor, middle class, and affluent that change into another class. The first graph shows the figures for 1967–1979, and the second for 1980–1991.* For each time period, assume the number of people who move directly from poor to affluent or from affluent to poor is essentially 0.

(a) Find the long-range percent of poor, middle class, and affluent people if the 1967–1979 trends were to continue.

(b) Find the long-range percent of poor, middle class, and affluent people if the 1980–1991 trends were to continue.

The poor are more likely to stay poor and the affluent are more likely to stay affluent.

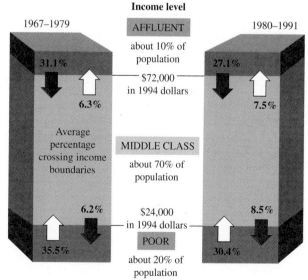

All figures are for household after-tax income, including wages, salaries, and some Government assistance programs like food stamps.

Sources: Greg Duncan, Northwestern University; Timothy Smeeding, Syracuse University

---

The New York Times, June 4, 1995, p. E4.

33. **Management** In 1992, many homeowners refinanced their mortgages to take advantage of lower interest rates. Most of the mortgages could be classified into three groups: 30-year fixed-rate, 15-year fixed-rate, and adjustable rate. (For this exercise, the small number of loans that are not of those three types will be classified with the adjustable rate loans.) Sometimes when a homeowner refinanced, the new loan was the same type as the old loan, and sometimes it was different. The breakdown of the percent in each category is shown in the following table.*

	NEW LOAN		
OLD LOAN	*30-year Fixed*	*15-year Fixed*	*Adjustable*
30-year fixed	.444	.479	.077
15-year fixed	.150	.802	.048
Adjustable	.463	.367	.170

If these conversion rates were to persist, find the long-range trend for the percent of loans of each type.

34. **Management** The probability that a complex assembly line works correctly depends on whether the line worked correctly the last time it was used. There is a .95 chance that the line will work correctly if it worked correctly the time before, and a .7 chance that it will work correctly if it did *not* work correctly the time before. Set up a transition matrix with this information and find the long-run probability that the line will work correctly. (See Example 5.)

35. **Management** Suppose improvements are made in the assembly line of Exercise 34, so that the transition matrix becomes

$$\begin{array}{c} \text{Works} \\ \text{Doesn't Work} \end{array} \begin{array}{cc} \text{Works} & \text{Doesn't Work} \\ \begin{bmatrix} .95 & .05 \\ .85 & .15 \end{bmatrix} \end{array}.$$

Find the new long-run probability that the line will work properly.

36. **Natural Science** In Exercises 46 and 47 of Section 9.4 we discussed the effect on flower color of cross-pollinating pea plants. As shown there, since the gene for red is dominant and the gene for white is recessive, 75% of these pea plants have red flowers and 25% have white flowers, because plants with 1 red and 1 white gene appear red. If a red-flowered plant is crossed with a red-flowered plant known to have 1 red and 1 white gene, then 75% of the offspring will be red and 25% will be white. Crossing a red-flowered plant that has 1 red and 1 white gene with a white-flowered plant produces 50% red-flowered offspring and 50% white-flowered offspring.

(a) Write a transition matrix using this information.
(b) Write a probability vector for the initial distribution of colors.
(c) Find the distribution of colors after 4 generations.
(d) Find the long-range distribution of colors.

37. **Natural Science** Snapdragons with 1 red gene and 1 white gene produce pink-flowered offspring. If a red snapdragon is crossed with a pink snapdragon, the probabilities that the offspring will be red, pink, or white are 1/2, 1/2, and 0, respectively. If 2 pink snapdragons are crossed, the probabilities of red, pink, or white offspring are 1/4, 1/2, and 1/4, respectively. For a cross between a white and a pink snapdragon, the corresponding probabilities are 0, 1/2, and 1/2. Set up a transition matrix and find the long-range prediction for the fraction of red, pink, and white snapdragons.

38. Many phenomena can be viewed as examples of a random walk. Consider the following simple example. A security guard can stand in front of any one of three doors 20 ft apart in front of a building, and every minute he decides whether to move to another door chosen at random. If he is at the middle door, he is equally likely to stay where he is, move to the door to the left, or move to the door to the right. If he is at the door on either end, he is equally likely to stay where he is or move to the middle door.

(a) Verify that the transition matrix is given by

$$\begin{array}{c} 1 \\ 2 \\ 3 \end{array} \begin{array}{ccc} 1 & 2 & 3 \\ \begin{bmatrix} \frac{1}{2} & \frac{1}{2} & 0 \\ \frac{1}{3} & \frac{1}{3} & \frac{1}{3} \\ 0 & \frac{1}{2} & \frac{1}{2} \end{bmatrix} \end{array}.$$

(b) Find the long-range trend for the fraction of time the guard spends in front of each door.

39. **Social Science** The probability that a homeowner will become a renter in five years is .03. The probability that a renter will become a homeowner in five years is .1. Suppose the proportions in the population are 64% homeowners (O), 35.5% renters (R), and .5% homeless (H), with the following transition matrix. Assume that these figures continue to apply.

$$\begin{array}{c} O \\ R \\ H \end{array} \begin{array}{ccc} O & R & H \\ \begin{bmatrix} .97 & .03 & 0 \\ .1 & .899 & .001 \\ 0 & .4 & .6 \end{bmatrix} \end{array}$$

Find the long-range probabilities for the three categories.

40. **Management** An insurance company classifies its drivers into three groups: G_0 (no accidents), G_1 (one accident), and G_2 (more than one accident). The probability that a driver in G_0 will stay in G_0 after 1 year is .85, that he will become a G_1 is .10, and that he will become a G_2 is .05. A driver in G_1 cannot move to G_0 (this insurance company has a long memory). There is an .80 probability that a G_1 driver

will stay in G_1 and a .20 probability that he will become a G_2. A driver in G_2 must stay in G_2.

(a) Write a transition matrix using this information.

Suppose that the company accepts 50,000 new policyholders, all of whom are in group G_0. Find the number in each group

(b) after 1 year;   (c) after 2 years;

(d) after 3 years;   (e) after 4 years.

(f) Find the equilibrium vector here. Interpret your result.

41. Management The difficulty with the mathematical model of Exercise 40 is that no "grace period" is provided; there should be a certain probability of moving from G_1 or G_2 back to G_0 (say, after 4 years with no accidents). A new system with this feature might produce the following transition matrix.

$$\begin{bmatrix} .85 & .10 & .05 \\ .15 & .75 & .10 \\ .10 & .30 & .60 \end{bmatrix}$$

Suppose that when this new policy is adopted, the company has 50,000 policyholders in group G_0. Find the number of these in each group

(a) after 1 year;   (b) after 2 years;

(c) after 3 years.

(d) Find the equilibrium vector here. Interpret your result.

42. Management Research done by the Gulf Oil Corporation* produced the following transition matrix for the probability that a person with one form of home heating would switch to another.

		Will Switch to	
	Oil	Gas	Electric
Oil	.825	.175	0
Now Has Gas	.060	.919	.021
Electric	.049	0	.951

The current share of the market held by these three types of heat is given by [.26   .60   .14]. Find the share of the market held by each type of heat after

(a) 1 year;   (b) 2 years;   (c) 3 years.

(d) What is the long-range prediction?

43. The results of cricket matches between England and Australia have been found to be modeled by a Markov chain.† The probability that England wins, loses, or draws

is based on the result of the previous game, with the following transition matrix.

	Wins	Loses	Draws
Wins	.443	.364	.193
Loses	.277	.436	.287
Draws	.266	.304	.430

(a) Compute the transition matrix for the game after the next one, based on the result of the last game.

(b) Use your answer from part (a) to find the probability that, if England won the last game, England will win the game after the next one.

(c) Use your answer from part (a) to find the probability that, if Australia won the last game, England will win the game after the next one.

44. Social Science At one liberal arts college, students are classified as humanities majors, science majors, or undecided. There is a 20% chance that a humanities major will change to a science major from one year to the next, and a 45% chance that a humanities major will change to undecided. A science major will change to humanities with probability .15, and to undecided with probability .35. An undecided will switch to humanities or science with probabilities of .5 and .3, respectively. Find the long-range prediction for the fraction of students in each of these three majors.

45. Management In the queuing chain, we assume that people are queuing up to be served by, say, a bank teller. For simplicity, let us assume that once two people are in line, no one else can enter the line. Let us further assume that one person is served every minute, as long as someone is in line. Assume further that in any minute, there is a probability of $\frac{1}{2}$ that no one enters the line, a probability of $\frac{1}{3}$ that exactly one person enters the line, and a probability of $\frac{1}{6}$ that exactly two people enter the line, assuming there is room. If there is not enough room for two people, then the probability that one person enters the line is $\frac{1}{2}$. Let the state be given by the number of people in line.

(a) Verify that the transition matrix is

$$\begin{array}{c} \\ 0 \\ 1 \\ 2 \end{array} \begin{array}{c} \begin{array}{ccc} 0 & 1 & 2 \end{array} \\ \begin{bmatrix} \frac{1}{2} & \frac{1}{3} & \frac{1}{6} \\ \frac{1}{2} & \frac{1}{3} & \frac{1}{6} \\ 0 & \frac{1}{2} & \frac{1}{2} \end{bmatrix} \end{array}.$$

(b) Find the transition matrix for a two-minute period.

(c) Use your result from part (b) to find the probability that a queue with no one in line has two people in line two minutes later.

Use a calculator or computer for Exercises 46 and 47.

46. Management A company with a new training program classified each employee in one of four states: s_1, never in the program; s_2, currently in the program; s_3, discharged;

(exercise continues)

*From "Forecasting Market Shares of Alternative Home-Heating Units by Markov Process Using Transition Probabilities Estimates from Aggregate Time Series Data" by Ali Ezzati, from *Management Science*, Vol. 21, No. 4, December 1974. Copyright © 1974 The Institute of Management Sciences. Reprinted by permission.

†From Derek Colwell, Brian Jones, and Jack Gillett, "A Markov Chain in Cricket," *The Mathematical Gazette*, June 1991.

s_4, completed the program. The transition matrix for this company is given below.

$$
\begin{array}{c c}
 & \begin{array}{cccc} s_1 & s_2 & s_3 & s_4 \end{array} \\
\begin{array}{c} s_1 \\ s_2 \\ s_3 \\ s_4 \end{array} &
\left[\begin{array}{cccc}
.4 & .2 & .05 & .35 \\
0 & .45 & .05 & .5 \\
0 & 0 & 1 & 0 \\
0 & 0 & 0 & 1
\end{array} \right]
\end{array}
$$

(a) What percent of employees who had never been in the program (state s_1) completed the program (state s_4) after the program had been offered five times?

(b) If the initial percent of employees in each state was [.5 .5 0 0], find the corresponding percents after the program had been offered four times.

47. **Management** Find the long-range prediction for the percent of employees in each state for the company training program from Exercise 46.

10.5 PROBABILITY DISTRIBUTIONS; EXPECTED VALUE

In this section we shall see that the *expected value* of a probability distribution is a kind of average. Probability distributions were introduced briefly in Chapter 9. A probability distribution depends on the idea of a *random variable,* so we begin with that.

Suppose a bank is interested in improving its services to the public. The manager decides to begin by finding the amount of time tellers spend on each transaction, rounded to the nearest minute. To each transaction, then, will be assigned one of the numbers 0, 1, 2, 3, 4, ... That is, if t represents an outcome of the experiment of timing a transaction, then t may take on any of the values from the list 0, 1, 2, 3, 4, ... The value that t takes on for a particular transaction is random, so t is called a random variable.

> A **random variable** is a function that assigns a real number to each outcome of an experiment.

PROBABILITY DISTRIBUTIONS Suppose that the bank manager in our example records the times for 75 different transactions, with results as shown in Table 1. As the table shows, the shortest transaction time was 1 minute, with 3 transactions of 1-minute duration. The longest time was 10 minutes. Only 1 transaction took that long.

In Table 1, the first column gives the 10 values assumed by the random variable t, and the second column shows the number of occurrences corresponding to each of these values, the *frequency* of that value. Table 1 is an example of a **frequency distribution,** a table listing the frequencies for each value a random variable may assume.

TABLE 1

Time	Frequency
1	3
2	5
3	9
4	12
5	15
6	11
7	10
8	6
9	3
10	1
Total:	75

Now suppose that several weeks after starting new procedures to speed up transactions, the manager takes another survey. This time she observes 57 transactions, and she records their times as shown in the frequency distribution in Table 2.

TABLE 2

Time	Frequency
1	4
2	5
3	8
4	10
5	12
6	17
7	0
8	1
9	0
10	0
	Total: 57

Do the results in Table 2 indicate an improvement? It is hard to compare the two tables, since one is based on 75 transactions and the other on 57. To make them comparable, we can add a column to each table that gives the *relative frequency* of each transaction time. These results are shown in Tables 3 and 4. Where necessary, decimals are rounded to the nearest hundredth. To find a **relative frequency,** divide each frequency by the total of the frequencies. Here the individual frequencies are divided by 75 or 57. Many of these frequencies have been rounded.

1 Complete the probability distribution below. (f represents frequency and p represents probability.)

x	f	p
1	3	
2	7	
3	9	
4	3	
5	2	

Answer:

x	f	p
1	3	.13
2	7	.29
3	9	.38
4	3	.13
5	2	.08
	Total: 24	

TABLE 3

Time	Frequency	Relative Frequency
1	3	$3/75 = .04$
2	5	$5/75 \approx .07$
3	9	$9/75 = .12$
4	12	$12/75 = .16$
5	15	$15/75 = .20$
6	11	$11/75 \approx .15$
7	10	$10/75 \approx .13$
8	6	$6/75 = .08$
9	3	$3/75 = .04$
10	1	$1/75 \approx .01$

TABLE 4

Time	Frequency	Relative Frequency
1	4	$4/57 \approx .07$
2	5	$5/57 \approx .09$
3	8	$8/57 \approx .14$
4	10	$10/57 \approx .18$
5	12	$12/57 \approx .21$
6	17	$17/57 \approx .30$
7	0	$0/57 = 0$
8	1	$1/57 \approx .02$
9	0	$0/57 = 0$
10	0	$0/57 = 0$

The results shown in the two tables are easier to compare using the relative frequencies. We will return to this problem later in this section.

The relative frequencies in Tables 3 and 4 can be considered probabilities. Such a table that lists all the possible values of a random variable, together with the corresponding probabilities, is called a **probability distribution.** The sum of the probabilities in a probability distribution must always be 1. (The sum in an actual distribution may vary slightly from 1 because of rounding.) **1**

EXAMPLE 1 Many plants have seed pods with variable numbers of seeds. One variety of green beans has no more than 6 seeds per pod. Suppose that examination of 30 such bean pods gives the results shown in Table 5. Here the random variable x tells the number of seeds per pod. Give a probability distribution for these results.

The probabilities are found by computing the relative frequencies. A total of 30 bean pods were examined, so each frequency should be divided by 30 to get the probability distribution shown as Table 6. Some of the results have been rounded to the nearest hundredth.

TABLE 5

x	Frequency
0	3
1	4
2	6
3	8
4	5
5	3
6	1
	Total: 30

TABLE 6

x	Frequency	Probability
0	3	$3/30 = .10$
1	4	$4/30 \approx .13$
2	6	$6/30 = .20$
3	8	$8/30 \approx .27$
4	5	$5/30 \approx .17$
5	3	$3/30 = .10$
6	1	$1/30 \approx .03$
	Total: 30	

2 In Example 1, what is
(a) $P(x = 0)$?
(b) $P(x = 1)$?

Answers:
(a) .10
(b) .13

As shown in Table 6, the probability that the random variable x takes on the value 2 is 6/30, or .20. This is often written as

$$P(x = 2) = .20.$$

Also, $P(x = 5) = .10$, and $P(x = 6) \approx .03$. ■ **2**

Instead of writing the probability distribution in Example 1 as a table, we could write the same information as a set of ordered pairs:

$$\{(0, .10), (1, .13), (2, .20), (3, .27), (4, .17), (5, .10), (6, .03)\}.$$

There is just one probability for each value of the random variable. Thus, a probability distribution defines a function, called a **probability distribution function** or simply a **probability function.** We shall use the terms "probability distribution" and "probability function" interchangeably.

The information in a probability distribution is often displayed graphically as a special kind of bar graph called a **histogram.** The bars of a histogram all have the same width, usually 1. The heights of the bars are determined by the frequencies. A histogram for the data in Table 3 is given in Figure 10.5. A histogram shows important characteristics of a distribution that may not be readily apparent in tabular form, such as the relative sizes of the probabilities and any symmetry in the distribution.

The area of the bar above $t = 1$ in Figure 10.5 is the product of 1 and .04, or $.04 \times 1 = .04$. Since each bar has a width of 1, its area is equal to the probability that corresponds to that value of t. The probability that a particular value will occur is thus given by the area of the appropriate bar of the graph. For example, the probability that a transaction will take less than 4 minutes is the sum of the areas for $t = 1$, $t = 2$, and $t = 3$. This area, shown in color in Figure 10.6, corresponds to 23% of the total area, since

$$P(t < 4) = P(t = 1) + P(t = 2) + P(t = 3)$$
$$= .04 + .07 + .12 = .23.$$

FIGURE 10.5

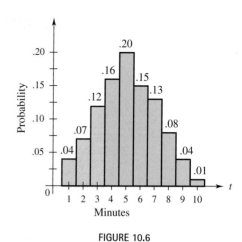

FIGURE 10.6

3 (a) Prepare a histogram for the probability distribution in Problem 1 at the side.

(b) Find $P(x \le 2)$.

Answers:
(a)

(b) .42

EXAMPLE 2   Construct a histogram for the probability distribution in Example 1. Find the area that gives the probability that the number of seeds will be more than 4.

A histogram for this distribution is shown in Figure 10.7. The portion of the histogram in color represents

$$P(x > 4) = P(x = 5) + P(x = 6)$$
$$= .10 + .03$$
$$= .13,$$

or 13% of the total area.   ■   **3**

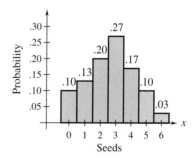

FIGURE 10.7

TECHNOLOGY TIP Virtually all graphing calculators can produce histograms. The procedures differ on various calculators, but usually require you to enter the outcomes in one list and the corresponding frequencies in a second list. Depending on the viewing window chosen, the histogram bars may not always be centered over the corresponding *x*-value. For specific details, check your instruction manual under "statistics graphs," or "statistical plotting." ✔

✳ **EXAMPLE 3** **(a)** Give the probability distribution for the number of heads showing when 2 coins are tossed.

Let x represent the outcome "number of heads." Then x can take on the values 0, 1, or 2. Now find the probability of each outcome. The results are shown in Table 7.

TABLE 7

x	$P(x)$
0	1/4
1	1/2
2	1/4

(b) Find the probability that at least one coin comes up heads.

$$P(x \geq 1) = P(x = 1) + P(x = 2)$$
$$= \frac{1}{2} + \frac{1}{4} = \frac{3}{4}$$

(c) Draw a histogram for the distribution in Table 7.

Two graphing calculator versions of the histogram are shown in Figure 10.8, both having bars of width 1. (See the Technology Tip below.) ■ ☐4

 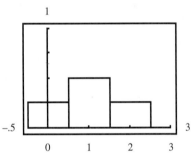

FIGURE 10.8

TECHNOLOGY TIP To obtain the histograms in Figure 10.8 on most graphing calculators, enter the outcomes 0, 1, and 2 on the first list and enter the probabilities .25, .5, and .25 on a second list. Use the second list as "frequencies" when graphing and use a window with $0 \leq y \leq 1$. On TI-81/82/85, however, frequencies must be integers. So make the entries of the second list 1, 2, and 1 (corresponding to 1/4, 2/4, and 1/4) and use a window with $0 \leq y \leq 4$. ✔

In working with experimental data, it is often useful to have a typical or "average" number that represents the entire set of data. For example, we compare our heights and weights to those of the typical or "average" person on weight charts. Students are familiar with the "class average" and their own "average" in a given course.

In a recent year, a citizen of the United States could expect to complete about 13 years of school, to be a member of a household earning \$35,400 per year, and to live in a household of 2.6 people. What do we mean here by "expect"? Many people have completed less than 13 years of school; many others have completed more. Many households have less income than \$35,400 per year; many others have more. The idea

---

☐4 **(a)** Give the probability distribution for the number of heads showing when 3 coins are tossed.

(b) Find the probability that no more than 1 coin shows heads.

Answers:

(a)

x	$P(x)$
0	1/8
1	3/8
2	3/8
3	1/8

(b) 1/2

of a household of 2.6 people is a little hard to swallow. The numbers all refer to *averages.* When the term "expect" is used in this way, it refers to *mathematical expectation,* which is a kind of average.

EXPECTED VALUE What about an average value for a random variable? As an example, let us find the average number of offspring for a certain species of pheasant, given the probability distribution in Table 8.

TABLE 8

Number of Offspring	Frequency	Probability
0	8	.08
1	14	.14
2	29	.29
3	32	.32
4	17	.17
	Total: 100	

We might be tempted to find the typical number of offspring by averaging the numbers 0, 1, 2, 3, and 4, which represent the numbers of offspring possible. This won't work, however, since the various numbers of offspring do not occur with equal probability; for example, a brood with 3 offspring is much more common than one with 0 or 1 offspring. The differing probabilities of occurrence can be taken into account with a **weighted average,** found by multiplying each of the possible numbers of offspring by its corresponding probability, as follows.

$$\textbf{Typical number of offspring} = 0(.08) + 1(.14) + 2(.29) + 3(.32) + 4(.17)$$
$$= 0 + .14 + .58 + .96 + .68$$
$$= \textbf{2.36}$$

Based on the data given above, a typical brood of pheasants includes 2.36 offspring.

It is certainly not possible for a pair of pheasants to produce 2.36 offspring. If the numbers of offspring produced by many different pairs of pheasants are found, however, the average of these numbers will be about 2.36.

We can use the results of this example to define the mean, or *expected value,* of a probability distribution as follows. (We will have more to say about the mean in Chapter 11.)

EXPECTED VALUE

Suppose the random variable x can take on the n values $x_1, x_2, x_3, \ldots, x_n$. Also, suppose the probabilities that these values occur are respectively $p_1, p_2, p_3, \ldots, p_n$. Then the **expected value** of the random variable is

$$E(x) = x_1 p_1 + x_2 p_2 + x_3 p_3 + \cdots + x_n p_n.$$

As in the example above, the expected value of a random variable may be a number that can never occur in any one trial of the experiment.

Physically, the expected value of a probability distribution represents a balance point. Figure 10.9 shows a histogram for the distribution of the pheasant offspring. If the histogram is thought of as a series of weights with magnitudes represented by the heights of the bars, then the system would balance if supported at the point corresponding to the expected value.

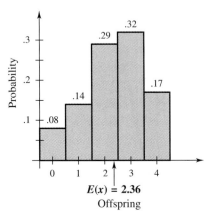

$E(x) = 2.36$
Offspring

FIGURE 10.9

EXAMPLE 4 Find the expected transaction time for each of the survey results from the bank example at the beginning of this section. What can we conclude from the results?

Refer to Tables 3 and 4. From Table 3, using the relative frequencies as probabilities, the expected transaction time is

$$1(.04) + 2(.07) + 3(.12) + 4(.16) + 5(.20) + 6(.15) + 7(.13)$$
$$+ 8(.08) + 9(.04) + 10(.01) = 5.09. \quad \boxed{5}$$

From Problem 5 at the side, the expected transaction time for the second survey is 4.4. The expected values give us a useful way to compare the results of the two surveys. The second survey had a smaller expected transaction time, so the new procedures may have sped up transactions. ■

EXAMPLE 5 Suppose a local church decides to raise money by raffling a microwave oven worth $400. A total of 2000 tickets are sold at $1 each. Find the expected value of winning for a person who buys 1 ticket in the raffle.

Here the random variable represents the possible amounts of net winnings, where net winnings = amount won − cost of ticket. The net winnings of the person winning the oven are $400 (amount won) − $1 (cost of ticket) = $399. The net winnings for each losing ticket are $0 − $1 = −$1.

The probability of winning is 1 in 2000, or 1/2000, while the probability of losing is 1999/2000. See Table 9.

5 Using Table 4, find the expected transaction time for the second survey.

Answer:
4.4

TABLE 9

Outcome (net winnings)	Probability
$399	$\dfrac{1}{2000}$
−$1	$\dfrac{1999}{2000}$

The expected winnings for a person buying 1 ticket are

$$399\left(\frac{1}{2000}\right) + (-1)\left(\frac{1999}{2000}\right) = \frac{399}{2000} - \frac{1999}{2000}$$

$$= -\frac{1600}{2000}$$

$$= -.80.$$

On the average, a person buying 1 ticket in the raffle will lose $.80, or 80¢.

It is not possible to lose 80¢ in this raffle—either you lose $1, or you win a $400 prize. If you bought tickets in many such raffles over a long period of time, however, you would lose 80¢ per ticket, on the average. ■ 6

[6] Suppose you buy 1 of 1000 tickets at 10¢ each in a lottery where the prize is $50. What are your expected net winnings? What does this answer mean?

Answer:
−5¢. On the average you lose 5¢ per ticket purchased.

EXAMPLE 6 Each day Donna and Mary toss a coin to see who buys the coffee (60¢ a cup). One tosses and the other calls the outcome. If the person who calls the outcome is correct, the other buys the coffee; otherwise the caller pays. Find Donna's expected winnings.

Assume that an honest coin is used, that Mary tosses the coin, and that Donna calls the outcome. The possible results and corresponding probabilities are shown below.

	Possible Results			
Result of toss	Heads	Heads	Tails	Tails
Call	Heads	Tails	Heads	Tails
Caller Wins?	Yes	No	No	Yes
Probability	1/4	1/4	1/4	1/4

Donna wins a 60¢ cup of coffee whenever the results and calls match, and she loses a 60¢ cup when there is no match. Her expected winnings are

$$(.60)\left(\frac{1}{4}\right) + (-.60)\left(\frac{1}{4}\right) + (-.60)\left(\frac{1}{4}\right) + (.60)\left(\frac{1}{4}\right) = 0.$$

[7] Find Mary's expected winnings.

Answer:
0

On the average, over the long run, Donna neither wins nor loses. ■ 7

A game with an expected value of 0 (such as the one in Example 6) is called a **fair game.** Casinos do not offer fair games. If they did, they would win (on the average) $0, and have a hard time paying the help! Casino games have expected winnings for the house that vary from 1.5¢ per dollar to 60¢ per dollar. Exercises 37–42 at the end of the section ask you to find the expected winnings for certain games of chance.

The idea of expected value can be very useful in decision making, as shown by the next example.

8 A person can take one of two jobs. With job A, there is a 50% chance of making $60,000 per year after 5 years and a 50% chance of making $30,000. With job B, there is a 30% chance of making $90,000 per year after 5 years and a 70% chance of making $20,000. Based strictly on expected value, which job should be taken?

EXAMPLE 7 At age 50, Earl Karn receives a letter from the Mutual of Mauritania Insurance Company. According to the letter, he must tell the company immediately which of the following two options he will choose: take $20,000 at age 60 (if he is alive, $0 otherwise) or $30,000 at age 70 (again, if he is alive, $0 otherwise). Based *only* on the idea of expected value, which should he choose?*

Life insurance companies have constructed elaborate tables showing the probability of a person living a given number of years in the future. From a recent such table, the probability of living from age 50 to age 60 is .88, while the probability of living from age 50 to 70 is .64. The expected values of the two options are given below.

Answer:
Job A has an expected salary of $45,000; job B, $41,000. Take job A.

First Option: $(20,000)(.88) + (0)(.12) = 17,600$
Second Option: $(30,000)(.64) + (0)(.36) = 19,200$

Based strictly on expected values, Karn should choose the second option. ■ **8**

*Other considerations might affect the decision, such as the rate at which Karn might invest the $20,000 at age 60 for 10 years.

10.5 EXERCISES

In Exercises 1–4 **(a)** *give the probability distribution, and* **(b)** *sketch its histogram. (See Examples 1–3.)*

1. The number of accidents at a certain intersection in a metropolitan area were counted each day for 7 days with the following results.

Number of Accidents	Frequency
0	2
1	3
2	1
3	0
4	1

2. A teacher polled a finite mathematics class each semester for 10 semesters to see how many students were business majors. The results are given in the table.

Number of Business Majors	Frequency
13	1
14	0
15	3
16	3
17	2
18	1

3. At a training program for police officers, each member of a class of 25 took 6 shots at a target. The numbers of bullseyes shot are shown in the table.

Number of Bullseyes	Frequency
0	0
1	1
2	0
3	4
4	10
5	8
6	2
Total:	25

4. Five people are given a pain reliever. After a certain time period, the number who experience relief is noted. The experiment is repeated 20 times with the results shown in the table.

Number Who Experience Relief	Frequency
0	3
1	5
2	6
3	3
4	2
5	1
Total:	20

For each of the experiments below, let x represent a random variable, and use your knowledge of probability to prepare a probability distribution.

5. The numbers 1 to 5 are written on cards, placed in a basket, mixed up, and then 2 cards are drawn. The number of even numbers are counted.

6. A bowl contains 5 yellow, 8 orange, and 3 white jelly beans. Two jelly beans are drawn at random and the number of white jelly beans is counted.

7. Two dice are rolled, and the total number of points is recorded.

8. Three cards are drawn from a deck. The number of aces is counted.

9. Two balls are drawn from a bag in which there are 4 white balls and 2 black balls. The number of black balls is counted.

10. Five cards are drawn from a deck. The number of black threes is counted.

Draw a histogram for each of the following, and shade the region that gives the indicated probability. (See Example 2.)

11. Exercise 7; $P(x \geq 11)$

12. Exercise 8; P(at least one ace)

13. Exercise 9; P(at least one black ball)

14. Exercise 10; $P(x = 1 \text{ or } x = 2)$

15. Exercise 5; $P(1 \leq x \leq 2)$

16. Exercise 10; $P(0 \leq x \leq 1)$

Find the expected value for each random variable. (See Example 4.)

17.

x	5	6	7	8
P(x)	.3	.1	.4	.2

18.

y	10	11	12	13	14
P(y)	.1	.05	.15	.4	.3

19.

z	22	23	24	25	26
P(z)	.15	.23	.34	.20	.08

20.

x	0	1	2	3	4
P(x)	.31	.35	.20	.11	.03

In Exercises 21–22, **(a)** *give a probability distribution,* **(b)** *sketch its histogram, and* **(c)** *find the expected value.*

21. Management According to officials of Mars, the makers of M & M Plain Chocolate Candies, 20% of the candies in

each bag are red.* 4 candies are selected from a bag and the number of red candies is recorded.

22. Social Science In 1992, the Big 10 collegiate sports conference moved to have women compose at least 40% of its athletes within five years.† Suppose they exactly achieved the 40% figure, and that 5 athletes are picked at random from Big 10 universities. The number of women is recorded.

23. Refer to Example 6. Suppose one day Mary brings a two-headed coin and uses it to toss for the coffee. Since Mary tosses, Donna calls.
(a) Is this still a fair game?
(b) What is Donna's expected gain if she calls heads?
(c) What is it if she calls tails?

Find the expected values for the random variables x having the probability functions graphed below.

24.

25.

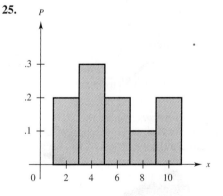

*NCTM News Bulletin, Feb. 1995, p. 5.

†Chicago Tribune, April 28, 1993, p. 19.

26.

27.

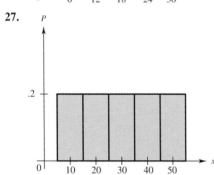

28. A raffle offers a first prize of $100, and two second prizes of $40 each. One ticket costs $1, and 500 tickets are sold. Find the expected winnings for a person who buys 1 ticket. (See Example 5.)

29. A raffle offers a first prize of $1000, two second prizes of $300 each, and twenty prizes of $10 each. If 10,000 tickets are sold at 50¢ each, find the expected winnings for a person buying 1 ticket.

Many of the following exercises use the idea of combinations discussed in Section 10.1.

30. If 3 marbles are drawn from a bag containing 3 yellow and 4 white marbles, what is the expected number of yellow marbles in the sample?

31. Management The shipping manager at a company receives a package of one dozen computer monitors, of which, unknown to him, three are broken. He checks four of the monitors at random to see how many are broken in his sample of four. What is the expected number of broken monitors in the sample?

32. Social Science According to the National Center for Education Statistics, 79% of the U.S. holders of Bachelor's degrees in Education in 1992–1993 were women.* What is the expected number of women in a random sample of five people holding U.S. Bachelor's degrees in Education?

*The New York Times, Jan. 7, 1996, Education Life, p. 24.

33. If 2 cards are drawn at one time from a deck of 52 cards, what is the expected number of diamonds?

34. Suppose someone offers to pay you $5 if you draw 2 diamonds in the game of Exercise 33. He says that you should pay 50¢ for the chance to play. Is this a fair game?

The following chart shows the number of times in a recent 52-week period that each possible number was chosen in any of the four digit positions in the Pick-4 game of the Illinois lottery.

No.	1st	2nd	3rd	4th	Any
0	38	29	32	34	133
1	32	35	39	40	146
2	35	37	39	39	150
3	46	43	35	42	166
4	33	50	38	40	161
5	31	35	28	37	131
6	46	35	36	32	149
7	41	37	42	33	153
8	26	28	43	35	132
9	35	34	31	31	131

Find each of the following expected values.

35. the expected number in first position

36. the expected number in any position

Find the expected winnings for the following games of chance.

37. A state lottery requires you to choose 1 heart, 1 club, 1 diamond, and 1 spade in that order from the 13 cards in each suit. If all four choices are correct, you win $5000. It costs $1 to play.

38. If exactly 3 of the 4 selections in Exercise 37 are correct, the player wins $200. (It still costs $1 to play.)

39. In one form of roulette, you bet $1 on "even." If 1 of the 18 even numbers comes up, you get your dollar back, plus another one. If 1 of the 20 noneven (18 odd, 0, and 00) numbers comes up, you lose.

40. In another form of roulette, there are only 19 noneven numbers (no 00).

41. Numbers is an illegal game in which you bet $1 on any 3-digit number from 000 to 999. If your number comes up, you get $500.

42. In Keno, the house has a pot containing 80 balls, each marked with a different number from 1 to 80. You buy a ticket for $1 and mark 1 of the 80 numbers on it. The house then selects 20 numbers at random. If your number is among the 20, you get $3.20 (for a net winning of $2.20).

43. Your friend missed class the day probability distributions were discussed. How would you explain to him what a probability distribution is?

44. How is the relative frequency of a random variable found?

45. The expected value of a random variable is
   (a) the most frequent value of the random variable;
   (b) the average of all possible values of the random variable;
   (c) the sum of the products formed by multiplying each value of the random variable by its probability;
   (d) the product of the sum of all values of the random variable and the sum of the probabilities.

46. Management   An insurance company has written 100 policies of $10,000, 500 of $5000, and 1000 policies of $1000 on people of age 20. If experience shows that the probability of dying during the 20th year of life is .001, how much can the company expect to pay out during the year the policies were written?

47. Jack must choose at age 40 to inherit $25,000 at age 50 (if he is still alive) or $30,000 at age 55 (if he is still alive). If the probabilities for a person of age 40 to live to be 50 and 55 are .90 and .85, respectively, which choice gives him the larger expected inheritance? (See Example 7.)

48. Management   A magazine distributor offers a first prize of $100,000, two second prizes of $40,000 each, and two third prizes of $10,000 each. A total of 2,000,000 entries are received in the contest. Find the expected winnings if you submit 1 entry to the contest. If it would cost you 35¢ in time, paper, and stamps to enter, would it be worth it?

49. Management   Levi Strauss and Company* uses expected value to help its salespeople rate their accounts. For each account, a salesperson estimates potential additional volume and the probability of getting it. The product of these gives the expected value of the potential, which is added to the existing volume. The totals are then classified as A, B, or C as follows: below $40,000, class C; between $40,000 and $55,000 inclusive, class B; above $55,000, class A. Complete the chart below for one of its salespeople.

50. Natural Science   One of the few methods that can be used in an attempt to cut the severity of a hurricane is to *seed* the storm. In this process, silver iodide crystals are dropped into the storm. Unfortunately, silver iodide crystals sometimes cause the storm to *increase* its speed. Wind speeds may also increase or decrease even with no seeding. The probabilities and amounts of property damage shown in the probability tree are from an article by R. A. Howard, J. E. Matheson, and D. W. North, "The Decision to Seed Hurricanes."*
   (a) Find the expected amount of damage under each option, "seed" and "do not seed."
   (b) To minimize total expected damage, what option should be chosen?

	Change in wind speed	Property damage (millions of dollars)
	0.038 +32%	335.8
	0.143 +16%	191.1
Seed	0.392 0	100.0
	0.255	
	0.172 −16%	46.7
	−34%	16.3
	0.054 +32%	335.8
	0.206 +16%	191.1
Do not seed	0.480 0	100.0
	0.206	
	0.054 −16%	46.7
	−34%	16.3

Account Number	Existing Volume	Potential Additional Volume	Probability of Additional Volume	Expected Value of Potential	Existing Volume + Expected Value of Potential	Class
1	$15,000	$10,000	.25	$2,500	$17,500	C
2	40,000	0	—	—	40,000	B
3	20,000	10,000	.20			
4	50,000	10,000	.10			
5	5,000	50,000	.50			
6	0	100,000	.60			
7	30,000	20,000	.80			

*This example was supplied by James McDonald, Levi Strauss and Company, San Francisco.

*Figure, "The Decision to Seed Hurricanes," by R. A. Howard from *Science,* Vol. 176, pp. 1191–1202, Copyright 1972 by the American Association for the Advancement of Science and the author.

51. Social Science Mr. Statistics (a feature in *Fortune* magazine) investigated the claim of the United States Postal Service that 83% of first class mail in New York City arrives by the next day.* (The figure is 87% nationwide.) He mailed a letter to himself on ten consecutive days; only four were delivered by the next day.

*Daniel Seligman, "Ask Mr. Statistics," *Fortune,* July 24, 1995, pp. 170–171.

(a) Find the probability distribution for the number of letters delivered by the next day if the overall probability of next-day delivery is 83%.

(b) Using your answer to part (a), find the probability that four or fewer out of 10 letters would be delivered by the next day.

(c) Based on your answer to part (b), do you think it is likely that the 83% figure is accurate? Explain.

(d) Find the number of letters out of 10 that you would expect to be delivered by the next day if the 83% figure is accurate.

10.6 DECISION MAKING

John F. Kennedy once remarked he had assumed that as president it would be difficult to choose between distinct, opposite alternatives when a decision needed to be made. Actually, however, he found that such decisions were easy to make; the hard decisions came when he was faced with choices that were not as clear-cut. Most decisions fall in this last category—decisions which must be made under conditions of uncertainty. In the previous section we saw how to use expected values to help make a decision. These ideas are extended in this section, where we consider decision making in the face of uncertainty. Let us begin with an example.

EXAMPLE 1 Freezing temperatures are endangering the orange crop in central California. A farmer can protect his crop by burning smudge pots—the heat from the pots keeps the oranges from freezing. However, burning the pots is expensive; the cost is $4000. The farmer knows that if he burns smudge pots he will be able to sell his crop for a net profit (after smudge pot costs are deducted) of $10,000, provided that the freeze does develop and wipes out other orange crops in California. If he does nothing he will either lose $2000 already invested in the crop if it does freeze, or make a profit of $9600 if it does not freeze. (If it does not freeze, there will be a large supply of oranges, and thus his profit will be lower than if there was a small supply.) What should the farmer do?

He should begin by carefully defining the problem. First, he must decide on the **states of nature,** the possible alternatives over which he has no control. Here there are two: freezing temperatures, or no freezing temperatures. Next, the farmer should list the things he can control—his actions or **strategies.** He has two possible strategies; to use smudge pots or not. The consequences of each action under each state of nature, called **payoffs,** are summarized in a **payoff matrix,** as shown below. The payoffs in this case are the profits for each possible combination of events.

1 Explain how each of the following payoffs in the matrix were obtained.

(a) −$2000

(b) $10,000

Answers:
(a) If it freezes and smudge pots are not used, the farmer's profit is −$2000 for labor costs.

(b) If it freezes and smudge pots are used, the farmer makes a profit of $10,000.

States of Nature
Freeze   No Freeze

Strategies of Farmer — Use Smudge Pots / Do Not Use Pots

$$\begin{bmatrix} \$10{,}000 & \$5600 \\ -\$2000 & \$9600 \end{bmatrix}$$

To get the $5600 entry in the payoff matrix, use the profit if there is no freeze, $9600, and subtract the $4000 cost of using the pots. **1**

Once the farmer makes the payoff matrix, what then? The farmer might be an optimist (some might call him a gambler); in this case he might assume that the best will happen and go for the biggest number of the matrix ($10,000). For this profit, he must adopt the strategy "use smudge pots."

On the other hand, if the farmer is a pessimist, he would want to minimize the worst thing that could happen. If he uses smudge pots, the worst thing that could happen to him would be profit of $5600, which will result if there is no freeze. If he does not use smudge pots, he might face a loss of $2000. To minimize the worst, he once again should adopt the strategy "use smudge pots."

Suppose the farmer decides that he is neither an optimist nor a pessimist, but would like further information before choosing a strategy. For example, he might call the weather forecaster and ask for the probability of a freeze. Suppose the forecaster says that this probability is only .1. What should the farmer do? He should recall the discussion of expected value from the previous section and work out the expected profit for each of his two possible strategies. If the probability of a freeze is .1, then the probability that there is no freeze is .9. This information leads to the following expected values.

2 What should the farmer do if the probability of a freeze is .6? What is his expected profit?

Answer:
Use smudge pots; $8240

If smudge pots are used: $10{,}000(.1) + 5600(.9) = 6040$

If no smudge pots are used: $-2000(.1) + 9600(.9) = 8440$

Here the maximum expected profit, $8440, is obtained if smudge pots are not used. ■ **2**

As the example shows, the farmer's beliefs about the probabilities of a freeze affect his choice of strategies.

EXAMPLE 2 A small manufacturer of Christmas cards must decide in February about the type of cards she should emphasize in her fall line. She has three possible strategies: emphasize modern cards, emphasize old-fashioned cards, or a mixture of the two. Her success is dependent on the state of the economy in December. If the economy is strong, she will do well with her modern cards, while in a weak economy people long for the old days and buy old-fashioned cards. In an in-between economy, her mixture of lines would do the best. She first prepares a payoff matrix for all three possibilities. The numbers in the matrix represent her profits in thousands of dollars.

		States of Nature		
		Weak Economy	In-between	Strong Economy
	Modern	40	85	120
Strategies	Old-fashioned	106	46	83
	Mixture	72	90	68

(a) What would an optimist do?

If the manufacturer is an optimist, she should aim for the biggest number on the matrix, 120 (representing $120,000 in profit). Her strategy in this case would be to produce modern cards.

(b) How would a pessimist react?

A pessimistic manufacturer wants to find the best of the worst of all bad things that can happen. If she produces modern cards, the worst that can happen is a profit of $40,000. For old-fashioned cards, the worst is a profit of $46,000. From a mixture, the worst is a profit of $68,000. Her strategy here is to use a mixture.

3 Suppose the manufacturer reads another article which gives the following predictions: 35% chance of a weak economy, 25% chance of an in-between economy, and a 40% chance of a strong economy. What is the best strategy now? What is the expected profit?

(c) Suppose the manufacturer reads in a business magazine that leading experts believe there is a 50% chance of a weak economy at Christmas, a 20% chance of an in-between economy, and a 30% chance of a strong economy. How might she use this information?

The manufacturer can now find her expected profit for each possible strategy.

Modern:	$40(.50) + 85(.20) + 120(.30) = 73$
Old-fashioned:	$106(.50) + 46(.20) + 83(.30) = 87.1$
Mixture:	$72(.50) + 90(.20) + 68(.30) = 74.4$

Answer:
Modern; $83,250

Here the best strategy is old-fashioned cards; the expected profit is 87.1, or $87,100. ■ 3

10.6 EXERCISES

1. **Management** A developer has $100,000 to invest in land. He has a choice of two parcels (at the same price), one on the highway and one on the coast. With both parcels, his ultimate profit depends on whether he faces light opposition from environmental groups or heavy opposition. He estimates that the payoff matrix is as follows (the numbers represent his profit).

	Opposition	
	Light	Heavy
Highway	$70,000	$30,000
Coast	$150,000	−$40,000

What should the developer do if he is
(a) an optimist? **(b)** a pessimist?
(c) Suppose the probability of heavy opposition is .8. What is his best strategy? What is the expected profit?
(d) What is the best strategy if the probability of heavy opposition is only .4?

2. Hillsdale College has sold out all tickets for a jazz concert to be held in the stadium. If it rains, the show will have to be moved to the gym, which has a much smaller seating capacity. The dean must decide in advance whether to set up the seats and the stage in the gym or in the stadium, or both, just in case. The payoff matrix below shows the net profit in each case.

	States of Nature	
	Rain	No Rain
Set up in Stadium	−$1550	$1500
Strategies Set up in Gym	$1000	$1000
Set up Both	$750	$1400

What strategy should the dean choose if she is
(a) an optimist? **(b)** a pessimist?

(c) If the weather forecaster predicts rain with a probability of .6, what strategy should she choose to maximize expected profit? What is the maximum expected profit?

3. **Management** An analyst must decide what fraction of the items produced by a certain machine are defective. He has already decided that there are three possibilities for the fraction of defective items: .01, .10, and .20. He may recommend two courses of action: repair the machine or make no repairs. The payoff matrix below represents the *costs* to the company in each case, in hundreds of dollars.

		States of Nature		
		.01	.10	.20
Strategies	Repair	130	130	130
	No Repair	25	200	500

What strategy should the analyst recommend if he is
(a) an optimist? **(b)** a pessimist?
(c) Suppose the analyst is able to estimate probabilities for the three states of nature as follows.

Fraction of Defectives	*Probability*
.01	.70
.10	.20
.20	.10

Which strategy should he recommend? Find the expected cost to the company if this strategy is chosen.

4. **Management** The research department of the Allied Manufacturing Company has developed a new process which it believes will result in an improved product. Management must decide whether or not to go ahead and

market the new product. The new product may be better than the old or it may not be better. If the new product is better and the company decides to market it, sales should increase by $50,000. If it is not better and they replace the old product with the new product on the market, they will lose $25,000 to competitors. If they decide not to market the new product they will lose $40,000 if it is better, and research costs of $10,000 if it is not.

(a) Prepare a payoff matrix.

(b) If management believes the probability that the new product is better to be .4, find the expected profits under each strategy and determine the best action.

5. Management A businessman is planning to ship a used machine to his plant in Nigeria. He would like to use it there for the next 4 years. He must decide whether or not to overhaul the machine before sending it. The cost of overhaul is $2600. If the machine fails when in operation in Nigeria, it will cost him $6000 in lost production and repairs. He estimates the probability that it will fail at .3 if he does not overhaul it, and .1 if he does overhaul it. Neglect the possibility that the machine might fail more than once in the 4 years.

(a) Prepare a payoff matrix.

(b) What should the businessman do to minimize his expected costs?

6. Management A contractor prepares to bid on a job. If all goes well, his bid should be $30,000, which will cover his costs plus his usual profit margin of $4500. However, if a threatened labor strike actually occurs, his bid should be $40,000 to give him the same profit. If there is a strike and he bids $30,000, he will lose $5500. If his bid is too high, he may lose they job entirely, while if it is too low, he may lose money.

(a) Prepare a payoff matrix.

(b) If the contractor believes that the probability of a strike is .6, how much should he bid?

7. Natural Science A community is considering an anti-smoking campaign.* The city council will choose one of three possible strategies: a campaign for everyone over age 10 in the community, a campaign for youths only, or no campaign at all. The two states of nature are a true cause-effect relationship between smoking and cancer and no cause-effect relationship. The costs to the community (including loss of life and productivity) in each case are as shown at the top of the next column.

	States of Nature	
Strategies	Cause-Effect Relationship	No Cause-Effect Relationship
Campaign for All	$100,000	$800,000
Campaign for Youth	$2,820,000	$20,000
No Campaign	$3,100,100	$0

What action should the city council choose if it is

(a) optimistic? **(b)** pessimistic?

(c) If the Director of Public Health estimates that the probability of a true cause-effect relationship is .8, which strategy should the city council choose?

8. Management An investor has $20,000 to invest in stocks. She has two possible strategies: buy conservative blue-chip stocks or buy highly speculative stocks. There are two states of nature: the market goes up or the market goes down. The following payoff matrix shows the net amounts she will have under the various circumstances.

	Market Up	Market Down
Buy Blue-Chip	$25,000	$18,000
Buy Speculative	$30,000	$11,000

What should the investor do if she is

(a) an optimist? **(b)** a pessimist?

(c) Suppose there is a .7 probability of the market going up. What is the best strategy? What is the expected profit?

(d) What is the best strategy if the probability of a market rise is .2?

Sometimes the numbers (or payoffs) in a payoff matrix do not represent money (profits or costs, for example), but utility. A utility *is a number that measures the satisfaction (or lack of it) that results from a certain action. The numbers must be assigned by each individual, depending on how he or she feels about a situation. For example, one person might assign a utility of +20 for a week's vacation in San Francisco, with −6 being assigned if the vacation were moved to Sacramento. Work the following problems in the same way as those above.*

9. Social Science A politician must plan her reelection strategy. She can emphasize jobs or she can emphasize the environment. The voters can be concerned about jobs or about the environment. A payoff matrix showing the utility of each possible outcome is shown below.

		Voters	
		Jobs	Environment
Candidate	Jobs	+25	−10
	Environment	−15	+30

The political analysts feel that there is a .35 chance that the voters will emphasize jobs. What strategy should the candidate adopt? What is its expected utility?

10. In an accounting class, the instructor permits the students to bring a calculator or a reference book (but not both) to an examination. The examination itself can emphasize either problems or definitions. In trying to decide which aid to take to an examination, a student first decides on the utilities shown in the following payoff matrix.

$$
\begin{array}{c}
\text{\textit{Exam Emphasizes}} \\
\begin{array}{cc}
\text{Numbers} & \text{Definitions}
\end{array}
\end{array}
$$

$$
\textit{Student Chooses}\quad
\begin{array}{l}
\text{Calculator} \\
\text{Book}
\end{array}
\begin{bmatrix}
+50 & 0 \\
+10 & +40
\end{bmatrix}
$$

(a) What strategy should the student choose if the probability that the examination will emphasize numbers is .6? What is the expected utility in this case?

(b) Suppose the probability that the examination emphasizes numbers is .4. What strategy should be chosen by the student?

CHAPTER 10 SUMMARY

Key Terms and Symbols

10.1
$n!$ n factorial
multiplication principle
permutations
combinations

10.3
Bernoulli trials (processes)
binomial experiment

10.4
stochastic processes
Markov chain
state
transition diagram
transition matrix
probability vector
equilibrium vector (fixed vector)
regular transition matrix
regular Markov chain

10.5
random variable
frequency distribution
relative frequency
probability distribution
probability function (probability
    distribution function)
histogram
weighted average
expected value
fair game

10.6
states of nature
strategies
payoffs
payoff matrix

Key Concepts

Multiplication Principle: If there are m_1 ways to make a first choice, m_2 ways to make a second choice, and so on, where none of the choices depend on any of the others, then there are $m_1 m_2 \cdots m_n$ different ways to make the entire sequence of choices.

The number of **permutations** of n elements taken r at a time is $\displaystyle {}_nP_r = \frac{n!}{(n-r)!}$.

The number of **combinations** of n elements taken r at a time is

$$
{}_nC_r = \binom{n}{r} = \frac{n!}{(n-r)!\,r!}.
$$

Binomial experiments have the following characteristics: The same experiment is repeated several times. There are only *two* outcomes, labeled success and failure. The trials are independent so that the probability of success is the same for each trial. If the probability of success in a single trial is p, then the probability of x successes in n trials is

$$
\binom{n}{x} p^x (1-p)^{n-x}.
$$

> **Markov Chains:**   A **transition matrix** must be square, with all entries between 0 and 1 inclusive, and the sum of the entries in any row must be 1. A Markov chain is *regular* if some power of its transition matrix P contains all positive entries. The long-range probabilities for a regular Markov chain are given by the **equilibrium** or **fixed vector** V, where for any initial probability vector v, the products vP^n approach V as n gets larger and larger, and $VP = V$. To find V, solve the system of equations formed by $VP = V$ and the fact that the sum of the entries of V is 1.

> **Expected Value of a Probability Distribution:**   For a random variable x with values $x_1, x_2, \ldots, x_n$ and probabilities $p_1, p_2, \ldots, p_n$, respectively, the expected value is
> $$E(x) = x_1 p_1 + x_2 p_2 + \cdots + x_n p_n.$$

> **Decision Making: A payoff matrix** which includes all available strategies and states of nature is used in decision making to define the problem and the possible solutions. The expected value of each strategy can help to determine the best course of action.

Chapter 10 Review Exercises

1. In how many ways can 5 shuttle vans line up at the airport?

2. How many variations in first-, second-, and third-place finishes are possible in a 100-yard dash with 7 runners?

3. In how many ways can a sample of 3 pears be taken from a basket containing a dozen pears?

4. If 2 of the pears in Exercise 3 are spoiled, in how many ways can the sample of 3 include the following?
(a) 1 spoiled pear   (b) No spoiled pears
(c) At most 1 spoiled pear

5. In how many ways can 3 pictures, selected from a group of 6 pictures, be arranged in a row on a wall?

6. In how many ways can the 6 pictures in Exercise 5 be arranged in a row, if a certain one must be first?

7. In how many ways can the 6 pictures in Exercise 5 be arranged if 3 are landscapes and 3 are puppies, and if
(a) like types must be kept together?
(b) landscapes and puppies are alternated?

8. A representative is to be selected from each of 3 departments in a large company. There are 7 people in the first department, 5 in the second department, and 8 in the third department.
(a) How many different groups of 3 representatives are possible?
(b) How many groups are possible, if any number (at least 1) up to 3 representatives can form a group?

9. Explain under what circumstances a permutation should be used in a probability problem, and under what circumstances a combination should be used.

10. Discuss under what circumstances the binomial probability formula should be used in a probability problem.

Suppose a family plans to have 4 children, and the probability that a particular child is a boy is 1/2. Find the probability that the family will have the following.

11. Exactly 2 boys

12. All girls

13. At least 2 boys

14. No more than 3 girls

Suppose 2 cards are drawn without replacement from an ordinary deck of 52 cards. Find the probabilities of the following results.

15. Both cards are black.

16. Both cards are hearts.

17. Exactly 1 is a face card.

18. At most 1 is an ace.

A collection of golf balls contains 4 yellow, 2 blue, and 6 white balls. A golfer selects 3 balls at random. Find the probability that the selection includes the following.

19. All white balls

20. All yellow balls

21. At least 1 blue ball

22. 1 ball of each color

Decide whether each of the following is a regular transition matrix.

23. $\begin{bmatrix} 0 & 1 \\ .8 & .2 \end{bmatrix}$

24. $\begin{bmatrix} -.1 & .4 \\ .3 & .7 \end{bmatrix}$

25. $\begin{bmatrix} .3 & 0 & .7 \\ .5 & .1 & .4 \\ 1 & 0 & 0 \end{bmatrix}$

26. $\begin{bmatrix} .2 & .3 & .5 \\ .4 & .4 & .2 \\ .5 & .1 & .4 \end{bmatrix}$

Management   *A bottle capping machine has an error rate of .01. A random sample of 20 bottles is selected. Find the following probabilities.*

27. Exactly 4 bottles are improperly capped.

28. No more than 3 bottles are improperly capped.

29. At least 1 bottle is improperly capped.

30. Management A credit card company classifies its customers in three groups: nonusers in a given month, light users, and heavy users. The transition matrix for these states is

	Nonuser	Light	Heavy
Nonuser	.8	.15	.05
Light	.25	.55	.2
Heavy	.04	.21	.75

Suppose the initial distribution for the three states is [.4   .4   .2]. Find the distribution after
(a) 1 month;   **(b)** 2 months.
(c) What is the long-range prediction for the distribution of users?

31. Management Savmor Investments starts a heavy advertising campaign. At the start of the campaign, Savmor sells 35% of all mutual funds sold in the area, while 65% are sold by the Highrate Company. The campaign produces the following transition matrix.

		After Campaign	
		Savmor	Highrate
Before Campaign	Savmor	.8	.2
	Highrate	.4	.6

(a) Find the market share for each company after the campaign.
(b) Find the share of the market for each company after three such campaigns.
(c) Predict the long-range market share for Savmor.

In Exercises 32–36, **(a)** *give a probability distribution, and* **(b)** *sketch its histogram.*

32.

x	6	7	8	9	10
Frequency	3	7	9	3	2

33.

x	1	2	3	4	5	6
Frequency	1	0	2	5	8	4

34. A coin is tossed 3 times and the number of heads is recorded.

35. A pair of dice are rolled and the sum of the results for each roll is recorded.

36. Natural Science Patients in groups of 5 are given a new cancer treatment. The experiment is repeated 10 times with the following results.

Number with Significant Improvement	Frequency
0	1
1	1
2	2
3	3
4	3
5	0
Total:	10

In Exercises 37 and 38, give the probability that corresponds to the shaded region of each histogram.

37.

38.

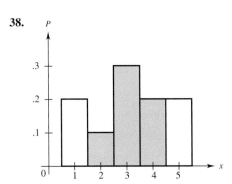

Solve the following problems.

39. Suppose someone offers to pay you $100 if you draw 3 cards from a standard deck of 52 cards and all the cards are clubs. What should you pay for the chance to win if it is a fair game?

40. You pay $6 to play in a game where you will roll a die, with payoffs as follows: $8 for a 6, $7 for a 5, and $4 for any other results. What are your expected winnings? Is the game fair?

41. A lottery has a first prize of $5000, two second prizes of $1000 each, and two $100 third prizes. A total of 10,000 tickets is sold, at $1 each. Find the expected winnings of a person buying 1 ticket.

42. Find the expected number of girls in a family of 5 children.

43. Three cards are drawn from a standard deck of 52 cards.
(a) What is the expected number of aces?
(b) What is the expected number of clubs?

44. Management In labor-management relations, both labor and management can adopt either a friendly or a hostile attitude. The results are shown in the following payoff matrix. The numbers give the wage gains made by an average worker.

$$\begin{array}{c} \qquad\qquad\quad \textit{Management} \\ \qquad\qquad \text{Friendly} \quad \text{Hostile} \\ \textit{Labor} \begin{array}{c} \text{Friendly} \\ \text{Hostile} \end{array} \left[\begin{array}{cc} \$600 & \$800 \\ \$400 & \$950 \end{array} \right] \end{array}$$

(a) Suppose the chief negotiator for labor is an optimist. What strategy should he choose?

(b) What strategy should he choose if he is a pessimist?

(c) The chief negotiator for labor feels that there is a 70% chance that the company will be hostile. What strategy should he adopt? What is the expected payoff?

(d) Just before negotiations begin, a new management is installed in the company. There is only a 40% chance that the new management will be hostile. What strategy should be adopted by labor?

45. Social Science A candidate for city council can come out in favor of a new factory, be opposed to it, or waffle on the issue. The change in votes for the candidate depends on what her opponent does, with payoffs as shown.

$$\begin{array}{c} \qquad\qquad\qquad\qquad \textit{Opponent} \\ \qquad\qquad \text{Favors} \quad \text{Waffles} \quad \text{Opposes} \\ \textit{Candidate} \begin{array}{c} \text{Favors} \\ \text{Waffles} \\ \text{Opposes} \end{array} \left[\begin{array}{ccc} 0 & -1000 & -4000 \\ 1000 & 0 & -500 \\ 5000 & 2000 & 0 \end{array} \right] \end{array}$$

(a) What should the candidate do if she is an optimist?

(b) What should she do if she is a pessimist?

(c) Suppose the candidate's campaign manager feels there is a 40% chance that the opponent will favor the plant, and a 35% chance that he will waffle. What strategy should the candidate adopt? What is the expected change in the number of votes?

(d) The opponent conducts a new poll which shows strong opposition to the new factory. This changes the probability he will favor the factory to 0 and the probability he will waffle to .7. What strategy should our candidate adopt? What is the expected change in the number of votes now?

*Exercises 46 and 47 are taken from actuarial examinations given by the Society of Actuaries.**

46. Management A company is considering the introduction of a new product that is believed to have probability .5 of being successful and probability .5 of being unsuccessful. Successful products pass quality control 80% of the time.

*Problem from "Course 130 Examination, Operations Research," of the *Education and Examination Committee of the Society of Actuaries.* Reprinted by permission of the Society of Actuaries.

Unsuccessful products pass quality control 25% of the time. If the product is successful, the net profit to the company will be $40 million; if unsuccessful, the net loss will be $15 million. Determine the expected net profit if the product passes quality control.

(a) $23 million (b) $24 million
(c) $25 million (d) $26 million
(e) $27 million

47. Management A merchant buys boxes of fruit from a grower and sells them. Each box of fruit is either Good or Bad. A Good box contains 80% excellent fruit and will earn $200 profit on the retail market. A Bad box contains 30% excellent fruit and will produce a loss of $1000. The a priori probability of receiving a Good box of fruit is .9. Before the merchant decides to put the box on the market, he can sample one piece of fruit to test whether it is excellent. Based on that sample, he has the option of rejecting the box without paying for it. Determine the expected value of the right to sample. (*Hint:* If the merchant samples the fruit, what are the probabilities of accepting a Good box, accepting a Bad box, and not accepting the box? What are these probabilities if he does not sample the fruit?)

(a) 0   (b) $16   (c) $34
(d) $72   (e) $80

48. Management The March 1982 issue of *Mathematics Teacher* included "Overbooking Airline Flights," an article by Joe Dan Austin. In this article, Austin developed a model for the expected income for an airline flight. With appropriate assumptions, the probability that exactly x of n people with reservations show up at the airport to buy a ticket is given by the binomial probability formula. Assume the following: 6 reservations have been accepted for 3 seats, $p = .6$ is the probability that a person with a reservation will show up, a ticket costs $100, and the airline must pay $100 to anyone with a reservation who does not get a ticket. Complete the following table.

Number Who Show Up (x)	0	1	2	3	4	5	6
Airline's Income							
P(x)							

(a) Use the table to find $E(I)$, the expected airline income from the 3 seats.

(b) Find $E(I)$ for $n = 3$, $n = 4$, and $n = 5$. Compare these answers with $E(I)$ for $n = 6$. For these values of n, how many reservations should the airline book for the 3 seats in order to maximize the expected revenue?

CASE 10

Optimal Inventory for a Service Truck

For many different items it is difficult or impossible to take the item to a central repair facility when service is required. Washing machines, large television sets, office copiers, and computers are only a few examples of such items. Service for items of this type is commonly performed by sending a repair person to the item, with the person driving to the location in a truck containing various parts that might be required in repairing the item. Ideally, the truck should contain all the parts that might be required. However, most parts would be needed only infrequently, so that inventory costs for the parts would be high.

An optimum policy for deciding on which parts to stock on a truck would require that the probability of not being able to repair an item without a trip back to the warehouse for needed parts be as low as possible, consistent with minimum inventory costs. An analysis similar to the one below was developed at the Xerox Corporation.*

To set up a mathematical model for deciding on the optimum truck-stocking policy, let us assume that a broken machine might require one of 5 different parts (we could assume any number of different parts—we use 5 to simplify the notation). Suppose also that the probability that a particular machine requires part 1 is p_1; that it requires part 2 is p_2; and so on. Assume also that failures of different part types are independent, and that at most one part of each type is used on a given job.

Suppose that, on the average, a repair person makes N service calls per time period. If the repair person is unable to make a repair because at least one of the parts is unavailable, there is a penalty cost, L, corresponding to wasted time for the repair person, an extra trip to the parts depot, customer unhappiness, and so on. For each of the parts carried on the truck, an average inventory cost is incurred. Let H_i be the average inventory cost for part i, where $1 \le i \le 5$.

Let M_1 represent a policy of carrying only part 1 on the repair truck, M_{24} represent a policy of carrying only parts 2 and 4, with M_{12345} and M_0 representing policies of carrying all parts and no parts, respectively.

For policy M_{35}, carrying parts 3 and 5 only, the expected cost per time period per repair person, written $C(M_{35})$, is

$$C(M_{35}) = (H_3 + H_5) + NL[1 - (1 - p_1)(1 - p_2)(1 - p_4)].$$

(The expression in brackets represents the probability of needing at least one of the parts not carried, 1, 2, or 4 here.) As further examples,

$$C(M_{125}) = (H_1 + H_2 + H_5) + NL[1 - (1 - p_3)(1 - p_4)],$$

while

$$C(M_{12345}) = (H_1 + H_2 + H_3 + H_4 + H_5) + NL[1 - 1]$$
$$= H_1 + H_2 + H_3 + H_4 + H_5,$$

and

$$C(M_0) = NL[1 - (1 - p_1)(1 - p_2)(1 - p_3)(1 - p_4)(1 - p_5)].$$

To find the best policy, evaluate $C(M_0)$, $C(M_1)$, . . . , $C(M_{12345})$, and choose the smallest result. (A general solution method is in the *Management Science* paper.)

EXAMPLE Suppose that for a particular item, only 3 possible parts might need to be replaced. By studying past records of failures of the item, and finding necessary inventory costs, suppose that the following values have been found.

p_1	p_2	p_3		H_1	H_2	H_3
.09	.24	.17		$15	$40	$9

Suppose $N = 3$ and L is $54. Then, as an example,

$$C(M_1) = H_1 + NL[1 - (1 - p_2)(1 - p_3)]$$
$$= 15 + 3(54)[1 - (1 - .24)(1 - .17)]$$
$$= 15 + 3(54)[1 - (.76)(.83)]$$
$$\approx 15 + 59.81$$
$$= 74.81.$$

*Reprinted by permission of Stephen Smith, John Chambers, and Eli Shlifer. "Optimal Inventories Based on Job Completion Rate for Repairs Requiring Multiple Items." *Management Science*, Vol. 26, No. 8, August 1980, copyright © 1980 by The Institute of Management Sciences.

Thus, if policy M_1 is followed (carrying only part 1 on the truck), the expected cost per repair person per time period is $74.81. Also,

$$C(M_{23}) = H_2 + H_3 + NL[1 - (1 - p_1)]$$
$$= 40 + 9 + 3(54)(.09)$$
$$= 63.58,$$

so that M_{23} is a better policy than M_1. By finding the expected values for all other possible policies (see the exercises), the optimum policy may be chosen. ■

EXERCISES

1. Refer to the example and find each of the following.
   (a) $C(M_0)$ (b) $C(M_2)$ (c) $C(M_3)$
   (d) $C(M_{12})$ (e) $C(M_{13})$ (f) $C(M_{123})$

2. Which policy leads to the lowest expected cost?

3. In the example, $p_1 + p_2 + p_3 = .09 + .24 + .17 = .50$. Why is it not necessary that the probabilities add up to 1?

4. Suppose an item to be repaired might need one of n different parts. How many different policies would then need to be evaluated?

CHAPTER 11

Introduction to Statistics

Mean! 8.8
Median! 8
Mode! 5, 0, 13, 10
Range! 13

Standard Deviation:
stuff

13, 5, 13, 10, 5, 3, 14, 6, 4
88, 16
19, 25
9, 15, 25, 25, 36, 64, 64, 100, 169, 169
3, 4, 5, 5, 6, 8, 8, 10, 13, 13, 14
16

8.9 | 1260-1213
         132

212 | N 8 5
        9 6
        9

23
8

31
8

39

49
13   2 56   9 38   100
62   196   64   1113
13   352   1000
75   169   36   1213
89   521   1030
16   169   25   108
105   800   1063   12
       25
      084   210
     1080   1050
430   16

47
____
132

12·105 - 1213
_____
   12·11

$S = \sqrt{\dfrac{12 \cdot 105 - 1213}{12 \cdot 11}}$

12   1256   d7
11   1213   ___
      ____   132
        47

Statistics is a branch of mathematics that deals with the collection and summarization of data. Methods of statistical analysis make it possible to draw conclusions about a population based on data from a sample of the population. Statistical models have become increasingly useful in manufacturing, government, agriculture, medicine, and the social sciences, and in all types of research. An Indianapolis race-car team is using statistics to improve performance by gathering data on each run around the track. They sample data 300 times a second, and use computers to process the data. In this chapter we give a brief introduction to some of the key topics from statistical theory.

In the previous chapter, we saw that a frequency distribution can be transformed into a probability distribution by using the relative frequency of each value of the random variable as its probability. Sometimes it is convenient to work directly with a frequency distribution.

11.1 FREQUENCY DISTRIBUTIONS; MEASURES OF CENTRAL TENDENCY

Often, a researcher wishes to learn something about a characteristic of a population, but because the population is very large or mobile, it is not possible to examine all of its elements. Instead, a limited sample drawn from the population is studied to determine the characteristics of the population. For example, a book by Frances Cerra Whittelsey, *Why Women Pay More,* published by Ralph Nader's Center for Responsive Law, documents how women are charged more than men for the same service. In the studies cited in this book, the population is U.S. women. The studies involved data collected from a sample of U.S. women.

For these inferences to be correct, the sample chosen must be a **random sample.** Random samples are representative of the population because they are chosen so that every element of the population is equally likely to be selected. For example, a hand dealt from a well-shuffled deck of cards is a random sample.

After a sample has been chosen and all data of interest are collected, the data must be organized so that conclusions may be more easily drawn. One method of organization is to group the data into intervals; equal intervals are usually chosen.

EXAMPLE 1 A survey asked 30 business executives how many college units in management each had. The results are shown below. Group the data into intervals and find the frequency of each interval.

3	25	22	16	0	9	14	8	34	21
15	12	9	3	8	15	20	12	28	19
17	16	23	19	12	14	29	13	24	18

The highest number in the list is 34 and the lowest is 0; one convenient way to group the data is in intervals of size 5, starting with 0–4 and ending with 30–34. This gives an interval for each number in the list and results in seven equal intervals of a convenient size. Too many intervals of smaller size would not simplify the data enough, while too few intervals of larger size would conceal information that the data might provide. A rule of thumb is to use from six to fifteen intervals.

First tally the number of college units falling into each interval. Then total the tallies in each interval, as in the table below. This table is an example of a **grouped frequency distribution.**

College Units	Tally	Frequency
0–4	III	3
5–9	IIII	4
10–14	JHI I	6
15–19	JHI III	8
20–24	JHI	5
25–29	III	3
30–34	I	1
	Total:	30

■ **1**

The frequency distribution in Example 1 shows information about the data that might not have been noticed before. For example, the interval with the largest number of units is 15–19, and 19 executives (more than half) had between 9 and 25 units. Also, the frequency in each interval increases rather evenly (up to 8) and then decreases at about the same pace. However, some information has been lost; for example, we no longer know how many executives had 12 units.

The information in a grouped frequency distribution can be displayed in a histogram similar to the histograms for probability distributions in the previous chapter. The intervals determine the widths of the bars; if equal intervals are used, all the bars have the same width. The heights of the bars are determined by the frequencies.

A **frequency polygon** is another form of graph that illustrates a grouped frequency distribution. The polygon is formed by joining consecutive midpoints of the tops of the histogram bars with straight line segments. Sometimes the midpoints of the first and last bars are joined to endpoints on the horizontal axis where the next midpoint would appear. (See Figure 11.1 on the next page.)

EXAMPLE 2 A grouped frequency distribution of college units was found in Example 1. Draw a histogram and a frequency polygon for this distribution.

First, draw a histogram, shown in black in Figure 11.1. To get a frequency polygon, connect consecutive midpoints of the tops of the bars. The frequency polygon is shown in color. ■ **2**

1 An accounting firm selected 24 complex tax returns prepared by a certain tax preparer. The number of errors per return were as follows.

8	12	0	6	10	8	0	14
8	12	14	16	4	14	7	11
9	12	7	15	11	21	22	19

Prepare a grouped frequency distribution for this data. Use intervals 0–4, 5–9, and so on.

Answer:

Interval	Frequency
0–4	3
5–9	7
10–14	9
15–19	3
20–24	2
	Total: 24

2 Make a histogram and a frequency polygon for the distribution found in Side Problem 1 above.

Answer:

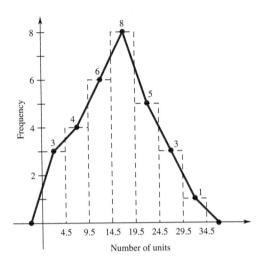

FIGURE 11.1

TECHNOLOGY TIP As noted in Section 10.5, most graphing calculators can display histograms. Many will also display frequency polygons (which are usually labeled LINE or xyLINE in calculator menus). When dealing with grouped frequency distributions, however, certain adjustments must be made on a calculator.

1. *A calculator list of outcomes must consist of single numbers, not intervals.* The table in Example 1, for example, cannot be entered as shown. To convert the first column of the table for calculator use, choose one number in each interval, say 2 in the interval 0–4, 7 in the interval 5–9, 12 in the interval 10–14, etc. Then use 2, 7, 12, . . . as the list of outcomes to be entered into the calculator. The frequency list (the last column of the table) remains the same.

2. *The histogram bar width affects the shape of the graph.* If you use a bar width of 4 in Example 1, the calculator may produce a histogram with gaps in it. To avoid this use the interval $0 \leq x < 5$ in place of $0 \leq x \leq 4$, and similarly for the other intervals and make 5 the bar width.

Following this procedure, we obtain the calculator-generated histogram and frequency polygon in Figure 11.2 for the data from Example 1. Note that the width of each histogram bar is 5. Some calculators cannot display both the histogram and the frequency polygon on the same screen as is done here. ✔

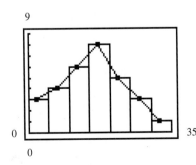

FIGURE 11.2

N O T E The remainder of this section deals with topics that are generally referred to as "measures of central tendency." Computing these various measures is greatly simplified by the statistical capabilities of most scientific and graphing calculators. Calculators vary considerably in how data is entered, so read your instruction manual to learn how to enter lists of data and the corresponding frequencies. On scientific calculators with statistical capabilities, there are keys for finding most of the measures of central tendency discussed below. On graphing calculators, most or all of these measures can be obtained with a single keystroke (look for *one-variable statistics,* which is often labeled 1-VAR, in the STAT menu or its CALC submenu). ◆

MEAN   The average value of a probability distribution is the expected value of the distribution. Three measures of central tendency, or "averages," are used with frequency distributions: the mean, the median, and the mode. The most important of these is the mean, which is similar to the expected value of a probability distribution. The **mean** (the arithmetic average) of a set of numbers is the sum of the numbers, divided by the total number of numbers. We write the sum of n numbers x_1, x_2, $x_3, \ldots, x_n$ in a compact way using **summation notation,** also called **sigma notation.** With the Greek letter Σ (sigma), the sum

$$x_1 + x_2 + x_3 + \cdots + x_n$$

is written

$$x_1 + x_2 + x_3 + \cdots + x_n = \sum_{i=1}^{n} x_i.$$

In statistics, $\sum_{i=1}^{n} x_i$ is often abbreviated as just Σx. The symbol $\bar{x}$ (read "x-bar") is used to represent the mean of a sample.

MEAN

The mean of the n numbers $x_1, x_2, x_3, \ldots, x_n$ is

$$\bar{x} = \frac{x_1 + x_2 + \cdots + x_n}{n} = \frac{\Sigma x}{n}.$$

EXAMPLE 3   The number of bankruptcy petitions (in thousands) filed in the United States in the fiscal years 1988–1993 are given in the table on the next page.* Find the mean number of bankruptcy petitions filed annually during this period.

*Administrative Offices of the U.S. Courts, *Annual Report of the Director.*

Year	Petitions Filed
1988	594
1989	643
1990	725
1991	880
1992	973
1993	919

3 Find the mean of the following list of sales at a boutique.

$25.12	$42.58
$76.19	$32
$81.11	$26.41
$19.76	$59.32
$71.18	$21.03

Answer:
$45.47

Let $x_1 = 594$, $x_2 = 643$, and so on. Here, $n = 6$, since there are 6 numbers in the list.

$$\bar{x} = \frac{594 + 643 + 725 + 880 + 973 + 919}{6} = 789$$

The mean number of bankruptcy petitions filed during the given years is 789,000. ■ **3**

TECHNOLOGY TIP The mean of the six numbers in Example 3 is easily found by using the $\bar{x}$ key on a scientific calculator or the one-variable statistics key on a graphing calculator. A graphing calculator will also display additional information, which will be discussed in the next section. ✔

The mean of data that have been arranged in a frequency distribution is found in a similar way. For example, suppose the following data are collected.

Value	Frequency
84	2
87	4
88	7
93	4
99	3
	Total: 20

The value 84 appears twice, 87 four times, and so on. To find the mean, first add 84 two times, 87 four times, and so on; or get the same result faster by multiplying 84 by 2, 87 by 4, and so on, and then by adding the results. Dividing the sum by 20, the total of the frequencies, gives the mean.

$$\bar{x} = \frac{(84 \cdot 2) + (87 \cdot 4) + (88 \cdot 7) + (93 \cdot 4) + (99 \cdot 3)}{20}$$

$$= \frac{168 + 348 + 616 + 372 + 297}{20}$$

$$= \frac{1801}{20}$$

$$\bar{x} = 90.05$$

Verify that your calculator gives the same result.

EXAMPLE 4 Find the mean for the data shown in the following frequency distribution.

Value	Frequency	Value × Frequency
30	6	30 · 6 = 180
32	9	32 · 9 = 288
33	7	33 · 7 = 231
37	12	37 · 12 = 444
42	6	42 · 6 = 252
	Total: 40	Total: 1395

4 Find $\bar{x}$ for the following frequency distribution.

Value	Frequency
7	2
9	3
11	6
13	4
15	1
17	4

Answer:
$\bar{x} = 12.1$

A new column, "Value × Frequency," has been added to the frequency distribution. Adding the products from this column gives a total of 1395. The total from the frequency column is 40. The mean is

$$\bar{x} = \frac{1395}{40} = 34.875. \quad \blacksquare \quad \boxed{4}$$

N O T E The mean in Example 4 is found in the same way as was the expected value of a probability distribution in the previous chapter. In fact, the words *mean* and *expected value* are often used interchangeably. ◆

The mean of grouped data is found in a similar way. For grouped data, intervals are used rather than single values. To calculate the mean, it is assumed that all these values are located at the midpoint of the interval. The letter x is used to represent the midpoints and f represents the frequencies, as shown in the next example.

EXAMPLE 5 Find the mean for the following grouped frequency distribution.

Interval	Midpoint, x	Frequency, f	Product, xf
40–49	44.5	2	89
50–59	54.5	4	218
60–69	64.5	7	451.5
70–79	74.5	9	670.5
80–89	84.5	5	422.5
90–99	94.5	3	283.5
		Total: 30	Total: 2135

A column for the midpoint of each interval has been added. The numbers in this column are found by adding the endpoints of each interval and dividing by 2. For the interval 40–49, the midpoint is (40 + 49)/2 = 44.5. The numbers in the product column on the right are found by multiplying frequencies and corresponding midpoints. Finally, we divide the total of the product column by the total of the frequency column to get

$$\bar{x} = \frac{2135}{30} = 71.2 \text{ (to the nearest tenth)}. \quad \blacksquare$$

The formula for the **mean of a grouped frequency distribution** is given below.

5 Find the mean of the following grouped frequency distribution.

Interval	Frequency
0–4	6
5–9	4
10–14	7
15–19	3

Answer:
8.75

> ## MEAN OF A GROUPED DISTRIBUTION
>
> The mean of a distribution where x represents the midpoints, f the frequencies, and $n = \Sigma f$, is
>
> $$\bar{x} = \frac{\Sigma(xf)}{n}.$$

CAUTION Note that in the formula above, n is the sum of the frequencies in the entire data set, not the number of intervals. ◆ **5**

MEDIAN Asked by a reporter to give the average height of the players on his team the Little League coach lined up his 15 players by increasing height. He picked out the player in the middle and pronounced this player to be of average height. This kind of average, called the **median,** is defined as the middle entry in a set of data arranged in either increasing or decreasing order. If there is an even number of entries, the median is defined to be the mean of the two center entries. The following table shows how to find the median for two sets of data: {8, 7, 4, 3, 1} and {2, 3, 4, 7, 9, 12}.

Odd Number of Entries	Even Number of Entries
8	2
7	3
Median = 4	4 ⎫
3	7 ⎬ Median = $\frac{4+7}{2}$ = 5.5
1	9 ⎭
	12

N O T E As shown in the table above, when there are an even number of entries, the median is not always equal to one of the data entries. ◆

The procedure for finding the median of a grouped frequency distribution is more complicated. We omit it here because it is more common to find the mean when working with grouped frequency distributions.

EXAMPLE 6 Find the median for the following lists of numbers.

(a) 11, 12, 17, 20, 23, 28, 29

The median is the middle number, in this case 20. (Note that the numbers are already arranged in numerical order.) In this list, three numbers are smaller than 20 and three are larger.

(b) 15, 13, 7, 11, 19, 30, 39, 5, 10

First arrange the numbers in numerical order, from smallest to largest.

$$5, \quad 7, \quad 10, \quad 11, \quad 13, \quad 15, \quad 19, \quad 30, \quad 39$$

The middle number can now be determined; the median is 13.

6 Find the median for each of the following lists of numbers.

(a) 12, 15, 17, 19, 35, 42, 58

(b) 28, 68, 7, 15, 47, 59, 13, 74, 32, 25

Answers:
(a) 19
(b) 30

(c) 47, 59, 32, 81, 74, 153

Write the numbers in numerical order.

$$32, \quad 47, \quad 59, \quad 74, \quad 81, \quad 153$$

There are six numbers here; the median is the mean of the two middle numbers, or

$$\text{Median} = \frac{59 + 74}{2} = \frac{133}{2} = 66.5. \quad \blacksquare$$

CAUTION Remember, the data must be arranged in numerical order before locating the median. ◆ **6**

TECHNOLOGY TIP Many graphing calculators (including TI-82/83/86, HP-38, Casio 9800/9850) display the median when doing one-variable statistics. You may have to scroll down to a second screen to find it. ✔

In some situations, the median gives a truer representative or typical element of the data than the mean. For example, suppose in an office there are 10 salespersons, 4 secretaries, the sales manager, and Ms. Daly, who owns the business. Their annual salaries are as follows: secretaries, $15,000 each; salespersons, $25,000 each; manager, $35,000; and owner, $200,000. The mean salary is

$$\bar{x} = \frac{(15,000)4 + (25,000)10 + 35,000 + 200,000}{16} = \$34,062.50.$$

However, since 14 people earn less than $34,062.50 and only 2 earn more, this does not seem very representative. The median salary is found by ranking the salaries by size: $15,000, $15,000, $15,000, $15,000, $25,000, $25,000, . . . , $200,000. There are 16 salaries (an even number) in the list, so the mean of the 8th and 9th entries will give the value of the median. The 8th and 9th entries are both $25,000, so the median is $25,000. In this example, the median is more representative of the distribution than the mean.

MODE   Sue's scores on ten class quizzes include one 7, two 8's, six 9's and one 10. She claims that her average grade on quizzes is 9, because most of her scores are 9's. This kind of "average," found by selecting the most frequent entry, is called the **mode.**

7 Find the mode for each of the following lists of numbers.

(a) 29, 35, 29, 18, 29, 56, 48

(b) 13, 17, 19, 20, 20, 13, 25, 27, 13, 20

(c) 512, 546, 318, 729, 854, 253

Answers:
(a) 29
(b) 13 and 20
(c) No mode

EXAMPLE 7   Find the mode for each list of numbers.

(a) 57, 38, **55**, **55**, 80, 87, 98

The number 55 occurs more often than any other, so it is the mode. It is not necessary to place the numbers in numerical order when looking for the mode.

(b) 182, **185**, 183, **185**, **187**, **187**, 189

Both 185 and 187 occur twice. This list has *two* modes.

(c) 10,708, 11,519, 10,972, 17,546, 13,905, 12,182

No number occurs more than once. This list has no mode. ■ **7**

The mode has the advantages of being easily found and not being influenced by data that are very large or very small compared to the rest of the data. It is often used in samples where the data to be "averaged" are not numerical. The major disadvantage of the mode is that there may be more than one, in case of ties, or there may be no mode at all when all entries occur with the same frequency.

The mean is the most commonly used measure of central tendency. Its advantages are that it is easy to compute, it takes all the data into consideration, and it is reliable—that is, repeated samples are likely to give very similar means. A disadvantage of the mean is that it is influenced by extreme values, as illustrated in the salary example above.

The median can be easy to compute and is influenced very little by extremes. Like the mode, the median can be found in situations where the data are not numerical. A disadvantage of the median is the need to rank the data in order; this can be tedious when the number of items is large.

11.1 EXERCISES

For Exercises 1–4, (a) group the data as indicated; (b) prepare a frequency distribution with columns for intervals and frequencies; (c) construct a histogram; (d) construct a frequency polygon. (See Examples 1 and 2.)

1. Use six intervals, starting with 0–24.

74	133	4	127	20	30
103	27	139	118	138	121
149	132	64	141	130	76
42	50	95	56	65	104
4	140	12	88	119	64

2. Use seven intervals, starting with 30–39.

79	71	78	87	69	50	63	51	60	46
65	65	56	88	94	56	74	63	87	62
84	76	82	67	59	66	57	81	93	93
54	88	55	69	78	63	63	48	89	81
98	42	91	66	60	70	64	70	61	75
82	65	68	39	77	81	67	62	73	49
51	76	94	54	83	71	94	45	73	95
72	66	71	77	48	51	54	57	69	87

3. Use 70–74 as the first interval.

79	84	88	96	102	104	110	108	106	106
104	99	97	92	94	90	82	74	72	83
84	92	100	99	101	107	111	102	97	94	92

4. Use 140–149 as the first interval.

174	190	172	182	179	186	171	152	174	185
180	170	160	173	163	177	165	157	149	167
169	182	178	158	182	169	181	173	183	176
170	162	159	147	150	192	179	165	148	188

5. How does a frequency polygon differ from a histogram?

6. Discuss the advantages and disadvantages of the mean as a measure of central tendency.

Find the mean for each list of numbers. Round to the nearest tenth.

7. 8, 10, 16, 21, 25

8. 44, 41, 25, 36, 67, 51

9. 21,900, 22,850, 24,930, 29,710, 28,340, 40,000

10. 38,500, 39,720, 42,183, 21,982, 43,250

11. 9.4, 11.3, 10.5, 7.4, 9.1, 8.4, 9.7, 5.2, 1.1, 4.7

12. 30.1, 42.8, 91.6, 51.2, 88.3, 21.9, 43.7, 51.2

Find the mean for each of the following. Round to the nearest tenth. (See Example 4.)

13.

Value	Frequency
3	4
5	2
9	1
12	3

14.

Value	Frequency
9	3
12	5
15	1
18	1

15.

Value	Frequency
12	4
13	2
15	5
19	3
22	1
23	5

16.

Value	Frequency
25	1
26	2
29	5
30	4
32	3
33	5

Find the median for each of the following lists of numbers. (See Example 6.)

17. 12, 18, 32, 51, 58, 92, 106

18. 596, 604, 612, 683, 719

19. 100, 114, 125, 135, 150, 172

20. 1072, 1068, 1093, 1042, 1056, 1005, 1009

21. 28.4, 9.1, 3.4, 27.6, 59.8, 32.1, 47.6, 29.8

22. .6, .4, .9, 1.2, .3, 4.1, 2.2, .4, .7, .1

Find the mode or modes for each of the following lists of numbers. (See Example 7.)

23. 4, 9, 8, 6, 9, 2, 1, 3

24. 21, 32, 46, 32, 49, 32, 49

25. 74, 68, 68, 68, 75, 75, 74, 74, 70

26. 158, 162, 165, 162, 165, 157, 163

27. 6.8, 6.3, 6.3, 6.9, 6.7, 6.4, 6.1, 6.0

28. 12.75, 18.32, 19.41, 12.75, 18.30, 19.45, 18.33

29. When is the median the most appropriate measure of central tendency?

30. Under what circumstances would the mode be an appropriate measure of central tendency?

For grouped data, the modal class *is the interval containing the most data values. Give the mean and modal class for each of the following collections of grouped data. (See Example 5.)*

31. The distribution in Exercise 3.

32. The distribution in Exercise 4.

For each set of ungrouped data, (a) Find the mean, median, and mode. (b) Discuss which of the three measures best represents the data and why.

33. The weight gains of 10 experimental rats fed on a special diet were −1, 0, −3, 7, 1, 1, 5, 4, 1, 4.

34. A sample of 7 measurements of the thickness of a copper wire were .010, .010, .009, .008, .007, .009, .008.

35. The times in minutes that 12 patients spent in a doctor's office were 20, 15, 18, 22, 10, 12, 16, 17, 19, 21, 23, 13.

36. The scores on a 10-point botany quiz were 6, 6, 8, 10, 9, 7, 6, 5, 6, 8, 3.

37. **Management**   A firm took a random sample of the number of days absent in a year for 40 employees, with results as shown below.

Days Absent	Frequency
0–2	23
3–5	11
6–8	5
9–11	0
12–14	1

Sketch a histogram and a frequency polygon for the data.

38. **Natural Science**   The size of the home ranges (in square kilometers) of several pandas were surveyed over a year's time, with the following results.

Home Range	Frequency
.1–.5	11
.6–1.0	12
1.1–1.5	7
1.6–2.0	6
2.1–2.5	2
2.6–3.0	1
3.1–3.5	1

Sketch a histogram and frequency polygon for the data.

39. **Social Science**   The histogram below shows the percent of the U.S. population in each age group in 1980.* What percent of the population was in each of the following age groups?
    (a) 10–19   (b) 60–69
    (c) What age group had the largest percent of the population?

40. **Social Science**   The histogram below shows the estimated percent of the U.S. population in each age group in the year 2000.* What percent of the population is estimated to be in each of the following age groups then?
    (a) 20–29   (b) 70+
    (c) What age group will have the largest percent of the population?
    (d) Compare the histogram in Exercise 39 with the histogram below. What seems to be true of the U.S. population?

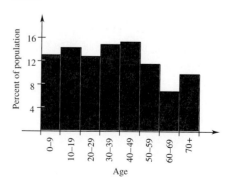

*Data from Census Bureau statistics.

Management *U.S. wheat prices and production figures for a recent decade are given below.*

Year	Price ($ per bushel)	Production (millions of bushels)
1982	2.70	2200
1983	2.30	2000
1984	2.95	1750
1985	3.80	2200
1986	3.90	2400
1987	3.60	2800
1988	3.55	2800
1989	3.50	2450
1990	3.35	2600
1991	3.20	2750

Find the mean for each of the following.

41. Price per bushel of wheat

42. Wheat production

43. Social Science The table shows the median age at first marriage in the U.S. over the last 50 years.*

Year	Male	Female
1990	26.1	23.9
1980	24.7	22.0
1970	23.2	20.8
1960	22.8	20.3
1950	22.8	20.3
1940	24.3	21.5

(a) Find the mean of the median ages for males.
(b) Find the mean of the median ages for females.

*U.S. National Center for Health Statistics of the United States, *Vital Statistics of the United States,* annual, *Monthly Vital Statistics Report.*

44. Social Science The number of nations participating in the winter Olympic games, from 1968 to 1992, is given below.*

Year	Nations Participating
1968	37
1972	35
1976	37
1980	37
1984	49
1988	57
1992	64

Find the following measures for the data.
(a) Mean **(b)** Median **(c)** Mode
(d) Which of these measures best represents the data? Explain your reasoning.

45. Social Science Washington Post writer John Schwartz pointed out that if Microsoft Corp. cofounder Bill Gates, who was reportedly worth $10 billion in 1995, lived in a town with 10,000 totally penniless people, the average personal wealth in the town would make it seem as if everyone were a millionaire.†
(a) Verify Schwartz' statement.
(b) What would be the median personal wealth in this town?
(c) What would be the mode for the personal wealth in this town?
(d) In this example, which average is most representative: the mean, the median, or the mode?

46. According to an article in the *Chance* electronic newsletter, the mean salary for National and American League baseball players in 1994 was $1,183,416, while the median salary was $500,000.‡ What might explain the large discrepancy between the mean and the median?

The Universal Almanac, John C. Wright, General Editor, Andrews and McMeel, Kansas City, p. 682.

†John Schwartz, "Mean Statistics: When is Average Best?" *The Washington Post,* Jan. 11, 1995, p. H7.

‡*Chance,* Aug. 13, 1994.

11.2 MEASURES OF VARIATION

The mean gives a measure of central tendency of a list of numbers, but tells nothing about the *spread* of the numbers in the list. For example, look at the following three samples.

I	3	5	6	3	3
II	4	4	4	4	4
III	10	1	0	0	9

Each of these three samples has a mean of 4, and yet they are quite different; the amount of dispersion or variation within the samples is different. Therefore, in addition to a measure of central tendency, another kind of measure is needed that describes how much the numbers vary.

The largest number in sample I is 6, while the smallest is 3, a difference of 3. In sample II this difference is 0; in sample III it is 10. The difference between the largest and smallest number in a sample is called the **range,** one example of a measure of variation. The range of sample I is 3, of sample II is 0, and of sample III is 10. The range has the advantage of being very easy to compute and gives a rough estimate of the variation among the data in the sample. However, it depends only on the two extremes and tells nothing about how the other data are distributed between the extremes.

EXAMPLE 1   Find the range for each list of numbers.

(a) 12,   27,   6,   19,   38,   9,   42,   15

The highest number here is 42; the lowest is 6. The range is the difference of these numbers, or

$$42 - 6 = 36.$$

(b) 74,   112,   59,   88,   200,   73,   92,   175

$$\text{Range} = 200 - 59 = 141 \quad \blacksquare \quad \boxed{1}$$

$\boxed{1}$ Find the range for the numbers   159,   283,   490, 390,   375,   297.

Answer:
331

TECHNOLOGY TIP Many graphing calculators list the largest and smallest numbers in a list when displaying one-variable statistics, usually on the second screen of the display. ✔

The most useful measure of variation is the *standard deviation.* Before defining it, however, we must find the **deviations from the mean,** the differences found by subtracting the mean from each number in a distribution.

EXAMPLE 2   Find the deviations from the mean for the numbers

$$32,\quad 41,\quad 47,\quad 53,\quad 57.$$

Adding these numbers and dividing by 5 gives a mean of 46. To find the deviations from the mean, subtract 46 from each number in the list. For example, the first deviation from the mean is $32 - 46 = -14$; the last is $57 - 46 = 11$.

$\boxed{2}$ Find the deviations from the mean for each set of numbers.

(a) 19,   25,   36,   41,   52,   61

(b) 6,   9,   5,   11,   3,   2

Answers:
(a) Mean is 39; deviations are -20, -14, -3,   2,   13,   22.

(b) Mean is 6; deviations from the mean are 0,   3, -1,   5, -3, -4.

Number	Deviation from Mean
32	-14
41	-5
47	1
53	7
57	11

To check your work, find the sum of these deviations. It should always equal 0. (The answer is always 0 because the positive and negative numbers cancel each other.) $\blacksquare$ $\boxed{2}$

To find a measure of variation, we might be tempted to use the mean of the deviations. However, as mentioned above, this number is always 0, no matter how widely the data are dispersed. To avoid the problem of the positive and negative deviations averaging 0, statisticians square each deviation (producing a list of nonnegative numbers) and then find the mean.

Number	Deviation from Mean	Square of Deviation
32	−14	196
41	−5	25
47	1	1
53	7	49
57	11	121

In this case the mean of the squared deviations is

$$\frac{196 + 25 + 1 + 49 + 121}{5} = \frac{392}{5} = 78.4.$$

This number is called the **population variance,** because the sum was divided by $n = 5$, the number of items in the original list.

In most applications, however, it isn't practical to use the entire population, so a sample is used instead to estimate the variance and other measures. Since a random sample may not include extreme entries from the list, statisticians prefer to divide the sum of the squared deviations in a sample by $n - 1$, rather than n, to create what is called an *unbiased estimator.* The informal idea behind this is that using $n - 1$ increases the value of the variance, as would be the case if extreme entries were used. Using $n - 1$ in the distribution above gives

$$\frac{196 + 25 + 1 + 49 + 121}{5 - 1} = \frac{392}{4} = 98.$$

This number 98 is called the **sample variance** of the distribution and is denoted s^2 because it is found by averaging a list of squares. In this case, the population and sample variances differ by quite a bit. But when n is relatively large, as is the case in real-life applications, the difference between them is rather small.

SAMPLE VARIANCE

The variance of a sample of n numbers $x_1, x_2, x_3, \ldots, x_n$, with mean $\bar{x}$, is

$$s^2 = \frac{\Sigma(x - \bar{x})^2}{n - 1}.$$

When computing the sample variance by hand, it is often convenient to use the following shortcut formula, which can be derived algebraically from the definition in the box above.

$$s^2 = \frac{\Sigma x^2 - n\bar{x}^2}{n - 1}$$

To find the sample variance, we square the deviations from the mean, so the variance is in squared units. To return to the same units as the data, we use the *square root* of the variance, called the **sample standard deviation,** denoted *s*.

SAMPLE STANDARD DEVIATION

The standard deviation of a sample of *n* numbers $x_1, x_2, x_3, \ldots, x_n$, with mean $\bar{x}$, is

$$s = \sqrt{\frac{\Sigma(x - \bar{x})^2}{n - 1}}.$$

Similarly, the **population standard deviation** is the square root of the population variance (just replace $n - 1$ by n in the denominator in the box above).

TECHNOLOGY TIP When a graphing calculator computes one-variable statistics for a list of data, it usually displays the following information (not necessarily in this order and sometimes on two screens), and possibly other information as well.

Information	Notation
Number of data entries	*n* or $N\Sigma$
Mean	$\bar{x}$ or mean Σ
Sum of all data entries	Σx or TOT Σ
Sum of the squares of all data entries	Σx^2
Sample standard deviation	Sx or sx or $x\sigma_{n-1}$ or SSDEV
Population standard deviation	σx or $x\sigma_n$ or PSDEV
Largest / smallest data entries	maxX/minX or MAXΣ/MINΣ
Median	Med or MEDIAN ✔

N O T E Hereafter we shall deal exclusively with the sample variance and the sample standard deviation. So whenever standard deviation is mentioned, it means "sample standard deviation." ◆

As its name indicates, the standard deviation is the most commonly used measure of variation. The standard deviation is a measure of the variation from the mean. The size of the standard deviation indicates how spread out the data are from the mean.

EXAMPLE 3 Find the standard deviation of the numbers

$$7, \quad 9, \quad 18, \quad 22, \quad 27, \quad 29, \quad 32, \quad 40$$

by hand, using the shortcut variance formula at the bottom of page 518.

Arrange the work in columns, as shown in the table.

Number	Square of the Number
7	49
9	81
18	324
22	484
27	729
29	841
32	1024
40	1600
184	5132

Now find the mean.

$$\bar{x} = \frac{\Sigma x}{8} = \frac{184}{8} = 23$$

The total of the second column gives $\Sigma x^2 = 5132$. The variance is

$$s^2 = \frac{\Sigma x^2 - n\bar{x}^2}{n - 1}$$

$$= \frac{5132 - 8(23)^2}{8 - 1}$$

$$= 128.6 \quad \text{(rounded)},$$

and the standard deviation is

$$s \approx \sqrt{128.6} \approx 11.3.$$

Use your calculator to verify this result. ■ 3

3 Find the standard deviation of each set of numbers. The deviations from the mean were found in Problem 2 at the side.

(a) 19, 25, 36, 41, 52, 61

(b) 6, 9, 5, 11, 3, 2

Answers:
(a) 15.9
(b) 3.5

One way to interpret the standard deviation uses the fact that, for many populations, most of the data are within three standard deviations of the mean. (See Section 11.3.) This implies that, in Example 3, most of the population from which this sample is taken is between

$$\bar{x} - 3s = 23 - 3(11.3) = -10.9$$

and

$$\bar{x} + 3s = 23 + 3(11.3) = 56.9.$$

This has important implications for quality control. If the sample in Example 3 represents measurements of a product that the manufacturer wants to be between 5 and 45, the standard deviation is too large, even though all the numbers are within these bounds.

For data in a grouped frequency distribution, a slightly different formula for the standard deviation is used.

STANDARD DEVIATION FOR A GROUPED DISTRIBUTION

The standard deviation for a distribution with mean $\bar{x}$, where x is an interval midpoint with frequency f, and $n = \Sigma f$, is

$$s = \sqrt{\frac{\Sigma fx^2 - nx^2}{n-1}}.$$

The formula indicates that the product fx^2 is to be found for each interval. Then these products are summed, n times the square of the mean is subtracted, and the difference is divided by one less than the total frequency; that is, by $n - 1$. The square root of this result is s, the standard deviation. The standard deviation found by this formula may (probably will) differ somewhat from the standard deviation found from the original data.

CAUTION In calculating the standard deviation for either a grouped or ungrouped distribution, using a rounded value for the mean or variance may produce an inaccurate value. ◆

EXAMPLE 4 Find s for the grouped data of Example 5, Section 11.1. Begin by including columns for x^2 and fx^2 in the table.

Interval	f	x	x^2	fx^2
40–49	2	44.5	1980.25	3,960.50
50–59	4	54.5	2970.25	11,881.00
60–69	7	64.5	4160.25	29,121.75
70–79	9	74.5	5550.25	49,952.25
80–89	5	84.5	7140.25	35,701.25
90–99	3	94.5	8930.25	26,790.75
Total:	30			Total: 157,407.50

4 Find the standard deviation for the grouped data that follows. (*Hint:* $\bar{x} = 28.5$)

Value	Frequency
20–24	3
25–29	2
30–34	4
35–39	1

Answer:
5.3

Recall from Section 11.1 that $\bar{x} = 71.2$. Use the formula above with $n = 30$ to find s.

$$s = \sqrt{\frac{\Sigma fx^2 - n\bar{x}^2}{n-1}}$$

$$= \sqrt{\frac{157,407.50 - 30(71.2)^2}{30 - 1}}$$

$$\approx 13.5 \quad \blacksquare \quad \boxed{4}$$

NOTE A calculator is almost a necessity for finding a standard deviation. With a non-graphing calculator, a good procedure to follow is to first calculate $\bar{x}$. Then for each x, square that number, then multiply the result by the appropriate frequency. If your calculator has a key that accumulates a sum, use it to accumulate the total in the last column of the table. With a graphing calculator, simply enter the midpoints and the frequencies, then ask for the 1-variable statistics. ◆

11.2 EXERCISES

1. How are the variance and the standard deviation related?
2. Why can't we use the sum of the deviations from the mean as a measure of dispersion of a distribution?

Find the range and standard deviation for each of the following sets of numbers. (See Examples 1 and 3.)

3. 6, 8, 9, 10, 12
4. 12, 15, 19, 23, 26
5. 7, 6, 12, 14, 18, 15
6. 4, 3, 8, 9, 7, 10, 1
7. 42, 38, 29, 74, 82, 71, 35
8. 122, 132, 141, 158, 162, 169, 180
9. 241, 248, 251, 257, 252, 287
10. 51, 58, 62, 64, 67, 71, 74, 78, 82, 93

Find the standard deviation for the grouped data in Exercises 11 and 12. (See Example 4.)

11. (From Exercise 1, Section 11.1)

College Units	Frequency
0–24	4
25–49	3
50–74	6
75–99	3
100–124	5
125–149	9

12. (From Exercise 2, Section 11.1)

Scores	Frequency
30–39	1
40–49	6
50–59	13
60–69	22
70–79	17
80–89	13
90–99	8

13. **Natural Science** Twenty-five laboratory rats, used in an experiment to test the food value of a new product, made the following weight gains in grams.

5.25	5.03	4.90	4.97	5.03
5.12	5.08	5.15	5.20	4.95
4.90	5.00	5.13	5.18	5.18
5.22	5.04	5.09	5.10	5.11
5.23	5.22	5.19	4.99	4.93

Find the mean gain and the standard deviation of the gains.

14. **Management** An assembly-line machine turns out washers with the following thicknesses (in millimeters).

1.20	1.01	1.25	2.20	2.58	2.19	1.29	1.15
2.05	1.46	1.90	2.03	2.13	1.86	1.65	2.27
1.64	2.19	2.25	2.08	1.96	1.83	1.17	2.24

Find the mean and standard deviation of these thicknesses.

*An application of standard deviation is given by **Chebyshev's theorem**. (P. L. Chebyshev was a Russian mathematician who lived from 1821 to 1894.) This theorem applies to any distribution of numerical data. It states:*

> *For any distribution of numerical data, at least $1 - 1/k^2$ of the numbers lie within k standard deviations of the mean.*

Example *For any distribution, at least*

$$1 - \frac{1}{3^2} = 1 - \frac{1}{9} = \frac{8}{9}$$

of the numbers lie within 3 standard deviations of the mean. Find the fraction of all the numbers of a data set lying within the following numbers of standard deviations from the mean.

15. 2
16. 4
17. 5

In a certain distribution of numbers, the mean is 50 with a standard deviation of 6. Use Chebyshev's theorem to tell what percent of the numbers are

18. between 32 and 68;
19. between 26 and 74;
20. less than 38 or more than 62;
21. less than 32 or more than 68;
22. less than 26 or more than 74.

23. **Management** The Britelite Company conducted tests on the life of its light bulbs and those of a competitor (Brand X) with the following results for samples of 10 bulbs of each brand.

	Hours of Use (in 100s)									
Britelite	20	22	22	25	26	27	27	28	30	35
Brand X	15	18	19	23	25	25	28	30	34	38

Compute the mean and standard deviation for each sample. Compare the means and standard deviations of the two brands and then answer the questions below.
(a) Which bulbs have a more uniform life in hours?
(b) Which bulbs have the highest average life in hours?

24. **Management** The weekly wages of the six employees of Harold's Hardware Store are $300, $320, $380, $420, $500, and $2000.

(a) Find the mean and standard deviation of this distribution.

(b) How many of the employees earn within one standard deviation of the mean? How many earn within two standard deviations of the mean?

25. **Social Science** The number of unemployed workers in the United States in 1988–1993 (in millions) is given below.*

Year	Number Unemployed
1988	6.70
1989	6.53
1990	6.87
1991	8.43
1992	9.38
1993	8.73

(a) Find the mean number unemployed (in millions) in this period. Which year has unemployment closest to the mean?

(b) Find the standard deviation for the data.

(c) In how many of these years is unemployment within 1 standard deviation of the mean?

26. **Social Science** In an article comparing national mathematics examinations in the U.S. and some European countries, a researcher found the results shown in the histogram for the number of minutes allowed for each open-ended question.†

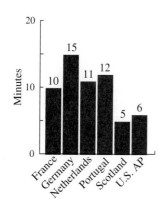

(a) Find the mean and standard deviation. (*Hint:* Do not use the rounded value of the mean to find the standard deviation.)

*U.S. Bureau of Labor Statistics, Bulletin 2307; and *Employment and Earnings,* monthly.

†Information from article comparing national mathematics examinations in the U.S. and some European countries as appeared in FOCUS, June 1993. Reprinted by permission of the Mathematical Association of America.

(b) How many standard deviations from the mean is the largest number of minutes?

(c) How many standard deviations from the mean is the U.S. AP examination?

27. **Social Science** For all questions, the researcher in Exercise 26 found the following results for Scotland and the United States, the countries with the least rigorous examinations.

(a) Find the mean and standard deviation.

(b) How many standard deviations from the mean is the AT (achievement test)?

(c) How many standard deviations from the mean is the Scottish test?

28. **Social Science** Recently Germany has endured rioting and disruption because of the increasing numbers of immigrants. The nine top countries of origin of German immigrants are shown in the histogram.

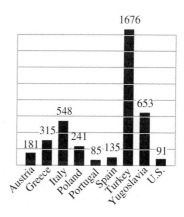

(a) Find the mean and standard deviation of these data.

(b) How many standard deviations from the mean is the largest number of immigrants? the smallest? Which country of origin is closest to the mean?

29. Social Science The numbers of immigrants to the United States (in thousands) from selected parts of the world in 1990 are shown in the table.*

Region	Immigrants
Europe	145.4
Asia	357.0
Canada	15.2
Mexico	213.8
Other America	210.3

(a) Find the mean number of immigrants from these regions. Which region produces the number of immigrants closest to the mean?

(b) Find the standard deviation. Are any of the data more than 2 standard deviations from the mean? If so, which ones? Are any of the data more than 1 standard deviation from the mean? If so, which ones?

30. Management The Quaker Oaks Company conducted a survey to determine if a proposed premium, to be included in their cereal, was appealing enough to generate new sales.† Four cities were used as test markets, where the cereal was distributed with the premium, and four cities as control markets, where the cereal was distributed without the premium. The eight cities were chosen on the basis of their similarity in terms of population, per capita income, and total cereal purchase volume. The results were as follows.

		Percent Change in Average Market Shares per Month
Test Cities	1	+18
	2	+15
	3	+7
	4	+10
Control Cities	1	+1
	2	-8
	3	-5
	4	0

(a) Find the mean of the change in market share for the four test cities.

(b) Find the mean of the change in market share for the four control cities.

(c) Find the standard deviation of the change in market share for the test cities.

(d) Find the standard deviation of the change in market share for the control cities.

(e) Find the difference between the mean of part (a) and the mean of part (b). This represents the estimate of the percent change in sales due to the premium.

(f) The two standard deviations from part (c) and part (d) were used to calculate an "error" of ± 7.95 for the estimate in part (e). With this amount of error what is the smallest and largest estimate of the increase in sales? (*Hint:* Use the answer to part (e).)

On the basis of the results of Exercise 30, the company decided to mass produce the premium and distribute it nationally.

31. Management The following table gives 10 samples, of three measurements, made during a production run.

	SAMPLE NUMBER									
1	*2*	*3*	*4*	*5*	*6*	*7*	*8*	*9*	*10*
2	3	-2	-3	-1	3	0	-1	2	0
-2	-1	0	1	2	2	1	2	3	0
1	4	1	2	4	2	2	3	2	2

(a) Find the mean $\bar{x}$ for each sample of three measurements.

(b) Find the standard deviation s for each sample of three measurements.

(c) Find the mean $\bar{X}$ of the sample means.

(d) Find the mean $\bar{s}$ of the sample standard deviations.

(e) The upper and lower control limits of the sample means here are $\bar{X} \pm 1.954\bar{s}$. Find these limits. If any of the measurements are outside these limits, the process is out of control. Decide if this production process is out of control.

32. Discuss what the standard deviation tells us about a distribution.

Social Science *The reading scores of a second-grade class given individualized instruction are shown below. The table also shows the reading scores of a second-grade class given traditional instruction in the same school.*

Scores	Individualized Instruction	Traditional Instruction
50–59	2	5
60–69	4	8
70–79	7	8
80–89	9	7
90–99	8	6

33. Find the mean and standard deviation for the individualized instruction scores.

*Source: U.S. Immigration and Naturalization Service, *Statistics Yearbook,* annual; and releases.

†This example was supplied by Jeffrey S. Berman, Senior Analyst, Marketing Information, Quaker Oats Company.

34. Find the mean and standard deviation for the traditional instruction scores.

35. Discuss a possible interpretation of the differences in the means and the standard deviations in Exercises 33 and 34.

36. Social Science In 1998, nineteen state governors earned $100,000 or more (not counting expense allowances) as listed in the following table. (Salaries are given in thousands of dollars and are rounded to the nearest $1000.)*
  (a) Find the mean salary of these governors. Which state has the governor with the salary closest to the mean?
  (b) Find the standard deviation for the data.
  (c) What percent of the governors have salaries within 1 standard deviation of the mean?
  (d) What percent of the governors have salaries within 2 standard deviations of the mean?

*World Almanac and Book of Facts, 1998.

State	Salary
California	131
Delaware	107
Florida	111
Georgia	103
Illinois	123
Maryland	120
Massachusetts	100
Michigan	127
Minnesota	115
Missouri	107
New York	130
North Carolina	107
Ohio	116
Pennsylvania	125
South Carolina	106
Texas	115
Virginia	110
Washington	121
Wisconsin	102

11.3   NORMAL DISTRIBUTIONS

The general idea of a probability distribution was first discussed in Section 10.5. In this section, we consider an important specific type of probability distribution.

Figure 11.3(a) shows a histogram and frequency polygon for the bank transaction example in Section 10.5. The heights of the bars are the probabilities. The transaction times in the example were given to the nearest minute. Theoretically, at least, they could have been timed to the nearest tenth of a minute, or hundredth of a minute, or even more precisely. In each case, a histogram and frequency polygon could be drawn. If the times are measured with smaller and smaller units, there are more bars in the histogram, and the frequency polygon begins to look more and more like the curve in Figure 11.3(b) instead of a polygon. Actually it is possible for the transaction times to take on any real number value greater than 0. A distribution in which the outcomes can take any real number value within some interval is a **continuous distribution.** The graph of a continuous distribution is a curve.

(a)

(b)

FIGURE 11.3

The distribution of heights (in inches) of college women is another example of a continuous distribution, since these heights include infinitely many possible measurements, such as 53, 58.5, 66.3, 72.666, . . . , and so on. Figure 11.4 shows the continuous distribution of heights of college women. Here the most frequent heights occur near the center of the interval shown.

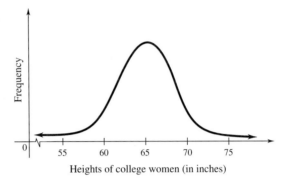

Heights of college women (in inches)

FIGURE 11.4

Another continuous curve, which approximates the distribution of yearly incomes in the United States, is given in Figure 11.5. The graph shows that the most frequent incomes are grouped near the low end of the interval. This kind of distribution, where the peak is not at the center, is called **skewed.**

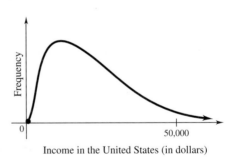

Income in the United States (in dollars)

FIGURE 11.5

Many natural and social phenomena produce continuous probability distributions whose graphs are approximated very well by bell-shaped curves, such as those shown in Figure 11.6. Such distributions are called **normal distributions** and their graphs are called **normal curves.** Examples of distributions that are approximately normal are the heights of college women and the errors made in filling 1-pound cereal boxes. We use the Greek letters μ (mu) to denote the mean and σ (sigma) to denote the standard deviation of a normal distribution.

Three normal distributions

FIGURE 11.6

There are many normal distributions. Some of the corresponding normal curves are tall and thin and others short and wide, as shown in Figure 11.6. But every normal curve has the following properties.

1. Its peak occurs directly above the mean μ.
2. The curve is symmetric about the vertical line through the mean (that is, if you fold the page along this line, the left half of the graph will fit exactly on the right half).
3. The curve never touches the x-axis—it extends indefinitely in both directions.
4. The area under the curve (and above the horizontal axis) is 1. (As is shown in calculus, this is a consequence of the fact that the sum of the probabilities in any distribution is 1.)

It can be shown that a normal distribution is completely determined by its mean μ and standard deviation σ.* A small standard deviation leads to a tall, narrow curve like the one in the center of Figure 11.6, because most of the data are close to the mean. A large standard deviation means the data are very spread out, producing a flat wide curve like the one on the right in Figure 11.6.

Since the area under a normal curve is 1, parts of this area can be used to determine certain probabilities. For instance, Figure 11.7(a) on the next page is the probability distribution of the annual rainfall in a certain region. The probability that the annual rainfall will be between 25 and 35 inches is the area under the curve from 25 to 35. The general case, shown in Figure 11.7(b), can be stated as follows.

> The area of the shaded region under the normal curve from a to b is the probability that an observed data value will be between a and b.

---

*As is shown in more advanced courses, its graph is the graph of the function

$$f(x) = \frac{1}{\sigma\sqrt{2\pi}}e^{-(x-\mu)^2/(2\sigma^2)},$$

where $e \approx 2.71828$ is the real number discussed in Section 5.3.

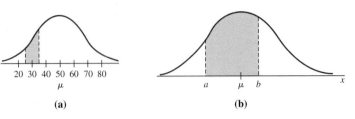

(a)   (b)

FIGURE 11.7

To use normal curves effectively we must be able to calculate areas under portions of these curves. These calculations have already been done for the normal curve with mean $\mu = 0$ and standard deviation $\sigma = 1$ (which is called the **standard normal curve**) and are available in Table 2 at the back of the book. The following examples demonstrate how to use Table 2 to find such areas. Later we shall see how the standard normal curve may be used to find areas under any normal curve.

The horizontal axis of the standard normal curve is usually labeled z. Since the standard deviation of the standard normal curve is 1, the numbers along the horizontal axis (the z-values) measure the number of standard deviations above or below the mean $z = 0$.

TECHNOLOGY TIP Some graphing calculators (such as the TI-83 and Casio 9800/9850) have the ability to graph a normal distribution given its mean and standard deviation and to find areas under the curve between two x-values. For an area under the curve, some calculators will give the corresponding z-value. For details, see your instruction book. (Look for "distribution" or "probability distribution.") A calculator-generated graph of the standard normal curve is shown in Figure 11.8. ✔

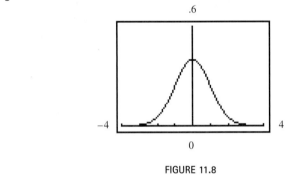

FIGURE 11.8

EXAMPLE 1   Find the following areas under the standard normal curve.

(a) The area between $z = 0$ and $z = 1$, the shaded region in Figure 11.9

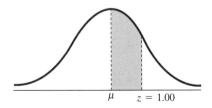

FIGURE 11.9

1 Find the percent of area between the mean and

(a) $z = 1.51$;

(b) $z = -2.04$.

(c) Find the percent of area in the shaded region.

Answers:
(a) 43.45%

(b) 47.93%

(c) 26.42%

2 If your calculator can graph probability distributions and find area, use it to find the areas required in Example 1.

Answers:
(a) 34.13%

(b) 49.25%

Find the entry 1 in the z-column of Table 2. The entry next to it in the A-column is .3413, which means that the area between $z = 0$ and $z = 1$ is .3413. Since the total area under the curve is 1, the shaded area in Figure 11.9 is 34.13% of the total area under the normal curve.

(b) The area between $z = -2.43$ and $z = 0$

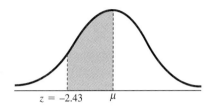

FIGURE 11.10

Table 2 lists only positive values of z. But the normal curve is symmetric around the mean $z = 0$, so the area between $z = 0$ and $z = -2.43$ is the same as the area between $z = 0$ and $z = 2.43$. Find 2.43 in the z-column of Table 2. The entry next to it in the A-column shows that the area is .4925. Hence, the shaded area in Figure 11.10 is 49.25% of the total area under the curve. ■ **1** **2**

EXAMPLE 2 Use a calculator or Table 2 to find the percent of the total area for the following areas under the standard normal curve.

(a) The area between 1.41 standard deviations *below* the mean and 2.25 standard deviations *above* the mean (that is, between $z = -1.41$ and $z = 2.25$)

First, draw a sketch showing the desired area, as in Figure 11.11. From Table 2, the area between the mean and 1.41 standard deviations below the mean is .4207. Also, the area from the mean to 2.25 standard deviations above the mean is .4878. As the figure shows, the total desired area can be found by *adding* these numbers.

$$\begin{array}{r} .4207 \\ + .4878 \\ \hline .9085 \end{array}$$

The shaded area in Figure 11.11 represents 90.85% of the total area under the normal curve.

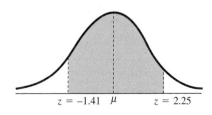

FIGURE 11.11

3 Find the following standard normal curve areas as percents of the total area.

(a) Between .31 standard deviations below the mean and 1.01 standard deviations above the mean

−.31 μ 1.01

(b) Between .38 and 1.98 standard deviations below the mean

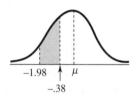

−1.98 μ
−.38

(c) To the right of 1.49 standard deviations above the mean

μ 1.49

(d) What percent of the area is within 1 standard deviation of the mean? within 2 standard deviations of the mean? within 3 standard deviations of the mean? What can you conclude from the last answer?

Answers:
(a) 46.55%

(b) 32.82%

(c) 6.81%

(d) 68.3%, 95.47%, 99.7%
Almost all the data lies within 3 standard deviations of the mean.

4 Find each z-score using the information in Example 3.

(a) $x = 36$

(b) $x = 55$

Answers:
(a) −3.5

(b) 1.25

(b) The area between .58 standard deviation above the mean and 1.94 standard deviations above the mean

Figure 11.12 shows the desired area. The area between the mean and .58 standard deviation above the mean is .2190. The area between the mean and 1.94 standard deviations above the mean is .4738. As the figure shows, the desired area is found by *subtracting* the two areas.

$$
\begin{array}{r}
.4738 \\
- .2190 \\
\hline
.2548
\end{array}
$$

The shaded area of Figure 11.12 represents 25.48% of the total area under the normal curve.

(c) The area to the right of 2.09 standard deviations above the mean

The total area under a normal curve is 1. Thus, the total area to the right of the mean is 1/2, or .5000. From Table 2, the area from the mean to 2.09 standard deviations above the mean is .4817. The area to the right of 2.09 standard deviations is found by subtracting .4817 from .5000.

$$
\begin{array}{r}
.5000 \\
- .4817 \\
\hline
.0183
\end{array}
$$

A total of 1.83% of the total area is to the right of 2.09 standard deviations above the mean. Figure 11.13 (which is not to scale) shows the desired area. ■ 3

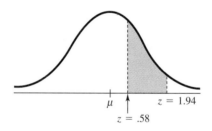

μ $z = 1.94$
$z = .58$

FIGURE 11.12

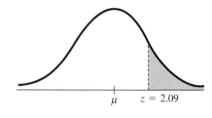

μ $z = 2.09$

FIGURE 11.13

The key to finding areas under *any* normal curve is to express each number x on the horizontal axis in terms of standard deviations above or below the mean. The **z-score** for x is the number of standard deviations that x lies from the mean (positive if x is above the mean, negative if x is below the mean).

EXAMPLE 3 If a normal distribution has mean 50 and standard deviation 4, find the following z-scores.

(a) The z-score for $x = 46$

Since 46 is 4 units below 50 and the standard deviation is 4, 46 is 1 standard deviation below the mean. So, its z-score is −1.

(b) The z-score for $x = 60$

The z-score is 2.5 because 60 is 10 units above the mean (since $60 - 50 = 10$) and 10 units is 2.5 standard deviations (since $10/4 = 2.5$). ■ 4

In Example 3(b) we found the z-score by taking the difference between 60 and the mean and dividing this difference by the standard deviation. The same procedure works in the general case.

If a normal distribution has mean μ and standard deviation σ, then the z-score for the number x is

$$z = \frac{x - \mu}{\sigma}.$$

The importance of z-scores is the following fact, whose proof is omitted.

AREA UNDER A NORMAL CURVE

The area under a normal curve between $x = a$ and $x = b$ is the same as the area under the standard normal curve between the z-score for a and the z-score for b.

Therefore, by converting to z-scores and using a graphing calculator or Table 2 for the standard normal curve, we can find areas under any normal curve. Since these areas are probabilities (as explained earlier), we can now handle a variety of applications.

Graphing calculators, computer programs, and CAS programs (such as DERIVE) can be used to find areas under the normal curve and hence, probabilities. The equation of the standard normal curve, with $\mu = 0$ and $\sigma = 1$, is $f(x) = (1/\sqrt{2\pi})e^{-x^2/2}$. A good approximation of the area under this curve (and above $y = 0$) can be found by using the x-interval $[-4, 4]$. See Chapter 14 for more information on finding such areas.

EXAMPLE 4   Dixie Office Supplies finds that its sales force drives an average of 1200 miles per month per person, with a standard deviation of 150 miles. Assume that the number of miles driven by a salesperson is closely approximated by a normal distribution.

(a) Find the probability that a salesperson drives between 1200 and 1600 miles per month.

Here $\mu = 1200$ and $\sigma = 150$, and we must find the area under the normal distribution curve between $x = 1200$ and $x = 1600$. We begin by finding the z-score for $x = 1200$.

$$z = \frac{x - \mu}{\sigma} = \frac{1200 - 1200}{150} = \frac{0}{150} = 0$$

The z-score for $x = 1600$ is

$$z = \frac{x - \mu}{\sigma} = \frac{1600 - 1200}{150} = \frac{400}{150} = 2.67.^*$$

---

*All z-scores here are rounded to two decimal places.

So the area under the curve from $x = 1200$ to $x = 1600$ is the same as the area under the standard normal curve from $z = 0$ to $z = 2.67$, as indicated in Figure 11.14. A graphing calculator or Table 2 shows that this area is .4962. Therefore, the probability that a salesperson drives between 1200 and 1600 miles per month is .4962.

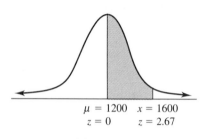

$\mu = 1200$ $x = 1600$
$z = 0$ $z = 2.67$

FIGURE 11.14

$\mu = 1200$
1000   1500

FIGURE 11.15

(b) Find the probability that a salesperson drives between 1000 and 1500 miles per month.

5 The heights of female sophomore college students at one school have $\mu = 172$ centimeters, with $\sigma = 10$ centimeters. Find the probability that the height of such a student is

(a) between 172 cm and 185 cm;

(b) between 160 cm and 180 cm;

(c) less than 165 cm.

Answers:
(a) .4032

(b) .6730

(c) .2420

As shown in Figure 11.15, z-scores for both $x = 1000$ and $x = 1500$ are needed.

For $x = 1000$,

$$z = \frac{1000 - 1200}{150}$$

$$= \frac{-200}{150}$$

$$= -1.33.$$

For $x = 1500$,

$$z = \frac{1500 - 1200}{150}$$

$$= \frac{300}{150}$$

$$= 2.00.$$

From the table $z = 1.33$ leads to an area of .4082, while $z = 2.00$ corresponds to .4773. A total of $.4082 + .4773 = .8855$, or 88.55%, of all drivers travel between 1000 and 1500 miles per month. From this, the probability that a driver travels between 1000 and 1500 miles per month is .8855. ■ **5**

EXAMPLE 5 A tire store finds that the tread life of its tires is normally distributed, with a mean of 26,640 miles and a standard deviation of 4000 miles. The store sold 9000 tires this month. How many of them can be expected to last more than 35,000 miles?

Here $\mu = 26,640$ and $\sigma = 4000$. The probability that a tire will last more than 35,000 miles is the area under the normal curve to the right of $x = 35,000$. The z-score for $x = 35,000$ is

$$z = \frac{x - \mu}{\sigma} = \frac{35,000 - 26,640}{4000} = \frac{8360}{4000} = 2.09.$$

Example 2(c) and Figure 11.13 show that the area to the right of $z = 2.09$ is .0183, which is 1.83% of the total area under the curve. Therefore, 1.83% of the tires can be expected to last more than 35,000 miles. Thus,

$$1.83\% \text{ of } 9000 = .0183 \cdot 9000 = 164.7,$$

or approximately 165 tires can be expected to last more than 35,000 miles. ■

As mentioned above, z-scores are standard deviations, so $z = 1$ corresponds to 1 standard deviation above the mean, and so on. As found in Side Problem 3(d) of this section, 68.3% of the area under a normal curve lies within 1 standard deviation of the mean. Also, 95.47% lies within 2 standard deviations of the mean, and 99.7% lies within 3 standard deviations of the mean. These results, summarized in Figure 11.16, can be used to get a quick estimate of results when working with normal curves.

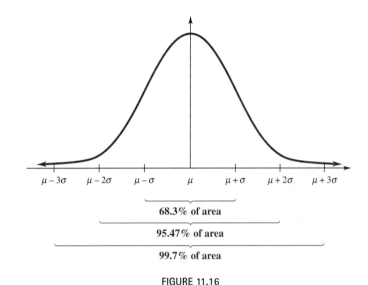

68.3% of area

95.47% of area

99.7% of area

FIGURE 11.16

11.3 EXERCISES

1. In your own words, discuss the characteristics of a normal curve, mentioning the area under the curve, the location of the peak, and the end-behavior.

2. Explain how z-scores are found for normal distributions where

$$\mu \neq 0 \quad \text{or} \quad \sigma \neq 1.$$

3. Explain how the standard normal curve is used to find probabilities for other normal distributions.

Find the percent of the area under a normal curve between the mean and the following number of standard deviations from the mean. (See Example 2.)

4. 1.75   5. .23   6. −.43

7. −2.1   8. 3.1

Find the percent of the total area under the standard normal curve between the following z-scores. (See Examples 1 and 2.)

9. $z = 1.41$ and $z = 2.83$

10. $z = .64$ and $z = 2.11$

11. $z = -2.48$ and $z = -.05$

12. $z = -1.74$ and $z = -1.02$

13. $z = -3.11$ and $z = 1.44$

14. $z = -2.94$ and $z = -.43$

Find a z-score satisfying each of the following conditions. (Hint: Use Table 2 backwards or a graphing calculator.)

15. 5% of the total area is to the right of z.

16. 1% of the total area is to the left of z.

17. 15% of the total area is to the left of z.

18. 25% of the total area is to the right of z.

Use a graphing calculator or a computer to find the following probabilities by finding the comparable area under a standard normal curve.

19. $P(1.372 \leq z \leq 2.548)$

20. $P(-2.751 \leq z \leq 1.693)$

21. $P(z > -2.476)$

22. $P(z < 1.692)$

Use a graphing calculator or a computer to find the following probabilities for a distribution with a mean of 35.693 and a standard deviation of 7.104.

23. $P(12.275 < x < 28.432)$

24. $P(x > 38.913)$

25. $P(x < 17.462)$

26. $P(17.462 \leq x \leq 53.106)$

Assume the following distributions are all normal, and use the areas under the normal curve given in Table 2 or a graphing calculator to answer the questions. (See Example 4.)

27. Management According to the label, a regular can of Campbell's soup holds an average of 305 grams with a standard deviation of 4.2 grams. What is the probability that a can will be sold that holds more than 306 grams?

28. Management A jar of Adams Old Fashioned Peanut Butter contains 453 grams with a standard deviation of 10.1 grams. Find the probability that one of these jars contains less than 450 grams.

29. Management A General Electric soft white 3-way bulb has an average life of 1200 hours with a standard deviation of 50 hours. Find the probability that the life of one of these bulbs will be between 1150 and 1300 hours.

30. Management A 100-watt light bulb has an average brightness of 1640 lumens with a standard deviation of 62 lumens. What is the probability that a 100-watt bulb will have a brightness between 1600 and 1700 lumens?

31. Social Science The scores on a standardized test in a suburban high school have a mean of 80 with a standard deviation of 12. What is the probability that a student will have a score less than 60?

32. Social Science The average time to complete the test in Exercise 31 is 40 minutes with a standard deviation of 5.2 minutes. Find the probability that a student will require more than 50 minutes to finish the test.

33. A 120-minute video tape has a standard deviation of .7 minute. Find the probability that a 118-minute movie will not be entirely copied.

34. Natural Science The distribution of low temperatures for a city has an average of 44° with a standard deviation of 6.7°. What is the probability that the low temperature will be above 55°?

Social Science *New studies by Federal Highway Administration traffic engineers suggest that speed limits on many thoroughfares are set arbitrarily and often are artificially low. According to traffic engineers, the ideal limit should be the "85th percentile speed." This means the speed at or below which 85 percent of the traffic moves. Assuming speeds are normally distributed, find the 85th percentile speed for roads with the following conditions.*

35. The mean speed is 50 mph with a standard deviation of 10 mph.

36. The mean speed is 30 mph with a standard deviation of 5 mph.

Social Science *One professor uses the following system for assigning letter grades in a course.*

Grade	Total Points
A	Greater than $\mu + \frac{3}{2}\sigma$
B	$\mu + \frac{1}{2}\sigma$ to $\mu + \frac{3}{2}\sigma$
C	$\mu - \frac{1}{2}\sigma$ to $\mu + \frac{1}{2}\sigma$
D	$\mu - \frac{3}{2}\sigma$ to $\mu - \frac{1}{2}\sigma$
F	Below $\mu - \frac{3}{2}\sigma$

What percent of the students receive the following grades?

37. A

38. B

39. C

40. Do you think the system in Exercises 37–39 would be more likely to be fair in a large freshmen class in psychology or in a graduate seminar of five students? Why?

Natural Science *In nutrition, the recommended daily allowance of vitamins is a number set by the government as a guide to an individual's daily vitamin intake. Actually, vitamin needs vary drastically from person to person, but the needs are very closely approximated by a normal curve. To calculate the recommended daily allowance, the government first finds the average need for vitamins among people in the population and then the standard deviation. The recommended daily allowance is defined as the mean plus 2.5 times the standard deviation.*

Find the recommended daily allowance for the following vitamins.

41. Mean = 500 units, standard deviation = 50 units

42. Mean = 1800 units, standard deviation = 140 units

43. Mean = 159 units, standard deviation = 12 units

44. Mean = 1200 units, standard deviation = 92 units

Social Science *The mean performance score of a large group of fifth-grade students on a math achievement test is 88. The scores are known to be normally distributed. What percent of the students had scores as follows?*

45. More than 1 standard deviation above the mean

46. More than 2 standard deviations above the mean

Social Science *A teacher gives a test to a large group of students. The results are closely approximated by a normal curve. The mean is 74, with a standard deviation of 6. The teacher wishes to give A's to the top 8% of the students and F's to the bottom 8%. A grade of B is given to the next 15%, with D's given similarly. All other students get C's. Find the bottom cutoff (rounded to the* nearest whole number) for the following grades. (Hint: Use Table 2 backwards.)

47. A **48.** B

49. C **50.** D

Social Science *Studies have shown that women are charged an average of $500 more than men for cars.* Assume a normal distribution of overcharges with a mean of $500 and a standard deviation of $60. Find the probability of a woman's paying the following additional amounts for a car.*

51. Less than $500

52. At least $600

53. Between $400 and $600

Social Science *Women earn an average of $.74 for every $1 earned by a man, a difference of $.26. Assume differences in pay are normally distributed with a mean of $.26 and a standard devi-*

**"From repair shops to cleaners, women pay more," by Bob Dart as appeared in The Chicago Tribune, May 27, 1993. Reprinted by permission of the author.

ation of $.08. Find the probability that a woman earns the following amounts less than a man in the same job.

54. Between $.20 and $.30 **55.** More than $.19

56. At most $.26

Management *At a ranch in the Sacramento valley, the tomato crop yields an average of 25 tons per acre with a standard deviation of 2.2 tons per acre. Find the probability of each of the following yields in tons per acre.*

57. At least 24 **58.** At most 25.7

59. Between 24.5 and 25.5

Management *According to the manufacturer, a certain automobile averages 28.2 miles per gallon of gasoline with a standard deviation of 2.1 miles per gallon. Find the probability of each of the following miles per gallon.*

60. Between 27 and 29 **61.** At least 28

62. At most 29

11.4 BINOMIAL DISTRIBUTIONS

Many practical experiments have only two possible outcomes, *success* or *failure*. Such experiments are called *binomial experiments* or *Bernoulli trials* and were first studied in Section 10.3. Examples of binomial experiments include flipping a coin (with heads being a success, for instance, and tails a failure) or testing TVs coming off the assembly line to see whether or not they are defective.

A **binomial distribution** is a probability distribution that satisfies the following conditions.

1. The experiment is a series of repeated independent trials.
2. There are only two possible outcomes for each trial, success or failure.
3. The probability of success is constant from trial to trial.

For example, suppose a fair die is tossed 5 times, with 1 or 2 considered a success and 3, 4, 5, or 6 a failure. Each trial is independent of the others and the probability of success in each trial is $p = \frac{2}{6} = \frac{1}{3}$. In 5 tosses there can be any number of successes from 0 through 5. But these 6 possible results are not equally likely, as shown in the chart at the side, which was obtained by using the following formula from Section 10.3.

$$P(x) = \binom{n}{x} p^x (1-p)^{n-x},$$

where n is the number of trials (here, $n = 5$), x is the number of successes, p is the probability of success in a single trial (here, $p = \frac{1}{3}$), and $P(x)$ is the probability that exactly x of the n trials result in success.

x	$P(x)$
0	$\binom{5}{0}\left(\frac{1}{3}\right)^0\left(\frac{2}{3}\right)^5 = \frac{32}{243}$
1	$\binom{5}{1}\left(\frac{1}{3}\right)^1\left(\frac{2}{3}\right)^4 = \frac{80}{243}$
2	$\binom{5}{2}\left(\frac{1}{3}\right)^2\left(\frac{2}{3}\right)^3 = \frac{80}{243}$
3	$\binom{5}{3}\left(\frac{1}{3}\right)^3\left(\frac{2}{3}\right)^2 = \frac{40}{243}$
4	$\binom{5}{4}\left(\frac{1}{3}\right)^4\left(\frac{2}{3}\right)^1 = \frac{10}{243}$
5	$\binom{5}{5}\left(\frac{1}{3}\right)^5\left(\frac{2}{3}\right)^0 = \frac{1}{243}$

TECHNOLOGY TIP The TI-83 computes binomial probabilities. See the Technology Tip on page 471. ✔

The expected value of this experiment (that is, the expected number of successes in 5 trials) is the sum

$$0 \cdot P(0) + 1 \cdot P(1) + 2 \cdot P(2) + 3 \cdot P(3) + 4 \cdot P(4) + 5 \cdot P(5)$$

$$= 0\left(\frac{32}{243}\right) + 1\left(\frac{80}{243}\right) + 2\left(\frac{80}{243}\right) + 3\left(\frac{40}{243}\right) + 4\left(\frac{10}{243}\right) + 5\left(\frac{1}{243}\right)$$

$$= \frac{405}{243} = 1\frac{2}{3} \quad \text{or} \quad \frac{5}{3}.$$

This result agrees with our intuition because there is, on average, 1 success in 3 trials; so we would expect 2 successes in 6 trials, and a bit less than 2 in 5 trials. From another point of view, since the probability of success is 1/3 in each trial and 5 trials were performed, the expected number of successes should be $5 \cdot (1/3) = 5/3$. More generally, we can show the following.

> The expected number of successes in n binomial trials is np, where p is the probability of success in a single trial.

The expected value of any probability distribution is actually its mean. To show this, suppose an experiment has possible numerical outcomes $x_1, x_2, \ldots, x_k$, which occur with frequencies $f_1, f_2, \ldots, f_k$, respectively. If n is the total number of items, then (as shown in Section 11.1) the mean of this grouped distribution is

$$\mu = \frac{x_1 f_1 + x_2 f_2 + \cdots + x_k f_k}{n} = \frac{x_1 f_1}{n} + \frac{x_2 f_2}{n} + \cdots + \frac{x_k f_k}{n}$$

$$= x_1\left(\frac{f_1}{n}\right) + x_2\left(\frac{f_2}{n}\right) + \cdots + x_k\left(\frac{f_k}{n}\right).$$

But in each case, the fraction f_i/n (frequency over total number) is the probability p_i of outcome x_i. Hence,

$$\mu = x_1 p_1 + x_2 p_2 + \cdots + x_k p_k,$$

which is the expected value.

We saw above that the expected value (mean) of a binomial distribution is np. It can be shown that the variance and standard deviation of a binomial distribution are also given by convenient formulas.

> $$\sigma^2 = np(1 - p) \quad \text{and} \quad \sigma = \sqrt{np(1 - p)}$$

Substituting the appropriate values for n and p from the example into this new formula gives

$$\sigma^2 = 5\left(\frac{1}{3}\right)\left(\frac{2}{3}\right) = \frac{10}{9},$$

and $\sigma = \sqrt{10/9} \approx 1.05$.   [1]

[1] Find μ and σ for a binomial distribution having $n = 120$ and $p = 1/6$.

Answer:
$\mu = 20$; $\sigma = 4.08$

A summary of these results follows.

> ## BINOMIAL DISTRIBUTION
>
> Suppose an experiment is a series of n independent repeated trials, where the probability of a success in a single trial is always p. Let x be the number of successes in the n trials. Then the probability that exactly x successes will occur in n trials is given by
>
> $$\binom{n}{x} p^x (1 - p)^{n-x}.$$
>
> The mean μ and variance σ^2 of a binomial distribution are, respectively,
>
> $$\mu = np \quad \text{and} \quad \sigma^2 = np(1 - p).$$
>
> The standard deviation is
>
> $$\sigma = \sqrt{np(1 - p)}.$$

EXAMPLE 1 The probability that a plate picked at random from the assembly line in a china factory will be defective is .01. A sample of 3 is to be selected. Write the distribution for the number of defective plates in the sample, and give its mean and standard deviation.

Since 3 plates will be selected, the possible number of defective plates ranges from 0 to 3. Here, n (the number of trials) is 3; p (the probability of selecting a defective on a single trial) is .01. The distribution and the probability of each outcome are shown below.

x	$P(x)$
0	$\binom{3}{0}(.01)^0(.99)^3 \approx .970$
1	$\binom{3}{1}(.01)(.99)^2 \approx .029$
2	$\binom{3}{2}(.01)^2(.99) \approx .0003$
3	$\binom{3}{3}(.01)^3(.99)^0 \approx .000001$

The mean of the distribution is

$$\mu = np = 3(.01) = .03.$$

The standard deviation is

$$\sigma = \sqrt{np(1 - p)} = \sqrt{3(.01)(.99)} = \sqrt{.0297} \approx .17. \quad \blacksquare \quad \boxed{2}$$

2 The probability that a can of soda from a certain plant is defective is .005. A sample of 4 cans is selected at random. Write a distribution for the number of defective cans in the sample, and give its mean and standard deviation.

Answer:

x	$P(x)$
0	.98
1	.0197
2	.00015
3	.0000005
4	6.25×10^{-10}

$\mu = .02, \sigma = .14$

Binomial distributions are very useful, but the probability calculations become difficult if n is large. If you do not have a calculator or computer that does binomial distributions, the standard normal distribution of the previous section (which shall be referred to as the normal distribution from now on) can be used to get a good approximation of binomial distributions.

As an example, to show how the normal distribution is used, consider the distribution for the expected number of heads if 1 coin is tossed 15 times. This is a binomial distribution with $p = 1/2$ and $n = 15$. The distribution is shown with Figure 11.17. The mean of the distribution is

$$\mu = np = 15\left(\frac{1}{2}\right) = 7.5.$$

The standard deviation is

$$\sigma = \sqrt{np(1 - p)} = \sqrt{15\left(\frac{1}{2}\right)\left(1 - \frac{1}{2}\right)}$$

$$= \sqrt{15\left(\frac{1}{2}\right)\left(\frac{1}{2}\right)} = \sqrt{3.75} \approx 1.94.$$

In Figure 11.17 the normal curve with $\mu = 7.5$ and $\sigma = 1.94$ has been superimposed over the histogram of the binomial distribution.

x	$P(x)$
0	.00003
1	.00046
2	.00320
3	.01389
4	.04166
5	.09164
6	.15274
7	.19638
8	.19638
9	**.15274**
10	.09164
11	.04166
12	.01389
13	.00320
14	.00046
15	.00003

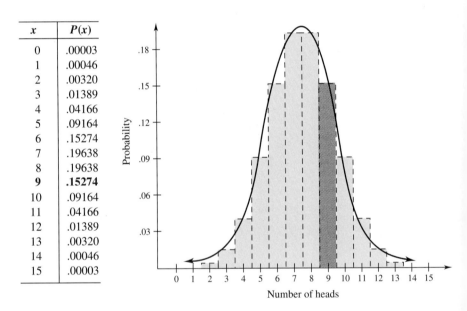

FIGURE 11.17

As shown in the distribution, the probability of getting exactly 9 heads in the 15 tosses is approximately .153. This answer is about the same fraction that would be found by dividing the area of the bar in color in Figure 11.17 by the total area of all 16 bars in the graph. (Some of the bars at the extreme left and right ends of the graph are too short to show up.)

As the graph suggests, the area in color is approximately equal to the area under the normal curve from $x = 8.5$ to $x = 9.5$. The normal curve is higher than the top of the bar in the left half but lower in the right half.

To find the area under the normal curve from $x = 8.5$ to $x = 9.5$, first find z-scores, as in the previous section. Use the mean and the standard deviation for the distribution, which we have already calculated, to get z-scores for $x = 8.5$ and $x = 9.5$.

For $x = 8.5$,

$$z = \frac{8.5 - 7.5}{1.94}$$

$$= \frac{1.00}{1.94}$$

$$z \approx .52.$$

For $x = 9.5$,

$$z = \frac{9.5 - 7.5}{1.94}$$

$$= \frac{2.00}{1.94}$$

$$z \approx 1.03.$$

3 Use the normal distribution to find the probability of getting exactly the following number of heads in 15 tosses of a coin.

(a) 7

(b) 10

Answers:
(a) .1985
(b) .0909

From Table 2, $z = .52$ gives an area of .1985, while $z = 1.03$ gives .3485. Find the required result by subtracting these two numbers.

$$.3485 - .1985 = .1500$$

This answer, .1500, is close to the answer, .153, found above.   **3**

TECHNOLOGY TIP If your graphing calculator does not give binomial distribution values, but does compute areas under the normal curve, it can be used to approximate binomial distribution values in the same way. Choose the option that gives the area under the normal curve between two z-values: enter .52 and 1.03. The result is .1500267505, which agrees with the answer found using the table. ✔

EXAMPLE 2   About 6% of the bolts of cloth produced by a certain machine have defects.

(a) Find the approximate probability that in a sample of 100 bolts, 3 or fewer have defects.

This problem satisfies the conditions of the definition of a binomial distribution, so the normal curve approximation can be used. First find the mean and the standard deviation using $n = 100$ and $p = 6\% = .06$.

$$\mu = 100(.06) \qquad \sigma = \sqrt{100(.06)(1 - .06)}$$
$$\mu = 6 \qquad\qquad = \sqrt{100(.06)(.94)}$$
$$= \sqrt{5.64}$$
$$\sigma \approx 2.37$$

As the graph of Figure 11.18 on the next page shows, the area to the left of $x = 3.5$ (since we want 3 or fewer bolts with defects) must be found. For $x = 3.5$,

$$z = \frac{3.5 - 6}{2.37} = \frac{-2.5}{2.37} \approx -1.05.$$

From Table 2, $z = -1.05$ corresponds to an area of .3531. Find the area to the left of z by subtracting .3531 from .5000 to get .1469. The approximate probability of 3 or fewer bolts with defects in a sample of 100 bolts is .1469, or 14.69%. (If you use a graphing calculator to approximate the area, use -4 as the lower bound. This works because the area under the normal curve between -4 and 4 is essentially equal to 1.)

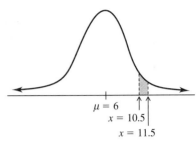

FIGURE 11.18

FIGURE 11.19

4 About 9% of the transistors produced by a certain factory are defective. Find the approximate probability that in a sample of 200, the following numbers of transistors will be defective. (*Hint:* $\sigma = 4.05$)

(a) Exactly 11

(b) 16 or fewer

(c) More than 14

Answers:

(a) .0226

(b) .3557

(c) .8051

(b) Find the probability of exactly 11 bolts with defects in a sample of 100 bolts. As Figure 11.19 shows, the area between $x = 10.5$ and $x = 11.5$ must be found.

$$\text{If } x = 10.5, \text{ then } z = \frac{10.5 - 6}{2.37} = 1.90.$$

$$\text{If } x = 11.5, \text{ then } z = \frac{11.5 - 6}{2.37} = 2.32.$$

Look in Table 2; $z = 1.90$ gives an area of .4713, while $z = 2.32$ yields, .4898. The final answer is the difference of these numbers, or

$$.4898 - .4713 = .0185.$$

The probability of having exactly 11 bolts with defects is .0185. ■ **4**

CAUTION The normal curve approximation to a binomial distribution is quite accurate *provided that n* is large and *p* is not close to 0 or 1. As a rule of thumb, the normal curve approximation can be used as long as both *np* and $n(1 - p)$ are at least 5. ◆

11.4 EXERCISES

1. What three conditions must be satisfied for a distribution to be binomial?

2. What must be known to find the mean and standard deviation of a binomial distribution?

In Exercises 3–6, binomial experiments are described. For each one (a) write the distribution; (b) find the mean; (c) find the standard deviation. (See Example 1.)

3. Management According to a recent survey, fifty percent of cereals are bought with a coupon. Six cereal purchases are selected randomly and checked for use of a coupon. Write the distribution for the number purchased with a coupon.*

4. Management Private labels (not nationally advertised) account for 7% of cold cereal sales. The label on each of six randomly selected purchases is noted. Write the distribution of the number with a private label.*

5. Social Science In the United States 14% of people ages five and older speak a language other than English at home. Five people are interviewed. Write the distribution of the number speaking a language other than English at home.

6. Management To maintain quality control on the production line, the Bright Lite Company randomly selects three light bulbs each day for testing. Experience has shown a defective rate of .02. Write the distribution for the number of defectives in the daily samples.

Natural Science *Work the following exercises involving binomial experiments.*

7. The probability that an infant will die in the first year of life is about .025. In a group of 500 babies, what are the mean and standard deviation of the number of babies who can be expected to die in their first year of life?

8. The probability that a particular kind of mouse will have a brown coat is 1/4. In a litter of 8, assuming independence, how many could be expected to have a brown coat? With what standard deviation?

9. The probability that a newborn infant will be a girl is .49. If 50 infants are born on Susan B. Anthony's birthday, how many can be expected to be girls? With what standard deviation?

10. What is the rule of thumb for using the normal distribution to approximate a binomial distribution?

11. Social Science According to the National Center for Education Statistics, 79% of the U.S. holders of Bachelor's degrees in Education in 1992–1993 were women.* Suppose five holders of Bachelor's degrees in Education are picked at random.
  (a) Find the expected value of the number of women selected.
  (b) Find the probability that four of the five are women.
  (c) Find the probability that at least two of the five are women.

For the remaining exercises, use the normal curve approximation to a binomial distribution. (See Example 2.)

Suppose 16 coins are tossed. Find the probability of getting exactly

12. 8 heads; **13.** 7 heads; **14.** 10 tails.

Suppose 1000 coins are tossed. Find the probability of getting each of the following.

15. Exactly 500 heads

16. Exactly 510 heads

17. 480 heads or more

18. Fewer than 470 tails

19. Fewer than 518 heads

20. More than 550 tails

A die is tossed 120 times. Find the probability of getting each of the following. (Hint: $\sigma = 4.08$)

21. Exactly 20 fives

22. Exactly 24 sixes

23. Exactly 17 threes

24. Exactly 22 twos

25. More than 18 threes

26. Fewer than 22 sixes

Management *Two percent of the quartz heaters produced in a certain plant are defective. Suppose the plant produced 10,000 such heaters last month. Find the probability that among these heaters, the following numbers were defective.*

27. Fewer than 170 **28.** More than 222

29. Natural Science For certain bird species, with appropriate assumptions, the number of nests escaping predation has a binomial distribution.† Suppose the probability of success (that is, a nest escaping predation) is .3. Find the probability that at least half of 26 nests escape predation.

30. Natural Science Under certain appropriate assumptions, the probability of a competing young animal eating x units of food is binomially distributed, with n equal to the maximum number of food units the animal can acquire, and p equal to the probability per time unit that an animal eats a unit of food.* Suppose $n = 120$ and $p = .6$.
  (a) Find the probability that an animal consumes 80 units of food.
  (b) Suppose the animal must consume at least 70 units of food to survive. What is the probability that this happens?

31. Social Science In a random sample of 100 drivers on Interstate 10 in Texas, 29% of the drivers exceeded the 70-mph speed limit. Find the probability of finding between 30 and 35 speeders in a random sample of 100.

32. Social Science A recent study of minimum wage earners found that 25.6% of them are teenagers.† Suppose a random sample of 60 minimum wage earners is selected. What is the probability that more than one-third of them are teenagers?

Use a graphing calculator or computer and the normal curve to approximate the following binomial probabilities. Compare these answers with those obtained by other methods in Exercises 42–45 of Section 10.3.

33. Natural Science A flu vaccine has a probability of 80% of preventing a person who is inoculated from getting the flu. A county health office inoculates 134 people. Find the probabilities of the following.
  (a) Exactly 10 of the people inoculated get the flu.
  (b) No more than 10 of the people inoculated get the flu.
  (c) None of the people inoculated get the flu.

34. Natural Science The probability that a male will be color-blind is .042. Find the probabilities that in a group of 53 men, the following will be true.
  (a) Exactly 5 are color-blind.
  (b) No more than 5 are color-blind.
  (c) At least 1 is color-blind.

35. Management The probability that a certain machine turns out a defective item is .05. Find the probabilities that in a run of 75 items, the following results are obtained.
  (a) Exactly 5 defectives
  (b) No defectives
  (c) At least 1 defective

36. Management A company is taking a survey to find out whether people like its product. Their last survey indicated that 70% of the population likes the product. Based on that, of a sample of 58 people, find the probabilities of the following.
  (a) All 58 like the product.
  (b) From 28 to 30 (inclusive) like the product.

---

*The New York Times, Jan. 7, 1996, Education Life, p. 24.

†From H. M. Wilbur, American Naturalist, vol. 111.

*From G. deJong, American Naturalist, vol. 110.

†The Chicago Tribune, Dec. 28, 1995, pp. 1, 14.

CHAPTER 11   SUMMARY

Key Terms and Symbols

11.1 Σ summation (sigma) notation
       $\bar{x}$ sample mean
       random sample
       grouped frequency distribution
       frequency polygon
       mean (arithmetic average)
       median
       mode

11.2 s^2 sample variance
       s sample standard deviation
       range
       deviations from the mean

population variance
population standard deviation

11.3 μ mean of a normal distribution
       σ standard deviation of a normal distribution
       continuous distribution
       skewed distribution
       normal distributions
       normal curves
       standard normal curve
       z-score

11.4   binomial distribution

Key Concepts

To organize the data from a sample, we use a **grouped frequency distribution,** a set of intervals with their corresponding frequencies. The same information can be displayed with a histogram, a bar graph with a bar for each interval. Each bar has width 1 and height equal to the probability of the corresponding interval. Another way to display this information is with a **frequency polygon,** which is formed by connecting the midpoints of consecutive bars of the histogram with straight line segments.

The **mean** $\bar{x}$ of a frequency distribution is the expected value.

For n numbers $x_1, x_2, \ldots, x_n$ For a grouped distribution

$$\bar{x} = \frac{\Sigma x}{n}. \qquad\qquad\qquad \bar{x} = \frac{\Sigma(xf)}{n}.$$

The **median** is the middle entry in a set of data arranged in either increasing or decreasing order.

The **mode** is the most frequent entry in a set of numbers.

The **range** of a distribution is the difference between the largest and smallest numbers in the distribution.

The **sample standard deviation** s is the square root of the **variance.**

For n numbers   For a grouped distribution

$$s = \sqrt{\frac{\Sigma x^2 - n\bar{x}^2}{n - 1}}. \qquad\qquad s = \sqrt{\frac{\Sigma fx^2 - nx^2}{n - 1}}.$$

A **normal distribution** is a continuous distribution with the following properties: The highest frequency is at the mean; the graph is symmetric about a vertical line through the mean; the total area under the curve, above the x-axis, is 1. If a normal distribution has mean μ and standard deviation σ, then the z-score for the number x is $z = \dfrac{x - \mu}{\sigma}$.

Area Under a Normal Curve   The area under a normal curve between $x = a$ and $x = b$ gives the probability that an observed data value will be between a and b.

> The **binomial distribution** is a distribution with the following properties: For n independent repeated trials, where the probability of success in a single trial is p, the probability of x successes is $\binom{n}{x}p^x(1-p)^{n-x}$. The mean is $\mu = np$ and the standard deviation is
>
> $$\sigma = \sqrt{np(1-p)}.$$

Chapter 11 Review Exercises

1. Discuss some reasons for organizing data into a grouped frequency distribution.

2. What is the rule of thumb for an appropriate interval in a grouped frequency distribution?

In Exercises 3 and 4, (a) write a frequency distribution; (b) draw a histogram; (c) draw a frequency polygon.

3. The following numbers give the sales in dollars for the lunch hour at a local hamburger store for the last twenty Fridays. (Use intervals 450–474, 475–499, and so on.)

   480 451 501 478 512 473 509 515 458 566
   516 535 492 558 488 547 461 475 492 471

4. The number of units carried in one semester by the students in a business mathematics class was as follows. (Use intervals of 9–10, 11–12, 13–14, 15–16.)

   10   9 16 12 13 15 13 16 15 11 13
   12 12 15 12 14 10 12 14 15 15 13

Find the mean for each of the following.

5. 41, 60, 67, 68, 72, 74, 78, 83, 90, 97

6. 105, 108, 110, 115, 106, 110, 104, 113, 117

7.
Interval	Frequency
10–19	6
20–29	12
30–39	14
40–49	10
50–59	8

8.
Interval	Frequency
40–44	2
45–49	5
50–54	7
55–59	10
60–64	4
65–69	1

9. What do the mean, median, and mode of a distribution have in common? How do they differ? Describe each in a sentence or two.

Find the median and the mode (or modes) for each of the following.

10. 32, 35, 36, 44, 46, 46, 59

11. 38, 36, 42, 44, 38, 36, 48, 35

The modal class is the interval containing the most data values. Find the modal class for the distributions of

12. Exercise 7 above;

13. Exercise 8 above.

14. What is meant by the range of a distribution?

15. How are the variance and the standard deviation of a distribution related? What is measured by the standard deviation?

Find the range and standard deviation for each of the following distributions.

16. 14, 17, 18, 19, 32

17. 26, 43, 51, 29, 37, 56, 29, 82, 74, 93

Find the standard deviation for the following.

18. Exercise 7 above

19. Exercise 8 above

20. Describe the characteristics of a normal distribution.

21. What is meant by a skewed distribution?

22. Find the percent of the area under a normal curve within 2.5 standard deviations of the mean.

Find the following areas under the standard normal curve.

23. Between $z = 0$ and $z = 1.27$

24. To the left of $z = .41$

25. Between $z = -1.88$ and $z = 2.10$

26. Find a z-score such that 8% of the area under the curve is to the right of z.

27. Why is the normal distribution not a good approximation of a binomial distribution that has a value of p close to 0 or 1?

Management   *The table gives the probability distribution of U.S. buyers of recorded music in 1995 by age.**

Age	Probability
10–14	.08
15–19	.171
20–24	.153
25–29	.121
30–34	.123
35–39	.108
40–44	.075
45 & up	.169

28. **(a)** Draw a histogram for this distribution.
    (b) From the shape of the histogram, is this a normal distribution?

---

*Recording Industry Association of America.

29. As shown in the distribution above, 16.9% of buyers of recorded music are age 45 or over. Assume a binomial distribution with $n = 6$ and $p = .169$.
 (a) Find the probability that 4 of 6 randomly selected recorded music buyers are age 45 or over.
 (b) Find the mean and standard deviation of the binomial distribution. Interpret your results.

30. Management The annual returns of two stocks for three years are given below.

	1997	*1998*	*1999*
Stock I	11%	−1%	14%
Stock II	9%	5%	10%

 (a) Find the mean and standard deviation for each stock over the three-year period.
 (b) If you are looking for security with an 8% return, which of these two stocks would you choose?

31. Natural Science The weight gains of two groups of 10 rats fed on two different experimental diets were as follows.

	Weight Gains									
Diet A	1	0	3	7	1	1	5	4	1	4
Diet B	2	1	1	2	3	2	1	0	1	0

Compute the mean and standard deviation for each group and compare them to answer the questions below.
 (a) Which diet produced the greatest mean gain?
 (b) Which diet produced the most consistent gain?

32. Social Science Between 1980 and 1990 HMO enrollment in the United States increased as shown below.

Location	*Gains*
South	7%
Midwest	10%
Northeast	13%
West	12%

 (a) Compute the mean and standard deviation of the gains.
 (b) How many standard deviations from the mean is the greatest gain? the smallest gain?

Social Science *On standard IQ tests, the mean is 100, with a standard deviation of 15. The results are very close to fitting a normal curve. Suppose an IQ test is given to a very large group of people. Find the percent of people whose IQ score is*

33. more than 130;

34. less than 85;

35. between 85 and 115.

36. Management A machine that fills quart orange juice cartons is set to fill them with 32.1 oz. If the actual contents of the cartons vary normally, with a standard deviation of .1 oz, what percent of the cartons contains less than a quart (32 oz)?

Management *About 60% of the blended coffee produced by a certain company contains cheap coffee. Assume a binomial distribution and use the normal distribution to find the probability that, in a sample of 500 bags of coffee, the following is true.*

37. 310 or fewer contain cheap coffee.

38. Exactly 300 contain cheap coffee.

Natural Science *An area infested with fruit flies is to be sprayed with a chemical which is known to be 98% effective for each application. A sample of 100 flies is checked. Assume a binomial distribution and use the normal distribution to approximate the following probabilities.*

39. 95% of the flies are killed in one application.

40. At least 95% of the flies are killed in one application.

41. At least 90% of the flies are killed in one application.

42. All the flies are killed in one application. Compare your answer with the probability found using the binomial distribution.

In Exercises 43–47, assume a binomial distribution and use the normal distribution approximation.

Natural Science *Twenty percent of third-world women would like to avoid pregnancy but are not using birth control. For 50 such women, use the normal curve approximation to find the probability of each of the following.*

43. Exactly 10 are not using birth control.

44. At least 20 are using birth control.

45. Social Science About 19% of the U.S. population older than 100 are African-Americans. Find the probability that in a group of 10 Americans older than 100, 4 are African-American.

46. Social Science A poll of 2000 teenagers found that 1 in 8 reported carrying a weapon for protection.* In a typical high school with 1200 students, what is the probability that more than 120 students, but fewer than 180, carry a weapon?

---

*The New York Times, Jan. 12, 1996, p. A6.

47. Social Science Only 1 out of 12 American parents requires that children do their homework before watching TV.* In a typical neighborhood, what is the probability that out of 51 parents, 5 or fewer require their children to do homework before watching TV?

▷ **48.** Much of our work in Chapters 10 and 11 is interrelated. Note the similarities in the following parallel treatments of a frequency distribution and a probability distribution.

Frequency Distribution

Complete the table below for the following data. (Recall that x is the midpoint of the interval.)

14, 7, 1, 11, 2, 3, 11, 6, 10, 13, 11, 11, 16, 12, 9, 11, 9, 10, 7, 12, 9, 6, 4, 5, 9, 16, 12, 12, 11, 10, 14, 9, 13, 10, 15, 11, 11, 1, 12, 12, 6, 7, 8, 2, 9, 12, 10, 15, 9, 3

Interval	Tally	x	f	x·f
1–3				
4–6				
7–9				
10–12				
13–15				
16–18				

*"Harper's Index," *Harper's*, Sept. 1996, p. 15.

Probability Distribution

A binomial distribution has $n = 10$ and $p = .5$. Complete the table below.

x	P(x)	x · P(x)
0	.001	
1	.010	
2	.044	
3	.117	
4		
5		
6		
7		
8		
9		
10		

(a) Find the mean (or expected value) for each distribution.
(b) Find the standard deviation for each distribution.
(c) For both distributions, use Chebyshev's theorem to find the interval that contains 75% of the distribution. (See Section 11.2 exercises.)
(d) Use the normal approximation of the binomial probability distribution to find the interval that contains 95.44% of that distribution.
(e) Why can't we use the normal distribution to answer probability questions about the frequency distribution?

CASE 11

Statistical Process Control

Statistical process control is a method of determining when a manufacturing process is out of control, producing defective items. The procedure involves taking samples of a measurement on a product over a production run and calculating the mean and standard deviation of each sample. These results are used to determine when the manufacturing process is out of control. For example, three sample measurements from a manufacturing process on each of four days are given in the table below. The mean $\bar{x}$ and standard deviation s are calculated for each sample.

DAY	1			2			3			4		
Sample Number	*1*	*2*	*3*	*1*	*2*	*3*	*1*	*2*	*3*	*1*	*2*	*3*
Measurements	−3	0	4	5	−2	4	3	−1	0	4	−2	1
	0	5	3	4	0	3	−2	0	0	3	0	3
	2	2	2	3	1	4	0	1	−2	3	−1	0
$\bar{x}$	−1/3	7/3	3	4	−1/3	11/3	1/3	0	−2/3	10/3	−1	4/3
s	2.5	2.5	1	1	1.5	.6	2.5	1	1.2	.6	1	1.5

Next, the mean of the 12 sample means, $\overline{X}$, and the mean of the 12 sample standard deviations, $\overline{s}$, are found (using the formula for $\overline{x}$). Here, these measures are

$$\overline{X} = 1.3 \quad \text{and} \quad \overline{s} = 1.41.$$

The control limits for the sample means are given by

$$\overline{X} \pm k_1 \overline{s},$$

where k_1 is a constant found from a manual. For samples of size 3, $k_1 = 1.954$, so the control limits for the sample means are

$$1.3 \pm (1.954)(1.41).$$

The upper control limit is 4.06, and the lower control limit is -1.46.

Similarly, the control limits for the sample standard deviations are given by $k_2 \cdot \overline{s}$ and $k_3 \cdot \overline{s}$, where k_2 and k_3 also are values given in the same manual. Here, $k_2 = 2.568$ and $k_3 = 0$, with the upper and lower control limits for the sample standard deviations equal to 2.568(1.41) and 0(1.41), or 3.62 and 0. As long as the sample means are between -1.46 and 4.06 and the sample standard deviations are between 0 and 3.62, the process is in control.

EXERCISES

1. The following table gives 10 samples, of three measurements, made during a production run.

SAMPLE NUMBER									
1	*2*	*3*	*4*	*5*	*6*	*7*	*8*	*9*	*10*
2	3	-2	-3	-1	3	0	-1	2	0
-2	-1	0	1	2	2	1	2	3	0
1	4	1	2	4	2	2	3	2	2

(a) Find the mean $\overline{x}$ for each sample of three measurements.
(b) Find the standard deviation s for each sample of three measurements.
(c) Find the mean $\overline{X}$ of the sample means.
(d) Find the mean $\overline{s}$ of the sample standard deviations.
(e) Using $k_1 = 1.954$, find the upper and lower control limits for the sample means.
(f) Using $k_2 = 2.568$ and $k_3 = 0$, find the upper and lower control limits for the sample standard deviations.

2. Given the following measurements from later samples on the process in Exercise 1, decide whether the process is out of control. (*Hint:* Use the results of Exercise 1(e) and (f).)

SAMPLE NUMBER					
1	*2*	*3*	*4*	*5*	*6*
3	-4	2	5	4	0
-5	2	0	1	-1	1
2	1	1	-4	-2	-6

Graphing Calculators

PART 1   INTRODUCTION

Graphing calculators are generally superior to ordinary scientific calculators, not only for graphing (which you would expect), but also for most multistep calculations. Their screens typically display 8 lines of text, so you can see what numbers and functions have been entered. You can easily edit or change these entries, without losing your place in the computation.

Although there are graphing calculators on the market for as little as $40, you would be well advised to use care when buying the less expensive ones, many of which lack useful features that are standard on more expensive models. Before buying a graphing calculator, you should consider which features you are most likely to need. The following chart may be helpful. Many of the features mentioned in the table are discussed in the next part of this appendix.

Feature	Used in
Table Making Capability	Chapters 1–6
Root Finder (Equation Solver)	Chapters 1–5, 12–14
Financial Functions	Chapter 6
Matrix Row Operations	Chapter 7–8
System of Linear Equations Solver	Chapter 7
Statistical Operations	Chapters 9–11
Maximum/Minimum Finder	Chapters 12–14
Intersection Finder	Chapters 3–5, 7–8, 12–14
Numerical Derivatives	Chapters 12–14
Numerical Integration	Chapter 14

Current models that have most or all of these features include TI-82/83/85/86/89/92, Casio 9800/9850, HP-38/48, and Sharp 9200/9600. Most graphing calculators have additional features that do not play a role in this book (such as polar coordinates, parametric graphing, and complex numbers). A few, such as the TI-92 and HP-48, can perform symbolic operations (such as factoring and finding derivatives).

ADVICE ON USING A GRAPHING CALCULATOR

1. **BASICS**   Graphing calculators have forty-nine or more keys. Most modern desk-top computers have 101 keys on their keyboards. With fewer keys, each key must be used for more actions, so you will find special mode-changing keys such as "**2nd,**" "**shift,**" "**alpha,**" and "**mode.**" Become familiar with the capabilities of the machine, the layout of the keyboard, how to adjust the screen contrast, and so on.

2. **EDITING**   When keying in expressions, you can pause at any time and use the **arrow keys,** located at the upper right of the keyboard, to move the cursor to any point in the text. You can then make changes by using the "**DEL**" key to delete and the "**INS**" key to insert material. ("INS" is the "second function" of "DEL.") After an expression has been entered or a calculation made, it can still be edited by using the **edit/replay** feature, available on almost every calculator model. On TI calculators (except TI-81), use "2nd, ENTER" to return to the previously entered expression. On Casio calculators, use the left or right arrow key, and on Sharp models and HP-38 use the up arrow key.

3. **SCIENTIFIC NOTATION**   Learn how to enter and read data in **scientific notation** form. This form is used when the numbers become too large or too small (too many zeros between the decimal point and the first significant digit) for the machine's display.

4. **FUNCTION GRAPHING**
   A. **Function Memory**   Learn how to enter functions in the "**function memory**" (labeled "**Y=**" on TI, "**SYMB**" on HP-38, and "**EQTN**" on Sharp), and to mark them to be graphed. Depending on the calculator, you can store from 4 to 99 functions in the function memory, so that you don't need to key them in each time you want to use them.
   B. **Viewing Window**   Use the "**RANGE**" key (labeled **WINDOW** or "**PLOT SETUP**" on some calculators) to determine what part of the coordinate plane will appear on the screen. You must key in the minimum and maximum values for x and y as well as the distance between tick marks on the axes (for instance, Xscl = 1 and Yscl = 2 means that tick marks will be one unit apart on the x-axis and 2 units apart on the y-axis).
   C. **Graphing Mode**   Normally your calculator is set to graph in "**connected mode,**" meaning that it plots the points and connects all of them with a continuous curve (essentially what you usually do when graphing by hand, except that the calculator plots more points). Sometimes—particularly when there should be breaks in the graph, such as a vertical asymptote—connected mode graphing produces misleading or inaccurate graphs. On these occasions, you may want to change the graphing mode to "**DOT,**" so that the calculator will plot the points, but won't connect them.
   D. **Trace**   With the "**Trace**" feature, the left/right arrow keys can be used to move the cursor along the last curve plotted, and the values of x and y will be displayed for each point plotted on the screen. On most calculators, if more than one graph was plotted, you can move the cursor vertically between the different graphs by using the up/down arrow keys.
   E. **Zoom**   The "**zoom**" feature allows a quick redrawing of your graph using smaller ranges of values for x and y ("**zoom in**") or larger ranges of values ("**zoom out**"). Thus one can easily examine the behavior of a function within the close vicinity of a particular point or the general behavior as seen from farther away.

F. Plot With "**Plot**" or "**Draw**" you can move the cursor to any point on the screen and either have the machine plot a single point or have it display the "screen coordinates" of the point. For example, on the Casio, the command "Plot 2,3" will cause this point to be plotted on the graph screen. However, the "screen coordinates" will usually be approximations of the specified values.

SOLVING EQUATIONS

1. EQUATIONS OF THE FORM $f(x) = 0$ A graphing calculator produces highly accurate approximations of the solutions of equations such as

$$x^3 - 2x^2 + x - 1 = 0.$$

To solve this equation, graph the function $f(x) = x^3 - 2x^2 + x - 1$ (see Figure 1). The places where the graph touches the x-axis (the x-intercepts of $y = f(x)$) are the values of x that make $f(x) = 0$, that is, the solutions of the equation. Figure 1 suggests that in this case, there is just one x-intercept, located between $x = 1$ and $x = 2$. On a typical calculator it can be precisely located in several ways.

Graphical Root Finder In the CALC or MATH or G-SOLVE or FCN menu, look for a key labeled "ZERO" or "ROOT." The syntax for this command varies with the calculator. On some calculators, you will be asked to specify a lower and upper bound (that is, numbers on either side of the x-intercept you are seeking) and possibly to make an initial guess. Check your instruction manual for details.* The root finder on a TI-83 shows that the solution is $x \approx 1.7548777$ (Figure 2).

Zoom-in Use the zoom-in feature of the calculator, or repeatedly change the range settings by hand, so that only a very tiny portion of the graph near the x-intercept is shown. Figure 3, in which the endpoints of the x-axis are 1.7 and 1.8, shows the final window in such a process. The tick marks on the x-axis are .01 unit apart and the desired x-intercept is between 1.75 and 1.76, at approximately 1.754 or 1.755. This is very similar to the approximation obtained by the Root Finder.

FIGURE 1

FIGURE 2

FIGURE 3

Algebraic Equation Solver Some calculators have an algebraic equation solver as well, which will solve an equation without the need to graph it first. Check your instruction manual for the correct syntax.

---

*Root finder programs for TI-81 and Casio 77/8700 are in the Program Appendix.

2. SYSTEMS OF TWO EQUATIONS IN TWO UNKNOWNS Graphing calculators can solve systems of equations in two variables by finding intersection points. For example to solve the system

$$y = x^3 - 2x^2 + x - 3$$
$$y = 4x^2 - 3x - 7,$$

graph both equations on the same screen, as in Figure 4. The points where the graphs intersect are the points (x, y) that satisfy both equations, that is, the solutions of the system. They can be found in two ways.

Graphical Intersection Finder In the CALC or MATH or G-SOLVE or FCN menu, look for a key labeled "INTERSECTION" or "ISCT." The syntax for this command varies with the calculator. On some calculators, you will be asked to specify the two curves and possibly to make an initial guess. Check your instruction manual for details. The intersection finder on a TI-83 (Figure 5) shows that the intersection point to the right of the y-axis has coordinates

$$x \approx 1.482696 \quad \text{and} \quad y \approx -2.654539.$$

This is one solution of the system.

Zoom-in Use the zoom-in feature of the calculator, or repeatedly change the range settings by hand, so that only a very tiny portion of the graph near one of the intersection points is shown. Figure 6 (graphed in "grid on" mode) shows the final window in such a process. The grid is determined by the tick marks on the axes, which are .01 unit apart. Using the trace feature we estimate that the intersection point has coordinates $(-.534, -4.257)$. So the other solution of the system is $x \approx -.534$ and $y \approx -4.257$.

FIGURE 4

FIGURE 5

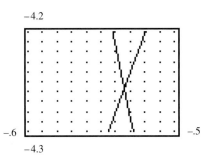

FIGURE 6

3. SYSTEMS OF LINEAR EQUATIONS Systems of two linear equations in two unknowns can be solved by the methods of the preceding paragraphs. Some calculators have a built-in solver for systems of linear equations and all graphing calculators have matrix capabilities that enable you to solve larger systems by using row operations or matrix inverses. These techniques are described in Chapter 7 of this book.

FINDING MAXIMUM AND MINIMUM FUNCTION VALUES

The graph of $f(x) = x^3 + x^2 - 3x - 2$ is shown in Figure 7. A graphing calculator can find the local maximum and minimum values of this function (which correspond graphically to the "top of the hill" to the left of the y-axis and the "bottom of the valley" to the right of the y-axis). As with roots and intersections, this can be done in several ways.

Maximum/Minimum Finder   To find the local maximum (top of the hill), look in the CALC or MATH or G-SOLVE or FCN menu for a key labeled "MAXIMUM" or "MAX" or "EXTREMUM." The syntax for this command varies with the calculator. On some calculators, you will be asked to specify upper and lower bounds and possibly to make an initial guess. Check your instruction manual for details. The maximum finder on a TI-83 (Figure 8) shows that the local maximum occurs when $x \approx -1.3874$ and $f(x) \approx 1.4165$.

Zoom-in   Use the zoom-in feature of the calculator, or repeatedly change the range settings by hand, so that only a very tiny portion of the graph near the local minimum point is shown. Figure 9 (in "grid on" mode) shows the final window in such a process. The grid is determined by the tick marks on the axes, which are .01 unit apart on the x-axis and .001 unit apart on the y-axis. Using the trace feature we estimate that the local minimum point has approximate coordinates $(.721, -3.268)$. So the local minimum occurs when $x \approx .721$ and $f(x) \approx -3.268$.

FIGURE 7   FIGURE 8   FIGURE 9

PROGRAMMING

Many relatively complicated formulas and processes must be used frequently (for example, the various financial formulas in Chapter 6 and the simplex method in Chapter 8). In such cases you can use the programming feature of a graphing calculator to automate this process. Once a program has been entered into the calculator, it can be called up whenever needed. Typically, the program asks the user to enter the necessary data (for instance, the amount of the loan, the interest rate, and the number of payments) and then quickly produces the desired information (for example, the monthly payment for the loan). Programming syntax and procedures vary from calculator to calculator, so check your instruction manual.

Part 2 of this appendix contains useful programs for many graphing calculators. These programs are of two types: programs to give older calculators some of the features that are built-in to newer ones (such as a table maker for TI-81/85 or a root finder for TI-81 and Casio 7700) and programs to do specific tasks (such as financial formulas, amortization tables, and the simplex method).

SOME SUGGESTIONS FOR REDUCING FRUSTRATION

We all find ways to make even the simplest machines do the wrong things without even trying. One of the more common problems with graphing calculators is getting a blank screen when a graph was expected. This usually results from not setting the "**Range**" values appropriately before graphing the function, although it could also easily result from incorrectly entering the function.

Another common error is having more right parentheses than left parentheses. To confuse us further, these same calculators think it is perfectly OK to have more left parentheses than right parentheses. For instance, the expression 5(3 − 4(2 + 7) has two left parentheses and one right parenthesis, but it will be evaluated as 5(3 − 4(2 + 7)).

The message "**Ma ERROR**" appears when a number is too large or when a number is not allowed. If you try to find the 1000th power of ten or divide a number by zero you will most certainly see some kind of error message. By pressing one of the "**cursor**" keys on the Casio you will see the cursor blinking at the location of the error in your expression. When the TI models detect an error, they display a special menu that lists a code number and a name for the type of error. For certain types of errors the choice "**Go to error**" is offered. The TI-85 will display the number 9.99999999 E999 but shows "ERROR 01 OVERFLOW" for 10 E999 (which means 10 times the 999th power of ten). The latter and other more advanced calculators display "(0,2)" when asked to find the square root of −4. The "(0,2)" represents the complex number $0 + 2i$.

When graphing **rational functions** you sometimes will see strange little "blips" in an otherwise smooth graph. This usually means there is a **vertical asymptote** at that location due to a division by zero, but you cannot see the actual behavior there because your "**window**" is too large. "**Zoom in**" on the graph in the region of the irregularity to obtain a better view. Also, as mentioned earlier in this section, one can switch to **dot mode** to eliminate unwanted lines in the graph. (See Section 4.4.)

SOME FINAL COMMENTS

While studying mathematics it is important to learn the mathematical concepts well enough to make intelligent decisions about when to use and when not to use "high tech" aids such as computers and graphing calculators. These machines make it easy to experiment with graphs of mathematical relations. One can learn much about the behavior of different types of functions by playing "what if" games with the formulas. However, in a timed test situation you may find yourself spending too much time working with the graphing calculator when a quick algebraic solution and a rough sketch with pencil and paper are more appropriate.

To get the most return on your investment, learn to use as many features of your calculator as possible. Of course, some of the features may not be of use to you, so feel free to ignore them. A first session of two or three hours with your graphing calculator and your user's manual is essential. Be sure to keep your manual handy, referring to it when needed.

A final word of caution: These calculators are fun to use, but they can be addictive. So set time limits for yourself, or you may find that your graphing calculator has been more of a detriment than a help!

PART 2   **PROGRAM APPENDIX**

CONTENTS

The programs listed here are of two types: programs to give older calculators some of the features that are built-in on newer ones (such as a root finder) and programs to do specific tasks discussed in this book (such as linear programming). Each program is preceded by a *Description,* which describes, in general terms, how the program operates and what it does. Some programs require that certain things be done before the program is run (such as entering a function in the function memory); these requirements are listed as *Preliminaries.* Occasionally, italic remarks appear in brackets after a program step; they are *not* part of the program, but are intended to provide assistance when you are entering the program into your calculator. A remark such as "[*MATH NUM menu*]" means that the symbols or commands needed for that step of the program are in the NUM submenu of the MATH menu.

FRACTION CONVERSION FOR TI-81, CASIO, AND SHARP

Description: Enter a repeating decimal; the program converts it into a fraction. The denominator is displayed on the last line and the numerator on the line above.

TI-81
:ClrHome
:Lbl 2
:Input N
:0 → D
:Lbl 1
:D + 1 → D
:If fPart round (ND, 7) ≠ 0   [*MATH NUM menu*]
:Goto 1

:Int (ND + .5) → N
:Disp N
:Disp D

CASIO

Fix 7
Lbl 2
"N =" ? → N
0 → D
Lbl 1
D + 1 → D
N × D
Rnd　[*MATH NUM menu*]
(Frc Ans) ≠ 0 ⇒ Goto 1　[*MATH NUM menu*]
(Ans + .5) → N
Norm　[*DISP menu*]
(Int N)◢
D

SHARP

Input n
d = 0
Label 1
d = d + 1
If fpart $(10^7 \cdot n \cdot d) \geq .5$ Goto 3　[*MATH menu*]
t = ipart $(10^7 \cdot n \cdot d)/10^7$　[*MATH menu*]
Goto 4
Label 3
t = (ipart $(10^7 \cdot n \cdot d) + 1)/10^7$
Label 4
If fpart t ≠ 0 Goto 1
n = ipart (n · d + .5)
Print n
Print d

GRAPHICAL ROOT FINDER, MAX/MIN FINDER, AND INTERSECTION FINDER FOR TI-81

Preliminaries: Enter the function whose roots or extrema are to be found in the function memory as Y_1. If an intersection point is to be found, enter the other function as Y_2. Set the RANGE so that the desired portions of the graph(s) will be displayed.
Description: Select root, intersection, or min/max. The appropriate function(s) are graphed. Use the arrow keys to move the cursor close to the desired root (x-intercept), intersection point, or extremum. Press ENTER, and its approximate coordinates are displayed. (If the calculator continues working for a long time without displaying an answer, press ON to quit the program. Then run it again, but pick a different point than before near the point you want to find.)

:ClrHome
:Disp "SELECT NUMBER, PRESS ENTER"

```
:Disp "1 = ROOT"
:Disp "2 = INTERSECTION"
:Disp "3 = MIN/MAX"
:Input C
:If C = 1
:Goto 1
:If C = 2
:Goto 2
:Goto 3
:Lbl 2
:"Y₁ − Y₂" → Y₄   [Y-VARS menu]
:Y₁ − On   [Y-VARS ON menu]
:Y₂ − On
:Y₃ − Off   [Y-VARS OFF menu]
:Y₄ − Off
:Input
:Goto 4
:Lbl 3
:"NDeriv(Y₁, .000001)" → Y₄   [MATH menu]
:Y₁ − On
:Y₂ − Off
:Y₃ − Off
:Y₄ − Off
:Input
:Goto 4
:Lbl 1
:Y₂ − Off
:Y₃ − Off
:"Y₁" → Y₄
:Input
:Lbl 4
:X − Y₄/NDeriv(Y₄, .000001) → G   [Use the X|T key for X.]
:If abs (X − G) < .0000000001
:Goto 5
:G → X
:Goto 4
:Lbl 5
:ClrHome   [Optional]
:G → X
:Disp "X = "
:Disp X
:Disp "Y = "
:Disp Y₁
```

ROOT FINDER FOR CASIO 7700/8700

Preliminaries: Enter the function whose roots are to be found in the equation memory as f_1.

Description: Enter two numbers; the program uses the bisection method to find a root (*x*-intercept) of the function that lies between the two numbers (if there is one).

"LOWER BOUND"? → A
"UPPER BOUND"? → B
0 → K
Lbl 1
B → X   [*Use the variable key for X.*]
f_1 → C   [*7700/8700 FMEM menu*]
C = 0 ⇒ Goto 5
((A + B) ÷ 2) → M
M → X
f_1 → D
CD < 0 ⇒ Goto 2
M → A
Goto 3
Lbl 2
1 → K
M → B
Lbl 3
abs(B − A) < .000000001 ⇒ Goto 4
Goto 1
Lbl 4
(B + A) ÷ 2 → M
K = 0 ⇒ Goto 7
Lbl 8
"ROOT = "
M
Goto 6
Lbl 7
"NO SIGN CHANGE"
Goto 6
Lbl 5
A → M
Goto 8
Lbl 9
B → M
Goto 8
Lbl 6

QUADRATIC FORMULA

Description: Enter the coefficients of the quadratic equation $ax^2 + bx + c = 0$; the program finds all real solutions.

TI-81/82/83
:ClrHome
:Disp "AX^2 + BX + C = 0"   [*Optional*]

```
:Prompt A†
:Prompt B†
:Prompt C†
:(B² − 4AC) → S
:If S < 0
:Goto 1
:Disp (−B + √S̄)/2A
:Disp (−B − √S̄)/2A
:Stop
:Lbl 1
:Disp "NO REAL ROOTS"
```

SHARP 9300

[Gives both real and complex solutions when complex mode is selected.]

```
Print "ax² + bx + c = 0"
x₁ = (1/(2a))(−b + √(b² − 4ac) )
Print x₁
x₂ = (1/(2a))(−b − √(b² − 4ac) )
Print x₂
```

CASIO 7700/8700

```
"AX² + BX + C = 0"   [Optional]
"A =" ? → A
"B =" ? → B
"C =" ? → C
B² − 4AC → D
D < 0 ⟹ Goto 1
(−B + √D̄) ÷ (2A)
(−B − √D̄) ÷ (2A)
Goto 2
Lbl 1
"NO REAL ROOTS"
Lbl 2
```

---

†*On TI-81, replace* Prompt A *by two lines:* :Disp "A = " *and* :Input A.

TABLE MAKER FOR TI-81 AND 85

Preliminaries: Enter the function to be evaluated in the function memory as Y_1. *Description:* Select a starting point and an increment (the amount by which adjacent x entries differ); the program displays a table of function values. To scroll through the table a page at a time, press "down" or "up" on TI-85, and "0 ENTER" or "1 ENTER" on TI-81. To end the program, press "quit" on TI-85 or "2 ENTER" on TI-81. (*Note:* An error message will occur if the calculator attempts to evaluate the function at a number for which it is not defined. In this case, change the starting point or increment to avoid the undefined point.)

TI-81
:ClrHome
:Disp "START AT"
:Input A
:Disp "INCREMENT"
:Input D
:6 → Arow   [*VARS DIM menu*]
:2 → Acol   [*VARS DIM menu*]
:A → X   [*Use the variable key for X.*]
:1 → J
:Lbl 1
:X → [A](J,1)   [*Second function of 1*]
:Y_1 → [A](J,2)   [*Y-VARS menu*]
:J + 1 → J
:If J > 6
:Goto 2
:X + D → X
:Goto 1
:Lbl 2
:Disp [A]
:Disp "0 = DN 1 = UP 2 = QT"
:Input K
:If K = 2
:Stop
:If K = 1
:Goto 4
:1 → J
:X + D → X
:Goto 1
:Lbl 4
:1 → J
:X − 11D → X
:Goto 1

TI-85
:Lbl SETUP
:ClLCD
:Disp "TABLE SETUP"
:Input "TblMin = ", tblmin
:Input "ΔTbl = ", dtbl   [*CHAR GREEK menu*]
:tblmin → x   [*Use the x-var key for x.*]
:Lbl CONTD
:ClLCD
:Outpt(1,1,"x")   [*I/O menu*]
:Outpt(1,10,"y_1")
:For (cnt,2,7.1)
:Outpt(cnt,1,x)
:Outpt(cnt,8," ")
:Outpt(cnt,9,y_1)
:x + dtbl → x

:End
:Menu(1,"Down", CONTD, 2, "UP", CONTU, 4, "SETUP", SETUP, 5, "quit",
   TQUIT)
:Lbl CONTU
:x − 12*dtbl → x
:Goto CONTD
:Lbl TQUIT
:ClLCD
:Stop

PRESENT AND FUTURE VALUE OF AN ANNUITY

Description: Enter the requested information about the ordinary annuity: the periodic payment, the interest rate per period, and the number of payments. The program displays the present value and future value of this annuity.

TI
ClrHome   [*labeled ClLCD on TI-85/86*]
Disp "PAYMENT"
Input R
Disp "RATE"
Input I
Disp "NUMBER OF PYMTS"
Input N
Disp "PRESENT VALUE"
Disp round (R(1 − (1 + I)^{-N})/I,2)†   [*MATH NUM menu*]
Disp "FUTURE VALUE"
Disp round(R((1 + I)^N − 1)/I,2)

CASIO
Fix 2   [*DISP or SETUP DISPLAY menu*]
"PAYMENT"? → R
"RATE"? → I
"NUMBER OF PYMTS"? → N
"PRESENT VALUE"
R(1 − (1 + I)^{-N})/I
"FUTURE VALUE"
R((1 + I)^N − 1)/I

SHARP
Print "p = payment"
Print "i = interest"
Print "n = number of pymts"
presentvalue = p(1 − (1 + i)^{-n})/i
Print presentvalue
futurevalue = p((1 + i)^n − 1)/i

---

†On TI-81, replace this line (and ones like it) by two lines: round(R(1 − (1 + I)^{-N})/I,2) → F *and* Disp F.

LOAN PAYMENT

Description: Enter the amount of the loan, the interest rate per payment period, and the number of payments; the program displays the periodic payment.

TI

ClrHome   [*labeled ClLCD on TI-85/86*]
Disp "AMOUNT"
Input P
Disp "RATE PER PERIOD"
Input I
Disp "NUMBER OF PYMTS"
Input N
Disp "PAYMENT"
Disp round $(P((1 - (1 + I)^{-N})/I)^{-1}, 2)$†   [*MATH NUM menu*]

CASIO

Fix 2   [*DISP or SETUP DISPLAY menu*]
"AMOUNT"? $\to$ P
"RATE"? $\to$ I
"NUMBER OF PYMTS"? $\to$ N
$P((1 - (1 + I)^{-N})/I)^{-1}$

SHARP

Print "p = amount"
Print "i = rate per period"
Print "n = number of pymts"
payment = $p((1 - (1 + i)^{-n})/i)^{-1}$

---

†On TI-81, replace this line by two lines: round $(P((1 - (1 + I)^{-N})/I)^{-1}, 2) \to$ F *and* Disp F.

LOAN BALANCE AFTER *n* PAYMENTS

Description: Enter the amount of the loan (starting balance), the interest rate per payment period, and the number of payments; the program displays the remaining balance. This is useful for calculating the payoff amount when a loan is repaid early.

TI

ClrHome   [*labeled ClLCD on TI-85/86*]
Disp "STARTING BALANCE"
Input b
Disp "RATE PER PERIOD"
Input I
Disp "PAYMENT"
Input P
Disp "NUMBER OF PYMTS"
Input N
$1 \to$ K
Lbl 1 [*Use P1 in place of 1 on TI-85/86.*]
$(B - P + \text{round}(B \cdot I, 2)) \to$ B   [*MATH NUM menu*]
$K + 1 \to$ K

If K $\leq$ N
Goto 1 [*Use P1 in place of 1 on TI-85/86.*]
Disp "ENDING BALANCE"
Disp B

CASIO
Fix 2   [*DISP or SETUP DISPLAY menu*]
"STARTING BALANCE"? $\to$ B
"RATE PER PERIOD"? $\to$ I
"PAYMENT"? $\to$ P
"NUMBER OF PYMTS"? $\to$ N
1 $\to$ K
Lbl 1
B $\cdot$ I
Rnd   [*MATH NUM or OPTN NUM menu*]
Ans $\to$ S
B $-$ P + S $\to$ B
K + 1 $\to$ K
K $\leq$ N $\Rightarrow$ Goto 1
"ENDING BALANCE"
B

SHARP
Print "starting balance"
Input b
Print "rate per period"
Input i
Print "payment"
Input p
Print "number of payments"
Input n
k = 1
Label 1
If fpart(100 $\cdot$ b $\cdot$ i) $\geq$ .5 Goto 2   [*MATH menu*]
t = ipart(100 $\cdot$ b $\cdot$ i)/100   [*MATH menu*]
Goto 3
Label 2
t = (ipart(100 $\cdot$ b $\cdot$ i) + 1)/100
Label 3
b = b $-$ p + t
k = k + 1
If k $\leq$ n Goto 1
Print "ending balance"
Print b

AMORTIZATION TABLE (IN MATRIX FORM) FOR TI (EXCEPT TI-81) AND CASIO 9850

Preliminaries: No preliminaries are needed for TI calculators. Before running the program on Casio 9850, open the matrix editor and change the dimensions of Mat A to $N \times 5$, where N is the number of payments.

Description: Enter the appropriate information about the loan; the program displays an amortization table for the loan in the form of a matrix, which you can scroll through by using the arrow keys. In this table, the five vertical columns are as follows.

payment number	amount of payment	interest for period	portion to principal	new balance

To complete the program, press ENTER, and the total payments made and the total interest paid will be displayed. When the program is finished, the table will be stored as matrix A, so that you can call it up and examine it further (until the program is run again and a new table takes its place).

When the regular payment is too small, the final payment may be quite large. When the regular payment is too large, the final payment may occur before N is reached, where N is the number of payments entered at the beginning of the program. In this case the matrix will either have fewer than N rows (TI) or will have all zero entries in the rows below row N (Casio).

TI
ClrHome   *[labeled ClLCD on TI-85/86]*
Disp "STARTING BALANCE"
Input B
Disp "RATE PER PERIOD"
Input I
Disp "PAYMENT"
Input P
Disp "NUMBER OF PYMTS"
Input N
$\{N, 5\} \to$ dim [A]
$1 \to K$
$0 \to W$
Lbl 1 *[Use P1, P2 in place of 1, 2 for label numbers on TI-85/86.]*
If $K = N + 1$
Then
Pause [A]
ClrHome
Disp "TOTAL PAYMENTS"
Disp $(J - 1) \cdot P + [A](J, 2)$
Disp "TOTAL INTEREST"
Disp $W + [A] (J, 3)$
Stop
End
$K \to [A](K, 1)$
$P \to [A](K, 2)$
round $(I \cdot B, 2) \to [A](K, 3)$
$[A](K, 3) + W \to W$
$P - [A](K, 3) \to [A](K, 4)$
$B - [A](K, 4) \to [A](K, 5)$
$[A](K, 5) \to B$
$K + 1 \to K$
If $K = N$

Goto 2 [*Use P1, P2 in place of 1, 2 on TI-85/86.*]
If B ≤ P
Goto 2
Goto 1
Lbl 2
K → [A](K, 1)
B + round(B · I, 2) → [A](K, 2)
round (B · I, 2) → [A](K, 3)
[A](K, 2) − [A](K, 3) → [A](K, 4)
B − [A](K, 4) → [A](K, 5)
{K, 5} → dim [A]
K → J
N + 1 → K
Goto 1

CASIO 9850
Fill (0, Mat A)
Fix 2
"STARTING BALANCE"? → B
"RATE PER PERIOD"? → I
"PAYMENT"? → P
"NUMBER OF PYMTS"? → N
1 → K
0 → W
Lbl 1
If K = N + 1
Then Mat A◢
Clr Text
"TOTAL PAYMENTS"
(J − 1) · P + Mat A[J, 2]◢
"TOTAL INTEREST"
W + Mat A[J, 3]
Norm
Stop
IfEnd
K → Mat A[K, 1]
P → Mat A[K, 2]
I · B
rnd
Ans → Mat A[K, 3]
Mat A[K, 3] + W → W
P − Mat A[K, 3] → Mat A[K, 4]
B − Mat A[K, 4] → Mat A[K, 5]
Mat A[K, 5] → B
K + 1 → K
K = N ⇒ Goto 2
B ≤ P ⇒ Goto 2
Goto 1
Lbl 2
K → Mat A[K, 1]

B · I
rnd
Ans → S
B + S → Mat A[K, 2]
S → Mat A[K, 3]
Mat A[K, 2] − Mat A[K, 3] → Mat A[K, 4]
B − Mat A[K, 4] → Mat A[K, 5]
K → J
N + 1 → K
Goto 1

AMORTIZATION TABLE (ONE LINE AT A TIME) FOR TI-81, CASIO, AND SHARP

Description: Enter the appropriate information about the loan; the program displays the following entries from the first line of the amortization table.

payment number   interest for period   portion to principal   new balance

On Casio and Sharp, only one or two pieces of information will appear at a time (press ENTER to see the remaining ones). Press ENTER and the same information is displayed for the second line of the amortization table. Continue in this fashion through the remaining lines of the table. Then press ENTER again to see the total payments made and the total interest paid.

TI-81
ClrHome
Disp "STARTING BALANCE"
Input B
Disp "RATE PER PERIOD"
Input I
Disp "PAYMENT"
Input P
Disp "NUMBER OF PYMTS"
Input N
1 → K
0 → W
Lbl 1
ClrHome
Disp K
round (I · B, 2) → R
Disp "INTEREST"
Disp R
B + W → W
P − R → S
Disp "PRINCIPAL"
Disp S
B − S → T
Disp "NEW BALANCE"
Disp T
Pause

$T \rightarrow B$

$K + 1 \rightarrow K$

If $K = N$

Goto 2

If $B \leq P$

Goto 2

Goto 1

Lbl 2

ClrHome

$B + \text{round}\,(B \cdot I, 2) \rightarrow Q$

Disp "LAST PAYMENT"

Disp Q

Disp "INTEREST"

$\text{round}\,(I \cdot B, 2) \rightarrow S$

Disp S

$Q - S \rightarrow T$

Disp "PRINCIPAL"

Disp T

Pause

ClrHome

Disp "TOTAL PAYMENTS"

Disp $(K - 1) \cdot P + Q$

Disp "TOTAL INTEREST"

Disp $W + S$

CASIO

Fix 2

"STARTING BALANCE"? $\rightarrow B$

"RATE PER PERIOD"? $\rightarrow I$

"PAYMENT"? $\rightarrow P$

"NUMBER OF PYMTS"? $\rightarrow N$

$1 \rightarrow K$

$0 \rightarrow W$

Lbl 1

"PAYMENT NUMBER"

K◢

"INTEREST"

$I \cdot B \rightarrow R$◢

$R + W \rightarrow W$

$P - R \rightarrow S$

"PRINCIPAL"

S◢

$B - S \rightarrow T$

"NEW BALANCE"

T◢

$T \rightarrow B$

$K + 1 \rightarrow K$

$K = N \Rightarrow$ Goto 2

$B \leq P \Rightarrow$ Goto 2

Goto 1

Lbl 2
"LAST PAYMENT"
B
rnd
Ans → B
B × I
rnd
Ans → S
B + S → Q◢
"INTEREST"
S◢
"PRINCIPAL"
Q − S → T◢
"TOTAL PAYMENTS"
(K − 1) · P + Q◢
"TOTAL INTEREST"
W + S

SHARP
Print "starting balance"
Input b
Print "rate per period"
Input i
Print "payment"
Input p
Print "number of pymts"
Input n
k = 1
w = 0
j = 0
Label 1
ClrT
Print "payment number"
Print K
Label 6
If fpart(100 · b · i) < .5 goto 3   [*MATH menu*]
r = (ipart(100 · b · i) + 1)/100   [*MATH menu*]
If j = 0 goto 4
If j = 1 goto 7
Label 3
r = ipart(100 · b · i)/100
If j = 0 goto 4
If j = 1 goto 7
Label 4
Print "interest"
Print r
Wait
w = r + w
s = p − r
Print "principal"

Print s
t = b − s
Print "new balance"
Print t
Wait
b = t
k = k + 1
If k = n goto 5
If b ≤ p goto 5
Goto 1
Label 5
j = 1
Goto 6
Label 7
ClrT
Print "last payment"
f = b + r
Print f
Print "interest"
Print r
Wait
Print "principal"
Print b
Wait
ClrT
Print "total payments"
t = (k − 1) · p + b + r
Print t
Print "total interest"
m = w + r
Print m

SIMPLEX METHOD FOR TI (EXCEPT TI-81) AND CASIO 9850

Preliminaries: Before running the program, enter the initial simplex matrix in the matrix memory as matrix A. It remains there, unchanged, when the program is run.
Description: When the initial simplex matrix is entered, the program does the first round of pivoting, and displays the resulting simplex matrix, which can be examined by using the arrow keys to scroll through it. When you press ENTER, the next round of pivoting is performed, and so on, until the final simplex matrix is reached. The final simplex matrix is stored in the matrix memory as matrix B.

TI

N O T E   The program is written for TI-82/83, in which the matrix memories are denoted [A], [B], etc. For TI-85/86/92, use A, B in place of [A], [B]. Similarly, on TI-82/83, the list memories are labeled L_1, L_2, etc. and are on the keyboard. For TI-85/86/92, simply type in L1, L2, etc. ◆

ClrHome   *[labeled ClLCD on TI-85/86]*
[A] → [B]

dim [B] → L$_1$ [*MATRIX MATH or MATRIX OPS menu*]

L$_1$(1) → R

L$_1$(2) → S

seq([B](R,I), I, 1, S − 1, 1) → L$_2$ [*LIST OPS menu*]

min(L$_2$) → T   [*LIST MATH or LIST OPS menu*]

If T ≥ 0   [*TEST menu*]

Goto 1   [*Here and below on TI-85/86, use P1, P2, etc. in place of 1, 2 for label numbers.*]

Lbl 13   [*P13 in place of 13 on TI-85/86; F in place of 13 on TI-82*]

1 → J

Lbl 2

If [B](R,J) = T   [*TEST menu; here and below, use = = in place of = on TI-85/86.*]

Goto 3

J + 1 → J

Goto 2

Lbl 3

1 → K

Lbl 6

If [B](K,J) ≤ 0

Goto 4

[[[B](K,S)/[B](K,J)]] → [E]

K + 1 → K

Lbl 7

If [B](K,J) ≤ 0

Goto 5

augment ([E], [[[B](K, S)/[B](K, J)]]) → [E]   [*"aug" on TI-85/86; in MATRIX MATH or MATRIX OPS menu*]

K + 1 → K

Goto 7

Lbl 8

dim ([E]) → L$_3$

seq([E] (1, I), I, 1, L$_3$ (2), 1) → L$_4$

min (L$_4$) → M

1 → I

Lbl 10 [*Use C in place of 10 on TI-82.*]

If [B](I, J) = 0

Goto 14 [*Use G in place of 14 on TI-82.*]

If [B](I, S)/[B](I, J) = M

Goto 11 [*Use D in place of 11 on TI-82.*]

I + 1 → I

Goto 10 [*Use G in place of 10 on TI-82.*]

Lbl 11 [*Use D in place of 11 on TI-82.*]

If I = 1

Goto 12 [*Use E in place of 12 on TI-82.*]

*row ([B](I, J)$^{-1}$, [B],I) → [B]   [*MATRIX MATH or MATRIX OPS menu; on TI-85/86 *row is labeled MultR.*]

For (K, 1, I − 1, 1)

*row + (−[B](K, J), [B], I, K) → [B]   [*MATRIX MATH or MATRIX OPS menu; on TI-85/86 *row + is labeled mRAdd.*]

End

For (K, I + 1, R, 1)

*row + (−[B](K, J), [B], I, K) → [B]

End
round ([B], 9) → [B]   *[MATH NUM menu; use 11 in place of 9 on TI-85/86.]*
seq ([B] (R, I), I, 1, S − 1, 1) → L_2
min (L_2) → T
If T ≥ 0
Goto 1
ClrHome
Pause [B]▶FRAC   *[MATH or MATH MISC menu]*
Goto 13 *[Use F in place of 13 on TI-82.]*
Lbl 12 *[Use E in place of 12 on TI-82.]*
*row ([B](I, J)$^{-1}$, [B], 1) → [B]
For (K, 2, R, 1)
*row+ (−[B](K, J), [B], 1, K) → [B]
End
round ([B], 9) → [B]
seq ([B] (R, I), I, 1, S − 1, 1) → L_2
min (L_2) → T
If T ≥ 0
Goto 1
ClrHome
Pause [B]▶FRAC
Goto 13 *[Use F in place of 13 on TI-82.]*
Lbl 4
K + 1 → K
If K > (R − 1)
Goto 9
Goto 6
Lbl 5
K + 1 → K
If K > (R − 1)
Goto 8
Goto 7
Lbl 1
ClrHome
Disp "FINAL SIMPLEX"
Disp "MATRIX"
Disp "   "
Pause [B]▶FRAC
Stop
Lbl 9
Disp "NO MAXIMUM SOLUTION"
Stop
Lbl 14 *[Use G in place of 14 on TI-82.]*
I + 1 → I
Goto 10 *[Use C in place of 10 on TI-82.]*

CASIO 9850
Mat A → Mat B
"INITIAL SIMPLEX"
"MATRIX"
"NUMBER OF ROWS"? → R

"NUMBER OF COLS"? → S
1 → J
1 → I
Lbl 2
Mat B[R, I] → T
Lbl 1
I + 1 → I
I > (S − 1) ⇒ Goto 3
Mat B[R,I] ≥ T ⇒ Goto 1
I → J
Goto 2
Lbl 3
T ≥ 0 ⇒ Goto 4
10^{10} → u
0 → K
Lbl 5
K + 1 → K
K > (R − 1) ⇒ Goto 6
Mat B[K, J] ≤ 0 ⇒ Goto 5
Mat B[K, S]/Mat B[K, J] < u ⇒ K → I
Mat B[K, S]/Mat B[K, J] < u ⇒ Mat B[K, S]/Mat B[K, J] → u
Goto 5
Lbl 6
u ≤ 10^{10} ⇒ Goto 8
I = 1 ⇒ Goto 7
*Row Mat B[I, J]$^{-1}$, B, I
For 1 → K to I − 1
*Row + −Mat B[K, J], B, I, K
Next
For I + 1 → K to R
*Row + −Mat B[K, J], B, I, K
Next
1 → J
1 → I
Mat B[R, I] → T
Lbl 9
I + 1 → I
I > (S − 1) ⇒ Goto 0
Mat B[R, I] ≥ T ⇒ Goto 9
I → J
Mat B[R, I] → T
Goto 9
Lbl 0
T ≥ 0 ⇒ Goto 4
Mat B ◢
Goto 3
Lbl 7
*Row Mat B[1, J]$^{-1}$, B, 1
For 2 → K to R
*Row + −Mat B[K, J], B, 1, K
Next

$1 \rightarrow I$
Mat B[R, I] $\rightarrow$ T
Goto 9
Lbl4
"FINAL SIMPLEX MATRIX"
Mat B ◢
STOP
Lbl 8
"NO MAXIMUM SOLUTION"

NUMERICAL DERIVATIVE FOR CASIO 7700/8700

Preliminaries: Before running the program, enter the function in the function memory as f_1.
Description: The program approximates the value of the derivative of a function at a specified value of x.

"X"? $\rightarrow$ A   [*Use the x, θ, τ key for X.*]
A + .0001 $\rightarrow$ X
$f_1 \rightarrow$ B
A $-$ .0001 $\rightarrow$ X
$f_1 \rightarrow$ C
(B $-$ C)/.0002

RECTANGLE APPROXIMATION OF $\int_a^b f(x)\ dx$
(USING LEFT ENDPOINTS)

Preliminaries: Enter the function $f(x)$ in the function memory as Y_1 or f_1.
Description: Enter the endpoints a and b and the number of rectangles to be used. The program approximates the value of the integral by using left endpoints.

TI
ClrHome   [*labeled ClLCD on TI-85/86*]
Disp "A = "
Input A
Disp "B = "
Input B
Disp "N = "
Input N
(B $-$ A)/N $\rightarrow$ D
$0 \rightarrow$ S
For (K, 0, N $-$ 1, 1)
(A + K $\cdot$ D) $\rightarrow$ X
Y_1 + S $\rightarrow$ S
End
Disp D $\cdot$ S

[*On TI-81, replace these five lines by:*
  K $\rightarrow$ 0
  Lbl 1
  A + K $\cdot$ D $\rightarrow$ X
  Y_1 + S $\rightarrow$ S
  K + 1 $\rightarrow$ K
  If K $<$ N
  Goto 1
  Disp D $\cdot$ S]

CASIO

"A = "? $\rightarrow$ A
"B = "? $\rightarrow$ B
"N = "? $\rightarrow$ N
(B − A)/N $\rightarrow$ D
0 $\rightarrow$ S
0 $\rightarrow$ K
Lbl 1
A + K · D $\rightarrow$ X
f_1 + S $\rightarrow$ S
K + 1 $\rightarrow$ K
K < N $\Rightarrow$ Goto 1
D · S

NUMERICAL INTEGRATION (SIMPSON'S RULE) FOR TI-81

Preliminaries: Enter the function to be integrated in the function memory as Y_1.
Description: Enter *a, b,* and *n* (the number of subdivisions to be used); the program calculates the approximate value of $\int_a^b f(x)\, dx$.

ClrHome
Disp "A = "
Input A
Disp "B = "
Input B
Disp "N = "
Input N
(B − A)/N $\rightarrow$ D
0 $\rightarrow$ S
A $\rightarrow$ X
$Y_1 \rightarrow$ S
B $\rightarrow$ X
(Y_1 + S) $\rightarrow$ S
For (K, 1, N/2, 1)
(A + (2K − 1) · D) $\rightarrow$ X
(4Y_1 + S) $\rightarrow$ S
End
For (K, 1, N/2 − 1, 1)
(A + 2K · D) $\rightarrow$ X
(2Y_1 + S) $\rightarrow$ S
End
Disp S · D/3

Tables

TABLE 1   Combinations $_nC_r = \binom{n}{r} = \dfrac{n!}{(n-r)!\,r!}$

n	$\binom{n}{0}$	$\binom{n}{1}$	$\binom{n}{2}$	$\binom{n}{3}$	$\binom{n}{4}$	$\binom{n}{5}$	$\binom{n}{6}$	$\binom{n}{7}$	$\binom{n}{8}$	$\binom{n}{9}$	$\binom{n}{10}$
0	1										
1	1	1									
2	1	2	1								
3	1	3	3	1							
4	1	4	6	4	1						
5	1	5	10	10	5	1					
6	1	6	15	20	15	6	1				
7	1	7	21	35	35	21	7	1			
8	1	8	28	56	70	56	28	8	1		
9	1	9	36	84	126	126	84	36	9	1	
10	1	10	45	120	210	252	210	120	45	10	1
11	1	11	55	165	330	462	462	330	165	55	11
12	1	12	66	220	495	792	924	792	495	220	66
13	1	13	78	286	715	1287	1716	1716	1287	715	286
14	1	14	91	364	1001	2002	3003	3432	3003	2002	1001
15	1	15	105	455	1365	3003	5005	6435	6435	5005	3003
16	1	16	120	560	1820	4368	8008	11440	12870	11440	8008
17	1	17	136	680	2380	6188	12376	19448	24310	24310	19448
18	1	18	153	816	3060	8568	18564	31824	43758	48620	43758
19	1	19	171	969	3876	11628	27132	50388	75582	92378	92378
20	1	20	190	1140	4845	15504	38760	77520	125970	167960	184756

For $r > 10$, it may be necessary to use the identity

$$\binom{n}{r} = \binom{n}{n-r}.$$

TABLE 2 Areas under the Normal Curve

The column under *A* gives the proportion of the area under the entire curve that is between $z = 0$ and a positive value of z.

z	A	z	A	z	A	z	A
.00	.0000	.49	.1879	.98	.3365	1.47	.4292
.01	.0040	.50	.1915	.99	.3389	1.48	.4306
.02	.0080	.51	.1950	1.00	.3413	1.49	.4319
.03	.0120	.52	.1985	1.01	.3438	1.50	.4332
.04	.0160	.53	.2019	1.02	.3461	1.51	.4345
.05	.0199	.54	.2054	1.03	.3485	1.52	.4357
.06	.0239	.55	.2088	1.04	.3508	1.53	.4370
.07	.0279	.56	.2123	1.05	.3531	1.54	.4382
.08	.0319	.57	.2157	1.06	.3554	1.55	.4394
.09	.0359	.58	.2190	1.07	.3577	1.56	.4406
.10	.0398	.59	.2224	1.08	.3599	1.57	.4418
.11	.0438	.60	.2258	1.09	.3621	1.58	.4430
.12	.0478	.61	.2291	1.10	.3643	1.59	.4441
.13	.0517	.62	.2324	1.11	.3665	1.60	.4452
.14	.0557	.63	.2357	1.12	.3686	1.61	.4463
.15	.0596	.64	.2389	1.13	.3708	1.62	.4474
.16	.0636	.65	.2422	1.14	.3729	1.63	.4485
.17	.0675	.66	.2454	1.15	.3749	1.64	.4495
.18	.0714	.67	.2486	1.16	.3770	1.65	.4505
.19	.0754	.68	.2518	1.17	.3790	1.66	.4515
.20	.0793	.69	.2549	1.18	.3810	1.67	.4525
.21	.0832	.70	.2580	1.19	.3830	1.68	.4535
.22	.0871	.71	.2612	1.20	.3849	1.69	.4545
.23	.0910	.72	.2642	1.21	.3869	1.70	.4554
.24	.0948	.73	.2673	1.22	.3888	1.71	.4564
.25	.0987	.74	.2704	1.23	.3907	1.72	.4573
.26	.1026	.75	.2734	1.24	.3925	1.73	.4582
.27	.1064	.76	.2764	1.25	.3944	1.74	.4591
.28	.1103	.77	.2794	1.26	.3962	1.75	.4599
.29	.1141	.78	.2823	1.27	.3980	1.76	.4608
.30	.1179	.79	.2852	1.28	.3997	1.77	.4616
.31	.1217	.80	.2881	1.29	.4015	1.78	.4625
.32	.1255	.81	.2910	1.30	.4032	1.79	.4633
.33	.1293	.82	.2939	1.31	.4049	1.80	.4641
.34	.1331	.83	.2967	1.32	.4066	1.81	.4649
.35	.1368	.84	.2996	1.33	.4082	1.82	.4656
.36	.1406	.85	.3023	1.34	.4099	1.83	.4664
.37	.1443	.86	.3051	1.35	.4115	1.84	.4671
.38	.1480	.87	.3079	1.36	.4131	1.85	.4678
.39	.1517	.88	.3106	1.37	.4147	1.86	.4686
.40	.1554	.89	.3133	1.38	.4162	1.87	.4693
.41	.1591	.90	.3159	1.39	.4177	1.88	.4700
.42	.1628	.91	.3186	1.40	.4192	1.89	.4706
.43	.1664	.92	.3212	1.41	.4207	1.90	.4713
.44	.1700	.93	.3238	1.42	.4222	1.91	.4719
.45	.1736	.94	.3264	1.43	.4236	1.92	.4726
.46	.1772	.95	.3289	1.44	.4251	1.93	.4732
.47	.1808	.96	.3315	1.45	.4265	1.94	.4738
.48	.1844	.97	.3340	1.46	.4279	1.95	.4744

Table 2   Areas under the Normal Curve   **803**

TABLE 2   *(continued)*

z	A	z	A	z	A	z	A
1.96	.4750	2.45	.4929	2.94	.4984	3.43	.4997
1.97	.4756	2.46	.4931	2.95	.4984	3.44	.4997
1.98	.4762	2.47	.4932	2.96	.4985	3.45	.4997
1.99	.4767	2.48	.4934	2.97	.4985	3.46	.4997
2.00	.4773	2.49	.4936	2.98	.4986	3.47	.4997
2.01	.4778	2.50	.4938	2.99	.4986	3.48	.4998
2.02	.4783	2.51	.4940	3.00	.4987	3.49	.4998
2.03	.4788	2.52	.4941	3.01	.4987	3.50	.4998
2.04	.4793	2.53	.4943	3.02	.4987	3.51	.4998
2.05	.4798	2.54	.4945	3.03	.4988	3.52	.4998
2.06	.4803	2.55	.4946	3.04	.4988	3.53	.4998
2.07	.4808	2.56	.4948	3.05	.4989	3.54	.4998
2.08	.4812	2.57	.4949	3.06	.4989	3.55	.4998
2.09	.4817	2.58	.4951	3.07	.4989	3.56	.4998
2.10	.4821	2.59	.4952	3.08	.4990	3.57	.4998
2.11	.4826	2.60	.4953	3.09	.4990	3.58	.4998
2.12	.4830	2.61	.4955	3.10	.4990	3.59	.4998
2.13	.4834	2.62	.4956	3.11	.4991	3.60	.4998
2.14	.4838	2.63	.4957	3.12	.4991	3.61	.4999
2.15	.4842	2.64	.4959	3.13	.4991	3.62	.4999
2.16	.4846	2.65	.4960	3.14	.4992	3.63	.4999
2.17	.4850	2.66	.4961	3.15	.4992	3.64	.4999
2.18	.4854	2.67	.4962	3.16	.4992	3.65	.4999
2.19	.4857	2.68	.4963	3.17	.4992	3.66	.4999
2.20	.4861	2.69	.4964	3.18	.4993	3.67	.4999
2.21	.4865	2.70	.4965	3.19	.4993	3.68	.4999
2.22	.4868	2.71	.4966	3.20	.4993	3.69	.4999
2.23	.4871	2.72	.4967	3.21	.4993	3.70	.4999
2.24	.4875	2.73	.4968	3.22	.4994	3.71	.4999
2.25	.4878	2.74	.4969	3.23	.4994	3.72	.4999
2.26	.4881	2.75	.4970	3.24	.4994	3.73	.4999
2.27	.4884	2.76	.4971	3.25	.4994	3.74	.4999
2.28	.4887	2.77	.4972	3.26	.4994	3.75	.4999
2.29	.4890	2.78	.4973	3.27	.4995	3.76	.4999
2.30	.4893	2.79	.4974	3.28	.4995	3.77	.4999
2.31	.4896	2.80	.4974	3.29	.4995	3.78	.4999
2.32	.4898	2.81	.4975	3.30	.4995	3.79	.4999
2.33	.4901	2.82	.4976	3.31	.4995	3.80	.4999
2.34	.4904	2.83	.4977	3.32	.4996	3.81	.4999
2.35	.4906	2.84	.4977	3.33	.4996	3.82	.4999
2.36	.4909	2.85	.4978	3.34	.4996	3.83	.4999
2.37	.4911	2.86	.4979	3.35	.4996	3.84	.4999
2.38	.4913	2.87	.4980	3.36	.4996	3.85	.4999
2.39	.4916	2.88	.4980	3.37	.4996	3.86	.4999
2.40	.4918	2.89	.4981	3.38	.4996	3.87	.5000
2.41	.4920	2.90	.4981	3.39	.4997	3.88	.5000
2.42	.4922	2.91	.4982	3.40	.4997	3.89	.5000
2.43	.4925	2.92	.4983	3.41	.4997		
2.44	.4927	2.93	.4983	3.42	.4997		

TABLE 3   Integrals

(C is an arbitrary constant.)

1. $\displaystyle\int x^n\,dx = \frac{1}{n+1}x^{n+1} + C \quad (n \neq -1)$

2. $\displaystyle\int e^{kx}\,dx = \frac{1}{k}e^{kx} + C$

3. $\displaystyle\int \frac{a}{x}\,dx = a\ln|x| + C$

4. $\displaystyle\int \ln|ax|\,dx = x(\ln|ax|-1) + C$

5. $\displaystyle\int \frac{1}{\sqrt{x^2+a^2}}\,dx = \ln\left|x+\sqrt{x^2+a^2}\right| + C$

6. $\displaystyle\int \frac{1}{\sqrt{x^2-a^2}}\,dx = \ln\left|x+\sqrt{x^2-a^2}\right| + C$

7. $\displaystyle\int \frac{1}{a^2-x^2}\,dx = \frac{1}{2a}\cdot\ln\left|\frac{a+x}{a-x}\right| + C \quad (a \neq 0)$

8. $\displaystyle\int \frac{1}{x^2-a^2}\,dx = \frac{1}{2a}\cdot\ln\left|\frac{x-a}{x+a}\right| + C \quad (a \neq 0)$

9. $\displaystyle\int \frac{1}{x\sqrt{a^2-x^2}}\,dx = -\frac{1}{a}\cdot\ln\left|\frac{a+\sqrt{a^2-x^2}}{x}\right| + C \quad (a \neq 0)$

10. $\displaystyle\int \frac{1}{x\sqrt{a^2+x^2}}\,dx = -\frac{1}{a}\cdot\ln\left|\frac{a+\sqrt{a^2+x^2}}{x}\right| + C \quad (a \neq 0)$

11. $\displaystyle\int \frac{x}{ax+b}\,dx = \frac{x}{a} - \frac{b}{a^2}\cdot\ln|ax+b| + C \quad (a \neq 0, b \neq 0)$

12. $\displaystyle\int \frac{x}{(ax+b)^2}\,dx = \frac{b}{a^2(ax+b)} + \frac{1}{a^2}\cdot\ln|ax+b| + C \quad (a \neq 0)$

13. $\displaystyle\int \frac{1}{x(ax+b)}\,dx = \frac{1}{b}\cdot\ln\left|\frac{x}{ax+b}\right| + C \quad (b \neq 0)$

14. $\displaystyle\int \frac{1}{x(ax+b)^2}\,dx = \frac{1}{b(ax+b)} + \frac{1}{b^2}\cdot\ln\left|\frac{x}{ax+b}\right| + C \quad (b \neq 0)$

15. $\displaystyle\int \sqrt{x^2+a^2}\,dx = \frac{x}{2}\sqrt{x^2+a^2} + \frac{a^2}{2}\cdot\ln\left|x+\sqrt{x^2+a^2}\right| + C$

16. $\displaystyle\int x^n\cdot\ln|x|\,dx = x^{n+1}\left[\frac{\ln|x|}{n+1} - \frac{1}{(n+1)^2}\right] + C \quad (n \neq -1)$

17. $\displaystyle\int x^n e^{ax}\,dx = \frac{x^n e^{ax}}{a} - \frac{n}{a}\cdot\int x^{n-1}e^{ax}\,dx + C \quad (a \neq 0)$

Answers to Selected Exercises

CHAPTER 1
Section 1.1 (page 9)

1. True   **3.** True   **5.** Identity property (addition)
7. Commutative property (addition)   **9.** Associative property
(multiplication)   **11.** Distributive property and commutative
property (multiplication)   **15.** -54 **17.** $-1/2$ **19.** $1/5$
21. 4   **23.** -4 **25.** -1 **27.** Irrational; 9.4248
29. Irrational; 1.7321   **31.** Rational   **33.** $5 < 7$ **35.** $y \le 8.3$
37. $t > 0$
39.

41.

43.

45.

47.

49.

51.

53. 42°   **55.** 36°   **57.** 4   **59.** -19 **61.** $=$ **63.** $=$
65. $=$ **67.** $=$ **69.** $=$ **71.** $-(a - 7) = -a + 7$
77. Let x represent the number of dollars contributed annually by
foreign students to the California economy. Then $x > 1,000,000,000$.
79. Let x represent the percent of foreign students now in the U.S.
from Middle Eastern countries. Then $0 \le x < 7.5$.   **81.** Let x
represent the percent of foreign students in the U.S. that are in
California. Then $x > 13$.   **83.** 100; 0   **85.** 67; 11

Section 1.2 (page 20)

1. 5   **3.** 7   **5.** $-19/52$ **7.** $-9/8$ **9.** 2   **11.** 40/7
13. 26/3   **15.** $-12/5$ **17.** $-59/6$ **19.** $-9/4$
21. $x = .72$ **23.** $r \approx -13.26$ **25.** $x = (5a - b)/6$
27. $x = 3b/(5 + a)$ **29.** $x = (-3a + 3)/(1 - a^2 - a)$ or
$x = (3a - 3)/(a^2 + a - 1)$ **31.** $V = k/P$ **33.** $g = (V - V_0)/t$
35. $B = (2A/h) - b$ or $B = (2A - bh)/h$ **37.** $R = (r_1 r_2)/(r_1 + r_2)$
39. $2, -3$ **41.** $-2, 8$ **43.** 5/2, 7/2   **45.** $-4/3, 2/9$
47. $-7, -3/7$ **49. (a)** .0352   **(b)** Approximately .015 or 1.5%
(c) Approximately one case   **51.** 13%   **53.** $4000
55. $205.41   **57. (a)** Let x represent the height.
(b) $16x + 288 + 36x = 496$ **(c)** 4 ft   **59.** 6 cm
61. About 6.42 hr   **63.** 20 min   **65.** $21,000

67. $70,000 for land that made a profit, $50,000 for land that
produced a loss   **69.** $800   **71.** 400/3 liters

Section 1.3 (page 26)

1. 118,587,876,497   **3.** 289.099133864   **7.** 2^7
9. $(-5)^7 = -5^7$ **11.** -3^9 **13.** $(2z)^{11}$ **15.** $-x^3 + x^2 + 3x$
17. $-6y^2 + 3y + 10$ **19.** $-10x^2 + 4x - 2$
21. $-18m^3 - 27m^2 + 9m$ **23.** $12z^3 + 14z^2 - 7z + 5$
25. $12k^2 - 20k + 3$ **27.** $6y^2 + 7y - 5$ **29.** $18k^2 - 7kq - q^2$
31. $4.34m^2 + 5.68m - 4.42$ **33.** $-k + 3$ **35.** $7x^2 - 4x - 12$
37. (a) 4.2 million   **(b)** 4.269 million   **39. (a)** 19.5 million
(b) 19.281 million   **41. (a)** $\frac{1}{3}hb^2$ (This is the correct formula.)
(b) Approximately 91.6 million cubic feet; slightly smaller
(c) Approximately 13.1 acres   **43. (a)** 4   **(b)** 4   **(c)** 7
47. $5x^3 - 4x^2 - 8x + 1$ **49.** $x^2 - x + 40$

Section 1.4 (page 32)

1. $12x(x - 2)$ **3.** $r(r^2 - 5r + 1)$ **5.** Cannot be factored
7. $6z(z^2 - 2z + 3)$ **9.** $5p^2(5p^2 - 4pq + 20q^2)$
11. $2(2y - 1)^2(5y - 1)$ **13.** $(x + 5)^4(x^2 + 10x + 28)$
15. $(2a + 5)(a - 1)$ **17.** $(x + 8)(x - 8)$ **19.** $(3p - 4)^2$
21. $(r + 2t)(r - 5t)$ **23.** $(m - 3n)^2$ **25.** $(2p + 3)(2p - 3)$
27. $3(x - 4z)^2$ **29.** Cannot be factored
31. $(-x + 4)(x - 3)$ or $(x - 4)(-x + 3)$ **33.** $(3a + 5)(a - 6)$
35. $(7m + 2n)(3m + n)$ **37.** $(4y - x)(5y + 11x)$
39. Cannot be factored   **41.** $(y - 7z)(y + 3z)$
43. $(n + 2)(n - 2)(m^2 + 3)$ **45.** $(11x + 8)(11x - 8)$
47. $2a^2(4a - b)(3a + 2b)$ **49.** $3x^3(3x - 5z)(2x + 5z)$
51. $m^3(m - 1)^2(m^2 + m + 1)^2(5 + 3m^2 - 3m^5)$
55. $(a - 6)(a^2 + 6a + 36)$ **57.** $(2r - 3s)(4r^2 + 6rs + 9s^2)$
59. $(4m + 5)(16m^2 - 20m + 25)$
61. $(10y - z)(100y^2 + 10yz + z^2)$

Section 1.5 (page 37)

1. $x/5$ **3.** $4/(7p)$ **5.** 5/4   **7.** $4/(w + 3)$
9. $(y - 4)/(3y^2)$ **11.** $2(x + 2)/x$ **13.** $(m - 2)/(m + 3)$
15. $(x + 4)/(x + 1)$ **17.** $2p/7$ **19.** $9a^3/4$ **21.** $15/(8c)$
23. 3/4   **25.** 2/9   **27.** 3/10   **29.** $2(a + 4)/(a - 3)$
31. $(k + 2)/(k + 3)$ **35.** $-1/(15z)$ **37.** 4/3
39. $(12 + x)/(3x)$ **41.** $(8 - y)/(4y)$ **43.** $137/(30m)$
45. $(3m - 2)/[m(m - 1)]$ **47.** $14/[3(a - 1)]$
49. $23/[20(k - 2)]$ **51.** $(7x + 9)/[(x - 3)(x + 1)(x + 2)]$
53. $y^2/[(y + 4)(y + 3)(y + 2)]$
55. $(k^2 - 13k)/[(2k - 1)(k + 2)(k - 3)]$ **57.** $(x + 1)/(x - 1)$
59. $-1/[x(x + h)]$ **61.** $(2 + b - b^2)/(b - b^2)$

Section 1.6 (page 45)

1. 1 **3.** 1/6 **5.** 1/32 **7.** $-1/64$ **9.** .00162 **11.** 9
13. 625/16 **17.** 7 **19.** 1.7 **21.** 9 **23.** 15.5
25. -16 **27.** 3/5 **29.** 27/64 **31.** 4/3 **33.** 1/9
35. 5 **37.** $1/12^{1/2}$ **39.** 5^3 **41.** $3^{7/6}$ **43.** k^3
45. $1/(5^2 x^{10})$ **47.** $z^6/(2^3 y^6)$ **49.** $5^3 \cdot 3^4/m^{14}$ **51.** $5^{9/4}k^{13/4}$
53. $6 + 2z$ **55.** (b) **57.** (d) **59.** (a) **61.** (e) **63.** 4
65. 5 **67.** -2 **69.** 9 **71.** $\sqrt{77}$ **73.** 7 **75.** 3

77. $5\sqrt{5} + 58$ **79.** $\dfrac{\sqrt{5}-1}{2}$ **81.** $-1 - \sqrt{3}$ **83.** (a) 14

(b) 85 **(c)** 58 **85.** 50 cm **87.** (a) $-60.9°$ **(b)** $-64.4°$
89. (a) $-63°$ **(b)** $-47°$ **(c)** The expression in Exercise 88
provides a better model.

Chapter 1 Review Exercises (page 49)

1. 0, 6 **3.** $-12, -6, -9/10, -\sqrt{4}, 0, 1/8, 6$ **5.** Commutative
multiplication **7.** Distributive **9.** $x \geq 6$ **11.** $-7, -3, -2,$
$0, \pi, 8$ **13.** $-|3 - (-2)|, -|-2|, |6 - 4|, |8 + 1|$
15. -3 **17.** -1
19.
─────●────────→
      -3

21.
←────○────────
      -2

23. -8 **25.** $-3/4$ **27.** 3 **29.** -2 **31.** No solution
33. $x = 1/(5a - 1)$ **35.** $x = (c - 3)/(3a - ac - 2)$ **37.** 12,
-6 **39.** $-38, 42$ **41.** 2, 1/5 **43.** 20 lb of hearts, 10 lb of
kisses **45.** $4x^4 - 4x^2 + 13x$ **47.** $2q^4 + 14q^3 - 4q^2$
49. $15z^2 - 4z - 4$ **51.** $16k^2 - 9h^2$ **53.** $36x^2 + 36xy + 9y^2$
55. $k(2h^2 - 4h + 5)$ **57.** $a^2(3a + 1)(a + 4)$
59. $(2y - 1)(5y - 3)$ **61.** $(2a - 5)^2$
63. $(12p + 13q)(12p - 13q)$ **65.** $(2y - 1)(4y^2 + 2y + 1)$
67. $(7x^2)/3$ **69.** 4 **71.** $[(y - 5)(4y - 5)]/(6y)$
73. $(m - 1)^2/[3(m + 1)]$ **75.** $1/(6z)$ **77.** $64/(35q)$
83. $1/10^2$ or $1/100$ **85.** $-1/3$ **87.** $-3^3/2^3$ or $-27/8$
89. $1/7^6$ **91.** $1/6^5$ **93.** $1/k^5$ **95.** 4/9 **97.** $2^3 = 8$
99. 7/12 **101.** 1 **103.** $2^{17/6}p^{31/6}$ **105.** 5
107. Not a real number **109.** $3\sqrt{7}$ **111.** $2a\sqrt[4]{4ab^3}$
113. $(x\sqrt{6xz})/(2z)$ **115.** $12\sqrt{7}$ **117.** 4 **119.** $82 - \sqrt{14}$
121. $(16 + 4\sqrt{2} + 4\sqrt{5} + \sqrt{10})/11$ **123.** $27,000,000
125. $6,704,420

Case 1 (page 52)

1. $c = 700 + 85x$ **3.** The \$700 refrigerator costs \$300 more over
10 years.

CHAPTER 2
Section 2.1 (page 59)

1. Yes **3.** No

5.

7.

9.

11. x-intercepts $-1, 3.5$; y-intercept 1 **13.** x-intercepts $-2, 0, 2$;
y-intercept 0 **15.** x-intercept 8, y-intercept 6 **17.** x-intercept 3,
y-intercept -2 **19.** x-intercepts 3 and -3, y-intercept -9
21.

23.

25.

27.

29.

31.

33. Approximately 12 seconds   **35.** Fargo, about 2 P.M.; Seattle, about 5 P.M.   **37.** From 11 A.M. until 6 P.M.
39. (a) About $1,250,000   **(b)** About $1,750,000   **(c)** About $4,250,000   **41. (a)** About $500,000   **(b)** About $1,000,000
(c) About $1,500,000   **43. (a)** From the graph, the answers are about $740, $595, and $280. The exact answers are $736.39, $595.89, and $281.64.   **(b)** During the 22nd year (that is, between $t = 21$ and $t = 22$)   **45.** From approximately mid-1990 through early 1992

Section 2.2 (page 71)

1. $-1/2$ **3.** -1 **5.** $-3/2$ **7.** Undefined   **9.** $y = 3x + 5$
11. $y = -2.3x + 1.5$ **13.** $y = -3x/4 + 4$ **15.** $m = 2$; $b = -7$ **17.** $m = 3$; $b = -2$ **19.** $m = 2/3$; $b = -14/9$
21. $m = 2/3$; $b = 0$ **23.** $m = 1$; $b = 5$ **25. (a)** C **(b)** B
(c) B **(d)** D
27.   **29.**

31.

33. Perpendicular   **35.** Parallel   **37.** Neither   **39. (a)** 2/5, 9/8, $-5/2$ **(b)** Yes   **41. (a)** The slope of $-.0221$ indicates that on the average from 1912 to 1992 the 5000-meter run is being run .0221 second faster every Olympic Game. It is negative because the times are generally decreasing as time progresses.   **(b)** World War II occurred during the years 1940 and 1944, and no Olympic Games were held.   **43.** $3y = -2x + 4$ **45.** $y = 2x + 2$ **47.** $y = 2$
49. $x = 6$ **51.** $3y = 4x + 7$ **53.** $2y = 5x - 1$ **55.** $y = 7x$
57. $x = 5$ **59.** $y = 2x - 2$ **61.** $y = x - 5$ **63.** $y = -x + 2$
65. (b)   **67.** $3308.33

69. (a) $135   **(b)** $205   **(c)** $275   **(d)** $345
(e)

71. 48 liters per second   **73.** 1989   **75. (a)** $17.96
(b) $20.13   **(c)** $21.21

Section 2.3 (page 81)

1. (a) 14.4°   **(b)** 122°   **(c)** 14°   **(d)** $-28.9°$ **3.** $-40°$
5. (a) $y = 82,500x + 850,000$ **(b)** $1,840,000   **(c)** 2003
7. $y = (-65/7)x + 504$; $439   **9.** $y = 4x + 120$; 4 ft
11. $y = 806,400x$; 24,192,000 gallons   **13.** $y = .0985x$; the time would be 1 hour, 9 minutes, 16.2 seconds, which is faster than the marathon record.   **15.** $y = 900x$; 27,000 cu ft
17. $y = 4.43x - 381.7$ **19.** $(3, -1)$ **21.** $(-(11/4, -61/4)$
23. Approximately $(-1.3593, -2.5116)$ **25.** $(0, -3)$ and $(20, 87)$
27. (a) 200,000 policies ($x = 200$)
(b)

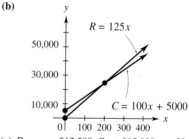

(c) Revenue: $12,500; Cost: $15,000   **29. (a)** $c = .126x + 1.5$
(b) $2.382 million   **(c)** About 17.857 units   **31.** Break-even point is about 467; do not produce the item.   **33.** Break-even point is about 1036; produce the item.   **35.** About 95 units in 1977
37. (a) Canada: $y = 1.56x + 12.1$; Japan: $y = 10.34x + 4.7$
(b)

(c) $(.84, 13.4)$; the investments of the Canadians and Japanese were both about $13.4 million near the end of 1980.   **39.** $140
41. 10 items, 10 items   **43. (a)** $16   **(b)** $11   **(c)** $6   **(d)** 8
(e) 4   **(f)** 0

(g)

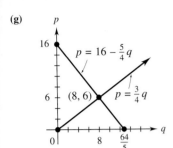

(h) 0   **(i)** 40/3   **(j)** 80/3   **(k)** See part (g).   **(l)** 8   **(m)** \$6

45. (a) p

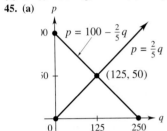

(b) 125   **(c)** 50¢   **(d)** [0, 125)

Section 2.4 (page 93)

1. $-3, 12$ **3.** $-5, 0$ **5.** $0, 2$ **7.** $-7, -8$ **9.** $3/2, 1$
11. $-1/2, 1/3$ **13.** $5/2, 4$ **15.** $5, 2$ **17.** $4/3, -4/3$
19. $0, 1$ **21.** $2 \pm \sqrt{7}$ **23.** $(1 \pm 2\sqrt{5})/4$
25. $(-5 \pm \sqrt{17})/4; -.219, -2.281$ **27.** $(-1 \pm \sqrt{5})/4; .309,$
$-.809$ **29.** $(-3 \pm \sqrt{19})/5; .272, -1.472$
31. No real number solutions   **33.** $5/2, 1$ **35.** $4/3, 1/2$
37. No real number solutions   **39.** 1   **41.** 2   **43.** $x \approx .4701$
or 1.8240   **45.** $x \approx -1.0376$ or .67196   **47.** $\pm\sqrt{5}$
49. $\pm\sqrt{6}/2$ **51.** $11/3, 3/2$ **53.** $-5, 3/2$ **55.** $-4 \pm 3\sqrt{2}$
57. (a) $(-r^2 + 3r - 12)/[r(r - 2)]$ **(b)** $4, -3$ **59. (a)** $150 - x$
or $5000/x$ **(b)** $x(150 - x) = 5000$ or $2(5000/x) + 2x = 300$
(c) Length: 100 m; width: 50 m   **61.** 1 ft   **63.** 176 mph
65. 30 bicycles   **67. (a)** 2 sec   **(b)** 1/2 sec or 7/2 sec
(c) It reaches the given height twice—once on the way up and once
on the way down.   **69.** $t = \pm\sqrt{2Sg}/g$ **71.** $h = (\pm d^2\sqrt{kL})/L$
73. $R = (-2Pr + E^2 \pm E\sqrt{E^2 - 4Pr})/(2P)$

Section 2.5 (page 101)

3. $[-4, \infty)$

5. $(-\infty, 0)$

7. $(-\infty, 8/3]$

9. $(-\infty, -7]$

11. $(-\infty, 3)$

13. $(-1, \infty)$

15. $(-\infty, 1]$

17. $(1/5, \infty)$

19. $(-5, 6)$

21. $[7/3, 4]$

23. $[-11/2, 7/2]$

25. $[-17/7, \infty)$

27. $(.1745, \infty)$

29. $(-\infty, 1.50)$

31. $(-\infty, .50]$

33. (a) Let x represent the number of mg per liter of lead in the water.
(b) $.038 \le x \le .042$ **(c)** Yes   **35. (a)** $.0015 \le R \le .006$
(b) $.00002 \le R/75 \le .00008$
37. $(-1, 1)$

39. No solution
41. $(-4, -1)$

43. $(-\infty, -8/3]$ or $[2, \infty)$

45. $(-3/2, 13/10)$

47. $-140 \le C \le -28$ **49.** $|F - 730| \le 50$
51. (a) $25.33 \le R_L \le 28.17; 36.58 \le R_E \le 40.92$
(b) $5699.25 \le T_L \le 6338.25; 8230.5 \le T_E \le 9207$ **53.** At most
14 min   **55.** At least 629 mi   **57.** $[500, \infty)$ **59.** $[45, \infty)$
61. Impossible to break even   **63.** $|x - 2| \le 4$
65. $|z - 12| \ge 2$ **67.** If $|x - 2| \le .0004$, then $|y - 7| \le .00001$.

Section 2.6 (page 108)

1. $[-5, 3/2]$

3. $(-\infty, -3)$ or $(-1, \infty)$

5. $[-2, 1/4]$

7. $(-\infty, -1)$ or $(1/4, \infty)$

9. $[-5, 5]$

11. $(-\infty, 0)$ or $(16, \infty)$

13. $[-3, 0]$ or $[3, \infty)$ **15.** $[-6, -1]$ or $[4, \infty)$ **17.** $(-\infty, -4)$ or $(-1, 3)$ **19.** $(-\infty, -1/2)$ or $(0, 4/3)$ **21.** No   **23.** $(-\infty, 1)$ or $[3, \infty)$ **25.** $(7/2, 5)$ **27.** $(-\infty, 2)$ or $(5, \infty)$ **29.** $(-\infty, -1)$ **31.** $(-\infty, -2)$ or $(0, 3)$ **33.** $(-\infty, -2.1196)$ or $(.7863, \infty)$ **35.** $(-.0806, 2.4806)$ **37.** $[-2.2635, .7556]$ or $[3.5079, \infty)$ **39.** $(.5, .8393)$ **41.** $[-1, 1/2]$ **43.** $10 < x < 35$ **45.** $(0, 5/3)$ or $(10, \infty)$ **47.** $(100, 150)$ **49.** $[4, 9.75]$

Chapter 2 Review Exercises (page 110)

1. $(-2, 3), (0, -5), (3, -2), (4, 3)$
3.

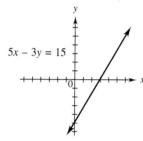

$5x - 3y = 15$

5.

$y + 3 = 0$

7.

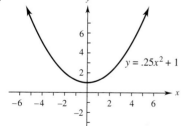

$y = .25x^2 + 1$

9. (a) About 11:30 A.M. to about 7:30 P.M.   **(b)** From midnight until about 5 A.M. and after about 10:30 P.M.   **13.** $-5/6$ **15.** $1/4$
17. 4   **19.** 0   **21.** -3

23.

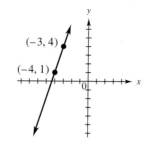

25. $3y = 2x - 13$ **27.** $4y = -5x + 17$ **29.** $x = -1$
31. $3y = 5x + 15$ **33. (a)** $y = .88x + 4.2$
(b)

(c) Positive; the percent is increasing, so the line moves upward from left to right.   **35. (a)** $r = 40x$ **(b)** 5 units   **(c)** $200
(d) $p = 20x - 100$ **37.** About 4.2   **39. (a)** $y = .57x$
(b) No, because there is 79.8 ppb remaining.   **(c)** About 87.7 ppb
41. (a) $p = .5q - 2$ **(b)** $5 for 14 units   **43.** 2   **45.** 0
47. $-7 \pm \sqrt{5}$ **49.** $1/2, -2$ **51.** $1 \pm \sqrt{3}$ **53.** $(6 \pm \sqrt{58})/2$
55. $5/2, -3$ **57.** $7, -3/2$ **59.** $\pm\sqrt{3}/3$ **61.** $\pm\sqrt{5}$
63. $5, -2/3$ **65.** $-5, -1/2$ **67.** $r = (-Rp \pm E\sqrt{Rp})/p$
69. $s = (a \pm \sqrt{a^2 + 4k})/2$ **71.** 5 ft   **73.** 50 m by 225 m or 112.5 m by 100 m   **75.** $(3/8, \infty)$ **77.** $(-\infty, 1/4]$
79. $[-1/2, 2]$ **81.** $[-8, 8]$ **83.** $(-\infty, 2]$ or $[5, \infty)$
85. $[-9/5, 1]$ **87.** $5.5 \le y \le 9.5$ **89.** $0 \le m < 500$, where m is the number of miles driven   **91.** $(-3, 2)$ **93.** $(-\infty, -5]$ or $[3/2, \infty)$ **95.** $(-\infty, -5]$ or $[-2, 3]$ **97.** $[-2, 0)$
99. $(-1, 3/2)$ **101.** $[-19, -5)$ or $(2, \infty)$

Case 2 (page 115)

1.

Year	Straight-Line	Sum-of-the-Years'-Digits	Double Declining Balance
1	$11,000	$18,333.33	$22,000
2	11,000	14,666.67	13,200
3	11,000	11,000.00	7,920
4	11,000	7,333.33	5,940
5	11,000	3,666.67	5,940

3. Either the sum-of-the-years'-digits method or the double declining balance method gives the largest early years' deductions.

CHAPTER 3
Section 3.1 (page 122)

1. Function   **3.** Function   **5.** Not a function   **7.** Function
9. $(-\infty, \infty)$ **11.** $(-\infty, \infty)$ **13.** $(-\infty, 0]$ **15.** All real
numbers except 1   **17.** $(-\infty, \infty)$ **19.** (a) 6 (b) 6 (c) 6
(d) 6   **21.** (a) 48 (b) 6 (c) 25.38 (d) 28.42
23. (a) $\sqrt{7}$ (b) 0 (c) $\sqrt{5.7} \approx 2.3875$ (d) Not defined
25. (a) -18.2 (b) -97.3 (c) -14.365 (d) -289.561
27. (a) 12 (b) 23 (c) 12.91 (d) 49.41
29. (a) $\sqrt{3}/15 \approx .1155$ (b) Not defined (c) $\sqrt{1.7}/6.29 \approx .2073$
(d) Not defined
31.

33. (a) $5 - p$ (b) $5 + r$ (c) $2 - m$ **35.** (a) $\sqrt{4 - p}$ $(p \le 4)$
(b) $\sqrt{4 + r}$ $(r \ge -4)$ (c) $\sqrt{1 - m}$ $(m \le 1)$ **37.** (a) $p^3 + 1$
(b) $-r^3 + 1$ (c) $m^3 + 9m^2 + 27m + 28$ **39.** Assume all
denominators are nonzero. (a) $3/(p - 1)$ (b) $3/(-r - 1)$
(c) $3/(m + 2)$ **41.** 2   **43.** $2x + h$ **45.** (a) About 89%
(b) About 47%   **47.** (a) About 50,000; about 55,700
(b) About 157,300   **49.** $f(t) = 2000 - 475t$

Section 3.2 (page 132)

1.

3.

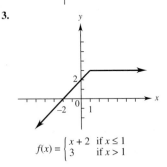

$f(x) = \begin{cases} x + 2 & \text{if } x \le 1 \\ 3 & \text{if } x > 1 \end{cases}$

5.

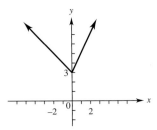

$y = \begin{cases} 3 - x & \text{if } x \le 0 \\ 2x + 3 & \text{if } x > 0 \end{cases}$

7.

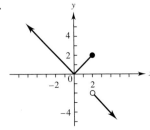

$f(x) = \begin{cases} |x| & \text{if } x \le 2 \\ -x & \text{if } x > 2 \end{cases}$

9.

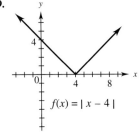

$f(x) = |x - 4|$

11.

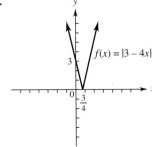

$f(x) = |3 - 4x|$

13.

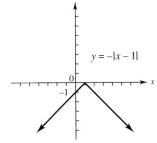

$y = -|x - 1|$

15.

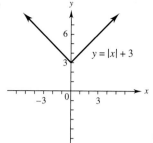

$y = |x| + 3$

17.

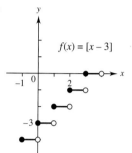

$f(x) = [x - 3]$

19.

$g(x) = [-x]$

21.

$f(x) = [x] + [-x]$

23.

$f(x) = 3 - 2x^2$

25.

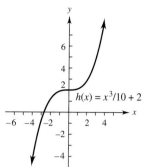

$h(x) = x^3/10 + 2$

27.

$g(x) = \sqrt{-x}$

29.

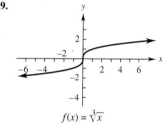

$f(x) = \sqrt[3]{x}$

31. Function   **33.** Not a function   **35.** Function

37.

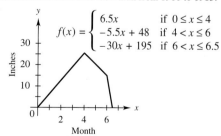

$$y = \begin{cases} .22x + 5.5 & \text{for 1960 to 1985} \\ .29x + 3.75 & \text{for 1985 to 1992} \end{cases}$$

They rose faster from 1985 to 1992 than from 1960 to 1985.

39. (a)

$$f(x) = \begin{cases} 6.5x & \text{if } 0 \le x \le 4 \\ -5.5x + 48 & \text{if } 4 < x \le 6 \\ -30x + 195 & \text{if } 6 < x \le 6.5 \end{cases}$$

(b) At the beginning of February; 26 in.   **(c)** Begins in early October; ends in mid-April

41. (a) $f(x) = \begin{cases} .012x + .05 & \text{if } 0 \le x \le 10 \\ .006x + .11 & \text{if } 10 < x \le 20 \end{cases}$

(b)

(c) 11% in 1975; 18.2% in 1982
43. (a) 26,300 BTUs per dollar   **(b)** $120 billion   **(c)** 23,500 BTUs per dollar, $85 billion; 1981   **(d)** None
45. (a)

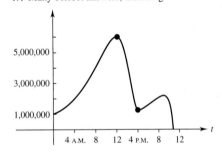

(b) $y = .04375x - 86.21875$ **(c)** $f(x) = .04375x - 86.21875$
(d) 4.125; yes; yes
47. Many correct answers, including:

49. (a) $29   **(b)** $29   **(c)** $33   **(d)** $33
(e)

51.

53.

Section 3.3 (page 142)

1. (a) $f(x) = -.15x + 31.2, 0 \le x \le 6$ **(b)** $-.15$ million
3. (a) The data appear linear.   **(b)** $f(x) = (1000/13)x$ **(c)** About 780 megaparsecs (approximately 1.5×10^{22} miles)   **(d)** About 12.4 billion years   **5. (a)** $f(x) = (4/75)x - (13/300)$
(b) Approximately 10.2 years   **7. (a)** $u(x) = 187 - .85x$;
$l(x) = 154 - .7x$ **(b)** 140 to 170 beats per minute   **(c)** 126 to 153 beats per minute   **(d)** 16 and 52; 143 beats per minute
9. (a) 2000   **(b)** 2900   **(c)** 3200   **(d)** Yes   **(e)** 300
11. (a) The number of new car dealers selling used cars was decreasing at the rate of 2720 per year.   **(b)** The slope of .3 indicates that the sales at superstores are increasing at the rate of 300,000 per year.   **(c)** As the number of new car dealers selling used cars decreased, the number of used cars sold in superstores increased, suggesting that used car buyers are changing to this new way of buying used cars.   **13. (a)** 3   **(b)** 3.5   **(c)** 1   **(d)** 0   **(e)** -2 **(f)** -2
(g) -3 **(h)** -1 **15.** Let $C(x)$ be the cost of renting a saw for x hours; $C(x) = 12 + x$.   **17.** Let $P(x)$ be the cost (in cents) of parking for x half-hours; $P(x) = 30x + 35$.   **19.** $C(x) = 30x + 100$
21. $C(x) = 120x + 3800$ **23. (a)** $5.25   **(b)** $66,250
25. (a) $C(x) = 10x + 500$ **(b)** $R(x) = 35x$
27. (a) $C(x) = 18x + 250$ **(b)** $R(x) = 28x$ **29. (c)**

Chapter 3 Review Exercises (page 146)

1. Not a function   **3.** Function   **5.** Not a function
7. (a) 23   **(b)** -9 **(c)** $4p - 1$ **(d)** $4r + 3$ **9. (a)** -28
(b) -12 **(c)** $-p^2 + 2p - 4$ **(d)** $-r^2 - 3$ **11. (a)** -13
(b) 3   **(c)** $-k^2 - 4k$ **(d)** $-9m^2 + 12m$ **(e)** $-k^2 + 14k - 45$
(f) $12 - 5p$

13.

$f(x) = |x| - 3$

15.

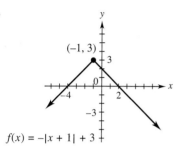

$(-1, 3)$

$f(x) = -|x + 1| + 3$

17.

$f(x) = [x - 3]$

19.

$f(x) = \begin{cases} -4x + 2 & \text{if } x \le 1 \\ 3x - 5 & \text{if } x > 1 \end{cases}$

21.

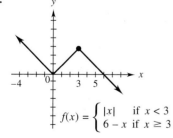

$f(x) = \begin{cases} |x| & \text{if } x < 3 \\ 6 - x & \text{if } x \ge 3 \end{cases}$

23.

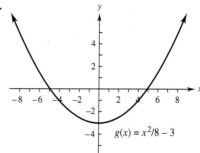

$g(x) = x^2/8 - 3$

25. (a) $C(x) = 30x + 60$ (b) \$30 (c) \$30.60
27. (a) $C(x) = 30x + 85$ (b) \$30 (c) \$30.85
29. (a) 1.4 (b) $-.75$ (c) .8 (d) 1.25
31. (a) Yes
(b)

(c) Domain: $(0, \infty)$; range: $\{47, 49, 51, 53, \ldots\}$
33. $y = -.8x + 187.7$ **35.** (a) $y = .18t + 14.0$ (b) 2020

Case 3 (page 148)

1. 4.8 million units   **3.** In the interval under discussion (3.1 to 5.7 million units), the marginal cost always exceeds the selling price.

CHAPTER 4
Section 4.1 (page 155)

1. $(5, 2)$; upward   **3.** $(1, 2)$; downward   **5.** $(-6, -35)$; upward
7. $(-1, -1)$; upward   **9.** x-intercepts 1, 3; y-intercept 9
11. x-intercepts $-1, -3$; y-intercept 6   **13.** $f(x) = -(x - 3)^2 - 5$
15. $f(x) = (1/4)(x - 1)^2 + 2$ **17.** $f(x) = (x + 1)^2 - 2$
19. $(-2, 0)$, $x = -2$

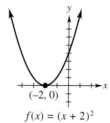

$(-2, 0)$

$f(x) = (x + 2)^2$

21. $(1, -3), x = 1$

$f(x) = (x - 1)^2 - 3$

23. $(2, 2), x = 2$

$f(x) = x^2 - 4x + 6$
$f(x) = (x - 2)^2 + 2$

25. $(1, 3), x = 1$

$f(x) = 2x^2 - 4x + 5$
$f(x) = 2(x - 1)^2 + 3$

27. $(3, 3), x = 3$

$f(x) = -x^2 + 6x - 6$
$f(x) = -(x - 3)^2 + 3$

29. Vertex $(7.2, 12)$ **31.** Vertex $(95, -2695.5)$
33. (a)–(d)

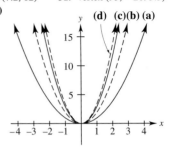

(e) The graph of $f(x) = ax^2$ is the graph of $k(x) = x^2$ stretched away from the x-axis by a factor of a.

35. (a)–(d)

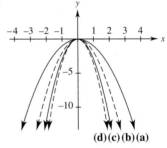

(e) Changing the sign of a reflects the graph in the x-axis.

37. (a)–(d)

(e) The graph of $f(x) = x^2 - c$ is the graph of $k(x) = x^2$ shifted c units downward.   **39.** $f(x) = 3x^2$

Section 4.2 (page 161)

1. (a) $16; $8; $32
(b)

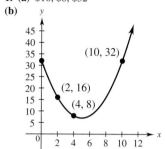

$$C(x) = x^2 - 10x + 32$$

(c) (5, 7) **(d)** 5; $7 per box   **3. (a)** $.60 **(b)** $800
5. 16 ft; 2 sec   **7. (a)** $R(x) = 150x - x^2/4$ **(b)** 300
(c) $22,500   **9. (a)** $R(x) = x(500 - x)$
(b)

(c) $250 **(d)** $62,500   **11. (a)** About 12 **(b)** 10   **(c)** About 7
(d) 0 **(e)** 5 **(f)** About 7 **(g)** 10 **(h)** About 12
(i)

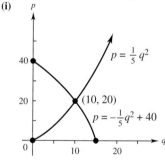

13. (a) $R(x) = (100 - x)(200 + 4x) = 20,000 + 200x - 4x^2$
(b)

(c) 25 **(d)** $22,500   **15.** 13 weeks; $96.10

Section 4.3 (page 170)

1.

$$f(x) = x^4$$

3.

$$h(x) = -.2x^5$$

5. (a) Yes **(b)** No **(c)** No **(d)** Yes   **7. (a)** Yes **(b)** Yes
(c) No **(d)** Yes   **9. (a)** Yes **(b)** No **(c)** No **(d)** No
11. (a) Yes **(b)** No **(c)** Yes **(d)** No
13.

$$f(x) = (x + 2)(x - 3)(x + 4)$$

15.

$$f(x) = x^2(x - 2)(x + 3)$$

17. (a) 250,000

(b) $g(x) = 1817.95(x - 2)^2 + 620$ **(c)** Approximately 589,636
19. 80 ft by 160 ft   **21.** 2.5 in.   **23. (a)** 2.3 or 17.7 **(b)** 15
(c) 10 **(d)** $120 **(e)** $0 < x < 2.3$ or $x > 17.7$ **(f)** $2.3 < x < 17.7$

17.

$f(x) = x^3 + x^2 - 6x$

19.

$f(x) = x^3 + 3x^2 - 4x$

21.

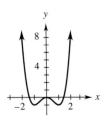

$f(x) = x^4 - 2x^2$

23. $-3 \le x \le 5$ and $-20 \le y \le 5$ **25.** $-3 \le x \le 4$ and $-35 \le y \le 20$ **27. (a)** 1.0 tenth of a percent, or .1%   **(b)** 2.0 tenths of a percent, or .2%   **(c)** 3.3 tenths of a percent, or .33% **(d)** 3.1 tenths of a percent, or .31%   **(e)** .8 tenths of a percent, or .08% **(f)**

$A(x) = -.015x^3 + 1.058x$

(g) Between 4 and 5 hours, closer to 5 hours   **(h)** From about 1.5 hours to about 7.5 hours   **29. (a)** 0; 108; 28; 10 **(b)**

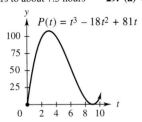

$P(t) = t^3 - 18t^2 + 81t$

(c) Increasing for years 0 to 3 and from the 9th year on; decreasing for the years 3 to 9   **31. (a)** About 106,875,000,000; about 401,500,000,000   **(b)** About 917,100,000,000; about 1,089,000,000,000   **(c)** About 109,469

Section 4.4 (page 179)

1. $x = -5, y = 0$

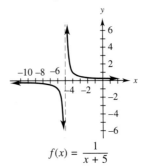

$f(x) = \dfrac{1}{x + 5}$

3. $x = -5/2, y = 0$

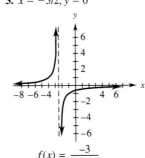

$f(x) = \dfrac{-3}{2x + 5}$

5. $x = 1, y = 3$

$f(x) = \dfrac{3x}{x - 1}$

7. $x = 4, y = 1$

$f(x) = \dfrac{x + 1}{x - 4}$

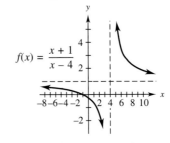

9. $x = 3, y = -1$

$f(x) = \dfrac{2 - x}{x - 3}$

11. $x = -1/2, y = 1/2$

$f(x) = \dfrac{2x - 1}{4x + 2}$

13. $x = -4, x = 1, y = 0$

$h(x) = \dfrac{x + 1}{x^2 + 3x - 4}$

15. $x = -1, x = 1, y = 1$

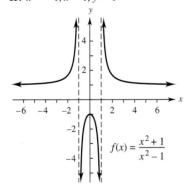

$f(x) = \dfrac{x^2 + 1}{x^2 - 1}$

17. $x = -2, x = 1$ **19.** $x = -1, x = 5$
21. **(a)** \$4300 **(b)** \$10,033.33 **(c)** \$17,200 **(d)** \$38,700
(e) \$81,700 **(f)** \$210,700 **(g)** \$425,700 **(h)** No

(i)

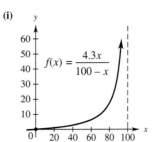

$f(x) = \dfrac{4.3x}{100 - x}$

23. **(a)** 125,000 **(b)** 187,500 **(c)** 200,000
(d)

$N(x) = \dfrac{250x}{x + 6}$

(e) The part in the first quadrant, because x and $N(x)$ are both positive
(f) $y = 250$; 250,000
25. **(a)**

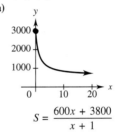

$$S = \dfrac{600x + 3800}{x + 1}$$

(b) Yes; \$600; the horizontal asymptote **(c)** After 7 days; the cost of
ads will be more than the sales they generate.
27. 30,000 reds, 10,000 blues

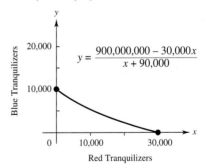

$$y = \dfrac{900,000,000 - 30,000x}{x + 90,000}$$

Blue Tranquilizers / Red Tranquilizers

29. **(a)** $C(x) = 2.6x + 40,000$ **(b)** $\overline{C}(x) = \dfrac{2.6x + 40,000}{x}$
(c) $y = 2.6$; the average cost may get close to, but will never equal
\$2.60.
31. **(a)**

(b) They appear almost identical because the parabola is an asymptote of the graph.

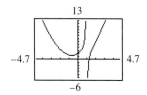

Chapter 4 Review Exercises (page 181)

1. Upward; $(2, 6)$ **3.** Downward; $(-1, 8)$
5.

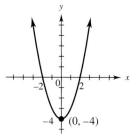

$f(x) = x^2 - 4$

7.

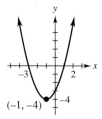

$f(x) = x^2 + 2x - 3$
$f(x) = (x + 1)^2 - 4$

9.

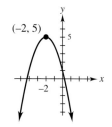

$f(x) = -x^2 - 4x + 1$
$f(x) = -(x + 2)^2 + 5$

11.

$f(x) = 2x^2 + 4x - 3$
$f(x) = 2(x + 1)^2 - 5$

13. Minimum value, -11 **15.** Maximum value, 7
17. 4 months after she began   **19.** Between 350 and 1850
21. 10 ft by 15 ft; 150 sq ft   **23. (a)** $f(x) = .6944(x - 1)^2 + 40$
(b) 265 billion; 291 billion

25.

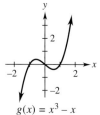

$g(x) = x^3 - x$

27.

$f(x) = x(x - 2)(x + 3)$

29.

$f(x) = x(2x - 1)(x + 2)$

31.

$f(x) = 2x^3 - 3x^2 - 2x$

33.

$f(x) = x^4 - 5x^2 - 6$

35. $x = 3, y = 0$

$f(x) = \dfrac{1}{x - 3}$

37. $x = 2, y = 0$

$$f(x) = \frac{-3}{2x - 4}$$

39. $x = -1/2, x = 3/2, y = 0$

$$g(x) = \frac{5x - 2}{4x^2 - 4x - 3}$$

41. (a) About \$10.83 **(b)** About \$4.64 **(c)** About \$3.61
(d) About \$2.71
(e)

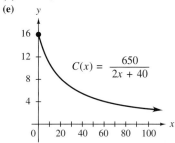

$$C(x) = \frac{650}{2x + 40}$$

43. (a) $(10, 50)$

(b) $(10, \infty)$ **(c)** $(0, 10)$

Case 4 (page 184)

1. $(2.9, 5.7, 8.5, 11.3, 14.1, 16.9)$
3. $(2.3, 1.1, 1.7, 2.4, 1.5, -2.7, -11.9, -27.8)$

CHAPTER 5
Section 5.1 (page 190)

1. Quadratic   **3.** Exponential   **5. (a)** The graph lies entirely above the x-axis and falls from left to right. It falls relatively steeply until it reaches the y-intercept 1 and then falls slowly, with the positive x-axis as a horizontal asymptote.   **(b)** $(0, 1), (1, .8)$ **7. (a)** The graph lies entirely above the x-axis and rises from left to right. The negative x-axis is a horizontal asymptote. The graph rises slowly until it reaches the y-intercept 1 and then rises quite steeply.
(b) $(0, 1), (1, 5^{.4})$

9.

$$f(x) = 3^x$$

11.

$$f(x) = 2^{x/2}$$

13.

$$f(x) = (1/5)^x$$

15. (a)–(c)

17. 2.3   **19.** .75   **21.** .31

23. (a) $a > 1$ **(b)** Domain: $(-\infty, \infty)$; range: $(0, \infty)$
(c)

$g(x) = -a^x$

(d) Domain: $(-\infty, \infty)$; range: $(-\infty, 0)$
(e)

$h(x) = a^{-x}$

(f) Domain: $(-\infty, \infty)$; range: $(0, \infty)$ **25. (a)** 3 **(b)** 1/3
(c) 9 **(d)** 1
27.

$f(x) = 2^{-x^2+2}$

29.

$f(x) = x \cdot 2^x$

31. 2   **33.** 3   **35.** 1.5   **37.** 1   **39.** 1/4   **41.** 1/9
43. 4, -4 **45.** 2
49. (a)

t	0	1	2	3	4	5	6	7	8	9	10
y	1	1.06	1.12	1.19	1.26	1.34	1.42	1.50	1.59	1.69	1.79

(b)

$y = (1.06)^t$

51. (a) About $141,892 **(b)** About $64.10   **53. (a)** 1.11 million
(b) 1.50 million **(c)** 2.01 million   **55. (a)** 11 **(b)** 53 **(c)** 576
57. About 56 years   **59.** $19,420.26   **63. (a)** About 520
(b) About 1559 **(c)** About 32.4 days

Section 5.2 (page 196)

1. 30.0 million   **3. (a)** 100 g **(b)** About 70.9 g **(c)** About
50.3 g **(d)** About 11.6 g   **5. (a)** 50,000 **(b)** About 47,561
(c) About 40,937 **(d)** About 30,327   **7.** 11.6 lb per square inch

9. (a) $21,665.74 **(b)** $29,836.49 **(c)** $44,510.82 **(d)** About
13.75 yr   **11. (a)** 12 **(b)** About 20 **(c)** About 27
(d) About 27   **13. (a)** 6 **(b)** About 23 **(c)** 25
15. (a) 446 (actual 383) billion **(b)** 1227 (actual 1000) billion
(c) 4075 (actual 4077) billion   **17.** 6.25°C   **19. (a)** 1000
(b) About 1882   **21. (a)** .13 **(b)** .23
23. (a) 100

(b) .182 (In many states the legal limit is .100.)

25. (a) The points fit a logistic model.

(b) Yes

(c) About 5970 **(d)** Yes; the limit is 10,000.

Section 5.3 (page 209)

1. a^y **3.** It is missing the value that equals b^y. If that value is x,
then it should read $y = \log_b x$.   **5.** $10^5 = 100,000$ **7.** $3^4 = 81$
9. $\log 75 = 1.8751$ **11.** $\log_3(1/9) = -2$ **13.** 3   **15.** 2
17. 3   **19.** -2 **21.** 1/2   **23.** 3.78   **25.** 1.672
27. -1.047 **31.** $\log 16$ **33.** $\ln 5$ **35.** $\log\left(\dfrac{u^2 w^3}{v^6}\right)$
37. $\ln\left(\dfrac{(x+1)^2}{x+2}\right)$ **39.** $\dfrac{1}{2}\ln 6 + 2\ln m + \ln n$
41. $\dfrac{1}{2}\log x - \dfrac{5}{2}\log z$ **43.** $2u + 5v$ **45.** $3u - 2v$
47. 3.5145   **49.** 2.4851   **51.** There are many correct answers,
including $b = 1, c = 2$.   **53.** 1/5   **55.** 3/2   **57.** 16
59. 8/5   **61.** 1/3   **63.** 10   **65.** 5.2378   **67.** 5

71.

$y = \ln(x + 2)$

73.

$y = \log(x - 3)$

77. $\ln 2.75 = 1.0116009$; $e^{1.0116009} = 2.75$ **79. (a)** 17.67
(b) 9.01 **(c)** 4.19 **(d)** 2.25 **(e)** $72/4 = 18$; $72/8 = 9$; $72/18 = 4$; $72/2 = 36$; So it takes about $72/k$ years for money to double at $k\%$ interest. **81. (a)** About 8 **(b)** About 12 **83. (a)** The percent of change in rents increased rapidly from 1992 to 1993, leveled off to a slow increase from 1993 to 1994, and then increased rapidly again from 1994 to 1996. From 1996 to 1999 (estimated) the percent of increase slowed and became constant at about 7%. **(b)** $f(92) = -3.4$; $f(99) = 7.1$; $f(92) = -3.4$ agrees closely with the y-value on the graph of about -3; $f(99) = 7.1$ is also reasonably close to the y-value on the graph of about 7.5.

Section 5.4 (page 217)

1. 1.465 **3.** 2.710 **5.** -1.825 **7.** .805 **9.** $-.123$
11. ± 2.141 **13.** 1.386 **15.** $\dfrac{\log d + 3}{4}$ **17.** $\dfrac{\ln b + 1}{2}$
19. 5 **21.** No solution **23.** 11 **25.** 3 **27.** No solution
29. $-2, 2$ **31.** 1, 10 **33.** $\dfrac{4 + b}{4}$ **35.** $\dfrac{10^{2-b} - 5}{6}$
39. (a) 25 g **(b)** About 4.95 yr **41.** About 3606 years old
43. (a) About 23 days **(b)** About 46 days **45.** $10^{1.2} \approx 15.85$
47. (a) 10^{10} times more **(b)** 10^5 times more **(c)** 10^5 times more
49. .629 **51. (a)** About 527 **(b)** About 3466 **(c)** About 6020
(d) About 1783 **(e)** $r \approx .8187$ **53.** In 2006 **55.** 14.2 hours
57. (a) About 1980 **(b)** Never; the function limits the percent to 25.

Chapter 5 Review Exercises (page 220)

1. (c) **3.** (d) **5.** $0 < a < 1$ **7.** $(0, \infty)$ **9.** -1 **11.** 2
13.

$f(x) = 4^x$

15.

$f(x) = \ln x + 5$

17. (a) 207 **(b)** 235 **(c)** 249
(d)

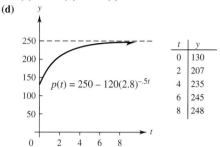

$p(t) = 250 - 120(2.8)^{-.5t}$

t	y
0	130
2	207
4	235
6	245
8	248

19. $\log 47 = 1.6721$ **21.** $\ln 39 = 3.6636$ **23.** $10^3 = 1000$
25. $e^{4.5581} = 95.4$ **27.** 3 **29.** 7.4 **31.** 4/3 **33.** $\log 20k^4$
35. $\log\left(\dfrac{b^2}{c^3}\right)$ **37.** 1.416 **39.** -2.807 **41.** -3.305
43. .747 **45.** 28.463 **47.** 8 **49.** 6 **51.** 2 **53.** 3
55. (a) \$15 million **(b)** \$15.6 million **(c)** \$16.4 million
57. (a) 10 g **(b)** About 140 days **(c)** About 243 days
59. 81.25° Celsius **61.** 1999

63. (a)

(b) Exponential
(c)

The graph fits the data very well. **(d)** 825 millibars; 224 millibars; Both predicted values are close to the actual values.

Case 5 (page 224)

1. 23.6; 47.7; 58.4; the estimates are a bit low.

CHAPTER 6
Section 6.1 (page 230)

1. Time and interest rate **3.** \$231.00 **5.** \$286.75
7. \$119.15 **9.** \$241.56 **11.** \$304.38 **13.** \$3654.10
17. \$46,265.06 **19.** \$28,026.37 **21.** \$8898.75
23. \$48,140.65 **25.** 10.7% **27.** 11.8% **29.** \$27,894.30
31. 6.8% **33.** \$34,438.29 **35.** \$6550.92 **37.** 11.4%
39. \$13,683.48; yes **41.** (c)

Section 6.2 (page 238)

3. \$1593.85 **5.** \$1515.80 **7.** \$13,213.16 **9.** \$10,466.35
11. \$2307.95 **13.** \$1968.48 **15.** \$26,545.91
17. \$45,552.97 **21.** 4.04% **23.** 8.16% **25.** 12.36%

27. $8954.58   **29.** $3255.55   **31.** $11,572.58   **33.** $1000 now   **35.** $13,693.34   **37.** $7743.93   **39.** $142,886.40   **41.** $123,506.50   **43.** 4.91, 5.20, 5.34, 5.56, 5.63   **45.** $8763.47   **47.** $30,611.30   **49.** About $1.946 million   **51.** $11,940.52   **53.** About 12 yr   **55.** About 18 yr   **57.** 10.00%   **59.** (a)

Section 6.3 (page 248)

1. 135   **3.** 6   **5.** 2315.25   **7.** 15   **9.** 6.24   **11.** 594.048   **13.** 15.91713   **15.** 21.82453   **17.** 22.01900   **21.** $4625.30   **23.** $327,711.81   **25.** $112,887.43   **27.** $1,391,304.39   **29.** $7118.38   **31.** $447,053.44   **33.** $36,752.01   **35.** $137,579.79   **37.** $3750.91   **39.** $236.61   **41.** $775.08   **43.** $4462.94   **45.** $148.02   **47.** $79,679.68   **49.** $66,988.91   **51.** $130,159.72   **53.** $284,527.35   **55.** $1349.48   **57.** 6.5%   **59.** (a) 7 yr   **(b)** 5 yr   **61. (a)** $120   **(b)** $681.83 except the last payment of $681.80

(c)

Payment Number	Amount of Deposit	Interest Earned	Total
1	$681.83	$ 0.00	$ 681.83
2	681.83	54.55	1418.21
3	681.83	113.46	2213.50
4	681.83	177.08	3072.41
5	681.80	245.79	4000.00

Section 6.4 (page 254)

1. (c)   **3.** 9.71225   **5.** 12.65930   **7.** 14.71787   **11.** $8415.93   **13.** $9420.83   **15.** $210,236.40   **17.** $103,796.58   **21.** $1603.01   **23.** $3109.71   **25.** $12.43   **27.** $681.02   **29.** $916.07   **31.** $86.24   **33.** $13.02   **35.** $48,677.34   **37. (a)** $158   **(b)** $1584   **39.** $573,496.06   **41.** About $77,790

The computer program, "Explorations in Finite Mathematics," was used for the following amortization schedules. Answers found using other computer programs or calculators may differ slightly.

43.

Payment Number	Amount of Payment	Interest for Period	Portion to Principal	Principal at End of Period
0	—	—	—	$72,000.00
1	$9247.24	$2160.00	$7087.24	64,912.76
2	9247.24	1947.38	7299.85	57,612.91
3	9247.24	1728.39	7518.85	50,094.06
4	9247.24	1502.82	7744.42	42,349.64

45.

Payment Number	Amount of Payment	Interest for Period	Portion to Principal	Principal at End of Period
0	—	—	—	$20,000.00
1	$2404.83	$700.00	$1704.83	18,295.17
2	2404.83	640.33	1764.50	16,530.68
3	2404.83	578.57	1826.25	14,704.42
4	2404.83	514.65	1890.17	12,814.25

47. $280.46; $32,310.40   **49. (a)** $2349.51; $197,911.80

(b) $2097.30; $278,352   **(c)** $1965.82; $364,746   **(d)** After 170 payments   **51. (a)** $1352.45   **(b)** $156,428.83   **(c)** $14,216.43   **(d)** 20.5 years (246 payments)   **53.** $432.53

Section 6.5 (page 257)

1. $45,401.41   **3.** $350.26   **5.** $5958.68   **7.** $2036.46   **9.** $8258.80   **11.** $47,650.97; $14,900.97   **13.** $16,876.98   **15.** $23,243.61   **17.** $32,854.98   **19.** $8362.60   **21.** $1228.20   **23.** $9843.39   **25.** $6943.65   **27.** $26,850.21   **29.**

Payment Number	Amount of Payment	Interest for Period	Portion to Principal	Principal at End of Period
0	—	—	—	$8500.00
1	$1416.18	$340.00	$1076.18	7423.82
2	1416.18	296.95	1119.23	6304.59
3	1416.18	252.18	1164.00	5140.59
4	1416.18	205.62	1210.56	3930.03
5	1416.18	157.20	1258.98	2671.05
6	1416.18	106.84	1309.34	1361.71
7	1416.18	54.47	1361.71	0

31. 5.1%   **33.** $8017.02   **35.** $2572.38   **37.** $89,659.63; $14,659.63   **39.** $6659.79   **41.** $4587.64   **43.** $212,026

Chapter 6 Review Exercises (page 260)

1. $426.88   **3.** $62.91   **7.** $78,717.19   **9.** $739.09   **13.** $70,538.38; $12,729.04   **15.** $5282.19; $604.96   **17.** $12,857.07   **19.** $1923.09   **21.** 4, 2, 1, 1/2   **23.** −32   **25.** 5500   **29.** $137,925.91   **31.** $25,396.38   **35.** $2619.29   **37.** $916.12   **39.** $31,921.91   **41.** $14,222.42   **43.** $3628.00   **45.** $546.93   **47.** $896.06   **49.** $2696.12   **51.** $10,550.54   **53.** 8.21%   **55.** $5596.62   **57.** $24,818.76; $2418.76   **59.** $3560.61   **61. (a)** $571.28   **(b)** $532.50   **(c)** Method 1: $56,324.44; method 2: $56,325.43   **(d)** $7100   **(e)** Method 1: $72,575.56; method 2: $72,574.57   **63.** $64,826.57

Case 6 (page 264)

1.

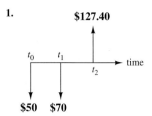

$50(1 + i)^2 + 70(1 + i) − 127.40 = 0$; 4.3%
3. (a) $5864.17

(b) 5.2%

5. (a)

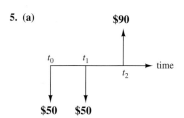

$50(1 + i)^2 + 50(1 + i) - 90 = 0; -.068, -2.932$ **(b)** $-.068$ is reasonable, -2.932 is not.

CHAPTER 7
Section 7.1 (page 276)

1. Yes **3.** $(11/5, -7/5)$ **5.** $(28, 22)$ **7.** $(2, -1)$ **9.** No solution **11.** $(4y + 1, y)$ for any real number y **13.** No solution **15. (a)** **17.** $(5, 10)$ **19.** $(-5, -3)$ **21.** Dependent **23.** Inconsistent **25.** Independent **27.** $(2, 1, 4)$ **29.** $(-1, 4, 2)$ **31.** No solution **33.** $(2, 0, 1)$ **35.** $((-7 - 2z)/5, (11z + 21)/5, z)$ for any real number z **37.** $(1, 2)$ **39. (a)** $y = x + 1$ **(b)** $y = 3x + 4$ **(c)** $(-3/2, -1/2)$ **43.** $W_1 = \dfrac{300}{1 + \sqrt{3}} = 110$ pounds; $W_2 = \dfrac{300 - W_1}{\sqrt{2}} = 134$ pounds **45.** Eastward: 7 hours; difference in time zones: 6 hours **47.** $27 **49.** 5 model 201, 8 model 301 **51.** $3000 at 8%, $6000 at 9%, $1000 at 5% **53.** 5 bags of brand A, 2 bags of brand B, 3 bags of brand C **55. (a)** $12; 5, 6, 7, \ldots, 46$ **(b)** The solution with 1 knife, 34 forks, and 5 spoons

Section 7.2 (page 284)

1. $\begin{bmatrix} 2 & 1 & 1 & | & 3 \\ 3 & -4 & 2 & | & -5 \\ 1 & 1 & 1 & | & 2 \end{bmatrix}$ **3.** $\begin{aligned} 2x + 3y + 8z &= 20 \\ x + 4y + 6z &= 12 \\ 3y + 5z &= 10 \end{aligned}$

5. $\begin{bmatrix} 1 & 2 & 3 & | & -1 \\ 2 & 0 & 7 & | & -4 \\ 6 & 5 & 4 & | & 6 \end{bmatrix}$ **7.** $\begin{bmatrix} -4 & -3 & 1 & -1 & | & 2 \\ 0 & -4 & 7 & -2 & | & 10 \\ 0 & -2 & 9 & 4 & | & 5 \end{bmatrix}$

9. $(-3, 4, 0)$ **11.** $(1, 0, -1)$ **13.** $((-9z + 5)/23, (10z - 3)/23, z)$ for any real number z **15.** $(-1, 23, 16)$ **17.** No solution **19.** $(-7, 5)$ **21.** $(-3, z - 17, z)$ for any real number z **23.** $(-1, 1, -3, -2)$ **25.** No solution **27.** Garcia 20 hr, Wong 15 hr **29.** 5 trucks, 2 vans, 3 station wagons **31. (a)** Let x_1 be the units purchased from the first supplier for Canoga Park, x_2 be the units from the first supplier for Wooster, x_3 be the units from the second supplier for Canoga Park, and x_4 be the units from the second supplier for Wooster. **(b)** $x_1 + x_2 = 75; x_3 + x_4 = 40; x_1 + x_3 = 40; x_2 + x_4 = 75;$ $70x_1 + 90x_2 + 80x_3 + 120x_4 = 10,750$ **(c)** $(40, 35, 0, 40)$ **33.** Three possibilities: 7 trucks, no vans, 2 station wagons; or 7 trucks, 1 van, 1 station wagon; or 7 trucks, 2 vans, no station wagons **35. (a)** $12,000 at 6%, $7000 at 7%, and $6000 at 10% **(b)** $10,000 at 6%, $15,000 at 7%, and $5000 at 10% **(c)** $20,000 at 6%, $10,000 at 7%, and $10,000 at 10% **37. (a)** $y = .01x^2 - .3x + 4.24$ **(b)** 15 platters; $1.99 **39. (a)** No **(b)** Yes; 150 acres for honeydews, 50 acres for onions, and 20 acres for lettuce **41. (a)** $C = .0378t^2 + .1t + 315$ **(b)** 2048

Section 7.3 (page 293)

1. $2 \times 3;$ $\begin{bmatrix} -7 & 8 & -4 \\ 0 & -13 & -9 \end{bmatrix}$ **3.** $3 \times 3;$ square matrix; $\begin{bmatrix} 3 & 0 & -11 \\ -1 & -\frac{1}{4} & 7 \\ -5 & 3 & -9 \end{bmatrix}$ **5.** $2 \times 1;$ column matrix; $\begin{bmatrix} -7 \\ -11 \end{bmatrix}$ **7.** B is a 5×3 zero matrix. **9.** $\begin{bmatrix} 9 & 12 & 0 & 2 \\ 1 & -1 & 2 & -4 \end{bmatrix}$

11. $\begin{bmatrix} 5 & 13 & 0 \\ 3 & 1 & 8 \end{bmatrix}$ **13.** $\begin{bmatrix} 3 & -7 & 7 \\ 4 & 4 & -8 \end{bmatrix}$ **15.** $\begin{bmatrix} -4 & 8 \\ 0 & 6 \end{bmatrix}$

17. $\begin{bmatrix} 24 & -8 \\ -16 & 0 \end{bmatrix}$ **19.** $\begin{bmatrix} -22 & -6 \\ 20 & -12 \end{bmatrix}$ **21.** $\begin{bmatrix} 4 & -\frac{7}{2} \\ 4 & \frac{21}{2} \end{bmatrix}$

23. $X + T = \begin{bmatrix} x & y \\ z & w \end{bmatrix} + \begin{bmatrix} r & s \\ t & u \end{bmatrix} = \begin{bmatrix} x + r & y + s \\ z + t & w + u \end{bmatrix}$ (a 2×2 matrix)

25. $X + (T + P) = \begin{bmatrix} x + (r + m) & y + (s + n) \\ z + (t + p) & w + (u + q) \end{bmatrix} =$ $\begin{bmatrix} (x + r) + m & (y + s) + n \\ (z + t) + p & (w + u) + q \end{bmatrix} = (X + T) + P$

27. $P + 0 = \begin{bmatrix} m + 0 & n + 0 \\ p + 0 & q + 0 \end{bmatrix} = \begin{bmatrix} m & n \\ p & q \end{bmatrix} = P$

29. $\begin{bmatrix} 18 & 2.7 \\ 10 & 1.5 \\ 8 & 1.0 \\ 10 & 2.0 \\ 10 & 1.7 \end{bmatrix}$; $\begin{bmatrix} 18 & 10 & 8 & 10 & 10 \\ 2.7 & 1.5 & 1.0 & 2.0 & 1.7 \end{bmatrix}$

31. (a) $\begin{bmatrix} 5.6 & 6.4 & 6.9 & 7.6 & 6.1 \\ 144 & 138 & 149 & 152 & 146 \end{bmatrix}$

(b) $\begin{bmatrix} 10.2 & 11.4 & 11.4 & 12.7 & 10.8 \\ 196 & 196 & 225 & 250 & 230 \end{bmatrix}$

(c) $\begin{bmatrix} 4.6 & 5.0 & 4.5 & 5.1 & 4.7 \\ 52 & 58 & 76 & 98 & 84 \end{bmatrix}$

(d) $\begin{bmatrix} 12.0 & 12.9 & 13.7 & 14.5 & 12.8 \\ 221 & 218 & 254 & 283 & 250 \end{bmatrix}$

33. (a) Chicago: $\begin{bmatrix} 4.05 & 7.01 \\ 3.27 & 3.51 \end{bmatrix}$; Seattle: $\begin{bmatrix} 4.40 & 6.90 \\ 3.54 & 3.76 \end{bmatrix}$

(b) $\begin{bmatrix} 4.24 & 6.95 \\ 3.42 & 3.64 \end{bmatrix}$ **(c)** $\begin{bmatrix} 4.42 & 7.43 \\ 3.38 & 3.62 \end{bmatrix}$ **(d)** $\begin{bmatrix} 4.41 & 7.17 \\ 3.46 & 3.69 \end{bmatrix}$

Section 7.4 (page 304)

1. $2 \times 2; 2 \times 2$ **3.** $3 \times 2;$ BA does not exist. **5.** AB does not exist; 3×2 **7.** Columns; rows **9.** $\begin{bmatrix} 13 \\ 25 \end{bmatrix}$ **11.** $\begin{bmatrix} -2 & 10 \\ 0 & 8 \end{bmatrix}$

13. $\begin{bmatrix} -4 & 1 \\ 2 & -3 \end{bmatrix}$ **15.** $\begin{bmatrix} 3 & -5 & 7 \\ -2 & 1 & 6 \\ 0 & -3 & 4 \end{bmatrix}$ **17.** $\begin{bmatrix} 16 & 11 \\ 37 & 32 \\ 58 & 53 \end{bmatrix}$

19. $AB = \begin{bmatrix} -30 & -45 \\ 20 & 30 \end{bmatrix}$, but $BA = \begin{bmatrix} 0 & 0 \\ 0 & 0 \end{bmatrix}$

21. $(A + B)(A - B) = \begin{bmatrix} -7 & -24 \\ -28 & -33 \end{bmatrix}$, but $A^2 - B^2 = \begin{bmatrix} -37 & -69 \\ -8 & -3 \end{bmatrix}$

23. $(PX)T =$

$\begin{bmatrix} (mx + nz)r + (my + nw)t & (mx + nz)s + (my + nw)u \\ (px + qz)r + (py + qw)t & (px + qz)s + (py + qw)u \end{bmatrix}$

$P(XT)$ is the same so $(PX)T = P(XT)$.

25. $k(X + T) = k\begin{bmatrix} x + r & y + s \\ z + t & w + u \end{bmatrix} = \begin{bmatrix} k(x + r) & k(y + s) \\ k(z + t) & k(w + u) \end{bmatrix}$

$= \begin{bmatrix} kx + kr & ky + ks \\ kz + kt & kw + ku \end{bmatrix} = \begin{bmatrix} kx & ky \\ kz & kw \end{bmatrix} + \begin{bmatrix} kr & ks \\ kt & ku \end{bmatrix} = kX + kT$

27. No   **29.** Yes   **31.** No   **33.** $\begin{bmatrix} 2 & -3 \\ -1 & 2 \end{bmatrix}$

35. No inverse   **37.** $\begin{bmatrix} 2 & -3 \\ -\frac{1}{2} & 1 \end{bmatrix}$ **39.** $\begin{bmatrix} 3 & 3 & -1 \\ -2 & -2 & 1 \\ -4 & -5 & 2 \end{bmatrix}$

41. $\begin{bmatrix} 2 & 1 & -1 \\ 8 & 2 & -5 \\ -11 & -3 & 7 \end{bmatrix}$ **43.** No inverse

45. $\begin{bmatrix} \frac{7}{4} & \frac{5}{2} & 3 \\ -\frac{1}{4} & -\frac{1}{2} & 0 \\ -\frac{1}{4} & -\frac{1}{2} & -1 \end{bmatrix}$ **47.** $\begin{bmatrix} \frac{1}{2} & \frac{1}{2} & -\frac{1}{4} & \frac{1}{2} \\ -1 & 4 & -\frac{1}{2} & -2 \\ -\frac{1}{2} & \frac{5}{2} & -\frac{1}{4} & -\frac{3}{2} \\ \frac{1}{2} & -\frac{1}{2} & \frac{1}{4} & \frac{1}{2} \end{bmatrix}$

49. (a) $\begin{bmatrix} 900 & 1500 & 1150 \\ 600 & 950 & 800 \\ 750 & 900 & 825 \end{bmatrix}$ **(b)** $[1.50 \quad .90 \quad .60]$

(c) $[1.50 \quad .90 \quad .60]\begin{bmatrix} 900 & 1500 & 1150 \\ 600 & 950 & 800 \\ 750 & 900 & 825 \end{bmatrix} = [2340 \quad 3645 \quad 2940]$

(d) $8925   **51. (a)** $\begin{array}{c} \\ S \\ C \end{array}\begin{array}{ccc} CC & MM & AD \\ \end{array}\begin{bmatrix} .5 & .4 & .3 \\ .2 & .3 & .3 \end{bmatrix}$ **(b)** $\begin{array}{c} \\ SD \\ MC \\ M \end{array}\begin{array}{cc} S & C \\ \end{array}\begin{bmatrix} 3 & 3 \\ 2 & 3 \\ 1 & 4 \end{bmatrix}$

(c) $\begin{array}{c} \\ SD \\ MC \\ M \end{array}\begin{array}{ccc} CC & MM & AD \\ \end{array}\begin{bmatrix} 2.1 & 2.1 & 1.8 \\ 1.6 & 1.7 & 1.5 \\ 1.3 & 1.6 & 1.5 \end{bmatrix}$ **(d)** $1.60   **(e)** $1200 in Managua

53. (a) $\begin{bmatrix} .027 & .009 \\ .030 & .007 \\ .015 & .009 \\ .013 & .011 \\ .019 & .011 \end{bmatrix}$; $\begin{bmatrix} 1596 & 218 & 199 & 425 & 214 \\ 1996 & 286 & 226 & 460 & 243 \\ 2440 & 365 & 252 & 484 & 266 \\ 2906 & 455 & 277 & 499 & 291 \end{bmatrix}$

(b) $\begin{array}{c} \\ 1960 \\ 1970 \\ 1980 \\ 1990 \end{array}\begin{array}{cc} \text{Births} & \text{Deaths} \\ \end{array}\begin{bmatrix} 62.208 & 24.710 \\ 76.459 & 29.733 \\ 91.956 & 35.033 \\ 108.283 & 40.522 \end{bmatrix}$ **55.** $X = \begin{bmatrix} 0 \\ \frac{3}{2} \end{bmatrix}$

Section 7.5 (page 314)

1. $\begin{bmatrix} 8 \\ 6 \end{bmatrix}$ **3.** $\begin{bmatrix} \frac{1}{2} & 1 \\ \frac{3}{2} & 1 \end{bmatrix}$ **5.** $\begin{bmatrix} 11 \\ -3 \\ 5 \end{bmatrix}$ **7.** $(-31, 24, -4)$

9. $(-31, -131, 181)$ **11.** $(15, 5, -1)$ **13.** $(-7, -34, -19, 7)$

15. $\begin{bmatrix} -6 \\ -14 \end{bmatrix}$ **17.** $\begin{bmatrix} \frac{32}{3} \\ \frac{25}{3} \end{bmatrix}$ **19.** $\begin{bmatrix} 6.43 \\ 26.12 \end{bmatrix}$ **21.** $\begin{bmatrix} 6.67 \\ 20 \\ 10 \end{bmatrix}$

23. Either 10 buffets, 5 chairs, no tables, or 11 buffets, 1 chair, 1 table   **25.** 2340 of the first species, 10,128 of the second species, 224 of the third species (rounded)   **27.** About 1073 metric tons of wheat, about 1431 metric tons of oil   **29.** Gas $98 million, electric $123 million   **31. (a)** 7/4 bushels of yams, 15/8 pigs   **(b)** 167.5 bushels of yams, 153.75 pigs   **33.** Agriculture 4100/23, manufacturing 4080/23, transportation 3790/23

35. $\begin{bmatrix} 23 \\ 51 \end{bmatrix}, \begin{bmatrix} 13 \\ 30 \end{bmatrix}, \begin{bmatrix} 45 \\ 96 \end{bmatrix}, \begin{bmatrix} 69 \\ 156 \end{bmatrix}, \begin{bmatrix} 87 \\ 194 \end{bmatrix}, \begin{bmatrix} 23 \\ 51 \end{bmatrix}, \begin{bmatrix} 51 \\ 110 \end{bmatrix}, \begin{bmatrix} 45 \\ 102 \end{bmatrix}, \begin{bmatrix} 69 \\ 157 \end{bmatrix}$

37. (a) 3   **(b)** 3   **(c)** 5   **(d)** 3   **39. (a)** $B = \begin{bmatrix} 0 & 2 & 3 \\ 2 & 0 & 4 \\ 3 & 4 & 0 \end{bmatrix}$

(b) $B^2 = \begin{bmatrix} 13 & 12 & 8 \\ 12 & 20 & 6 \\ 8 & 6 & 25 \end{bmatrix}$ **(c)** 12   **(d)** 14

41. (a) $\begin{array}{c} \\ \text{Dogs} \\ \text{Rats} \\ \text{Cats} \\ \text{Mice} \end{array}\begin{array}{cccc} \text{Dogs} & \text{Rats} & \text{Cats} & \text{Mice} \\ \end{array}\begin{bmatrix} 0 & 1 & 1 & 1 \\ 0 & 0 & 0 & 1 \\ 0 & 1 & 0 & 1 \\ 0 & 0 & 0 & 0 \end{bmatrix} = C$

(b) $C^2 = \begin{bmatrix} 0 & 1 & 0 & 2 \\ 0 & 0 & 0 & 0 \\ 0 & 0 & 0 & 1 \\ 0 & 0 & 0 & 0 \end{bmatrix}$;

C^2 gives the number of food sources once removed from the feeder. Thus, since dogs eat rats and rats eat mice, mice are an indirect as well as a direct food source for dogs.

Chapter 7 Review Exercises (page 318)

1. $(-5, 7)$ **3.** $(0, -2)$ **5.** No solution; inconsistent system   **7.** $(-35, 140, 22)$ **9.** 8000 standard, 6000 extra large   **11.** $7000 in the first; $11,000 in the second   **13.** 50 from source I, 150 from source II, 100 from source III   **15.** $(0, 3, 3)$ **17.** No solution   **19.** $(-79, 99, -8)$ **21.** 2×2; square   **23.** 1×4; row   **25.** 2×3 **27.** $\begin{bmatrix} 8 & 8 & 8 \\ 10 & 5 & 9 \\ 7 & 10 & 7 \\ 8 & 9 & 7 \end{bmatrix}$ **29.** $\begin{bmatrix} -2 & -3 & 2 \\ -2 & -4 & 0 \\ 0 & -1 & -2 \end{bmatrix}$

31. $\begin{bmatrix} 2 & 30 \\ -4 & -15 \\ 10 & 13 \end{bmatrix}$ **33.** Not possible

35. Next day $\begin{bmatrix} 2310 & -\frac{1}{4} \\ 1258 & -\frac{1}{4} \\ 5061 & \frac{1}{2} \\ 1812 & \frac{1}{2} \end{bmatrix}$; two-day total $\begin{bmatrix} 4842 & -\frac{1}{2} \\ 2722 & -\frac{1}{8} \\ 10{,}035 & -1 \\ 3566 & 1 \end{bmatrix}$

37. $\begin{bmatrix} 18 & 80 \\ -7 & -28 \\ 21 & 84 \end{bmatrix}$ **39.** $\begin{bmatrix} 13 & 43 \\ 17 & 46 \end{bmatrix}$ **41.** $\begin{bmatrix} 222 & 632 \\ -77 & -224 \\ 231 & 672 \end{bmatrix}$

43. (a) $\begin{bmatrix} \frac{1}{4} & \frac{1}{2} \\ \frac{1}{3} & \frac{1}{3} \end{bmatrix}$ (b) Cutting 34 hr; shaping 46 hr **45.** There are

many correct answers, including $\begin{bmatrix} 1 & 2 \\ 3 & 4 \end{bmatrix}$.

47. $\begin{bmatrix} -\frac{1}{4} & \frac{1}{6} \\ 0 & \frac{1}{3} \end{bmatrix}$ **49.** No inverse **51.** $\begin{bmatrix} \frac{1}{4} & \frac{1}{2} & \frac{1}{2} \\ \frac{1}{4} & -\frac{1}{4} & \frac{1}{2} \\ \frac{1}{8} & -\frac{1}{4} & -\frac{1}{4} \end{bmatrix}$

53. No inverse **55.** $\begin{bmatrix} -\frac{2}{3} & -\frac{17}{3} & -\frac{14}{3} & -3 \\ \frac{1}{3} & \frac{1}{3} & \frac{1}{3} & 0 \\ -\frac{1}{3} & -\frac{10}{3} & -\frac{7}{3} & -2 \\ 0 & 2 & 1 & 1 \end{bmatrix}$

57. $\begin{bmatrix} -\frac{7}{19} & \frac{4}{19} \\ \frac{3}{19} & \frac{1}{19} \end{bmatrix}$ **59.** $\begin{bmatrix} 1 & 1 \\ -2 & -3 \end{bmatrix}$ **61.** No inverse **63.** $\begin{bmatrix} 18 \\ -7 \end{bmatrix}$

65. $\begin{bmatrix} -22 \\ -18 \\ 15 \end{bmatrix}$ **67.** $(2, 2)$ **69.** $(2, 1)$ **71.** $(-1, 0, 2)$

73. No inverse; no solution for the system   **75.** 18 liters of the 8%; 12 liters of the 18%   **77.** 30 liters of 40% solution, 10 liters of 60% solution   **79.** 80 bowls, 120 plates   **81.** $12,750 at 8%, $27,250 at 8 1/2%, $10,000 at 11%   **83.** $\begin{bmatrix} 218.09 \\ 318.27 \end{bmatrix}$ **85.** (a) $\begin{bmatrix} 1 & -\frac{1}{4} \\ -\frac{1}{2} & 1 \end{bmatrix}$

(b) $\begin{bmatrix} \frac{8}{7} & \frac{2}{7} \\ \frac{4}{7} & \frac{8}{7} \end{bmatrix}$ (c) $\begin{bmatrix} 2800 \\ 2800 \end{bmatrix}$ **87.** Agriculture $140,909,

manufacturing $95,455   **89.** (a) $\begin{bmatrix} 54 \\ 32 \end{bmatrix}, \begin{bmatrix} 134 \\ 89 \end{bmatrix}, \begin{bmatrix} 172 \\ 113 \end{bmatrix}, \begin{bmatrix} 118 \\ 74 \end{bmatrix}, \begin{bmatrix} 208 \\ 131 \end{bmatrix}$

(b) $\begin{bmatrix} 2 & -3 \\ -\frac{1}{2} & 1 \end{bmatrix}$

Case 7 (page 323)

1. (a) $A = \begin{bmatrix} .245 & .102 & .051 \\ .099 & .291 & .279 \\ .433 & .372 & .011 \end{bmatrix}$; $D = \begin{bmatrix} 2.88 \\ 31.45 \\ 30.91 \end{bmatrix}$; $X = \begin{bmatrix} x_1 \\ x_2 \\ x_3 \end{bmatrix}$

(b) $I - A = \begin{bmatrix} .755 & -.102 & -.051 \\ -.099 & .709 & -.279 \\ -.433 & -.372 & .989 \end{bmatrix}$ (d) $\begin{bmatrix} 18.2 \\ 73.2 \\ 66.8 \end{bmatrix}$

$18.2 billion of agriculture, $73.2 billion of manufacturing, and $66.8 billion of household would be required (rounded to three significant digits).

CHAPTER 8
Section 8.1 (page 331)

1.

3.

5.

7.

9.

11.

13.

15.

17.

19.

23.

25.

27.

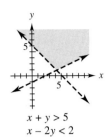

$x + y > 5$
$x - 2y < 2$

29.

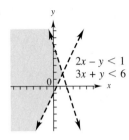

$2x - y < 1$
$3x + y < 6$

31.

$-x - y < 5$
$2x - y < 4$

33.

$3x + y \geq 6$
$x + 2y \geq 7$
$x \geq 0$
$y \geq 0$

35.

$-2 < x < 3$
$-1 \leq y \leq 5$
$2x + y < 6$

37.

$2y - x \geq -5$
$y \leq 3 + x$
$x \geq 0$
$y \geq 0$

39.

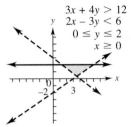

$3x + 4y > 12$
$2x - 3y < 6$
$0 \leq y \leq 2$
$x \geq 0$

41. $2 < x < 7$
$-1 < y < 3$

43. (a)

	Number	Hours Spinning	Hours Dyeing	Hours Weaving
Shawls	x	1	1	1
Afghans	y	2	1	4
Maximum number of hours available		8	6	14

(b) $x + 2y \leq 8; x + y \leq 6; x + 4y \leq 14; x \geq 0; y \geq 0$

(c)

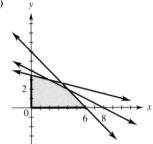

45. $x \geq 3000; y \geq 5000; x + y \leq 10{,}000$

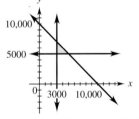

47. $x + y \geq 3.2; .5x + .3y \leq 1.8; .16x + .20y \leq .8; x \geq 0; y \geq 0$

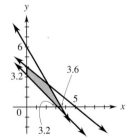

Section 8.2 (page 338)

1. Maximum of 65 at (5, 10); minimum of 8 at (1, 1)
3. Maximum of 9 at (0, 12); minimum of 0 at (0, 0)   **5. (a)** No
maximum; minimum of 16 at (0, 8)   **(b)** No maximum; minimum of
18 at (3, 4)   **(c)** No maximum; minimum of 21 at (13/2, 2)   **(d)** No
maximum; minimum of 12 at (12, 0)   **7.** Maximum of 42/5 at
$x = 6/5, y = 6/5$ or (6/5, 6/5)   **9.** Minimum of 13 at $x = 5, y = 3$
or (5, 3)   **11.** Maximum of 235/4 at $x = 105/8, y = 25/8$ or
(105/8, 25/8)   **13.** No maximum; minimum of 9 at (1, 3/2)
15. Maximum 22; no minimum   **17. (a)** (18, 2) **(b)** (12/5, 39/5)
(c) No maximum

Section 8.3 (page 344)

1. $6x + 4y \le 90$, where x = number of canoes, y = number of
rowboats.   **3.** $150x + 750y \le 8500$, where x = number of radio
spots, y = number of TV ads.   **5.** $x + y \ge 30$, where x = number
of pounds of cashews, y = number of pounds of peanuts.   **7.** Send
35 systems from Meadville to Superstore, 45 from Meadville to
ValueHouse, 15 from Cambridge to ValueHouse, and none from
Cambridge to Superstore for a minimum cost of $1015.   **9. (a)** 6
units of policy A and 16 units of policy B for a minimum cost of $940
(b) 30 units of policy A and no units of policy B for a minimum cost
of $750   **11.** 6.4 million gal of gasoline and 3.05 million gal of
fuel oil for maximum revenue of $11,050,000   **13.** 800 Type 1 and
1600 Type 2 for maximum revenue of $272   **15.** 250,000 hectares
to each crop for a maximum profit of $132,500,000   **17.** From
warehouse I ship 60 boxes to San Jose and 300 boxes to Memphis.
From warehouse II ship 290 boxes to San Jose and none to Memphis,
for a minimum cost of $147.70.   **19.** 8/7 units of Species I and
10/7 units of Species II will meet the requirements with a minimum
expenditure of 6.57 units of energy; however, a predator probably can
catch and digest only whole numbers of prey.   **21.** 5 humanities,
12 science   **23.** (b)   **25.** (c)

Section 8.4 (page 357)

1. (a) 3   **(b)** x_3, x_4, x_5 **(c)** $4x_1 + 2x_2 + x_3 = 20$
$5x_1 + x_2 + x_4 = 50$
$2x_1 + 3x_2 + x_5 = 25$

3. (a) 3   **(b)** x_4, x_5, x_6 **(c)** $3x_1 - x_2 + 4x_3 + x_4 = 95$
$7x_1 + 6x_2 + 8x_3 + x_5 = 118$
$4x_1 + 5x_2 + 10x_3 + x_6 = 220$

5.

	x_1	x_2	x_3	x_4	x_5	z	
	2	3	1	0	0	0	6
	4	1	0	1	0	0	6
	5	2	0	0	1	0	15
	-5	-1	0	0	0	1	0

7.

	x_1	x_2	x_3	x_4	x_5	x_6	z	
	1	2	3	1	0	0	0	10
	2	1	1	0	1	0	0	8
	3	0	2	0	0	1	0	6
	-1	-5	-10	0	0	0	1	0

9. 3 in row 2, column 1   **11.** 6 in row 1, column 1

13.

	x_1	x_2	x_3	x_4	x_5	z	
	-1	0	3	1	-1	0	16
	1	1	$\frac{1}{2}$	0	$\frac{1}{2}$	0	20
	2	0	$-\frac{1}{2}$	0	$\frac{3}{2}$	1	60

15.

	x_1	x_2	x_3	x_4	x_5	x_6	z	
	$-\frac{1}{2}$	$\frac{1}{2}$	0	1	$-\frac{1}{2}$	0	0	10
	$\frac{3}{2}$	$\frac{1}{2}$	1	0	$\frac{1}{2}$	0	0	50
	$-\frac{7}{2}$	$\frac{1}{2}$	0	0	$-\frac{3}{2}$	1	0	50
	2	0	0	0	1	0	1	100

17. (a) Basic: x_3, x_5; nonbasic: x_1, x_2, x_4 **(b)** $x_1 = 0, x_2 = 0$,
$x_3 = 16, x_4 = 0, x_5 = 29, z = 11$ **(c)** Not maximum
19. (a) Basic: x_1, x_2, x_5; nonbasic: x_3, x_4, x_6 **(b)** $x_1 = 6, x_2 = 13$,
$x_3 = 0, x_4 = 0, x_5 = 21, x_6 = 0, z = 18$ **(c)** Maximum
21. Maximum is 30 when $x_1 = 0, x_2 = 10, x_3 = 0, x_4 = 0, x_5 = 16$
23. Maximum is 8 when $x_1 = 4, x_2 = 0, x_3 = 8, x_4 = 2, x_5 = 0$
25. No maximum   **27.** No maximum   **29.** Maximum is 34
when $x_1 = 17, x_2 = 0, x_3 = 0, x_4 = 0, x_5 = 14$ or when $x_1 = 0$,
$x_2 = 17, x_3 = 0, x_4 = 0, x_5 = 14$.   **31.** Maximum is 26,000 when
$x_1 = 60, x_2 = 40, x_3 = 0, x_4 = 0, x_5 = 80, x_6 = 0$.   **33.** Maximum
is 64 when $x_1 = 28, x_2 = 16, x_3 = 0, x_4 = 0, x_5 = 28, x_6 = 0$.
35. Maximum is 250 when $x_1 = 0, x_2 = 0, x_3 = 0, x_4 = 50, x_5 = 0$,
$x_6 = 50$.   **37. (a)** Maximum is 24 when $x_1 = 12, x_2 = 0, x_3 = 0$,
$x_4 = 0, x_5 = 6$. **(b)** Maximum is 24 when $x_1 = 0, x_2 = 12, x_3 = 0$,
$x_4 = 0, x_5 = 18$. **(c)** The unique maximum value of z is 24, but this
occurs at two different basic feasible solutions.

Section 8.5 (page 363)

1.

	x_1	x_2	x_3	x_4	x_5	
	2	1	1	0	0	90
	1	2	0	1	0	80
	1	1	0	0	1	50
	-12	-10	0	0	0	0

where x_1 is the number of Siamese cats and x_2 is the number of Persian
cats.

3.

	x_1	x_2	x_3	x_4	x_5	x_6	
	-2	3	1	0	0	0	0
	1	0	0	1	0	0	6700
	0	1	0	0	1	0	5500
	1	1	0	0	0	1	12,000
	-8.5	-12.10	0	0	0	0	0

where x_1 is the number of trucks and x_2 the number of fire engines.
5. Make no 1-speed or 3-speed bicycles; make 2295 10-speed
bicycles; maximum profit is $55,080.   **7.** 48 loaves of bread and 6
cakes for a maximum gross income of $168   **9.** 100 Basic Sets, no
Regular Sets, and 200 Deluxe Sets for a maximum profit of $15,000
11. 4 radio ads, 6 TV ads, and no newspaper ads for a maximum
exposure of 64,800   **13. (a)** (3) **(b)** (4) **(c)** (3)   **15.** 40
Siamese and 10 Persian for a maximum gross income of $580
17. 6700 trucks, 4466 fire engines for a maximum profit of $110,989

Section 8.6 (page 372)

1. $\begin{bmatrix} 3 & 1 & 0 \\ -4 & 10 & 3 \\ 5 & 7 & 6 \end{bmatrix}$ **3.** $\begin{bmatrix} 3 & 4 \\ 0 & 17 \\ 14 & 8 \\ -5 & -6 \\ 3 & 1 \end{bmatrix}$

5. Maximize $z = 4x_1 + 6x_2$
subject to: $3x_1 - x_2 \le 3$
$x_1 + 2x_2 \le 5$
$x_1 \ge 0, x_2 \ge 0.$

7. Maximize $z = 18x_1 + 15x_2 + 20x_3$
subject to: $x_1 + 4x_2 + 5x_3 \le 2$
$7x_1 + x_2 + 3x_3 \le 8$
$x_1 \ge 0, x_2 \ge 0, x_3 \ge 0.$

9. Maximize $z = -8x_1 + 12x_2$
subject to: $3x_1 + x_2 \le 1$
$4x_1 + 5x_2 \le 2$
$6x_1 + 2x_2 \le 6$
$x_1 \ge 0, x_2 \ge 0.$

11. Maximize $z = 5x_1 + 4x_2 + 15x_3$
subject to: $x_1 + x_2 + 2x_3 \le 8$
$x_1 + x_2 + x_3 \le 9$
$x_1 + 3x_3 \le 3$
$x_1 \ge 0, x_2 \ge 0, x_3 \ge 0.$

13. $y_1 = 0, y_2 = 100, y_3 = 0$; minimum is 100.   **15.** $y_1 = 0,$
$y_2 = 12, y_3 = 0$; minimum is 12.   **17.** $y_1 = 0, y_2 = 0, y_3 = 2$;
minimum is 4.   **19.** $y_1 = 10, y_2 = 10, y_3 = 0$; minimum is 320.
21. $y_1 = 0, y_2 = 0, y_3 = 4$; minimum is 4.   **23.** 4 servings of A, 2
servings of B for a minimum cost of $1.76   **25.** Make 15 tables
and 45 chairs for a minimum cost of $4080.   **27. (a)** 1 bag of Feed
1, 2 bags of Feed 2   **(b)** $6.60 per day for 1.4 bags of Feed 1 and 1.2
bags of Feed 2

29. (a) Minimize $w = 100y_1 + 20,000y_2$
subject to: $y_1 + 400y_2 \ge 120$
$y_1 + 160y_2 \ge 40$
$y_1 + 280y_2 \ge 60$
$y_1 \ge 0, y_2 \ge 0.$

(b) $6300   **(c)** $5700

Section 8.7 (page 383)

1. (a) Maximize $z = 5x_1 + 2x_2 - x_3$
subject to: $2x_1 + 3x_2 + 5x_3 - x_4 = 8$
$4x_1 - x_2 + 3x_3 + x_5 = 7$
$x_1 \ge 0, x_2 \ge 0, x_3 \ge 0, x_4 \ge 0, x_5 \ge 0.$

(b)

x_1	x_2	x_3	x_4	x_5	
2	3	5	-1	0	8
4	-1	3	0	1	7
-5	-2	1	0	0	0

3. (a) Maximize $z = 2x_1 - 3x_2 + 4x_3$
subject to: $x_1 + x_2 + x_3 + x_4 = 100$
$x_1 + x_2 + x_3 - x_5 = 75$
$x_1 + x_2 - x_6 = 27$
$x_1 \ge 0, x_2 \ge 0, x_3 \ge 0, x_4 \ge 0, x_5 \ge 0, x_6 \ge 0.$

(b)

x_1	x_2	x_3	x_4	x_5	x_6	
1	1	1	1	0	0	100
1	1	1	0	-1	0	75
1	1	0	0	0	-1	27
-2	3	-4	0	0	0	0

5. Maximize $z = -2y_1 - 5y_2 + 3y_3$
subject to: $y_1 + 2y_2 + 3y_3 \ge 115$
$2y_1 + y_2 + 2y_3 \le 200$
$y_1 + y_3 \ge 50$
$y_1 \ge 0, y_2 \ge 0, y_3 \ge 0.$

y_1	y_2	y_3	y_4	y_5	y_6	
1	2	3	-1	0	0	115
2	1	1	0	1	0	200
1	0	1	0	0	-1	50
2	5	-3	0	0	0	0

7. Maximize $z = -y_1 + 4y_2 - 2y_3$
subject to: $7y_1 - 6y_2 + 8y_3 \ge 18$
$4y_1 + 5y_2 + 10y_3 \ge 20$
$y_1 \ge 0, y_2 \ge 0, y_3 \ge 0.$

y_1	y_2	y_3	y_4	y_5	
7	-6	8	-1	0	18
4	5	10	0	-1	20
1	-4	2	0	0	0

9. Maximum is 480 when $x_1 = 40, x_2 = 0$.   **11.** Maximum is 114
when $x_1 = 38, x_2 = 0, x_3 = 0$.   **13.** Maximum is 90 when $x_1 = 12,$
$x_2 = 3$ or when $x_1 = 0, x_2 = 9$.   **15.** Minimum is 40 when $y_1 = 10,$
$y_2 = 0$.   **17.** Minimum is 26 when $y_1 = 6, y_2 = 2$.

19.

y_1	y_2	y_3	y_4	y_5	y_6	y_7	y_8	
1	0	1	0	-1	0	0	0	32
0	1	0	1	0	-1	0	0	20
1	1	0	0	0	0	1	0	25
0	0	1	1	0	0	0	1	30
14	22	12	10	0	0	0	0	0

21.

y_1	y_2	y_3	y_4	y_5	y_6	y_7	
1	1	1	-1	0	0	0	10
1	1	1	0	1	0	0	15
1	$-\frac{1}{4}$	0	0	0	-1	0	0
-1	0	1	0	0	0	-1	0
.30	.09	.27	0	0	0	0	0

23. Ship 5000 barrels of oil from supplier 1 to distributor 2; ship 3000
barrels of oil from supplier 2 to distributor 1. Minimum cost is
$175,000.   **25.** Make $3,000,000 in commercial loans and
$22,000,000 in home loans for a maximum return of $2,940,000.
27. Use 1,060,000 kg for whole tomatoes and 80,000 kg for sauce for
a minimum cost of $4,500,000.   **29.** Order 1000 small test tubes
and 500 large ones for a minimum cost of $210.

Chapter 8 Review Exercises (page 386)

1.

3.

$3x + 4y \geq 12$

5.

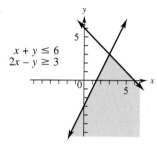

$x + y \leq 6$
$2x - y \geq 3$

7.

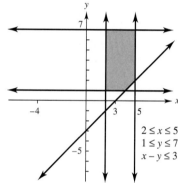

$2 \leq x \leq 5$
$1 \leq y \leq 7$
$x - y \leq 3$

9. Let x represent the number of batches of cakes, and let y represent the number of batches of cookies. Then

$2x + (3/2)y \leq 15$
$3x + (2/3)y \leq 13$
$x \geq 0$
$y \geq 0$.

11. Minimum of 8 at $(2, 1)$; maximum of 40 at $(6, 7)$
13. Maximum is 15 when $x = 5, y = 0$. **15.** Minimum is 60 when $x = 0, y = 20$. **17.** Make 3 batches of cakes and 6 of cookies for a maximum profit of $210. **19. (a)** Let $x_1 =$ number of item A, $x_2 =$ number of item B, $x_3 =$ number of item C.
(b) $z = 4x_1 + 3x_2 + 3x_3$ **(c)** $2x_1 + 3x_2 + 6x_3 \leq 1200$
$x_1 + 2x_2 + 2x_3 \leq 800$
$2x_1 + 2x_2 + 4x_3 \leq 500$
$x_1 \geq 0, x_2 \geq 0, x_3 \geq 0$

21. (a) Let $x_1 =$ number of gallons of Fruity and $x_2 =$ number of gallons of Crystal. **(b)** $z = 12x_1 + 15x_2$ **(c)** $2x_1 + x_2 \leq 110$
$2x_1 + 3x_2 \leq 125$
$2x_1 + x_2 \leq 90$
$x_1 \geq 0, x_2 \geq 0$

23. When there are more than two variables
25. Any standard minimization problem
27. (a) $3x_1 + 5x_2 + x_3 \qquad\qquad\quad = 47$
$x_1 + x_2 \qquad + x_4 \qquad\quad = 25$
$5x_1 + 2x_2 \qquad\qquad + x_5 \quad = 35$
$2x_1 + x_2 \qquad\qquad\qquad + x_6 = 30$

(b)

$$\begin{array}{cccccc} x_1 & x_2 & x_3 & x_4 & x_5 & x_6 \\ \end{array}$$
$$\left[\begin{array}{cccccc|c} 3 & 5 & 1 & 0 & 0 & 0 & 47 \\ 1 & 1 & 0 & 1 & 0 & 0 & 25 \\ 5 & 2 & 0 & 0 & 1 & 0 & 35 \\ 2 & 1 & 0 & 0 & 0 & 1 & 30 \\ \hline -2 & -7 & 0 & 0 & 0 & 0 & 0 \end{array}\right]$$

29. (a) $x_1 + x_2 + x_3 + x_4 \qquad\qquad = 100$
$2x_1 + 3x_2 \qquad\qquad + x_5 \qquad = 500$
$x_1 \qquad + 2x_3 \qquad\qquad + x_6 = 350$

(b)

$$\begin{array}{cccccc} x_1 & x_2 & x_3 & x_4 & x_5 & x_6 \\ \end{array}$$
$$\left[\begin{array}{cccccc|c} 1 & 1 & 1 & 1 & 0 & 0 & 100 \\ 2 & 3 & 0 & 0 & 1 & 0 & 500 \\ 1 & 0 & 2 & 0 & 0 & 1 & 350 \\ \hline -4 & -6 & -3 & 0 & 0 & 0 & 0 \end{array}\right]$$

31. Maximum is 80 when $x_1 = 16, x_2 = 0, x_3 = 0, x_4 = 12, x_5 = 0$.
33. Maximum is 35 when $x_1 = 5, x_2 = 0, x_3 = 5, x_4 = 35, x_5 = 0$, $x_6 = 0$. **35.** Maximize $z = -18y_1 - 10y_2$ with the same constraints. **37.** Maximize $z = -6y_1 + 3y_2 - 4y_3$ with the same constraints. **39.** The simplex method gives a minimum of 40 at $(10, 0)$. (The graphical method gives $(34, 12)$ as another solution.)
41. Minimum of 1957 at $(47, 68, 0, 92, 35, 0, 0)$
43. $(9, 5, 8, 0, 0, 0)$; minimum is 62. **45.** Get 250 of A, none of B, none of C for a maximum profit of $1000. **47.** Make 36.25 gal of Fruity and 17.5 gal of Crystal for a maximum profit of $697.50.
49. Build 3 atlantic and 3 pacific models for a minimum cost of $21,000.

CHAPTER 9
Section 9.1 (page 398)

1. False **3.** True **5.** True **7.** True **9.** False **13.** $\subseteq$
15. $\not\subseteq$ **17.** $\subseteq$ **19.** $\subseteq$ **21.** 8 **23.** 64 **27.** $\cap$
29. $\cap$ **31.** $\cup$ **33.** $\cup$ **35.** $\{3, 5\}$ **37.** $\{7, 9\}$ **39.** $\emptyset$
41. U or $\{2, 3, 4, 5, 7, 9\}$ **43.** All students in this school not taking this course **45.** All students in this school taking accounting and zoology **47.** C and D, A and E, C and E, D and E
49. B' is the set of all stocks in the list with a last price less than $75 or greater than $100; $B' = \{$ATT, GnMill, PepsiCo$\}$. **51.** $(A \cup B)'$ is the set of all stocks with a high price less than $100 and a last price less than $75 or more than $100; $(A \cup B)' = \{$ATT, GnMill, PepsiCo$\}$.
53. $M \cap E$ is the set of all male, employed applicants.
55. $M' \cup S'$ is the set of all female or married applicants.
57. $\{1989, 1993, 1995\}$ **59.** $\{$HBO, The Disney Channel$\}$
61. $\{$Spice$\}$ **63.** $\{$The Disney Channel, Spice$\}$ **65.** $\{s, d, c\}$
67. $\{g\}$ **69.** $\{s, d, c\}$

Section 9.2 (page 406)

1.

$B \cap A'$

3.

$A' \cup B$

5.

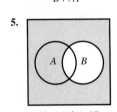

$B' \cup (A' \cap B')$

7. ∅

9.

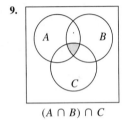

$(A \cap B) \cap C$

11.

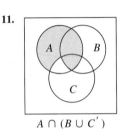

$A \cap (B \cup C')$

13.

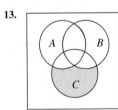

$(A' \cap B') \cap C$

15.

$(A \cap B') \cap C$

17. (a) 342 **(b)** 192 **(c)** 72 **(d)** 86 **19. (a)** 89% **(b)** 66%
(c) 4% **(d)** 1% **(e)** 4% **21. (a)** 50 **(b)** 2 **(c)** 14
23. (a) 54 **(b)** 17 **(c)** 10 **(d)** 7 **(e)** 15 **(f)** 3 **(g)** 12
(h) 1 **25. (a)** 124.5 million **(b)** 74.2 million **(c)** 53.8 million
(d) 37.1 million **(e)** 228.6 million **(f)** 30.7 million **29.** 9
31. 27

33.

35.

37.

39.

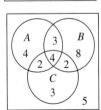

Section 9.3 (page 414)

3. {1, 2, 3, 4, . . . , 30} **5.** {buy, lease}
7. {January, February, March, . . . , December}
11. $\{y_1, y_2, y_3, w_1, w_2, w_3, w_4, b_1, b_2, b_3, b_4, b_5, b_6, b_7, b_8\}$
(a) $\{y_1, y_2, y_3\}$ **(b)** $\{b_1, b_2, b_3, b_4, b_5, b_6, b_7, b_8\}$
(c) $\{w_1, w_2, w_3, w_4\}$ **(d)** ∅
13. {ccc, ccw, cwc, cww, wcc, wcw, wwc, www} **(a)** {www}
(b) {ccw, cwc, wcc} **(c)** {cww}
15. {(1, 1), (1, 2), (1, 3), (1, 4), (1, 5), (2, 1), (2, 2), (2, 3), (2, 4),
(2, 5), (3, 1), (3, 2), (3, 3), (3, 4), (3, 5), (4, 1), (4, 2), (4, 3), (4, 4),
(4, 5)}
(a) {(2, 1), (2, 2), (2, 3), (2, 4), (2, 5), (4, 1), (4, 2), (4, 3), (4, 4), (4, 5)}
(b) {(1, 2), (1, 4), (2, 2), (2, 4), (3, 2), (3, 4), (4, 2), (4, 4)}
(c) {(1, 4), (2, 3), (3, 2), (4, 1)} **(d)** ∅ **17.** 1/6
19. 1/3 **21.** 1/13 **23.** 1/26 **25.** 1/52 **27.** 1/13
29. 1/5 **31.** 9/25 **33.** 21/25 **35.** 9/25
37. (a) .197 **(b)** .228 **(c)** .229

Section 9.4 (page 423)

1. No **3.** No **5.** Yes **7.** .143 **9. (a)** 1/2 **(b)** 1/2
(c) 7/10 **(d)** 0 **11. (a)** 3/5 **(b)** 7/10 **(c)** 3/10
13. (a) 3/13 **(b)** 7/13 **(c)** 3/13 **15. (a)** .72 **(b)** .70
(c) .79 **17.** 0 **19.** .90 **21.** .84 **23. (a)** 1 to 5
(b) 1 to 2 **(c)** 1 to 1 **(d)** 1 to 5 **25.** 13 to 37 **27.** 8/25
31. Possible **33.** Not possible because the sum of the probabilities
is less than 1 **35.** Not possible because one probability is negative
37. (a) .187 **(b)** .540 **(c)** .769 **39. (a)** .45 **(b)** .75 **(c)** .35
(d) .46 **41.** .527 **43.** .466 **45.** .515 **47.** 1/4
49. 1/2 **51.** .90
**In Exercises 53–57, answers will vary. Theoretical probabilities
are given.**
53. The theoretical probability is 5/32 = .15625. **55.** The
theoretical probability is 1/221 = .004525. **57.** The theoretical
probability is 7/8.

Section 9.5 (page 437)

1. 1/3 **3.** 1 **5.** 1/6 **7.** 4/17 **9.** 11/51 **11.** .012
13. .245 **19.** .815 **21.** .875 **23. (a)** .94 **(b)** .1282
25. .7456 **27.** .154 **29.** .43 **31.** The probability that a
person in that group is white, given that the person is male is .956.
33. The probability that a person in that group is black, given that the
person is a woman is .083. **35.** .527 **37.** .042 **39.** 6/7 or
.857 **41.** Yes **43.** .05 **45.** .25 **47.** .287 **49.** .248
51. .084 **53.** .4955 **55.** .0361 **57. (a)** $(.98)^3 \approx .94$
(b) Not very realistic **59. (a)** 3 backups **(b)** Yes **61.** No

Section 9.6 (page 444)

1. 1/3 **3.** 2/41 ≈ .0488 **5.** 21/41 ≈ .5122 **7.** 8/17 ≈ .4706
9. .0135 **11.** .342 **13.** .006 **15.** .091 **17.** .556
19. .706 **21.** 82.4% **23. (c)** **25.** .481 **27. (a)** .161
29. .043 **31. (a)** .53 **(b)** .1623

Chapter 9 Review Exercises (page 447)

1. False **3.** False **5.** True **7.** True **9.** {New Year's
Day, Martin Luther King's Birthday, Washington's Birthday
(President's Day), Memorial Day, Independence Day, Labor Day,
Columbus Day, Veteran's Day, Thanksgiving, Christmas}

11. {1, 2, 3, 4}   **13.** {B_1, B_2, B_3, B_6, B_{12}}   **15.** {A, C, E}
17. {A, B_3, B_6, B_{12}, C, D, E}   **19.** The set of all students in the class who are redheaded males   **21.** The set of all students in the class who are male or younger than 21   **23.** The set of all students in the class who are not A students and do not have red hair
25.

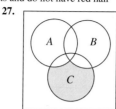

27.

29. 5   **31.** {1, 2, 3, 4, 5, 6}   **33.** {(red, 10), (red, 20), (red, 30), (blue, 10), (blue, 20), (blue, 30), (green, 10), (green, 20), (green, 30)}
35. {(2, blue), (4, blue), (6, blue), (8, blue), (10, blue)}   **37.** No
39. $E \cup F$ **45.** 3/13   **47.** 1/2   **49.** 1   **51.** 1 to 25
53. .86

55.

	N_2	T_2
N_1	$N_1 N_2$	$N_1 T_2$
T_1	$T_1 N_2$	$T_1 T_2$

57. 1/2   **59.** 5/36 ≈ .139   **61.** 5/18 ≈ .278   **63.** 1/3
65. .66   **67.** .71   **69. (a)** 3/11 **(b)** 7/39   **73.** 3/22
75. 2/3   **77.** 5/11   **79. (a)** Row 1: 400; row 2: 150; row 3: 750, 1000 **(b)** 1000 **(c)** 300 **(d)** 250 **(e)** 600 **(f)** 150
(g) Those who purchased a used car, given that the buyer was not satisfied **(h)** 3/5 **(i)** 1/4

Case 9 (page 450)

1. .076   **3.** .051

CHAPTER 10
Section 10.1 (page 460)

1. 12   **3.** 56   **5.** 8   **7.** 24   **9.** 792   **11.** 156
13. 6,375,600   **15.** 240,240   **17.** 352,716   **19.** 2,042,975
21. 10^9 = 1,000,000,000; yes   **23.** $10^9 \neq$ 1,000,000,000
25. 840   **27. (a)** 17,576,000 **(b)** 17,576,000 **(c)** 456,976,000
29. 800   **31.** 72,000,000   **33.** 60,949,324,800   **35.** 132
37. 210   **39.** 5814   **41.** 2,118,760   **43.** 1,621,233,900
45. (a) 9 **(b)** 6 **(c)** 3; yes, from both   **49. (a)** 84 **(b)** 10
(c) 40 **(d)** 74   **51. (a)** 961 **(b)** 29,791   **53.** 5040
55. 81   **57.** 1   **59.** 55,440   **61. (a)** 7,059,052
(b) $3,529,526 **(c)** No   **63.** 5.524 × 10^{26} **65. (a)** 840
(b) 180 **(c)** 420   **67. (a)** 362,880 **(b)** 6 **(c)** 1260

Section 10.2 (page 468)

1. .424   **3.** 7/9   **5.** 5/12   **7.** .147   **9.** .020   **11.** .314
13. 1326   **15.** .851   **17.** .765   **19.** .941

23. $\dfrac{\binom{4}{2}\binom{4}{3}\binom{44}{8}}{\binom{52}{13}}$ **25. (a)** 8.9 × 10^{-10} **(b)** 1.2 × 10^{-12}

27. 3.3 × 10^{-16} **29. (a)** .218 **(b)** .272 **(c)** .728

31. (a) .003 **(b)** .005 **(c)** .057 **(d)** .252
33. $(1 - _{365}P_{100})/(365)^{100}$ **35.** .350   **37.** There were 10 balls with 9 of them blue.   **39.** Answers will vary. The theoretical answers are: **(a)** .0399 **(b)** .5191 **(c)** .0226.

Section 10.3 (page 473)

1. .067   **3.** .066   **5.** .934   **7.** .0000000005   **9.** .269
11. .875   **13.** 1/32   **15.** 13/16   **19.** .129   **21.** .549
23. .337   **25.** .349   **27.** .736   **29.** .218   **31.** .000729
33. .118   **35.** .121   **37. (a)** .001 **(b)** .088   **39.** .6930
41. (a) .199 **(b)** .568 **(c)** .133 **(d)** .794   **43. (a)** .047822
(b) .976710 **(c)** .897110   **45. (a)** 0 **(b)** .002438

Section 10.4 (page 480)

1. No   **3.** Yes   **5.** No   **7.** No
9. Yes

11. No   **13.** Not a transition diagram

15. $\begin{array}{c} \\ A \\ B \\ C \end{array} \begin{bmatrix} .6 & .2 & .2 \\ .9 & .02 & .08 \\ .4 & 0 & .6 \end{bmatrix}$ with columns labeled A B C

17. Yes   **19.** Yes   **21.** No   **23.** $\begin{bmatrix} \frac{1}{3} & \frac{2}{3} \end{bmatrix}$ **25.** $\begin{bmatrix} \frac{3}{11} & \frac{8}{11} \end{bmatrix}$
27. [.4633 .1683 .3684]   **29.** $\begin{bmatrix} \frac{1}{4} & \frac{1}{4} & \frac{1}{2} \end{bmatrix}$
31. The first power is the given transition matrix;

$$\begin{bmatrix} .23 & .21 & .24 & .17 & .15 \\ .26 & .18 & .26 & .16 & .14 \\ .23 & .18 & .24 & .19 & .16 \\ .19 & .19 & .27 & .18 & .17 \\ .17 & .20 & .26 & .19 & .18 \end{bmatrix};$$

$$\begin{bmatrix} .226 & .192 & .249 & .177 & .156 \\ .222 & .196 & .252 & .174 & .156 \\ .219 & .189 & .256 & .177 & .159 \\ .213 & .192 & .252 & .181 & .162 \\ .213 & .189 & .252 & .183 & .163 \end{bmatrix};$$

$$\begin{bmatrix} .2205 & .1916 & .2523 & .1774 & .1582 \\ .2206 & .1922 & .2512 & .1778 & .1582 \\ .2182 & .1920 & .2525 & .1781 & .1592 \\ .2183 & .1909 & .2526 & .1787 & .1595 \\ .2176 & .1906 & .2533 & .1787 & .1598 \end{bmatrix};$$

$$\begin{bmatrix} .21932 & .19167 & .25227 & .17795 & .15879 \\ .21956 & .19152 & .25226 & .17794 & .15872 \\ .21905 & .19152 & .25227 & .17818 & .15898 \\ .21880 & .19144 & .25251 & .17817 & .15908 \\ .21857 & .19148 & .25253 & .17824 & .15918 \end{bmatrix}; .17794$$

33. 24.0% 30-year fixed rates, 69.7% 15-year fixed rates, 6.3% adjustable   **35.** 17/18

37.
$$\begin{array}{c} \\ R \\ P \\ W \end{array}\begin{array}{c} \begin{array}{ccc} R & P & W \end{array} \\ \begin{bmatrix} \frac{1}{2} & \frac{1}{2} & 0 \\ \frac{1}{4} & \frac{1}{2} & \frac{1}{4} \\ 0 & \frac{1}{2} & \frac{1}{2} \end{bmatrix} \end{array}; \begin{bmatrix} \frac{1}{4} & \frac{1}{2} & \frac{1}{4} \end{bmatrix}$$

39. Homeowners: .769; renters: .231; homeless: 0   **41. (a)** 42,500 in G_0; 5000 in G_1; 2500 in G_2 **(b)** 37,125 in G_0; 8750 in G_1; 4125 in G_2 **(c)** 33,281 in G_0; 11,513 in G_1; 5206 in G_2 **(d)** $\begin{bmatrix} \frac{28}{59} & \frac{22}{59} & \frac{9}{59} \end{bmatrix}$ or [.475 .373 .152]; in the long run, the probabilities of no accidents, one accident, and more than one accident are .475, .373, and .152, respectively.   **43. (a)** $\begin{bmatrix} .348 & .379 & .273 \\ .320 & .378 & .302 \\ .316 & .360 & .323 \end{bmatrix}$ **(b)** .348

(c) .320

45. (b) $\begin{bmatrix} \frac{5}{12} & \frac{13}{36} & \frac{2}{9} \\ \frac{5}{12} & \frac{13}{36} & \frac{2}{9} \\ \frac{1}{4} & \frac{5}{12} & \frac{1}{3} \end{bmatrix}$ **(c)** 2/9   **47.** [0 0 .102273 .897727]

Section 10.5 (page 492)

1. (a)

Number of accidents	0	1	2	3	4
Probability	$\frac{2}{7}$	$\frac{3}{7}$	$\frac{1}{7}$	0	$\frac{1}{7}$

(b)

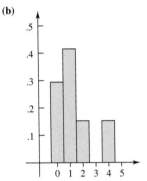

3. (a)

Number	0	1	2	3	4	5	6
Probability	0	.04	0	.16	.40	.32	.08

(b)

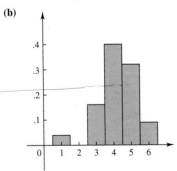

5.

Number of even numbers	0	1	2
Probability	$\frac{3}{10}$	$\frac{3}{5}$	$\frac{1}{10}$

7.

Number of points	2	3	4	5	6	7	8	9	10	11	12
Probability	$\frac{1}{36}$	$\frac{1}{18}$	$\frac{1}{12}$	$\frac{1}{9}$	$\frac{5}{36}$	$\frac{1}{6}$	$\frac{5}{36}$	$\frac{1}{9}$	$\frac{1}{12}$	$\frac{1}{18}$	$\frac{1}{36}$

9.

Number of black balls	0	1	2
Probability	$\frac{2}{5}$	$\frac{8}{15}$	$\frac{1}{15}$

11.

13.   **15.**

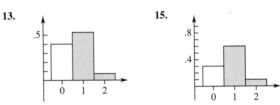

17. 6.5   **19.** 23.83

21. (a)

Number	0	1	2	3	4
Probability	.4096	.4096	.1536	.0256	.0016

(b)   **(c)** .8

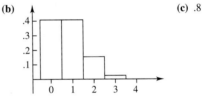

23. (a) Yes; the probability of a match is still 1/2.   **(b)** 60¢
(c) −60¢   **25.** 5.6   **27.** 30   **29.** −32¢   **31.** 1
33. 1/2   **35.** 4.4435   **37.** −82¢   **39.** −5¢
41. −50¢   **45. (c)**   **47.** $30,000 at age 55

49.

Account Number	Expected Value	Total	Class
3	2,000	22,000	C
4	1,000	51,000	B
5	25,000	30,000	C
6	60,000	60,000	A
7	16,000	46,000	B

51. (a)

Number	0	1	2	3	4	5
Probability	.0000	.0000	.0000	.0003	.0024	.0141

Number	6	7	8	9	10
Probability	.0573	.1600	.2929	.3178	.1552

(b) .0027   **(d)** 8, 3

Section 10.6 (page 498)

1. (a) Choose the coast.   **(b)** Choose the highway.   **(c)** Choose the highway; $38,000.   **(d)** Choose the coast.   **3. (a)** Make no repairs.   **(b)** Make repairs.   **(c)** Make no repairs; $10,750.

Fails   Doesn't

5. (a) Overhaul $\begin{bmatrix} -8600 & -2600 \\ -6000 & 0 \end{bmatrix}$
Don't

(b) Do not overhaul before shipping.
7. (a) No campaign   **(b)** Campaign for all   **(c)** Campaign for all
9. Emphasize the environment; 14.25.

Chapter 10 Review Exercises (page 501)

1. 120   **3.** 220   **5.** 120   **7. (a)** 72   **(b)** 72   **11.** 3/8
13. 11/16   **15.** .245   **17.** .362   **19.** .091   **21.** .455
23. Yes   **25.** No   **27.** .000041   **29.** .182
31. (a) [.54 .46]   **(b)** [.6464 .3536]   **(c)** 2/3
33. (a)

x	1	2	3	4	5	6
$P(x)$	.05	0	.1	.25	.4	.2

(b)

35. (a)

Number	2	3	4	5	6	7	8	9	10	11	12
Probability	$\frac{1}{36}$	$\frac{1}{18}$	$\frac{1}{12}$	$\frac{1}{9}$	$\frac{5}{36}$	$\frac{1}{6}$	$\frac{5}{36}$	$\frac{1}{9}$	$\frac{1}{12}$	$\frac{1}{18}$	$\frac{1}{36}$

(b)

37. .6   **39.** $1.29   **41.** -28¢   **43. (a)** 3/13 or .23   **(b)** 3/4
or .75   **45. (a)** Oppose   **(b)** Oppose   **(c)** Oppose; 2700
(d) Oppose; 1400   **47. (c)**

Case 10 (page 505)

1. (a) $69.01   **(b)** $79.64   **(c)** $58.96   **(d)** $82.54   **(e)** $62.88
(f) $64.00   **3.** The events of needing parts 1, 2, and 3 are not the
only events in the sample space.

CHAPTER 11
Section 11.1 (page 514)

1. (a)–(b)

Interval	Frequency
0–24	4
25–49	3
50–74	6
75–99	3
100–124	5
125–149	9

(c)–(d)

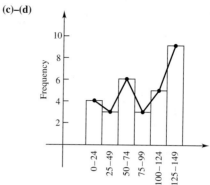

3. (a)–(b)

Interval	Frequency
70–74	2
75–79	1
80–84	3
85–89	2
90–94	6
95–99	5
100–104	6
105–109	4
110–114	2

(c)–(d)

7. 16   **9.** 27,955   **11.** 7.7   **13.** 6.7   **15.** 17.2   **17.** 51
19. 130   **21.** 29.1   **23.** 9   **25.** 68 and 74   **27.** 6.3
31. 94.9; 90–94 and 100–104   **33. (a)** Mean = 1.9; median = 1;
mode = 1   **35. (a)** Mean = 17.2; median = 17.5; no mode
37.

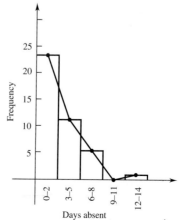

Days absent

39. (a) About 17.5% **(b)** About 8% **(c)** 20–29 **41.** $3.29
43. (a) 23.98 **(b)** 21.47 **45. (a)** The mean wealth is
$999,900.01 or roughly $1 million. **(b)** $0 **(c)** $0 **(d)** Median or
mode

Section 11.2 (page 522)

3. 6; 2.2 **5.** 12; 4.7 **7.** 53; 21.8 **9.** 46; 16.1 **11.** 45.2
13. Mean: 5.0876; standard deviation: .1087 **15.** At least 3/4
17. At least 24/25 **19.** At least 93.8% **21.** No more than
11.1% **23.** Britelite: $\bar{x} = 26.2$, $s = 4.4$; Brand X: $\bar{x} = 25.5$,
$s = 7.2$ **(a)** Britelite **(b)** Britelite **25. (a)** 7.77; 1991
(b) 1.22 **(c)** 4 **27. (a)** 1.83; .76 **(b)** 1.09 **(c)** .88
29. (a) 188.3; Mexico **(b)** 123.9; No; Yes, Canada and Asia
31. (a) 1/3; 2; −1/3; 0; 5/3; 7/3; 1; 4/3; 7/3; 2/3 **(b)** 2.1; 2.6; 1.5;
2.6; 2.5; .6; 1.0; 2.1; .6; 1.2 **(c)** 1.13 **(d)** 1.68 **(e)** 4.41; −2.15;
the process is out of control. **33.** Mean = 80.17; standard
deviation = 12.2

Section 11.3 (page 533)

5. 9.10% **7.** 48.21% **9.** 7.7% **11.** 47.35%
13. 92.42% **15.** 1.64 or 1.65 **17.** −1.04 **19.** .0796
21. .9933 **23.** .1527 **25.** .0051 **27.** .4052 **29.** .8186
31. .0475 **33.** .0021 **35.** 60.4 mph **37.** 6.68%
39. 38.3% **41.** 625 **43.** 189 **45.** 15.87% **47.** 82
49. 70 **51.** .5000 **53.** .9050 **55.** .8106 **57.** .6736
59. .1820 **61.** .5398

Section 11.4 (page 540)

3. (a)

x	0	1	2	3	4	5	6
$P(x)$	.016	.094	.234	.313	.234	.094	.016

(b) 3 **(c)** 1.2
5. (a)

x	0	1	2	3	4	5
$P(x)$	.470	.383	.125	.020	.002	5×10^{-5}

(b) .7 **(c)** .78 **7.** 12.5; 3.49 **9.** 24.5; 3.53 **11. (a)** 3.95
(b) .409 **(c)** .9919 **13.** .1747 **15.** .0240 **17.** .9032
19. .8665 **21.** .0956 **23.** .0760 **25.** .6443 **27.** .0146
29. .0222 **31.** .3798 **33. (a)** .0001 **(b)** .0002 **(c)** .0000
35. (a) .1686 **(b)** .0304 **(c)** .9575

Chapter 11 Review Exercises (page 543)

3. (a)

Interval	Frequency
450–474	5
475–499	6
500–524	5
525–549	2
550–574	2

(b)–(c)

5. 73 **7.** 34.9 **11.** 38; 36 and 38 **13.** 55–59 **17.** 67;
23.9 **19.** 6.2 **23.** .3980 **25.** .9521 **29. (a)** .008
(b) $\mu = 1.014$; $\sigma = .9180$ **31. (a)** Diet A (mean = 2.7)
(b) Diet B ($s = .95$) **33.** 2.27% **35.** 68.26% **37.** .8315
39. .0305 **41.** 1.000 **43.** .1428 **45.** .0806 **47.** .7357

Case 11 (page 546)

1. (a) 1/3; 2; −1/3; 0; 5/3; 7/3; 1; 4/3; 7/3; 2/3 **(b)** 2.1; 2.6; 1.5;
2.6; 2.5; .6; 1.0; 2.1; .6; 1.2 **(c)** 1.13 **(d)** 1.68 **(e)** 4.41; −2.15
(f) 4.31; 0

CHAPTER 12
Section 12.1 (page 555)

1. (a) 3 **(b)** 0 **3. (a)** Does not exist **(b)** −1/2 **5. (a)** 2
(b) Does not exist **7. (a)** 1 **(b)** −1 **11.** 1 **13.** 2
15. 0 **17.** 8 **19.** Does not exist **21.** 8 **23.** 2
25. 4 **27.** 3/2
29. (a)

$$f(x) = \begin{cases} 3 - x & \text{if } x < -2 \\ x + 2 & \text{if } -2 \leq x < 2 \\ 1 & \text{if } x \geq 2 \end{cases}$$

(b) Does not exist **(c)** 3 **(d)** Does not exist **31.** 41
33. 9/7 **35.** 6 **37.** −5 **39.** 1/2 **41.** Does not exist
43. $\sqrt{5}$ **45.** Does not exist **47.** −1/9 **49.** 1/10
51. $1/(2\sqrt{5})$ **53. (a)** 25 **(b)** 5.2 **(c)** 7
55. (a) 3 **(b)** Does not exist **(c)** 2 **(d)** 16 months
57. (a) Approx. .0444 **(b)** Approx. .0667 **(c)** Approx. .0667

Section 12.2 (page 565)

1. 7 **3.** 8 **5.** 1/3 **7.** −1/3 **9. (a)** 2 **(b)** 3 **(c)** 2
(d) 1/2 **(e)** 1/2 **11. (a)** 3 **(b)** 0 **(c)** −9/5 **(d)** Sales
increase in years 0–4, stay constant until year 7, and then decrease.
13. (a) 73.7/yr **(b)** 261.4/yr **(c)** 362/yr **(d)** 398.1/yr
(e) 273.8/yr **(f)** It is increasing. **(g)** Average the answers to parts
(a)–(d). **15. (a)** $11 million, −$1 million **(b)** −$1 million,
$9 million **(c)** Civil penalties were increasing from 1987 to 1988
and decreasing from 1988 to 1989. Criminal penalties decreased

slightly from 1987 to 1988, then increased from 1988 to 1989. This indicates that criminal penalties began to replace civil penalties in 1988. (d) About a \$3.5 million increase; the general trend was upward. More is being done to impose penalties for polluting.
19. 20 ft per sec **21.** 60 ft per sec **23.** 7 ft per sec
25. (a) $x_0^2 + 2x_0h + h^2 + 1$ **(b)** $2x_0 + h$ **(c)** 10
27. (a) $x_0^2 + 2x_0h + h^2 - x_0 - h - 1$ **(b)** $2x_0 + h - 1$ **(c)** 9
29. (a) $x_0^3 + 3x_0^2h + 3x_0h^2 + h^3$ **(b)** $3x_0^2 + 3x_0h + h^2$ **(c)** 75
31. (a) \$5998 **(b)** \$6000 **(c)** \$5998 **(d)** They are the same.

Section 12.3 (page 578)

1. 35 **3.** 1/8 **5.** 1/8 **7.** $y = 6x - 4$ **9.** $y = 5 - 1.25x$
11. $y = (2/3)x + 6$ **13.** $-6; 6$ **15.** $-5; -3; 0; 2; 4$
17. (a) x_5 **(b)** x_4 **(c)** x_3 **(d)** x_2
19.

21. (a) Distance **(b)** Velocity **23.** $f'(x) = -8x + 11; -5; 11; 35$ **25.** $f'(x) = 8; 8; 8; 8$ **27.** $f'(x) = 2/x^2; 1/2$; does not exist; 2/9 **29.** $f'(x) = -4/(x - 1)^2; -4; -4; -1/4$ **31. (a)** \$16 per table **(b)** \$15.998 or \$16 **(c)** The marginal revenue found in part (a) approximates the actual revenue from the sale of the 1001st item found in part (b). **33. (a)** $-4p + 4$ **(b)** -36; demand is decreasing at the rate of about 36 items for each increase in price of \$1. **35. (a)** 20 **(b)** 0 **(c)** -10 **(d)** At 5 hr **37. (a)** Just past 3/4 of an hour (at $x = .775$) and just before 3 hours (at $x = 2.975$) **(b)** 1000; the oven temperature is increasing at 1000° per hour. **(c)** 0; the oven temperature is not changing. **(d)** -1000; the oven temperature is decreasing at 1000° per hour. **39. (a)** The derivative is positive because the tangent line is rising from left to right at $x = 100$ and thus has positive slope. **(b)** The derivative is negative because the tangent line is falling from left to right at $x = 200$ and thus has negative slope.
41. (a)

(b)

(c) The graphs appear identical, which suggests that $f'(x) = g(x)$.
43. (d)

Section 12.4 (page 591)

1. $f'(x) = 8x - 6$ **3.** $y' = 6x^2 + 6x - 5$
5. $g'(x) = 4x^3 + 9x^2 - 6$ **7.** $f'(x) = 9x^{-5} - 2x^{-.5}$ or $9x^{-5} - 2/x^{.5}$
9. $y' = -48x^{2.2} + 3.8x^9$ **11.** $y' = 36t^{1/2} + 2t^{-1/2}$ or $36t^{1/2} + 2/t^{1/2}$ **13.** $y' = 4x^{-1/2} + (9/2) x^{-1/4}$ or $4/x^{1/2} + 9/(2x^{1/4})$
15. $g'(x) = -30x^{-6} + x^{-2}$ or $-30/x^6 + 1/x^2$
17. $y' = -20x^{-3} - 12x^{-5} - 6$ or $-20/x^3 - 12/x^5 - 6$
19. $f'(t) = -6t^{-2} + 16t^{-3}$ or $-6/t^2 + 16/t^3$
21. $y' = -36x^{-5} + 24x^{-4} - 2x^{-2}$ or $-36/x^5 + 24/x^4 - 2/x^2$
23. $g'(x) = -4/x^{3/2} - 5/(2x^{1/2}) + 1$ **25.** $y' = -6/x^{5/2} - 4/x^{3/2} + 2x$
27. $y' = -3/(2x^{5/4})$ **29.** $y' = -5/(3t^{2/3})$
31. $dy/dx = -40x^{-6} + 36x^{-5}$ or $-40/x^6 + 36/x^5$
33. $(-9/2)x^{-3/2} - 3x^{-5/2}$ or $-9/(2x^{3/2}) - 3/x^{5/2}$ **35.** -28
37. 11/16 **39.** (b) **41.** 0; $y = 0$ **43.** 7; $y = 7x - 3$
45. (d) **47. (a)** .4824 **(b)** 2.216 **49. (a)** $C'(x) = 1 + 28x^{-1/3}$ **(b)** About \$9190 **(c)** About \$9150 **(d)** Yes
51. (a) $P(x) = 201\sqrt[3]{x} - .1x^2 - 3x - 40$ **(b)** About \$353; about \$406; about \$405; about \$300 **(c)** $P'(x) = 67x^{-2/3} - .2x - 3$ **(d)** About \$9.43; about \$2.09; about $-\$2.06$; about $-\$8.06$
53. (a) 100 **(b)** 1 **(c)** $-.01$; the percent of acid is decreasing at the rate of .01 per day after 100 days.
55. (a) $f'(x) = .8036x + 2.039$ **(b)** 6.8606 **(c)** Increasing because $f'(6)$ is positive. **57. (a)** 30 **(b)** 0 **(c)** -10
59. (a) 264 **(b)** 510 **(c)** 97/4 or 24.25 **(d)** The number of matings per hour is increasing by about 24.25 matings at a temperature of 16°C. **61. (a)** $v(t) = 16t + 3$ **(b)** 3; 83; 163
63. (a) $v(t) = 6t^2 + 12t$ **(b)** 0; 210; 720
65. (a) -32 ft per sec; -64 ft per sec **(b)** In 3 sec
(c) -96 ft per sec **69. (a)** None **(b)** None **(c)** $(-\infty, \infty)$
(d) The derivative is always negative, so the graph of $g(x)$ is always decreasing.

Section 12.5 (page 599)

1. $y' = 9x^2 + 2x - 6$ **3.** $y' = 120x^3 - 54x^2 + 10$
5. $y' = 24x^5 - 35x^4 - 20x^3 + 42x^2 + 4x - 6$
7. $y' = 144x^3 + 144x^2 + 32x$ **9.** $y' = 54x^5 + 30x^4 + 4x^3$
11. $y' = -7/(x - 4)^2$ **13.** $f'(t) = (t^2 + 6t - 12)/(t + 3)^2$
15. $g'(x) = (-6x^4 - 4x^3 - 6x - 1)/(2x^3 - 1)^2$
17. $y' = (x^2 + 6x - 14)/(x + 3)^2$
19. $r'(t) = [-\sqrt{t} + 3/(2\sqrt{t})]/(2t + 3)^2$ or $(3 - 2t)/[2\sqrt{t}(2t + 3)^2]$
21. $y' = (9\sqrt{x}/2 + 4/\sqrt{x})/x$ or $(9x + 8)/(2x\sqrt{x})$
23. $y' = 42x^5 - 112x^3 + 6x^2 - 8$ **25.** $y' = 3\sqrt{x} - 3/(2\sqrt{x}) - 2$
27. $y' = -12 + 15/\sqrt{x}$ **29.** $y' = \dfrac{2}{(1 - x)^2}$
31. $f'(p) = (24p^2 + 32p + 29)/(3p + 2)^2$
33. $g'(x) = (10x^4 + 18x^3 + 6x^2 - 20x - 9)/[(2x + 1)^2(5x + 2)^2]$
35. In the first step, the numerator should be $(x^2 - 1)2 - (2x + 5)(2x)$.
37. $y = -2x + 9$ **39. (a)** \$22.86 per unit **(b)** \$12.92 per unit **(c)** $(3x + 2)/(x^2 + 4x)$ hundred dollars per unit **(d)** $\overline{C}'(x) = (-3x^2 - 4x - 8)/(x^2 + 4x)^2$ **41. (a)** $N'(t) = \dfrac{4200t}{(30 + t^2)^2}$
(b) 8.28 words per min; 6.94 wpm; 4.71 wpm; 2.49 wpm; .97 wpm
43. (a) -5 ft/sec **(b)** -1.7 ft/sec **45. (a)** .1173 **(b)** 2.625

Section 12.6 (page 609)

1. 275 **3.** 107 **5.** $24x + 4; 24x + 35$ **7.** $-64x^3 + 2; -4x^3 + 8$ **9.** $1/x^2; 1/x^2$ **11.** $\sqrt{8x^2 - 4}; 8x + 10$

In Exercises 13–19 other answers are possible.

13. If $f(x) = x^5$ and $g(x) = 4x + 3$, then $y = f[g(x)]$.

15. If $f(x) = \sqrt{x}$ and $g(x) = 6 + 3x$, then $y = f[g(x)]$.

17. If $f(x) = (x + 3)/(x - 3)$ and $g(x) = \sqrt{x}$, then $y = f[g(x)]$.

19. If $f(x) = x^2 + x + 5$ and $g(x) = x^{1/2} - 3$, then $y = f[g(x)]$.

21. $y' = 9(3x - 4)^2$ **23.** $y' = 72(3x + 2)^3$

25. $y' = -128x(8x^2 + 6)^3$ **27.** $y' = 36(2x + 5)^{1/2}$

29. $y' = (-21/2)(8x + 9)(4x^2 + 9x)^{1/2}$ **31.** $y' = 16/\sqrt{4x + 7}$

33. $y' = \dfrac{-2x - 4}{\sqrt{x^2 + 4x}}$ or $-\dfrac{2x + 4}{\sqrt{x^2 + 4x}}$

35. $y' = 2(x + 1)(x - 3) + (x - 3)^2$

37. $y' = 70(x + 3)(2x - 1)^4(x + 2)$

39. $y' = [(3x + 1)^2(21x + 1)]/(2\sqrt{x})$

41. $y' = -2(x - 4)^{-3}$ or $-2/(x - 4)^3$

43. $y' = [2(4x + 3)(4x - 7)]/(2x - 1)^2$

45. $y' = (-5x^2 - 36x + 8)/(5x + 2)^4$

47. $y' = (3x^{1/2} - 1)/[4x^{1/2}(x^{1/2} - 1)^{1/2}]$

49. (a) -2 **(b)** $-24/7$ **51. (d)**

53. $D(c) = (-c^2 + 10c - 25)/25 + 500$

55. $A[r(t)] = A(t^2) = \pi t^4$; this function represents the area of the oil slick as a function of time t after the beginning of the leak.

57. $C'(x) = \dfrac{15x}{\sqrt{50 + 15x^2}}$ **59.** \$77.46 per additional employee

61. (a) $R'(x) = \dfrac{5(300 - 4x)}{\sqrt{300x - 2x^2}}$ **(b)** \$10.61; \$2.89; $-$\$2.89; $-$\$10.61

63. (a) $F'(t) = \dfrac{150t}{(8 + t^2)^{3/2}}$ **(b)** About 3.96; 1.34; .36; .09

65. (a) $-.5$ **(b)** $-1/54 \approx -.02$ **(c)** $-.011$
(d) $-1/128 \approx -.008$ **67. (a)** $(-1, 1)$ **(b)** $x = -1$
(c) $(-\infty, -1), (1, \infty)$ **(d)** The derivative is 0 when the graph of $G(x)$ is at a low point. It is positive where $G(x)$ is increasing and negative where $G(x)$ is decreasing.

Section 12.7 (page 616)

1. $y' = 3e^{3x}$ **3.** $f'(x) = 10e^{2x}$ **5.** $g'(x) = 20e^{-5x}$
7. $y' = 2xe^{x^2}$ **9.** $f'(x) = xe^{x^2/2}$ **11.** $y' = -18xe^{3x^2+5}$
13. $y' = (-8x + 3)/(-4x^2 + 3x)$ **15.** $y' = 1/(2x + 1)$

17. $f'(x) = \dfrac{6x^2 - 6x + 8}{(2x - 3)(x^2 + 4)}$ **19.** $y' = 2x(1 - x)e^{-2x}$

21. $y' = (-9x^2 + 18x - 4)e^{-3x}$ **23.** $y' = \dfrac{2x}{1 + x^2}$

25. $y' = -23/[(6 - x)(3x + 5)]$
27. $y' = 3(15x^2 - 2)/[2(5x^3 - 2x)]$
29. $y' = -2x^2/(2 - x^2) + \ln(2 - x^2)$
31. $y' = (2x - 1)e^x/(2x + 1)^2$ **33.** $y' = (1 - 3\ln|x|)/x^4$
35. $y' = \dfrac{-20/x + 8 - 8\ln|x|}{(5 - 2x)^2}$
37. $y' = [3x^3 \ln|x| - (x^3 - 1)]/[2x(\ln|x|)^2]$
39. $y' = 1/[2(x - 3)\sqrt{\ln(x - 3)}]$
41. $y' = (xe^x \ln|x| - e^x + 1)/[x(\ln|x|)^2]$
43. $y' = (xe^x + xe^{-x} - e^x + e^{-x})/x^2$ or $[e^x(x - 1) + e^{-x}(x + 1)]/x^2$
45. $f'(x) = \dfrac{4e^{3x+2}}{4x - 5} + 3e^{3x+2}\ln(4x - 5)$ **47.** $y' = \dfrac{2000e^{4x}}{(7 - 10e^{4x})^2}$
49. $y' = 1250e^{-.5x}/(12 + 5e^{-.5x})^2$ **51. (d)**   **53.** 1/8
55. (a) $dR/dx = 100 + 50(\ln x - 1)/(\ln x)^2$ **(b)** \$112.48
57. (a) $C'(x) = 100$ **(b)** $P(x) = (50x/\ln x) - 100$ **(c)** \$12.48

59. (a) $P'(t) = 50{,}000[(1 + .2t)(-.04)e^{-.04t} + .2e^{-.04t}]$
(b) Gaining in 1985 and 1995; losing in 2005 and 2015
61. (a) About 817 vehicles/hr; -41.2 vehicles/hr per ft
(b) About 522 vehicles/hr; -20.9 vehicles/hr per ft
63. (a) Increase at \$54,065 per year   **(b)** Increase at \$13,116 per year   **(c)** Decrease at \$7935 per year   **(d)** Decrease at \$19,517 per year   **(e)** Between years 10 and 15   **65. (a)** 7.42   **(b)** 21.22
(c) .79   **67. (a)** .005   **(b)** .0007   **(c)** .000013   **(d)** $-.022$
(e) $-.0029$ **(f)** $-.000054$

Section 12.8 (page 624)

1. $x = 0, 2$ **3.** $x = 1$ **5.** $x = -3, 0, 3$ **7.** $x = 0, 6$
9. Yes, no   **11.** No, no, yes   **13.** Yes, no, yes
15. Yes, no, yes   **17.** No, yes, yes   **19.** Yes, no   **21.** 1
23. Discontinuous at $t = m$; differentiable everywhere except $t = m$ and the endpoints   **25. (a)** 6 cents   **(b)** Does not exist   **(c)** Any three of the years 1935, 1943, 1949, 1967, 1973, 1989, or 1991
27. (a) \$15   **(b)** \$25   **(c)** \$50   **(d)** \$92.50   **(e)** At $t = 20$
29. $(-4, 6)$ **31.** $(-\infty, -2]$ **33.** $[-25, 0]$ **35.** $(\pi, 10)$
37. $(-4, \infty)$ **39.** $(-\infty, 0)$ **41.** $(-3, 0), (0, 3)$

Chapter 12 Review Exercises (page 628)

1. 4   **3.** -1.2 **5.** .3536   **7.** -1 **9.** 13/2   **11.** 4
13. -13 **15.** 1/6   **17.** 1/4   **19.** 21   **21.** About $-.3846$
23. $y' = 2$ **25.** $y' = 4x - 1$ **27.** $-2; y + 2x = -4$
29. 2/9; $2x - 9y = 2$ **31. (a)** \$150   **(b)** \$187.50   **(c)** \$189
(d)

(e) At $x = \$125$ **33. (a)** -25.33 per dollar   **(b)** -16.75 per dollar   **(c)** -31.12 per dollar   **35.** $y' = 10x - 7$
37. $y' = 14x^{4/3}$ **39.** $f'(x) = -3x^{-4} + (1/2)x^{-1/2}$ or $-3/x^4 + 1/(2x^{1/2})$ **41.** $y' = 15t^4 + 12t^2 - 7$
43. $y' = 16x^{3/4} + 6x^{-1/4}(2x + 3)$ **45.** $f'(x) = \dfrac{-2x^2 + 4}{(x^2 + 2)^2}$
47. $y' = (2 - x + 2\sqrt{x})/[2\sqrt{x}(x + 2)^2]$
49. $y' = (x^2 - 2x)/(x - 1)^2$ **51.** $f'(x) = 12(3x - 2)^3$
53. $y' = 1/\sqrt{2t - 5}$ **55.** $y' = 18x(3x - 4)^2 + 2(3x - 4)^3$
57. $f'(u) = \dfrac{(2u + 3)(6u - 4) - 6(3u^2 - 4u)}{(2u + 3)^4}$
59. $y' = -6x^2e^{-2x^3}$ **61.** $y' = 10xe^{2x} + 5e^{2x}$ or $5e^{2x}(2x + 1)$
63. $y' = \dfrac{2x + 4}{x^2 + 4x + 1}$ **65.** $y' = \dfrac{x - 1/x - 2x\ln 4x}{(x^2 - 1)^2}$
67. $y' = \dfrac{x^2 - 6x + 1}{(x - 3)^2}$ **69.** $y' = -12e^{2x}$
71. $-1/[x^{1/2}(x^{1/2} - 1)^2]$
73. $dy/dt = (1 + 2t^{1/2})/[4t^{1/2}(t^{1/2} + t)^{1/2}]$ **75.** $-2/3$

77. None   **79.** No, yes, yes   **81.** No, yes, no   **83.** Yes, no, yes
85. (a) $367 per hundred   **(b)** $219.50 per hundred   **(c)** $-$528 per
hundred   **87. (a)** 55/3; sales will increase by 55 million dollars
when 3 thousand more dollars are spent on research.   **(b)** 65/4; sales
will increase by 65 million dollars when 4 thousand more dollars are
spent on research.   **(c)** 15; sales will increase by 15 million dollars
when 1 thousand more dollars are spent on research.   **(d)** As more is
spent on research, the increase in sales is decreasing.
89. (a) -9.5; costs will decrease by $9500 for the next $100 spent on
training.   **(b)** -2.375; costs will decrease by $2375 for the next $100
spent on training.   **(c)** Decreasing   **91.** 0 mph per sec for the
hands and approximately 640 mph per sec for the bat; this represents
the acceleration of the hands and the bat at the moment when their
velocities are equal.

Case 12 (page 636)

1. (a) .17   **(b)** Inelastic   **3.** The elasticity is negative.

CHAPTER 13
Section 13.1 (page 644)

1. Increasing on $(1, \infty)$; decreasing on $(-\infty, 1)$; local minimum at 1
3. Increasing on $(-\infty, -2)$; decreasing on $(-2, \infty)$; local maximum at
-2 **5.** Increasing on $(-\infty, -4)$ and $(-2, \infty)$; decreasing on
$(-4, -2)$; local maximum at -4; local minimum at -2
7. Increasing on $(-7, -4)$ and $(-2, \infty)$; decreasing on $(-\infty, -7)$ and
$(-4, -2)$; local maximum at -4; local minima at -7 and -2
9. Increasing on $(-\infty, -1/3)$ and $(2, \infty)$; decreasing on $(-1/3, 2)$
11. Increasing on $(-\infty, -4)$ and $(-4, \infty)$ **13.** Decreasing on
$(-\infty, 5)$ **15.** Increasing on $(-\infty, -1)$ and $(2, \infty)$; decreasing on
$(-1, 2)$ **17.** $-2, -1, 2$ **19.** Local maximum at 0; local
minimum at 2   **21.** Local maximum at -3; local minimum at -1
23. Local maximum at $-1/4$; local minimum at -5 **25.** Local
maximum at -2; local minimum at 3   **27.** Local maximum at 0;
local minimum at 2   **29.** Local maximum at 0; $f(0) = 1$; local
minimum at 6/11; $f(6/11) \approx .7804$ **31.** Local maximum at 3/4;
$f(3/4) = 4$ **33.** Local minimum at 0; $f(0) = 0$ **35.** Local
maximum at -1; $f(-1) = 1/e$ **37.** Local maximum at $-1/e$;
$f(-1/e) = 1/e$; local minimum at $1/e$; $f(1/e) = -1/e$ **39.** Local
minimum at $-1/3$; $f(-1/3) \approx -2.1226$ **41.** Local minimum at
$x = 0$; $f(0) = 2$ **43.** Approximate local maximum: $(3.35, 161.98)$;
approximate local minima: $(-6.77, -709.87)$ and $(10.92, -240.08)$
45. Approximate local maxima: $(-80.7064, 8671701.6)$ and
$(.3982, 5.0017)$; approximate local minima: $(-1.6166, -8.8191)$ and
$(1.9249, -1.7461)$ **49. (a)** 6000   **(b)** $324,000   **51. (a)** The
derivative is undefined at any point where the graph shows a sharp
change from increasing to decreasing.   **(b)** From April to July; from
July to November; from January to April and from November to
December   **(c)** From January to April and for November and
December; air pollution is greatly reduced when the temperature is
low, as is the case during these months.   **53.** $(0, 8000)$ **55.** After
10 days   **57. (a)** $(0, 3)$ **(b)** $(3, \infty)$ (*Note:* x must be at least 0.)
59. Local maxima of 4900 in 1985, 6900 in 1992, 6580 in 1995; local
minima of 4700 in 1986 and 6500 in 1994
61. (a) [1945, 1965], [1970, 1974]   **(b)** [1945, 1986]
(c) 1965–1967, 1974; 1987   **63.** 100 units
65. 1,000,000 tires; $700,000   **67.** About 12.6°C   **69.** 10

Section 13.2 (page 657)

1. $f''(x) = 6x - 10$; -10; 2; -28 **3.** $f''(x) = 12(x + 2)^2$; 48; 192;
12   **5.** $f''(x) = 2/(1 + x)^3$; 2; 2/27; $-1/4$ **7.** $f''(x) =$
$-(x + 4)^{-3/2}/4$ or $-1/[4(x + 4)^{3/2}]$; $-1/32$; $-1/(4 \cdot 6^{3/2})$; $-1/4$
9. $f''(x) = (-6/5)x^{-7/5}$ or $-6/(5x^{7/5})$; $f''(0)$ does not exist;
$-6/(5 \cdot 2^{7/5})$; $-6/[5(-3)^{7/5}]$ **11.** $f''(x) = 2e^x$; 2; $2e^2$; $2e^{-3}$ or $2/e^3$
13. $f''(x) = 20e^{2x}$; 20; $20e^4$; $20e^{-6}$ **15.** $f''(x) = -1/x^2$; not
defined; $-1/4$; $-1/9$ **17.** $f''(x) = 1/x$; not defined; 1/2; $-1/3$
19. $f'''(x) = 72x + 12$; $f^{(4)}(x) = 72$ **21.** $f'''(x) = 6/x^4$;
$f^{(4)}(x) = -24/x^5$ **23.** $v(t) = 12t + 2$; $a(t) = 12$; $v(0) = 2$ cm/sec;
$v(4) = 50$ cm/sec; $a(0) = a(4) = 12$ cm/sec^2
25. $v(t) = 9t^2 - 8t + 8$; $a(t) = 18t - 8$; $v(0) = 8$ cm/sec; $v(4) = 120$
cm/sec; $a(0) = -8$ cm/sec^2; $a(4) = 64$ cm/sec^2
29. Concave upward on $(-\infty, \infty)$; no points of inflection
31. Concave upward on $(-4/3, \infty)$; concave downward on $(-\infty, -4/3)$;
point of inflection at $(-4/3, 425/27)$ **33.** Concave upward on $(3, \infty)$;
concave downward on $(-\infty, 3)$; no points of inflection
35. Concave upward on $(-\infty, -5)$ and $(1, \infty)$; concave downward on
$(-5, 1)$; points of inflection at $(-5, -1248)$ and $(1, 0)$
37. $(14, 26688)$ **39.** No critical numbers; no local extrema
41. Local maximum at $x = -5$; local minimum at $x = 4$
43. Local maximum at $x = 0$; local minima at $x = 2$ and $x = -2$
45. Local maximum at $x = -\sqrt{3}$; local minimum at $x = \sqrt{3}$
47. Local maximum at $x = -3$; local minimum at $x = 3$
49. No critical numbers; no local extrema   **51.** Local maximum at
2; local minima at 1 and 4; points of inflection at $x = (7 + \sqrt{7})/3$ and
$x = (7 - \sqrt{7})/3$ **53.** Local minimum at 1; points of inflection at
$x = 4/3$ and $x = 2$ **55.** Many correct answers, including $f(x) = \sqrt{x}$
57. (a) -96 ft/sec   **(b)** -160 ft/sec   **(c)** -256 ft/sec
(d) -32 ft/sec^2 **59.** $(15, 86250)$ **61.** 8   **63.** 10,000
65. (a) 15 days   **(b)** 16.875%

Section 13.3 (page 669)

1. Absolute maximum at 4; absolute minimum at 1   **3.** Absolute
maximum at 2; absolute minimum at -2 **5.** Absolute maximum at
-4; absolute minima at -7 and -2 **7.** Absolute maximum at
$x = 6$; absolute minima at $x = -4$ and $x = 4$ **9.** Absolute
maximum at $x = 6$; absolute minimum at $x = 4$ **11.** Absolute
maximum at $x = \sqrt{2}$; absolute minimum at $x = 0$ **13.** Absolute
maxima at $x = -3$ and $x = 3$; absolute minimum at $x = 0$
15. Absolute maximum at $x = 0$; absolute minima at $x = -1$ and
$x = 1$ **17. (a)** $200 + 20x$ **(b)** $80 - x$ **(c)** $R(x) =$
$16,000 + 1400x - 20x^2$ **(d)** 35   **(e)** $40,500   **19. (a)** 90
(b) $405   **21.** 8 in. by 8 in.   **23. (a)** 54.8 mph   **(b)** $73.03
25. (a) $1200 - 2x$ **(b)** $A(x) = 1200x - 2x^2$ **(c)** 300 m
(d) 180,000 m^2 **27.** 20,000 ft^2 **29.** $1000
31. (a) Day 12   **(b)** 50 per ml   **(c)** Day 1   **(d)** 81.365
33. 4 in. by 4 in. by 2 in.   **35.** 10 cm; 10 cm
37. Width $= 2\sqrt{6} + 2$ in. ≈ 6.90 in.; length $= 3\sqrt{6} + 3$ in. $\approx$
10.35 in.   **39.** 1 mile from A **41.** About 269; about $577
43. (a) 600   **(b)** 958   **45.** 10   **47.** 60   **49.** 5000   **51. (c)**

Section 13.4 (page 679)

1. No points of inflection

$f(x) = -x^2 - 10x - 25$

3. Point of inflection at $x = 1/3$

$f(x) = 3x^3 - 3x^2 + 1$

5. Point of inflection at $x = -3/2$

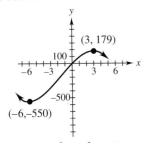

$f(x) = -2x^3 - 9x^2 + 108x - 10$

7. Point of inflection at $x = -7/12$

$f(x) = 2x^3 + \frac{7}{2}x^2 - 5x + 3$

9. No points of inflection

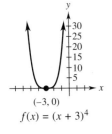

$f(x) = (x + 3)^4$

11. Points of inflection at $x = -\sqrt{3}$ and $x = \sqrt{3}$

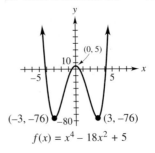

$f(x) = x^4 - 18x^2 + 5$

13. No points of inflection

$f(x) = x - \frac{1}{x}$

15. No points of inflection

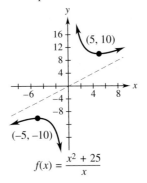

$f(x) = \frac{x^2 + 25}{x}$

17. No points of inflection

$f(x) = \frac{x - 1}{x + 1}$

For Exercises 19–23, there are many correct answers, including the following.

19.

21.

23.

25. (a) The first derivative is always positive because the function is increasing. **(b)** The second derivative is positive because the graph is concave upward. This means that the risk is increasing at a faster and faster rate. **27.** Local maximum at $(-.0241, 1.0000)$; local minimum at $(.6908, .9818)$; point of inflection at $(1/3, 5351/5400)$ **29.** Local maxima at $(-80.7064, 8671701.5635)$ and $(.3982, 5.0017)$; local minima at $(-1.6166, -8.8191)$ and $(1.9249, -1.7461)$; points of inflection at $(-60.4796, 5486563.4612)$, $(-.7847, -2.7584)$, and $(1.2643, 1.2972)$

Chapter 13 Review Exercises (page 682)

5. Increasing on $(-7/2, \infty)$; decreasing on $(-\infty, -7/2)$
7. Increasing on $(-\infty, -2/3)$ and $(1, \infty)$; decreasing on $(-2/3, 1)$
9. Decreasing on $(-\infty, 3)$ and $(3, \infty)$ **11.** Local maximum of 101 at $x = -3$; local minimum of -24 at $x = 2$ **13.** Local maximum of 1 at $x = 0$; local minima of -44 at $x = -3$ and $-4/3$ at $x = 1$
15. Local maximum of $3/e$ at $x = 1$ **17.** $f''(x) = 40x^3 - 24x$; 16; -272 **19.** $f''(x) = -80e^{4x}$; $-80e^4$; $-80e^{-8}$
21. Local maximum at $x = 1/3$; local minimum at $x = -1/2$; point of inflection at $x = -1/12$; increasing on $(-1/2, 1/3)$; decreasing on $(-\infty, -1/2)$ and $(1/3, \infty)$; concave upward on $(-\infty, -1/12)$; concave downward on $(-1/12, \infty)$

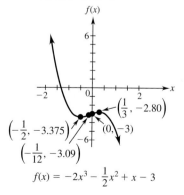

$$f(x) = -2x^3 - \frac{1}{2}x^2 + x - 3$$

23. Local maximum at $x = 0$; local minima at $x = -1$ and $x = 2$; points of inflection at $x = (1 - \sqrt{7})/3$ and $x = (1 + \sqrt{7})/3$; increasing on $(-1, 0)$ and $(2, \infty)$; decreasing on $(-\infty, -1)$ and $(0, 2)$; concave upward on $(-\infty, (1 - \sqrt{7})/3)$ and $((1 + \sqrt{7})/3, \infty)$; concave downward on $((1 - \sqrt{7})/3, (1 + \sqrt{7})/3)$

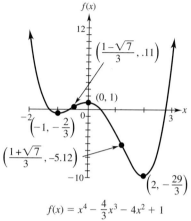

$$f(x) = x^4 - \frac{4}{3}x^3 - 4x^2 + 1$$

25. No local extrema or points of inflection; increasing on $(-\infty, -1/2)$ and $(-1/2, \infty)$; concave upward on $(-\infty, -1/2)$; concave downward on $(-1/2, \infty)$

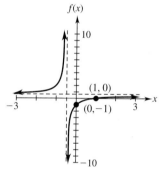

$$f(x) = \frac{x - 1}{2x + 1}$$

27. Local maximum at $x = 1/2$; local minimum at $x = -2/3$; point of inflection at $x = -1/12$; increasing on $(-2/3, 1/2)$; decreasing on $(-\infty, -2/3)$ and $(1/2, \infty)$; concave upward on $(-\infty, -1/12)$; concave downward on $(-1/12, \infty)$

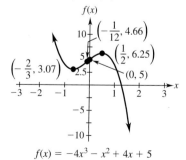

$$f(x) = -4x^3 - x^2 + 4x + 5$$

29. Local minimum at $x = 0$; concave upward on $(-\infty, \infty)$

$f(x) = x^4 + 2x^2$

31. Local maximum at $x = -2$; local minimum at $x = 2$; no point of inflection; increasing on $(-\infty, -2)$ and $(2, \infty)$; decreasing on $(-2, 0)$ and $(0, 2)$; concave upward on $(0, \infty)$; concave downward on $(-\infty, 0)$

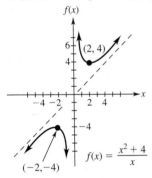

$f(x) = \dfrac{x^2 + 4}{x}$

33. Absolute maximum of $29/4$ at $x = 5/2$; absolute minima of 5 at $x = 1$ and $x = 4$ **35.** Absolute maximum of 39 at $x = -3$; absolute minimum of $-319/27$ at $x = 5/3$ **37. (a)** $2000
(b) $400   **39.** 3 m by 3 m by 3 m
41. 225 m by 450 m   **43.** 126   **45.** 225 m by 225 m

Case 13 (page 684)

1. $Z'(m) = -C_1 m^2 + DC_3/2$ **3.** About 3.33
5. $Z(m^+) = Z(4) = \$11,400$; $Z(m^-) = Z(3) = \$11,300$

CHAPTER 14
Section 14.1 (page 693)

1. They differ only by a constant.
5. $5x^2 + C$ **7.** $8p^3/3 + C$ **9.** $100x + C$ **11.** $5z^2/2 - z + C$
13. $z^3/3 - 2z^2 + 2z + C$ **15.** $x^4/4 - 14x^3/3 + 10x^2 + 3x + C$
17. $4y^{3/2} + C$ **19.** $12t^{5/2}/5 + 2t^{3/2} + C$
21. $112t^{3/2}/3 + 4t^{9/2} + C$ **23.** $-12/x^2 + C$
25. $-1/y - 4\sqrt{y} + C$ **27.** $-3x^{-2} + 4\ln|x| + C$
29. $4e^{3u}/3 + C$ **31.** $-15e^{-.2x} + C$ **33.** $3\ln|x| - 8e^{-.5x} + C$
35. $\ln|t| + 2t^3/3 + C$ **37.** $e^{2u}/2 + u^2/8 + C$
39. $x^3/3 + x^2 + x + C$ **41.** $6x^{7/6}/7 + 3x^{2/3}/2 + C$
43. $f(x) = 2x^3 - 2x^2 + 3x + 1$ **45.** $C(x) = .2x^3/3 + 5x^2/2 + 10$
47. $C(x) = 2x^{3/2}/3 + 52/3$ **49.** $C(x) = x^3/3 - x^2 + 3x + 6$
51. $C(x) = .000375x^4 + .011x^3 + .022x^2 + .25x + 5.55$
53. $C(x) = 3e^{.01x} + 5$ **55.** $p = 50 - 3x/2 - x^2/3$
57. (a) $f(t) = -e^{-.01t} + k$ **(b)** .095 unit

Section 14.2 (page 700)

3. $(12x - 1)^3/12 + C$ **5.** $-2/[3(3t + 1)] + C$
7. $-1/\sqrt{x^2 + 2x - 4} + C$ **9.** $2(r^3 + 3)^{3/2}/9 + C$
11. $-3e^{5k}/5 + C$ **13.** $2e^{2w^3}/3 + C$ **15.** $e^{4t-t^2}/2 + C$
17. $2e^{\sqrt{y}} + C$ **19.** $-4\ln|2 + 5x|/5 + C$
21. $(1/2)\ln|e^{2t} + 1| + C$ **23.** $-1/[8(2x^2 + 8x)^2] + C$
25. $\left(\dfrac{1}{r} + r\right)^2/2 + C$ **27.** $(x^3 + 3x)^{1/3} + C$
29. $(1/6)\ln|3x^2 + 12x + 8| + C$ **31.** $(x^2 + 1)^4/4 + C$
33. $(x^2 + 12x)^{3/2}/3 + C$ **35.** $(1 + \ln x)^3/3 + C$
37. $2(u - 1)^{3/2}/3 + 2(u - 1)^{1/2} + C$
39. $2(5t - 1)^{5/2}/125 + 2(5t - 1)^{3/2}/75 + C$
41. (a) $f(x) = .029x^4 - .601x^3 + 3.43x^2 + 3.24x + 26$ **(b)** 82.5%
43. (a) $f(x) = 1110\ln|x| - 3994.8$ **(b)** 1060 per 1000 people

Section 14.3 (page 709)

3. 28; 32   **5.** 8; 10   **7.** 12.8; 27.2   **9.** 2.67; 2.08
11. (a) 3   **(b)** 3.5   **(c)** 4   **13.** 124.5833   **15.** 74.0439
17. 1.8719   **19.** About \$83,000   **21.** About 35 liters
23. About 1300 ft   **25. (a)** About 690 BTUs
(b) About 180 BTUs   **27. (a)** 9 ft   **(b)** 2 sec   **(c)** 4.62 ft
(d) Between 3 and 3.5 seconds

Section 14.4 (page 719)

1. 52   **3.** 13   **5.** 1/3   **7.** 76   **9.** 4/3
11. $5\ln 5 + (12/25) \approx 8.527$
13. $20e^{-.2} - 20e^{-.3} - 3\ln 3 - 3\ln 2 \approx 2.775$
15. $e^{10}/5 - e^5/5 - 1/2 \approx 4375.1$ **17.** 91/3
19. $447/7 \approx 63.857$ **21.** $(\ln 2)^2/2 \approx .24023$ **23.** 49
25. $1/4 - 1/(3 + e) \approx .075122$ **27.** $(6/7)(128 - 2^{7/3}) \approx 105.39$
29. A loss for the first year and a half; $\displaystyle\int_0^2 (6x^2 - 7x - 3)\,dx$
31. 23/3   **33.** 9.5   **35.** $e^2 + e^{-1} - 3 \approx 4.757$
37. 1   **39.** 23/6   **41.** $(e + e^3 - 2e^2)/2 \approx 4.01$
43. 19   **45. (a)** \$22,000   **(b)** \$62,000   **(c)** About 15.32 days
47. (a) $(9000/8)(17^{4/3} - 2^{4/3}) \approx \$46,341$
(b) $(9000/8)(26^{4/3} - 17^{4/3}) \approx \$37,477$ **(c)** It is slowly increasing
without bound.   **49. (a)** $f(x) \approx 109.0e^{.0198x} + 1.6$ **(b)** 119.6,
which differs from the actual value by 1.0   **51. (a)** About 414
barrels   **(b)** About 191 barrels   **(c)** Decreasing to 0
53. (a) $\displaystyle\int_0^{10} 1.2e^{.04t}\,dt$ **(b)** $30e^{.4} - 30 \approx 14.75$ billion   **(c)** About
12.8 years   **(d)** About 14.4 years

Section 14.5 (page 729)

1. \$280; \$11.67   **3.** \$2500; \$30,000   **5.** No   **7. (a)** 8 yr
(b) About \$148   **(c)** About \$771   **9. (a)** 39 days   **(b)** \$3369.18
(c) \$484.02   **(d)** \$2885.16   **11. (a)** 5   **(b)** About 15.6 tons
13. 13 yr

15. (a)

$S(q) = \frac{7}{5}q$

$D(q) = -\frac{3}{5}q + 10$

(b) (5, 7)   **(c)** $7.50   **(d)** $17.50   **17.** $1999.54   **19.** $40.50
21. (a)

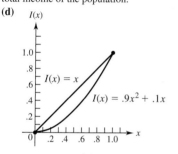

$S(q) = q^2 + \left(\frac{11}{4}\right)q$

$D(q) = 150 - q^2$

(b) (8, 86)   **(c)** $341.33   **(d)** $429.33
23. $5.76 million
25. (a) .019; the lower 10% of the income producers earn 1.9% of the total income of the population.   **(b)** .275; the lower 50% of the income producers earn 27.5% of the total income of the population. **(c)** .819; the lower 90% of the income producers earn 81.9% of the total income of the population.
(d)

$I(x)$

$I(x) = x$

$I(x) = .9x^2 + .1x$

(e) .15; the amount of inequality of income distribution
27. (a) 10 items   **(b)** 10 hr
The answers in Exercises 31 and 33 may vary slightly.
31. 161.2   **33.** 5.516

Section 14.6 (page 733)

1. $-4 \ln|x + \sqrt{x^2 + 36}| + C$
3. $\ln|(x-3)/(x+3)| + C$ $(x^2 > 9)$
5. $(4/3)\ln[(3 + \sqrt{9 - x^2})/x] + C$ $(0 < x < 3)$
7. $-2x/3 + 2\ln|3x + 1|/9 + C$
9. $(-2/15)\ln|x/(3x - 5)| + C$
11. $\ln|(2x - 1)/(2x + 1)| + C$ $(x^2 > 1/4)$
13. $-3\ln|(1 + \sqrt{1 - 9x^2})/(3x)| + C$ $(0 < x < 1/3)$
15. $2x - 3\ln|2x + 3| + C$
17. $1/[25(5x - 1)] - (\ln|5x - 1|)/25 + C$
19. $x^5[(\ln|x|)/5 - 1/25] + C$
21. $(1/x)(-\ln|x| - 1) + C$
23. $-xe^{-2x}/2 - e^{-2x}/4 + C$ **25.** $2094.71
27. About 15,100 microbes

Section 14.7 (page 740)

1. $y = -x^2 + x^3 + C$ **3.** $y = 3x^4/8 + C$ **5.** $y^2 = x^2 + C$
7. $y = Me^{x^2}$ **9.** $y = Me^{x^3-x^2}$ **11.** $y = Mx$
13. $y = Me^x + 5$ **15.** $-1/(e^x + C)$ **17.** $y = x^3 - x^2 + 2$
19. $y^2 = \dfrac{2\ln|x^3 + 28|}{3} + 6$ **21.** $y^2 = 2x^3/3 + 9$
23. $y = e^{x^2+3x}$ **25.** $y^2/2 - 3y = x^2 + x - 4$ **29. (a)** 108.4
(b) 121.2   **(c)** No; if $x = 8$, the denominator becomes 0.
31. $y = 26.7e^{.02336t}$ **33.** About 1.35×10^5 **35. (a)** About 178.6 thousand   **(b)** About 56.4 thousand   **(c)** About 50.0 thousand
37. 54.6 g   **39.** About 10   **41. (a)** $T = 38 + 60.6e^{-.343t}$
(b) About 68.5°   **(c)** About 1.44 hours   **(d)** About 25.4 hours

Chapter 14 Review Exercises (page 743)

1. $x^3/3 - 3x^2/2 + 2x + C$ **3.** $2x^{3/2} + C$ **5.** $2x^{3/2}/3 + 9x^{1/3} + C$ **7.** $2x^{-2} + C$ **9.** $-3e^{2x}/2 + C$
11. $2\ln|x - 1| + C$ **13.** $e^{3x^2}/6 + C$
15. $(3\ln|x^2 - 1|)/2 + C$ **17.** $-(x^3 + 5)^{-3}/9 + C$
19. $\ln|2x^2 - 5x| + C$ **21.** $-e^{-3x^4}/12 + C$ **23.** $2e^{-5x}/5 + C$
27. (a) About 136.97   **(b)** 148.8   **29.** 28   **33.** $72/25 \approx 2.88$
35. $2\ln 3$ or $\ln 9 \approx 2.1972$ **37.** $2e^4 - 2 \approx 107.1963$
39. $(2/3)(22^{3/2} - 2^{3/2}) \approx 66.907$ **41.** $e^2 - 1 \approx 6.3891$
43. $C(x) = 10x - x^2 + 4$ **45.** $C(x) = (2x - 1)^{3/2} + 145$
47. 36,000   **49.** About 26.3 yr   **53.** About 142
55. $-\ln|(5 + \sqrt{25 + x^2})/x| + C$
57. $15[x/2 + 5\ln|2x - 5|/4] + C$ **59.** $y = x^4/2 + 3x^2 + C$
61. $y^2 = 3x^2 + 2x + C$ **63.** $y = -5e^{-x} - 5x + 22$
65. $y = 2e^{5x-x^2}$ **67. (a)** $dA/dt = .05A - 1000$ **(b)** $9487.29
69. (a) About 111 items   **(b)** Not exactly, but for practical purposes, yes
71. $.2\ln y - .5y = -.3\ln x + .4x + C$; $x = 3/4$ unit, $y = 2/5$ unit

Case 14 (page 746)

1. About 102 years   **3.** About 45.4 years

CHAPTER 15
Section 15.1 (page 755)

3. 4; -22; -20; 12   **5.** 3; $\sqrt{33}$; $\sqrt{17}$; 8
9.

11.

13.

15.

17.

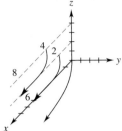

19. $y = 500^{10/3}x^{-7/3} \approx 10^9/x^{7/3}$

21. 1.12; the IRA account grows faster. **23.** $f(L, W, H) = L + 2H + 2W$ **25. (a)** $18x + 9h$ **(b)** $-6y - 3h$ **27. (c)**
29. (e) **31. (b)**

Section 15.2 (page 763)

1. (a) $24x^2 - 8xy$ **(b)** $-4x^2 + 18y$ **(c)** 48 **(d)** -40
3. $f_x = -2xy + 12x^3; f_y = -x^2; 100; -16$ **5.** $f_x = 2e^{2x+y};$
$f_y = e^{2x+y}; 2e^3; e^{-5}$ **7.** $f_x = 2/(e^{x+2y}); f_y = 4/(e^{x+2y}); 2; 4/e^2$
9. $f_x = (y^3 - x^2 - 6xy^2)/(x^2 + y^3)^2;$
$f_y = (6x^2y - 3xy^2 - 3y^4)/(x^2 + y^3)^2; -17/9;$ about .083
11. $f_x = \dfrac{2 - 2xy}{2x - x^2y}; f_y = \dfrac{-x}{2 - xy}; 3/4; 2/7$
13. $f_x = 2x^2ye^{2xy} + 2xe^{2xy}; f_y = 2x^3e^{2xy}; -4e^{-4}; -128e^{-24}$
15. $f_{xx} = 20y^3 - 30x; f_{xy} = f_{yx} = 60xy^2; f_{yy} = 60x^2y$
17. $h_{xx} = -8y^2; h_{xy} = h_{yx} = -16xy + 14y; h_{yy} = -6 - 8x^2 + 14x$
19. $R_{xx} = \dfrac{24y}{(2x + y)^3}; R_{xy} = R_{yx} = \dfrac{-12x + 6y}{(2x + y)^3}; R_{yy} = \dfrac{-12x}{(2x + y)^3}$
21. $z_{xx} = 0; z_{yy} = 4xe^y; z_{xy} = z_{yx} = 4e^y$ **23.** $r_{xx} = -1/(x + y)^2;$
$r_{yy} = -1/(x + y)^2; r_{xy} = r_{yx} = -1/(x + y)^2$ **25.** $z_{xx} = 1/x;$
$z_{yy} = -x/y^2; z_{xy} = z_{yx} = 1/y$ **27.** $1; -1/4$ **29.** $x = -4, y = 2$
31. $x = 0, y = 0$ or $x = 3, y = 3$ **33.** $f_x = 2x; f_y = z;$
$f_z = y + 4z^3; f_{yz} = 1; 3; 1$ **35.** $f_x = 6/(4z + 5); f_y = -5/(4z + 5);$
$f_z = -4(6x - 5y)/(4z + 5)^2; f_{yz} = 20/(4z + 5)^2$
37. $f_x = (2x - 5z^2)/(x^2 - 5xz^2 + y^4); f_y = 4y^3/(x^2 - 5xz^2 + y^4);$
$f_z = -10xz/(x^2 - 5xz^2 + y^4); f_{yz} = 40xy^3z/(x^2 - 5xz^2 + y^4)^2$
41. (a) 2.04% **(b)** 1.26% **(c)** .05%: the rate of change of the
probability for an additional semester of high school math; .003%: the
rate of change of the probability per unit of change in the SAT score
43. (a) $206,800 **(b)** $f_p = 132 - 2i - .02p; f_i = -2p$; the rate at
which sales revenue is changing per unit of change in price (f_p) or
interest rate (f_i) **(c)** A sales revenue drop of $18,800

45. (a) 46.656 **(b)** $f_x(27, 64) = .6912$ and is the rate at which
production is changing when labor changes by 1 unit from 27 to 28
and capital remains constant; $f_y(27, 64) = .4374$ and is the rate at
which production is changing when capital changes by 1 unit from 64
to 65 and labor remains constant. **(c)** Production would be increased
by about 69 units when labor is increased by 1 unit.
47. $.7x^{-.3}y^{.3}$ or $.7y^{.3}/x^{.3}; .3x^{.7}y^{-.7}$ or $.3x^{.7}/y^{.7}$
49. (a) About 5.71 **(b)** About 6.84 **(c)** $1/(a - v)$ **(d)** $b/(a - v)^2$
51. (a) $F_m = (gR^2)/r^2$; the rate of change in force per unit change in
mass; $F_r = (-2mgR^2)/r^3$; the rate of change in force per unit change
in distance

Section 15.3 (page 771)

3. Local minimum of -22 at $(3, -1)$
5. Local minimum of 0 at $(-2, -2)$ **7.** Saddle point at $(-3, 4)$
9. Local maximum of 17 at $(2/3, 4/3)$ **11.** Saddle point at $(2, -2)$
13. Saddle point at $(1, 2)$ **15.** Saddle point at $(0, 0)$; local
minimum of -201 at $(6, 18)$ **17.** Saddle point at $(0, 0)$; local
minimum of -5157 at $(72, 12)$ **19.** Saddle point at $(0, 0)$
21. Local maximum of 9/8 at $(-1, 1)$; saddle point at $(0, 0)$; (a)
23. Local minima of $-33/16$ at $(0, 1)$ and $(0, -1)$; saddle point at
$(0, 0)$; (b) **25.** Local maxima of 17/16 at $(1, 0)$ and $(-1, 0)$; local
minima of $-15/16$ at $(0, 1)$ and $(0, -1)$; saddle points at $(0, 0), (-1, 1)$,
$(1, -1), (1, 1)$, and $(-1, -1)$; (e) **27.** $P(12, 40) = 2744$
29. $P(12, 72) = \$2,528,000$ **31.** $C(12, 25) = 2237$
33. 3 m by 3 m by 3 m **35.** 18 inches by 18 inches by 36 inches
37. 1000 tons of grade A ore and 2000 tons of grade B ore for
maximum profit of $1400

Chapter 15 Review Exercises (page 774)

1. 32; 156 **3.** $-6/5; -8/3$
7.

9.

11. (a) $-4x + 5y$ **(b)** 3 **(c)** 5
15. $f_x = 12x^2y + 10y^4; f_y = 4x^3 + 40xy^3$
17. $f_x = (12y - 3x^2 + 4xy^2)/(x^2 + 4y)^2;$
$f_y = (-4x^2y - 8y^2 - 12x)/(x^2 + 4y)^2$ **19.** $f_x = 2(y + 1)^2e^{2x+y};$
$f_y = (y + 1)(y + 3)e^{2x+y}$ **21.** $f_x = 3x^2y^2/(1 + x^3y^2);$
$f_y = 2x^3y/(1 + x^3y^2)$ **23.** $f_{xx} = 24xy^2; f_{xy} = 24x^2y - 8$
25. $f_{xx} = 8y/(x - 2y)^3; f_{xy} = (-4x - 8y)/(x - 2y)^3$
27. $f_{xx} = 2e^y; f_{xy} = 2xe^y$ **29.** $f_{xx} = (-2x^2y^2 - 4y)/(2 - x^2y)^2;$
$f_{xy} = -4x/(2 - x^2y)^2$ **31.** Local minimum at $(0, 1)$
33. Saddle point at $(0, 2)$ **35.** Local minimum at $(1, 1/2)$; saddle
point at $(-1/3, 11/6)$ **37.** $F(L, W, H) = 2LW + 2WH + 2LH$
39. (a) Local minimum at $(11, 12)$ **(b)** $431

Index of Applications

Index

Average speed, 558
Axis of a parabola, 151

B
Back substitution, 270
Bank discount, 228
Base, 22, 200, 201, 202
Basic feasible solutions, 355, 375
   steps to solve, 377
Basic probability principle, 412
Basic variables of a simplex tableau, 353, 375
Bayes, Thomas, 442
Bayes' formula, 439, 441
   application of, 442
   special case for, 440
Bernoulli, Jakob, 469, 687
Bernoulli processes, 469
Bernoulli trials, 469, 535
Binomial distribution, 535, 537
Binomial experiment, 469, 535
Binomial probability, 471
Binomials, 24
   factoring of, 30
Boundary of a linear inequality, 325
Bounded region, 334
Break-even analysis, 76
Break-even point, 77

C
Calculator, programming, 779
Calculator graphing, 775
Calculus, Fundamental Theorem of, 711
Carbon-14 dating, 214
Cartesian coordinate system, 53
Celsius-Fahrenheit conversion, 75, 97
Central tendency, measures of, 506
Certain event, 412
Chain rule for derivatives, 602, 603
   alternate form of, 604
Change in x, 61
Change in y, 61
Change of base theorem for logarithms, 205
Closed interval, 621
Cobb-Douglas production function, 754
Code theory, 311
Coefficient matrix, 307
Coefficients of a polynomial, 24
Column matrix, 288

Column of a matrix, 271
Column vector, 288
Combinations, 455
   distinguishing from permutations, 457
   formula for, 456
   notation for, 455
   table of, 801
Common denominator, 13
Common logarithms, 200
Common ratio of a geometric sequence, 241
Commutative properties, 3
Complement of a set, 395
Complement rule for an event, 419
Completing the square, 88
Complex fractions, 36
   simplifying, 36
Composition of functions, 601
   chain rule for derivatives of, 602, 603
Compound amount, 231
Compound interest, 231
   formula for, 232
   periods of, 232
   present value with, 236
Compounding, continuously, 234
Concave downward, 651, 652
Concave upward, 651, 652
Conditional probability, 427
Conjugate, 45
Constant, Hubble, 142
Constant
   of antiderivatives, 686
   growth rate, 738
   of integration, 686
Constant multiple rule for antiderivatives, 689
Constant rule for derivatives, 583
Constant term of a polynomial, 24
Constant times a function rule for derivatives, 585
Constraints, 332
Consumers' surplus, 726
Continuity
   on an open interval, 621
   at a point, 618, 619
Continuous compounding, 234
Continuous distribution, 525
Coordinate system, Cartesian, 53
Corner point of a region, 334
Corner point theorem, 335
Cost
   fixed, 139
   marginal, 564, 587, 588